Reproductive and Developmental Toxicology

edited by

KENNETH S. KORACH
National Institute of Environmental Health Sciences
Research Triangle Park, North Carolina

MARCEL DEKKER, INC.   NEW YORK · BASEL · HONG KONG

Library of Congress Cataloging-in-Publication Data

Reproductive and developmental toxicology/edited by Kenneth S. Korach.
    p. cm.
  Includes bibliographical references and index.
  ISBN 0-8247-9857-0 (alk. paper)
  1. Reproductive toxicology. 2. Developmental toxicology.
I. Korach, Kenneth S.
RA1224.2.R46 1998
615.9--dc21
                                           98-5182
                                              CIP

The publisher offers discounts on this book when ordered in bulk quantities. For more information, write to Special Sales/Professional Marketing at the address below.

This book is printed on acid-free paper.

MARCEL DEKKER, INC.
270 Madison Avenue, New York, New York 10016
http://www.dekker.com
Current printing (last digit):
10 9 8 7 6 5 4 3 2 1

PRINTED IN THE UNITED STATES OF AMERICA

Reproductive and Developmental Toxicology

Foreword

Thanks to the insight of the editor and the scholarly contributions of the authors, this book captures in one volume the essence of the threshold of science that is currently being experienced in the areas of biology and toxicology. It is not very often in science and medicine that scientific breakthroughs occur at a time when the public and the community of scientists and clinicians are primed to accept them and use them to move the field forward. A scientific breakthrough may not cause much of a stir if it happens at a time when there is no critical context in which to place the finding. On the other hand, a human health or environmental catastrophe may present a critical context for good science, but if new scientific understanding is not forthcoming, the incident passes without a step forward in science. When thalidomide and 1,2-dibromo-3-chloropropane (DBCP) produced chemical-induced human malformations and infertility, respectively, in neither case did we learn much about the biology of development or reproduction beyond the establishment of causality. Science was not prepared to capitalize on these situations when the tragedies occurred. Thus, the exposure was stopped and further damage was prevented, but the knowledge gained was chemical-specific and did not have broad application in toxicology or medicine.

The compilation of data in this book represents integration of a new level of biological understanding into the fields of reproductive and developmental toxicology. This new influx of mechanistic understanding comes at a time when extrapolation of animal data to humans has reached a scientific plateau. Importantly, these scientific advancements that represent threshold referred to earlier have occurred at a time when the public and scientific communities have been sensitized to the importance of this area by questions about endocrine disrupters in the environment. Scientists are poised today to evaluate the significance of new findings through the use of research tools that were unavailable during the crises with thalidomide and DBCP. Thus, the question of whether exposure to low levels of hormonally active substances in the environment poses a threat to human health has surfaced at a time when new laboratory tools, such as transgenic animals and molecular techniques to better understand the underlying basis for toxicological findings, can be used to provide mechanism-based information to support decisions for protection of human and environmental health.

All three sections of this book successfully demonstrate the interface between new biological understanding and its application to toxicological problems. The authors of individual chapters are leaders in their respective disciplines. The findings they present are an objective view of the leading edge of science in the areas of development and male and female reproduction. Perhaps more importantly, their research approaches and the underlying biological concepts help to form the future basis for identifying potential toxicological hazards and for enhancing risk assessment procedures in the areas of reproductive and developmental toxicology.

Bernard A. Schwetz
Director, National Center for
Toxicological Research
Jefferson, Arkansas, and
Interim Chief Scientist of the FDA
Rockville, Maryland

Preface

Reproduction and reproductive sciences encompass an area that spans a number of scientific disciplines, including endocrinology, cell biology, physiology, anatomy, developmental biology, biochemistry, and molecular biology. Because the reproductive process is critical for perpetuation of any species of organisms, factors or agents that alter or disrupt this process can have devastating consequences.

We now realize that such agents can arise from varying sources—which can be pharmacological, environmental, and natural—having extensive chemical structural diversity. In addition, the effects can be through a single action, or in combination, and can influence either individual or multiple cellular signaling pathways in a tissue or organ system. As such, it is difficult to ascribe a single action or effect to certain agents. This is why studies, such as those described in this volume, which incorporate the latest scientific approaches and findings, provide the scientific basis from experimental studies to determine the extent, magnitude, and severity of exposure to certain compounds. Such experimental observations will have direct relevance toward human exposure and its resultant consequences for fertility and normal development.

Most important in recent toxicology studies is the number of examples showing that exposure during specific periods of development results in long-term effects that occur following sexual maturity and adulthood. Certain organ systems are more susceptible to toxicants during these developmental periods. It is extremely important to realize that not only adults, but also children, infants, and the developing fetus, are potential targets for toxicological insult. Exposure during these sensitive periods either alters normal development, resulting in immediate or acute effects, or may subsequently compromise normal physiology and function later in life.

Some biological processes are common to both sexes; however, some are unique to a certain sex, and these require even more precise understanding in order to evaluate their sex-specific actions. Reproduction is a complex physiological process involving several organ systems and encompassing neuroendocrine, reproductive tract, accessory sex organs, and gonads. All of these systems act in a coordinated manner to produce successful fertility of both sexes. Infertility and compromised reproductive function can occur at a specific organ, such as the gonads, or at a number of points or organ systems, such as neuroendocrine centers, which ultimately affect reproduction through an alternative mechanism. Parts II and III of this volume directly address the experimental evidence illustrating toxic effects on male and female reproduction at several levels. Even though reproduction is a hormonally regulated endocrine response, agents besides those having hormonal or antihormonal activities are also described. Our knowledge within the scientific disciplines governing reproduction continues to unravel, and understanding the basic mechanisms involved in reproductive processes is crucial to determining the activities and effects of certain reproductive toxicants. This volume attempts to provide some of the latest findings to help address these important questions.

Kenneth S. Korach

Contents

Part I. Development and Neurotoxicity

Contents   ix

Contributors

Herman Adlercreutz, M.D., Ph.D. Department of Clinical Chemistry, University of Helsinki and Folkhälsan Research Center, Helsinki, Finland

Linda S. Birnbaum, Ph.D. Experimental Toxicology Division, National Health and Environmental Effects Laboratory, U.S. Environmental Protection Agency, Research Triangle Park, North Carolina

Jack B. Bishop, Ph.D. National Institute of Environmental Health Sciences, Research Triangle Park, North Carolina

James H. Clark, Ph.D. Department of Cell Biology, Baylor College of Medicine, Houston, Texas

Ralph L. Cooper, Ph.D. Developmental Toxicology Division, U.S. Environmental Protection Agency, Research Triangle Park, North Carolina

William R. Crowley, Ph.D. Department of Pharmacology, University of Tennessee, College of Medicine, Memphis, Tennessee

Audrey M. Cummings, Ph.D. Reproductive Toxicology Division, U.S. Environmental Protection Agency, Research Triangle Park, North Carolina

Gerald R. Cunha, Ph.D. Departments of Anatomy, Obstetrics and Gynecology and Urology, University of California, San Francisco, California

Barbara J. Davis, V.M.D., Ph.D. Laboratory of Experimental Pathology, Environmental Toxicology Program, National Institute of Environmental Health Sciences, Research Triangle Park, North Carolina

Patricia K. Donahoe, M.D. Pediatric Surgical Services, Massachusetts General Hospital, and Department of Surgery, Harvard Medical School, Boston, Massachusetts

Annemarie A. Donjacour, Ph.D. Department of Anatomy, University of California, San Francisco, California

Rama S. Dwivedi, Ph.D. Department of Pediatrics, Northwestern University Medical School, Chicago, Illinois

Elaine M. Faustman, Ph.D. , M.D. Department of Environmental Health, University of Washington, Seattle, Washington

Walderico M. Generoso, Ph.D. Biology Division, Oak Ridge National Laboratory, Oak Ridge, Tennessee

Jerome M. Goldman, Ph.D. Reproductive Toxicology Division, U.S. Environmental Protection Agency, Research Triangle Park, North Carolina

Gaylia Jean Harry, Ph.D. Laboratory of Toxicology, National Institute of Environmental Health Sciences, Research Triangle Park, North Carolina

Simon W. Hayward, Ph.D. Department of Anatomy, University of California, San Francisco, California

Jerrold J. Heindel, Ph.D. Division of Extramural Research and Training, National Institute of Environmental Health Sciences, Research Triangle Park, North Carolina

Rex A. Hess, Ph.D. Department of Veterinary Biosciences, College of Veterinary Medicine, University of Illinois, Urbana, Illinois

Claude L. Hughes, Jr., M.D., Ph.D. Department of Obstetrics and Gynecology and the Integrated Toxicology Program, Duke University Medical Center, Durham, North Carolina

Philip M. Iannaccone, M.D., D.Phil. Department of Pediatrics, Northwestern University Medical School, Chicago, Illinois

Robert J. Kavlock, Ph.D. Developmental Toxicology Division, U.S. Environmental Protection Agency, Research Triangle Park, North Carolina

Gary R. Klinefelter, Ph.D. National Health and Environmental Effects Research Laboratory, Reproductive Toxicology Division, U.S. Environmental Protection Agency, Research Triangle Park, North Carolina

Ling-Hong Li, Ph.D. Laboratory of Toxicology, Environmental Toxicology Program, National Institute of Environmental Health Sciences, Research Triangle Park, North Carolina

David T. MacLaughlin, Ph.D. Department of GYN-Vincent Research Laboratory, Massachusetts General Hospital, and Harvard Medical School, Boston, Massachusetts

Michael J. McPhaul, M.D. Department of Internal Medicine, Division of Endocrinology, University of Texas Southwestern Medical Center, Dallas, Texas

Retha R. Newbold, M.A., CT(ASCP) National Institute of Environmental Health Sciences, Research Triangle Park, North Carolina

Sonoko Ogawa, Ph.D. The Rockefeller University, New York, New York

Roger A. Pedersen, Ph.D. Department of Obstetrics, Gynecology, and Reproductive Sciences, University of California, San Francisco, California

Sally D. Perreault, Ph.D. Reproductive Toxicology Division, U.S. Environmental Protection Agency, Research Triangle Park, North Carolina

Richard E. Peterson, Ph.D. School of Pharmacy and Environmental Toxicology Center, University of Wisconsin, Madison, Wisconsin

Donald W. Pfaff, Ph.D. Laboratory of Neurobiology and Behavior, The Rockefeller University, New York, New York

Rafael A. Ponce, Ph.D. Department of Environmental Health, University of Washington, Seattle Washington

John M. Rogers, Ph.D.   Reproductive Toxicology Division, U.S. Environmental Protection Agency, Research Triangle Park, North Carolina

Beth L. Roman, Ph.D.   School of Pharmacy and Environmental Toxicology Center, University of Wisconsin, Madison, Wisconsin

Ellen A. Rorke, Ph.D.   Department of Environmental Health Sciences, Case Western Reserve University, Cleveland, Ohio

Joe C. Rutledge, M.D.   Department of Laboratory Medicine, Children's Hospital and University of Washington, Seattle, Washington

Stephen H. Safe, Ph.D.   Department of Veterinary Physiology and Pharmacology, Texas A&M University, College Station, Texas

Richard M. Sharpe, Ph.D. MRC Reproductive Biology Unit, Edinburgh, Scotland

Kimberly Silvers, Ph.D. Case Western Reserve University, Cleveland, Ohio

Rebecca Z. Sokol, M.D.   Departments of Obstetrics and Gynecology and Medicine, University of Southern California, Los Angeles, California

Ginger Tansey, D.V.M.   Department of Comparative Medicine, Wake Forest University School of Medicine, Winston-Salem, North Carolina

Robert S. Tebbs, Ph.D. Department of Dermatology, University of California, San Francisco, California

Lee Tyrey, Ph.D. Duke University Medical Center, Durham, North Carolina

Timothy Zacharewski, Ph.D. Department of Biochemistry, Michigan State University, East Lansing, Michigan

Reproductive and Developmental Toxicology

1

Müllerian-Inhibiting Substance Activity and Normal Male Sex Determination

David T. MacLaughlin and Patricia K. Donahoe
*Massachusetts General Hospital and Harvard Medical School,
Boston, Massachusetts*

INTRODUCTION

It is well established that for normal mammalian reproductive tract development to occur, a number of specific molecular events must occur in an orderly sequence at specific gestational stages. Since the gonadal ridge of undifferentiated embryos possesses both male and female reproductive tract anlagen, Wolffian and Müllerian ducts, respectively, the manifestation of chromosomal or genetic sex determination must include the stimulation of the appropriate gender-specific ductal system and regression of the other. Significant deviation from this program results in abnormal reproductive tract organization in males and females and, thus, varying degrees of ambiguous genitalia in the newborn.

As the result of his pioneering work in the late 1940s, Professor Alfred Jost tested the hypothesis that the fetal testis produced a nonsteroidal factor that is responsible for the regression of the Müllerian ducts (Jost, 1947), the anlagen of the female reproductive tract. His research into the gonadal control of reproductive tract development in rabbit fetuses revolutionized the then contemporary thinking that testosterone played a dual role in the early stages of normal male phenotypic development (i.e., that testosterone stimulated Wolffian duct development while simultaneously blocking Müllerian duct growth). Professor Jost unequivocally proved the validity of the former idea and disproved the latter, which at the same time provided a partial explanation for the clinical syndrome of testicular feminization.

We now understand that the normal process of producing the male phenotype requires two active mechanisms, both of which depend upon testicular differentiation. Using a gonadectomized fetal rabbit experimental model, Jost showed that testosterone, synthesized and released from the newly formed testis, stimulates development of the Wolffian ducts to mature into epididymis, vas deferens, and seminiferous tubules. At the same time, a testicular nonsteroidal compound, which he called the "Müllerian inhibitor," causes the regression of the Müllerian ducts, precursors to the uterus, fallopian tubes, cervix, and upper third of the vagina. Androgen action alone is insufficient to attain this development hallmark, since treatment of gonadectomized fetal animals with testosterone crystals produced newborns with retained Müllerian structures. More than three decades after this initial observation was published, several investigators identified this compound as a 140 kDa glycoprotein homodimer (Picard et al., 1978; Swann et al., 1979; Budzik et al., 1983; Donahoe et al., 1984; Picard and Josso, 1984) produced by Sertoli cells (Blanchard and Josso, 1974) of the fetal, neonatal, and adult testis (Donahoe et al., 1977b). The gene for the protein, now called Müllerian inhibiting substance (MIS) or anti-Müllerian hormone (AMH), has been cloned from a variety of mammalian species (Cate et al., 1986a; Picard et al., 1986; Münsterberg and Lovell-Badge, 1991; Haqq et al., 1992), and much work has been done to elucidate its molecular mechanism of action. Most recently candidate type I and II receptors for MIS have been cloned from rabbit (di Clemente et al., 1994), human (Imbeaud et al., 1995), and rat (He et al., 1993; Baarends et al., 1994; Teixeira et al., 1996), thus opening up an exciting new era in the examination of the action of this naturally occurring growth suppressor. Using specific MIS receptor probes it will also be possible to test the hypothesis, generated from a large body of preliminary in vitro and in vivo data (Donahoe et al., 1979, 1981, 1984; Fuller et al., 1982, 1985; Chin et al., 1991; Parry et al., 1992; Boveri et al., 1993), that MIS may be a tissue-specific, nontoxic therapeutic agent for tumors arising from Müllerian precursors (Cate et al., 1986b).

Although early workers believed MIS to be a male-specific protein, later studies demonstrated production of the protein by granulosa cells in the ovary (Vigier et al., 1984; Takahashi et al., 1986 a,b; Ueno et al., 1988, 1989a,b; Seifer, et al., 1993; Hirobe et al., 1994). Furthermore, evidence exists that MIS in the female is one of the regulators of oocyte maturation (Takahashi et al., 1986b; Ueno et al., 1988) and that, in vitro at least, MIS inhibits granulosa cell division and progesterone synthesis (Hirobe et al., 1992; Kim, et al., 1992). Another non-Müllerian target for MIS action in the fetus is suggested by the work of Catlin and colleagues, who demonstrated MIS inhibition of surfactant production in developing lung (Catlin et al., 1988, 1991), providing a possible explanation for the male preponderance of respiratory distress in the premature infant (Catlin et al., 1992).

A summary of the current understanding of MIS activity will be provided below, with an emphasis upon relevant clinical correlations whenever possible.

MÜLLERIAN-INHIBITING SUBSTANCE PROTEIN STRUCTURE

Experiments in our laboratories at the Massachusetts General Hospital and in Paris, at the Enfant Malade, were successful in identifying a 140 kDa glycoprotein disulfide–linked homodimer from bovine testes, called Müllerian inhibiting substance (MIS) by our group (Donahoe et al., 1977a,b; Swann et al., 1979; Budzik et al., 1983) and anti-Müllerian hormone (AMH) by the investigators in Paris (Blanchard and Josso, 1974; Picard et al., 1978; Picard and Josso, 1984). The protein is secreted by fetal Sertoli cells of the newly formed testes (Hayashi et al., 1984) and causes Müllerian duct regression in normal male embryos as measured in elegant fetal rat urogential ridge in vitro bioassays developed by each group (Donahoe et al., 1977a) based on the original assay of Picon (1969).

Subsequently, the bovine (Cate et al., 1986a), mouse (Münsterberg and Lovell-Badge, 1991), and rat cDNAs (Haqq et al., 1992) and the human genomic sequences were identified and cloned (Cate et al., 1986a). The deduced amino acid sequence showed the protein to be divided into two structural domains: an amino-terminal region with nearly 70% homology among species, and the essentially invariant 108-amino-acid carboxy-terminal region that can be produced by proteolytic processing with plasmin (Pepinsky et al., 1988). Recent studies show that the biological activity of human recombinant MIS resides in the carboxy-terminal dimer (MacLaughlin et al., 1992a). Two consensus N-linked glycosylation sites are found in the amino-terminal domain, and although the structures of the carbohydrate side chains of MIS remain to be solved, it is known that they comprise nearly 15% of the total molecular weight of the protein. Since isolated MIS carboxy-terminus causes complete regression of Müllerian ducts in vitro (MacLaughlin et al, 1992a), the amino-terminus is not absolutely required for biological activity. There is evidence, however, that the amino-terminus can augment the activity of the carboxy terminus, suggesting it plays a role in maintaining conformation or bioavailability of the bioactive domain (Wilson et al., 1993). The importance of the cleavage of MIS for bioactivity was confirmed by producing mutants of the MIS molecule. Using site-directed mutagenesis, a noncleavable species of the recombinant human protein was prepared by replacing a threonine residue for the arginine at the basic cleavage site (Cate et al., 1990). The resulting protein was inactive in the standard MIS bioassay (Cate et al., 1990) and in in vitro antiproliferation studies (Boveri et al., 1993). Conversely, if proteolytic cleavage is made more efficient by creating a dibasic ARG-ARG site in place of the existing ARG-SER sequence, MIS activity is enhanced (Kurian et al., 1995). The protease responsive for the bioactivation of MIS in vivo has not been identified, nor has the site of cleavage been established. A reasonable case can be made that cleavage occurs in the testes or ovaries due to the action of a biosynthetic protease of the Kex family of enzymes originally purified from yeast (Leibowitz and Wickner, 1976) and shown to be in the testis (Seidah et al., 1992; Torii et al., 1993) and to possess substrate specificities similar to those needed for the MIS cleavage. These proteins, also called pro-hormone convertases, have been shown in a number of in vitro settings to cleave and thereby activate members of the transforming growth factor beta (TGF-β) family (Dubois et al., 1995) and other proteins with similar mono- or dibasic cleavage sites (Oda et al., 1992; Yanagita et al., 1992; Alacron et al., 1994; Bravo et al., 1994; Liu et al., 1995).

MIS is synthesized with a 25-amino-acid leader sequence, the first two thirds of which constitutes a signal peptide that when cleaved produces a pro-sequence of MIS. Current data suggest that MIS is not stored in the Sertoli or granulosa cells but is secreted by a constitutive mechanism.

Solving the primary structure of MIS allowed comparisons to be made with other proteins and identified significant sequence homology with members of the TGF-β family (Cate et al., 1986a) particularly in the carboxy-terminal domains of mature TGF-βs (Massague, 1990). A remarkable feature of all proteins in the TGF-β family is the presence of an ordered array of seven to nine cysteines and a consensus basic proteolytic cleavage site, which separates the amino- from the carboxy-terminus (Sha et al., 1989). X-ray crystallographic analysis of a TGF-β carboxy-terminus by two groups (Daopin et al., 1992; Schlunegger and Grutter, 1993) and more recently of OP-1 (Griffith et al., 1994) clearly demonstrated that cysteine residues form a "knot" of intrachain disulfide bonds similar to those detected in tumor neurosis factor (TNF) (Banner et al., 1993) and hCG (Lapthorn et al., 1994). A single interchain disulfide bond holds the dimeric, biologically active carboxy-terminus together. Although the MIS crystal structure has not yet been solved, predictions of its conformation based on homology with TGF-β suggest a similar three-dimensional configuration.

MIS GENE STRUCTURE

Purification of the bovine MIS molecule from neonatal testes was accomplished by a combination of conventional biochemical (Budzik et al., 1983; Picard and Josso, 1984) and immunoaffinity techniques (Shima et al., 1984; Cate et al., 1986a; Ragin et al., 1992). Sequence analysis of tryptic fragments of the purified protein provided investigators with the necessary information to construct degenerate oligonucleotide probes to screen cDNA libraries for the MIS mRNA (Cate et al., 1986a). These experiments, conducted more than a decade ago, culminated in the cloning of bovine and human cDNAs and the human genomic sequence of MIS and its transfection into CHO cells for the production of the recombinant protein now in use for current studies. The rat mouse and human genes, the former of which is found on chromosome 10 (King et al., 1991) and the latter on chromosome 19 (Cohen-Haguenauer et al., 1987), each contain five exons and four introns and are GC rich (65–75%) in both the introns and exons. The 5' untranslated pro-region of the human and bovine genes is 10 bases in length and the 3' UTR distal to the poly A region is 30 bases in length. Of the genes sequenced thus far only the human lacks a TATA box in the 5' of the transcription start site.

SEXUALLY DIMORPHIC MIS GENE ACTIVATION AND EXPRESSION

While it was appreciated that MIS is produced in male embryos just as the testis differentiates, thus allowing Müllerian duct regression to subsequently occur, it was perhaps not expected that MIS continues to be produced after the Müllerian ducts have regressed. Biological, immunohistochemical, quantitative immunological assays as well as northern and in situ hybridization assays confirm that MIS is synthesized, secreted, and measurable in the circulation well after birth and into adulthood in rats and humans (Necklaws et al., 1987; Hudson et al., 1990; Baker et al., 1990; Josso et al., 1990; Lee et al., 1996). Roles for postnatal MIS activity in the male remain unclear, particularly since recent MIS knockout studies conducted in mice showed that while the affected animals had retained Müllerian ducts (Behringer et al., 1994) other abnormalities were more subtle. Leydig cell hyperplasia and rare tumors were also found, and this phenotype was amplified in the MIS and inhibin double knockout, suggesting a Leydig cell suppressor function for both of these protein factors (Matzuk et al., 1995). The fact that serum MIS does not significantly drop in concentration until the peripubertal period, coincident with the resumption of spermatogenesis, suggests that MIS may be an inhibitor of that process.

Indirect evidence implies that MIS in the embryo is involved in testicular development and not merely as a regressor of Müllerian ducts. Taketo et al. (1991, 1993), analyzing a genetic mouse model of gonadal sex reversal, showed that in the ovotestes of these animals, the MIS expressed in the central testicular tissue correlates with meiotic arrest of the gonocytes, while in the more polar ovarian structures, loss of MIS expression correlates with advancement of gonocyte development and eventual loss of the germ cell pool. Thus, in the testes MIS may cause protective germ cell arrest as was previously seen in the immature ovary (Takahashi et al., 1986a,b; Ueno et al., 1988, 1989a,b).

It is now unequivocal that MIS is produced by females after birth (Vigier et al., 1984; Takahashi et al., 1986a,b; Ueno et al., 1988, 1989a,b; Seifer et al., 1993; Hirobe et al., 1994; Lee, et al., 1996). In humans MIS can be detected at extremely low levels in the serum of normal females by the age of 3 or 4 years but is not routinely measurable until after puberty (Lee et al., 1996). It is teleological that MIS can play a role in females during fetal development as Müllerian duct regression would not support procreation of the species, but a possible function for MIS in adult females was elucidated in oocytes harvested from immature rats. MIS added to cultures of denuded or cumulus enclosed oocytes reversibly inhibited premature germinal vesicle breakdown by holding the oocyte in meiotic arrest (Takahashi et al., 1986a,b; Ueno et al., 1988, 1989a,b).

With the development of human specific MIS serum assays (Hudson et al., 1990; Baker et al., 1990; Josso et al., 1990; Lee et al., 1996) it was possible to document the sexually dimorphic expression of the MIS gene. The levels of MIS in males actually shows a significant rise in the first 18 postnatal months, suggesting a release from maternal endocrine influences. Thereafter, the MIS levels remain consistently high until the males approach puberty, and testosterone production increases MIS levels fall to basal levels. Factors leading to the downregulation of MIS production remain to be elucidated. In females, MIS is not produced by granulosa cells until well after birth, ultimately reaching male levels in serum by puberty. Because of the differences in MIS production in males and females during the neonatal period, serum MIS measurements can be useful in the management of intersex anomalies in the newborn (Gustafson et al., 1993). In the adult, serum MIS may prove to be an important cell-type specific marker of granulosa cell and sex cord tumors (Gustafson et al., 1992).

The findings that MIS is produced by females as well as males and that the MIS gene is located on an autosome and not the X or Y chromosomes raises interesting questions about the role for MIS in females as well as how the gene is regulated. Our investigations have focused on identifying transcription factors and cis-acting elements specific for the MIS gene.

The promoter region of the MIS gene has a number of evolutionary conserved regions in the first 300 base pairs. Since the polyadenylation region of an snRNA gene, SPA 62, is found between −300 and −400 of the MIS start site (Hacker et al., 1995), it is likely that most of the promoter regulation resides in this 300-base-pair region immediately upstream of the start site. An evolutionary conserved region, which we originally called M1, has a GATA-containing motif. An M2 region, also conserved, has multiple half-sites related to the steroid hormone response element and binds truncated SF1 (Shen et al., 1994). A region called SRYe because of the HMG box of SRY footprints that area, suggested that SRY may be directly involved in the regulation of expression of MIS (Haqq et al., 1993). Developmentally this makes good sense, since there is overlap in the expression of SRY and MIS during early urogenital ridge development, and MIS is an early protein produced by the differentiated testis. Cotransfection of human SRY with the promoter region of the MIS gene demonstrated activation of a gene regulatory pathway leading to MIS expression (Haqq et al., 1994). In addition, SRY constructs containing mutations associated with human sex reversal failed in cotransfection assays to induce MIS transcription and HMG proteins with the same mutations showed abnormal interactions in band shift assays with ATTGTT, a high affinity binding site for SRY. Experiments are underway to determine whether SRY acts directly or indirectly on the MIS promoter to regulate expression of MIS. Because of the cyclic nature of ovarian expression and the abrupt decrease of expression in the pubertal testis, a search for specific proteins that suppress expression of MIS is also underway.

MIS RECEPTORS AND A MODEL MECHANISM OF MIS ACTION

The fact that essentially all of the actions of MIS in Müllerian and non-Müllerian tissues can be inhibited by epidermal growth factor (EGF) provides a framework in which to define the molecular mechanism of action of MIS. Experimental data shows the net effect of co-incubation of tissue, cells, or plasma membrane fractions of MIS-responsive cell types with MIS is the reduction of the autophosphorylation of the EGF receptor (Coughlin et al., 1987; Cigarroa et al., 1989; Maggard et al., 1996). Since the action of EGF begins with this catalytic event, it can be argued that MIS action is manifested by the inhibition of EGF activity. It remains to be established by what molecular mechanism the inhibition is accomplished. Recent data from our laboratory demonstrating that a specific tyrosine phosphatase is stimulated by the interaction of MIS with the plasma membrane of MIS responsive cells (Maggard et al., 1996) suggest a possible mechanism, which is being further investigated in our laboratory. It is our hypothesis that MIS interacts with its own receptor on Müllerian and other non-Müllerian MIS-responsive structures as evident from MIS binding studies (Catlin et al., 1992, 1993; MacLaughlin et al., 1992b). An early response is to block EGF activity via stimulation of tyrosine phosphatase. Subsequent MIS receptor downstream interactions with cytosolic proteins have begun to unravel the molecular mechanism that elicits the myriad responses attributed to this molecule.

Recently, cytosolic proteins that interact with the putative type I receptor for MIS (R1) have been uncovered using a protein-interaction trap technology in yeast and confirmed in mammalian cells with immunoprecipitation techniques. FKBP12 was found to bind to all type I receptors of the TGF-β family of proteins (Wang et al., 1994) and to be released by ligand binding, indicating that it functions as a growth inhibitor in the unbound state (Wang et al., unpublished). Since macrolides such as FK506 and it derivatives bind to FKBP12 at a binding site presumably shared with the type I receptors of the TGF-β receptor, they were used to study the functional significance of the downstream pathway. Interaction of these compounds with FKBP12 potentiates the activity of the ligand, suggesting that FKBP12 release is permissive for the activation of other downstream transducers. The putative MIS type I receptor cytosolic domain (He et al., 1993) also binds specifically to a novel Myc-like protein, termed Myx, indicating that the Myc pathway, known to be involved in apoptosis, a characteristic of Müllerian duct regression, may be part of the downstream signaling mechanism of MIS. It will be challenging to determine how these molecular events will coordinate with the observed MIS-mediated dephosphorylation of the EGF receptor via a specific phosphatase.

SUMMARY

Normal male phenotypic development is the product of the coordinated activity of the androgenic gonadal steroid hormone testosterone and its 5-α reduced metabolite, dihydrotestosterone, and the fetal Sertoli cell product Müllerian inhibiting substance. Following the initiation of testicular organization by SRY, the androgens are responsible for directing the maturation and differentiation of the Wolffian duct–derived structures while MIS ensures that the precursors of the female reproductive tract, the Müllerian ducts, are ablated. In this setting, the term activity refers to the end result of the properly timed synthesis and secretion of sufficient quantities of the hormones, their transport to target tissues, proper high affinity interaction with hormone-specific receptors, and the elaboration of downstream metabolic events. The androgen receptor, which must be present in the target tissues, is a ligand transcription regulator in the nucleus that induces (or inhibits) expression of particular genes. The specifics of MIS activity, however, are less well characterized. It is known that MIS is produced by and released from the newly differentiated Sertoli cells and that the MIS must be proteolytically cleaved to be bioactive. MIS then binds to a cell-surface macromolecule(s) and thereby reduces the EGF-induced tyrosine-phosphorylation state of the EGF receptor, thus diminishing the subsequent signaling pathways flowing from the kinase activity of the EGF-receptor complex. Current evidence suggests the receptor for MIS is also a protein kinase but with a specificity for serine and/or threonine, although substrates have not been detected. Since the discovery of cytoplasmic protein interactors with the putative MIS type I receptors, such as FKBP-12 and Myx, it is probable that MIS does more than simply block EGF-receptor signaling. The molecular nature of these MIS-specific downstream signaling pathway(s) is the subject of continuing research in our laboratory.

Cataloging the sequence of events required for normal phenotypic development in the male reproductive tract provides a logical framework for understanding the pathophysiology that could result from altering the subsequent pathways from MIS ligand binding. Since specific serum MIS assays are available and molecular probes exist for MIS, the putative type I and type II receptors, and the cytosolic partners of the type II receptors, it is possible to examine patients with retained Müllerian duct structures or ambiguous genitalia for reasonable explanations, as evidenced by variations in MIS levels or significant mutations in any of the pathway genes. More problematic is explaining the observed abnormal phenotypes when blood levels of MIS are within normal limits and genotypic analyses are normal. Presumably, defects in such cases exist downstream from the MIS-receptor interaction or from improper proteolytic processing of MIS, which could provide clues to further downstream interactors and regulated genes. The interplay of environmental toxins on this genetic pathway has yet to be examined, particularly in the ovary and testes, where cyclic variations leave windows for defects to occur in stressed repair pathways.

REFERENCES

Alarcon, C., Cheatham, B., Lincoln, B., Kahn, C. R., Siddle K., and Rhodes, C. J. (1994). *Biochem. J.*, *301*: 257.

Baarends, W. M., vanHelmond, M. J. L., vander Schoot, P. J. C. M., Hoogrtbrugge, J. W., de Winter, J. P., Uilenbroek, J. T. L., Karels, B., Wilming, L. G., Meijers, J. H. C., Themmen, A. P. N., and Grootegoed, J. A. (1994). *Development*, *120*: 189.

Baker, M. L., Metcalf, S. A., and Hutson, J. M. (1990). *J. Clin. Endocrinol. Metab.*, *70*: 11.

Banner, D. W., D'Arcy, A., Janes, W., Gentz, R., Schoenfeld, H. J., Broger, C., Loetscher, H., and Lesslauer, W. (1993). *Cell*, *73*: 431.

Behringer, R. R., Finegold, M. J., and Cate, R. (1994). *Cell*; *79*: 415.

Blanchard, M. G. and Josso, N. (1974). *Pediatr Res.*, *8*: 968.

Boveri, J. F., Parry, R. L., Ruffin, W. K., Gustafson, M. L., Lee, K. W., He, W. W., and Donahoe, P. K. (1993). *Int. J. Oncol.*, *2*: 135.

Bravo, D. A., Gleason, J. B., Sanchez, R. I., Roth, R. A., and Fuller R. S. (1994). *J. Biol. Chem.*, *269*: 25830.

Budzik G. P., Powell, S. M., Kamagata, S., and Donahoe, P. K. (1983). *Cell*, *34*: 307.

Cate, R. L., Mattaliano, R. J., Hession, C., Tizard, R., Farber, N. M., Cheung, A., Ninfa, E. G., Frey, A. Z., Gash, D. J., Chow, E. P., Fisher, R. A., Bertonis, J. M., Torres, G., Wallner, B. P., Ramachandran, K. L., Ragin, R. C., Manganaro, T. F., MacLaughlin, D. T., and Donahoe, P. K. (1986a). *Cell*, *45*: 685.

Cate, R. L., Ninfa, E. G., Pratt, D. J., MacLaughlin, D. T., and Donahoe, P. K. (1986b). *Cold Spring Harbor Symp.*, *51*: 641.

Catlin, E. A., Manganaro, T. F., and Donahoe, P. K. (1988). *Am. J. Obstet. Gynecol.*, *159*: 1299.

Catlin, E. A., Powell, S. M., Manganaro, T. F., Hudson, P. L., Ragin, R. C., Epstein, J., and Donahoe, P. K. (1990). *Ann. Rev. Respir. Dis.*, *141*: 466.

Catlin, E. A., Uitvlugt, N. D., Powell, D. M., Donahoe, P. K., and MacLaughlin, D. T. (1991). *Metabolism*, *40*: 1178.

Catlin, E. A., Ezzell, R., Donahoe, P. K., Manganaro, T. F., Ebb, R. G., and MacLaughlin, D. T. (1992). *Dev. Dynam.* (*formerly Am. J. Anat.*), *193*: 295.

Catlin, E. A., Ezzell, R. M., Donahoe, P. K., Gustafson, M. L., Son, E. V., and MacLaughlin, D. T. (1993). *Endocrinology*, *133*: 3007.

Chin, T., Parry, R. L., and Donahoe, P. K. (1991). *Cancer Res.*, *51*: 2101.

Cigarroa, F. G., Coughlin, J. P., Donahoe, P. K., White, M., Uitvlugt, N., and MacLaughlin, D. T. 1989). *Growth Factors*, *1*: 179.

Cohen-Haguenauer, H. O., Picard, J. Y., Mattei, M. G., Serero, S., Nguyen, V. C., de Trand, M. F., Guerrier, D., Hors-Cayla, M. D., Josso, N., and Frezal, J. (1987). *Cytogenet. Cell Genet.*, *44*: 2.

Coughlin, J. P., Donahoe, P. K., Budzik, G. P., and MacLaughlin, D. T. (1987). *Mol. Cell Endocrinol.*, *49*: 75.

Daopin, S., Piez, K. A., Ogawa, Y., and Davies, D. R. (1992). *Science.*, *257*: 369.

di Clemente, N., Wilson, C., Faure, E., Boussin, L., Carmillo, P., Tizard, R., Picard, J. Y., Vigier, B., Josso, N., and Cate, R. (1994). *Mol. Endocrinol.*, *8*: 1006.

Donahoe, P. K., Ito, Y., Marfatia, S., and Hendren, W. H. (1977a). *J. Surg. Res.*, *23*: 141.

Donahoe, P.K., Ito, Y., Price, J. M., and Hendren, W. H. (1977b). *Biol. Reprod.*, *16*: 238.

Donahoe, P. K., Swann, D. A., Hayashi, A., and Sullivan, M. D. (1979). *Science*, *205*: 913.

Donahoe, P. K., Fuller, A. F. Jr., Scully, R. E., Guy, S. R., and Budzik, G. P. (1981). *Ann. Surg.*, *194*: 472.

Donahoe, P. K., Budzik, G. P. Kamagata, S., Hudson, P. L., and Mudgett-Hunter, M. (1984). *Hybridoma*, *3*: 201.

Donahoe, P. K., Krane, I., Bodgen, A. E., Kamagata, S., and Budzik, G. P. (1984). *J. Pediatr. Surg.*, *19*: 863.

Dubois, C. M., Laprise, M. H., Blanchette, F., Gentry, L. E., and Leduc, R. (1995). *J. Biol. Chem.*, *270*: 10618.

Fuller, A. F. Jr., Guy, S. R., Budzik, G. P., and Donahoe, P. K. (1982). *J. Clin. Endo. Metab.*, *54*: 1051.

Fuller, A. F. Jr., Krane, I. M., Budzik, G. P., and Donahoe, P. K. (1985). *Gynecol. Oncol.*, *22*: 135.

Griffith, D. L., Oppermann, H., Rueger, D. C., Sampath, T. K., Tucker, R. F., and Carlson, W. D. (1994). *J. Mol. Biol.*, *16*: 657.

Gustafson, M. L., Lee, M. M., Scully, R. E., Moncure, A. C., Hirakawa, T., Goodman, A. K., Muntz, H. G., Donahoe, P. K., MacLaughlin, D. T., and Fuller, A. F. (1992). *N. Engl. J. Med.*, *326*: 466.

Gustafson, M. L., Lee, M. M., Asmundson, L., MacLaughlin, D. T., and Donahoe, P. K. (1993). *J. Pediatr. Surg.*, *28*: 439.

Hacker, A., Capel, B., Goodfellow, P., and Lovell-Badge, R. (1995). *Development*, *121*: 1603.

Haqq, C., Lee, M. M., Tizard, R., Wysk, M., DeMarinis, J., Donahoe, P. K., and Cate, R. (1992). *Genomics*, *12*: 665.

Haqq, C. M., King, C-Y., Donahoe, P. K., and Weiss, M. A. (1993). *Proc. Natl. Acad. Sci. USA*, *90*: 1097.

Haqq, C. M., King, C-Y., Ukiyama, E., Falsafi, S., Haqq, T. N., and Donahoe, P. K., and Weiss, M. A. (1994). *Science*, *266*: 1494.

Hayashi, M., Shima, H., Hayashi, K., Trelstad, R. L., and Donahoe PK. (1984). *J. Histochem. Cytochem.*, *32*: 649.

He, W. W., Gustafson, M. L., Hirobe, S., and Donahoe, P. K. (1993). *Dev. Dynam.*, *196*: 133.

Hirobe, S., He, W. W., Lee, M. M., and Donahoe, P. K. (1992). *Endrocrinology*, *131*: 854.

Hirobe, S., He, W. W., Gustafson, M. L., MacLaughlin, D. T. and Donahoe, P. K. (1994). *Biol. Reprod.*, *50*: 1238.

Hudson, P. L., Dougas, I., Donahoe, P. K., Cate, R. L., Epstein, J., Pepinsky, R. B., and MacLaughlin, D. T. (1990). *J. Clin. Endocrinol. Metab.*, *70*: 16.

Imbeaud, S., Faure, E., Lamarre, I., Mattie, M-G., di Clemente, N., Tizard, R., Carre-Eusebe, D., Belville, C., Tragethon, L., Tonkin, C., Nelson, J., McAuliffe, M., Bidart, J-M., Lababidi, A., Josso, N., Cate, R. L., and Picard, J. Y. (1995). *Nature Genet.*, *11*: 382.

Josso, N., Picard, J. Y., and Tran, D. (1977). *Recent Prog. Hormone Res.*, *33*: 117.

Josso, N., Legeai, L., Forest, M. G., Chaussain, J. L., and Brauner, R. (1990). *J. Clin. Endocrinol. Metab.*, *70*: 23.

Jost A. (1947). *Arch. Anat. Microsc. Mrophol. Exp.*, *36*: 271.

Kim, J. H., Seibel, M. M., MacLaughlin, D. T., et al. (1992). *J. Clin. Endocrinol. Metab.*, *7*: 911.

King, T. R., Lee, B. K., Behringer, R. R., and Eicher, E. M. (1991). *Genomics*, *11*: 273.

Kurian, M. S., de la Cuesta, R. S., Waneck, G. L., MacLaughlin, D. T., Manganaro, T. F., and Donahoe, P. K. (1995). *Clin. Cancer Res.*, *1*: 343.

Lapthorn, A. J., Harris, D. C., Littlejohn, A., Lustbader, J. W., Canfield, R. E., Machin, K. J., Morgan, F. J., and Issacs, N. W. (1994). *Nature*, *369*: 455.

Lee, ML, Donahoe, P. K., Hasegawa, T., Silverman, B., Crist, G. B., Best, S., Hasegawa, Y., Noto, R. A., Schoenfeld, D., MacLaughlin, D. T. (1996). *J. Clin. Endocrinol. Metab. 81*: in press.

Leibowitz, M. J., and Wickner, R. B. (1976). *Proc. Natl. Acad. Sci. USA*, *89*: 73.

Liu B, Goltzman D, Rabbani SA. (1995). *Am. J. Physiol. 268*: E832.

MacLaughlin, D. T., Hudson, P. L., Graciano, A. L., Kenneally, M. K., Ragin, R. C., Manganaro, T. F., and Donahoe, P. K. (1992a). *Endocrinology*, *131*: 291.

MacLaughlin, D. T., Levin, R. K., Catlin, E. A., Taylor, L. A., Preffer, F. I., and Donahoe, P. K. (1992b. *Hormone Metab. Res.*, *24*: 570.

Maggard, M. A., Catlin, E. A., Hudson, P. L., Donahoe, P. K., and MacLaughlin, D. T. (1996). *Metabolism*, *45*: 190.

Massague, J. (1990). *Ann. Rev. Cell Biol.*, *6*: 597.

Matzuk, M. M., Finegold, M. J., Mishina, Y., Bradley, A., and Behringer, R. R. (1995). *Mol. Endocrinol.*, *9*: 1337.

Münsterberg, A., and Lovell-Badge, R. (1991). *Development*, *113*: 613.

Oda, K., Misumi, Y., Ikehara, Y., Brennan, S. O., Hatsuzawa, K., and Nakayama, K. (1992). *Biochem. Biophys. Res. Commun.*, *189*: 1353.

Parry, R. L., Chin, T. W., Epstein, J., Powell, D. M., Hudson, P. L., and Donahoe, P. K. (1992). *Cancer Res.*, *52*: 1182.

Pepinsky, R. B., Sinclair, L. K., Chow, E. P., Mattaliano, R. J., Manganaro, T. F., Donahoe, P. K., and Cate, R. L. (1988). *J. Biol. Chem.*, *263*: 18961.

Picard, J. Y., Tran, D. and Josso, N. (1978). *Mol. Cell Endocrinol.*, *12*: 17.

Picard, J. Y., and Josso, N. (1984). *Mol. Cell Endocrinol.*, *34*: 23.

Picon, R. (1969). *Arch. Anat. Microsc. Morphol. Exp.*, *58*: 1.

Ragin, R. C., Donahoe, P. K., Kenneally, M. K., Ahmad, M., and MacLaughlin, D. T. (1992). *Prot. Expr. Purif.*, *3*: 236.

Schlunegger, M. P. and Grutter, M. G. (1993). *J. Mol. Biol.*, *231*: 445.

Seidah, N. G., Day, R., Hamelin, J., Gaspar, A., Collard, M. W., and and Chretien, M. (1992). *Mol. Endocrinol.*, *6*: 1559.

Seifer, D. B., MacLaughlin, D. T., Penzias, A. S., Behrman, H. R., Asmundson, L., Donahoe P. K., Hanning, Jr R. V., and Flynn, S. D. (1993). *J. Clin. Endocrinol. Metab.*, *76*: 711.

Sha, X., Brunner, A. M., Purchio, A. F., and Gentry, L. E. (1989). *Mol. Endocrinol.*, *3*: 1090.

Shen, W. H., Moore, C. C., Ikeda, Y., Parker, K. L., and Ingraham, H. A. (1994). *Cell*, *77*: 651.

Swann, D. A., Donahoe, P. K., Ito, Y., Morikawa, Y., and Hendren, W. H. (1979). *Dev. Biol.*, *69*: 73.

Takahashi, M., Hayashi, M., Manganaro, T. F. and Donahoe, P. K. (1986a). *Biol Reprod.*, *35*: 447.

Takahashi, M., Koide, S. S., and Donahoe, P. K. (1986b). *Mol Cell Endocrinol.*, *47*: 225.

Taketo, T., Saeed, J., Nishioka, Y., and Donahoe, P. K. (1991). *Dev. Biol.*, *146*: 386.

Taketo, T., Saeed, J., Manganaro, T. F., Takahashi, M., and Donahoe, P. K. (1993). *Biol. Reprod.*, *49*: 13.

Teixeira, J., He, W. W., Shah, P. C., Morikawa, N., Lee, M. M., Catlin, E. A., Hudson, P. L., Wing, J., MacLaughlin, D. T., and Donahoe, P. K. (1996). *Endocrinology*, *137*: 160.

Torii, S., Yamagishi, T., Murakami, K., and Nakayama, K. (1993). *FEBS Lett.*, *316*: 12.

Ueno, S., Manganaro, T. F., and Donahoe, P. K. (1988). *Endocrinology*, *123*: 1652.

Ueno, S., Takahashi, M., Manganaro, T. F., Ragin, R. C., and Donahoe, P. K. (1989a). *Endocrinology*, *124*: 1000.

Ueno, S., Kuroda, T., MacLaughlin, D. T., Ragin, R. C., Manganaro, T. F., and Donahoe, P. K. (1989b). *Endocrinology*, *125*: 1060.

Vigier, B., Picard, J. Y., Tran, D., Legeai, L., and Josso, N. (1984). *Endocrinology*, *114*: 1315.

Wang, T., Donahoe, P. K., and Zervos, A. J. (1994). *Science*, *265*: 674.

Wilson, C. A., di Clemente, N., Ehrenfels, C., Pepinsky, R. B., Josso, N., Vigier, B., and Cate, R. L. (1993). *Mol. Endocrinol.*, *7*: 247.

Yanagita, M., Nakayama, K., and Takeuchi, T. (1992). *FEBS Lett.*, *311*: 55.

2

Effects of Environmental Chemicals on Early Development

Rama S. Dwivedi and Philip M. Iannaccone
Northwestern University Medical School, Chicago, Illinois

INTRODUCTION

Exposure of the pregnant woman and her unborn child to environmental chemicals is a major public health concern. There are more than 3.5 million births every year in the USA (Conover, 1994), and approximately 4–6% of birth defects are estimated to be due to chemical agents in our environment (Wilson, 1977). In 1988 about 250,000 children in the United States were born with birth defects; 600,000 women experienced a miscarriage or fetal death (Chelimsky, 1991), and many young children were exposed to chemicals that will reduce their ability to develop the intellectual skills necessary to function by the twenty-first century. Apart from the voluminous information available regarding the effects of environmental chemicals on neonatal, perinatal (Mirkin and Singh, 1976), and postnatal periods of development (Hood, 1990; Schardein, 1993; Needleman and Bellinger, 1994; Sastry, 1995), very little information is available concerning chemical exposure in the preimplantation period of the embryo, when fundamental decisions effecting future development are made (Watson, 1992).

In the majority of cases when embryos are exposed to toxicants during the preimplantation period, they are refractory to the toxic insult. They either die or recover and develop normally. This type of all-or-none response towards the toxic insult is due to the unique regulatory ability of embryos to survive. The totipotency of the embryo's cells during this period of development plays a very important role in this adaptive response. It has been assumed that under such circumstances any adverse effects would be likely to damage either most or all of the embryonic cells before implantation to the same degree. If the damage is too severe, the embryo will be killed; if not, it can recover from the insult and may replace the damaged cells with other cells that have the potential to develop normally. The embryo can recover its normal size during the process of subsequent development. This all-or-none assumption does not seem justified now since the blastocyst stage of development has been demonstrated to be sensitive to drugs such as cyclophosphamide (Spielmann et al., 1982), heavy metals, and trypan blue (Lin and Monie, 1973). Studies from our laboratory (Iannaccone et al., 1982, 1987; Iannaccone, 1984; Bossert and Iannaccone, 1985) and elsewhere (Brunstrom, 1991;

Kaufmann and Armant, 1992; Kholkute et al., 1994a,b) have provided evidence that toxic insult during the preimplantation period can have a deleterious effect on the normal development of the embryo leading to fetal, neonatal death, dysmorphogenesis, or indeed poor outcomes after birth. The toxic effects of alkylnitrosoureas on pregnant rodents have also been demonstrated to lead to failure to implant, embryonic death, and decreased birth rate (Napalkov et al., 1968).

Embryos in the preimplantation period have therefore been shown to be susceptible to chemical insult (Spielmann et al., 1982; Generoso et al., 1988) depending on their developmental stage. It has been demonstrated that mouse embryos at the early blastocyst stage (day 3 following conception) are much more sensitive to the deleterious effects of actinomycin D compared with other stages of development. Cleavage of the developing embryo is prevented by actinomycin D at other stages, an effect not due to cell death (Epstein and Smith, 1978). This differential response is not the result of the rate or extent of chemical accumulation, since actinomycin D accumulates more rapidly and to a higher level in the less sensitive late blastocyst stage than during the more susceptible early blastocyst stage (Iannaccone et al., 1987). It has also been reported that the effects of methyl mercury are more deleterious on the blastocyst than on the morula stage of the embryo development. A higher concentration of methyl mercury exerts its acute effect by arresting the implantation of the embryo, while a lower concentration prevents the development of the inner cell mass (Matsumoto and Spindle, 1982). Exposure to methylnitrosourea does not stop the progression of the morula to the blastocyst stage at a concentration that causes blastocyst dysfunction (Iannaccone, 1984).

A differential response toward chemical insult within the blastocyst itself was also noticed between the inner cell mass and the trophectoderm cells. The inner cell mass as a whole was found more drastically affected following the parenteral administration of various substances such as diethylstilbestrol, antimitotic agents, polyfunctional alkylating agents, and purine and pyrimidine analogs to experimental animals during an early period of development (Adams et al., 1961). The distinctive susceptibility of the inner cell mass and trophectoderm cells is presumably the expression of differences in both type and rate of metabolic processes in these tissues. The differential susceptibility might also reflect the difference in survival potential of these two tissues. In this chapter we will discuss the adverse effects and mechanisms of action of environmental chemicals on the development of the embryo during the preimplantation period.

EARLY DEVELOPMENT

We will refer to early embryonic development as the period in mammals from fertilization to implantation of the embryos in the uterus, sometimes called the preimplantation period. The length of this period varies with different species: about 7 days in humans, 4 days in mice, and 5 days in rats. In many mammalian species, at this stage of development, the embryo is 100 μm in diameter, is associated with two polar bodies, and is surrounded by a chemically complex structure called the zona pellucida. The zona pellucida is a highly refractile, noncellular coat consisting of mucopolysaccharide and scialic acid residues. The zona pellucida is important to early development for a variety of reasons, although its removal does not prevent normal development. The zona provides a constraint to cleavage and ensures that as the blastomeres divide they remain together and in proper orientation. The zona also prevents the embryos from adhering to the wall of the oviduct.

Three phases of preimplantation development will be discussed. The earliest phase is the development from the fertilized egg to the two-cell zygote, which is regulated mainly by maternal message. The second phase is the cleavage stage from the two-cell stage to the formation of a blastocoel (cavitation), including the morula stage, and marks the beginning of the embryonic genome expression. The last phase is the formation of an expanded blastocyst with two distinct cell populations—the inner cell mass and the outer trophectoderm (Fig. 1). A brief outline of early development of the mouse embryo is presented next.

Oogenesis

The primordial germ cells of the female appear at the 8th day of gestation and migrate along the dorsal mesentery to the mesenteric folds and reach the germinal ridges by the 9th to 10th days. The characteristic features of the ovary are seen at the 11th day of gestation, and meiotic division by day 13 results in ovum formation. At the 14th and 15th days of gestation, leptotene and zygotene stages may be seen in oocytes with the beginning of oogonial division. By the 16th and 17th days of gestation, numerous pachytene stages are seen. Some primordial ova acquire follicle cells at the time of birth, and shortly afterward all of them have follicle cells. Primary oocytes are distinguished from other cells by their relatively large size and large spherical nucleus. About 3 days after birth the oocytes acquire a static state known as the dictyate stage, and by 5 days after birth all oocytes are in the diplotene stage of the prophase of the first meiotic division (Rugh, 1991).

Each oocyte is contained within a follicle and is surrounded by multiple layers of follicle cells, which play an important role in oocyte growth and differentiation. The follicle cells and oocytes are gradually separated by the deposition of the zona pellucida, a layer of extracellular material synthesized and deposited by growing oocytes. More than half of the primordial follicles present in the mouse ovary at birth degenerate by 3–5 weeks of age. The female mouse reaches sexual maturity at about 6–8 weeks of age, depending on the strain and associated environmental factors. By this time each ovary contains approximately 10,000 oocytes at different stages of maturity (Hogan et al., 1986). Once oogenesis has started, it continues regularly on 4.5- to 5-day cycles throughout the reproductive life of the mouse until about 12–14 months of age. The mature ovum is slightly polarized due to the presence of a cortical zone of RNA on the presumptive dorsal side and some vacuoles on the presumptive ventral side. At each estrous approximately 1000 or more ova are released for oogenesis, but only 1% of these ever mature. This means that in an average reproductive life of 2–12 months of age, some 60,000 ova are available, of which 500 mature and only about 100 can result in offspring (Rugh, 1991).

Ovulation

Ovulation is the release of mature ova from their follicles into the periovarial space following the coordinated response of both the follicle cells and the oocytes in each cycle. Only a few follicles respond to the increase in the level of follicle stimulating

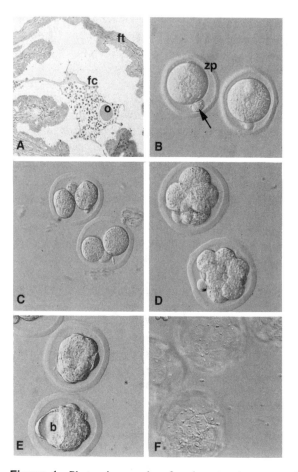

Figure 1 Photomicrographs of early rat and mouse embryos. (A) Rat oviduct immediately postovulation. The ovum (o) is surrounded by follicular cells (fc), which have aided in egg obtainment by the oviduct (Fallopian tube, ft) through interactions with the epithelial lining. The ovum (o) is about 100 μm in diameter. H+E stained fixed tissue section. (B) The mouse one-cell stage about 14 h after fertilization. The arrow indicates one of the polar bodies, which is in the perivitelline space defined by the zona pellucida (zp). A pronucleus is apparent in the center of the fertilized egg which is about 100 μm in diameter. (C) The two-cell, four-cell (D, top), and eight-cell (D, bottom) cleavage stages are all about 100 μm in diameter. (E) The blastocyst stage shows the formation of a blastocoel (b). The embryos at this stage are still about 100 μm in diameter, although they become somewhat larger at the late blastocyst stage (F), when they are expanded and the inner cell mass has become small and hard to see. The outer trophectoderm is most visible in the embryos in F as pavemented epithelial-like cells. B–F were living embryos as photographed with Hoffman modulation contrast optics. (Modified from Iannaccone, 1990).

hormone (FSH), which is produced by the pituitary. The stimulated follicle cells break contact with the oocyte and increase their synthesis and secretion of high molecular weight proteoglycan and tissue plasminogen activator (TPA). At the same time the follicle accumulates fluid, swells and moves toward the periphery of the ovary, ready for the final maturation and release. Ovulation occurs in response to a surge in the level of leutinizing hormone (LH), which is produced by the pituitary. Followed by LH stimulation the oocyte undergoes nuclear maturation and meiotic division. One set of chromosomes, surrounded by a small amount of cytoplasm, is extruded as the first polar body. It is in this stage that the oocyte is finally released from the follicle (Hogan et al., 1986).

Each ovulated oocyte is surrounded by its zona and a mass of follicle cells (cumulus mass). The eggs are swept toward the open end of the oviduct, or infundibulum, by the coordinated action of numerous cilia on the surface of the epithelium. In a natural ovulation, 8–12 eggs are released over a period of 2–3 h. Prepubertal mice can be induced to ovulate by treatment with hormones and superovulation can occur producing up to 100 ova. After ovulation the follicle cells differentiate into steroid-secreting cells, which help to maintain the pregnancy. The mature ovum has a diameter of about ~95 μm and a volume of about 200,000 μm³ (Rugh, 1991).

Fertilization

Fertilization is the union of male and female pronuclei followed by the initiation of a series of cellular transformations (Ziomek, 1987). Sperm are released into the reproductive tract of the female and reach to the ampulla of the oviduct within 5 minutes; however, these are not capable of fertilization for about an hour. This process of maturation is known as capacitation, which induces hypermotility and prepares the sperm for the acrosomal reaction. Activation stimulates the ovum to complete its interrupted second meiotic division, by which it becomes haploid. To reach to the surface of the ovum the sperm first penetrates the cumulus mass and then actively binds to the zona pellucida. An ~83 kDa glycoprotein, ZP3, isolated from the zona pellucida (Saling, 1989) is believed to bind the mouse sperm with the help of three adhesive proteins on the sperm cell membrane and initiates the acrosomal reaction. The molecular mechanism (s) by which zona and sperm recognize each other is not clearly known. Following the acrosomal reaction at its head, sperm release a variety of hydrolytic enzymes from the space between the membrane of the sperm head and sperm nucleus. It is believed that the acrosomal reaction is a very important trigger of fertilization. Fusion of the additional sperm to the ovum is probably controlled by a change in the surface of the ovum and the release of some substances that reduce the penetrability of other sperm to the zona pellucida. Another process that prevents the polyspermy is the Ca^{2+}-dependent release of cortical granules.

An active spermatozoa in the ampulla of the oviduct retains its capacity to fertilize the egg for 8 h, and generally all eggs of a female are fertilized within 6 h after mating. After fertilization the zona pellucida expands and the ovum shrinks resulting a perivitelline space between the ovum and the zona pellucida. The first polar body is visible in this space. Both male and female pronuclei may be seen as early as 6 h after mating. Fusion of the posterior part of the sperm head with egg membrane triggers a cascade of reactions. During fertilization, the head, midpiece, and a large part of the tail of sperm are incorporated into egg cytoplasm. Fertilization triggers the second meiotic division and the release of the second polar body. Haploid male and female pronuclei move toward the center of the egg, and DNA replication takes place during this migration. The pronuclei do not fuse, but the membrane breaks down and the chromosomes assemble on the spindle. Fertilization is considered to be complete when the pronuclei are almost contiguous and the first cleavage spindle is forming.

Cleavage

DNA replication in the one-cell zygote and the first cleavage leading to the two-cell stage seems to be under the control of the maternal genome in the mouse embryo (Telford et al., 1990). Either the embryonic genome is inactive or its activity is irrelevant to the next developmental stage. Experimental evidence has demonstrated that a switch from maternal to embryonic control occurs at the two-cell stage. The activity of RNA polymerase II has not been found in the one-cell zygote, while it is present in two-cell embryos (Moore, 1975). In other mammals the change in control varies from 8- to 16-cell morula. The cleaving embryo is held in the zona pellucida, which prevents it from adhering to the oviduct wall. Cleavage in mammals is a relatively slow process, varying from 12 to 24 h following fertilization (Gilbert, 1994). The first cleavage producing two cells begins in the ampulla of the oviduct of most mammals, including humans, and continues until the embryo moves into the uterus. The dividing embryo moves through the oviduct as a result of a combined ciliary motion of the oviductal epithelium and direction neutral muscular motion. While maternal factors might affect cleavage, the process is inherently under the control of the embryo.

In vitro experimentation suggests that environmental factors in the oviduct significantly contribute to the initial process of cleavage. Arrangement of the spindle and asters followed by an elongation of the fertilized ovum lead towards the process of early cleavage. The chromatin contents of the two pronuclei are not mixed before the first cleavage, but soon after they combine with each other. The amount of DNA in each nucleus doubles before each division, and RNA and protein synthesis is enhanced significantly during cleavage. The spermatozoa provides the kinetic center in its proximal centriole and might affect the time for the cleavage. The time lapse between the first and second cleavages is variable and depends on the strain of the mice (Rugh, 1991). Successive cleavages occur after shorter and shorter intervals and become asynchronous. The second cleavage to produce four cells occurs about 37 h following fertilization, the third to produce eight cells about 47 hour after fertilization.

Compaction

During the process of early cleavage, the embryo undergoes compaction normally at the late eight-cell morula stage and results in a spherical embryo. Individual blastomeres change shape and flatten against each other, increasing cell-to-cell contact. Compaction appears to involve three fundamental events: (1) cell surface recognition and the attachment of a ring of lateral microvilli to adjacent blastomeres, (2) subsequent shortening of the microfilaments in these microvilli, and (3) the maintenance of the compacted, polarized state, which results in apical localization of microvilli on each blastomere. Compaction seems to be reversible and can be reinduced. The polarization of microvilli can occur only once, however (Johnson and Ziomek, 1981).

Prior to compaction, sister blastomeres of mouse embryos exhibit ionic coupling and possess cytoplasmic bridges that admit both fluorescein and horseradish peroxidase, a large molecule unable to cross through gap junctions. Blastomeres from different mitotic pairs do not have demonstrable intercellular pathways. Intercellular communication mediated by gap junctions is first detected in the compacted eight-cell mouse embryos. At this stage, ionic coupling is observed and fluorescein injected into one cell is distributed to all eight cells of the embryo. Gap junction formation does not appear to require concurrent protein synthesis since the inhibition of protein synthesis by cyclohexamide treatment fails to block gap junction formation. The fourth cleavage to produce 16 cells 60 h after fertilization results in a solid ball of cells (Kalthoff, 1996). The increase in cell number from the 8 cell to the 16 cell stage necessitates that a certain number of cells occupy the interior of the morula and are completely separated from the outside by other cells. The cells create a distinct inside and outside microenvironment (Ducibella and Anderson, 1975) to which the cells respond. These microenvironmental differences due to geographical location are the basis for the differentiation of blastomeres to trophoblast and inner cell mass (ICM) according to the inside-out side hypothesis (Tarkowski and Wroblewska, 1967). Each blastomere of the four-cell embryo or early eight-cell embryo can form both trophectoderm and ICM of the blastocyst. According to the polarization hypothesis (Johnson and Maro, 1986), a positional signal is transmitted between blastomeres at the time of compaction to determine the axis of polarity involving the calcium-dependent cell adhesion glycoprotein "ovomorulin" or E-cadherin (Winkel et al., 1990). Those polarized cells on the outside will form trophectoderm and its derivatives, while the apolar cells on the inside will form ICM and its derivatives including the fetus.

Blastulation (Cavitation)

The late morula stage acquires an eccentrically placed, silt-like fluid-filled structure, the blastocoel. The process of blastulation begins with the 16- to 32-cell morula stage embryo. Following cycles of contraction and expansion, zonular tight junctions are formed between peripheral blastomeres resulting in a barrier that allows for the development of the blastocoel within the embryo. The following models have been proposed to account for this process:

1. The secretion cavitation model (Wiley and Eglitis, 1980) explains cavitation by the appearance of cytoplasmic droplets at the basolateral borders of the outer blastomeres, which contain the nascent blastocoel fluid.
2. The transporting cavitation model (Ducibella and Anderson, 1975) proposes that the transport of ions and water across the outer blastomeres is responsible for the origin of nascent blastomeric fluid, however, this model does not explain the appearance of cytoplasmic droplets at the beginning of blastocoel formation.
3. The metabolic activation model (Wiley, 1984) takes account of the two previous models and explains that the juxtaposition of mitochondria, lipid droplets, and the Na^+/K^+ ATPase located on the basolateral membrane is responsible for the production of blastocoel fluid. ATP produced by β-oxidation is utilized by ATPase to pump Na^+ out into the intercellular spaces. The intercellular space left by the exit of Na^+ is occupied by passive diffusion of water. At the end of the cavitation the blastocoel is formed with about 64 cells (mouse embryo).

The blastocoel enlarges as cells degenerate in this area, until the embryo resembles a hollow sphere. It comprises a single layer of large flat cells on the periphery called the trophectoderm or trophoblast and an eccentric aggregate known as the inner cell mass (ICM). The trophectoderm will develop into extraembryonic tissue, including the placenta, and is responsible for implantation, while the ICM will develop into the fetus. The blastocyst exhibits a high rate of respiratory metabolism, resulting in the process of the hatching and shedding of the zona pellucida. This action includes an enlargement and a rhythmic undulating movement. Single large contractions of the blastocyst can result in nearly complete disappearance of the zona pellucida. Large contractions occur every 6–8 h, and they are interspersed with smaller contractions every 20–100 min. Some contractions are slow (5–6 min), while others are relatively fast (15–20 s). At the time of hatching the blastocyst is no longer spherical but slightly ovoid and varies considerably in size (96–108 μm). At this point the blastocyst can contact the uterine epithelium for the attachment phase of implantation. Nucleaolar activity of the embryonic cells is accelerated during the blastocyst stage, synthesizing ribosomal RNA with a high guanine-cytosine–rich component.

Implantation

Implantation begins in the late blastocyst stage (64-cell stage). Following hatching from the zona pellucida, the blastocyst is ready for implantation. Hatching is brought about by a trypsin-like enzyme (strypsin), which is synthesized by cells in the mural trophoblast and by the expansion and contraction of the blastocoel as described above. There appears to be an estrogen surge at this time, just before the time of the implantation, initiated by the secretion of LH. Progesterone from the corpora lutea prepares the endometrium for implantation, while the levels of progesterone and estrogen are balanced for the maintenance of implantation (Rugh, 1991). Embryos will not be implanted if the uterus is not ready, and there is only very short time during which a transplanted blastocyst can survive in the uterus.

Implantation consists of the following steps: (1) adhesion and attachment of the trophoblast to the uterine mucosa where receptors on the trophectoderm are responsible for tight binding to endometrial epithelium, (2) invasion and spread of the embryonic components in the mucosa, and (3) endometrial response to the implanted embryo. In the case of humans, the trophectoderm cells invade through the basement membrane of the uterine epithelium to establish an implantation site in the stroma of the endometrium. In the mouse and rat, the embryo hatches out of the zona pellucida and then the blastocyst is engulfed by endometrium, resulting in an interstitial implantation. In a few Eutherian species (including humans, primates, and murine rodents), the uterine stromal cells undergo decidualization. The cells of the decidual swelling may be important in support of the pregnancy and in suppression of the immune rejection of the implanted embryo. Decidual prolactin may also be involved in the regulation of the amniotic fluid volume, and decidual luteotropin (a prolactin-like hormone in the rat) has been implicated in the maintenance of the corpus luteum.

The developing embryo lies within a crypt in the uterine mucosa on the ventral or antimesometrial side of the lumen. It is the uterus and not the blastocyst that determines the site for implantation. The factors that control the positioning of embryo implantation are not known. The onset of the degenerative changes in nearby embryos is coincidental with the rupture of the blood vessels in the area of the implanting blastocyst. The beginning of the implantation appears to be the release of a secretion from the trophoblast, which adheres to the mucosa. Leukocytes of the uterus rush to the site of the contact, and mucosal cells loosen and seem to degenerate or are digested away.

It is evident that the preimplantation embryo may be affected by environmental chemicals since exposures can occur via the mother's genital fluid, oviductal fluids, or uterine melieu (Agostoni, 1993). The preimplantation period is where fundamental developmental decisions are made that culminate in the formation of the blastocyst and lead to further development of the embryo. Important gene products such as growth factors, e.g., TGF-α and epidermal growth factor (EGF), tight junction protein-Zol, alpha and beta subunits of Na^+/K^+ ATPase, and uvomorulin, all contribute to the development of an early embryo (Watson, 1992) during the preimplantation stage and are potential targets of toxic injury.

Gastrulation

Gastrulation (Gk. *gaster*, "stomach") is the formation of three primary germ layers-ectoderm, endoderm, and mesoderm. Gastrulation involves delamination, cell differentiation, and cell movement. Following implantation, rapid growth and differentiation of tissues ensure. There is an expansion of the trophoblast population and an expansion of the endoderm to surround distally expanding embryonic ectoderm. Even at this point in development, major axes of asymmetry are not apparent. The two tissues—trophectoderm and inner cell mass—have become embryonic and extraembryonic ectoderm, embryonic and extraembryonic visceral endoderm, parietal endoderm, and trophoblast. The embryo expands distally and the embryonic ectoderm remains totipotent. About 6 days following fertilization in the mouse, presumptive dorsal-ventral and anterior-posterior axes are established. Gastrulation begins on the posterior aspect of the embryonic ectoderm where embryonic ectoderm and endoderm delineate with extrusion of mesoderm from ectoderm into the resulting space (Fig. 2) following instructive induction as with the gene *nodal* (Iannaccone et al., 1992; Zhou et al., 1993).

GENE EXPRESSION DURING PREIMPLANTATION DEVELOPMENT OF THE MOUSE EMBRYO

Earlier expression of three complexes of phosphoproteins has been demonstrated in the mouse zygote from maternal mRNA. Heat shock proteins, e.g., Hsp68, appear early in the two-cell embryo (Bensaude et al., 1983) when the major activation of the zygotic genome occurs. HSP70. 1 is among the earliest genes found to be expressed in

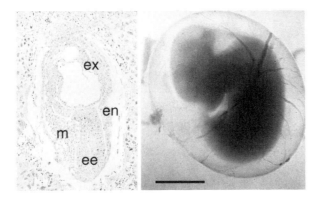

Figure 2 Photomicrograph of a postimplantation mouse embryo (left). The mesoderm (m) is developing from embryonic ectoderm (ee) and is surrounded by endoderm (en). The embryo quickly develops into a fetus with recognizable features (right). The fetus was photographed at day 12 of development. Bar = 2.5 mm. (Modified from Iannaccone et al., 1992.)

the two-cell stage mouse embryo (Thompson et al., 1995). A number of other proteins such as lamin A and B, laminin, uvomorulin, and gap-junction proteins have also been shown to be expressed during the preimplantation development of the embryo (Glover and Hames, 1989). An in vitro investigation of the growing blastocyst has shown the presence of EGF receptor on the surface of trophoblast. However, EGF is not produced by the mouse embryo until midgestation or late in development (Adamson and Meek, 1984). The mRNA for platelet-derived growth factor (PDGF) A and B forms are reported to be present during the gastrulation stage (Mercola and Stiles, 1988).

The development of asymmetry leading to body plan seems to involve coordinated expression of a large number of genes, many of which are highly conserved. Retroviral mutational studies (Iannaccone et al., 1992) have suggested the involvement of certain gene(s) in the regulation of growth and differentiation of cells, in particular, in the early egg cylinder stage of development of the mouse embryo. Subsequent experiments (Zhou et al., 1993) have identified a novel gene, *nodal*, which encodes a new member of the TGF-β superfamily. This plays a crucial role in mesoderm formation and subsequent organization of axial structures of the mouse embryo (Zhou et al., 1993). This gene was found to be essential for the formation of mesoderm, and the retroviral insertion within the nodal locus stops embryonic mesoderm formation in the mouse (Iannaccone et al., 1992). Later nodal seems important in a patterning leading to right-left asymmetry.

Patterns of gene expression for several antioxidative enzymes have also been examined using RT-PCR techniques during the preimplantation period of development (Harvey et al., 1995). Recent studies have reported an increase of $\sim$20-fold in thymidine kinase (TK) gene expression during the development of mouse embryo from two-cell to late blastocyst stage. RT-PCR analysis of the TK-mRNA have demonstrated a very low level of detectable TK-mRNA in one-cell and two-cell embryo, which sharply increased in day 5 embryo suggesting the regulation of TK gene expression at the transcriptional level during the preimplantation period of development (Lee et al., 1994). Likewise, the Na^+/K^+-ATPase (α-subunit) gene expression has also been reported in late morula and early blastocyst stages during the preimplantation development of the mouse embryo (MacPhee et al., 1994). Transcripts of catalase, CuZn superoxide dismutase (CuZn-SOD), Mn-SOD, glutathione peroxidase (GPX), and glutamyl cysteine synthetase have been detected in mouse embryos at all stages of preimplantation development both in vitro and in vivo. Expression of the genes encoding these antioxidative enzymes during the preimplantation period suggest the existence of a potential oxidative defense mechanism against oxidative stress during early development of the embryo.

Highly conserved genes from the *Drosophila* developmental hierarchy are known to be important in mammalian development. For example, homeobox (Hox) genes Hox-1.3(a5) and -2.1(b5) are detected by RNA analysis and Hox-1.5(a3) and -3.1(c8) by in situ hybridization at the gastrulation stage of development between $\sim$7.5 and 8.5 days postcoitus (Jackson et al., 1985). The homeobox is characterized by a 183 bp sequence, encoding a 61-amino-acid domain, shared between several homeotic genes of *Drosophila* (Gehring, 1987). The initial homeobox genes were confined to only two major homeotic gene clusters: the Antennapedia and Bithorax complexes (ANTP-C and BX-C). However, more divergent homeobox clusters containing numerous genes have been described (Rauskolb et al., 1993; Lints et al., 1996) during murine embryogenesis. Most of the members of the homeobox family share ~70% amino acid identity in the homeodomain, which is the DNA-binding domain of these transcription factors.

More than 20 mouse Hox genes in four clusters, such as Hox-1.4(a4), Hox-2.6(b4), and Hox-4.2/5.1(d4), have now been isolated that belong to the Antennapedia family. Transcripts are seen in embryonic ectoderm, mesoderm, and allantois. Each of the genes is expressed in a different domain along the anterior posterior (A/P) axis of the embryo where the anterior extent of the expression is strictly regulated. This type of pattern is very similar to the expression of *Drosophila* homeotic genes, where different members of ANTP-C and BX-C are expressed in the ectoderm and endoderm in discrete but overlapping domains. The mechanism of the regulation of Hox gene expression in mouse embryos is not clearly understood. For example, is heterodimerization of various gene products required for effective DNA binding? Site-directed mutagenesis experiments with murine *evx1* gene (Spyropoulos and Capecchi, 1994), a homolog of the *Drosophila* even-skipped (*eve*) gene and *Hlx* (Lints et al., 1996), a divergent homeobox gene, suggest the critical function of these genes in initiating the process of gastrulation or regulating the patterning/growth through the cell-cell signaling mechanism.

The "paired box," which also contains a homeobox, is found in a number of *Drosophila* segmentation genes and encodes a 126-amino-acid domain, with 80–90% identity between the different members of this family of mouse genes. *Pax-1*, for example (Deutsch et al., 1988), in mouse has 70% identity with *Drosophila* and is first detected at the 9th day postcoitus (dpc) and not involved in regulatory function. Pax-1 has not been detected during the early development of the mouse. Similarly, the zinc finger encoding genes of the Krüppel and hunchback family, which regulate the early development of the *Drosophila* embryo, are not detected in mouse during the preimplantation period. Recent in vitro experiments reveal that the disruption of GATA-4 genes (a family of zinc finger encoding genes) by targeted mutagenesis (Soudais et al., 1995) results in a specific block in visceral endoderm formation. The conserved genes of developmental hierarchies are obviously potential targets for environmentally based mutagenesis, and as gene networks are discovered in preimplantation development (e.g., *fgf* and its receptors), their susceptibilities to environmentally based damage will be a critical object of investigation.

EFFECTS OF ENVIRONMENTAL CHEMICALS

According to an estimation of the U.S. Environmental Protection Agency (EPA), there are approximately 55,000 chemicals in use with an estimated 1000 added every year (Schardein, 1993). This number could reach more than 80,000 by the end of this century. About 12,800 chemicals are produced in a quantity of more than 1 million lb/yr (U.S. Congress, 1985) and 50 in a quantity of over 1 billion lb/yr (Bergin and Grandon, 1984). More than 35,000 pesticides are registered with the EPA, and 3600 food additives are approved by the U.S. Food and Drug Administration (FDA) for routine use. More than 1500 chemicals are being incorporated in cosmetics (Lowrence, 1976), and over 1200 are contained in numerous household products. More than 700 contaminants, including pesticides, solvents, metals, and many others, have been found in our drinking water (Jacobson et al., 1991). Thus, an average American is consuming about 9 lb of different chemical additives every year apart from sugar and salt (Epstein et al., 1982) and is exposed to a variety of chemicals through regular household products, occupational involvement, and environmental contamination. The following information gives an account of the effects of environmental chemicals on the early stages of the developing embryo of mammalian species.

Metals

Metals are present in our environment because of their release from a wide variety of sources such as smelting of metallic ores, industrial fabrication, commercial applications of metals as well as burning of fossil fuels (Nriagu and Pacyna, 1988). Each year millions of tons of metals are produced from mines and subsequently redistributed in our biosphere, being dispersed in air, food, soil, and water (Landrigan, 1982). Environmental exposure to heavy metals is associated with a wide variety of toxicological manifestations with profound effects on the preimplantation embryos and in particular to the blastocysts (Storeng and Jansen, 1980; Matsumoto and Spindle, 1982; Izquierdo and Becker, 1982). Elements such as arsenic, cadmium, lead, and mercury are proven developmental toxicants with significant effects on embryonic and fetal development (Domingo, 1994).

Arsenic

Arsenic is widely present in the natural environmental and is found in two valence forms. Naturally occurring arsenic is comparatively nontoxic compared to trivalent arsenic, which is an extremely toxic metal. Exposure to arsenic compounds may occur from sources such as herbicides, insecticides, rodenticides, paint pigments, wool preservatives, hair dyes, and tonics. Seafoods, pork, and liver have a high arsenic content. Environmental arsenic exposure has received attention because of the complications resulting following the ingestion of food or water containing arsenic (Franzblau and Lilis, 1989).

The effects of arsenic on the preimplantation stage of mouse embryo were assessed by exposing embryos to 100 μM of sodium arsenate. This dose of sodium arsenate was found to kill an embryo instantaneously (Muller et al., 1986). A lower dose of 1 μM almost completely inhibited the blastocyst formation. Concentrations below 0.1 μM did not show any appreciable effects on mouse embryos with respect to the endpoints examined. Other fetal anomalies resulting from i.p. administration of arsenate to pregnant mice have been reported (Hood et al., 1988), however, very little information is available with regard to its effect on preimplantation embryos. One case of human neonatal death due to maternal ingestion of arsenic has been described (Lugo et al., 1969).

Cadmium

Some effects related to cadmium toxicity were found to enhance the levels of lipid peroxidation indices and lower enzymatic activity of enzymes, such as superoxide dismutase and glutathione peroxidase, which prevent oxidative damage and maintain cellular defense mechanisms against oxidative stress (Manca et al., 1991). Alterations in membrane structure, function (Muller, 1986), nucleic acids (Coogan et al., 1992), energy metabolism (Pederson and Lin, 1978), and interference with metalloenzymes have also been suggested as results of cadmium-induced toxicity. Cadmium has also been shown to exert its toxic effects on preimplantation embryos by interfering with EGF metabolism and calcium mobilization, which are important for mouse preimplantation embryo development (Wiley et al., 1992).

Cadmium exposure is of particular interest since it is a component of cigarette smoke, which is found to be accumulated in the follicular fluid of pregnant women smokers (Kuhnert et al., 1988; Maximovich and Beyler, 1995). It has been proven to be embryotoxic (Yu and Chan, 1986; Andrews et al., 1991) and to have adverse effects on fertility, the early embryo, and the fetus. Animal model studies have demonstrated that cadmium is a developmental toxicant and produces defects in lung, brain, testes, eye, and palate, as well as intrauterine growth retardation and fetal death (Holt and Webb, 1987). Exposure of mouse preimplantation embryos at the two-cell stage of development to cadmium concentrations of 1–2.5 μM in vitro results in an adverse effect on blastocyst formation (Pederson and Lin, 1978; Peters et al., 1995a). The frequency of blastocyst formation and mean embryo cell number was decreased when exposed to a cadmium concentration of more than 2.5 μM in culture. A cadmium concentration of more than 2.5 μM stops blastocyst formation and inhibits embryonic development to less than two to three cell cycles (Peters et al., 1995a). The morula stage embryos were found to be less sensitive to cadmium treatment than later stages of development when cultured with 3.0–6.0 μM cadmium (Peters et al., 1995a). These results are consistent with previous studies, which also demonstrated that cadmium is likely to have more severe effects on the later stages of embryonic development (De et al., 1993). Cadmium administration on day 2 postcoitus resulted in a transient delay in implantation by increasing the vascular permeability at the site of implantation, which in turn hampered embryo identification of implantation sites. In contrast to this cadmium administration at day 4 caused pregnancy failure in all the mice examined. No implantation sites were detected following treatment with cadmium on day 5 postcoitus. A rapid accumulation and efflux of cadmium by blastocyst stage embryos compared to two-cell embryos partially explains increased cadmium induced toxicity at later preimplantation stages. The possible mechanism(s) to account for cadmium intoxication include its interference with metalloenzymes (Pederson and Lin, 1978), alteration in thiol proteins, alterations in DNA structure, altered membrane structure/function (Muller, 1986), and excessive oxidative damage (Manca et al., 1991).

Pretreatment of embryos for 24 h with antioxidant glutathione (GSH), prior to cadmium exposure was found to eliminate the cadmium toxicity at lower concentrations (Peters et al., 1995a). Amelioration of cadmium-induced toxicity by intracellular GSH has also been shown by many other laboratories (Chubatsu et al., 1992; Li et al., 1993). A significant improvement in the process of differentiation and proliferation of the embryos by prior treatment with GSH followed by cadmium intoxication clearly suggests the role of intracellular GSH in preventing cadmium-induced impairment during the early stages of embryo development. Conversely, an increased sensitivity to cadmium toxicity was observed in cells with low intracellular GSH concentration exposed to cadmium (Chan and Cherian, 1992).

The exact mechanism of GSH protection against cadmium-induced toxicity in early embryo development is not known, although it has been suggested that GSH could alter the cellular uptake of cadmium (Chan and Cherian, 1992) or inhibit cadmium binding to intracellular sulfhydryl and subsequent alteration in the cadmium-binding proteins and their enzymatic activity (Li et al., 1993). GSH could also be involved in preventing oxidative damage produced by cadmium intoxication and other specific effects on cadmium metabolism (Peters et al., 1995a). Recent studies have shown a high expression of metallothionein (MT) mRNA during the early stages of preimplantation embryo development (Peters et al., 1995b), although knockout mice lacking both MT-I and MT-II were found to reproduce normally suggesting that MT may not have the same important role in metal detoxification during early development (Master et al., 1994) as it does in the adult.

Lead

Lead poisoning is a serious pediatric health problem in many countries, including the United States. More than 800,000 people in this country have lead exposure in their workplace. The principal route of lead exposure is through food and beverages. Virtually all food contains lead (Schardein, 1993). Drinking water is the another important source of lead ingestion. Other sources include lead-based interior paints (banned since 1977), air from lead-containing auto exhausts, waterproofing, varnishes, lead dryers, gold processing, insecticides, wood preservatives, and as chemical catalysts in industry (Winter, 1979). The total lead intake for an average American is approximately 100–200 µg/day (NAS, 1972).

The effects of lead on growth and development were documented by the mid-1900s and are summarized by Cantarow and Trumper (1944): "It is generally agreed that if pregnancy does occur it is frequently characterized by miscarriage, intrauterine death of the fetus, premature birth and if living children are borne, they are usually smaller, weaker, slower in development and have a higher infant mortality." Recent studies confirm these observations (Needleman and Bellinger, 1994) in relation to lead exposure to pregnant women with blood lead levels exceeding 50 µg/dl.

The Clean Air Scientific Advisory Committee of the EPA in its review of the proposed National Ambient Air Quality Standards for Lead (1990) concluded that "blood lead levels above 10 µg/dl clearly warrant avoidance, especially for development of the adverse health effects in sensitive populations." The Agency for Toxic Substances and Disease Registry (1988) has also concluded recently that "the available evidence for a potential risk of developmental toxicity from lead exposure of the fetuses in pregnant women points towards a Pb-B level of 10 to 15 µg/dl, and perhaps even lower."

In vitro studies with mouse embryos have demonstrated that treatment of two-cell embryos with lead chloride in late G2-phase at 32 h postcoitus impaired the formation and hatching of blastocysts. When embryos were X-irradiated 1 hour after the lead treatment, more profound effects were noticed than with lead alone: cell proliferation was disturbed and a significant reduction in cell number was found on a per embryo basis due to increased cell death (Molls et al., 1983). Long-term exposure of male mice to inorganic lead (lead chloride) in the drinking water reduced their fertility and decreased implantation of embryos. It was found that spermatozoa from lead-exposed males had a significantly lower ability to fertilize mouse eggs than from unexposed males (Johansson et al., 1987). A decreased frequency of inner cell mass development was also noticed along with an increased frequency of delayed hatching of the blastocysts from the zona pellucida in the lead-treated group.

Reduced fertility, spontaneous fetal loss, and increased stillbirths have been reported among women with a relatively high dose of occupational lead exposure (Nordstorm et al., 1978). Lead has been shown to be teratogenic at high doses in a variety of rodents and avian model systems (Gerber et al., 1980), but data on human teratogenicity are limited. An association between increased prenatal lead levels and preterm delivery or reduced length of gestation has been reported in several studies (McMichael et al., 1986; Rothenberg et al., 1989).

Mercury

Sources of mercury contamination in the environment include waste discharges of chlorine and caustic soda-manufacturing plants, mercury catalyst used in industries, fungicides used in pulp and paper industry, pharmaceutical manufacturing by-products, and the burning of fossil fuels. These account for approximately 70% of the mercury contamination to the American environment. Other sources include medical and scientific wastes, naturally occurring geological formations, and the processing of raw ores containing mercury.

The most toxic form of mercury is methyl mercury. Human exposure to methyl mercury occurs because of its use as a fungicidal seed dressing (now banned) and consumption of fish contaminated with methyl mercury due to effluent discharge into the sea. Inorganic mercury discharged into rivers, lakes, and oceans undergoes microbial conversion to the methyl form of organic mercury by methanogenic bacteria present at the bottom of bodies of water. This form tends to accumulate in the aquatic food chain. The highest concentration of methyl mercury was found in predatory species such as snapper, pike, swordfish, tuna, and shark.

Methyl mercury crosses the placenta and has proven to be a potent teratogen. It induces a number of abnormalities, especially of the central nervous system. Administration of methyl mercury compounds to pregnant mice induced brain and jaw defects, cleft palate, and postbehavioral alterations. Methyl mercury is the cause of birth defects and neurological deficits in Minamata disease (Takeuchi, 1966). It has also been shown to be embryotoxic and teratogenic in golden hamsters (Harris et al., 1972; Hoskins and Hupp, 1978), rats (Hoskins and Hupp, 1978), and mice (Sanchez et al., 1993). A high incidence of resorptions and dead fetuses was noticed, whereas cleft palate was the most frequent malformation. In vitro experiments with methyl mercury have also demonstrated that it impedes the assembly of microtubules, a principal element of the mitotic spindle.

To assess the effects of methyl mercury on preimplantation embryos, late blastocysts were exposed to various concentrations of methyl mercury or mercuric chloride for 24 h followed by another 24 h of exposure in a mercury-free medium. The results demonstrate a significant inhibition in cell proliferation and protein synthesis of the blastocyst stage embryo following the exposure of mercury or methyl mercury chloride (Katayama et al., 1984).

Cobalt

There is no substantial information available implicating other metals in the induction of serious malformations as a result of exposure during the preimplantation period. Preimplantation losses have recently been reported following a 10-week administration of cobalt to male mice (Pedigo and Vernon, 1993). These losses were not due to an adverse effect on the preimplantation development of the embryo, but rather to decreases in sperm concentration and function in cobalt-treated male mice. This study concludes that cobalt has detrimental effects on the preimplantation embryo possibly through its toxicological effects on the male reproductive system. Enhanced deleterious and synergetic effects of cadmium, lead, arsenic, or mercuric compounds were found when preimplantation embryos were exposed to these metals in combination with radiation. Mercury had the greatest potential for enhanced radiation risk to aberrant morphological development and cell proliferation of the preimplantation embryo.

Radiation Effects

The effects of irradiation on preimplantation embryos have been investigated in several in vivo (Russell and Russell, 1954; Michel et al., 1979; Muller and Streffer, 1990) and in vitro (Streffer and Molls, 1987; Jung and Streffer, 1992; Streffer, 1993) studies. Most of these data demonstrate an impairment of the morphological development of the mouse embryo as a result of the irradiation. Prenatal doses of ionizing radiation between 50 and 500 rad were found to induce testicular hypoplasia and sterility. In vitro studies suggest that mouse oocytes are most sensitive to irradiation. A rapid change in germ cell structure characterized by the condensation of the chromosomes and damage to the nuclear envelope was noticed following the irradiation. Impaired gestation due to the reduced distensibility of the irradiated uterine musculature and abdominal cavity as well as uterine vascular insufficiency was observed among the survivors of Wilm's tumor where pregnant women were exposed to radiation doses of ~900–1100 rad (Lione, 1987).

It has been reported that the human embryo is most sensitive to the lethal effects of radiation during the preimplantation period of gestation (0–2 weeks) (Brent, 1972; Benture, 1990), probably due to the genetic damage (Lione, 1987). A dose-dependent effect of irradiation of two-cell mouse embryos (36 h postcoitus) was noted. The ED_{50} for blastocyst formation was 2.8 Gy, and for hatching 1.9 Gy. At 59 h postcoitus a significant accumulation of four- and eight-cell stage embryos can be seen in experimental groups compared to the control nonirradiated group indicating significant delay of development. The percentage of morula stage embryos was lower (40–62%) in the experimental irradiated group than in the control group (82%), indicating significant embryo lethality. Experiments with pregnant animals have shown that lethality is the major outcome of the exposure of the preimplantation embryos to ionizing radiation. A variety of malformations (Brent, 1989) have been described in fetuses that have aborted following irradiation of the mother during the early stages of the development, the first 2 months of the pregnancy (Jankowski, 1986).

Progressive degenerative morphological changes such as enlargement of the perivitelline space and reduction of embryonic dimension were observed in a dose-dependent manner when the blastocyst stage of the embryo was irradiated (Kohler et al., 1994). Effects of different fluorochromes, acridine orange (AO), ethidium bromide (EB), and propidium iodide (PI) were also investigated on early developmental stages of the embryo to determine whether they have any synergistic effects with radiation. ED_{50} values for blastocoel formation and hatching rate were (1) for AO, 0.49 and 0.23 µg/ml, and (2) for EB, 6.63 and 5.30 µg/ml. Hatching rates seem to be more severely affected by the combined exposure of x-rays and AO staining than blastocyst formation. However, synergistic modification of radiogenic effects were not significant. No effects were observed for PI on blastocoel formation and hatching.

Oxygen Toxicity and Oxidative Stress

The success rate of pregnancies in humans following in vitro fertilization and embryo transfer is extremely low (10–30% worldwide) (Droesch et al., 1988; Scott et al., 1989).

One reason for this low success rate could be the poor in vitro culture conditions due to a high atmospheric oxygen (224 μM) (Jones, 1985) for embryo culture compared to the physiological concentration (28–42 μM) (Schuchhardt, 1973). Oxygen intoxication is considered to be mediated by superoxide radicals (O_2-) produced as intermediates during enzymatic metabolic reactions catalyzed by oxidases such as xanthine oxidase, monoamine oxidase, and aldehyde oxidase (Fridovich, 1972). Superoxide radicals are spontaneously dismutated to produce hydrogen peroxide (H_2O_2) by superoxide dismutase and extremely toxic hydroxyl radicals (OH^-) in the Haber-Weiss reaction (Haber and Weiss, 1932). Both of these radicals O_2^- and OH^- have been reported to enhance cellular lipid peroxidation, enzyme inactivation, and DNA damage (Ding and Chan, 1984).

Experimental studies have recently demonstrated that of embryos cultured in 5% oxygen after 1-h transient exposure to atmospheric oxygen, 20% showed a significantly lower rate of blastulation than those cultured continuously at a 5% oxygen environment (Umaoka et al., 1992). Pabon et al. (1989) found the blastulation rate of mouse embryos under standard conditions (1.5%) was greatly increased to 28.5% if embryos were cultured under low-oxygen conditions. This rate was further accelerated to 75.2% when embryos were cultured under low-oxygen conditions in addition to superoxide dismutase (SOD; 500 μg/ml). In previous investigations we have also successfully demonstrated that in vitro damage of the mouse blastocyst caused by liver microsomes and NADPH, presumably because the superoxide radicals, could partially be prevented by butyl hydroxytoluene (BHT), an antioxidant treatment (Iannaccone, 1986). We also found that an addition of as little as 0.06 IU/ml of SOD to the incubation medium significantly prevented damage caused by superoxide radicals to the blastocysts. These studies, therefore, demonstrate that early stages of development of mouse embryos can be affected to a great extent by insults caused by an oxidative stress.

Oxidative stress has also been implicated in the "two-cell blockage" of morula development of the mouse embryo in vitro due to an increase in hydrogen peroxide levels and a subsequent decrease in intracellular GSH concentration (Nasr-esfahani et al., 1992). To understand further the involvement of oxidative stress and GSH during preimplantation developmental stages, two-cell and morula/blastocyst embryos were cultured for 15 min in the presence of 13.3 μM tertiary butyl hydroperoxide (tBH)—an oxidant. This was found to decrease the GSH concentration of the two-cell embryo by 75% and the blastocyst stage by 25% (Gardiner and Reed, 1994), with a concomitant increase in the equimolar concentration of oxidized GSH and protein mixed disulfides, indicating an oxidative stress on the developing embryos. Addition of GSH to the culture medium at the two-cell stage significantly improved the later stages of embryo development. An increase in protein mixed disulfides could also be correlated to the loss of functionality of some important proteins such as Na^+/K^+ ATPase, which might have detrimental effects on preimplantation embryo viability and development (Gardiner and Menino, 1993). Early development of the embryo is therefore very susceptible to reactive oxygen species capable of producing an oxidative stress.

Drug Metabolism During Early Development

The overall toxicity for early embryonic development may depend on the relative levels of formation and detoxification of toxic metabolites from xenobiotics (Juchau, 1989, 1990). Bioactivation of numerous xenobiotic agents into toxic metabolites in the tissues of a conceptus is a prerequisite for abnormal embryonic development. Therefore, the extent of activation into toxic metabolites and detoxification efficiency in these tissues may in part determine normal embryonic development during the preimplantation period in the face of toxic insult. Reports from our laboratory have demonstrated that the endometrial lining cells of the mouse uterus contain cytochrome-dependent mixed function oxidase activity (Iannaccone et al., 1983). Cytochrome P-450 is known to be the enzyme system principally responsible for bioactivating polycyclic aromatic hydrocarbons (PAHs) to reactive electrophilic species capable of covalent interaction with DNA. The stromal cells of the mouse uterine lining have also been recently reported to express PAH-inducible cytochrome P-450 activity (Savas et al., 1993). Several stromal cell populations that are responsive and nonresponsive to steroid hormones, estradiol, diethylstilbestrol, or progesterone have been isolated (Iannaccone, 1984).

Polychlorinated biphenyls

Polychlorinated biphenyls (PCBs) are a family of polycyclic synthetic hydrocarbons whose production in the United States started around 1929 as industrial chemicals for stable fluids, plasticizers, adhesives, and dielectric fluids in capacitors and transformers. Chemically PCBs are a mixture of approximately 209 possible chlorinated biphenyls congeners and traces of toxic impurities such as polychlorodibenzofurans (PCDF). It has been estimated that each year some 10 million lb of PCB escape into our environment through dumping, vaporization, spills, and leaks. PCB production was banned in the United States in 1976, since they were recognized as environmental pollutants (Miller, 1977). Up to this time approximately 400,000 tons of PCBs have been produced, 80% of which were released into the environment. Current use is restricted to insulating electrical material (Rogan, 1982) and is under the strict control of the Toxic Substances Control Act (TSCA) of 1976. Like DDT and other slowly degraded environmental contaminants, PCBs are concentrated in food chains. Because of their high lipid solubility PCBs tend to accumulate in adipose tissue once they have been absorbed. Detectable levels of PCBs can now be found in tissues from most fish, birds, and mammals, including humans.

Human exposure to PCBs usually occurs from aquatic or terrestrial food chains. Massive accidental exposure of the human population to PCBs has occurred on three occasions. According to available reports, about 3000–5000 people were exposed to PCBs in 1968 from the consumption of contaminated rice oils in Kyushu, Japan, and 40–60 babies were born showing a characteristic intrauterine intoxication. Severely intoxicated mothers had stillbirths. Those who ingested the oil developed a peculiar acne-like skin eruption (chloracne) known as rice oil disease or "Yusho." The mothers' intake was estimated to be as high as 839 μg/kg/day (Yamaguchi et al., 1971), and all the babies had dark brown ("cola") staining of the skin. An accidental outbreak occurred in 1973 in Michigan because of a packaging error at the Michigan Chemical Corporation. The food additive magnesium oxide under the trade name of "Nutrimaster" was replaced by hexabrominated biphenyl manufactured by the same company under the trade name of "Firemaster." A very similar outbreak as that in Japan occurred in Taiwan in the Taichung area in 1979, and the PCB exposure known as "Yucheng" was characterized by hyperpigmentation. Developmental delays and neurobehavioral problems were reported in children born with Yusho disease when offspring were reexamined at 7–9 years of age (Rogan, 1982).

Maternal consumption of PCB-contaminated fish in humans has been reported to cause reduced birth weight, diminished head circumference, and reduced gestational age (Swain, 1991). A series of reports have been published on neural developmental deficits in children exposed to PCBs in utero. Subacute or chronic exposures to PCBs pre- or postnatal or during the pregnancy have been found to be detrimental to reproductive functions in laboratory and wild mammalian species (Fuller and Hobson, 1990). A substantial amount of PCBs and related contaminants were reported in human follicular fluids, ovarian tissues, embryos, and fetuses. Hydroxylated and methyl sulfone derivatives of PCBs have also been found in uterine fluid of pregnant mice following exposure (Fuller and Hobson, 1990).

Aroclor

Recent studies have demonstrated that Aroclor-1254, an important congener of PCBs, adversely affected in vitro fertilization in mouse (Kholkute et al., 1994a). Cumulus masses containing oocytes and capacitated epididymal sperm were placed in a medium containing Aroclor-1254. In vitro fertilization rates were dramatically decreased 20–24 h following insemination. Sperm motility was, however, not affected. Decreased sperm counts and smaller male reproductive organs, decreased litter size, and embryo/fetal loss in experimental animals exposed with PCBs have been demonstrated.

Prolonged estrous cycles were noted in mice fed 0.8 mg/kg/day PCB for a 10-week period and in rats administered Aroclor-1254 (i.p., 10 mg/kg/day) for 6 weeks (Golub et al., 1991). Addition of Aroclor-1254 (10 µg/ml) to the embryo culture medium significantly decreased the probability of embryo growth to the four-cell stage assessed at 48 h and affected blastocoel development assessed at 96 h. PCBs therefore have the potential to affect the early developing embryo by direct toxic effects (Golub et al., 1991).

PCBs have also been reported in the past to decrease the frequency of implantation and litter size in mice and rats (Lindner et al., 1974; Orberg, 1978) and to cause reproductive failure in mink (Platonow and Karstad, 1973). An oral treatment of 375–500 mg/kg body weight of 2,2'-2CB to mice on gestation days 1–3 significantly delayed the implantation of embryos (Torok, 1978). Mortality of pups was markedly increased when female rats were fed varying doses of Aroclor-1254 (maximum dose, 269 ppm) from the time of mating to the time of weaning of the pups. Dose-related reduction of pup weight gain and delayed reflux ontogeny were observed at all doses tested. According to the studies performed on rodents, PCBs may induce general toxic effects in the offspring, however, they do not result in birth defects (Mantovani et al., 1993). Reduced and delayed implantation with impaired menstrual cycle have been observed following PCB exposures in rats.

2, 3, 7, 8-Tetrachlorodibenzo-p-dioxine

2, 3, 7, 8-Tetrachlorodibenzo-*p*-dioxine (TCDD) is one member of a family of structurally related polyhalogenated aromatic hydrocarbons and is considered a widespread environmental contaminant. TCDD is produced as a byproduct during the manufacturing of chlorinated phenols and their derivatives. It is also produced as a pyrolysis product during high-temperature combustion processes and chemical bleaching of pulp. TCDD is one of the most toxic compounds within its class of halogenated hydrocarbons, and its high toxicity is related to the unique position of the chlorine atoms at four lateral positions of the planar TCDD molecule (Environmental Protection Agency, 1989).

Two cohorts of occupational exposure to human populations, the Dow Chemical Midland workforce and the Monsanto Nitro workforce, have been described in the past (Silbergeld and Mattison, 1987). Offsite contamination by TCDD has also been reported in the Nitro and Midland plants by the U.S. EPA (Environmental Protection Agency, 1989). At high levels of exposure to humans, a skin disorder known as chloracne involving hyperplasia of the epithelial cells was observed. Human studies have failed to identify an association between TCDD exposure and evidence of developmental toxicity or resulting teratogenicity (Kimbrough et al., 1984).

Biological effects of TCDD in animals include carcinogenesis, immune dysfunction, hepatotoxicity, teratogenesis, and reproductive toxicity. Exposure of male rats to TCDD has been shown to alter the androgenic status of the animals, which could be responsible for a decrease in testis and epididymal weights and consequently a decreased number of sperm (Mably et al., 1992) in the cauda epididymis. However, these studies could not be reproduced by other investigators and in any other strains of rats. Very little is known about TCDD exposure effects in female rats. Reduced fertility, decreased litter size, gonadal effects, and some changes in estrous/menstrual cycle was noticed in female rats exposed to TCDD. In a recent study where rhesus monkeys were exposed to TCDD (5 and 25 ppt) in the diets for 4 years, a dose-dependent increase in the prevalence and severity of endometriosis was induced (Eskenazi and Kimmel, 1995) 10 years following the initiation of the experiments (Rier et al., 1993). TCDD exposure of mice during pregnancy resulted in severe alterations in embryos and fetal development including hydronephrosis, cleft palate, and thymic hypoplasia. Pre- and postimplantation embryo loss and other impairments resulting in differentiation of cultured blastocysts have also been demonstrated following TCDD exposure in mice (Blankenship et al., 1993).

The underlying mechanism of TCDD-induced alterations during the early developmental period is thought to be mediated by its binding to a cytosolic aryl hydrocarbon receptor (Ahr) since the presence of Ahr-mRNA and protein in mouse preimplantation embryos has recently been described (Peters and Wiley, 1995). This study also provides evidence that the preimplantation embryos have the capacity to transcribe the gene rather than inheriting the Ahr-mRNA maternally. Ahr is presumed to play an important role in embryonic cell differentiation and proliferation. The mechanism by which TCDD-like compounds cause adverse effects on development of the embryo is not clearly understood. TCDD binds to the Ah-receptor and the ligand receptor complex is then translocated to the nucleus where it binds to DNA, which in turn results in alterations in gene expression (Jones et al., 1985), and this may be responsible for TCDD-induced alterations during preimplantation development of the embryo.

Hexachlorobenzene

Hexachlorobenzene (HCB) was used as a seed grain antifungal from the 1940s until it was banned in 1970. Its production has been continued in large quantities as a byproduct in the manufacture of chlorinated solvents and as impurities in several pesticides. HCB continues to pose a risk to human health due to its continued entry into our environment. Data from animal studies demonstrated that HCB is both a developmental and a reproductive toxicant. It has been shown to cross to the placenta and to affect the developing embryos of mice, rats, and rabbits (Courtney et al., 1985). HCB treatment (10 mg/kg/day) of cynomolgus monkey for 90 days resulted in depressed ovulatory levels of estradiol with degenerative changes in the primordial follicles and germinal epithelium of the ovary, suggesting that HCB is a ovarian toxicant and induces premature ovarian failure (Foster, 1995). Apart from systemic toxicity and impairment during in vitro fertilization, a dose-dependent affect on early embryo cleavage was reported recently following HCB exposure in sexually mature cynomolgus monkeys (Jarrell et al., 1993).

Effects of AHH Induction

The growth of progesterone-responsive cell strains can be inhibited by estrone or estriol. These cells contain inducible aryl hydrocarbon hydroxylase (AHH) activity that catalyzes the bioactivation of benzo(*a*)pyrene to 3-hydroxy-benzo(*a*)pyrene. Several other cell strains that were isolated in the presence or absence of estradiol vary in response to estradiol with respect to cell growth. Estradiol can reduce the basal or the induced levels of AHH activity by TCDD. The pattern of induction by benzanthracene in these cells is different from TCDD, which implies that AHH induction depends on the type of inducer and may have different mechanism (s) (Okey and Vella, 1982; Stols and Iannaccone, 1985). The modulation of cytochrome P-450 activity by estrogen is associated with alterations in the pattern of microsomal proteins as revealed by denaturing SDS-PAGE electrophoresis (Iannaccone et al., 1987). The electrophoretic patterns of the microsomal proteins have also revealed that induction of AHH activity in mouse liver occurs via different isozymes than in the endometrium. Different isoforms of cytochrome P-450 were also found to be induced following benzanthracene and TCDD induction in endometrial cells. In addition to this, a 52.4 kDa protein induced by benzanthracene in the absence of estradiol could not be induced in the presence of estradiol. There are several possible explanations regarding this inducibility. One is that the induction of cytochrome P-450 by benzanthracene is altered in some cells by estrogen as a result of an interaction between their activated receptors, which direct a different form of isozyme induction. Another explanation is that the expression of the gene for the respective isoforms is directly affected by estrogen, or the proteins are post-transcriptionally modified in the presence of estrogen. Induction of uterine aromatase in the presence of the steroids may be viewed similarly (Tseng, 1984). Other laboratories have also subsequently reported that the human conceptus expresses cytochrome P-450–associated mixed function oxidase (MFO) and other related enzymes that participate in a xenobiotic biotransformation processes (Savas et al., 1993; Peters and Wiley, 1995).

Cigarette Smoke

Smoking has been known for many years to produce deleterious effects on the developing fetus. Frequent direct nicotine effects on fetal development were suspected for many decades. Abortion was reported in women working tobacco factories in Europe during 1868 (Ballantyne, 1902). Spontaneous abortions, stillbirths, premature births, low birth weight, and sudden infant death syndrome are associated with maternal cigarette smoking during pregnancy and are well documented in the literature (Streissguth et al., 1994). Cigarette smoke is a colloidal suspension of more than 2000 known compounds including nicotine, nitrosamine, tar, heavy metals (Cu, Cd, Hg, Ni) (Sastry and Janson, 1995), carbon particulates, carbon monoxide, hydrogen cyanide, nitrosamines, and polynuclear aromatic hydrocarbons (PAHs) (Claxton et al., 1989). Cigarette smoke has also been implicated in the premature rupture of the amniotic membrane, intrauterine growth retardation, congenital anomalies, and immature lung development. Nicotine is generally regarded as the principal alkaloid responsible for these pharmacodynamic effects. Nicotine in tobacco smoke particulates is about 100–2500 μg/cigarette. Nicotine has been shown to affect the development of the preimplantation embryo in rabbits by inhibiting DNA synthesis (Balling and Beier, 1985).

Administration of 7.5 mg of nicotine twice each day from pro-estrous through pregnancy in rats delayed the entry of ova into the uterus, implantation, and the subsequent developmental processes (Yoshinaga et al., 1979). It has been suggested that nicotine acts by delaying progesterone secretion, which is necessary to prepare the uterus for implantation. Tobacco-specific nitrosamines such as N-nitrosonornicotine (NNN), 4-(N-methyl-N-nitrosamino)4-(3-pyridyl)-1-butanal (NNA), and 4-(N-methyl-N-nitrosamino)1-(3-pyridyl)-1-butanone (NNK) have been reported to produce DNA damage by forming covalent DNA adducts in fetal tissues of hamsters (Rossignol et al. 1989). Smoking-induced cancer susceptibility has recently been suggested to be due to germline polymorphisms of the human cytochrome P-4501A1(CYP1A1) and glutathione S-transferase Mu-1 (GSTM1) genes (Weir, 1987). The genetic polymorphism of these two genes is considered as a risk factor for cigarette smoking associated with p53 mutations in lung cancer patients as a result of PAH from cigarette smoke metabolites.

N-Methyl-N-Nitrosourea

N-Methyl-N-nitrosourea (MNU) is a chemical analogue of the nicotine metabolites (tobacco-specific nitrosamines) and is a direct-acting mutagen (further metabolism is not required for its actions). We have studied the effects of in vitro exposure to MNU on mouse blastocyst development (Iannaccone et al., 1982). The results demonstrate that MNU exposure altered the rate of incorporation of precursors of macromolecule synthesis in the blastocysts. A dose-dependent decrease in incorporation of thymidine, uridine, and leucine was observed following MNU exposure. The resorption rate of MNU-exposed blastocysts when transferred to the uterine horns of the surrogate mothers was found to be significantly increased compared to untreated controls. A dose-dependent decrease in birth rate following exposure of blastocysts to MNU was correlated with the decreased incorporation of the macromolecule precursors in blastocysts. It has been shown that MNU treatment of the preimplantation stage embryo in vitro leads to other deleterious effects such as increased intrauterine mortality (Murray et al., 1979), abnormal implantations (Iannaccone et al., 1982), fetal growth retardation in the midgestational period, gross malformations, and an increased perinatal mortality of the exposed offspring (Iannaccone, 1984). Malformations include rotational malformation, limb deformities, microcephaly, hydrocephaly, and nasopharyngeal dysgenesis (Fig. 3). The effects of MNU exposure are embryo specific since the exposures were in vitro.

Results of our recent studies reveal that exposure of embryos to MNU during the preimplantation period leads to protein alterations in midgestational fetuses (Bossert et al., 1990). Two-dimensional gel electrophoresis of fetal proteins at day 12 following transfer of blastocysts exposed to MNU on day 3 postcoitus demonstrates that six proteins of blastocysts are altered following the MNU treatment. Five of these proteins

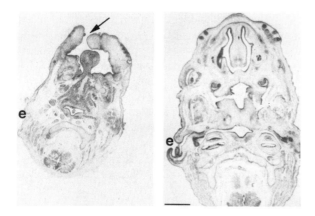

Figure 3 Photomicrographs of mouse fetus derived from blastocyst treated with 50 μg/ml MNU (right) or solvent (left) and examined at day 15 of gestation. Sections were taken from equivalent levels of the head. The external ear is present (e) in both sections. There are obvious abnormalities including failure of midline fusion (arrow) in the fetus that was derived from a blastocyst exposed in vitro to MNU. (Modified from Bossert and Iannaccone, 1985.)

show an alteration in pI. The pI alterations in proteins could be the result of either base changes resulting in amino acid substitution or posttranslational modification such as phosphorylation, glycosylation, acetylation, and methylation (Seeley and Faustman, 1995). Alterations at the protein level could be the consequence of mutational changes in regulatory genes at critical points during the process of development, and these alterations then lead to a poor fetal outcome (Iannaccone et al., 1987). These results suggest that MNU treatment of preimplantation stage embryos can alter the development program of the embryo with long-term effects extending even after birth (Bossert and Iannaccone, 1985). Recent studies with other alkylating agents on in vitro rat embryo differentiation and development suggest that the severity of the developmental toxicity of these agents is dose dependent (Seeley and Faustman, 1995). The dysmorphogenetic and mutagenic effects of MNU on the developing embryos are mediated by direct alkylation of DNA to form DNA adducts. This process is analogous to that of nicotine-specific nitrosamines (Hecht and Hoffmann, 1988) and thus the MNU model provides a useful system to study poor pregnancy outcome of smoking mothers.

Polynuclear Aromatic Hydrocarbons

PAHs such as benzo(a)pyrene are abundantly present in cigarette smoke and other environmental sources. They induce AHH (CYP1A1) activity favoring bioactivation and generation of toxic metabolites during the early phase of embryonic development when embryos are very sensitive to toxic insult. Spontaneous abortions and pregnancy loss have been reported in cigarette smoking pregnant women (Streissguth et al., 1994).

Results from our laboratory have demonstrated that endometrial stromal cells possess the ability to bioactivate benzo(a)pyrene to four enantiomeric forms by a P-450–dependent oxidation mechanism(s). These enantiomers are more chemically active than the parent benzo(a)pyrene (BP) compound and are capable of binding to DNA (Iannaccone et al., 1984). High-pressure liquid chromatography (HPLC) was used to demonstrate the site specificity of oxidation by cytochrome P-450 in microsomes isolated from estradiol-responsive mouse uterine cells. Induced microsomes from mouse endometrial stromal cells were incubated with tritium-labeled BP and their metabolites separated with HPLC. The resulting profiles demonstrated the predominant formation of *anti*-BP-7,8-dihydrodiol-9,10-epoxide (~80%) and smaller amounts of *syn*-BP-7,8-dihydrodiol-9,10-epoxide (~20%). The two most active enantiomers, *anti*- and *syn*-BP-7,8-dihydrodiol-9,10-epoxides (BPDE), when incubated with blastocysts, were found to disrupt blastocyst function (i.e., frequency of normal implantation rate and subsequent development of the preimplantation embryo) in a dose-dependent manner (Iannaccone et al., 1984).

These results show that endometrial stromal cells are capable of generating benzo(a)pyrene-diol epoxides from benzo(a)pyrene dihydrodiol. When embryos exposed to BP in vitro with endometrial cell microsomes were transferred to surrogate mothers and allowed to come to term, a dose-dependent decrease in live birth rate was observed. The metabolic specificity of this effect was determined by exposing the embryos to a racemic mixture of *anti*-BP-7, 8-dihydrodiol-9,10-epoxides. Following the insertion of embryos into the uteri of surrogate mothers, there was a significant dose-dependent decrease in implantation and birth rate and a dose-dependent increase in resorption rate. This particular metabolite has been demonstrated to have cytotoxic and mutagenic effects in mammalian cells but not in bacteria.

In order to establish the cytotoxic or mutagenic effects of these two enantiomers, blastocyst stage preimplantation embryos were exposed to racemic mixtures of either *anti*- or *syn*-BPDE at varying concentrations and transferred to the surrogate mothers. The data demonstrate that the *anti*-BPDEs were more effective in causing the failure of implantation, resorption of the fetus, and reduced live births when compared with *syn*-BPDE–exposed blastocysts. No significant alterations were noted on decidual swelling size of the blastocysts following their exposure to *anti*- or *syn*-BPDEs under equitoxic conditions, suggesting the mutational events are more likely responsible for these effects (Figs. 4, 5).

Figure 4 Effects of enantiomers of BPDE on mouse embryo development following brief exposure at the blastocyst stage. Open circles represent data from a 30-min exposure of blastocysts to *anti*-BPDE and subsequent transfer to surrogate mothers, while closed circles represent similar exposure to *syn*-BPDE. Top left: The percentage of exposed embryos that resulted in live births following the transfer to surrogate mothers. Each point represents data from 96–102 embryos. Top right: The percentage of exposed embryos that resulted in normal implantation sites determined 7 days following the transfer to surrogate mothers. Each point represents data from 95–103 embryos. Lower left: The percentage of exposed embryos that resulted in abnormal implantation sites determined 7 days following transfer to surrogate mothers. Each point represents data from 4 to 81 embryos. Lower right: The decidual swelling size in mm^3 determined by calculation from two axes measured 7 days following transfer of exposed embryos to surrogate mothers using the formula for the volume of an oblate spheroid. The BPDEs were prepared from stocks obtained from the NCI Chemical Repository. The BPDEs were stored in the dark in tetrahydrofuran:triethylamine (19:1) at $-20°C$, and the tetrol contamination in all preparations was determined by fluorescence (Geacintov et al., 1980) before each experiment. Dilutions were made under yellow light from BPDEs dissolved in acetone:NH_2OH (1000:1) and added to dishes containing the blastocysts in PB-1 as described previously (Iannaccone et al., 1984; Iannaccone, 1984; Bossert and Iannaccone, 1985; Bossert et al., 1990). *anti*-BPDE is a racemic mixture of 7β,8α-dihydroxy-9α,10α-epoxy-7,8,9,10-tetrahydrobenzo(a)pyrene and of 7α,8β-dihydroxy-9β,10β-epoxy-7,8,9,10 -tetrahydrobenzo(a)pyrene. *syn*-BPDE is a racemic mixture of r-7,*t*-8-dihydroxy-*c*-9,10-epoxy-7,8,9,10-tetrahydrobenzo(a)pyrene.

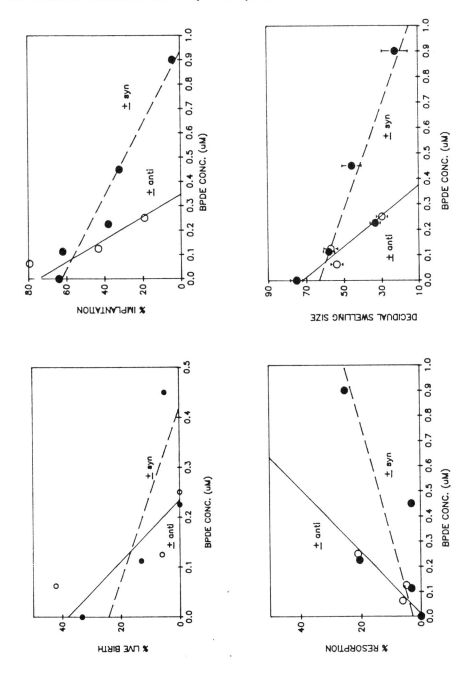

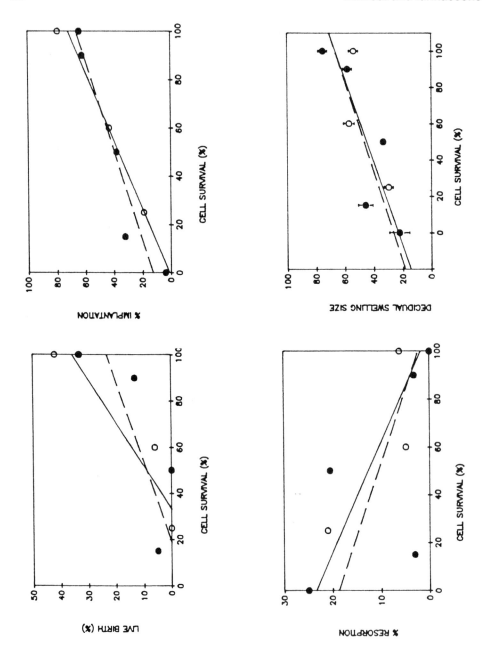

AHH (CYP1A1) activity, which is required for the bioactivation of PAHs, has been shown to be present in human placental tissues and is induced in human term placental tissues by exposure to cigarette smoke or other xenobiotics during pregnancy. It has also been shown that AHH activity is almost absent in nonsmoking women during pregnancy (Sanyal et al., 1993). The upregulation of AHH activity by PAH exposure in cigarette-smoking pregnant women produces toxic metabolites, which in turn are responsible for abnormal development of the pre- or postimplantation embryos. Deleterious effects of such toxic metabolites and possible mutations (Iannaccone et al., 1987) in dividing and differentiating embryonic cells significantly affect pregnancy outcome.

Pesticides

Pesticides are the most widely used chemicals in our environment. More than one billion pounds of pesticides are produced annually in the United States alone (Environmental Protection Agency, 1988). More than 1400 formulated products are registered by the EPA (Environmental Protection Agency, 1992). Eighty percent of all pesticides are used for agricultural purposes and pose a potential health risk problem to human populations. Occupational exposures in manufacturing these pesticides are a great concern. More than 200,000 persons were exposed, with 1754 deaths and 17,000 permanently disabled (Weir, 1987), following the accidental release of methyl isocyanate (MIC) from a carbamate pesticide plant in Bhopal, India, on December 2, 1984. The drinking water is another source of pesticide contamination. In the recent past The California State Water Resources Control Board verified 441 cases of groundwater contamination by 54 different pesticides in 28 counties. Additionally 2500 wells were found to be contaminated with 1,2-dibromo-3-chloropropane (DBCP), which has been banned from use in California since 1977 (Heindel et al., 1994). Effects of pre- or postconception exposure to pesticides includes sterility, embryo and fetal toxicity, abortion, stillbirth, teratogenicity, and developmental deficits (Shane, 1989). Prior-to-conception damage to germ cells has been reported in people exposed to pesticides in a case control study conducted from 1981 to 1986. A high prevalence of ancephaly and neural tube defects occurred from 1989 to 1991 in several Texas counties bordering Mexico (Canfield et al., 1996), presumably due to pesticide exposure.

Figure 5 Comparison of the effects of *anti-* or *syn-*BPDE at equitoxic concentrations on mouse blastocyst function following transfer to surrogate mothers. Legend as in Figure 4. At each concentration, the effect of treatment on the developmental parameter shown (live birth rate, implantation rate, resorption rate, or decidual swelling size) is plotted against the effect of that concentration on fibroblast survival [cell survival (%)] determined as previously described (Stevens et al., 1985). The lines are linear regression curves and the slopes of the curves from *anti* and *syn* BPDE for live birth rate, implantation rate, and resorption rate are significantly different, indicating that the effects of the stereoisomers on the parameter are different.

Stereoisomer of BPDE	Birth rate	Implantation rate	Resorption rate	Decidual swelling size
anti	0.53	0.70	−0.22	0.43
syn	0.29	0.53	−0.16	0.40

Toluene

Toluene is an aromatic hydrocarbon used in paints, lacquer, and the shoe industry. Exposure of human populations to toluene was first noticed through municipal water supplies where it was found at levels ranging from 0.1 to 11 µg/liter. Its metabolites, benzaldehyde and benzoic acids, were also detected in municipal water at a concentration of up to 19 µg/liter (Environmental Protection Agency, 1980). There is very little information available on toluene exposure to pregnant women or its effects on the growing fetus. Pregnant women who were exposed to toluene while working as shoemakers during the soling process gave birth to children with multiple malformations (Euler, 1967). Moreover, children born to mothers who intentionally inhaled a high dose of toluene during pregnancy were found to have congenital defects (Goodwin, 1988).

Impaired growth of the fetus and fetal skeletal anomalies were demonstrated following exposure of pregnant mice and rats to a high dose of toluene (Shigeta, 1982). Behavioral abnormalities have also been reported in mice exposed to toluene during fetal development. Significant embryonic mortality has been noticed in previous studies where toluene was administered to experimental mice by gavage. In a recent investigation, it was demonstrated that exposure of the developing embryos to toluene during the pre-implantation phase is very deleterious (Yelian and Dukelow, 1992). Significant degenerative changes were found in preimplantation embryos exposed to toluene in vitro at a dose higher than 8.67 µg/ml. Higher doses of toluene decreased the fertilization rate of exposed eggs and enhanced embryo degeneration resulting in increased embryonic lethality.

Nitrous Oxide

Epidemiological studies indicate that there is a strong correlation between occupational exposure to anesthetic gases, such as nitrous oxide, and spontaneous abortion (Rowland et al., 1995). Nitrous oxide exposure levels are higher in dental offices than in hospital operating rooms. In dental offices nitrous oxide is usually mixed with oxygen and is administered as a sedative agent, not as an anesthetic. Because of the operating inconvenience and nonavailability of appropriate scavenging equipment, nitrous oxide levels found in this setting are very high, up to 1000 ppm, approximately 40 times higher than the standard recommended by NIOSH (McGlothlin et al., 1988). It has been reported that exposure to nitrous oxide of more than 5 h per week is associated with reduced fertility in female dental associates.

Continuous exposure to nitrous oxide at a 1000 ppm or higher in pregnant laboratory animals (rats/mice) resulted in malformations, increased fetal resorptions, and decreased litter size (Fink et al., 1967). The exact mechanism for nitrous oxide–associated malformations during preimplantational development of the embryo is not understood, however, it has been hypothesized that nitrous oxide interferes with methionine synthetase, which is a vitamin B_{12}–dependent enzyme that plays an important role in DNA synthesis. It was demonstrated that DNA synthesis is affected in rat embryos treated with 50% nitrous oxide for 24 h (Hansen and Billing, 1986). This effect may cause impairment in embryonic development during the preimplantation stage.

Methanol

Methanol is generally produced during the manufacture of formaldehyde and is used in varnishes, paints, and stain remover as a solvent. Methanol may also be used as an alternate fuel. The major concern of methanol toxicity is its effect on the visual system of primates due to the formation of formic acid in an aqueous environment. Postnatal behavioral abnormalities are noticed in the pups of Evans rats following exposure on days 15–17 of gestation. Teratogenic effects were described at 20,000 ppm when administered to rats by inhalation during days 7–15 of pregnancy. Data from a recent in vivo investigation in which methanol was administered to female rats on days 1–8 of pregnancy demonstrate adverse effect on decidual growth (Cummings, 1993). On either side of the implantation site, extravasation of blood was clearly seen in methanol treated rats at day 9 of pregnancy. Thus, exposure to methanol during an early period of development impaired uterine decidualization as demonstrated by effects on the decidual cell response (DCR).

Chlorophenols

The chlorophenols (CPs) are environmental toxicants most often found in industrial and sewage effluents, soil, groundwater, and chlorinated drinking water (Zhao et al., 1995). These are either consumed directly by human beings via food ingestion or are formed by intermediary metabolism (e.g., from chlorobenzenes). CPs have been identified in the urine and blood of the American population. It is estimated that an average daily intake of CPs in the U.S. population is approximately 6 μg (National Research Council of Canada, 1982). Among the CPs, pentachlorophenols (C_5P) are the most toxic and were used up to the early 1970s as a component of wood preservatives. C_5Ps have been reported to have adverse effects on reproduction, embryonic, and fetal development and neonatal survival in rats, but there is not much information regarding its effects on preimplantation development.

CONCLUSIONS

The preimplantation period from fertilization until attachment to the uterine wall is a period of critical genetic determination. The interaction of extracellular and intracellular signals conspires to isolate an inside population of cells in the embryo, which by virtue of their position alone commits them to produce the fetus several days later. The feedback systems that sense the geographic location of blastomeres within the embryo and respond by creating polarized cells on the outside and apolar cells on the inside can compensate for some accidents, but clearly are not completely plastic. Such is also true of those signals that lead to compaction, the first visually apparent manifestation of this commitment. Among those cells inside and committed to the fetal lineages, there are certainly redundancies which must form the basis of adaptive response in mammals. The extent of plasticity with respect to cell number in this compartment is unknown. In fact, the entire preimplantation period is greatly understudied in this regard. The extent to which these cells are an "at-risk population" is essentially an unknown. This period is greatly underrepresented in studies of exposure risk from environmental compounds. This is indeed a pity, since, in addition to the critical nature of genetic determination in this period, in the human population the mother is largely unaware that a pregnancy exists in the preimplantation period. Thus, if significant risk from environmental exposures is documented, it is entirely plausible that by the time the pregnancy is known and the maternal environment is refined, significant damage to the offspring may have already occurred.

Most reproductive loss occurs during the preimplantation period, usually as a result of nondysjunctional events or other chromosomal aberrations. This undoubtedly represents a reproductive strategy among species where females are reproductively limiting and a pregnancy that fails early, if it is to fail, will reset the reproductive function quickly. Depending on the species more than 25 % of conceptions may be lost in this period and as a result exposure-based loss may be very hard to distinguish from "normal" wastage. This can be expected to make risk assessment in early developmental periods extremely difficult.

Our work has shown that mutagenic exposure in this period can have consequences to the development of the fetus derived from the exposed embryo long after the exposure, and indeed after birth. The possibility that subtle developmental defects may manifest themselves in behavioral or learning deficits is particularly alarming and again requires highly sophisticated epidemiological approaches to establish with certainty in human populations. The use of data emerging from heavily polluted environments, particularly in Eastern Europe, may provide important, if tragic, opportunities to study this issue.

Combined in vitro/in vivo methodologies in which exposures are carried out with isolated embryos under controlled conditions, with subsequent transfer to unexposed surrogate mothers for evaluation of development, are necessary to study the preimplantation stages since bioavailability to the embryo is so hard to assess when drugs are administered to the pregnant mother. Moreover, injury to the embryo cannot be ascribed as an embryonic effect if it could be the result of maternal effects secondary to injury to important maternal reproductive tissues. This is an important problem when one is trying to establish the mechanism of developmental dysfunction resulting from specific exposure.

The combined in vitro/in vivo development model makes it clear that implantation is not a gate through which the normal embryo may pass while the damaged embryo may not, but rather is like a threshold over which the embryo, if alive, will pass with both its accumulated genetic accomplishments and failures. The notion that implantation is an "all-or-nothing" phenomenon is particularly well rooted in the teratology literature, where following extensive radiation studies it took hold. Low-dose radiation exposure in the preimplantation period, i.e., doses low enough to avoid killing the embryo outright, may well fail to reveal a higher-resolution dose-response relationship in which mutagenetic effects do not prevent implantation yet do manifest as developmental defects later as the affected genes are required. Obviously, from the perspective of a teratologist if no offspring are born following an exposure to manifest a teratological result, the exposure cannot be considered teratological. In that context, low-dose radiation in the preimplantation period may well not be teratogenic, while mutagenic chemical exposure is.

An important hallmark of mammalian reproduction is adaptive response to injury. In the preimplantation period this is manifest as a robust ability to compensate for cell loss. It is established that in the mouse blastocyst no more than three to four ICM cells are required to give rise to the entire fetus. Since these cells are totipotent, it may not matter which three to four cells. There is clear evidence that the loss of large numbers of cells in early stages of development, particularly immediately following implantation, can be compensated for with rapid and accurate replacement of cell numbers. It has been argued that this compensatory growth is itself teratogenic. The issue has not been explored in the preimplantation period, although we do know that the embryo can compensate its size in the other direction very accurately by an unknown mechanism. That is, from chimeric embryos that have too many cells, a normal-sized fetus develops. Attempts to use stereoisomers with disproportionate mutagentic vs. cytotoxic attributes suggest that developmental deficits following exposure to PAHs are the result of mutations and not cell loss.

The new challenge is to determine which specific genes are involved in the risk of poor pregnancy outcome. The available evidence suggests that a relatively small number of genes are adversely affected under conditions that produce serious malformation at high frequency. In the end, though, an evaluation of the importance of understanding the relationship between exposure, specific mutations, and specific developmental deficits needs to be done. Perhaps exposure to toxic substances in the preimplantation period is a trivial issue in the broader consequence of toxic exposure worldwide. We suppose that this will not be the case particularly when we consider that specific exposures seem to carry the risk of serious health outcomes not only long after exposure but after birth as well. The ability to choose between the various exposure sets with respect to consequences of societal importance will allow for rational environmental refinement with respect to reproductive outcomes.

ACKNOWLEDGMENT

The data represented in Figures 4 and 5 were produced in collaboration with Dr. W. E. Fahl (Madison, WI).

REFERENCES

Adams, C. E., Hay, M. F., and Lutwak-Mann, C. (1961). *J. Embryol. Exp. Morphol. 9*: 468.
Adamson, E. D., and Meek, J. (1984). *Dev. Biol., 103*: 62.
Agostoni, E. (1993). *Ann. Ist. Super. Sanita, 29*: 15-25
Andrews, G. K., Huet-Hudson, Y. M., Paria, B. C., McMaster, M. T., De, S. K., and Dey, S. K. (1991). *Dev. Biol., 145*: 13.
Ballantyne, J. W. (1902). *The Fetus* (Willium, G., ed.), William Green and Sons, Edinburgh.
Balling, R. and Beier, H. M. (1985). *Toxicology, 34*: 309.
Bensaude, O., Babinet, C., Morange, M., and Jacob, F. (1983). *Nature, 305*: 331.
Benture, Y. (1990). *Maternal-Fetal Toxicology; A Clinicians' Guide* (Koren, G., ed.), Marcel Dekker, Inc., New York, p. 205.
Bergin, E. J., and Grandon, R. E. (1984). *How to Survive in Your Toxic Environment*. The American Survival Guide, Avon, New York.
Blankenship, A. L., Suffia, M. C., Matsumura, F., Walsh, K. J., and Wiley, L. M. (1993). *Reprod. Toxicol., 7*: 255
Bossert, N. L., and Iannaccone, P. M. (1985). *Proc. Natl. Acad. Sci. USA, 82*: 8757.
Bossert, N. L., Hitselberger, M. H., and Iannaccone, P. M. (1990). *Teratology, 42*: 147.
Brent, R. L. (1972). *Davis' Gynecology and Obstetrics*, Vol. 2 (Sciarra, J. J., ed.). Harper & Row, New York, p. 1.
Brent, R. L. (1989). *Semin. Oncol. 16*: 347.
Brunstrom, B. (1991). *Chem. Biol. Interact., 81*: 69.
Canfield, M. A., Annegers, J. F., Brender, J. D., Cooper, S. P., and Greenberg, F. (1996). *Am. J. Epidemiol. 143*: 1.
Cantarow, A., and Trumper, M. (1944). *Lead Poisoning*. Williams and Wilkins, Baltimore.

Chan, H. M., and Cherian, M. G. (1992). *Toxicology*, *72*: 281.

Chelimsky, E. (1991). *Reproductive and Developmental Toxicants*. Report to the Chairman, Committee on Governmental Affairs, U.S. Senate, United States General Accounting Office, Washington, DC.

Chubatsu, L. S., Gennari, M., and Meneghini, R. (1992). *Chem. Biol. Interact.*, *82*: 99.

Claxton, L. D., Morin, R. S., Hughes, T. J., and Lewtas, J. (1989). *Mutat. Res.*, *222*: 81.

Conover, E. (1994). *Gen. Birth Def. Environ. Hazards*, *23*: 524.

Coogan, T. P., Bare, R. M., and Waalkes, M. P. (1992). *Toxicol. Appl. Pharmacol. 113*: 227.

Courtney, K. D., and Andrews, J. E. (1985). *Fundam. Appl. Toxicol. 5*: 265.

Cummings, A. M. (1993). *Toxicology*, *79*: 205.

De, S. K., Paria, B. C., Dey, S. K., and Andrews, G. K. (1993). *Toxicology*, *80*: 13.

Deutsch, U., Dressler, G. R., and Gruss, P. (1988). *Cell*, *53*: 617.

Ding, A., and Chan, P. C. (1984). *Lipids*, *19*: 278.

Domingo, J. L. (1994). *J. Toxicol. Environ. Health*, *42*: 123.

Droesch, K., Jones, G. S., and Rosenwaks, Z. R. (1988). *Fertil. Steril.*, *50*: 451.

Ducibella, T., and Anderson, E. (1975). *Dev. Biol.*, *47*: 45.

Environmental Protection Agency, United States. (1980). *Ambient Water Quality Criteria for Toluene*, National Technical Information Service, Springfield, VA, A-2.

Environmental Protection Agency, United States. (19889). *Interim Procedures for Estimating Risks Associated with Exposures to Mixtures of Chlorinated Dibenzo-p-dioxins and Dibenzofurans (CDDs and CDFs)*, U.S. Environmental Protection Agency, Washington, DC.

Environmental Protection Agency, U.S. (1992). Press Advisory, R-257, U.S. Environmental Protection Agency, Washington, DC.

Environmental Protection Agency, U.S., and Office of Pesticide Program. (1988). *Pesticide Industry Sales and Usage: 1987 Market Estimates*, Environmental Protection Agency, Washington, DC.

Epstein, C. J., and Smith, S. A. (1978). *Exp. Cell Res.*, *111*: 117.

Epstein, S. S., Brown, L. O., and Pope, C. (1982). *Hazardous Waste in America*. Sierra Club Books, San Francisco.

Eskenazi, B., and Kimmel, G. (1995). *Environ. Health Perspect.*, *103* (Suppl. 2): 143.

Euler, H. H. (1967). *Arch. Gynakol.*, *204*: 258.

Fink, B. R., Shepard, T. H., and Blandau, R. J. (1967). *Nature*, *214*: 146.

Foster, W. G. (1995). *Environ. Health Perspect. 103* (Suppl. 9): 63.

Franzblau, A., and Lilis, R. (1989). *Arch. Environ. Health*, *44*: 385.

Fridovich, I. (1972). *Acct. Chem. Res.*, *5*: 321.

Fuller, G. B., and Hobson, W. C. (1990). *PCBs and the Environment* (Weid, J. S., ed.). CRC Press, Boca Raton, FL, p. 101.

Gardiner, C. S., and Menino, A. R. Jr. (1993). *Preimplantation Embryo Development* (Bavister, B. D., ed.). Springer-Verlag, New York, p. 200.

Gardiner, C. S., and Reed, D. (1994). *Biol. Reprod.*, *51*: 1307.

Geacintov, N. E., Ibanez, V., Gagliano, A. G., Yoshida, H., and Harvey R. G. (1980). *Biochem. Biophys. Res. Commun.*, *92*: 1335.

Gehring, W. J. (1987). *Science*, *236*: 1245.

Generoso, W. M., Rutledge, J. C., Cain, K. T., Hughes, L. A., and Downing, D. J. (1988). *Mutat. Res.*, *199*: 175.

Gerber, G., Leonard, A., and Jacquet, P. (1980). *Mutat. Res.*, *76*: 115.

Gilbert, S. F. (1994). *Developmental Biology*. Sinauer Associates, Inc., Sunderland, MA.

Glover, D. M., and Hames, B. D. (1989). *Genes and Embryos*. IRL Press, Oxford.

Golub, M. S., Donald, J. M., and Reyes, J. A. (1991). *Environ. Health Perspect.*, *94*: 245.

Goodwin, T. M. (1988). *Obstet. Gynecol.*, *71*: 715.

Haber, F., and Weiss, J. (1932). *Proc. R. Soc. Lond. A*, *147*: 332.

Hansen, D. K., and Billing, R. E. (1986). J. Pharmacol. Exp. Ther., 238: 985.

Harris, S. B., Wilson, J. G., and Printz, R. H. (1972). *Teratology*, 6: 139.

Harvey, M. B., Arcellana-Panlilio, M. Y., Zhang, X., Schultz, G. A., and Watson, A. J. (1995). *Biol. Reprod.*, 53: 532.

Hecht, S. H., and Hoffmann, D. (1988). *Carcinogenesis*, 9: 875.

Heindel, J. J., Chapin, R. E., Gulati, D. K., George, J. D., Price, C. J., Marr, M. C., myers, C. B., Barnes, L. H., Fail, P. A., Grizzle, T. B., Schwetz, B. A., and Yang, R. S. H. (1994). *Fundam. Appl. Toxicol.*, 22: 605.

Hogan, B., Costantini, F., and Lacy, E. (1986). *Manipulating the Mouse Embryo: A Laboratory Manual* (Hogan, B., Costantini, F., and Lacy, E., eds.). Cold Spring Harbor Laboratory, Cold Spring Harbor, New York, p. 17.

Holt, D., and Webb, M. (1987). *Arch. Toxicol.*, 59: 443.

Hood, R. D. (1990). *Developmental Toxicology: Risk Assessment and Future*, U.S. Environmental Protection Agency, Washington, DC.

Hood, R. D., Vedel, G. C., Zaworotko, M. J., Tatum, F. M., and Meeks, R. G. (1988). *J. Toxicol. Environ. Health*, 25: 423.

Hoskins, B. B., and Hupp, E. W. (1978). *Environ. Res.*, 15: 5.

Iannaccone, P. M. (1984). *Cancer Res.*, 44: 2785.

Iannaccone, P. M. (1986). *Terat. Carcinogen. Mutagen.*, 6: 237.

Iannaccone, P. M. (1990). *Principals and Practice of Endocrinology and Metabolism* (Becker, K. L., ed.). J. B. Lippincott Co., New York, p. 880.

Iannaccone, P. M., Tsao, T. Y., and Stols, L. (1982). *Cancer Res.*, 42: 864.

Iannaccone, P. M., Stols, L., Hollenberg, P. F., and Gurka, D. (1983). *J. Cell Physiol.*, 116: 227.

Iannaccone, P. M., Fahl, W. E., and Stols, L. (1984). *Carcinogenesis*, 5: 1437.

Iannaccone, P. M., Bossert, N. L., and Connelly, C. S. (1987). *Am. J. Obstet. Gynecol.*, 157: 476.

Iannaccone, P. M., Zhou, X., Khokha, M., Boucher, D., and Kuehn, M. R. (1992). *Dev. Dynam.*, 194: 198.

Izquierdo, L., and Becker, M. I. (1982). *J. Embryol. Exp. Morphol.*, 67: 51.

Jackson, I. J., Schofield, P., and Hogan, B. L. M. (1985). *Nature*, 317: 745.

Jacobson, M. F., Lefferts, L. Y., and Garland, A. W. (1991). *Safe Foods. Eating Wisely in a Risky World*. Living Planet Press, Los Angeles.

Jankowski, C. B. (1986). *Am. J. Nurs.*, 86: 260.

Jarrell, J. F., McMahon, A., Villeneuve, D., Franklin, C., Singh, A., Valli, V. E., and Bartlett, S. (1993). *Reprod. Toxicol.*, 7: 41.

Johansson, L., Sjoblom, P., and Wide, M. (1987). *Environ. Res.*, 42: 140.

Johnson, M. H., and Maro, B. (1986). *Experimental Approaches to Mammalian Embryonic Development* (Rossant, J., and Pederson, R., eds.). Cambridge University Press, Cambridge.

Johnson, M. H., and Ziomek, C. A. (1981). *Dev. Biol.*, 190: 287.

Jones, D. P. (1985). *Oxidative Stress* (Sies, H. ed.). Academic Press, London, p. 167.

Jones, P. B. C., Galeazzi, D. R., Fisher, J. M., and Whitlock, J. P. (1985). *Science*, 227: 1499.

Juchau, M. R. (1989). *Ann. Rev. Pharmacol. Toxicol.*, 29: 165.

Juchau, M. R. (1990). *Drug Toxicity and Metabolism in Pediatrics* (Kacew, S., ed.). CRC Press, Inc., Boca Raton, FL, p. 15.

Jung, T. H., and Streffer, C. (1992). *Int. J. Radiat. Biol.*, 62: 161.

Kalthoff, K. (1996). *Analysis of Biological Development*. McGraw-Hill, Inc., New York.

Katayama, S., Kubo, H., and Matsumoto, N. (1984). *Nippon Sanka Fujinka Gakk. Zasshi-Acta Ob. Gyn. Japonica*, 36: 1957.

Kaufmann, R. A., and Armant, D. R. (1992). *Teratology*, 46: 85.

Kholkute, S. D., Rodriguez, J., and Dukelow, W. R. (1994a). *Reprod. Toxicol.*, 8: 69.

Kholkute, S. D., Rodriguez, J., and Dukelow, W. R. (1994b). *Reprod. Toxicol.*, 8: 487.

Kimbrough, R. D., Falk, H., Stehr, P., and Fries, G. (1984). *J. Toxicol. Environ. Health*, 14: 47.

Kohler, M., Kundig, A., Reist, H. -W., and Michel, C. (1994). *Radiat. Environ. Biophys.*, *33*: 341.

Kuhnert, B. R., Kuhnert, P. J., and Zarlingo, T. J. (1988). *Obstet. Gynecol.*, *71*: 67.

Landrigan, P. J. (1982). *West. J. Med.*, *137*: 531.

Lee, D. K., Sun, W., Rhee, K., Cho, H., Lee, C. C., and Kim, K. (1994). Mol. Reprod. Dev., 39: 259.

Li, W., Zhao, Y., and Chou, I. -N. (1993). *Toxicology*, *77*: 65.

Lin, T. P., and Monie, I. W. (1973). *J. Reprod. Fertil.*, *32*: 149.

Lindner, R. E., Gaines, T. B., and Kimbrough, R. D. (1974). *Food Cosmet. Toxicol.*, *12*: 63.

Lints, T. J., Hartley, L., Parsons, L. M., and Harvey, R. P. (1996). *Dev. Dynam.*, *205*: 457.

Lione, A. (1987). Reprod. Toxicol., 1: 3.

Lowrence, W. W. (1976). *Of Acceptable Risk. Science and the Determination of Safety.* William Kaufman, Los Altos, CA.

Lugo, G., Cassady, G., and Palmisano, P. (1969). *Am. J. Dis. Child.*, *117*: 328.

Mably, T. A., Moore, R. W., Goy, R. W., and Peterson, R. E. (1992). *Toxicol. Appl. Pharmacol.*, *114*: 108.

MacPhee, D. J., Barr, K. J., De Sousa, P. A., Todd, S. D., and Kidder, G. M. (1994). *Dev. Biol.*, *162*: 259.

Manca, D., Ricard, A. C., Trottier, B., and Chevalier, G. (1991). *Toxicology*, *67*: 303.

Mantovani, A., Ricciardi, C., Macri, C., and Stazi, A. V. (1993). *Ann. Ist Super. Sanita.*, *29*: 47.

Masters, B. A., Kelly, E. J., Quaife, C. J., Brinster, R. L. and Palmiter, R. D. (1994). *Proc. Natl. Acad. Sci. USA*, *91*: 584.

Matsumoto, N., and Spindle, A. (1982). *Toxicol. Appl. Pharmacol.*, *64*: 108.

Maximovich, A., and Beyler, S. A. (1995). *J. Assist. Reprod. Gene.*, *12*: 75.

McGlothlin, J. D., Jensen, P. A., Todd, W. F., et al. (1988). *Study Protocol: Control of Anesthetic Gases in Dental Operatories.* National Institute for Occupational Safety and Health, Cincinnati, OH.

McMichael, A., Vimpani, G., Robertson, E., Baghurst, P., and Clark, P. (1986). *J. Epidemiol. Commun. Health*, *40*: 18.

Mercola, M., and Stiles, C. D. (1988). *Development*, *102*: 451.

Michel, C., Blattmann, H., Cordt-Reihle, I., and Fritz-Niggli, H. (1979). *Radiat. Environ. Biophys.*, *16*: 299.

Miller, R. W. (1977). *J. Pediatr.*, *90*: 510.

Mirkin, B. L., and Singh, S. (1976). *Perinatal Pharmacology and Therapeutics* (Mirkin, B. L., ed.). Academic Press, Inc., New York, p. 1.

Molls, M., Pon, A., Streffer, C., van Beuningen, D., and Zamboglou, N. (1983). *Int. Rad. Biol. Related Stud. Phys. Chem. Med.*, *43*: 57.

Moore, G. P. M. (1975). *J. Embryol. Exp. Morphol.*, *34*: 291.

Muller, L. (1986). *Toxicology*, *40*: 285.

Muller, W. U., and Streffer, C. (1990). *Teratology*, *42*: 643.

Muller, W. U., Streffer, C., and Fischer-Lahdo, C. (1986). *Arch. Toxicol.*, *59*: 172.

Murray, F. J., Schwetz, B. A., McBride, J. G., and Staples, R. E. (1979). *Toxicol. Appl. Pharmacol.*, *50*: 515.

Napalkov, N. P., and Alexandrov, V. A. (1968). *Z. Krebsforsch.*, *71*: 32.

Nasr-esfahani, M. H., Aitken, J. R., and Johnson, M. H. (1992). *Development*, *7*: 1281.

National Academy of Science, Committee on Medical and Biological Effects of Atmospheric Pollutants. (1972). *Lead. Airborne lead in perspective.* National Academy of Science, Washington, DC.

National Research Council of Canada. (1982). *Chlorinated Phenols: Criteria for Environmental Quality.* NRCC No. 18578,

Needleman, H. L., and Bellinger, D. (1994). *Prenatal Exposure to Toxicants. Developmental Consequences.* The John Hopkins University Press, Baltimore.

Nordstrom, S., Beckman, L., and Nordenson, I. (1978). *Hereditas*, *88*: 51.

Nriagu, J. O., and Pacyna, J. M. (1988). *Nature*, *333*: 134.

Okey, A. B., and Vella, L. M. (1982). *Br. J. Biochem.*, *127*: 39.

Orberg, J. (1978). *Acta Pharmacol. Toxicol.*, *42*: 323.

Pabon, W. E., Findley, W. E., and Gibbons, W. E. (1989). *Fertil. Steril.*, *51*: 896.

Pederson, R. A., and Lin, T. P. (1978). *Developmental Toxicology of Energy-Related Pollutants* (Mahlum, P. L., Hachett, P. L., and Andrew, F. D., eds.). Department of Energy, Washington, DC, p. 600.

Pedigo, N. G., and Vernon, M. W. (1993). *Reprod. Toxicol.*, *7*: 11.

Peters, J. M. and Wiley, L. M. (1995) *Toxicol. Appl. Pharmacol.*, *134*: 214.

Peters, J. M., Duncan, J. M., Wiley, L. M., and Keen, C. L. (1995a). *Toxicology*, *99*: 11.

Peters, J. M., Duncan, J. R., Wiley, L. M., Rucker, B. R., and Keen, C. L. (1995b). *Reprod. Toxicol.*, *9*: 123.

Platonow, N. S., and Karstad, L. H. (1973). *Can. J. Comp. Med.*, *37*: 391.

Rauskolb, C., Peifer, M., and Wieschaus, E. (1993). *Cell*, *74*: 1101.

Rier, S. E., Martin, D. C., Bowman, R. E., Dmowski, W. P., and Becker, J. L. (1993). *Fundam. Appl. Toxicol.*, *21*: 433.

Rogan, W. J. (1982). *Teratology*, *26*: 259.

Rossignol, G., Alaoui-Jamali, M. A., Castonguay, A., and Schuller, H. M. (1989). *Cancer Res.*, *49*: 5671.

Rothenberg, S., Schnaas, L., Cansino-Ortiz, S., Perroni-Hernandez, E., de la Torre, P., Neri-Mendeze, C., Ortega, P., Hidalgo-Loperena, H., and Svendsgaard, D. (1989). *Neurotoxicol. Teratol.*, *11*: 85.

Rowland, A. S., Baird, D. D., Shore, D. L., Weinberg, C. R., Savitz, D. A., and Wilcox, A. J. (1995). *Am. J. Epidemiol.*, *141*: 531

Rugh, R. (1991). *The Mouse*: *Its Reproduction and Development* (Rugh, R., ed.). Oxford University Press, New York, p. 7.

Russell, L. B., and Russell, W. L. (1954). *J. Cell Comp. Physiol.*, *43* (Suppl.): 103.

Saling, P. M. (1989). *Oxford Rev. Reprod. Biol.*, *11*: 339.

Sanchez, D. J., Gomez, M., Llobet, J. M., and Doming, J. L. (1993). *Ecotoxicol. Environ. Safety*, *26*: 33.

Sanyal, M. K., Li, Y. L., Biggers, W. J., Satish, J., and Barnes, E. R. (1993). *Am. J. Obstet. Gynecol.*, *168*: 1587.

Sastry, B. V. R., ed. (1995). *Placental Toxicology*. CRC Press Inc., Boca Raton, FL.

Sastry, B. V. R., and Janson, V. E. (1995). *Placental Toxicology* (Sastry, B. V. R. ed.). CRC Press Inc., Boca Raton, FL, p. 45.

Savas, U., Christou, M., and Jefcoate, C. R. (1993). *Carcinogenesis*, *14*: 2013.

Schardein, J. L. (1993). *Chemically Induced Birth Defects*. Marcel Dekker, New York.

Schuchhardt, S. (1973). *Oxygen Supply* (Lubbers, D. W., and Kessler, M., eds.). Urban & Schwarzenberg, Munich, p. 223.

Scott, R. T., Oehninger, S., and Rosenwaks, Z. (1989). *Fertil. Steril.*, *51*: 651.

Seeley, M. R., and Faustman, E. M. (1995). *Fundam. Appl. Toxicol.*, *26*: 136.

Shane, B. S. (1989). *Environ. Sci. Technol.*, *23*: 1187.

Shigeta, S. (1982). *Tokai J. Exp. Clin.* Med., *7*: 265.

Silbergeld, E. K., and Mattison, D. R. (1987). *Am. J. Ind. Med.*, *11*: 131.

Soudais, C., Bielinska, M., Heikinheimo, M., McArthur, C. A., Narita, N., Saffitz, J. E., Simon, M. C., Leiden, J. M., and Wilson, D. B. (1995). *Development*, *121*: 3877.

Spielmann, H., Eibs, H. G., Habenicht, U., and Jacob-Miller, U. (1982). *Exp. Biol. Med.*, *7*: 162.

Spyropoulos, D. C., and Capecchi, M. R. (1994). *Genes Dev.*, *8*: 1949.

Stevens, C. W., Bouck, N., Burgess, J. A., and Fahl, W. E. (1985). *Mutat. Res.*, *152*: 5.

Stols, L., and Iannaccone, P. M. (1985). *J. Cell Physiol.*, *123*: 395.

Storeng, R., and Jonsen, J. (1980). *Toxicology*, *17*: 183.

Streffer, C. (1993). *Mutat. Res., 299*: 313.

Streffer, C., and Molls, M. (1987). *Advances in Radiation Biology* (Lett, J., ed.). Academic Press, New York, p. 169.

Streissguth, A. P., Sampson, P. D., Barr, H. M., Bookstein, F. L., and Olson, H. C. (1994). *Prenatal Exposure to Toxicants: Developmental Consequen*ces (Needleman, H. L, and Bellinger, D., eds.). The Johns Hopkins University Press, Baltimore, MD, p. 148.

Swain, W. R. (1991). *J. Toxicol. Environ. Health, 33*: 587.

Takeuchi, T. (1966). *Minamata Disease: Study Group of Minamata Disease* (Katsuna, M., ed.). Kumamoto University, Japan, p. 141.

Tarkowski, A. K., and Wroblewska, J. (1967). *J. Embryol Exp. Morphol, 18*: 155.

Telford, N. A., Watson, A. J., and Schlutz, G. A. (1990). *Mol. Reprod. Dev., 26*: 90.

Thompson, E. M., Legouy, E., Christians, E., and Renard, J. P. (1995). *Development, 121*: 3425.

Torok, P. (1978). *Arch. Toxicol., 40*: 249.

Tseng, L (1984). *Endicrinology, 115*: 833.

U. S. Congress. (1985) Reproductive Health Hazards in the Workplace. Office of Technology Assessment, OTA-BA-266, Government Printing Office, Washington, DC.

Umaoka, Y., Noda, Y., Narimoto, K., and Mori, T. (1992). *Mol. Reprod. Dev., 31*: 28.

Watson, A. J. (1992). Mol. Reprod. Dev., 33: 492.

Weir, D. (1987). *The Bhopal Sybdrome. Pesticides, Environment, and Health.* Sierra Club Books, San Francisco.

Wiley, L. M. (1984). *Dev. Biol., 105*: 330.

Wiley, L. M., and Eglitis, M. A. (1980). *Exp. Cell Res., 127*: 89.

Wiley, L. M., Wu, J., Harari, I., and Adamson, E. D. (1992). *Dev. Biol., 149*: 247.

Wilson, J. G. (1977). *Fed. Proc., 36*: 1698.

Winkel, G. K., Ferguson, J. E., Takeichi, M., and Nuccitelli, R. (1990). *Dev. Biol., 138*: 1.

Winter, R. (1979). *Cancer-Causing Agents. A Preventive Guide.* Crown Publishers, New York.

Yamaguchi, A., Yoshimura, T., and Kuratsune, M. (1971). *Fukuoka Acta Med., 62*: 117.

Yelian, F. D., and Dukelow, R. (1992). *Arch. Toxicol., 66*: 443.

Yoshinaga, K., Rice, C., Krenn, J., and Pilot, R. L. (1979). *Biol. Reprod., 20*: 294.

Yu, H. S., and Chan, S. Y. H. (1986). *Toxicology, 34*: 323.

Zhao, F., Mayura, K., Hutchinson, R. W., Lewis, R. P., Burghardt, R. C., and Phillips, T. D. (1995). *Toxicol. Lett., 78*: 35.

Zhou, X., Saski, H., Linda, L., Brigid, L. M., and Kuehn, M. R. (1993). *Nature, 11*: 543.

Ziomek, C. A. (1987). *The Mammalian Preimplantation Embryo: Regulation of Growth and Differentiation In Vitro.* Plenum Press, New York.

3

Developmental Toxicology

John M. Rogers and Robert J. Kavlock
U.S. Environmental Protection Agency, Research Triangle Park, North Carolina

INTRODUCTION

The first reports of experimentally induced birth defects in mammals were published in the 1930s and were concerned with maternal nutritional deficiencies. Hale (1935) produced malformations including anophthalmia and cleft palate in offspring of sows fed a diet deficient in vitamin A. Beginning in 1940, Warkany and coworkers began a series of experiments in which they demonstrated that maternal dietary deficiencies and other environmental factors could affect intrauterine development in rats (Warkany and Nelson, 1940; Warkany and Schraffenberger, 1944; Warkany, 1945; Wilson et al., 1953). These experiments were followed by many other studies in which chemical and physical agents, e.g., nitrogen mustard, trypan blue, hormones, antimetabolites, alkylating agents, hypoxia, and x-rays, to name a few, were clearly shown to cause malformations in mammalian species (Warkany, 1965).

Gregg (1941) first reported human malformations induced by an environmental agent, linking an epidemic of rubella virus infection to an elevation in the incidence of congenital eye, heart, and ear defects, as well as mental retardation. Heart and eye defects were associated with infection in the first or second months of pregnancy, whereas hearing and speech defects and mental retardation were more common after infection in the third month. Later, the risks of congenital anomalies associated with maternal rubella infection were estimated to be 61% in the first 4 weeks of pregnancy, decreasing thereafter to 26% in weeks 5–8 and 8% in weeks 9–12 (Sever, 1967).

The finding that embryos of mammals, including humans, were susceptible to common external influences such as nutritional deficiencies and intrauterine infections did not attract wide attention at the time (Wilson, 1973). However, reports in 1961 (McBride, 1961; Lenz, 1961) of severe birth defects in infants following thalidomide ingestion by pregnant women had a major impact on the field of developmental toxicology. As a result of the thalidomide catastrophe, regulatory agencies in many countries began developing animal testing requirements for evaluating the effects of drugs on pregnancy outcome separate from chronic toxicity studies. In the United States, the Segment I, II, and III testing protocols were developed by the Food and Drug Administration (FDA) (Kelsey, 1988). Design of the Segment II study, which covers the period of organogenesis, will be presented later in this chapter, along with other tests and screens developed to evaluate environmental agents and drugs for potential developmental toxicity.

Today, estimates of adverse pregnancy outcomes in humans include postimplantation pregnancy loss, 31%; major birth defects, 4% at birth increasing to 6–7% at one year as more manifestations are diagnosed; minor birth defects, 14%; low birth weight, 7%; infant mortality (prior to one year of age), 1.4%; and abnormal neurological function, 16–17% (Schardein, 1993). Thus, less than half of all human pregnancies result in the birth of a completely normal, healthy infant. Reasons for the adverse outcomes are largely unknown. Brent and Beckman (1990) attributed 15–25% of human birth defects to genetic causes, 4% to maternal conditions, 3% to maternal infections, 1–2% to deformations (i.e., mechanical problems such as umbilical cord limb amputations), <1% to chemicals and other environmental influences, and 65% to unknown etiologies. These estimates are not dramatically different from those suggested by Wilson (1977). Regardless of the etiology, the sum total represents a significant health burden in light of the 2 million annual births in the United States.

Principles of teratology were proposed by James Wilson in 1959 (Wilson, 1959) and in a 1973 monograph (Wilson, 1973) (Table 1). Wilson stated that "very likely with increased understanding of teratogenic mechanisms, it will be necessary to formulate additional principles and to revise those now formulated." However, these principles can be restated largely unchanged over 20 years later. In the past two decades, animal experimentation has led to greater understanding of mechanisms and pathogenesis, and more human developmental toxicants have been identified. The study of developmental functional deficits, including neurobehavioral effects, has largely emerged during this time. As we rapidly increase our understanding of how genes direct development, we are presented with exciting new opportunities to apply powerful tools to understand mechanisms of teratogenesis. As has been true for the past two centuries, it is likely that progress in developmental biology and teratology will continue to be intimately linked.

In its broadest sense, human development extends from gametogenesis to adulthood. In this chapter we will concentrate on embryogenesis and organogenesis, the interval from gastrulation to the beginning of the fetal period. The emphasis will be

Table 1 Wilson's General Principles of Teratology

1. Susceptibility to teratogenesis depends on the genotype of the conceptus and the manner in which this interacts with adverse environmental factors.
2. Susceptibility to teratogenesis varies with the developmental stage at the time of exposure to an adverse influence.
3. Teratogenic agents act in specific ways (mechanisms) on developing cells and tissues to initiate sequences of abnormal developmental events (pathogenesis).
4. The access of adverse influences to developing tissues depends on the nature of the influence (agent).
5. The four manifestations of deviant development are death, malformation, growth retardation, and functional deficit.
6. Manifestations of deviant development increase in frequency and degree as dosage increases, from the no-effect to the totally lethal level.

Source: Wilson, 1959, 1973.

on the special vulnerabilities presented by the rapid and complex changes occurring during this life stage, the current status of test methods and rapid screens to detect potential developmental toxicity, and mechanisms of abnormal development. Manifestations of developmental toxicity include structural malformations, growth retardation, functional impairment, and/or death of the organism. Developmental toxicology is the study of pharmacokinetics, mechanisms, pathogenesis, and outcome following exposures to agents or conditions that cause abnormal development. Current status of the field and emerging issues and techniques are reviewed in this chapter.

THE CRITICAL PERIOD CONCEPT

Prerequisite to understanding abnormal development is a firm grasp of normal development. An overview of normal mammalian development is provided in this volume by Donahoe and MacLaughlin, and the dynamic changes occurring in the embryo/fetus are reflected by equally dynamic changes in sensitivity to environmental insults. Timing of some key developmental events in humans and experimental animal species is presented in Table 2.

Following implantation into the uterine endometrium, the embryo undergoes gastrulation, the process by which the three primary germ layers are formed: ectoderm, mesoderm, and endoderm. The three germ layers will give rise to all the organs and tissues of the body, as well as extraembryonic structures that are vital to intrauterine survival. During gastrulation, cells proliferate and migrate through a structure called the primitive streak, and their movements set up basic morphogenetic fields in the embryo (Smith et al., 1994). Gastrulation is a period that is quite susceptible to teratogenesis. A number of toxicants administered during gastrulation produce malformations

Table 2 Timing of Selected Developmental Events in Lab Species and Humans

	Day(s) of Gestation in:			
	Rat	Rabbit	Monkey	Human
Blastocyst formation	3–5	2.6–6	4–9	4–6
Implantation	5–6	6	9	6–7
Organogenesis	6–17	6–18	20–45	21–56
Primitive streak	9	6.5	18–20	16–18
Neural plate	9.5	—	19–21	18–20
First somite	10	—	—	20–21
First pharyngeal arch	10	—	—	20
Ten somites	10–11	9	23–24	25–26
Upper limb buds	10.5	10.5	25–26	29–30
Lower limb buds	11.2	11	26–27	31–32
Forepaw rays	13.4	14.5	34	35
Testes differentiation	14.5	20	—	43
Heart septation	15.5	—	—	46–47
Palate closure	16–17	19–20	45–47	56–58
Length of gestation	22	32	165	267

Source: Adapted from Shepard, 1992.

of the eye, brain, and face due to deficiencies in the anterior neural plate, one of the regions defined by the cellular movements of gastrulation.

During the period of organogenesis, the rudiments of most bodily structures are established, beginning with the formation of the neural plate. This is a period of high susceptibility to malformations, and it extends from approximately the third to the eighth week of gestation in humans. Within this short period, the embryo undergoes rapid and dramatic changes. At 3 weeks of gestation, the human conceptus is in most ways indistinguishable from other vertebrate embryos at the gastrula stage, consisting of only a few cell types in a trilaminar arrangement. Only 5 weeks later, at 8 weeks of gestation, the conceptus, which can now be termed a fetus, has a recognizable human form. Cell proliferation, cell migration, cell-cell interactions, and morphogenetic tissue remodeling occur during organogenesis. These processes are exemplified by the development of the neural crest cells. These cells originate at the edges of the neural plate along most of its length and migrate to form a wide variety of structures throughout the embryo, including facial bones, sensory ganglia, sheaths of Schwann, pigment cells, and various other cell types. Their survival, correct migration, and differentiation depend on many as yet poorly understood factors, including extracellular matrix components and cellular signaling. Neural crest cells appear to be sensitive targets for developmental toxicants. One insult to which they are very sensitive is maternal zinc deficiency. Figure 1 illustrates apoptosis induced in premigratory neural crest cells by only 4 days of maternal zinc deficiency in the rat (Rogers et al., 1995)

Within organogenesis, there are periods of maximum susceptibility for each forming structure. The work of Shenefelt (1972), who studied the developmental toxicity of carefully timed exposures to retinoic acid in the hamster, will be used to illustrate this concept. The incidence of defects seen after maternal retinoic acid treatment at different times during pregnancy are shown in Figure 2. Periods of sensitivity for spe-

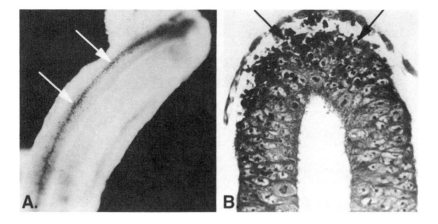

Figure 1 Apoptosis (programmed cell death) induced in premigratory neural crest cells by maternal zinc deficiency in the rat. Pregnant rats were fed a zinc-deficient diet from gestation days 7–11 and embryos were examined on gestation day 11. (A) Dorsal midline view of the tail region of a gestation day 11 rat embryo stained with the vital dye Nile blue sulfate, which stains areas where apoptosis is occurring (stippling in midline; arrows). (B) Histological transverse section through the same region, showing cell debris and apoptotic bodies (arrows). (Modified from Rogers et al., 1995.)

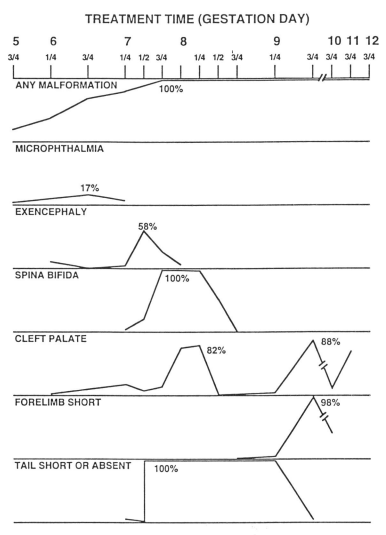

Figure 2 Critical periods of susceptibility to malformations in hamster fetuses following maternal treatment with retinoic acid on different days of gestation. Each selected malformation shows a different temporal pattern of sensitivity. Note that the total incidence of malformations increases to 100% during organogenesis. Incidences are estimates of malformations in live fetuses at a maternal dosage producing 50% embryo/fetal lethality. (Modified after Shenefelt, 1972.)

cific malformations coincide with the timing of key developmental events in affected structures. For example, the process of eye development starts quite early, and microphthalmia has an early critical period. Rudiments of the long bones of the limbs appear later, as does susceptibility to shortened limbs. The palate shows two peaks of susceptibility to retinoic acid. The first corresponds to the establishment of the palatal folds and the second corresponds to the events of palatal closure. Note also that the total incidence of malformations at the fetal LD_{50} is lower prior to organogenesis, but increases to 100% by gestation day 7¾, demonstrating the sensitivity of organogenesis to induced malformations.

A toxicant may affect a single or multiple developmental events, so the temporal pattern of susceptibility of a structure depends on both the developmental program of the structure and the nature of the toxic insult. Cleft palate occurs in mouse fetuses following maternal exposure to methanol as early as gestation day 5, peaks around day 7, and little sensitivity is observed after day 9 (Rogers and Mole, 1997). The peak critical period for induction of cleft palate in the mouse by most toxic agents lies between gestation days 11 and 13. Neubert's group reported that in NMRI mice the day of peak sensitivity to cleft palate induction was day 11 for TCDD, day 12 for 2,4,5-trichlorophenoxyacetic acid, and day 13 for dexamethasone (Neubert et al., 1973). It is likely that these various toxicants are preferentially affecting different developmental events in palatogenesis.

The closure of the palatal shelves is the landmark typically used to denote the end of organogenesis and the end of the embryonic period. The fetal period (from about day 56 through birth in humans) follows, characterized primarily by tissue differentiation, growth, and physiological maturation. Development of organs proceeds during the fetal period to attain requisite functionality prior to birth, including fine structural morphogenesis (e.g., neural outgrowth and synaptogenesis, branching morphogenesis of the bronchial tree and renal cortical tubules, further development of the reproductive tract) as well as biochemical maturation (e.g., induction of tissue-specific enzymes and structural proteins, synthesis of various hormones).

Chemical exposure during the fetal period is most likely to result in effects on growth and function. Functional anomalies of the central nervous system and reproductive organs including behavioral, mental, and motor deficits and decreases in fertility are among the possible adverse outcomes. These manifestations are not apparent prenatally and require postnatal observation and testing of offspring. Such postnatal functional manifestations can be sensitive indicators of in utero toxicity, and recent reviews of postnatal functional deficits of the central nervous system (Rodier, Cohen and Buelke, 1994; Chap. 10, this volume), immune system (Holladay and Luster, 1994), and heart, lung, and kidneys (Lau and Kavlock, 1994) are available.

DETECTION OF POTENTIAL FOR CHEMICAL OR PHYSICAL AGENTS TO CAUSE DEVELOPMENTAL TOXICITY

The Risk-Assessment Process

Characterization of the potential for an agent to cause developmental toxicity is carried out somewhat differently for pharmaceuticals, where exposure is voluntary and usually to relatively high dosages, and for environmental agents, where exposure is generally involuntary and to low levels. For drugs, a Use-in-Pregnancy rating is utilized (U.S. FDA, 1979). In this system, the letters A, B, C, D, and X are used to classify drugs based on the evidence concerning the risk that a chemical poses to the human conceptus. For example, drugs are placed in Category A if well-controlled studies in pregnant humans have not demonstrated a risk, and in Category X (contraindicated for pregnancy) if studies in animals or humans, or investigational or postmarketing reports, have shown fetal risk that outweighs any benefit to the patient. The default is Category C (risk cannot be ruled out), which is used when there is a lack of human studies and animal studies are either lacking or are positive for fetal risk, but the benefits may outweigh the potential risk. Categories B and D represent areas of relatively lesser, or greater concern for risk, respectively. Manson (1994) reviewed the 1992 *Physician's Desk Reference* and found that 7% of the 1033 drugs listed were in Category X, 66% in Category C, and only 0.7% in Category A. The FDA categorization procedure has recently been criticized (Teratology Society, 1994) as being too reliant on risk/benefit comparisons, since the level of risk is often poorly characterized, or benefits of exposure are not an issue (e.g., after the drug has been taken during early pregnancy and the question is then directed at the management of the exposed pregnancy).

For environmental agents, the risk-assessment process for developmental toxicity is carried out to define the dose, route, timing, and duration of exposures that induce an adverse effect at the lowest exposure level in the most relevant laboratory animal model (U.S. EPA, 1991a). In a given study, the highest test exposure level not producing this "critical effect" is termed the No Observed Adverse Effect Level (NOAEL). The NOAEL is then divided by a variety of safety or uncertainty factors in order to derive an exposure level for humans that is presumed to be safe. The uncertainty factors include one for extrapolation from the test species to the human and another for variation in susceptibility within the human population. The default value for each of these factors is 10, and additional uncertainty factors can be included based on data quality. The use of the NOAEL in the risk-assessment process has been criticized for several reasons. For example, since it is dependent on statistical power to detect pairwise differences between a treated and a control group, the use of larger sample sizes and more dosage groups (which might better characterize the dose-response relationship) can only yield lower NOAELs. Thus, using better experimental design may actually penalize a manufacturer of a chemical by lowering allowable exposure limits. In addition, the NOAEL is constrained to be one of the experimental dosage levels, and an experiment might need to be repeated to develop a NOAEL for risk assessment.

Crump (1984) proposed using a mathematical model to estimate the lower confidence bounds on a predetermined level of risk (the "benchmark dose") as a means of avoiding many of the disadvantages of the NOAEL. The application of this approach to a large compilation of Segment II type data sets (Faustman et al., 1994; Allen et al., 1994a,b; Kavlock et al., 1995) demonstrated that a variety of mathematical models can be applied to standard test results. On average, benchmark doses based on a 5–10% added risk of effect were approximately equivalent to traditionally determined NOAELs.

Certain default assumptions are generally made in extrapolating developmental toxicity data from animals to humans, including the following: (1) an agent that produces an adverse developmental effect in experimental animals can pose a hazard to humans following sufficient exposure during development; (2) all four manifestations of developmental toxicity (death, structural abnormalities, growth alterations, and functional deficits) are of concern; (3) the types of developmental effects seen in animal studies are not necessarily the same as those that might be produced in humans; (4) in general, a threshold is assumed for the dose-response curve for agents that produce developmental toxicity (see below for discussion of thresholds for developmental toxicity). The most appropriate species is used to estimate human risk when data are available to indicate that a particular species is most relevant, but in the absence of such data (which is generally the case), the most sensitive species is used.

One of the more difficult and controversial aspects of risk assessment for developmental toxicants is distinguishing in the animal model between adverse effects (defined as an unwanted effect that is detrimental to health) and lesser effects, which, while different from those observed in control groups, are of questionable significance to human health. The interpretation of reduced fetal growth observed in developmental toxicity studies is a case in point. There are accepted definitions of low birth weight in humans, and we know that intrauterine growth retardation is associated with an elevated risk of infant mortality, mental retardation, and other lifelong health effects. In contrast, we do not have similar knowledge of the effects of low fetal weight in rodents, and we seldom examine whether the reduced fetal weight at term in Segment II studies persists beyond birth. Although considered an adverse effect, regulating an agent on the basis of low fetal weight in the absence of other endpoints of developmental toxicity has been contentious.

In Vivo Regulatory Guidelines

Prior to the thalidomide tragedy, safety evaluations for reproductive effects were limited in both the types of chemicals evaluated and the sophistication of the endpoints. Subsequently, FDA issued the Segments I, II, and III testing protocols (U.S. FDA, 1966). These protocols were adopted with few changes by a variety of regulatory agencies around the world and have remained in place for over 30 years. The accumulated knowledge and experience of these many years of testing, increased knowledge of basic reproductive processes, and the need for consistent data requirements in different countries have recently resulted in new and streamlined international testing guidelines (U.S. FDA, 1994; U.S. EPA, 1997). These guidelines, the result of the International Conference on Harmonization of Technical Requirements for Registration of Pharmaceuticals for Human Use (ICH), specifically incorporate flexibility in their implementation depending on the circumstances surrounding the agent under evaluation. The ICH guidelines give the investigator the responsibility of meeting the primary goal of detecting and documenting any indication of developmental toxicity. Palmer (1993) has reviewed issues relevant to implementing the new ICH guidelines. In each protocol, guidance is provided on species/strain selection, route of administration, number and spacing of dosage levels, exposure duration, experimental sample size, observational techniques, statistical analysis, and reporting requirements. Details are available in the original publications and in several recent reviews (e.g., Manson, 1994; Francis, 1994). Variations of these protocols include extension of exposure to earlier or later time points in development and/or extension of the observation period to postnatal ages and inclusion of more sophisticated endpoints. For example, the U.S. Environmental Protection Agency's Developmental Neurotoxicity Protocol for the rat includes maternal exposure from gestation day 6 through lactation day 10 as well as observation of postnatal growth, developmental landmarks of puberty (balanopreputial separation, vaginal opening), motor activity, auditory startle, learning and memory, and neuropathology at various ages through postnatal day 60 (U.S. EPA, 1991b).

Information pertaining to developmental toxicity can also be obtained from studies in which animals are exposed to the test substance continuously over one or more generations, as discussed elsewhere in this volume.

Concordance of Data Between Animal Models and Humans

Reviews of the similarity of responses of laboratory animals and humans for developmental toxicants generally support the assumption that results from laboratory test animals are predictive of human effects. Concordance is strongest when there are positive data from more than one test species, yet it remains that specific types of effects do not extrapolate well across species. The predictiveness of animal data for developmental toxicants presumed to be negative in humans is less than that for positive agents, most likely because of the difficulty in ascertaining a negative response in humans as well as issues of inappropriate design or interpretation of animal studies. The comparisons that have been made suggest that humans tend to be more sensitive to developmental toxicants than the most sensitive test species.

Frankos (1985) reviewed data for 38 compounds having demonstrated or suspected toxicity in humans. All except tobramycin, which caused otological defects that may be difficult to assess on standard animal tests, were positive in at least one species, and 76% were positive in more than one test species. Predictiveness was highest in the mouse (85%) and rat (80%) compared to the rabbit (60%) or hamster (40%). While Frankos identified 165 chemicals with no evidence of human toxicity, only 29% of these were negative in all test species, while 51% were negative in more than one species.

Schardein and Keller (1989) examined 51 potential human developmental toxicants deemed to have adequate animal data for comparison (3 human developmental toxicants did not). Across all chemicals, the most common adverse outcomes in humans, rabbits, and monkeys were spontaneous abortion and fetal/neonatal death, followed by malformations and then growth retardation. In the rat, prenatal death, growth retardation, and then malformations was the typical pattern. The concordance of results is presented in Table 3. With one exception, all of the potential human developmental toxicants showed a positive response in at least one test species. The exception was formaldehyde, which was negative in both the rat and mouse, but was included as a potential human developmental toxicant based on a single occupational study. The agreement of test species data to the human data (i.e., matching without specifying endpoint of toxicity) was rat, 98%; mouse, 91%; hamster, 85%; monkey, 82%; and rabbit, 77%.

Jelovsek et al. (1989) reviewed the predictiveness of animal data for 84 negative human developmental toxicants, 33 with unknown activity, 26 considered suspicious,

Table 3 Predictiveness of Animal Data for 51 Potential Human Developmental Toxicants

		Mouse	Rat	Monkey	Rabbit	Hamster
Potential human developmental toxicants tested (%)		86	96	33	61	26
Concordance by class	G	61	57	65	39	39
	D	75	71	53	52	54
	M	71	67	65	65	62
	All	91	98	82	77	85
False positives	G	25	33	6	19	8
	D	11	16	18	10	0
	M	14	12	6	7	15
False negatives	G	10	14	29	39	54
	D	14	12	29	39	46
	M	11	25	29	29	23

G = growth retardation; D = death of conceptus; M = malformation; All = either growth, death, or malformations.
Source: Adapted from Schardein and Keller, 1989.

and 32 considered positive. Multivariate analysis using the human effect as the classification variable was used to examine predictiveness of the animal database. The animal data correctly classified 63–91% of the compounds, depending upon how the suspicious and unknown human toxicants were considered. The various models had a
sensitivity of 62–75%, a positive predictive value of 75–100%, and a negative predictive value of 64–91%.

Several attempts at comparing potencies across species have been made, although
these have been based upon administered dosage and have not considered pharmacokinetic differences. Schardein and Keller (1989) estimated the human and animal "threshold" dosages for 21 chemicals and found only two cases, aminopterin and carbon disulfide, for which developmental effects were seen at lower dosages in animals than
those believed to cause effects in humans. Ratios of the "threshold" dosages in the most
sensitive animals to those in humans ranged from 1.2 to 200.

Among the well-characterized human developmental toxicants valproic acid,
isotretinoin, thalidomide, and methotrexate, Newman et al. (1993) found the monkey
to be the most sensitive species for the first three chemicals, while the rabbit was the
most sensitive to methotrexate. Based upon the NOAEL of the most sensitive test species, human embryos were 0.9 to approximately 10 times more sensitive.

Dose-Response Patterns and the Threshold Concept

The major effects of prenatal exposure observed at birth in developmental toxicity studies
are embryolethality, malformations, and growth retardation. The relationship between
these endpoints varies with the type of agent, the time of exposure, and the dosage.
These endpoints may represent a continuum of increasing toxicity, with lower dosages
producing growth retardation and increasing dosages producing malformations and then
lethality. However, malformations and/or death can also occur in the absence of any
effect on intrauterine growth, and growth retardation and embryolethality can occur
without malformations. Agents producing the latter pattern of response would be considered embryotoxic or embryolethal, but not teratogenic (unless it was subsequently
established that death was due to a structural malformation).

A key element of the dose-response relationship is the shape of the dose-response
curve at low exposure levels. Mammalian developmental toxicity has typically been
considered a threshold phenomenon because of the high restorative growth potential of
the embryo, cellular homeostatic mechanisms, and maternal/placental metabolic defenses. In this context, threshold means that there is a maternal dosage below which
developmental toxicity is not elicited. Daston (1993) evaluated two approaches for establishing the existence of a threshold. The first, a "brute force" approach exemplified by a large teratology study on 2,4,5-T (Nelson and Holson, 1978), suggests that
no study is capable of evaluating the dose-response at low response rates (e.g., it was
found that 805 liters per dose would be necessary to detect a 5% increase in resorptions).
The second approach is to attempt to determine whether a threshold exists for the mechanism responsible for the observed effect. While mechanisms of developmental toxicity
are known with any certainty for few agents, cellular and embryonic repair mechanisms
and dose-dependent kinetics both suggest the plausibility of a mechanistic threshold.
Lack of a threshold infers that maternal exposure to any amount of a toxic chemical,
even one molecule, has the potential to cause developmental toxicity. One mechanism
of abnormal development for which this might be the case is gene mutation. A point

mutation in a critical gene could theoretically be induced by a single hit or single molecule, leading to a deleterious change in a gene product and consequent abnormal development. Another mechanism by which a single molecule might have an effect is in the case in which the toxic molecule serves as a ligand for a cellular receptor. A single molecule of the toxicant could have an incremental effect on a receptor-mediated cellular process if the toxic molecule acted in concert with an existing level of the endogenous ligand.

In the context of human health risk assessment, it is important to consider the distinction between individual thresholds and population thresholds. There is wide variability in the human population, and the population threshold is defined by the threshold of the most sensitive individual (Gaylor et al., 1988). Although the target of a developmental toxicant may exhibit a biological threshold, factors such as health status or concomitant exposures may render an individual at or even above the threshold for failure of that biological process. Any further toxic impact on that process, even one molecule, would be expected to increase risk.

Alternative Tests for Developmental Toxicity

A variety of alternative test systems have been devised to prescreen chemicals and reduce reliance on the standard regulatory tests for assessing prenatal toxicity (Table 4). These assay systems include those based on cell cultures, cultures of embryos in vitro (including submammalian species), and short-term in vivo tests. Validation of these alternative tests has been a contentious and as yet unresolved issue (reviewed in Neubert, 1989; Welsch, 1990). Assessing the sensitivity and specificity of results from such tests has been problematic, as there is no agreed-upon standard of comparison. It is generally accepted that a validated test would have to be predictive of results of full in vivo developmental toxicity protocols (Schwetz et al., 1991). The existing alternative test methods have not been shown to be generally applicable to all chemicals, although many of the tests have shown utility within a chemical class or exposure scenario. Given the complexity of development, it is unrealistic to expect a single test, or even a small battery, to accurately prescreen the activity of chemicals in general.

An exception to the poor acceptance of alternative tests for prescreening for developmental toxicity is the in vivo test developed by Chernoff and Kavlock (1982). In this test, pregnant females are exposed during organogenesis to dosages near those inducing maternal toxicity, and offspring are evaluated during the neonatal period for external malformations, growth and viability. This test has the advantage of keeping the complexity of the maternal/embryonal system intact, and it has proven reliable for a large number of chemical agents and classes (Hardin et al., 1987). A regulatory testing guideline for this assay has been developed (U.S. EPA, 1985).

RELATIONSHIPS OF MATERNAL AND DEVELOPMENTAL TOXICITY

All developmental toxicity must ultimately result from an insult to the conceptus at the cellular level. However, this insult may occur through a direct effect on the embryo/fetus, indirectly through toxicity of the agent to the mother and/or the placenta, or, probably most commonly, a combination of direct and indirect effects. Maternal conditions capable of adversely affecting the developing organism in utero include decreased

Table 4 Brief Survey of Alternate Developmental Toxicity Tests

Assay	Brief description and endpoints evaluated	Concordance[a]	Ref.
Mouse ovarian tumor	Labeled mouse ovarian tumor cells added to culture dishes with Concanavalin A–coated disks for 20 minutes. Endpoint is inhibition of attachment of cells to disks.	Sensitivity: 19/31; 19/30 Specificity: 7/13; 5/13	Steele et al., 1988 (results from two labs)
Human embryonic palate mesenchyme	Human embryonic palatal mesenchyme cell line growth in culture. Assess cell number after 3 days.	Sensitivity: 21/31; 21/30 Specificity: 7/13; 5/13	Steele et al., 1988 (results from two labs)
Micromass culture	Midbrain and limb bud cells dissociated from day 13 rat embryos and grown in micromass culture for 5 days. Endpoints include cell proliferation and biochemical markers of differentiation.	Sensitivity: 25/27; 20/33; 11/15 Specificity: 17/19; 18/18; 8/10	Flint and Orton, 1984; Renault et al., 1989; Uphill et al., 1991
Chick embryo neural retina cell culture	Neural retinas of day 6.5 chick embryos dissociated and grown in rotating suspension culture for 7 days. Endpoints include cellular aggregation, growth, differentiation, and biochemical markers.	Sensitivity: 36/41 Specificity: 14/17	Daston et al., 1991; Daston et al., 1995 (concordances combined)
Drosophila	Larva grown in treated media from egg to hatching of adults. Adult flies examined for specific structural defects (bent bristle and notched wing).	Sensitivity: 10/13 Specificity: 4/5	Lynch et al., 1991
Hydra	*Hydra attenuata* cells are aggregated to form an "artificial embryo" and allowed to regenerate. Dose response compared to that for adult *Hydra* toxicity.	Sensitivity: n/a Specificity: n/a	Johnson and Gabel, 1982
FETAX[b]	*Xenopus* mid-blastula embryos exposed for 96 hours, then evaluated for viability, growth, and morphology.	Sensitivity: n/a Specificity: n/a	Bantle, 1995
Chernoff/Kavlock assay	Pregnant mice or rats exposed during organogenesis and allowed to deliver. Postnatal growth, viability, and gross morphology of litters recorded.	Sensitivity: 49/58 Specificity: 28/34	Hardin et al., 1987

[a]Author's interpretation. Sensitivity: correct identification of "positive" chemicals. Specificity: correct identification of "negative" compounds.
[b]*Frog Embryo Teratogenesis Assay-Xenopus*
Source: Adapted from Rogers and Kavlock, 1996.

uterine blood flow, maternal anemia, altered nutritional status, toxemia, altered organ function, autoimmune states, diabetes, and electrolyte or acid-base disturbances (Daston, 1994). Induction or exacerbation of such maternal conditions by toxic agents and the degree to which they manifest in abnormal development are dependent on maternal genetic background, age, parity, size, nutrition, disease, stress, and other health parameters and exposures (DeSesso, 1987; Chernoff et al., 1989).

The distinction between direct and indirect developmental toxicity is important for interpreting safety assessment tests in pregnant animals, as the highest dosage level in these experiments is chosen based on its ability to produce some maternal toxicity (e.g., decreased food or water intake, weight loss, clinical signs). However, maternal toxicity defined only by such manifestations gives little insight to the toxic actions of a xenobiotic. When developmental toxicity is observed only in the presence of maternal toxicity, the developmental effects may be indirect, but they are nonetheless important. Understanding the physiological changes underlying the observed maternal toxicity and the association with developmental effects is needed before one can address the relevance of the observations to human safety assessment. Many known human developmental toxicants, including ethanol and cocaine, adversely affect the embryo/fetus predominately at maternally toxic levels, and part of their developmental toxicity may be ascribed to secondary effects of maternal physiological disturbances. For example, the nutritional status of alcoholics is generally inadequate, and effects on the conceptus may be exacerbated by effects of alcohol on placental transfer of nutrients. Effects of chronic alcohol abuse on maternal folate and zinc metabolism may be particularly important in the induction of fetal alcohol syndrome (Dreosti, 1993).

A retrospective analysis of relationships between maternal toxicity and specific types of prenatal effects in test animals found some associations between maternal toxicity and specific adverse developmental effects. Yet, among rat, rabbit, and hamster studies, 22% failed to show any developmental toxicity in the presence of significant maternal toxicity (Khera, 1984, 1985). In a study designed to test the potential of maternal toxicity to affect development, Kavlock et al. (1985) administered 10 structurally unrelated compounds to pregnant mice at maternotoxic dosages. Developmental effects were agent-specific, ranging from complete resorption to lack of effect. An exception was an increased incidence of supernumerary ribs (ribs on the first lumbar vertebra), which occurred with 7 of the 10 compounds. Chernoff et al. (1990) dosed pregnant rats for 10 days with a series of compounds chosen because they exhibited little or no developmental toxicity in previous studies. When these compounds were administered at dosages high enough to produce maternal toxicity (weight loss or lethality), a variety of developmental outcomes was noted. The lack of a consistent fetal outcome led these authors to conclude that maternal weight loss or mortality per se were not associated with any consistent syndrome of developmental effects in the rat.

Diverse forms of maternal toxicity may have in common the induction of a physiological stress response. Understanding potential effects of maternal stress on development may help to interpret the developmental toxicity observed in experimental animals at maternally toxic dosages. Various forms of physical stress have been applied to pregnant animals in attempts to isolate the effects of stress on development. Subjecting pregnant rats or mice to noise stress throughout gestation can affect development (Geber, 1966; Geber and Anderson, 1967; Kimmel et al., 1976; Nawrot et al., 1980, 1981). Restraint stress produces increased fetal death in rats (Euker and Riegle, 1973) and cleft palate (Barlow et al., 1975), fused and supernumerary ribs, and encephaloceles in mice (Beyer and Chernoff, 1986; Chernoff et al., 1988). Objective data on effects of stress in humans are difficult to obtain. Nevertheless, studies investigating the relationship of maternal stress and pregnancy outcome have indicated a positive correlation between stress and adverse developmental effects, including low birth weight and congenital malformations (Stott, 1973; Gorsuch and Key, 1974).

The genotype of the pregnant female has been well documented as a determinant of developmental outcome in both humans and animals. The incidence of cleft lip and/ or palate [CL(P)], which occurs more frequently in whites than in blacks, has been investigated in offspring of interracial couples in the United States (Khoury et al., 1983). Offspring of white mothers had a higher incidence of CL(P) than offspring of black mothers after correcting for paternal race, while offspring of white fathers did not have a higher incidence of CL(P) than offspring of black fathers after correcting for maternal race.

There are numerous examples in experimental animals of differential response to developmental toxicants depending on maternal genotype. Two related mouse strains, A/J and CL/Fr, produce spontaneous CL(P) at incidences of 8–10% and 18–26%, respectively. The incidence of CL(P) in offspring depends on the genotype of the mother rather than that of the embryo (Juriloff and Fraser, 1980). The response to vitamin A of murine embryos heterozygous for the curly-tail mutation also depends on the genotype of the mother (Seller et al., 1983). The teratogenicity of phenytoin has been compared in several inbred strains of mice. The susceptibility of offspring of crosses between susceptible A/J mice and resistant C57BL/6J mice was determined by the maternal, but not the embryonic genome (Hansen and Hodes, 1983).

Maternal disease can be a cause of or contributor to developmental toxicity. Chronic hypertension is a risk factor for the development of preeclampsia, eclampsia, and toxemia of pregnancy, and hypertension is a leading cause of pregnancy-associated maternal deaths. Uncontrolled maternal diabetes mellitus is a significant cause of prenatal morbidity. Certain maternal infections can adversely affect the conceptus (e.g., rubella virus, discussed earlier), either through indirect disease-related maternal alterations or direct transplacental infection. Cytomegalovirus infection is associated with fetal death, microcephaly, mental retardation, blindness, and deafness (MacDonald and Tobin, 1978), while maternal infection with *Toxoplasma gondii* is known to induce hydrocephaly and chorioretinitis in infants (Alford et al., 1974). One factor common to many disease states is hyperthermia. Hyperthermia is a potent experimental animal teratogen (Edwards, 1986), and there is a body of evidence associating maternal febrile illness during the first trimester of pregnancy with birth defects in humans, most notably malformations of the central nervous system (Warkany, 1986; Milunsky et al., 1992).

Dietary insufficiencies ranging from protein-calorie malnutrition to deficiencies of vitamins or trace elements are known to adversely affect pregnancy (Keen et al., 1993). Among the most significant findings related to human nutrition and pregnancy outcome in recent years are those from studies in which pregnant women at risk for having infants with neural tube defects (NTDs) were supplemented with folate (Wald, 1993). In one study, supplementation with 4 mg/d folic acid reduced NTD recurrence by over 70% (MRC, 1991). Results of these studies have prompted the U.S. Centers for Disease Control and Prevention to recommend folate supplementation for women of childbearing age.

PHARMACOKINETICS AND METABOLISM IN PREGNANCY

The maternal, placental, and embryonic compartments are independent, yet interacting systems that undergo striking changes during the course of pregnancy. Changes in maternal physiology during pregnancy involve the gastrointestinal tract, cardiovascular system, excretory system, and respiratory system (Hytten, 1984; Krauer, 1987; Mattison et al., 1991). While these physiological changes are necessary to support the growing needs of the conceptus in terms of energy supply and waste elimination, they can also have significant impact on the uptake, distribution, metabolism, and elimination of xenobiotics. For example, decreases in intestinal motility and prolongation of gastric emptying result in longer retention of ingested chemicals in the upper gastrointestinal tract. Cardiac output increases by 50% during the first trimester in humans and remains elevated throughout pregnancy, while blood volume increases and plasma protein concentration and peripheral vascular resistance decrease. The relatively greater increase in blood volume over red cell volume may lead to mild anemia and a generalized edema consisting of a 70% elevation of extracellular space. Thus, the volume of distribution of a chemical and the amount bound by plasma proteins may change considerably during pregnancy. Renal blood flow and glomerular filtration also increase in many species during pregnancy. Increases in tidal volume, minute ventilation, and minute O_2 uptake can result in increased pulmonary distribution of gasses and decreased time to reach alveolar steady state.

In addition to changes in maternal physiology, evidence suggests that drug-metabolizing capacity also changes during pregnancy (Juchau, 1981; Juchau and Faustman-Watts, 1983). In rats, liver weight increases by 40% during pregnancy. This increase in mass, however, is nearly balanced by a lowered metabolic activity, so that the hepatic metabolic capability of the pregnant rodent is comparable to that of the nonpregnant female. The decreased level of hepatic monooxygenase activity observed in maternal liver during pregnancy in rats has been attributed to decreased enzyme levels and to competitive inhibition by circulating steroids (Neims, 1976). Another factor contributing to reduced monooxygenase activities is that pregnant rats appear to be less responsive to phenobarbital (but not 3-methylcholanthrene) induction of hepatic cytochrome monooxygenase systems than are nonpregnant females (Guenther and Mannering, 1977). Although the literature on this topic is not comprehensive, there appears to be an overall decrease in hepatic xenobiotic metabolism during pregnancy. It is clear that maternal handling of a chemical bears considerable weight in determining the extent of developmental toxicity. Kimmel and Young (1983) showed that a linear combination of the 45-minute maternal blood concentration and the 24-hour maternal blood concentration was able to predict (within the 95% confidence interval) the litter response of pregnant rats dosed with 500 mg/kg sodium salicylate on gestation day 11. The predictive value of these two kinetic parameters probably reflects the influence of the peak drug concentration as well as the cumulative area under the concentration-time curve in inducing developmental disturbances.

The placenta mediates embryonic exposure by helping to regulate blood flow, by serving as a transport barrier, and by metabolizing chemicals (Slikker and Miller, 1994). Functionally, the placenta acts as a lipid membrane that permits bidirectional transfer of substances between maternal and fetal compartments. Although there are marked species differences in types of placentae, orientation of blood vessels, and numbers of intervening cell layers, these do not seem to play a dominant role in placental transfer of most chemicals.

The passage of most drugs across the placenta seems to occur by simple diffusion, the rate of which is proportional to the diffusion constant of the drug, the concentration gradient between the maternal and embryonic plasma, the area of exchange, and the inverse of the membrane thickness (Nau, 1992). Important factors modifying the rate of transfer include lipid solubility, molecular weight, protein binding, the type of transfer (passive diffusion, facilitated or active transport), the degree of ionization, and placental metabolizing enzymes. Weak acids appear to be rapidly transferred across the placenta, due in part to a pH gradient between the maternal and embryonic plasma, which can trap ionized forms of a drug in the slightly more acidic embryonic compartment (Nau and Scott, 1986). Blood flow probably constitutes the major rate-limiting step for more lipid-soluble compounds.

Quantitating the form, amount, and timing of chemical delivery to the embryonic compartment relative to concurrent developmental processes is an important component of understanding mechanisms of embryotoxicity and species differences in embryonic sensitivity (Nau, 1986). The small size of the conceptus during organogenesis and the fact that the embryo is changing at a rapid rate during this period makes assessment of toxicokinetics difficult. Nevertheless, there has been considerable progress in this area (Nau and Scott, 1987; Clark, 1993). Recent evidence suggests that the early embryo has the ability to metabolize some xenobiotics. Using an embryo culture system, Juchau and coworkers (1992) demonstrated that the rat conceptus was able to generate sufficient amounts of metabolites of the proteratogen 2-acetylaminofluorene (2-AAF) to induce dysmorphogenesis. Prior exposure of the dams to 3-methylcholanthrene increased the sensitivity of the cultured embryos to 2-AAF, demonstrating the inducibility of at least some cytochromes in the conceptus. These investigators later showed that the embryos could further metabolize the 7-hydroxy metabolite (the proximate teratogen) to an even more toxic catechol. No previous induction was necessary for this activation to occur, demonstrating the presence of constitutive metabolizing enzymes in the embryo.

Physiologically based pharmacokinetic models integrate what is known about physiological changes during pregnancy, both within and between species, with aspects of drug metabolism and embryonic development. Gabrielson and coworkers (Gabrielson and Paalkow, 1983; Gabrielson and Larsson, 1990) were among the first to develop physiologically based models of pregnancy, and others have added to their comprehensiveness (Fisher et al., 1989; O'Flaherty et al., 1992; Clark et al., 1993; Luecke et al., 1994). The pregnancy model of O'Flaherty and coworkers describes the entire period of gestation, and includes the uterus, mammary tissue, maternal fat, kidney, liver, other well-perfused maternal tissues, embryo/fetal tissues and yolk sac and chorioallantoic placentas. It takes into account the growth of various compartments during pregnancy (including the embryo itself), as well as changes in blood flow to the tissues and the stage-dependent pH gradients between maternal and embryonic plasma. Transfer across the placenta in the model is diffusion limited. The utility of the model was evaluated using 5,5'-dimethyloxazolidine-2,4-dione (DMO), a weak acid that is eliminated by excretion in the urine. The model demonstrated that the whole body disposition of DMO, including distribution to the embryo, can be accounted for solely on the basis of its pK_a and of the pH and volumes of body fluid spaces. Observed differences in disposition of DMO by the pregnant mouse and rat can be accounted for by differences in fluid pH.

MECHANISMS OF DEVELOPMENTAL TOXICITY

The term mechanisms is used here to refer to cellular level events that initiate the process leading to abnormal development. Pathogenesis comprises the ensuing cellular, tissue, and organ-level sequelae, which are ultimately manifest in abnormality. Mechanisms of teratogenesis listed by Wilson (1977) include mutations, chromosomal breaks, altered mitosis, altered nucleic acid integrity or function, diminished supplies of precursors or substrates, decreased energy supplies, altered membrane characteristics, osmolar imbalance, and enzyme inhibition. While these cellular insults are not unique to development, they may relatively quickly trigger unique pathogenetic responses in the embryo, such as reduced cell proliferation, cell death, altered cell-cell interactions, reduced biosynthesis, inhibition of morphogenetic movements, or mechanical disruption of developing structures.

If we are to continue making progress in preventing birth defects, it is imperative that we begin to understand the molecular and cellular events underlying abnormal development. Rapid advances are being made in our knowledge of normal development, and this knowledge can be applied to developmental toxicity. Modern molecular biological techniques are being used to address problems of abnormal development in an increasing number of laboratories, and the prospects for understanding toxicant effects on the embryo/fetus at the molecular and cellular levels are increasingly hopeful. It is also critical for us to be able to follow toxicant-induced molecular and cellular alterations through the resultant pathogenesis leading to a birth defect.

Linkages of mechanisms to pathogenesis and abnormal pregnancy outcome are essential for the development of biologically based dose-response (BBDR) models. Our current dose-response paradigm consists principally of comparing dosage administered to the pregnant animal to pregnancy outcome at term. BBDR models, in contrast, seek to understand the sequence of events intervening between administration of the test agent and final outcome. Physiologically based pharmacokinetic models (as described earlier) can be used to predict dose to the target (e.g., the litter, the embryo, or part of the embryo), and then events including interaction of the delivered dose with cellular or molecular targets, cellular response to the toxicant, and downstream pathogenetic events leading to abnormal development are modeled.

The BBDR concept is relatively new and still somewhat vague in its detail. There is consensus that our current outcome-drive risk-assessment approach is fraught with uncertainties and that we need to understand mechanisms to progress. However, how mechanistic information at the molecular/cellular level might be used in risk assessment has not been thoroughly explored. Qualitative differences in pharmacokinetics or in cellular responses to a toxicant have obvious value in choosing appropriate test species, but data on such differences between species are hard to obtain, especially in the human. Even more difficult is defining the quantitative relationships of events in the mechanistic pathway in a dose-response fashion. It is hoped that such efforts will reduce uncertainties in predicting safe exposure levels for humans. If molecular/biochemical endpoints are found to be more sensitive than outcome at term, they may provide data at a lower range of the dose-response curve. Such data may be useful for examining the question of the existence of a threshold, as discussed previously, or perhaps for helping to choose a biologically relevant mathematical model to use in the dose-response assessment.

A prototype BBDR model has been constructed for the chemotherapeutic agent 5-fluorouracil (5-FU) (Shuey et al., 1994). This well-studied animal teratogen probably works through a number of mechanisms, for the most part the same as those through which it exerts its antitumor activity. A proposed BBDR model of 5-FU developmental toxicity is presented in Figure 3. It is known that 5-FU inhibits the enzyme thymidylate synthetase, and this inhibition and downstream events form the core of this model. However, to different extents 5-FU may also be incorporated into DNA and/or RNA, and at high maternal dosages, fetal anemia is also produced and may contribute to other manifestations of toxicity.

Shuey and coworkers modeled the actions of 5-FU on thymidylate synthetase and attempt to assess the pathological sequence leading to digit defects in fetuses after maternal exposure on gestation day 14 in the rat. Dose-response relationships were examined for the effects of 5-FU on thymidylate synthetase activity, DNA synthesis, cell cycle distributions, early dysmorphogenesis of the hind limbs, and limb defects at term. The relationships of dose to thymidylate synthetase activity, thymidylate synthetase activity to altered cell cycle (represented by percent of cells in S-phase), percent of cells in S-phase to early digit outgrowth, and early digit outgrowth to limb defects at term were expressed quantitatively using Hill-type equations and linked to produce a predicted dose-response for hind limb defects (Fig. 4).

The study of Shuey and coworkers serves as a first step in formulating a BBDR model, but also illustrates some of the problems inherent in such an approach. First, for most chemicals undergoing a risk assessment we will have little or no idea of puta-

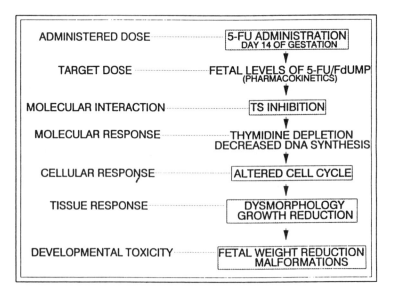

Figure 3 Proposed biologically based dose response (BBDR) model for the developmental toxicity of 5-fluorouracil (5-FU). The goal of BBDR models is to examine the sequence of events between administered dose and adverse outcome. For 5-FU, parameters in boxes in the right column were quantified, and relationships between these parameters are illustrated in Figure 4.

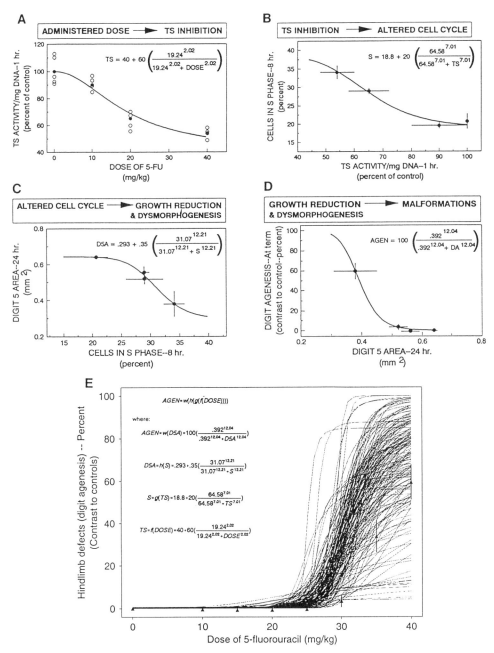

Figure 4 Biologically based dose-response model for the production of hindlimb defects by 5-fluorouracil (5-FU). The mechanistic sequence hypothesized for production of defects is: 5-FU → thymidylate synthetase inhibition → altered cell cycle → growth reduction and dysmorphogenesis → malformations (digit agenesis at term), as illustrated in Figure 3. The relationships among these sequential endpoints are shown in panels A through D, with the relationships expressed mathematically with Hill-type equations. These equations are then combined to produce the predicted dose response for hindlimb defects, as shown in panel E. Here, Monte-Carol simulation was used to incorporate variability in response for each endpoint and 200 simulations were run, represented by the individual dose-response data are represented by the solid triangles. The simulation results indicate that variability in the intermediate endpoints can account for differences between the predicted and actual dose response. **AGEN**: Digit agenesis at term; **D5A**: digit 5 area; **S**: percent of cells in S-phase. (Adapted from Shuey et al., 1994.)

tive mechanisms, so even proposing a BBDR model may require a great deal of preliminary research. Second, in addition to dose-response there is also a time course for the mechanistic events in the BBDR model. In the 5-FU model, the investigators chose fixed time points for assessment of each endpoint based on time of maximal effect (i.e., thymidylate synthetase activity at one hour, percent of cells in S-phase at 8 hours, digit outgrowth at 24 hours, and defects at term). However, this approach fails to incorporate the time-course profile of these events, so it loses some information and is somewhat arbitrary. Third, this model examines toxic events specific to 5-FU, and therefore may have limited applicability to other toxicants. A more comprehensive and generally applicable approach might be to model the normal biological processes and then input the putative effect(s) of the toxicant under study. In the case of 5-FU, nucleotide metabolism, cell cycle, and limb bud morphogenesis could be modeled (including the role of thymidylate synthetase), and then the effects of 5-FU on thymidylate synthetase could be input as the toxic mechanism. The value of such a model of normal biology would be that toxicants affecting other parts of the pathway could be assessed with the same model. Such efforts are obviously extremely complicated and may require vast amounts of data.

The most notorious human teratogen is thalidomide. Thalidomide is still in use in South America for the treatment of leprosy, and, predictably, pregnant women have been exposed and birth defects have occurred (Castilla et al., 1997). It is both ironic and telling that the chemical largely responsible for the advent of modern regulatory authority over potential developmental toxicants presents a very complex pattern of effects in various animal species used for safety evaluations.

There has been a 30-year search for the mechanism(s) underlying thalidomide teratogenesis. In 1988, Stephens reviewed 24 proposed mechanisms that had been studied or proposed up to that point. These mechanisms include antagonism of vitamins, chelation of trace elements, effects on nucleic acid synthesis, DNA intercalation, inhibition of oxidative metabolism or protein synthesis, inhibitor of cell-cell interactions, trophoblastic degeneration, nerve or blood vessel degeneration, hormonal disruption, and mutagenesis. Some of these mechanisms have been tested and rejected, while others have not been rigorously examined. Convincing evidence is lacking for most.

Recently, Neubert and coworkers (Neubert et al., 1996) have examined the effects of the thalidomide analog EM-12 on cell surface adhesion receptors. Thalidomide and EM-12 strongly altered the expression of numerous adhesion receptors on white blood cells in both nonhuman primates (Neubert et al., 1991, 1992a, 1993) and humans (Neubert et al., 1991, 1992b, 1993). These investigators have reported that the expression of several surface adhesions receptors is dramatically altered by thalidomide in limb bud cells and other cells during early organogenesis in the marmoset, a nonhuman primate. These results are summarized in Table 5. Such alterations would be expected to alter cell-cell interactions as well as cell migration and differentiation. Thus, for the first time, a primary mechanism of thalidomide teratogenesis may have been discovered.

SUMMARY AND CONCLUSIONS

Birth defects are not a single disease, but rather a broad spectrum of adverse outcomes representing both morphological and physiological or other functional deficits. Thus,

Table 5 Effects of Teratogenic Exposures In Utero to the Thalidomide Derivative EM-12 on the Expression of Cell Surface Adhesion Receptors in Marmoset (a primate) Embryos[a]

Receptor	Description	Limb		Heart		Head		Body	
		Small cells	Large cells	Small cells	Large cells	Small cells	Large cells	Small cells	Large cells
CD11a	LFA-1α	→	→	(↓)	(↓)	0	(↓)	0	(↓)
CD11b	Mac-1	?	?	?	?	0	0	?	?
CD18	LFA-β-chain	→	0	0	0	0	0	0	(↓)
CD49d	VLA-4α	→	→	→	(↓)	0	0	0	0
CD29	VLA-β-chain	→	→	(↓)	→	(↓)	0	→	0
CD61	β₃-Integrin β-chain	←	←	→	→	0	→	0	→
CD54	ICAM-1	←	←	?	0	0	?	0	(↓)
CD8		0	0	0	0	0	0	0	0

[a]Embryonic cells were analyzed by flow cytometry at development stages 11 and 12.
↓ = Drastically reduced percentage of cells with this receptor in treated embryos; (↓) = clearly reduced percentage of cells with this receptor in treated embryos; ↑ = clearly increased percentage of cells with this receptor in treated embryos; 0 = no clear-cut change in expression of this receptor; ? = questionable changes in treated embryos and/or insufficient data.

Source: Adapted from Neubert et al., 1996.

understanding and preventing birth defects is a vastly complicated and arduous undertaking. We can claim some success over the three decades since the thalidomide tragedy, in that there has been no recurrence of such an event. Yet it is impossible to know whether or to what extent this lack of recurrence is due to our diligence in testing drugs and chemicals, and much uncertainty remains in the testing and risk-assessment process. A panoply of modern techniques is now being brought to bear on the problems of normal and abnormal development, and as more light is shed on developmental mechanisms we will have opportunities to further reduce uncertainties in extrapolating results from test animals to humans.

Predicting toxic effects on a widely divergent human population will always be fraught with difficulties. Understanding how maternal health and social factors affect pregnancy and response to environmental exposures is critical to protect our whole society. Poverty is perhaps the worst teratogen we have, working insidiously through social mechanisms including poor health care before and during pregnancy, increased exposure to pollution, poor living and working conditions, poor nutrition, stress, smoking, and alcoholism and other drug abuse. These are causes of birth defects we know about, yet they persist widely.

Birth defects are a lifelong malady that result in a financial and quality-of-life burden on affected families as well as an economic impact on all of us. Continuously improving our ability to prevent birth defects, as well as increased advocacy for support for birth defect research, are goals for which we should all strive.

REFERENCES

Alford, C. A., Stagno, S., and Reynolds, D. W. (1974). *Clinical Perinatology* (S. Aldjem and A. K. Brown, eds.). C.V. Mosby, St. Louis, p. 31.

Allen, B. C., Kavlock, R. J., Kimmel, C. A., and Faustman, E. M. (1994a). *Fundam. Appl. Toxicol.*, *23*: 487.

Allen, B. C., Kavlock, R. J., Kimmel, C. A., and Faustman, E. M. (1994b). *Fundam. Appl. Toxicol.*, *23*: 496.

Bantle, J. A. (1995). *Fundamendals of Aquatic Toxicology. Effects, Environmental Fate, Risk Assessment*, 2nd ed. (G. Rand, ed.). Taylor and Francis, Washington, DC, p. 207.

Barlow, S. M., McElhatton, P. R., and Sullivan, F. M. (1975). *Teratology*, *12*: 97.

Beyer, P. E., and Chernoff, N. (1986). *Teratogen. Carcinogen. Mutagen.*, *6*: 419.

Brent, R. L., and Beckman, D. A. (1990). *Bull. NY Acad. Med.*, *66*: 123.

Castilla, E. E., Ashton-Prolla, P., Barreda-Mejia, E., Brunoni, D., Cavalcanti, D. P., Correa-Neto, J., Delgadillo, J. L., Dutra, M. G., Felix, T., Giraldo, A., Juarez, N., Lopez-Camelo, J. S., Nazer, J., Oriolo, I. M., Paz, J. E., Pessoto, M. A., Pina-Neto, J. M., Quadrelli, R., Rittler, M., Rueda, S., Saltos, M., Sánchez, O., and Schüler, L. (1997). *Teratology*, *54*: 273.

Chernoff, N., and Kavlock, R. J. (1982). *J. Environ. Toxicol. Health, 10*: 541.

Chernoff, N., Miller, D. B., Rosen, M. B., and Mattscheck, C. L. (1988). *Toxicology*, *51*: 57.

Chernoff, N., Rogers, J. M., and Kavlock, R. J. (1989). *Toxicology*, *59*: 111.

Chernoff, N., Setzer, R. W., Miller, D. B., Rosen, M. B., and Rogers, J. M. (1990). *Teratology*, *42*: 651.

Clark, D. O. (1993). *Toxicol. Meth.*, *3*: 223.

Clark, D. O., Elswick, B. A., Welsch, F., and Conolly, R. (1993). *Toxicol. Appl. Pharmacol.*, *121*: 239.

Crump, K. (1984). *Fundam. Appl. Toxicol.*, *4*: 854.

Daston, G. P. (1993). *Issue and Review in Teratology* (H. Kalter, ed.). Plenum Press, New York, p. 169.

Daston, G. P. (1994). *Developmental Toxicology, 2nd ed.* (C. A. Kimmel and J. Buelke-Sam, eds.). Raven Press, New York, p. 189.

Daston, G. P., Baines, D., and Yonker, J. E. (1991).*Toxicol. Appl. Pharmacol.*, *109*: 352.

Daston, G. P., Baines, D., Elmore, E., Fitzgerald, M. P., and Sharma, S.(1995). *Fundam. Appl. Toxicol.*, *26*: 203.

DeSesso, J. M. (1987). *Teratogen. Carcinogen. Mutagen.*, 7: 225.

Dreosti, I. E. (1993). *Ann. NY Acad. Sci.*, *678*: 193.

Edwards, M. J. (1986). *Teratogen. Carcinogen. Mutagen.*, 6: 563.

Euker, J. S., and Riegle, G. D. (1973). *J. Reprod. Fertil.*, *34*: 343.

Faustman, E. M., Allen, B. C., Kavlock, R. J., and Kimmel, C. A. (1994). *Fundam. Appl. Toxicol.*, *23*: 478.

Flint, O. P., and Orton, T. C. (1984). *Toxicol. Appl. Pharmacol.*, *76*: 383.

Fisher, J. W., Whitaker, T. A., Taylor, D. H., Clewell III, H. J., and Andersen, M. E. (1989). *Toxicol. Appl. Pharmacol.*, *99*: 395.

Francis, E. Z. (1994). *Developmental Toxicology.*, 2nd ed. (C. A. Kimmel and J. Buelke-Sam, eds.). Raven Press, New York, p. 403.

Frankos, V. H. (1985). *Fundam. Appl. Toxicol.*, *5*: 615.

Gabrielson, J. L., and Larson, K. S. (1990). *Pharmacol. Toxicol.*, *66*: 10.

Gabrielson, J. L., and Paalkow, L. K. (1983). *J. Pharmacokinet. Biopharmacol.*, *11*: 147.

Gaylor, D. W., Sheehan, D. M., Young, J. F., and Mattison, D. R. (1988). *Teratology*, *38*: 389.

Geber, W. F. (1966). *J. Embryol. Morphol.*, *16*: 1.

Geber, W. F., and Anderson, T. A. (1967). *Biol. Neonatol.*, *11*: 209.

Gorsuch, R. L., and Key, M. K. (1974). *Psychosom. Med.*, *36*: 362.

Gregg, N. M. (1941). *Tr. Ophthalmol. Soc. Aust.*, *3*: 40.

Guenther, T. M., and Mannering, G. T. (1977). *Biochem. Pharmacol.*, *26*: 577.

Hale, F. (1935). *J. Hered.*, *27*: 105.

Hansen, D. K., and Hodes, M. E. (1983). *Teratology*, *28*: 175.

Hardin, B. D., Becker, R. J., Kavlock, R. J., Seidenberg, J. M., and Chernoff, N. (1987). *Teratogen. Carcinogen. Mutagen.*, 7: 119.

Hytten, F. E. (1984). *Drugs and Pregnancy* (B. Krauer, F. Hytten, and E. del Pozo, eds.). Academic Press, New York, p. 7.

Jelovsek, F. R., Mattison, D. R., and Chen, J. J. (1989). *Obstet. Gynecol.*, *74*: 624.

Johnson, E. M., and Gabel, B. E. G. *J. Am. Coll. Toxicol.*, *1*: 57.

Juchau, M. R. (1981). *The Biochemical Basis of Chemical Teratogenesis* (M. R. Juchau, ed.). Elsevier/North Holland, New York, p. 63.

Juchau, M. R., and Faustman-Watts, E. M. (1983). *Clin. Obstet. Gynecol.*, *26*: 379.

Juchau, M. R., Lee, Q. P., and Fantel A. G. (1992). *Drug Metab. Rev.*, *24*: 195.

Juriloff, D. M., and Fraser, F. C. (1980). *Teratology*, *21*: 167.

Kavlock, R. J., Chernoff, N., and Rogers, E. H. (1985). *Teratogen. Carcinogen. Mutagen.*, 5: 3.

Kavlock, R. J., Allen, B. C., Faustman, E. M., and Kimmel, C. A. (1995). *Fundam. Appl. Toxicol.*, *26*: 211.

Keen C. L., Bendich, A., and Willhite, C. C. (eds.) (1993). *Ann. NY Acad. Sci.*, *678*: 372.

Kelsey, F. O. (1988). *Teratology*, *38*: 221.

Khera, K. S. (1984). *Teratology*, *29*: 411.

Khera, K. S. (1985). *Teratology*, *31*: 129.

Khoury, M. J., Erickson, J. D., and James, L. M. (1983). *Teratology*, *27*: 351.

Kimmel, C. A., and Young, J. F. (1983). *Fundam. Appl. Toxicol.*, *3*: 250.

Kimmel, C. A., Cook, R. O., and Staples, R. E. (1976). *Toxicol. Appl. Pharmacol.*, *36*: 239.

Krauer, B. (1987). *Pharmacokinetics in Teratogenesis,* Vol. 1 (H. Nau and W. J. Scott, eds.). CRC Press, Boca Raton, FL, p. 3.

Lau, C., and Kavlock R. J. (1994). *Developmental Toxicology,* 2nd ed. (C. A. Kimmel and J. Buelke-Sam, eds.). Raven Press, New York, p. 119.

Lenz, W. (1961). *Deutsche Med. Wochenschr.,* 86: 2555.

Lenz, W. (1994). *Fortschr. Med.,* 81: 148.

Luecke, R. H., Wosilait, W. D., Pearce, B. A., and Young, J. F. (1994). *Teratology,* 49: 90.

Lynch, D. W., Schuler, R. L., Hood, R. D., and Davis, D. G. (1991). *Teratogen. Carcinogen. Mutag.,* 11: 147.

MacDonald, H., and Tobin, J. O. H. (1978). *Dev. Med. Child. Neurol.,* 20: 271.

Manson, J. M. (1994). *Developmental Toxicology,* 2nd ed. (C. A. Kimmel and J. Buelke-Sam, eds.). Raven Press, New York, p. 379.

Mattison, D. R., Blann, E., and Malek, A. (1991). *Fundam. Appl. Toxicol.,* 16: 215.

McBride, W. G. (1961). *Lancet, ii:* 1358.

Milunsky, A., Ulcickas, M., Rothman, K. J., Willett, W., Jick, S. S., and Jick, H. (1992). *JAMA,* 268: 885.

MRC Vitamin Study Research Group (prepared by N. Wald with assistance from J. Sneddon, J. Densem, C. Frost, and R. Stone). (1991). *Lancet,* 338: 132.

Nau, H. (1986). *Environ. Health Perspect.,* 70: 113.

Nau, H. (1992). *Fetal and Neonatal Physiology,* Vol. 1. (R. A. Polin and W. W. Fox, eds.). Saunders, New York, p. 130.

Nau, H., and Scott, W. J. (1986). *Nature,* 323: 276.

Nau, H., and Scott, W. J. (1987). *Pharmacokinetics in Teratogenesis,* Vols. 1 and 2. CRC Press, Boca Raton, FL.

Nawrot, P. S., Cook, R. O., and Staples, R. E. (1980). *Teratology,* 22: 279.

Nawrot, P. S., Cook, R. O., and Hamm, C. W. (1981). *J. Toxicol. Environ. Health,* 8: 151.

Neims, A. H., Warner, M., Loughnan, P. M., and Aranda, J. V. (1976). *Ann. Rev. Pharmacol. Toxicol.,* 16: 427.

Nelson, C. J., and Holson, J. F. (1978). *J. Environ. Pathol. Toxicol.,* 2: 187.

Neubert, D. (1989). *Advances in Applied Toxicology* (A. D. Dayan and A. J. Paine, eds.). Taylor and Francis Ltd, London, p. 191.

Neubert, D., Zens, P., Rothenwallner, A., and Merker, H.-J. (1973). *Environ. Health Perspect.,* 5: 63.

Neubert, R., Nogueira, A. C., and Neubert, D. (1991). *Naunyn-Schmiedeberg's Arch. Pharmacol.,* 344 (suppl. 2): R123 (77).

Neubert, R., Nogueira, A. C., Helge, H., Stahlmann, R., and Neubert, D. (1992a). *Risk Assessment of Prenatally-Induced Adverse Health Effects* (D. Neubert, R. J. Kavlock, and H.-J. Merker, eds.). Springer-Verlag, Berlin, p. 313.

Neubert, R., Nogueira, A. C., and Neubert, D. (1992b). *Life Sci.,* 51: 2107.

Neubert, R., Nogueira, A. C., and Neubert, D. (1993). *Arch. Toxicol.,* 67: 1.

Neubert, R., Hinz, N., Thiel, R., and Neubert, D. (1996). *Life Sci.,* 58: 295.

Newman, L. M., Johnson, E. M., and Staples, R. E. (1993). *Reprod. Toxicol.,* 7: 359.

O'Flaherty, E. J., Scott, W. J., Shreiner, C., and Beliles, R. P. (1992). *Toxicol. Appl. Pharmacol.,* 112: 245.

Palmer, A. K. (1993). *Current Issues in Drug Development. II.* Huntington Research Centre, Ltd, UK, p. 1.

Renault, J.-Y., Melcion, C., and Cordier, A. (1989). *Teratogen. Carcinogen. Mutagen.,* 9: 83.

Rodier, P. M., Cohen, I. R., and Buelke-Sam, J. (1994). *Developmental Toxicology,* 2nd ed. (C. A. Kimmel and J. Buelke-Sam, eds.). Raven Press, New York, p. 65.

Rogers, J. M., and Kavlock, R. J. (1996). *Casarett and Doull's Toxicology* (C. D. Klaassen, ed.). McGraw-Hill, New York, p. 301.

Rogers, J. M., and Mole, M. L. (1997). *Teratology,* 55: 364.

Rogers, J. M., Taubeneck, M. W., Daston, G. P., Sulik, K. K., Zucker, R. M., Elstein, K. H., Jankowski, M. A., and Keen, C. L. (1995). *Teratology, 52:* 149.

Schardein, J. L. (1993). *Chemically Induced Birth Defects,* 2nd ed. Marcel Dekker, Inc., New York.

Schardein, J. L., and Keller, K. A. (1989). *CRC Crit. Rev. Toxicol., 19:* 251.

Schwetz, B. A., Morrissey, R. E., Welsch, F., and Kavlock, R. J. (1991). *Environ. Health Perspect., 94:* 265

Seller, M. J., Perkins, K. J., and Adinolfi, M. (1983). *Teratology, 28:* 123.

Sever, J. L. (1967). *Adv. Teratol., 2:* 127.

Shenefelt, R. E. (1972). *Teratology, 5:* 103.

Shepard, T. H. (1992). *Catalog of Teratogenic Agents,* 7th ed. The Johns Hopkins University Press, Baltimore.

Shuey, D. L., Lau, C., Logsdon, T. R., Zucker, R. M., Elstein, K. H., Narotsky, M. G., Setzer, R. W., Kavlock, R. J., and Rogers, J. M. (1994). *Toxicol. Appl. Pharmacol., 126:* 129.

Slikker, W., and Miller, R. K. (1994). *Developmental Toxicology,* 2nd ed. (C. A. Kimmel and J. Buelke-Sam, eds.). Raven Press, New York, p. 245.

Smith, J. L., Gesteland, K. M., and Schoenwolf, G. C. (1994). *Dev. Dynam., 201:* 279.

Stephens, T. D. (1988). *Teratology, 38:* 229.

Stott, D. H. (1973). *Dev. Med. Child. Neurol., 15:* 770.

Teratology Society. (1994). *Teratology, 49:* 446.

Uphill, P. F., Wilkins, S. R., and Allen, J. A. (1990). *Toxicol. in Vitro, 4:* 623.

U.S. Environmental Protection Agency. (1985). *Fed. Reg., 50:* 39429.

U.S. Environmental Protection Agency. (1991a). *Pesticide Neurotoxicity Assessment Guidelines, Subdivision F, Hazard Evaluation: Human and Domestic Animals, Addendum.* Series 81, 82, and 83. EPA 540/09-91-123 PB 91-154617.

U.S. Environmental Protection Agency (1991b). *Fed. Reg., 56:* 63798.

U.S. Food and Drug Administration (1979). *Fed. Reg., 44:* 37434.

U.S. Food and Drug Administration. (1966). *Guidelines for Reproduction Studies for Safety Evaluation of Drugs for Human Use.* Rockville, MD.

U.S. Food and Drug Administration (1994). *Fed. Reg., 59:* 48746.

Wald, N. (1993). *Ann. NY Acad. Sci., 678:* 112.

Warkany, J. (1945). *Vit. Horm., 3:* 73.

Warkany, J. (1965). *Teratology: Principles and Techniques* (J. G. Wilson and J. Warkany, eds.). University of Chicago Press, Chicago, p. 1.

Warkany, J. (1986). *Teratology, 33:* 365.

Warkany, J., and Nelson, R. C. (1940). *Science, 92:* 383.

Warkany, J., and Schraffenberger, E. (1944). *Am. J. Roentgenol. Radium Ther., 57:* 455.

Welsch, F. (1990). *Issues and Review in Teratology,* Vol. 5 (H. Kalter, ed.). Plenum Press, New York, p. 115.

Wilson, J. G. (1959). *J. Chron. Dis., 10:* 11.

Wilson, J. G. (1973). *Environment and Birth Defects.* Academic Press, New York.

Wilson, J. G. (1977). *Handbook of Teratology* (J. G. Wilson and F. C. Fraser, eds.). Plenum Press, New York, p. 309.

Wilson, J. G., Roth, C. B., and Warkany, J. (1953). *Am. J. Anat., 92:* 189.

4

Malformations in Pregastrulation Developmental Toxicology

Joe C. Rutledge
Children's Hospital and University of Washington, Seattle, Washington

Walderico M. Generoso
Oak Ridge National Laboratory, Oak Ridge, Tennessee

INTRODUCTION

The phenomenon in which the pregastrulation stage in mouse embryogenesis is a window in experimental induction of developmental anomalies has gained wide acceptance. Depending upon the agent used, this window may be restricted to the early zygotic period or it may be wide, involving embryos from cleavage through blastocyst stages. A common feature among xenobiotic agents that are toxic to pregastrulation stages is the induction of embryonic lethality that occurs around the time of implantation. Certain agents also induce mid-to-late gestation lethality and anatomical malformations in live fetuses, while others induce anatomical malformations in live fetuses with no fetal lethality associated. While this field of developmental toxicology is still in the formative period, it is already clear that many anatomical defects induced are unique to the pregastrulation stages and may have significance to the (1) specific interactions between teratogens and the conceptus, (2) existence of determinative developmental process already at pregastrulation stages, which is contrary to the present dogma, and (3) origin of a large class of human anatomical malformations. Thus, this chapter will focus on the characterization of pregastrulation-derived developmental anomalies with respect to their relevance to clinical observations and as they compare with malformations induced during organogenesis.

GENERAL PROCEDURE

The main bulk of results discussed in this chapter were generated in this laboratory. Fertilization was synchronized in naturally ovulating females by caging them with males

Research jointly sponsored by the National Toxicology Program under National Institute of Environmental Health Sciences Interagency Agreement Y01-ES-20085 and the Office of Health and Environmental Research, U.S. Department of Energy, under contract DE-AC05-96OR22464 with Lockheed Martin Energy Research Corporation.

for a period of 30 minutes early in the morning when freshly ovulated eggs are known to be already in the ampulla. In most cases fetuses were examined only for external anomalies, at gestational day (GD) 17 or GD 18. Skeletal analysis of GD-18 fetuses was performed in two cases, and a follow-up characterization of defects at midgestation was conducted in another.

For the purpose of relating a specific stage in embryogenesis treated to the developmental outcome, it is important to keep in mind that the results summarized in this chapter involved treatments by single intraperitoneal injection, by single exposure to acute ionizing radiation, or by inhalation for one hour. Therefore, unless a given agent or its active metabolite is persistent, the developmental anomalies induced may be attributed to the response of the specific embryonic stage at the time of treatment. The latter is presumed to be the case.

CLASSES OF DEVELOPMENTAL DEFECTS PRODUCED FROM TREATMENT OF PREGASTRULATION STAGES

The intent of this chapter is to provide a characterization of the classes of developmental defects that have been induced, so far, from treatment of pregastrulation stages, from zygote to blastocyst, and compare them to those induced in conventional teratology and to those that are observed clinically. Agents that are active during the pregastrulation stages vary with respect to the (1) window of susceptibility, (2) classes of developmental anomalies induced, and (3) rates at which the anomalies were induced. However, in order not to distract attention from this goal, little consideration will be paid to the comparison between specific agents and between the rates at which various classes of defects were induced, as these have been partially summarized in other reports (Rutledge et al., 1992).

Lethality

Peri-Implantation

A common feature among pregastrulation toxicants is the induction of lethality around the time of implantation. In fact, it may be stated that the first indication that a given agent is active in pregastrulation stages is the increase in the incidence of resorption bodies. This is true regardless of whether the agent is effective only in early zygotes or only in embryonic stages. At relatively high doses, treatment of zygotes may result in reduced implantation rates presumably due to death so early in embryonic development as to be lost prior to implantation. It may also be stated that, for all agents, increased incidence of external defects in living fetuses has always been associated with significant increases in resorption bodies, although the reverse may not necessarily be true.

Mid-to-Late Gestation

One of the most striking observations in pregastrulation teratology has been the narrow window of susceptibility during zygote development in which chemicals, such as ethylene oxide and ethyl methanesulfonate, induce varied developmental anomalies, including mid- and late-gestation deaths. These effects were induced by treatment 1.5

or 6 h postmating, but essentially not by treatment at 9 or 24 h. Such deaths are rare sequelae to exposure of germ cells to mutagens. Observation at varying gestational ages showed a higher rate of hydrops in midgestation than in late, suggesting that some gestational deaths occurred with hydropia (Rutledge and Generoso, 1989).

Hydropia and Associated Defects

Like mid-to-late fetal death, the high incidence of hydrops, generalized fetal edema, is also a feature of zygote exposure and distinguishes this period of development from organogenesis (Fig. 1). While hydrops is observed after teratogen exposure during organogenesis, the high rate and reproducibility among different mutagens is a distinguishing feature of zygotic exposure. Hydrops is first seen as pericardial edema on gestational day 13 and becomes more widely distributed on subsequent gestational days as it manifests as ascites and then subcutaneous edema. Hydrops may be transient in some fetuses, since its incidence early in gestation exceeds that occurring later. Many, but not all, fetuses with other anomalies have edema.

Other malformations occur together with hydropia, but not in an absolutely concordant fashion. Detailed studies of the skeletons of fetuses treated as zygotes with ethylene oxide showed a high incidence of sternal clefting and of bifurcation of the second cervical vertebrae (Polifka et al., 1996). Neither of these are common in control animals, neither were they observed when fetuses were exposed during organogenesis. Because hydropia begins before the sternum forms, it has been postulated that pressure created by the expansion of the pericardial sac may be responsible for the failure of the sternal plates to fuse. Similarly, the abdominal wall defects and opened eyelids

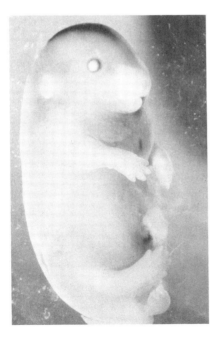

Figure 1   Eighteen-gestational-day fetus with hydrops after zygotic treatment with EMS. Note the diffuse subcutaneous edema, open eyelids, and abnormally positioned hind limbs.

observed in the zygote experiments may also be attributable to the inability of the bowel to return to the abdomen or failure of eyelid fusion to occur due to edema. Thus, hydropia could be a unifying idea for some of the observed anomalies.

Limb and Tail Abnormalities

A peculiar observation regarding the limbs is that while defects were observed, they were restricted to the hind limbs. Basically, there have been two types of hind limb and tail abnormalities induced in pregastrulation stages: bent or malpositioned limbs and duplication. The latter type is discussed separately later.

Bent or malpositioned hind limbs are a feature of exposed zygotes. One or both limbs, of normal length and constitution, are malpositioned, being bent proximally upward and toward the body or, in some instances, caudally with rotation of the plantar surface (Fig. 1). The limb and bent tail anomalies may be deformations rather than malformations, since skeletal staining did not reveal intrinsic defects of the bone. Moreover, in a few live-born affected mice, the limb defects resolved with time.

Abdominal Wall Defects

Abdominal wall defects vary from minute, skin-covered umbilical hernia with slight splaying of abdominal wall musculature to fully opened abdominal wall with extrusion of the intestines and liver (Fig. 2). This class of defect seems to be omnipresent, ap-

Figure 2 Eighteen-gestational-day fetus with liver and intestines protruding through a nonclosed anterior abdominal wall. This most severe end of a spectrum of abdominal wall defects occurred after ethylene oxide treatment of the zygote. Lower limb is malpositioned and left eyelid is not fused.

pearing in most experiments in which malformations have been significantly increased. It has been observed following treatment of zygotes, early cleavage embryos, and blastocysts. Abdominal wall defects were the most common type of malformation induced by exposure of zygotes to x-irradiation and neutrons (Pampfer and Streffer, 1988; Rutledge et al., 1992). The high rate of abdominal wall defect occurs in a stock with a high background rate (3%) in the laboratory of Pampfer and Streffer (1988). Gastroschisis has been associated with an increased rate of confined protein alterations in affected fetuses as examined by two-dimensional electrophoresis (Hillebrandt and Streffer, 1994). These nonspecific alterations point to the possible epigenetic origin of the induced malformations.

In certain cases the occurrence of abdominal wall defects may be associated with other defects. For example, those derived from chemical exposure of zygotes may have hydropia underlying the pathogenesis, as discussed earlier. Another example, based on a limited number of cases, is the presence of abdominal wall defect in many fetuses with exencephaly derived from treatment of blastocysts with *trans*-retinoic acid (Rutledge et al., 1994). The significance of this observation is not clear.

Eye Defects

Eye defects ranged from open eyelids to micro-ophthalmia and coloboma (Figs. 1–3). Like abdominal wall defects, they were also omnipresent, having been observed following treatment of different pregastrulation stages. Moreover, the eyes were affected to variable degrees in association with exencephaly when the absence of skull extended inferiorly to involve the orbits and eyes. In these instances, they are considered to be

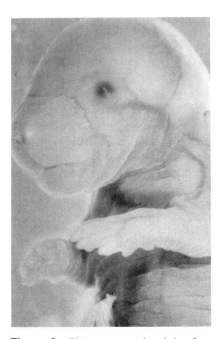

Figure 3 Eighteen-gestational-day fetuses with microophthalmia and coloboma (left) occurring after EMS treatment of the zygote.

part of the spectrum of the anterior neural tube defect. Absence of one or both eyes was also part of the spectrum of the severe craniofacial anomalies observed after treatment of pregastrulation stages with 5-azacytidine (Dellarco, personal communication). Similar defects, when occurring in combination with other defects, may well point to alternative pathogenetic mechanisms.

Exencephaly

Exencephaly, absence of the skull resulting from lack of anterior neural tube closure, is a common anomaly induced during organogenesis and observed in humans as absence of the brain, but termed anencephaly. Treatment of the zygote stages with most mutagens does not appreciably or consistently increase the incidence of exencephaly. Treatment with the genetic modifiers retinoic acid and 5-azacytidine, though, is very different. Exencephaly is induced at both zygotic stages and other pregastrulation stages with 5-azacytidine (Dellarco, personal communication), while retinoic acid produces the defect after the zygotic stage (Rutledge et al., 1994).

Neural tube closure is an event susceptible to teratogens during organogenesis and during the preimplantation period. The neural tube is one of the earliest structures defining the body plan, so it would not be unexpected that agents that disrupt the embryonic program would have adverse impacts on neural tube closure. The impact is, however, only on anterior neural tube closure, as the posterior spina bifida defect, myelomeningocele, is much less common in the pregastrulation experimental system in contrast to their relatively frequent occurrence in humans.

Craniofacial Defects

General

The spectrum of cranial facial anomalies is broad, in part due to the complexity of the structures involved. Isolated defects include ocular defects outlined above, exencephaly and associated structural deficiencies, cleft palate formation, and micrognathia. Cleft palate occurs after treatment of the zygote with a variety of mutagens and is usually an isolated defect. The rate is low. Micrognathia, small mandible, is uncommon and unassociated with other facial defects. In general, craniofacial defects are not induced at high rates, with the exception of those induced later in the pregastrulation period by certain agents.

Frontonasal Dysplasia

In addition to a high rate of exencephaly and ocular defects, 5-azacytidine produces throughout pregastrulation an unusual craniofacial malformation (Dellarco, personal communication). Affected fetuses have exencephaly with absent eyes and the nasal processes with maxilla are splayed laterally with the frontal lobes juxtaposed to tongue (Fig. 4). The mandible may be present, or in severe cases absent, with poorly defined nasal processes. Variations in positioning of the nasal processes from projecting lateral to midline fusion into an unrecognizable mass with no nares are intermixed with similar variations from anophthalmia to rudimentary eyes. The brain is occasionally diminutive and rarely the entire head is markedly reduced in size. It is unclear if the exencephaly and eye defects observed in these treatment groups might be part of the

Figure 4 Eighteen-gestational-day severe frontonasal dysplasia after treatment of GD 4 embryo with 5-azacytidine. Nasal processes are unfused and exencephalic brain protrudes anteriorly over the tongue in absence of cranial bones.

same spectrum. This syndrome is relatively unique among agents studied thus far in the pregastrulation stages.

Caudal Duplications

In contrast to the above-detailed anomalies with loss of body structures, retinoic acid induced duplications of body parts, but only when administered on gestational days 4.5–5.5. The syndrome was almost exclusively confined to the caudal region of the body and ranged from hindlimb digital duplications to an entire extracaudal region of the body (Figs. 5 and 6). In the former, partial extra foot structures were demonstrable by skeletal staining of cleared fetuses. The extreme examples included an accessory caudal segment with both supranumerary hindlimbs and rudimentary tails. The external genitalia were often duplicated, and internal examinations revealed some instances of renal duplication (J. E. Polifka, unpublished). The skeletal examinations showed that the vertebral column was not duplicated, but rather the pelvic bones were partially duplicated with an aberrant set of supranumeray limbs. This type of anomaly is unique in mammalian teratogenic studies.

COMPARISON BETWEEN PREGASTRULATION- AND ORGANOGENESIS-DERIVED DEVELOPMENTAL DEFECTS

General Differences

The main bulk of research in the field of experimental teratology has concentrated on the effects of exogenous agents on embryos during the period of organogenesis and

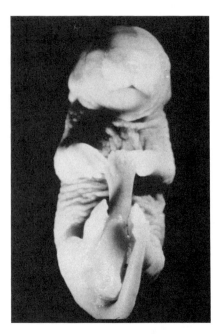

Figure 5 Eighteen-gestational-day fetus with caudal duplications including single supranumerary limb and micro-ophthalmia. Retinoic acid treatment at GD 5.

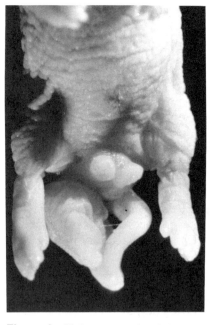

Figure 6 Eighteen-gestational-day caudal duplication with remnants of extra limbs and duplication of external genitalia. Retinoic acid treatment on GD 5.

growth. Depending upon the agent, dose, time of administration, genetic factors, etc., teratologists can induce a wide variety of fairly well-defined anomalies, often in the same susceptible organ system. Those induced during organogenesis are generally classified as malformations in which, in most cases, whole or part of organ systems fail to develop properly: syndactyly, hydrocephaly, spina bifida, exencephaly, ventricular septal defects, tail shortening, cleft palate, and hydronephrosis (LaBorde and Kimmel, 1980; Takeuchi, 1984; Daston et al., 1987; Wada et al., 1994). While certain of these malformations may be seen in pregastrulation experiments, they are much more restricted and many organs are relatively spared—for example, the genitourinary, pulmonary, cardiac, and gastrointestinal systems.

Hydrops

Certain types of developmental anomalies distinguish pregastrulation from organogenesis. Hydrops is a common anomaly after zygotic treatment; we have not yet established its pathogenesis. It occurs in association with other anomalies, especially cleft sternum and to a lesser degree bent limbs. It is possible that expansion of the pericardial sac by fluid, the earliest gestational site of hydrops, may prevent the sternum from forming properly. An analogous, but more speculative association is that expansion of the abdomen with fluid may distort limb positioning and may prevent the abdominal wall from closing fully, leading to abdominal wall defects. Because fewer of the fetuses with the later two defects had hydrops, such an association would depend on resolution of the fetal edema by gestational day 18. It is possible that hydrops in the fetus represents the basis for several anomalies.

The pathogenesis of hydrops is recognized to be multifactorial but usually is dependent on events late in gestation. In certain situations in humans (e.g., fetal anemia), the fluid accumulation is secondary to cardiac failure and reduced intravascular oncotic pressure. In Mirex-induced fetal edema in rats, the pathogenesis is related to heart failure from rhythm disturbances (Grabowski and Payne, 1983). In the case of hydrops with human sex chromosome loss, the pathogenesis is unclear, but hydrops may be transient in utero. It may also attend fetuses treated with a variety of teratogens, although it is not commonly induced. Focal subcutaneous edema is induced in rat fetuses as subectodermal blisters by ethylenethiourea treatment, but this localized cutaneous edema differs from the more generalized edema after zygotic treatment (Daston et al., 1987). Fetal hydrops is a nonspecific disorder caused in midgestation by hypoxia, for example, but it occurs commonly after chemical zygotic treatment with a variety of agents. This frequency distinguishes the pregastrulation period from organogenesis.

Craniofacial

Frontonasal Dysplasia

The severity and extent of certain pregastrulation-induced craniofacial syndromes further distinguish this developmental period from organogenesis. The severe frontonasal dysplasia induced by 5-azacytidine represents the extreme of a spectrum of midfacial malformations. During a set period of exposure, however, a continuum of malformations (syndrome) should be considered, and this concept confuses the issue as it relates to the singularity of a defect. Some components (e.g., cleft palate), are common after many different teratogens employed during organogenesis. The absence of midfacial structures is decidedly unusual. At a given dose and exposure, malformations of the same developmental field should be considered part of a continuum. In teratogenesis involving earlier stages, we may be more likely to observe a continuum of pathology. We may also expect an overlap of some features of a pregastrulation syndrome with those induced during organ formation.

Frontonasal malformations similar to those induced by 5-azacytidine have been produced by methotrexate administration on gestational day 9 in mice (Darab et al., 1987) and related to blood-filled blebs in the nasal placode region and distension of the neural tubes occurring within hours after administration. This possible mechanism must be different from that involved with the 5-azacytidine effects.

Eye

The ocular malformations observed are macroscopically identical to those induced by teratogens during organogenesis. X-ray irradiation of mice on gestational days 8 and 9 but not 6 or 7 induce anophthalmia, micro-ophthalmia, and coloboma (Kuno et al., 1994). Induced hyperthermia (42–23°C) on mouse gestational day 6.5 produces micro-ophthalmia in association with growth retardation.

Open Neural Tube Defect

In most of our experiments involving zygotic treatment, we did not induce anencephaly, oxycephaly, or spina bifida. This stands in contrast to the frequent occurrence of these defects accompanying treatment during organogenesis. Treatment of postzygotic pregastrulation stages, however, results in an increased incidence, sometimes with a broader set of craniofacial malformations.

Duplication Syndrome

An induced syndrome unique to pregastrulation teratogenesis is the caudal duplication syndrome induced by retinoic acid. This syndrome has a broad range of expression ranging from duplication of portions of hindlimbs to entire caudal duplication. This unique set of malformations differs from those induced with organogenesis treatment by both retinoic acid and most teratogens, which usually results in limb reduction defects rather than duplication disorders (Sanders and Stephens, 1991). Among the few duplication anomalies resulting from teratogens is polydactyly, which is induced at low rates after methynitrosourea treatment of gestational day 2 or 3 in mouse embryos (Nagao et al., 1991). Duplication of a limb or a digit could be induced during the limb bud stage and later, but duplication of a segment of the body must relate to the earlier stages of the formation of the body plan. The one-day gestational period of susceptibility suggests that the plan can be altered prior to gastrulation. Malformations that develop as a consequence of alteration in the body program seem to mark the post-zygotic/pregastrulation period as a special window in teratogenesis.

RELEVANCE TO HUMAN BIRTH DEFECTS

The etiology of most human birth defects remains unestablished. Estimates indicate that about 50% of human fertilizations do not result in viable offspring, and among those surviving to birth, 3–5% have major congenital anomalies exclusive of prematurity, growth problems, and late-onset diseases of genetic or nongenetic origin. The etiology of these anatomical anomalies is unknown in 65%, genetic (including aneuploidy) in 25%, environmental in 10% (mainly maternal or infectious), and due to exogenous agents in only about 1% (Brent, 1987). The group of birth defects with unknown etiology harbors a broad spectrum of developmental defects, and the syndromes associated with pregastrulation teratology may have relevance to the origin of some of these defects. The interactions and contributions of genetic factors and environmental factors is at best unclear (Rutledge, 1994).

Sporadic Defects

A large class of human anomalies manifest as defects in embryogenesis similar to those observed after exposure to the zygote: abdominal wall defects, hydrops, clubfoot, cleft palate, growth retardation. In the human instances, these common defects may occur in isolation or with other malformations. Most seem to be sporadic unless combined with a known inheritance pattern. This group of developmental defects is without established etiology. The experiments perturbing development in the one-cell stage and some of those producing defects after pregastrulation treatment suggest alternatives in our search for causation.

In general, we have not, in human studies, concentrated on events around the time of conception. One study in a well-defined population could not find periconceptional events leading to malformations in humans (Siffel and Czeizel, 1995). We have also not given much thought to the notion that human gestation might be alterable in the pregastrulation period. While it seems likely that common anomalies may have a genetic component, our multifactorial model has not considered that a laxity in preorganogenesis development might also predispose to anomalies that manifest as errors during organogenesis. While the studies to date do not assign the origin of human anomalies to the pregastrulation period (much less, implicating environmental exposures), they raise the possibility that new alternative mechanisms may exist in poorly studied mammalian early developmental systems.

Specific Defects

Some of the defects observed after pregastrulation treatment of mouse embryos are unique, and most dramatic is the caudal duplication syndrome induced by retinoic acid. Caudal duplication syndromes are present rarely in humans (Dominguez et al., 1993). Some affected human have supranumerary malformations that closely resemble those described above; because of their uniqueness they have been illustrated over the centuries in medical texts as curiosities. These types of anomalies in humans overlap with entities termed parasitic twins and with some teratomas. More recently, these limb duplications have been compared to those observed in mice with the semi-dominant gene *Disorganization* (Petzel and Erickson, 1991; Crosby et al., 1992). The experimental caudal duplication also resembles the malformations in mice with *Disorganization*, and it seems likely that the genetic defect in that animal may be traced to the early body plan program.

The less severe hindlimb defect of malpositioning is much more common in the human analogy: clubfoot. This sporadic condition is a common deformation that can be corrected by orthopedic casting and surgery. In some patients, it occurs in association with decreased amniotic fluid (oligohydramnios) and is hence derived from fetal constraint. In our studies, the amount of amniotic fluid was normal; the analogy then is to instances of clubfoot of unknown etiology.

Hydrops in humans is relatively common, especially accompanying fetal death, and is multifactorial (Machin, 1989). While the pathogenesis can be ascertained in many cases, in others it is idiopathic. The frequency of this finding after preimplantation treatment of mice opens another avenue of investigation.

Most occurrences of micro-ophthalmos and colobomas in humans have a genetic basis (Warburg, 1992). Cases do occur in association with other anomalies (CHARGE or VATER associations), which are sporadic. They have also been attributed to teratogenic exposure including maternal infection and fever. The neural crest is a likely target of these exposures. Sporadically occurring colobomas are associated with mental retardation and other anomalies (Leppig and Pagon, 1993). In particular, several cases of anophthalmos and choroidal colobomas are reported in association with a syndrome of limb deformities and open abdominal wall (Jensen et al., 1993). Some cases of these eye disorders have an unknown etiology and are associated with other nonrare malformation syndromes or associations.

The median facial cleft induced by 5-azacytidine may be a severe form of frontonasal dysplasia, which in humans is a nonspecific, usually sporadic developmental field defect (Gorlin et al., 1990). As with the mouse, this is among the most severe of human anomalies and may be accompanied by other anomalies. When combined with micro- or anophthalmia it is classified as ophthalmofrontonasal dysplasia. It seems reasonable that global craniofacial malformations may originate in early embryogenesis (Webster et al., 1988).

PERSPECTIVES

The pregastrulation-derived developmental anomalies have been characterized, for the most part, only at the level of external morphology. Nevertheless, for the purpose of relating them to embryological mechanisms and human birth defects, it is useful to describe them in terms of the accepted system of nomenclature. Anatomical anomalies form several conceptional groups: malformations, disruptions, syndromes, and associations (Stevenson et al., 1993). Malformations imply intrinsic errors in embryogenesis. In the case of deformations, the developmental program is unaltered, but growth after organogenesis is aberrant due to exerting mechanical forces. Thus, deformed body parts would grow normally in organ culture. Destruction of formed structures is classified as a disruption. Groups of anomalies related to a common etiology are termed syndromes (although some syndromes may have multiple etiologies), while those anomalies that often occur together without regard to etiology are classified as associations.

Developmental anomalies produced from treatment of pregastrulation stages are clearly malformations. As such, they are presumed to arise from alterations in certain embryological processes. If, indeed, the primary damage responsible for the malformations is caused directly by the chemicals administered and not by late-appearing metabolites that linger through gastrulation and later stages, it may be argued that the embryological processes that were disrupted were already in existence at the time of embryonic development when cells have been thought to still be totipotent. Note that chemicals such as ethylene oxide and ethyl methanesulfonate are very reactive with half-lives of less than one hour.

Two aspects of the pregastrulation effects may be important in this context. One is the narrow window within which certain anomalies were induced, e.g., hydropia, in the case of ethylene oxide or ethyl methanesulfonate, and limb and caudal duplications, in the case of retinoic acid. Another is the seemingly syndromatic manifestation of many of the malformations, e.g., patterns of anterior malformations, hydropia and related malformations, and duplication of entire limb or lower body portion. These syndromes have similarities to mutations in major genes with respect to pleiotropism and variable expression. Thus, it is tempting to speculate that chemical perturbations during pregastrulation stages may cause alterations in primary embryological areas, which, with development, migration, growth, etc., result in dispersed anomalies.

The developmental field defect concept explains some instances of difficult-to-reconcile anatomical defects (Opitz et al., 1986). More recently, Opitz (1993) has extended the concept of the developmental field to include blastogenesis. Malformations originating at these stages are presumed to be severe, complex, involving multiple systems, and lethal and appear like the above-described syndromes. It seems reasonable that damage occurring very early in embryogenesis would have these characteristics and would involve genes important in organization (Bamforth, 1994).

The developmental field concept was originally explained in terms of cell populations and corresponding anatomical regions. Damage to a progenitor cell population prior to its dispersal in the embryo may result in widespread malformations in the surviving embryo when the anatomical distribution of the malformations may be difficult to relate to damage occurring during organogenesis. Moreover, the developmental programs for establishing the body plan may begin in the pregastrulation stage, thus allowing for sets of malformations that, while not specific to damage at that stage, may be somewhat unique compared to those induced later in gestation.

Since agents that have been shown to induce pregastrulation–derived malformations are either mutagens (e.g., methylnitrosourea) or known modulators of gene expression (e.g., retinoic acid and 5-azacytidine), the pregastrulation malformations were hypothesized to arise from disruption in gene expression either through chromosomal damage (Streffer, 1993) or through epigenetic mechanisms (Katoh et al., 1989) that lead to cascades of temporally and spatially altered gene expression and of cellular proliferation and distribution.

Just a few years ago, the idea that the pregastrulation stages of mammalian development are a window in experimental induction of anatomical defects was not in the mainstream. The field of pregastrulation teratology is now not only well established, but it also begs research attention because of (1) its likely relevance to the etiology of many unexplained human birth defects, (2) its importance to our understanding of embryological processes during very early stages of mammalian development, and (3) the question it raises regarding current procedures used in screening developmental toxicants. The first two points have been touched on repeatedly throughout this chapter. The third is concerned with the issue of whether the pregastrulation stages should be specifically targeted in the detection and risk evaluation of developmental toxicants. It should be fair to say that all three are both exciting as well as important areas of research.

REFERENCES

Bamforth, J. S. (1994). *Reprod. Toxicol.,* 8: 455.

Brent, R. L. (1987). *Bradbury Report 26: Developmental Toxicology: Mechanisms and Risk* (J. A. McLachlan, R. M. Pratt, and C. L. Markert, eds.), p. 287.

Crosby, J. L., Varnum, D. S., Washburn, L. L., and Nadeau J. H. (1992). *Mechan. Dev., 37*: 121.

Darab, D. J., Minkoff, R., Sciote, J., and Sulik, K. K. (1987). *Teratology, 35*: 77.

Daston, G. P., Ebron, M. T., Carver, B., and Stefanadis, J. G. (1987). *Teratology, 35*: 239.

Dominquez R., Rott, J., Castillo, M. Pittaluga, R. R., and Corriere, J. N. Jr. (1993). *Am. J. Dis. Child., 147*: 1048.

Gorlin, R. J., Cohen, M. M., Jr., and Levin, L. S. (1990). *Syndromes of the Head and Neck*, 3rd ed. Oxford University Press, New York, p. 785.

Grabowski, C. T., and Payne, D. B. (1983). *Teratology, 27*: 7.

Hillebrandt, S., and Streffer, C. (1994). *Mutat. Res., 308*: 11.

Jensen, O. A., Hagerstrand, I., Brun, A., and Lofgren, O. (1993). *Pediatr. Pathol., 13*(4): 505.

Kalter, H. (1992). *Teratology, 46*: 207.

Katoh, M., Cacheiro, N. L. A., Cornett, C. V., Cain, K. T., Rutledge, J. C., and Generoso, W. M. (1989). *Mutat. Res., 210*: 337.

Kuno, H., Kemi, M., and Matsumoto, H. (1994). *Exp. Animals, 43*(1): 115.

LaBorde, J. B., and Kimmel, C. A. (1980). *Toxicol. Appl. Pharmacol., 56*: 16.

Leppig, K. A., and Pagon, R. A. (1993). *Clin. Dysmorph.*, 2(4): 322.

Machin, G. A. (1989). *Am. J. Med. Genet., 34*: 366.

Nagao, T., Morita, Y., Ishizuka, Y., Wada, A., and Mizutani, M. (1991). *Teratogen. Carcinogen. Mutagen., 11*: 1.

Optiz, J. M. (1993). *Birth Defects: Original Article Series, 29*: 3.

Optiz, J. M., Reynolds, J. F., and Spano, L. M., eds. (1986). *The Developmental Field Concept.* Alan R. Liss, Inc., New York.

Pampfer, S., and Streffer, C. (1988). *Teratology, 37*: 599.

Petzel, M. A., and Erickson, R. P. (1991). *J. Med. Genet., 28*: 712.

Polifka, J. E., Rutledge, J. C., Kimmel, G. L., Dellarco V., and Generoso, W. M. (1996). *Teratology, 53*: 1.

Rutledge, J. C. 1994. *Developmental Toxicology* (C. A. Kimmel and J. Buelke-Sam, eds.) Raven Press, New York, p. 333.

Rutledge, J. C., and Generoso, W. M. (1989). *Teratology, 39*: 563.

Rutledge, J. C., Generoso, W. M., Shourbaji, A., Cain, K. T., Gans M., and Oliva, J. (1992). *Mutat. Res., 296*: 167.

Rutledge, J. C., Shourbaji, A. G., Hughes, L. A., Politka, J. E., Cruz, Y. P., Bishop, J. B., and Generoso, W. M. (1994). *Proc. Natl. Acad. Sci., 91*: 5436.

Sanders, D.D., and Stephens, T. D. (1991). *Teratology, 44*: 335.

Siffel, C., and Czeizel, A. E. (1995). *Mutat. Res., 334*: 293.

Stevenson, R. E., Hall, J. G., and Goodman, R. M. (1993). *Human Malformations and Related Anomalies*. Oxford University Press, New York, p. 21.

Streffer, C. (1993). *Mutat. Res., 299*: 313.

Takeuchi, I. K. (1984). *Experientia, 40*: 879.

Wada, A., Sato, M., Takashima, H., and Nagao, T. (1994). *Teratogen. Carcinogen. Mutagen., 14*: 271.

Warburg, M. (1992). *Ophthal. Paed. Genet., 13*(2): 111.

Webster, W. S., Lipson, A. H., Sulik, K. K. (1988). *Am. J. Med. Genet., 31*: 505.

5
Developmental Effects of Dioxins

Linda S. Birnbaum

U.S. Environmental Protection Agency, Research Triangle Park, North Carolina

INTRODUCTION

Dioxin is the generic name for a broad group of chemical compounds that have a similar structure, common mechanism of action, and common spectrum of biochemical and toxicological effects (for recent reviews see Birnbaum, 1994a; Devito and Birnbaum, 1995). The prototypical compound for this group is the most toxic and best-studied member, 2,3,7,8-tetrachlorodibenzo-*p*-dioxin (TCDD). Initially identified as a toxic contaminant during production of certain herbicides, it is now known to be an unwanted byproduct of many types of combustion and industrial processes, as well as natural events such as volcanos and forest fires. Prior to the introduction of the large-scale industrial use of chlorine, core sediment samples failed to detect the presence of TCDD. However, since the 1920s, there has been a sharp rise in the presence of the dioxins, which appeared to peak in the late 1960s and 1970s. Recent surveys indicate that the levels in both environmental and biological samples have begun to decline, largely as a result of regulations both in the United States and in other industrialized countries. Although production of biocides and chlorine bleaching of paper and pulp products were major sources of these compounds during the midportion of this century, the predominant sources in the United States today are incineration of medical and municipal waste. Iron sintering, copper smelting, and production of PVC plastics have also been reported to be significant contributors to global contamination by dioxins. The importance of recycling of environmental loading is unclear. What does appear clear, however, is that the major route of exposure to the general population is via microcontamination of food. Because of its resistance to both physical and biological degradation and its lipophilicity, dioxin persists in the environment and bioaccumulates up the food chain (reviewed in U.S. EPA, 1994).

Disclaimer: The research described in this article has been reviewed by the National Health and Environmental Effects Research Laboratory, U.S. Environmental Protection Agency, and approved for publication. Approval does not signify that the contents necessarily reflect the views and policies of the Agency nor does mention of trade names or commercial products constitute endorsement or recommendation for use.

Dioxins are well absorbed following both oral and inhalation exposure (Diliberto et al., 1996). Because of their high lipophilicity, they partition into fats, resulting in preferential deposition in adipose tissue (for review of pharmacokinetics, see Van den Berg et al., 1994; Buckley, 1995). However, the distribution of these compounds is dose dependent (Diliberto et al., 1995). Dioxins induce a hepatic binding protein, tentatively identified as CYP1A2, which results in sequestration of these compounds in the liver (Andersen et al., 1993). Therefore, high-dose experimental studies, or environmental poisoning episodes, can lead to underprediction of concentrations in extrahepatic tissues present in background situations. It also raises the question whether estimation of body burdens based on lipid-adjusted serum levels will give an accurate estimation of total dioxin content.

MECHANISM OF ACTION

TCDD is but one of 75 chlorinated dibenzo-*p*-dioxins. There are 135 chlorinated dibenzofurans, 209 polychlorinated biphenyls, and hundreds of polychlorinated naphthalenes, azo- and azoxy-benzenes, as well as terphenyls, quarterphenyls, and biphenylenes. In addition, there are brominated congeners and mixed bromo-chloro compounds in these structural classes. Only a small subset of these compounds are approximate isostereomers of TCDD. However, those that are fully laterally substituted and can assume a planar conformation, can act as full or partial agonists, bringing about the same spectrum of responses. The first step in the action of all dioxin-like compounds is interaction with a cellular protein known as the Ah receptor (reviewed in Birnbaum, 1994a). Halogenated aromatics which do not bind to this receptor with high affinity do not bring about the same effects as TCDD. A rank order of binding to the Ah receptor with numerous effects such as enzyme induction, thymic atrophy, and even lethality has been well demonstrated. However, it is important to note that compounds that do not bind to the Ah receptor may have their own spectrum of adverse effects via independent mechanisms, that is, because a compound is not TCDD-like does not mean it is nontoxic.

The Ah receptor, so named because it was first associated with enzyme induction by aryl hydrocarbons, is a member of the basic helix-loop-helix family of regulatory proteins (Swanson and Bradfield, 1993). It is the first member of this family known to be ligand-activated. It is normally present in the cell complexed to heat shock protein 90 as well as additional proteins. Upon binding of TCDD or another dioxin-like chemical, the ligand-binding subunit undergoes a conformational change leading to dissociation of the complex and release of the other proteins. The ligand-bound Ah receptor then associates with its heterodimeric partner, ARNT (aryl hydrocarbon nuclear translocase). ARNT is another basic helix-loop-helix protein that has significant sequence similarity to the Ah receptor (Reyes et al., 1992). Other members of this family that share a common domain known as PAS include two regulatory proteins from *Drosophila*, *per* and *sim*, as well as the recently identified hypoxia-inducible factor 1 (Wang et al., 1995). Recent studies indicate that there may in fact be a family of ARNT-like proteins that can dimerize with the Ah receptor (Carrier et al., 1994). The ligand-bound heterodimer then binds to a specific DNA sequence, functioning as a transcriptional enhancer (Whitlock, 1993). Additional proteins may also play a role in this activation function of the Ah receptor. Functional dioxin response elements (DREs) have been found in the regulatory region of several genes involved in biotransformation reactions, such as CYP1A1, CYP1A2, glutathione-S-transferase, menadione reductase, uridine diphosphoglucuronylsyl transferase, and aldehyde dehydrogenase. In addition, DREs have been detected in the genes for several cytokines, growth factors, and steroid receptors (White and Gasiewicz, 1993).

Both the Ah receptor and ARNT have been shown to be present in the embryo. Using in situ hybridization to detect mRNA and immunohistochemical staining to recognize the presence of the expressed protein, Abbott and coworkers have demonstrated the tissue specific localization of both the Ah receptor (Abbott et al., 1995) and ARNT (Abbott and Probst, 1995) in the mouse embryo from gestation day (gd) 10–16. The presence of both proteins is developmentally regulated, being specific for both cell type, organ/tissue, and developmental stage. The expression of these heterodimeric partners is generally coordinated. High levels, which decrease toward the end of organogenesis, are expressed in the developing brain and heart. Low levels are present in the liver on gd 10, which then increase to represent the highest levels in the fetus by gd 16. A similar pattern is present in bone. Low-to-moderate levels are present in kidney, muscle, and lung by the end of organogenesis, whereas high levels appear for skin and adrenal by gd 16. Noncoordinate expression of the Ah receptor and ARNT occurs in the adrenal with levels of ARNT being much higher than Ah receptor. The Ah receptor is also present in the renal tubules and differentiating glomeruli, whereas ARNT is present in the developing ureter and collecting ducts. Expression in other organs appeared after gd 13 and increased with gestational age. In the developing human palate, costaining studies (Abbott et al., 1994d) demonstrated that although most cells that expressed the Ah receptor also expressed ARNT and vice versa, there were some cells that expressed only one of the two proteins. The developmental regulation of Ah receptor and ARNT not only occurred at the levels of the tissue and organ, but involved the subcellular localization. In neural tissue and bone, both proteins were predominantly nuclear throughout organogenesis. In contrast, these proteins were both nuclear and cytoplasmic in muscle and skin. In the adrenals, the Ah receptor was strongly nuclear, whereas ARNT was highly cytoplasmic. Liver Ah receptor and ARNT were both nuclear and cytoplasmic on gd 10, but by gd 16 both had become strongly nuclear.

Other investigators have detected the presence of the Ah receptor and/or ARNT earlier in development. Peters and Wiley (1995) used RT-PCR to determine whether the Ah receptor was expressed during murine preimplantation development. Their results suggest that the embryo transcribes, rather than maternally inherits, Ah receptor mRNA. They first detected Ah receptor mRNA at the compacted eight-cell stage, suggesting that the Ah receptor begins to function in preimplantation embryos that are well into the eight-cell stage. Ah receptor protein was detected in the blastocyst. As seen in multiple other systems where the Ah receptor functions in both proliferation and differentiation, in the early embryo the Ah receptor plays a major role in the differentiation of the trophectoderm as well as influences embryonic growth by enhancing cell proliferation. At the end of organogenesis, the levels of Ah receptor mRNA are low compared to those for ARNT (Carver et al., 1994). However, in human placenta at term, Ah receptor levels are high compared to those found in most adult tissues (Dolwick et al., 1993). The critical importance of this signal transduction system in normal development has been supported by the decreased viability and immune deficiencies of transgenic mice in which the Ah receptor gene has been knocked out (Fernandez-Salguero et al., 1995).

Although the direct transcriptional activation of the Ah receptor has been well demonstrated, recent studies suggest that the Ah receptor has an alternative mode of action involving direct activation of second messenger systems. Phosphorylation of tyrosine residues has been shown to be a rapid and sensitive response to dioxin exposure both in vivo and in vitro. Activation of c-src and the EGF receptor have been documented. Based on analogy to the steroid hormone receptors, which have been recently shown to have nonnuclear functions, both Birnbaum (1993, 1996) and Matsumura (1994) have suggested that additional proteins present in the multimeric protein complex present in the absence of ligand may be tyrosine kinases or proteins involved in activation of tyrosine kinases. Upon binding dioxin, these other proteins are released from the Ah receptor complex and from its inhibitory control. A negative role has been shown for other helix-loop-helix proteins such as Id which blocks the transcriptional activation function of NF-kappa (Beg et al., 1995).

EFFECTS OF DIOXINS

The receptor-mediated action of dioxin, as well as many functional analogies between the steroid receptors and the Ah receptor, have led to the consideration of TCDD and related compounds as potent environmental hormones (Birnbaum, 1994a,c; 1996). Dioxins have been shown to modulate the activity of multiple endocrine systems, including both steroid and peptide hormones (DeVito and Birnbaum, 1995). For example, dioxin can affect the serum concentrations of estrogens, androgens, corticosterone, melatonin, gastrin, and thyroid hormones. It can also either increase or decrease the amount of estrogen and glucocorticoid receptors. Growth factors such as insulin growth factor, EGF, TGF, TGF, and retinoic acid have been shown to be affected by exposure to dioxins. Many of their receptors have also been shown to be either up- or downregulated in response to dioxin exposure. As is true for any hormone or growth factor, the response is tissue specific and depends upon the integrated action of multiple regulatory molecules in the entire tissue. However, understanding that dioxin can perturb multiple hormonal and growth factor pathways, it is easy to comprehend how dioxins function as potent growth dysregulators, altering both proliferation and differentiation.

Although alterations in biochemical processes involving hormones, growth factors, and metabolizing enzymes are often extremely sensitive responses, clearly adverse and toxic effects may occur at similar and higher dose levels. TCDD has often been called the most toxic manmade chemical, because of the small amounts that can cause death in certain species. (For recent reviews of TCDD toxicity, see Birnbaum, 1994a; Schecter, 1994; Okey et al., 1994; Pohjanvirta and Tuomisto, 1994.) Lethality, which is preceded by a severe wasting syndrome, is delayed, and the time-to-death is species specific. Wasting appears to be a result of alteration of the hypothalamic setpoint for body weight (Seefeld et al., 1984). Hypoplasia of the lymphoid tissues and gonads also occurs. In contrast, hyperplastic responses are seen in the liver and gastrointestinal and urinary epithelia. Metaplasia occurs in the Meibomian and ceruminous glands, leading to excessive and inappropriate wax production. The tissue-specific nature of the dioxin response is characteristic of a hormone-disrupting compound. Chloracne, the "hallmark of dioxin toxicity" in people, involves both hyperplastic and metaplastic changes, altered proliferation and differentiation.

Dioxins are associated with a variety of "-icities." TCDD is clearly carcinogenic in multiple animal species, at multiple sites, and in both sexes. The recent human epidemiology studies are consistent with this response (Flesch-Janys et al., 1995; Kogevinas et al., 1995). Although there is little evidence that dioxin or its related isosteromers are directly mutagenic, exposure can lead to secondary genotoxicity. Dioxins are clearly immunotoxic in several species, affecting both T-cell and B-cell responses. Several studies have indicated that dioxins are neurotoxic, causing both central and peripheral effects, as well as affecting learning and memory. The liver, GI and genitourinary tracts, and cardiovascular system have all been shown to be targets for dioxins' adverse effects. Dioxins have also been shown to be both developmental and reproductive toxicants in multiple species. Both structural and functional defects have been noted, as well as altered fertility.

OVERVIEW OF DEVELOPMENTAL EFFECTS

Numerous reviews concerning the developmental effects of dioxin and related compounds have appeared in the past few years (Couture et al., 1990a; Birnbaum, 1991,1995a,1996; Peterson et al., 1993; Battershill, 1994; Sauer et al., 1994; Brouwer et al., 1995; Lindstrom et al., 1995; Birnbaum and Abbott, 1996; Abbott, 1996). Some have focused on TCDD alone; others have looked holistically at all the compounds that act via the Ah receptor; still others have focused on the polychlorinated biphenyls (PCBs). Certain reviews have concentrated on the animal data; others on the effects in both people and animals; and still others have emphasized the epidemiological literature. Certain analyses have been concerned only with the teratogenic effects of dioxin exposure; others with the functional developmental toxicities. In this chapter, both the structural and functional deficits induced by prenatal exposure to dioxin and related compounds in both animals and people will be critically examined, with emphasis on some of the extremely recent findings that have not been included in previous reviews.

Developmental toxicity associated with exposure to dioxin has been known for more than 20 years. High levels of exposure to industrial chemicals contaminated with TCDD and/or related chemicals resulted in fetal wastage in multiple species, both in the laboratory and in the wild. We now know that the lack of reproduction in Lake Trout in Lake Ontario has been caused by the presence of these chemicals in the lake (U. S. EPA, 1993). Although adult populations of lake trout can be reestablished, successful recruitment of juveniles fails to occur because of significant mortality during early life stage development (Walker et al., 1991). Mink, a species extremely sensitive to the acute lethality of dioxins, have failed to reproduce in the wild around Lake Michigan because of the elevated presence of PCBs (Heaton et al., 1995 a,b). Embryo lethality and birth defects in colonial fish-eating water birds around the Great Lakes correlate with the presence of dioxin-like compounds (Jones et al., 1994; Williams et al., 1995). Great blue heron hatchlings exposed to polychlorinated dibenzo-p-dioxins (PCDDs) and polychlorinated dibenzofurans (PCDFs) off the coast of British Columbia have morphometric abnormalites in their brains (Henschel et al., 1995).

Experimental studies with multiple animal species soon revealed that TCDD exposure during organogenesis resulted in fetotoxicity in chickens, rats, mice, rabbits, guinea pigs, hamsters, and monkeys. At dose levels below those where overt toxicity occurred in the dam, growth retardation was detected in the offspring. Although subcutaneous edema was observed in several species, gastrointestinal hemorrhage was seen in others. Higher levels of exposure were associated with fetal death and resorptions. Thymic and splenic atrophy were also noted. One of the most important observations was that the dose levels associated with fetal toxicity are similar across species, regardless of the dose associated with lethality in the adult. For example, although the adult hamster is relatively resistant to dioxin-induced lethality, with an LD_{50} ~5000 µg/kg, and the adult rat is more sensitive (LD_{50} ~50 µg/kg), the dose to the dam needed to cause fetotoxicity in the pups of either species is approximately the same (Olson et al., 1990). Doses needed to cause cleft palate in the rat and the guinea pig, which has an LD_{50} ~ <2 µkg, are essentially the same (Olson and McGarrigle, 1992). Han/Wistar (Kupio) rats, which are uniquely resistant to dioxin-induced lethality in the adult, exhibit similar developmental toxicity to Long-Evans (Turku) rats, which are extremely susceptible to the lethal effects of TCDD (Huuskonen et al., 1994).

CLEFT PALATE AND HYDRONEPHROSIS

In the early 1970s, several groups of investigators demonstrated that the teratogenic effects of 2,4,5-T in mice were due to contamination with TCDD (Courtney and Moore, 1971). Follow-up work with the purified compound revealed that TCDD exposure of the dam resulted in a pathognomonic syndrome of effects in the offspring at doses that resulted in no overt toxicity to the mother (Couture et al., 1990a). Either divided doses throughout organogenesis or a single higher dose during the midpoint of that developmental stage resulted in a similar spectrum of structural malformations in the pups. These consisted of clefting of the secondary palate and hydronephrosis. Considering gd 0 as the day on which the vaginal plug was detected, the peak window for sensitivity for induction of cleft palate was gd 11–12 (Couture et al., 1990b). Earlier treatment, from gd 6–10, required higher doses of TCDD; dosing on gd 13 was much less effective at inducing clefting. From gd 14 on, cleft palate could not be induced by prenatal dioxin exposure. This appears to be due to the fact that palatal fusion occurs on gd 14. Treatment at that time or later cannot block the fusion process.

Although induction of cleft palate has a clear-cut window of sensitivity, this is not the case for hydronephrosis. The same dose is associated with both the same incidence and severity of the renal lesion whenever treatment occurs during organogenesis (Couture et al., 1990b). In fact, hydronephrosis can also be brought about following lactational exposure to dioxin, although this is less efficient than transplacental treatment (Couture-Haws et al., 1991). Hydronephrosis is also a more sensitive response than cleft palate and can be detected at doses below that where cleft palate occurs. At low doses, mild hydronephrosis can be observed, predominantly in the right kidney. As the dose is raised, both incidence and severity of response increases in both kidneys. Hydronephrosis is often accompanied by hydroureter. Both malformations have been shown to have a common cause: inappropriate proliferation of the ureteric epithelium (Abbott et al., 1987). The thickening of the luminal lining results in a narrowing of the lumen that can be blocked by epithelial cells sloughing off. Once the fetal kidney begins to function, urinary outflow is blocked, leading to a build-up of pressure on the kidney resulting in destruction of the renal parenchyma. The pressure also leads to the production of hydroureter, especially at the portion of the ureter proximal to the kidney.

Inappropriate proliferation of epithelial cells appears to be a major causative agent in the induction of cleft palate. Pratt et al. (1984) had demonstrated that dioxin-induced clefting of the secondary palate was due to a failure of fusion of the opposing shelves. In the mouse, the palatal shelves extend from the maxillary process, grow vertically, orient to a horizontal position over the tongue, and grow together. The only step that dioxin blocks in this process is that of fusion. The shelves actually make contact but are unable to fuse. Normally, the peridermal cells covering the medial edge epithelium slough off, exposing the underlying tissues. During normal development the medial epithelial cells transform into mesenchyme (Fitchett and Hay, 1989; Shuler et al., 1992), allowing fusion to occur between the underlying mesenchymal cores of the two opposing shelves. Dioxin treatment blocks epithelial-mesenchymal transformation. The peridermal cells also fail to slough off. For several years, it was suggested that these cells underwent apoptosis and that dioxin blocked this programmed cell death (Pratt et al., 1984). Whether or not these cells actually undergo apoptosis or just slough off when the underlying epithelium becomes mesenchyme remains to be determined. However, the medial epithelium transforms under the influence of dioxin into an oral-like stratified squamous epithelium, complete with desmosomes, tonofilaments, and keratins (Abbott and Birnbaum, 1989a).

ROLE OF GROWTH FACTORS AND HORMONES

Although a recent study of Hassoun et al. (1995) suggested that reactive oxygen species may participate in the teratogenic effects of TCDD, it is clear that the alterations in proliferation and differentiation that lead to palatal clefting can be correlated with changes in various growth factors and receptors in the palatal tissue. During normal development, EGF levels decrease as the palate is fusing and those of TGF-α, an alternative ligand for the EGF receptor, slowly increase. Overall, however, the levels of these growth factors are extremely low in the developing palate. Dioxin has little effect on the levels of EGF, but it blocks the increase in TGF-α (Abbott and Birnbaum, 1990a, 1991). Dioxin also affects the expression of the EGF receptor. This receptor is normally present in the epithelium, but decreases during the time of palatal fusion (Abbott and Birnbaum, 1989a). Dioxin exposure leads to an increase in the expression of this receptor (Abbott and Birnbaum, 1989b; Abbott et al., 1994c). It is possible that the increase in this receptor is in compensation for the decrease in the presence of its ligands. These interactions likely play a role in the continued proliferation of the epithelial cells due to dioxin. Changes are also noted in several of the TGF-βs. Although these growth factors often cause inhibition of epithelial cell proliferation, they are often stimulatory to cells of mesenchymal origin. Dioxin appears to cause an increase in the levels of TGF-β1 and 2 in the medial epithelium (Abbott and Birnbaum, 1990a, 1991; Abbott et al., 1994c). In addition, there are increases noted in the expression of these growth factors in the underlying mesenchyme.

Glucocorticoids can also cause cleft palate; however, the mechanism is different from that of dioxin. Glucocorticoid-induced cleft palate is due to the growth inhibition caused by these hormones, resulting in shelves that are small in size, and therefore are unable to make contact at the time when palatal fusion should occur (Pratt, 1985). Cotreatment of mice with hydrocortisone and dioxin leads to a synergistic increase in the incidence of cleft palate (Birnbaum et al., 1986). However, the size of the cleft suggests that the glucocorticoid response predominates. Analysis of the growth factor changes also suggests that the changes in EGF, TGF-α, TGF-β, and the EGF receptor are more like that of hydrocortisone alone than of TCDD alone (Abbott, 1995). The effects of glucocorticoids are mediated by binding to the glucocorticoid receptor, which is downregulated following exposure to hydrocortisone (Abbott et al., 1994b). Sensitivity to glucocorticoid-induced cleft palate is, in fact, related to the numbers of glucocorticoid receptors. Dioxin exposure leads to an increase in the numbers of glucocorticoid receptors in the developing palate, suggesting that the synergism of glucocorticoid-induced cleft palate by TCDD may be associated with an increase in the number of steroid receptors.

However, there is an additional complexity present in this interaction. Hydrocortisone can increase the expression of the Ah receptor (Abbott et al., 1994b), which is required for the induction of cleft palate by dioxin. Mouse strains with a lower-affinity Ah receptor require higher doses of dioxin for the induction of cleft palate, as well as all other responses (reviewed in Birnbaum, 1994b). In the developing mouse palate, exposure to dioxin results in a decrease in the level of Ah receptor expression in the palate on gd 14, the time of fusion (Abbott et al., 1994a). This is blocked by the presence of hydrocortisone. Thus, the synergistic induction of cleft palate by dioxin and the steroid may involve multiple interactions.

Dioxin can also interact synergistically with retinoic acid in the induction of cleft palate (Birnbaum et al., 1989). The effect on growth factors in this combination treatment resembles those seen more with TCDD than with the retinoid (Abbott and Birnbaum, 1989b). Whether or not retinoic acid upregulates the dioxin receptor in the palate, as has been demonstrated for the glucocorticoid receptor, is unknown. However, a recent report has shown that dioxin exposure can block the retinoic acid increase in the retinoic acid receptor (Weston et al., 1995) in cultured murine embryonic palatal mesenchyme cells. In cultured keratinocytes, retinoic acid exposure suppressed the differentiation-induced increase in the Ah receptor (Wanner et al., 1995). Thus, there appears to be the potential for multiple opportunities for crosstalk between the Ah receptor and receptors in the steroid family.

Changes in certain growth factors and receptors have also been seen in the developing urinary tract in response to dioxin. The induction of hydronephrosis is associated with an increase in EGF receptors in the ureteric epithelium (Abbott and Birnbaum, 1990b), similar to that observed with the enhanced proliferation of the medial epithelium in the palate. However, while dioxin decreased Ah receptor expression in the developing palate at the time of fusion (Abbott et al., 1994a), there were no changes detected in the levels of Ah receptor in the developing urinary tract at the same time (Bryant et al., 1995). To date, none of the other growth factors or receptors have been examined in this target tissue. However, neither retinoids nor glucocorticoids cause hydronephrosis, either alone or in combination with dioxin.

Although the majority of studies examining the mechanism of dioxin teratogenicity in mice have been done with TCDD, there have been many investigations of the induction of cleft palate and hydronephrosis by other dioxin-like compounds (see cited reviews). Several higher chlorinated dioxins as well as 2,3,7,8-tetrabromodibenzo-*p*-dioxin (Birnbaum, 1991) have been shown to cause the same spectrum of birth defects as TCDD. The relative potency of these compounds follows a similar rank order as their binding to the Ah receptor. In addition, both chlorinated and brominated dibenzofurans (Birnbaum et al., 1991) cause cleft palate and hydronephrosis, as do some of the laterally substituted brominated naphthalenes (Miller and Birnbaum, 1986). The coplanar PCBs have also been shown to be dioxin-like teratogens (Marks et al., 1989; Zhao et al., 1995), as have several of the complex commercial PCB mixtures.

OTHER DEVELOPMENT EFFECTS

Although cleft palate and hydronephrosis occur following prenatal exposure in mice at doses that are not overtly toxic to the dam, no other species shows the same responses at nontoxic doses (reviewed in Couture et al., 1990a). Cleft palate has been reported in rats exposed during organogenesis, but the doses that have been needed result in fetotoxicity, fetal wastage, and adverse effects on the dam. Although hydronephrosis has been reported, in no case has the increased incidence been statistically significant. Cleft palate has not been reported in hamsters or guinea pigs prenatally exposed. Renal abnormalites have been seen in hamsters, again at doses where fetotoxicity is evident. Gastrointestinal hemorrhage has also been seen in guinea pigs and rats. This suggests a problem with development of the vasculature and is interesting in light of the cardiac abnormalities that have been observed in chick embryos (Henshel et al., 1993) and fish (Spitsbergen et al., 1991). High levels of expression of the Ah receptor have been seen in the developing heart (Abbott et al., 1995) compatible with the hypothesis that the developing cardiovascular system may be a potential target tissue. The requirement for a subclass of bHLH proteins for normal cardiac morphogenesis is of special interest in this regard (Srivastava et al., 1995).

The majority of the studies examining the developmental toxicity of dioxins have focused on exposure during organogenesis. Early studies in rats did expose animals early in gestation and observed increased fetotoxicity (Giavini et al., 1982). This was not unexpected because treatment early in development often fails to produce viable offspring; however, recent studies have indicated that dioxin can accelerate differentiation of the preimplantation embryo (Blankenship et al., 1993). This supports the functionality of the Ah receptor, which is present in the developing embryo from the eight-cell stage (Peters and Wiley, 1995).

Exposure toward the end of organogenesis has been used to examine the effects on the developing immune system (reviewed in Birnbaum, 1995b). Exposure during organogenesis does result in atrophy of the lymphoid tissues, but the mice will not survive because of the cleft palate. Therefore, mice were treated on gd 14 in order to examine the results of prenatal exposure on immune functions. Fine and coworkers noted that dioxin appeared to affect the differentiation of prothymocytes, the actual target being the bone marrow rather than the thymus (Fine et al., 1989, 1990a,b). Terminal differentiation of these cells required an enzyme, terminal deoxynucleotidyl transferase, which is decreased following dioxin exposure. Such treatment alters the ratio of T lymphocytes in the pups. Similar results were noted by Holladay et al. (1991), who observed changes in the surface marker of lymphocytes following late prenatal exposure. It is interesting that exposure of rats on gd 15, developmentally slightly earlier than gd 14 in the mouse, leads to similar changes in T-cell subset ratios in the rat offspring (Gehrs and Smialowicz, 1994), with permanent suppression in delayed-type hypersensitivity (Gehrs et al., 1995), an immunological function that is highly correlated with altered host resistance and increased disease sensitivity. The dose levels that are associated with immune suppression in the rat are even lower than those needed in the mouse, which is extremely interesting given the apparent resistance of the adult rat to immunosup-pression by dioxins.

FUNCTIONAL DEVELOPMENTAL TOXICITY

Male Reproductive System Effects

Examination of functional sequelae to prenatal dioxin exposure has resulted in key find-ings over the past several years. Not only have dioxin and related compounds been shown to cause functional developmental toxicity in multiple species, but additional structural abnormalites have been noted that are delayed in their appearance. Studies conducted by Peterson and coworkers (Mably et al., 1992 a,b,c,) began the intensive investigation into this field. Their paradigm involved exposure of the pregnant Holtzman (Sprague-Dawley) rat on GD 15 to a single dose of 1 µg TCDD/kg. This dose was chosen based on previous developmental toxicity studies in Sprague-Dawley rats where higher doses caused excessive fetal mortality. The entire focus of this research group was on effects in male offspring, based on their studies of dioxin effects in adult males. They examined the male pups from birth through middle age ($\sim$ 1 year). At birth and postnatal day (pnd) 4, the dioxin-treated pups had a decreased anogenital distance. At adulthood, the weights of the accessory sex organs of the prenatally exposed male off-spring were decreased. Both testicular and epididymal sperm counts were permanently down. Reproductive behaviors were also altered. The males took longer to mount re-ceptive females, had more difficulty in achieving intromission, and took more thrusts to achieve ejaculation. These investigators hypothesized that the effects of dioxin might be due to decreased levels of androgens in the pups, a response noted following high doses to adult males. In their early studies, they in fact reported a decrease in circulat-ing androgen levels in the male offspring at birth. However, in more recent studies they were not able to replicate these early findings (Roman et al., 1995). They also sug-gested that dioxin exposure not only demasculinized the male offspring, but feminized them as well. Castration of the male pups, followed by treatment with estrogens and progesterone, led to an increased lordotic response in the dioxin-exposed pups as com-pared to controls. This suggested that dioxin might have "feminized" the central ner-vous system. They also examined the sexually dimorphic nuclei of the preoptic area of the hypothalamus to see if this was "demasculinized." However, they have not been able to observe any change in this brain region (Bjerke et al., 1994).

Peterson's group also conducted dose-response studies in order to determine what are the most sensitive developmental responses seen in the rats (Mably et al., 1992 a,b,c). They gave doses from 0.64 to 1.0 µg TCDD/kg to the pregnant rat on gd 15. The decrease in epidydimal sperm count was present even at the lowest dose at pnd 63 and later. The sensitivity of this response to TCDD on the developing reproductive system is similar to that seen with biochemical alterations in the rat as well as immu-nological endpoints in other species.

Gray and colleagues attempted to replicate and extend the findings of Peterson's laboratory because of their importance and implications (reviewed in Gray et al., 1995b). They examined the effects of prenatal TCDD exposure on other strains of rats as well as hamsters, and studied the female offspring as well as the males. Rats were exposed on either gd 8 or 15; hamsters on gd 11 (which is similar developmentally to gd 15 in the rat). Initially studies were conducted at doses to the dam of 1 µg TCDD/kg (Gray et al., 1995a; Gray and Ostby, 1995). More recently (Gray et al., 1995b), dose-response studies have been conducted in the rat with doses of 0.8, 0.2, and 0.05 µg TCDD/kg maternal weight. In general, the results of Gray and coworkers confirmed the major findings of the Wisconsin studies: prenatal exposure results in decreased sperm counts, anogenital distance, and altered male mating behavior. However, there are some significant differences. The North Carolina group noted that the decrease in anogenital distance in the male pups is associated with a decrease in body weight; in fact, if the anogenital distance is normalized to pup weight, the difference between control and treated disappears (Gray et al., 1995a). They also found that the decrease in testicular sperm (∼6%) is unlikely to explain the much larger decrease in epididymal (∼35%) and ejaculated sperm counts (∼60%). In fact, recent studies using the dioxin-like PCB #77 have also indicated that prenatal exposure in mice leads to a transient drop in germ cell numbers, which recovers by adulthood (Rönnbäck and de Rooij, 1994).

Although male mating behavior was altered, Gray's group suggested that these effects may not be centrally mediated. The male pups appeared to become aroused as rapidly as controls, but had more difficulty in achieving intromission and ejaculation. Whether this is associated with subtle changes in the penis remains to be determined. Gray et al. (1995a) also failed to observe any change in androgen status, including the lack of effect on androgen receptors. Gray's studies have recently been confirmed by Peterson's laboratory (Roman et al., 1995). Using castrated Long-Evans rats, no effects were seen on feminization of behavior. Whether this is a strain difference remains to be resolved; however, the enhanced lordosis in the castrated TCDD-exposed male offspring, even when observed in Holtzman rats (Mably et al., 1992a; Bjerke et al., 1994), is not a robust response. The delay in puberty, however, as measured by preputial separation, appears to be a consistent finding.

Differences in the sensitivity of various responses was also noted between the two groups. Although Peterson's team observed a decrease in epididymal sperm count at doses as low as 64 ng/kg, Gray's group only noted a significant decrease in epidydimal sperm at 800 ng/kg or higher. This parameter was reduced 10% at 200 ng/kg. Few effects were noted in the male Long-Evans rat below 200 ng/kg. Gray et al. (1995a) also noted that exposure to 1 µg TCDD/kg on gd 8 was less effective than on gd 15 in regard to developmental reproductive effects in male offspring. This suggests that the window of sensitivity for the male effects occurs late in organogenesis.

Female Reproductive System Effects

Multiple developmental effects were also observed in the prenatally exposed female rats (Gray and Ostby, 1995). No abnormalities were noted at birth or pnd 4. However, vaginal opening was delayed. In some cases, it was never completed. A thread of tissue persisted across the vaginal opening. In addition, clefting of the external genitalia was present in the female pups. The persistence of the vaginal thread may also reflect a lack of appropriate differentiation. Targeting the genitourinary tracts in the rat and hamster is not surprising given the demonstrated morphological effects on the developing urinary tract in the mouse. There appeared to be no change in cyclicity, suggesting that a normal hormonal profile was present. The structural abnormalities of the external genitourinary tract occurred at a higher incidence following exposure on gd 15 than on gd 8. In contrast, exposure at the beginning of organogenesis resulted in premature reproductive senescence in the female pups. Exposure on gd 8 caused many of the female offspring to stop estrus cycling before 6 months of age. The mechanism of this response remains to be examined, although Rönnback (1991) has shown that exposure of of mice to the dioxin-like PCB #77 reduced the number of germ cells in the ovaries.

The effects of prenatal exposure to TCDD on the developing reproductive system of the rat are not unique to that species (Gray et al., 1995a,b). Exposure of the Syrian hamster on gd 11 to 2 µg TCDD/kg (remember that the adult LD_{50} for this species is ~5000 µg/kg) results in some effects similar to those observed in the rat. In TCDD-treated male hamster offspring, puberty is delayed and epididymal and ejaculated sperm count are permanently decreased. Female pups exhibit delays in vaginal opening. In some of the hamster pups, vaginal opening cannot be detected. All of the female pups exhibit clefting of the external genitalia. Fertility is also decreased in the female hamster offspring, likely due to the structural problems in the external genitalia.

CNS Effects

Although most of the effects of low levels of prenatal dioxin exposure appear to directly target the developing genitourinary tract, there is some evidence for involvement of the central nervous system. The strongest data for a direct effect on the brain are those of Gordon and coworkers (Gordon et al., 1995, 1996). They examined the effects of prenatal dioxin exposure on core body temperature in both rats and hamsters. In 18-month-old rats who had been exposed to 1 µg TCDD/kg on gd 15, the hypothalamic set point for body temperature appeared to be permanently depressed. The treated rats could respond to a cold or heat stress by raising or lowering their body temperature as required. However, the body temperature was always decreased relative to the controls. A similar effect has been recently reported in 1-year-old hamsters who had been exposed to 2 µg/kg on gd 11.

Persistence of Effects

One of the most important findings of all of these studies is the permanent nature of the response following prenatal exposure. This suggests the irreversibility of some of these effects, which is in contrast to the reversible nature of some of the biochemical outcomes. Peterson reported that induction of hepatic biotransformation enzymes, readily detectable in the dam following exposure and in the pups shortly after birth, was nondetectable by 4 months of age (Mably et al., 1992a). Given the pharmacokinetics of TCDD, it is likely that the tissue levels have dropped to background by this time. In fact, recent studies by Hurst and colleagues (Hurst et al., 1996) have demonstrated that 1 µg/kg exposure to the pregnant rate on gd 8 results in embryonic concentrations of ~44 ng/kg on gd 9 and fetal concentrations of ~100 ng/kg on gd 16. This low concentration is clearly high enough to result in the spectrum of adverse effects observed. The half-life in the pregnant rat is approximately 7 days (Hurst et al., 1996), in contrast to the nonpregnant rat of 26 days (Li et al., 1995). Previous studies had indicated a similar difference in half-life between the pregnant (Weber and Birnbaum, 1985) and nonpregnant (Birnbaum, 1986) mouse.

Even though the highest mass of dioxin is transferred lactationally, the concentration in the neonatal pup is not much higher than during in utero development. By 3 months after weaning, the concentration in the pups is decreased to a level where enzyme induction is no longer detectable. However, the changes in the reproductive system and in core body temperature resulting from prenatal exposure persist.

Related Compounds

While most of the studies of functional developmental toxicology have used TCDD, there have been a few looking at related compounds, especially PCBs. Smits-van Prooije et al. (1992) exposed pregnant rats on gd 1 to 1.8 mg/kg of PCB #169 (3,4,5,3′,4′,5′-hexachlorobiophenyl), a coplanar, dioxin-like PCB. They noted low fertility, longer gestation times, and effects on male sexual behavior. Gray and Kelce (1996) conducted similar studies, but exposed the ams on gd 8 instead of gd 1. PCB #169 produced a spectrum of effects in the offspring similar to that observed with TCDD, including delays in puberty and decreased fertility. More dramatic decreases were noted in testis and accessory sex gland weights. Bouwman et al. (1995) exposed pregnant rats on gd 1 to several dioxin-like compounds, including 2,3,4,7,8-pentachlorodibenzofuran, the dioxin-like PCB #126 (3,3′,4,4′,5-pentachlorobiphenyl), PCB #153 (2,2′,4,4′5,5′-hexachlorobiphenyl), which is a diortho, nondioxin-like PCB,, and PCB #118 (2,3′,4,4′,5-pentachlorobiphenyl), which has both dioxin-like and non–dioxin-like activity. TCDD-like vaginal anomalies were seen in the female pups. In the male offspring, puberty was delayed and caudal sperm count was decreased. The relative potency was PCB #126 > 2,3,4,7,8-pentachlorodibenzofuran > PCB #118. PCB #153 had no effect developmentally. The effects seen were similar to those observed both by Peterson's and Gray's laboratories following TCDD exposure on gd 15 and the structure-activity relationship suggests involvement of the Ah receptor. Early studies by Lucier et al. (1978) demonstrated that prenatal exposure to PCB #77 (3,3′,4,4′-tetrachlorobiphenyl), a coplanar dioxin-like PCB, may have been associated with cryptorchidism in mice. No effects were seen with the nondioxin-like, diortho PCB #153 or the similar nondioxin-like brominated compound, 2,2′4,4′,5,5′-hexabromobiphenyl.

A number of studies have also looked at complex mixtures of PCBs. These studies have been conducted in rats, mice, guinea pigs, and monkeys. Lundkvist (1990) exposed guinea pigs both prenatally and lactationally to Clophen A50, a commercial mixture containing approximately 50% chlorine by weight. This mixture contains a relatively high percentage of dioxin-like PCBs, especially PCB #126. As in rats exposed to TCDD and dioxin-like PCBs, exposure resulted in a delay in sexual maturity in both male and female offspring. Testes weight was decreased in male offspring and vaginal opening was delayed in female pups. In agreement with the recent studies of Gray et al. (1995b) and Roman et al. (1995), no effects were seen in plasma testosterone levels. Exposure of rats to Aroclor 1254, which is 54% chlorine by weight and contains significant concentrations of dioxin-like congeners, causes delayed puberty and decreased fertility in female offspring (Sager and Girard, 1994) and decreased fertility and decreased accessory sex organ weights in males (Sager, 1983). However, although neither Gray's nor Peterson's laboratories reported dramatic changes in testes weight following perinatal TCDD exposure, Sager (1983) noted a significant decrease in testes weight, similar to what was observed with Clophen 50 in guinea pigs. In contrast to these effects of in utero and lactational exposure, Cooke et al. (1996) recently reported that early postnatal exposure to Aroclor 1254 or 1242 increased adult testis size as well as daily sperm production. Whether these effects are due to the nondioxin-like PCBs in the different commercial mixtures or to different windows of sensitivity remain to be determined. However, exposure of rats prenatally and lactationally to Aroclor 1254 resulted in a decrease in core body temperature in the offspring (Seo and Meserve, 1995), similar to that seen following TCDD exposure (Gordon et al., 1995). Aroclor 1254 was much more effective than Aroclor 1221 at reducing in vitro fertilization in the mouse (Kholkute et al., 1994a). Aroclor 1254 directly targeted the oocyte with no effect on sperm motility and also inhibited embryonic growth in vitro (Kholkute et al., 1994b).

Exposure of female rhesus monkeys for several years to Aroclor 1254 in their diet led to decreased fertility, even at the lowest dose of 5 μg Aroclor 1254/kg maternal body weight ($p = .059$) (Arnold et al., 1995). The surviving offspring exhibited many of the classical symptoms of TCDD poisoning such as ectodermal dysplasia (effects on teeth, hair, and nails), inflamed Meibomian glands, and immunotoxicity (suppression of the primary antibody response to sheep red blood cells). There was also a decrease in fetal and postnatal survival and a decrease in size.

ENDOCRINE DISRUPTION AS MECHANISM FOR FUNCTIONAL DEVELOPMENTAL TOXICITY

Estrogen

Because dioxin disrupts multiple endocrine systems (reviewed in Birnbaum, 1994a), the adverse effects on the reproductive, immune, and hypothalamic systems, such as those involved in temperature control, have been hypothesized to have a hormonal basis. High doses of TCDD have been shown to have antiestrogenic action, causing a decreased number of estrogen receptors, in immature uterus and breast tissues. Bjerke et al. (1994) failed to show any effect of prenatal dioxin exposure on brain estrogen receptors in rats, although they did note an increase in hippocampal glucocorticoid receptors. Mink are extremely sensitive to dioxin-induced reproductive toxicity. Reduced fertility occurs following exposure to less than 1 ng dioxin equivalents per gram of liver (Heaton et al., 1991). Patnode and Curtis (1994) compared the role of steroid receptors in the developmental toxicity of two hexachlorobiphenyls, #169, which is dioxin-like, and #153, which is non–dioxin-like. The dioxin-like congener was more developmentally toxic than the non–dioxin-like congener. Although both congeners exhibit antiestrogenicity as defined by their ability to block the estradiol-stimulated increase in nuclear estrogen receptors in immature mink, estrogenic effects were also observed. However, during pregnancy, both 153 and 169 exposure led to a decrease in the binding affinity of progesterone for the progesterone receptors, events normally thought of as being mediated by estradiol. Thus, the reproductive toxicity of PCBs may have multiple mechanisms, only one of which is Ah receptor mediated. This is not surprising given that the effects of many PCBs, including PCB #153, are not dioxin-like. It is also interesting to note that many of the malformations observed following in utero and lactational exposure to TCDD in female rats and hamsters are "estrogen-like."

Thyroid

Alteration of thyroid hormone status is another potential mechanism by which dioxin and related compounds may induce developmental toxicity (Porterfield and Hendrich,

1993). TCDD has been shown to decrease levels of circulating thyroid hormones by multiple mechanisms. One involves the induction of UDP-glucuronosyltransferase, which leads to conjugation and elimination of both thyroxine (T4) and triiodothyronine (T3). Another involves alterations in plasma protein binding, which can alter thyroid hormone transport. A third mechanism involves the induction of deiodinases that convert T4 to T3 and T3 to T2. In many investigations in adult animals, exposure to TCDD or to PCB mixtures reduces the concentration of serum T4, without having a significant effect on serum T3 (reviewed in Kohn et al., 1996). In certain studies, this has led to an increase in thyroid-stimulating hormone (TSH) levels, which may play a role in the reported thyrotoxicity of these chemicals (i.e., goiter). Maternal ingestion of Aroclor 1254 increases neonatal thyroid weight (Seo and Meserve, 1995).

The effects of prenatal exposure have been recently explored. Seo et al. (1995) demonstrated that female pups who had been prenatally exposed to TCDD (100 ng/kg/day on gd 10–16) had moderately depressed levels ($\sim 20\%$) of plasma T4 at weaning. However, this T4 response was not as sensitive as enzyme induction. The decrease in T4 was accompanied by an increase in glucuronosyl transferase activity. As in adults, no effects were seen on T3 or TSH concentrations. Exposure to toxicologically equivalent doses of the dioxin-like PCBs #77 and #126 had similar effects as TCDD. Treatment of rats prenatally with the dioxin-like congeners PCB #77 or #169 increased peripheral T4 metabolism (Morse et al., 1993). Exposure to non–dioxin-like PCBs such as #153 also reduced serum T4 levels (Ness et al., 1993). Prenatal exposure to #118 (2,3′,4,4′,5-pentachlorobiphenyl), which has both dioxin-like and non–dioxin-like properties, was the most effective. This is interesting in light of the dramatic reduction in circulating T4 concentrations that have been reported following adult exposure to commercial PCB mixtures (Gray et al., 1993). Prenatal and lactational exposure of rats to Aroclor 1254 at doses where few overt effects were detected resulted in decreased circulating T4 concentrations during the neonatal period and through weaning, when modest decreases in T3 were also seen (Goldey et al., 1995). The depression in circulating T3 has been observed in certain studies (Meserve et al., 1992) but not in others (Juarez de Ku et al., 1994). T4 supplementation could overcome the PCB-induced deficit in choline acetyltransferase activity in the hippocampus and basal forebrain of prenatally exposed neonatal rats (Juarez de Ku et al., 1994).

NEUROBEHAVIORAL EFFECTS OF DEVELOPMENTAL EXPOSURES

The thyrotoxic effects of PCBS have been suggested to form the basis for the developmental neurotoxicity attributed to prenatal PCB exposure (Porterfield and Hendrich, 1993). Prenatal exposure of rodents to PCBs has been reported to result in cholinergic deficits and a host of behavioral alterations including impaired learning and memory, altered activity levels, delayed development of reflexes, impaired acquisition of active avoidance tasks, and delayed development of negative geotaxis, auditory startle, and air righting reflex (reviewed in Juarez de Ku and Meserve, 1994). Prenatally exposed monkeys have been retarded in the development of learning and object alternation tasks (Schantz et al., 1991). Mice exposed in utero to the dioxin-like PCB #77 have demonstrated spinning behavior, hyperactivity, and impaired acquisition of avoidance response (Tilson et al., 1979), and mice exposed during the early neonatal period exhibited a depression of spontaneous motor behavior (Eriksson et al., 1991). Recent studies by

Schantz and coworkers (Schantz et al., 1995) have demonstrated that non–dioxin-like congeners, such as #153, and those that have mixed activity, such as #118, had no effect on working or reference memory following prenatal exposure. Delays in spatial alternation, similar to those seen in monkeys (Schantz and Bowman, 1989), were observed in the female offspring at adulthood.

The mechanism of the PCB developmental neurotoxicity may involve both dioxin-like (and, thus, Ah receptor-mediated) and non–dioxin-like effects. Some of the most exciting findings, such as the induction of hearing deficits in rats, due to an increase in the auditory threshold (Goldey et al., 1995), can also result from exposure to TCDD alone (Goldey et al., 1995). Effects on core body temperature, which may involve an alteration in the hypothalamic set point, can result from exposure to TCDD alone (Gordon et al., 1995, 1996) or complex mixtures of PCBs (Seo and Meserve, 1995). However, prenatal dioxin exposure has not been shown to alter sexually dimorphic behaviors in the Morris water maze or saccharin preference in male offspring (Gray et al., 1995a). A recent study has reported that late gestational exposure did result in changes in locomotor activity and rearing behavior (Thiel et al., 1994). Prenatal and lactational exposure of monkeys to dioxin caused changes in object learning (Schantz and Bowman, 1989); however, these were different from the delays in spatial alternation reported following exposure to the low chlorinated PCB mixture, Aroclor 1016, which has few dioxin-like congeners (Schantz et al., 1989).

HUMAN DEVELOPMENTAL TOXICITY TO DIOXINS

Extrapolation from Animal Data

One of the major questions regarding dioxin toxicity is whether or not humans are a sensitive species to its developmental effects. Given the wide range of effects that dioxin causes (as reviewed above) in multiple species and tissues, it is unlikely that people would be immune to the entire spectrum of adverse effects of dioxin following prenatal exposure. In fact, a recent investigation has demonstrated that the dose-response relationships for dioxin responses are similar for humans and animals (DeVito et al., 1995). However, one major point for dioxin is that as a result of its hormonal nature, responses are tissue and stage specific. Therefore, a given species can be an outlier for a given response, but similar to other organisms for other effects (Birnbaum, 1994a). This has already been discussed in terms of the relative resistance of the adult hamster to lethality in contrast to the sensitivity of its fetus. The mouse seems to be uniquely sensitive to the induction of cleft palate and hydronephrosis by dioxin at doses that are not overtly maternally or fetally toxic (Birnbaum, 1991). Using organ cultures of developing mouse, rat, and human palates, Abbott and coworkers demonstrated that the concentration of dioxin needed to bring about clefting was approximately 200–1000 times greater in rat (Abbott and Birnbaum, 1990c) and human (Abbott and Birnbaum, 1991) palatal shelves than in mouse (Abbott et al., 1989). The concentration used in vitro resulted in palatal tissue concentrations similar to those measured following in vivo exposure resulting in all of the exposed fetuses having cleft palate (Abbott et al., 1989). However, the doses that result in this high incidence of cleft palate in mouse, for example, 24 µg TCDD/kg on gd 12 (Birnbaum et al., 1989), and lead to a palatal concentration of approximately 400 ng/kg 24 h after exposure of the dam (Abbott et al., 1996) would kill all of the rat fetuses. Couture et al. (1989) demonstrated that exposure of the rat to 2,3,4,7,8-pentachlorodibenzofuran, a potent dioxin-like congener, can result in cleft palate, but only at doses where fetotoxicity occurs.

Negative (?) Findings

What adverse effects have been observed in humans developmentally exposed to dioxin? The answer at this point is very few. There are at least three possible explanations for this generally negative finding. The first is that humans are not sensitive. The second is that exposure has not been very high. And the third possibility is that investigations have not targeted the appropriate response endpoints. In fact, there have been few epidemiological studies looking at health outcomes for children prenatally exposed to high levels of dioxin. The majority of epidemiological studies have focused on occupationally exposed men. There are few women in these cohorts, making it impossible to examine their offspring. Following the industrial accident in Seveso, Italy, in 1976, extremely high dioxin levels were measured in a small number of people living in the area (reviewed in Bertazzi and di Domenico, 1994). Standard birth defects were examined for babies born to women living in the contaminated area at the time of the explosion. No increased abnormalities were noted at birth. However, no studies have been conducted to follow these children to determine whether or not they exhibited latent functional developmental effects or problems associated at puberty or even later. The children of Ranch Hand veterans from the Vietnam conflict have been examined for birth defects, and no association has been found (Wolfe et al., 1994). This finding, however, is not unexpected given that there is no evidence from animal studies of a male-mediated adverse effect on developmental outcomes. The dioxin exposure to the Ranch Hand veterans was also quite low.

Yusho and Yu-cheng

Adverse developmental effects from exposure to dioxin-like chemicals, however, have been reported in several poisoning episodes as well as several studies of less highly exposed populations. Two incidents occurred in the Far East in which rice oil was contaminated with PCBs and polychlorinated dibenzofurans: one in 1968 in Japan ("Yusho") (reviewed in Masuda, 1994), and the other in Taiwan in 1979 ("Yu-cheng") (reviewed in Hsu et al., 1994). Several thousand people were exposed in their diet for many months. In both cases, there was an increased incidence of stillbirths and early infant death. The Yu-cheng cohort has been followed intensively. Babies born to exposed women from shortly after the poisoning was discovered until even 10 years later have been examined and their physical and behavioral development followed. The children born to exposed women exhibited a syndrome of ectodermal dysplasia, involving defects in teeth, hair, and nails. Alterations in pigmentation earned them the name "cola-colored" babies (Rogan, 1982). They were small in size, exhibited developmental delays, and had a lowered IQ (Rogan et al., 1988). None of these effects have been reversible. In fact, there appears to be a global cognitive delay. In addition, there appear to be some effects on sex-related behaviors, such that the prenatally exposed boys have spatial learning characteristics more typical of girls, and the prenatally exposed girls preferred activities more commonly selected by boys (Guo et al., 1995). The prenatally exposed boys not only have unusually high serum levels of estrogen, but have penises that are significantly smaller at puberty than those of unexposed boys (Guo et al., 1993, 1995). Urinary porphyrins are also elevated in the children transplacentally exposed (Gladen et al., 1988b). The children also have a hearing deficit associated with cognitive processing (Chen and Hsu, 1994). The entire spectrum of developmentally induced defects are clearly associated with the dietary exposure of the mothers. However, the causative agent cannot be uniquely identified because the poisoning involved a complex mixture of PCBs and PCDFs, as well as other related polyhalogenated aromatic hydrocarbons. Many of the structural abnormalities correlate better with the PCDF concentration, than that of the total PCBs, suggesting that these effects are due to the dioxin-like compounds. However, some of the placental effects correlate better with the total PCB exposure (Sunahara et al., 1987). Whether these are due to the dioxin-like PCBs, the non–dioxin-like PCBs, or an interaction between the different classes of compounds may never be able to be determined.

Michigan and North Carolina Cohorts

Several groups have investigated exposures that are not as high as Yusho or Yu-cheng, but are higher than those occurring in the general population. These may be associated with dietary habits, such as elevated fish consumption. A series of studies by the Jacobsons and coworkers (reviewed in Tilson et al., 1990; Jacobson, 1995) have looked at the children born to women who consumed fish from Lake Michigan as compared to women who did not eat fish. The children born to the more highly exposed women exhibited a spectrum of developmental delays and behavioral deficits that appear to be irreversible. There are some problems associated with the exposure assessment in these studies, because the method used to measure total PCBs was relatively insensitive. Rogan and coworkers (Gladen et al., 1988a) examined a cohort of North Carolina women and their babies. There was no reason, a priori, to expect these women to have been exposed to more PCBs than the general population. However, they observed that in the babies whose mothers had the highest PCB concentrations, there was significant evidence of developmental delay and neurobehavioral deficits. In general, their results are in agreement with those of the Jacobsons. They have similar problems of insensitivity of total PCB measurements. Neither study looked at the amounts of dioxins and dibenzofurans in the women or did congener-specific analysis of the PCBs. Thus, whether the observed effects, which are clearly associated with prenatal as opposed to lactational exposure (Tilson et al., 1990), are due to dioxin-like compounds, to the nondioxin-like PCBs, or to a combination of the two classes of chemicals is unknown.

Inuits

A cohort of more highly exposed women and their babies is currently being studied among Inuit women in Arctic Quebec who have elevated levels of PCBs, PCDDS, and PCDFs (Dewailly et al., 1992b, 1993a). The infants are small (Dewailly et al., 1992a) and appear to have a higher incidence of infectious disease and otitis (Dewailly et al., 1993b). This may be indicative of immune suppression based on a low immunization take rate (Birnbaum, 1995b). Enhanced respiratory disease had previously been noted in the prenatally exposed Yu-cheng cohort (Rogan et al., 1988). There also appear to be some alterations in the lymphocyte subsets. However, no neurobehavioral or sexual assessment has yet been reported in these children.

Dutch Study

Among the most insightful investigations concerning potential developmental effects following prenatal as well as both prenatal and lactational exposure to dioxins, dibenzofurans, and PCBs in the general population are being conducted in the Netherlands (Koopman-Esseboom et al., 1995a). Four hundred and eighteen mothers and their children are being studied. Approximately half of the cohort nursed their babies. All seventeen of the laterally substituted PCDDs and PCDFs as well as the coplanar (dioxin-like) PCB congeners #77, #126, and #169, plus 23 additional PCB congeners were measured by high-resolution gas chromatography/mass spectroscopy in milk taken in the second week after birth. Milk concentrations are an indirect measure of prenatal exposure. In addition, the concentrations of four major PCBs, three of which have some dioxin-like activity (#118, #138, and #180) as well as the prototypical non–dioxin-like PCB, #153, were determined in both maternal plasma and cord blood. The concentrations of these four PCB congeners in human milk were highly significantly correlated with their levels in maternal plasma and with total dioxin-like TEQ in human milk (Koopman-Esseboom et al., 1994a). The mean dioxin-like toxic equivalence (TEQ) in human milk in this study (65 pg/g lipid) is similar to that in other highly industrialized countries. Nearly half of the TEQ is due to the dioxins, and one fourth to the coplanar PCBs. The TEQ in human milk did not change between 2 and 6 weeks after birth.

Immunological parameters were examined in plasma samples from both the umbilical cord and the child (Weiglas-Kuperus et al., 1995). Although there was no relationship between developmental PCB/dioxin exposure and respiratory infections or humoral antibody production, a higher prenatal TEQ exposure was associated with an increase in the number of TcR gamma delta plus T cells at birth and 18 months of age. Total T cells, cytotoxic T cells, and TcR plus T cells were also increased at 18 months of age in the more highly exposed children within the general population. This is the first study to demonstrate that background levels of dioxin/PCB exposure can influence the human immune system. Although the clinical implications of these alterations in T-cell populations are unknown at this time, they are reminiscent of the recent results from Gehrs and Smialowicz (1994) in which increases in changes in T-lymphocyte subsets following prenatal exposure of rats are correlated with a permanent suppression of delayed-type hypersensitivity (Gehrs et al., 1995).

Effects of dioxins and PCBs on thyroid functions were assessed both in the mother and in the offspring (Koopman-Esseboom et al., 1994b). Infants exposed prenatally to higher TEQ levels had elevated levels of TSH both 2 weeks and 3 months after birth, and lower levels of free and total T4 2 weeks after birth. The elevation in plasma TSH is in agreement with an earlier report from another group of healthy Dutch newborns 11 weeks after birth (Pluim et al., 1993). In this case, the total TEQ was based only on dioxins and dibenzofurans. However, this earlier study observed an increase in total T4 levels at 1 and 11 weeks after birth. Overall, however, the results of these human studies resemble the effects observed following prenatal exposure in experimental animals. The thyroid hormone alterations observed in the infants are, however, within the normal range of clinical values. Whether these changes may have influenced fetal development or have a clinical significance in a segment of the population remains to be determined.

A series of neurodevelopmental parameters were also measured in the large Dutch study involving general population exposure. Huisman et al. (1995a) investigated whether exposure to PCDDs, PCDFs, and PCBs was correlated with the neurological condition of the neonate. Higher TEQ levels in maternal breast milk, as an index of prenatal and early postnatal exposure, were associated with reduced neonatal neurological optimality, as evaluated with the Prechtl neurological examination, which measures subtle neurological dysfunction. Elevated levels of coplanar PCBs were also associated with an increased incidence of hypotonia. These results are consistent with previous reports from the North Carolina (Rogan et al., 1986) and Michigan (Jacobson et al., 1984) cohorts concerning the neurotoxic effects of these chemicals on the developing human brain. However, in contrast to the earlier studies, the Dutch group found no effects at either 3 or 7 months on the infants' cognitive development assessed by means of the visual recognition memory (Fagan) test (Koopman-Esseboom et al., 1995b). Whether this is due to differences in levels of exposure are unknown. However, prenatal as well as early postnatal exposure to PCBs and dioxins was correlated with a small negative effect on early psychomotor development (Koopman-Esseboom et al., 1995c). Postnatal elevated TEQ exposure was related to a delay in psychomotor development at 7 months of age. A decrease in neurological optimality was still present at 18 months of age (Huisman et al., 1995b) in the toddlers who had been more highly exposed prenatally.

CONCLUSIONS

Taken together, the human studies involving complex mixtures of dioxin-like compounds, including both dioxin and non–dioxin-like PCBs, suggest that levels present in the general population may be associated with subtle signs of neurological dysfunction, delays in psychomotor development, alterations in thyroid hormone status, and changes in immunological functions. The newer data are consistent with earlier reports of decreased birth weight and gestational age associated with prenatal PCB exposure of occupationally exposed women (Taylor et al., 1989), as well as the generally adverse effects of PCBs on infant health (Swain, 1991). Reduced birth weight and length in the offspring of women chronically exposed to low levels of PCDDs/PCDFs have also been recently reported (Karmaus and Wolf, 1995). Many of these changes fall within the normal range. However, they do suggest a shift in the distribution of the population, potentially putting more children in an "at-risk" situation due to functional developmental toxicity. Higher levels of exposure clearly are associated with adverse outcomes in multiple systems. Whether the effects observed are due to dioxin itself, all dioxin-like compounds, non–dioxin-like PCBs, or a combination of the two classes in some ways is a moot question. Exposure to one rarely occurs without the other. The issue for the general population is whether or not the low levels of exposure to this chemical mix is associated with the potential for adverse effects from prenatal, and possibly early postnatal, exposure, even though some of these effects may not be evidence for many years.

REFERENCES

Abbott, B. D. (1995). *Toxicology, 105*: 365.
Abbott, B. D. (1996). *Drug Toxicity in Embryonic Development* (in press).

Abbott, B. D., and Birnbaum, L. S. (1989a). *Toxicol. Appl. Pharmacol., 9*: 276.

Abbott, B. D., and Birnbaum, L. S. (1989b). *Toxicol. Appl. Pharmacol., 99*: 287.

Abbott, B. D., and Birnbaum, L. S. (1990a). *Toxicol. Appl. Pharmacol., 106*: 418.

Abbott, B. D., and Birnbaum, L. S. (1990b). *Teratology, 41*: 71.

Abbott, B. D., and Birnbaum, L. S. (1990c). *Toxicol. Appl. Pharmacol., 103*: 441.

Abbott, B. D., and Birnbaum, L. S. (1991). *Teratology, 43*: 119.

Abbott, B. D., and Probst, M. R. (1995). *Dev. Dyn., 204*: 144.

Abbott, B. D., Birnbaum, L. S., and Pratt, R. M. (1987). *Teratology, 35*: 329.

Abbott, B. D., Diliberto, J. J., and Birnbaum, L. S. (1989). *Toxicol. Appl. Pharmacol., 100*: 119.

Abbott, B. D., Perdew, G. H., and Birnbaum, L. S. (1994a). *Toxicol. Appl. Pharmacol., 126*: L16.

Abbott, B. D., Perdew, G. H., Buckalew, A. R., and Birnbaum, L. S. (1994b). *Toxicol. Appl. Pharmacol., 128*: 138.

Abbott, B. D., Perdew, G. H., Fantel, A. G., and Birnbaum, L. S. (1994c). *Teratology, 49*: 378.

Abbott, B. D., Probst, M. R., and Perdew, G. H. (1994d). *Teratology, 50*: 361.

Abbott, B. D., Birnbaum, L. S., and Perdew, G. H. (1995). *Dev. Dyn., 204*: 133.

Abbott, B. D., Birnbaum, L. S., and Diliberto, J. J. (1996). *Toxicol. Appl. Pharmacol.*, in press.

Andersen, M. E., Mills, J. J., Gargas, M. L., Kedderis, L., Birnbaum, L. S., Neubert, D., and Greenlee, W. F. (1993). *Risk Anal., 13*: 25.

Arnold, D. L., Bryce, F., McGuire, P. F., Stapley, R., Tanner, J. R., Wrenshall, E., Mes, J., Fernie, S., Tryphonas, H., Hayward, S., and Malcolm, S. (1995). *Food Chem. Toxicol., 33*: 457.

Battershill, J. M. (1994). *Hum. Exp. Toxicol., 13*: 581.

Beg, A. A., Sha, W. C., Bronson, R. T., Ghosh, S., and Baltimore, D. (1995). *Lett. Nature, 376*: 167.

Bertazzi, P. A., and di Domenico, A. (1994). *Dioxins and Health* (A. Schecter, ed). Plenum Press, New York, p. 587.

Birnbaum, L. S. (1986). *Drug Metab. Dispos., 14*: 34.

Birnbaum, L. S. (1991). *Banbury Report 35: Biological Basis for Risk Assessment of Dioxins and Related Compounds.* Cold Spring Harbor Laboratory Press, Cold Spring Harbor, New York, p. 51.

Birnbaum, L. S. (1993). *VIP-32 Municipal Waste Combustion.* Proceedings of an International Specialty Conference, p. 598.

Birnbaum, L. S. (1994a). *Environ. Health Perspect., 102* (Suppl. 9): 157–167.

Birnbaum, L. S. (1994b). *Receptor-Mediated Biological Processes: Implications for Evaluating Carcinogens. Progress in Clinical and Biological Research*, Vol. 387 (H. L. Spitzer, T. L. Slaga, W. F. Greenlee, and M. McClain, eds.). Wiley-Liss, New York, p. 139.

Birnbaum, L. S. (1994c). *Environ. Health Perspect., 102*: 676.

Birnbaum, L. S. (1995a). *Environ. Health Perspect., 103* (Suppl. 7): 89.

Birnbaum, L. S. (1995b). *Environ. Health Perspect., 103* (Suppl. 2): 157.

Birnbaum, L. S. (1996). *Toxicol. Lett.* (in press).

Birnbaum, L. S., and Abbott, B. D. (1996). *Methods in Developmental Toxicology/Biology.* Blackwell Wissenchafts-Verlag, Berlin *82/83*: 743.

Birnbaum, L. S., Harris, M. W., Miller, C. P., Pratt, R. M., and Lamb, J. C. (1986). *Teratology, 33*: 29.

Birnbaum, L. S., Harris, M. W., Stocking, L., Clark, A. M., and Morrissey, R. E. (1989). *Toxicol. Appl. Pharmacol., 98*: 487.

Birnbaum, L. S., Morrissey, R. E., and Harris, M. W. (1991). *Toxicol. Appl. Pharmacol., 107*: 141.

Bjerke, D. L., Brown, T. J., MacLusky, N. J., Hochberg, R. B., and Peterson, R. E. (1994). *Toxicol. Appl. Pharmacol., 127*: 258.

Blankenship, A. L., Suffia, M. C., Matsumura, F., Walsh, K. J., and Wiley, L. M. (1993). *Reprod. Toxicol., 7*: 255.

Bouwman, C. A., Fase, K. M., Waalkens-Berendsen, I. D. H., Smits-van Prooije, A. E., Seinen, W., and Van den Berg, M. (1995). *Organohal. Compd., 25*: 39.

Brouwer, A., Ahlborg, U. G., Van den Berg, M., Birnbaum, L. S., Boersma, E. R., Bosveld, B., Denison, M. S., Gray, L. E., Hagmar, L., Holene, E., Huisman, M., Jacobson, S. W., Jacobson, J. L., Koopman-Esseboom, C., Koppe, J. G., Kulig, B. M., Morse, D. C., Muckle, G., Peterson, R.E., Sauer, P. J. J., Seegal, R. F., Smits-Van Prooije, A. E., Touwen, B. C. L., Weisglas-Kuperus, N., and Winneke, G. (1995). *Environ. Toxicol. Pharmacol., 239*: 1.

Bryant, P. L., Clark, G., Probst, M. R., and Abbott, B. D. (1995). *Toxicologist, 15*: 349.

Buckley, L. A. (1995). *Toxicology, 102*: 125.

Carrier, F., Chang, C., Duh, J., Nebert, D. W., and Puga, A. (1994). *Biochem. Pharmacol., 48*: 1767.

Carver, L. A., Hogenesch, J. B., and Bradfield, C. A. (1994). *Nucleic Acids Res., 22*: 3038.

Chen Y-J., and Hsu, C-C. (1994). *Dev. Med. Child. Neurol., 36*: 312.

Cooke, P. S., Zhao, Y-D., and Hansen, L. G. (1996). *Toxicol. Appl. Pharmacol., 136*: 112.

Courtney, K. D., and Moore, J. A. (1971). *Toxicol. Appl. Pharmacol., 20*: 396.

Couture, L. A., Harris, M. W., and Birnbaum, L. S. (1989). *Fundam. Appl. Toxicol., 12*: 358.

Couture, L. A., Abbott, B. D., and Birnbaum, L. S. (1990a). *Teratology, 43*: 619.

Couture, L. A., Harris, M. W., and Birnbaum, L. S., (1990b). *Fundam. Appl. Toxicol., 15*: 142.

Couture-Haws, L., Harris, M. W., Clark, A. M., and Birnbaum, L. S. (1991). *Toxicol. Appl. Pharmacol., 107*: 402.

DeVito, M. J., and Birnbaum, L. S. (1995). *Toxicology, 102*: 115.

DeVito, M. J., Birnbaum, L. S., Farland, W. H., and Gasiewicz, T. A. (1995). *Environ. Health Perspect., 103*: 820.

Dewailly, E., Bruneau, S., Laliberté, C., Bélanger, D., Gingras, S., Ayotte, P., and Nantel, A. (1992a). *Organohal. Compd., 10*: 257.

Dewailly, E., Nantel, A., Bruneau, S., Laliberté, C., Ferron, L., and Gingras, S. (1992b). *Chemosphere, 25*(7–10): 1245.

Dewailly, E., Ayotte, P., Bruneau, S., Laliberte, C., Muir, D. C. G., and Norstrom, R. J. (1993a). *Environ. Health Perspect., 101*: 618.

Dewailly, E., Bruneau, S., Laliberté, C., Belles-Iles, M., Weber, J. P., Ayotte, P., and Roy, R. (1993b). *Organohal. Compd., 13*: 403.

Diliberto, J. J., Akubue, P. I., Luebke, R. W., and Birnbaum, L. S. (1995). *Toxicol. Appl. Pharmacol., 130*: 197.

Diliberto, J. J., Jackson, J. A., and Birnbaum, L. S. (1996). *Toxicol. Appl. Pharmacol. 138*: 158.

Dolwick, K. M., Schmidt, J. V., Carver, L. A., Swanson, H. I., and Bradfield, C. A. (1993). *Mol. Pharmacol., 44*: 911.

Eriksson, P., Lundkvist, U., and Frederiksson, A. (1991). *Toxicology, 69*: 27.

Eskenazi, B., and Kimmel, G. (1995). *Environ. Health Perspect., 103*: 143.

Fernandez-Salguero, P., Pineau, T., Hilbert, D. M., McPhail, T., Lee, S. S. T., Kimura, S., Nebert, D. W., Rudikoff, S., Ward, J. M., and Gonzalez, F. J. (1995). *Science, 268*: 722.

Fine, J. S., Gasiewicz, T. A., and Silverstone, A. E. (1989). *Pharmacol., 35*: 18.

Fine, J. S., Gasiewicz, T. A., Fiore, N. C., and Silverstone, A. E. (1990a). *J. Exp. Pharmacol. Ther., 255*: 1.

Fine, J. S., Silverstone, A. E., and Gasiewicz, T. A. (1990b). *J. Immunol., 144*: 1169.

Fitchett, J. E., and Hay, E. D. (1989). *Dev. Biol., 131*: 455.

Flesch-Janys, D., Berger, J., Gurn, P., Manz, A., Nagel, S., Waltsgott, H., and Dwyer, J. H. (1995). *Am. J. Epidemiol., 142*: 1165.

Gehrs, B., and Smialowicz R. (1994). *Toxicologist, 14*: 382.

Ghers, B. D., Riddle, M. M., Williams, W. C., and Smialowicz, R. J. (1995). *Toxicologist, 15*: 551.

Giavini, E. M., Prati, M., and Vismara, C. (1982). *Environ. Res., 29*: 185.

Gladen, B. C., Rogan, W. J., Hardy, P., Thullen, J., Tingelstaad, J., and Stully, M. (1988a). *J. Pediatr., 113*: 991.

Gladen, B. C., Rogan, W. J., Ragan, N. B., and Spierto, F. W., (1988b). *Arch. Environ. Health, 43*: 54.

Goldey, E. S., Kehn, L. S., Lau, C., Rehnberg, G. L., and Crofton, K. M. (1995). *Toxicol. Appl. Pharmacol., 135*: 77.

Goldey, E. S., Lau, C., Kehn, L. S., and Crofton, K. M. (1996). *Fundam. Appl. Toxicol., 30*: 225.

Gordon, C. J., Gray, L. E., Jr., Monteiro-Riviere, N. A., and Miller, D. B. (1995). *Toxicol. Appl. Pharmacol., 133*: 172.

Gordon, C. J., Ying, Y., and Gray, L. E. (1996). *Toxicol. Appl. Pharmacol., 137*: (in press).

Gray, L. E., and Kelce, W. R. (1996). *Toxicol. Ind. Health* (in press).

Gray, L. E., Jr., and Ostby, J. S. (1995). *Toxicol. Appl. Pharmacol., 133*: 285.

Gray, L. E., Ostby, J., Marshall, R., and Andrews, J. (1993). *Fundam. Appl. Toxicol., 20*: 288.

Gray, L. E., Kelce, W. R., Monosson, E., Ostby, J. S., and Birnbaum, L. S. (1995a). *Toxicol. Appl. Pharmacol., 131*: 108.

Gray, L. E. Ostby, J., Wolf, C., Miller, D. B., Kelce, W. R., Gordon, C. J., and Birnbaum, L. (1995b). *Organohal. Compd., 25*: 33.

Guo, Y. L., Lai, T. J., Ju, S. H., Chen, Y. C., and Hsu, C. C. (1993). *Chemosphere, 14*: 235.

Guo, Y. L., Yu, M. L., Hsu, C. C., and Lambert G. H. (1995). Neuro-endocrine developmental effects in children exposed in utero to PCBs: Studies in Taiwan. Thirteenth International Neurotoxicology Conference, 5.

Hassoun, E. A., Bagchi, D., and Stohs, S. J. (1995). *Toxicol. Lett., 76*: 245.

Heaton, S. N., Aulerich, R. J., Bursian, S. J., Giesy, J. P., Render, J. A., Tillitt, D. E., and Kubiak,T. J. (1991). Proceedings of the Annual Meeting of the Society of Environmental Toxicology and Chemistry, p. 85.

Heaton, S. N., Bursian, S. J., Giesy, J. P., Tillitt, D. E., Render, J. A., Jones, P., Verbrugge, D., Kubiank, T. J., and Aulerich, R. J. (1995a). *Arch. Environ. Contam. Toxicol., 28*: 334.

Heaton, S. N., Bursian, S. J., Giesy, J. P., Tillitt, D. E., Render, J. A., Jones, P. D., Verbrugge, D. A., Kubiak, T. J., and Aulerich, R. J. (1995b). *Arch. Environ. Contam. Toxicol., 29*: 411.

Henshel, D. S., Hehn, B. M., Vo, M. T., and Steeves, J. D. (1993). *Environmental Toxicology and Risk Assessment*: Vol. 2, *ASTM STP 1173* (J. W. Gorsuch, F. J. Dwyer, C. G. Ingersoll, and T. W. La Point, eds.). American Society for Testing and Materials, Philadelphia.

Henshel, D. S., Martin, J. W., Norstrom, R., Whitehead, P., Steeves, J. D., and Cheng, K. M. (1995). *Environ. Health Perspect., 103*: 61.

Holladay, S. D., Linstrom, P., Blaylock, B. L., Comment, C. E., Germolec, D. R., Heindell, J. J., and Luster, M. I. (1991). *Teratology, 44*: 385.

Hsu, C-C., Yu, M-L., Chen, Y-C.J., Guo, Y-L.L., and Rogan W. J. (1994). *Dioxins and Health* (A. Schecter, ed.). Plenum Press, New York, p. 661.

Huisman, M., Koopman-Esseboom, C., Fidler, V., Hadders-Algra, M., Van Der Paauw, C. G., Tuinstra, L. G. M. Th., Weisglas-Kuperus, N., Sauer, P. J. J., Touwen, B. C. L., and Boersma, E. R. (1995a). *Early Hum. Dev., 41*: 111.

Huisman, M., Koopman-Esseboom, C., Lanting, C. I., Van Der Paauw, C. G., Tuinstra, L. G. M. Th., Fidler, V., Weisglas-Kuperus, N., Sauer, P. J. J., Boersma, R., and Touwen, B. C. L. (1995b). *Early Hum. Dev.* (in press).

Hurst, C., DeVito, M., Abbott, B., and Birnbaum, L. (1996). *Fundam. Appl. Toxicol., 30*: 198.

Huuskonen, H., Unkila, M., Pohjanvirta, R., and Tuomisto, J. (1994). *Toxicol. Appl. Pharmacol., 124*: 174.

Jacobson, J. L. (1995). Thirteenth International Neurotoxicology Conference, p. 5.

Jacobson, J. L., Jacobson, S. W., Schwartz, P. M., Fein, G. G., and Dowler, J. K. (1984). *Dev. Psychol., 10*: 523.

Jones, P. D., Giesy, J. P., Newsted, J. L., Verbrugge, D. A., Ludwig, J. P., Ludwig, M. E., Auman, H. J., Crawfford, R., Tillitt, D. E., Kubiak, T. J., and Best, D. A. (1994). *Ecotoxicol. Environ. Safety, 27*: 192.

Juarez de Ku, L. M., and Meserve, L. A. (1994). *Toxin-Induced Models of Neurological Disorders* (M. L. Woodruff and A. J. Nonneman, eds.). Plenum Press, New York, p. 281.

Juarez Ku, L. M., Sharma-Stokkermans, M., and Meserve, L. A. (1994). *Toxicology, 94*: 19.

Karmaus, W., and Wolf, N. (1995). *Environ. Health Perspect., 103*: 1120.

Kholkute, S. D., Rodriguez, J., and Dukelow, W. R. (1994a). *Reprod. Toxicol., 8*: 69.

Kholkute, S. D., Rodriguez, J., and Dukelow, W. R. (1994b). *Reprod. Toxicol., 8*: 487.

Kogevinas, M., Kauppinen, T., Winkelmann, R., Becher, H., Bertazzi, P. A., Bueno-de-Mesquita, H. B., Coggon, D., Green, L., Johnson, E., Littorin, M., Lynge, E., Marlow, D. A., Mathews, J. D., Neuberger, M., Benn, T., Pannett, B., Pearce, N., and Saracci, R. (1995). *Epidemiology, 6*: 396.

Kohn, M. C., Sewall, C. H., Lucier, G. W., and Portier, C. J. (1996). *Toxicol. Appl. Pharmacol., 165*: 29.

Koopman-Esseboom, C., Huisman, M., Weisglas-Kuperus, N., Van Der Paauw, C. G., Tuinstra, L. G. M. T., Boersma, E. R., and Sauer, P. J. J. (1994a). *Chemosphere, 28*: 1721.

Koopman-Esseboom, C., Morse, D. C., Weisglas-Kuperus, N., Lutkeschipholt, I. J., Van Der Paauw, C. G., Tuinstra, L. G. M. T., Brouwer, A., and Sauer, P. J. J. (1994b). *Pediatr. Res., 36*: 468.

Koopman-Esseboom, C., Huisman, M., Weisglas-Kuperus, N., Boersma, E. R., Touwen, B. C. L., and Sauer, P. J. J. (1995a). Results of the Dutch study on PCB and dioxin induced neurotoxicity in children. Thirteenth International Neurotoxicology Conference.

Koopman-Esseboom, C., Weisglas-Kuperus, N., de Ridder, M. A. J., Van Der Paauw, C. G., Tuinstra, L. G. M. T., and Sauer, P. J. J. (1995b). *Pediatr. Res.* (in press).

Koopman-Esseboom, C., Weisglas-Kuperus, N., de Ridder, M. A. J., Van Der Paauw, C. G., Tuinstra, L. G. M. T., and Sauer, P. J. J. (1995c). *Pediatr. Res.* (in press).

Li, X., Weber L. W. D., and Rozman, K. K. (1995). *Fundam. Appl. Toxicol., 27*: 70.

Lindstrom, G., Hopper, K., Petreas, M., Stephens, R., and Gilman, A. (1995). *Environ. Health Perspect., 103*: 135.

Lucier, G. W., Davis, G. J., and McLachlan, J. A. (1978). 17th Hanford Biology Symposium-Developmental Toxicology of Energy-Related Pollutants. Technical Information Center, Department of Energy, p. 188.

Lundkvist, U. (1990). *Toxicology, 61*: 249.

Mably, T. A., Moore, R. W., Goy, R. W., and Peterson, R. E. (1992a). *Toxicol. Appl. Pharmacol., 114*: 108.

Mably, T. A., Bjerke, D. L., Moore, R. W., Gendron-Fetzpatrick, A., and Peterson, R. E. (1992b). *Toxicol. Appl. Pharmacol., 114*: 118.

Mably, T. A., Moore, R. W., and Peterson, R. E. (1992c). *Toxicol. Appl. Pharmacol., 114*: 97.

Marks, T. A., Kimmel, G. L., and Staples, R. E. (1989). *Toxicol. Appl. Pharmacol., 13*: 681.

Masuda, Y. (1994). *Dioxins and Health* (A. Schecter, ed.). Plenum Press, New York, p. 633.

Matsumura, F. (1994). *Biochem. Pharmacol., 48*: 215.

Meserve, L. A., Murray, B. A., and Landis, J. A. (1992). *Bull. Environ. Contam. Toxicol., 48*: 715.

Miller, C. P., and Birnbaum, L. S. (1986). *Fundam. Appl. Toxicol., 7*: 398.

Morse, D. C., Groen, D., Veerman, M., Van Amerongen, C. J., Koëter, H. B. W. M., Smits-van Prooije, A. E., Visser, T. J., Koeman, J. H., and Brouwer, A. (1993). *Toxicol. Appl. Pharmacol., 122*: 27.

Ness, D. K., Schantz, S. L., Moshtaghian, J., and Hansen, L. G. (1993). *Toxicol. Lett., 68*: 311.

Okey, A. B., Riddick, D. S., and Harper, P. A. (1994). *Toxicol. Lett., 70*: 1.

Olson, J. R., and McGarrigle, B. P. (1992). *Chemosphere, 25*: 71.

Olson, J. R., McGarrigle, B. P., Tonucci, D. A., Schecter, A., and Eichelberger, H. (1990). *Chemosphere, 20*: 1117.

Patnode, K. A., and Curtis, L. R. (1994). *Toxicol. Appl. Pharmacol., 127*: 9.

Peters, J. M., and Wiley, L. M. (1995). *Toxicol. Appl. Pharmacol., 134*: 214.

Peterson, R. E., Theobald, H. M., and Kimmel, G. L. (1993). *Crit. Rev. Toxicol., 23*: 283.

Pluim H. J., de Vijder, J. J. M., Olie, K., Kok, J. H., Vulsma, T., van Tijn, D. A., van der Slikke, J. R., and Koppe, J. R. (1993). *Environ. Health Perspect., 101*: 504.

Pohjanvirta, R., and Tuomisto, J. (1994). *Pharmacol. Rev., 46*: 483.

Porterfield, S. P., and Hendrich, C. E. (1993). *Endocrinol. Rev., 14*: 94.

Pratt, R. M. (1985). Receptor-dependent mechanisms of glucocorticoid and dioxin-induced cleft palate. *Environ. Health Perspect. 61*: 35.

Pratt, R. M., Dencker, L., and Diewert, V. M. (1984). *Teratogen. Carcinogen. Mutgen., 4*: 427.

Reyes, H., Reisz-Porszasz, S., and Hankinson, O. (1992). *Science, 256*: 1193.

Rogan, W. J. (1982). *Teratology, 26*: 259.

Rogan, W. J., Gladen, B. C., McKinney, J. D., Carreras, N., Hardy, P., Thullen, J., Tinglestaad, S., and Tully, M. (1986). *J. Pediatr., 109*: 335.

Rogan, W. J., Gladen, B. C., Hung, K. L., Koong, S. L., Shih, L. Y., Taylor, J. S., Wu, Y. C., Yang, D., Rogan, N. B., and Hsu, C. C. (1988). *Science, 241*: 334.

Roman, B. L., Sommer, R. J., Shinomiya, K., and Peterson, R. E. (1995). *Toxicol. Appl. Pharmacol., 134*: 241.

Rönnbäck, C. (1991). *Pharmacol. Toxicol., 68*: 340.

Rönnbäck, C., and de Rooij, D. G. (1994). *Pharmacol. Toxicol., 74*: 287.

Sager, D. B. (1983). *Environ. Res., 31*: 76.

Sager, D. B., and Girard, D. M. (1994). *Environ. Res., 66*: 52.

Sauer, P. J. J., Huisman, M., Koopman-Esseboom, C., Morse, D. C., Smits-van Prooije, A. E., van de Berg, K. J., Tuinstra, L. G. M. Th., van der Paauw, C. G., Boersma, E. R., Weisglas-Kuperus, N., Lammers, J. H. C. M., Kulig, B. M., and Brouwer, A. (1994). *Hum. Exp. Toxicol., 13*: 900.

Schantz, S. L., and Bowman, R. E. (1989). *Neurotoxicol. Teratol., 11*: 13.

Schantz, S. L., Levin, E. D., Bowman, R. E., Heironimus, M. P., and Laughlin, N. K. (1989). *Neurotoxicol. Teratol., 11*: 243.

Schantz, S. L., Levin, E. L., and Bowman, R. E. (1991). *Environ. Toxicol. Chem., 10*: 747.

Schantz, S. L., Moshtaghian, J., and Ness, D. K. (1995). *Fundam. Appl. Toxicol., 26*: 117.

Schecter, A., ed. (1994). *Dioxins and Health*. Plenum Press, New York.

Seefeld, M. D., Corbett, S. W., Keesey, R. E., and Peterson, R. E. (1984). *Toxicol. Appl. Pharmacol., 73*: 311.

Seo, B-W., and Meserve, L. A. (1995). *Bull. Environ. Contam. Toxicol., 55*: 22.

Seo, B-W., Li, M-H., Hansen, L. G., Moore, R. W., Peterson, R. E., and Schantz, S. (1995). *Toxicol. Lett., 78*: 253.

Shuler, C. T., Halpern, D. E., Guo, Y., and Sank, A. C. (1992). *Dev. Biol., 154*: 318.

Smits-van Prooije, A. E., and Lammers, J. H. C. M., Waalkens-Berendsen, D. H., and Kulig, B. M. (1992). *Organohal. Compd., 10*: 217.

Spitsbergen, J. M., Walker, M. K., Olson, J. R., and Peterson, R. E. (1991). *Aquat. Toxicol., 19*: 41.

Srivastava, D., Cserjesi, P., and Olson, E. N. (1995). *Science, 270*: 1995.

Sunahara, G. I., Nelson, K. G., Wong, T. K., and Lucier, G. W. (1987). *Mol. Pharmacol., 32*: 572.

Swain, W. R. (1991). *J. Toxicol. Environ. Health, 33*: 587.

Swanson, H. I., and Bradfield, C. A. (1993). *Pharmacogenetics, 3*: 213.

Taylor, P. R., Stelma, J. M., and Lawrence, C. E. (1989.) *Am. J. Epidemiol., 129*: 395.

Thiel, R., Koch, E., Ulbrich, B., and Chahoud, I. (1994). *Arch. Toxicol., 69*: 79.

Tilson, H. A., Davis, G. J., KmcLachlan, J. A., and Lucier, G. W. (1979). *Environ. Res., 18*: 466.

Tilson, H. A., Jacobson, J. L., and Rogan, W. J. (1990). *Neurotoxicol. Teratol., 12*: 239.

U.S. Environmental Protection Agency. (1993). Interim report on data and methods for assessment of 2,3,7,8-tetrachlorodibenzo-*p*-dioxin risks to aquatic life and associated wildlife. Office of Research and Development, Environmental Research Laboratory at Duluth, MN. EPA/600/R-93/055.

U.S. Environmental Protection Agency. (1994). Health assessment document for 2,3,7,8-tetrachlorodibenzo-*p*-dioxin (TCDD) and related compounds. Office of Research and Development, Washington, DC. EPA/600/BP-92/001c.

Van den Berg, M., De Jongh, J., Poiger, H., and Olson, J.R. (1994). *CRC Crit. Rev. Toxicol., 24*: 1.

Walker, M. K., Spitsbergen, J. M., Olson, J. R., and Peterson, R. E. (1991). *Can. J. Fish. Aquat. Sci., 48*: 875.

Wang, G. L., Jiang, B-H., Rue, E. A., and Semenza, G. L. (1995). *Proc. Natl. Acad. Sci., 92*: 5510.

Wanner, R., Brömmer, S., Czarnetzki, B. M., and Rosenbach, T. (1995). *Biochem. Biophys. Res. Commun., 209*: 706.

Weber, H., and Birnbaum, L. (1985). *Arch. Toxicol., 57*: 159.

Weisglas-Kuperus, N., Sas, T. C. J., Koopman-Esseboom, C., Van Der Zwan, C. W., De Ridder, M. A. J., Beishuizen, A., Hooijkaas, H., and Sauer, P. J. J. (1995). *Pediatr. Res., 38*: 404.

Weston, W. M., Nugent, P., and Greene, R.M. (1995). *Biochem. Biophys. Res. Commun., 207*: 690.

White, T. E. K., and Gasiewicz, T. A. (1993). *Biochem. Biophy. Res. Comm., 193*: 956.

Whitlock Jr., J. P. (1993). *Chem. Res. Toxicol., 6*: 754.

Williams, L. L., Giesy, J. P., Verbrugge, D. A., Jurzyista, S., Heinz, G., and Stromborg, K. (1995). *Arch. Environ. Contam. Toxicol., 29*: 52.

Wolfe, W. H., Michalek, J. E., Miner, J. C., Rahe, A. J., Moore, C. A., Needham, L. L., and Patterson, D. G., Jr. (1994). *Epidemiology, 6*: 17.

Zhao, F. Mayura, K., Safe, S., and Phillips, T. D. (1995). *Toxicologist, 838*: 157.

6

Transgenic and Knockout Mice: Genetic Models for Environmental Stress

Robert S. Tebbs and Roger A. Pedersen
University of California, San Francisco, California

Jack B. Bishop
National Institute of Environmental Health Sciences, Research Triangle Park, North Carolina

ENVIRONMENTAL STRESS AND MAMMALIAN DEVELOPMENT

All living organisms have mechanisms for dealing with environmental stress, probably because of their evolution in environments encompassing natural radiation, oxygen, or dietary toxicants. This adaptive legacy is shared by organisms ranging from bacteria to humans. Because of the evolutionary relatedness of molecular mechanisms for dealing with environmental stress, studies in prokaryotic as well as eukaryotic systems have provided insight into how organisms cope with oxidative stress, radiation, and chemical toxicants.

Numerous transgenic and knockout mice have been designed to evaluate the role of distinct enzymes in protecting the organism against excessive molecular damage. These genetic models are useful for assessing toxicological potential by increasing or by decreasing the capacity of the organism to carry out specific molecular processes associated with the metabolism or detoxification of environmental agents or repair of damage resulting from exposure to them. Accordingly, these genetic models can be instrumental in defining the normal limits of organisms for dealing with environmental stress. For example, a mouse with augmented capacity for metabolizing reactive intermediates may reveal additional roles for such metabolic pathways. Conversely, a mouse lacking the ability to repair a particular type of DNA damage can reveal the consequences of exceeding the normal repair capacity, or amplify the consequences of long-term exposure. Because the mammalian conceptus develops within the maternal environment, it is protected from certain types of environmental stress, such as ultraviolet light; nonetheless, the embryo and fetus are susceptible to other radiation modalities and to many of the chemicals and metabolites present in the maternal environment. Moreover, the consequences of exposure to environmental stress during early developmental stages are likely to be more severe than exposure at postnatal or adult stages and may include developmental anomalies as well as developmental delays or learning deficiencies. Therefore, it is crucial to understand the molecular mechanisms by which the developing individual copes with environmental stresses, and mammalian genetic models are invaluable resources for achieving this objective.

TRANSGENIC AND GENE-TARGETING APPROACHES

Since their introduction in 1980, transgenic mice have provided valuable genetic models for understanding the role of individual gene products during the entire life cycle of the organism (see Brinster, 1993, for review), as well as the organisms response to environmental toxicants. By now, many transgenes—too numerous to review—have been introduced into the mouse germ line by zygotic DNA injection and embryo transfer to foster mothers, making this approach the most highly utilized means of genetic modification in mammals (reviewed by Brinster, 1993; Pinkert, 1994). The principal outcome achieved from using this "standard" transgenic approach is overexpression, or tissue-specific expression of the transgene. Such studies have provided useful models for determining the developmental or physiological role of specific gene products (Nizielski et al., 1996). Mouse transgenic models have been particularly valuable for environmental studies because the high frequency of insertional mutagenesis (Rijkers et al., 1994) has provided numerous target genes for assessing mutation frequency (Gahlmann, 1993) and tumorigenesis studies (Christofori and Hanahan, 1994; Viney, 1995). Finally, recent refinements have enabled standard transgenic approaches to achieve more reliable tissue-specific gene expression (Bonifer et al., 1996), and regulatable systems for gene expression (Kistner et al., 1996) or for tissue- or stage-specific gene ablation (Kuhn et al., 1995; Betz et al., 1996; Rickert et al., 1997). When used in conjunction with gene targeting in embryonic stem cells (see below), transgenic systems provide an integrated approach for genetic modification in mammals.

Pluripotent stem cells are unique embryonic or fetal cells that are capable of differentiating into all cell types upon exposure to the appropriate embryonic environments. In the mouse, where embryonic stem (ES) cells have been studied most extensively, they have been derived by culturing preimplantation embryos at the morula or blastocyst stage and isolating colonies of proliferating cells that maintain their capacity for differentiation (Martin, 1981; Evans and Kaufman, 1981). Pluripotent mouse ES cells have profoundly changed mammalian developmental genetics. The remarkable capacity of pluripotent stem cells to proliferate in culture, yet resume normal development and contribute to all tissues of the organism gave birth to mouse gene-targeting technology (for experimental approaches, see also Robertson, 1987; Pedersen et al., 1993a; Joyner, 1993; Wassarman, 1993; Hogan et al., 1994). The extensive use of ES cells as vectors for ablating the function of specific genes through homologous recombination, or gene targeting (reviewed by Joyner, 1993), has revealed the developmental and physiological roles of many genes (reviewed by Joyner, 1991; Bradley et al., 1992; Fung-Leung and Mak, 1992; Huang, 1993; Koller and Smithies, 1992; Pedersen et al., 1993; Pinkert, 1993; Ramirez-Solis et al., 1993; Robbins, 1993; Smithies, 1993; Hogan et al., 1994; Joyner, 1993; Copp, 1995; Rinkenberger et al., 1997). Such studies have used micromanipulation to generate chimeras that transmit the mutant genes through the germline of the resulting mouse (Robertson, 1987; Pedersen et al., 1993; Joyner, 1993; Wassarman and DePamphilis, 1993; Hogan et al., 1994; Camper et al., 1995).

Because they are pluripotent stem cells, mouse ES cells have provided valuable in vitro models for studying differentiation of the cell lineages that form during early embryonic development. Such studies have provided insight into mechanisms that regulate differentiation and proliferation of these early lineages. Pluripotent cells have also been derived by culturing mouse fetal gonads, giving rise to embryonic germ (EG) cell lines that have similar properties to mouse ES cells, including germline transmission in chimeras (reviewed in Donovan, 1994). To date, limited studies of pluripotent stem cells in other species, including bovine, hamster, mink, pig, sheep, rabbit, rat, and human, have been reported (reviewed in Pedersen, 1994). While germline transmission in chimeras derived from pluripotent stem cells has still only been accomplished in the mouse, it could, in principle, eventually be extended to other species.

Our purpose in this chapter is to describe the phenotypes of mouse mutant strains created by transgenic and gene-targeting approaches that are relevant to understanding the developmental effects of environmental stress. Two main areas have been explored: enzymes that detoxify reactive oxygen metabolites and enzymes that repair damage incurred to DNA.

GENETIC MODELS FOR OXIDATIVE STRESS

Nonvertebrate Models

The first line of defense against reactive molecules accumulating inside the cell involves molecular detoxification. The cell has evolved a collection of enzymes which effectively detoxify a wide variety of hazardous compounds. One of the most extensively studied detoxification systems in the cell is that involved in eliminating reactive oxygen species. Highly reactive oxygen species arise intracellularly as byproducts of metabolism in aerobic organisms. The univalent reduction of oxygen generates reactive oxygen intermediates; specifically, superoxide radical ($O_2\cdot^-$), hydrogen peroxide (H_2O_2), and hydroxyl radical ($OH\cdot$) (Fridovich, 1978). Superoxide radicals can escape from many naturally occurring reactions in the cell, including the mitochondrial electron transport chain, monooxygenases such as P450s, and can also be liberated from hemoglobin and myoglobin. Oxygen free radicals have also been implicated in the toxicity associated with exposures to exogenous agents and other pathophysiological situations, such as radiation, chemical agents (i.e., paraquat), hyperoxia, and ischemia. Superoxide dismutase (SOD), catalase (CAT), and glutathione peroxidase (GSHPx) are three well-known cellular enzymes, which constitute a major defense mechanism against reactive oxygen intermediates (see Fantel, 1996, for review).

Superoxide dismutase is found in almost all aerobic organisms from bacteria to humans (Fridovich, 1975, 1978), suggesting that superoxide is a universal hazard to all oxygen-metabolizing organisms. Three distinct enzymes have been discovered: an iron-containing enzyme (Fe-SOD), a manganese-containing enzyme (Mn-SOD), and an enzyme that contains both copper and zinc (CuZn-SOD). They all catalyze the same reaction with similar rates—the dismutation of $O_2\cdot^-$ into O_2 and H_2O_2.

A number of organisms have been studied that contain null-mutations in one or several SOD genes, as well as organisms that overproduce superoxide dismutase. *Escherichia coli* contains two SOD enzymes: Mn-SOD (the product of *sodA*), which is induced by oxidative stress, and FeSOD (product of *sodB*), expressed constitutively. *SodA*, *sodB*, and *sodA/sodB* (*sod⁻*) mutants were first constructed by Carlioz and Touati and found to be sensitive to oxidative stress induced by administration of H_2O_2 or the superoxide-generating drug paraquat (Carlioz and Touati, 1986). *SodA*, lacking the inducible form of the gene, is more sensitive to oxidative stress than *sodB*. However, only the mutant disrupted in both SOD genes demonstrates abnormal sensitivity under standard aerobic culture conditions.

Surprisingly, the overexpression of superoxide dismutase in *E. coli* increased toxicity to oxygen stress, rather than reducing it (Bloch and Ausubel, 1986; Scott et al., 1987; Liochev and Fridovich, 1991). This unexpected result seems to reflect the need to coordinate the detoxification of superoxide with other genes involved in the total response to elevated O_2^-, including downstream reactive intermediates such as H_2O_2 and $OH\cdot$.

Saccharomyces cerevisiae, like most eukaryotes, contains CuZn-SOD (product of the *SOD1* gene) in the cytosol and Mn-SOD (product of the *SOD2* gene) in the mitochondria. *S. cerevisiae* single and double mutants with null-mutations in *SOD1* and *SOD2* have been constructed and, like *E. coli*, demonstrate oxygen-associated growth sensitivity (Bilinski et al., 1985; von Loon et al., 1986; Longo et al., 1996). In addition, the viability of *SOD* single and double mutants in stationary phase (no nutrients provided) decreased over time at a faster rate than wild type (Longo et al., 1996). *SOD1* was more sensitive to hyperoxygenated conditions than *SOD2*. The viability of *SOD1* was completely restored to wild type when the *SOD1* gene was overexpressed (Longo et al., 1996).

Null mutations in *Drosophila melanogaster* have added another dimension to SOD, demonstrating systemic affects when intercellular regulation of superoxide is altered. As expected from studies with other organisms, adult and larvae *Drosophila* nullizygous for CuZn-SOD were highly sensitive to the lethal effects of paraquat compared to normal insects (Phillips et al., 1989). Heterozygote larvae demonstrated a slight increase in sensitivity to paraquat, but adult heterozygotes showed no increased sensitivity to paraquat above wild type (Phillips et al., 1989). Interestingly, homozygous mutants have a dramatic reduction in adult life expectancy (life span reduced approximately 80%) and are completely male sterile (females have reduced sterility) (Phillips et al., 1989). Histopathology of the eye revealed atrophy and necrosis in the retina of 7-day-old adults, primarily in the photoreceptor neurons (Phillips et al., 1995). Other defects in the nervous system have been reported, although not documented (Phillips et al., 1995). Partial restoration of CuZn-SOD activity (approximately 40% of wild type) in SOD-null *Drosophila* by transformation with a bovine CuZn-SOD transgene rescues male infertility and resistance to paraquat, but only partially rescues adult life expectancy and resistance to hyperoxia (Reveillaud et al., 1994).

Since reduced SOD activity decreases life expectancy, the next obvious question is whether life expectancy increases with increased SOD activity. A number of investigators have overexpressed Cu-ZN superoxide dismutase, and all have generally found a slight increase in mean life expectancy, although no increase in maximum life span (Staveley et al., 1990; Seto et al., 1990; Reveillaud et al., 1991; Orr and Sohal, 1993). Sensitivity to paraquat varied, with some investigators showing a decrease in sensitivity (Reveillaud et al., 1991), while others found increased sensitivity (Staveley et al., 1990; Seto et al., 1990). It is worth noting that no transgenic lines were recovered that contained twofold or greater increase in CuZn-SOD activity, leading to the suggestion that too high a level of SOD activity is deleterious. Indeed, strains overexpressing SOD were vulnerable to death during the process of eclosion (Reveillaud et al., 1991).

The study of multigene effects allows one to better appreciate the networking of gene products in vivo. *Drosophila* strains with a null mutation for the rosy (*ry*) gene are unable to synthesize urate, a proposed of oxygen radical scavenger, and *ry*-null flies, like CuZn-SOD–null flies, are hypersensitive to paraquat and hyperoxic conditions (Hilliker et al., 1992). *Drosophila* compound mutants doubly deficient for uric acid and Cu/Zn-SOD die before completion of metamorphosis (Hilliker et al., 1992). This indicates that Cu/Zn-SOD is part of a larger system (including urate) responsible for protection against reactive oxygen. Overexpression of catalase in *Drosophila* demonstrated characteristics similar to overexpression of CuZn-SOD—a slight increase in average life span, but not maximum life span (Orr and Sohal, 1992). Overexpression of both CuZn-SOD and catalase in the same strain (SOD activity 26% greater, and catalase activity 73% greater than control levels) demonstrated an increase in maximum life span as great as 34% over control strains (Orr and Sohal, 1994). Age-matched adult flies overexpressing CuZn-SOD and catalase show improved physical fitness based on movement analysis and demonstrate increased metabolic activity determined by oxygen consumption (Orr and Sohal, 1994, 1995).

It is clear that SOD protects against oxygen-induced stress in *E. coli, S. cerevisiae*, and *D. melanogaster*. All three nonvertebrates demonstrate increased lethality in hyperoxic conditions in the absence of SOD. It is also apparent that SOD is part of a complex system involving other enzymes, hence overexpression of SOD alone leads to an unbalanced response to oxygen stress and is toxic. Overexpressing SOD and catalase, on the other hand, can increase total metabolic activity during an average lifespan, leading to increased life expectancy. Altering SOD activity in *D. melanogaster* shows that multicellular organisms can manifest a more varied and complex effect, with some physiological systems (i.e., eye, nervous system, and gonads) more prone to developmental defects.

Mouse Transgenic and Knockout Models

Transgenic Models for SOD

Transgenic approaches have been valuable in revealing pathological consequences of overexpressing gene products (see Fantel, 1996, for review). Several stocks of transgenic mice overexpressing human CuZn-SOD have been constructed, each of which varies in CuZn-SOD activity; the level of activity varies between mice as well as between tissues within each mouse, ranging, for example, between 1.6- to 6.0-fold in the brain of transgenic mice compared to control mice (Epstein et al., 1987). hCuZn-SOD transgenic mice demonstrate no gross physical or behavioral abnormalities (Epstein et al., 1987). However, pathological analysis demonstrated premature degeneration of the neuromuscular junctions of the tongue of transgenic mice at 2–4 months, which typically arise in control mice at ~1 year (Avraham et al., 1988). When the levels of endogenous Mn-SOD, catalase, and glutathione peroxidase activity were measured in the brains of hCuZn-SOD transgenic mice, Mn-SOD and catalase activity increased, indicating an effort to try to balance the total response (Przedborski et al., 1992). In a separate experiment, a mouse was constructed overexpressing human CuZn-SOD, and although all tissues examined overexpressed hCuZn-SOD, expression was most predominant in the brain (1.93-fold) where it was found that hCuZn-SOD expression localized in neurons, similar to endogenous mouse transcripts (Ceballos-Picot et al., 1991). In agreement with Przedborski et al. (1992), these transgenic mice also demonstrate a slight increase in endogenous Mn-SOD activity compared to controls, but no increase in glutathione peroxidase activity (catalase activity was not measured in this study) (Cellabos-

Picot et al., 1992b) In addition, these hCuZn-SOD–transgenic mice contain significantly enhanced levels of malondialdehyde in whole brain homogenate, but not liver homogenate, relative to control mice, indicating elevated lipid peroxidation in the brain (Ceballos-Picot et al., 1991, 1992b). Evidence has indicated that overexpression of CuZn-SOD could be deleterious to the cell, by nonenzymatically catalyzing the production of hydroxyl radical from hydrogen peroxide (for review see Ceballos-Picot et al., 1992a). Although transgenic mice do demonstrate slight defects in the nervous system, they appear better able to adapt to the overexpression of CuZn-SOD compared to *D. melanogaster* and other invertebrates, possibly by the compensatory upregulation of other genes.

Transgenic mice overexpressing superoxide dismutase have been exploited for the purpose of understanding more clearly the effect of SOD on oxidative stress in lung and brain tissue. White et al. (1991) found that transgenic mice overexpressing human CuZn-SOD (2- to 2.5-fold increased activity in lung) have a lower mortality rate when subjected to hyperoxia ($>99\%$ O_2) compared to control mice. The protective effect of hCuZn-SOD is further shown by histopathology, where decreased edema and hyaline membrane formation is seen in lung tissue of transgenic mice vs. control mice after exposure to hyperoxia (White et al., 1991). Transgenic mice expressing human Mn-SOD specifically in the distal epithelial cells of the lung were constructed by utilizing transcriptional elements from the human pulmonary surfactant protein C gene (Wispe et al., 1992). hMn-SOD protein is increased approximately 4-fold in alveolar Type II cells in these transgenic mice and show no associated increase in CuZn-SOD, glutathione peroxidase, or catalase activity (Wispe et al., 1992). hMn-SOD–transgenic mice demonstrate no detectable edema, hyaline membrane formation, or hemorrhage in the lungs after exposure to 95% O_2 for 5 days, at a time when nontransgenic mice demonstrate tissue damage (Wispe et al., 1992). hMn-SOD-transgenic mice survive nearly twice as long as control mice when continuously exposed to 95% O_2 (Wispe et al., 1992). Thus, increased SOD activity can protect against lung-tissue toxicity during hyperoxic stress.

The brain is particularly vulnerable to free radicals based on its high consumption of oxygen and its high concentration of phospholipids, which can propagate free radical–generated reactions. The role of CuZn-SOD in focal cerebral ischemia was evaluated through the use of transgenic mice overexpressing human CuZn-SOD (2.6-fold increased level of SOD enzymatic activity). A focal cerebral ischemia was created surgically by closing off the blood flow to a select set of arteries (left middle cerebral artery and left common carotid artery) and temporarily closing off the right common carotid artery. After 1 hour of temporary occlusion of the right common carotid artery followed by 24 hours of reperfusion, the brains of hCuZn-SOD–transgenic mice demonstrate a 30% reduction in total infarct volume and a significant reduction in brain edema compared to nontransgenic controls (Kinouchi et al., 1991). In another study, focal ischemia was created by middle cerebral artery occlusion for 3 hours, followed by 3 hours of reperfusion (Yang et al., 1994). Under these conditions, the infarct volume is reduced by 26% in hCuZn-SOD–transgenic mice compared with nontransgenic mice, and neurological deficits measured by gross observation of physical abilities were also significantly reduced in transgenic mice (Yang et al., 1994). Brain injury induced by a 30-second exposure to a −50°C brass probe applied directly to the skull was reduced, and cold-induced cerebral infarction was reduced 52% in transgenic mice overexpressing human CuZn-SOD compared with nontransgenic mice (Chan et al.,

1991). In addition there was a reduction in brain edema in hCuZn-SOD–transgenic mice relative to control mice (Chan et al., 1991). In conclusion, genetically increasing SOD activity in mice can protect the lung and the brain against oxygen toxicity. In humans, oxygen toxicity is a real threat for stroke patients, and therefore increasing SOD activity could be protective therapy for patients at risk (see also Fantel, 1996).

SOD Gene-Targeting Models

Mice deficient for CuZn-SOD develop normally, show no detectable motor deficiencies and no neuropathological abnormalities through 6 months of age (Reaume et al., 1996). No significant differences were observed in either lipid peroxidation or protein carbonyl content in CuZn-SOD (–/–) mice relative to control mice (Reaume et al., 1996). However, CuZn-SOD (–/–) mice demonstrate increased motor neuron loss following facial axotomy relative to CuZn-SOD (+/+) mice, while CuZn-SOD (–/+) mice show an intermediate degree of neuronal death (Reaume et al., 1996). In contrast to CuZn-SOD (–/–) mice, mice lacking Mn-SOD die within 10 days of birth (Li et al., 1995). Mn-SOD (–/–) neonates are pale, less energetic, hypertonic, and hyperthermic (Li et al., 1995). Histological analysis demonstrates that Mn-SOD (–/–) animals contain an enlarged heart and show lipid deposition in liver and muscle tissue (Li et al., 1995). The severe toxicity of Mn-SOD deficiency compared to CuZn-SOD deficiency in the mouse is curious since they catalyze the same reaction. Mn-SOD, unlike CuZn-SOD, is inducible, and Mn-SOD is localized to the mitochondria, whereas CuZn-SOD is in the cytosol and extracellular spaces. The highest concentration of internally generated superoxide is likely to be in the mitochondria, the site of ATP production by oxidative phosphorylation. The accumulation of oxidative damage during embryogenesis is probably the cause of neonatal lethality in Mn-SOD mutants.

GENETIC MODELS FOR DNA DAMAGE AND REPAIR

Numerous enzymes and multiprotein complexes exist in the cell for the purpose of repairing practically all possible forms of DNA damage, to maintain the integrity of the genetic blueprint, and preserve efficient cellular biochemistry (for extensive discussions on DNA repair, see Freidberg et al., 1995; McEntyre, 1995). There are at least five DNA repair pathways requiring multiple proteins: nucleotide excision repair, base excision repair, mismatch repair, double-strand break repair, and recombinational repair. A mutation in any protein associated with a particular repair pathway can inactivate the whole pathway. Although there appears to be some degree of redundancy and interaction between the different repair pathways, a defect in any pathway is typically reflected by an increased sensitivity to a particular class of genotoxic agents. Organisms from bacteria to humans have been studied to determine the effects of defective DNA repair. Organisms compromised for DNA repair often demonstrate an associated decrease in survival and increase in mutation frequency and cancer incidence following exposure to selected mutagens and carcinogens. Nonetheless, there are conspicuous phenotypic trends among transgenics and knockouts of the diverse DNA repair pathways, which we will address at the conclusion of this review. We focus here on mammalian transgenic and knockout models affecting repair of DNA damage (Table 1).

Table 1 Transgenic Mice Deficient for DNA Repair Genes

Gene	Repair defect	Predominant physiological characteristic
XPA	NER	UV-sensitive skin; elevated spontaneous tumors, and UV-/DMBA-induced skin cancer
XPC	NER	UV-sensitive skin; increase in UV-induced skin cancer
XPC/p53	NER	UV-sensitivity to skin is worsened; UV-induced skin cancer has shorter latency period
ERCC1	NER	Severe growth defect; premature death due to liver failure
CSB	NER	UV-sensitive skin; increase in UV-/DMBA-induced skin cancer
XPB	NER	Embryonic lethal (very early)
XPD	NER	Embryonic lethal (very early)
Ref-1/Ape	BER	Embryonic lethal (defect begins ~E5.5)
polβ	BER	Embryonic lethal (midgestation)
Xrcc1	BER	Embryonic lethal (defect begins ~E6.5)
PARP	BER	Predisposition to developing skin disease with age
Aag	BER	No physiological abnormalities reported
Scid	DSB	Immunodeficient; small lymphoid tissues; higher incidence of lymphomas; radiosens
Ku80	DSB	Immunodeficient; small lymphoid tissues; growth defect; mothers don't nurse pups
Rad51	REC	Embryonic lethal (defect begins ~E5.5)
Rad51/p53	REC	Embryonic lethal (defect begins ~E7.5); p53 deficiency delays embryonic death
Brca1	REC	Embryonic lethal (defect begins ~E4.5)
Brca2	REC	Embryonic lethal (defect begins ~E6.5)
RAD54	REC	No physiological abnormalities reported
MSH2	MMR	High incidence and early onset of lymphomas; Intestinal tumor frequency increases with age
MSH2/Min	MMR	Intestinal tumor onset time shortened and multiplicity increased
PMS2	MMR	Male infertility (abnormal spermatozoa structure); predisposition to lymphoma
MLH1	MMR	Male and female infertility; no data on tumor susceptibility available
MGMT	Other	Growth defects; hypersensitive to MNU-induced lethality

BER, Base excision repair; DSB, double strand break; MMR, mismatch repair; NER, nucleotide excision repair; REC, recombinational repair.

Nucleotide Excision Repair

Nucleotide excision repair (NER) is a versatile repair mechanism, recognizing a wide variety of DNA adducts, and is the predominant repair pathway involved in the removal of UV photoproducts and bulky chemical adducts bound to DNA. Over 25 polypeptides are required for NER in mammalian cells. Mutations in NER genes are associated with the human diseases xeroderma pigmentosum (XP), Cockayne syndrome (CS), and trichothiodystrophy (TTD). Two NER genes, *XPB* and *XPD*, are also essential components of the RNA polymerase II basal transcription factor TFIIH. The dual function of TFIIH in transcription and repair has been evoked as the basis for the finding of three distinct human diseases, each resulting from mutations affecting one or both functions of the protein (Schaeffer et al., 1993).

There are two distinguishable forms of NER: global genome repair, which removes DNA damage without respect to gene activity, and transcription-coupled repair, which selectively removes damage from the template strand of actively transcribed genes. Although both NER subpathways use many of the same proteins, select proteins are unique to each system, and therefore NER can be studied generally or each subpathway can be studied independently. Knockout mice have been constructed such that each of the three possible NER-deficient scenarios are represented (*XPA* (–/–) and *ERCC1* (–/–) mice are deficient in both global genome and transcription-coupled repair, *XPC* (–/–) mice are deficient in global genome repair, and *CSB* (–/–) mice are deficient in transcription-coupled repair).

XPA Knockout Mice

The XP-A protein in a component of nucleotide excision repair and is presumably involved in DNA damage recognition. There are eight XP complementation groups, XP-A through G plus a variant demonstrating normal excision repair. Human patients with a defect in the *XPA* gene usually represent the most severe clinical form of XP (Cleaver and Kraemer, 1995). *XPA* (–/–) mice have been constructed and found to be physically and pathologically normal in their first year of life (Nakane et al., 1995; de Vries et al., 1995). Approximately 20% of older *XPA* (–/–) mice spontaneously develop tumors, mostly hepatocellular adenomas (de Vries et al., 1997a). Like human XP patients, *XPA* (–/–) mice show a heightened intolerance to UV-B, evidenced by increased edema, erythema, and hyperkeratosis upon exposure to UV, and an enhanced immunosuppressive response after UV-B radiation (Nakane et al., 1995; de Vries et al., 1995; Miyauchi-Hashimoto et al., 1996). Moreover, *XPA* (–/–) mice demonstrate a marked elevation in incidence of skin tumors following exposure to UV-B (Nakane et al., 1995; de Vries et al., 1995). *XPA* (–/–) mice are also more prone to developing skin tumors induced by the topical administration of 7,12-dimethylbenz[a]anthracene (Nakane et al., 1995; de Vries et al., 1995) and to developing T-cell lymphomas following the oral administration of benzo[a]pyrene (de Vries et al., 1997a). Mice administered benzo-[a]pyrene orally three times a week demonstrate an increased level of DNA adducts in the liver, lung, and spleen, eventually reaching a steady-state level at ~5 weeks of treatment in wild-type mice, however, in *XPA* (–/–) mice the number of adducts continue to increase in most tissues through 13 weeks of treatment (de Vries et al., 1997b). Benzo[a]pyrene-induced mutation frequencies (determined by a *lacZ* transgene) were elevated only in the spleen of *XPA* (–/–) mice compared to wild-type mice, and not elevated in the lung or liver (de Vries et al., 1997b). It was suggested that the proliferative state of the spleen, being a lymphoid organ involved in the development of effector T lymphocytes, is better able to fix premutagenic lesions into permanent mutations, which might then lead to an increased incidence of lymphomas (de Vries et al., 1997b). Thus, *XPA* (–/–) mice are hypersensitive to the mutagenic and carcinogenic affects of environmental carcinogens and are more prone to developing spontaneous tumors in adult life.

XPC Knockout Mice

Mice disrupted at *XPC* have been constructed and develop into normal, fertile adults, showing no signs of tumor formation in the first year of life (Sands et al., 1995; Cheo et al., 1997). Mouse embryonic fibroblasts derived from *XPC* (–/–) embryos are defective in global genome NER, but retain a significant capacity for transcription-coupled NER (Cheo et al., 1997). Like the *XPA* (–/–) mice described above, *XPC* (–/–) mice display a wide range of UV-induced histopathological damage to the skin and eyes (Sands et al., 1995). *XPC* (–/–) mice exposed daily to ultraviolet light developed skin carcinomas with 100% penetrance ($n = 9$), whereas no tumors were detected in mice heterozygous or wild type for functional *XPC* under the same conditions (Sands et al., 1995).

    XPC mutant mice were crossed to *p53* mutant mice to generate double-mutant animals (Cheo et al., 1996). Late-stage *XPC* (–/–) *p53* (–/–) embryos contain a high number of neural tube defects, mostly found in females, which result in neonatal death of more than half of female double-knockouts, a characteristic previously found in *p53* (–/–) embryos at moderate levels (Sah et al., 1995; Armstrong et al., 1995). UV-induced skin disorders increase as a function of *p53* status in *XPC* (–/–) mice, becoming progressively worse when the *p53* genotype in *XPC* (–/–) mice changes from (+/+) to (–/+) to (–/–) (Cheo et al., 1996). Furthermore, *XPC* (–/–) *p53* (–/+) mice develop UV-induced skin tumors with a latency period that is significantly reduced compared to *XPC* (–/–) *p53* (+/+) mice (data for *XPC-p53* double-knockout mice is not available) (Cheo et al., 1996). Thus, *XPC* (–/–) mice are prone to developing tumors induced by environmental carcinogens, and added deficiencies in *p53* exacerbate the disease.

ERCC1 Knockout Mice

ERCC1 is another component of NER, not found to correct any of the 8 XP complementation groups. Mice deficient in *ERCC1* were the first DNA repair-deficient mice constructed (McWhir et al., 1993). *ERCC1* (–/–) mice are runted at birth and die within the first month of life from liver failure (McWhir et al., 1993). *ERCC1* (–/–) mouse liver cells contain low levels of perinatal nuclear polyploidy, progressing to severe aneuploidy by 3 weeks of age (McWhir et al., 1993). There was also an indication that p53 levels were elevated in liver, brain, and kidney (McWhir et al., 1993).

    In a recent report from a workshop on DNA damage, another mouse containing deletion of the C-terminal seven amino acids of *ERCC1* was described which also exhibited a severe growth defect; however, they survive to early adulthood, dying by 6 months of age presumably from liver failure (Lehman et al., 1996).

CSB Knockout Mice

Mice containing a knockout mutation in the Cockayne syndrome group B gene (*ERCC6*) develop normally and are fertile, however, they do demonstrate a slight but significant growth defect, more pronounced in males than in females (van der Horst et al., 1997). Primary embryonic fibroblasts derived from *CSB* (–/–) embryos demonstrate no transcription-coupled NER, but show normal global genome NER (van der Horst et al., 1997). Like human CS patients which show mental deficiencies, *CSB* (–/–) mice show some neurological abnormalities, detected as circling behavior and poor balance on the rotarod (van der Horst et al., 1997). *CSB* (–/–) mice develop UV-induced skin pathologies including severe erythema and hyperplasia, which were not found in similarly treated (–/+) or (+/+) mice (van der Horst et al., 1997). Chronic exposure of *CSB* (–/–) mice to UV-B light resulted in skin tumor formation with a shorter latency period and higher multiplicity than heterozygous and wild-type control animals (van der Horst et al., 1997). In addition, *CSB* (–/–) mice are also slightly more sensitive to DMBA (7,12-dimethylbenz[a]anthracene)-induced tumor formation compared to controls (van der Horst et al., 1997). Thus *CSB* (–/–) mice are predisposed to carcinogen-induced skin cancer and demonstrate a mild growth and neurological deficiency. A direct comparison for the relative cancer susceptibility between *XPA* (–/–), *XPC* (–/–), and *CSB* (–/–) mice await a formal study, but preliminary evidence indicates *XPA* (–/–) mice are more prone to turmorigenesis than *CSB* (–/–) mice (van der Horst et al., 1997).

Other NER-Deficient Mice

As cited by Lehman et al. (1996) and Friedberg et al. (1997), Hoeijmakers and co-workers have created knockout mice deficient in *XPB* and in *XPD*, and they have found that both mice are embryonic lethal. The embryonic lethality of *XPB*- and *XPD*-deficient mice is likely due to their more essential role in basal transcription.

Base Excision Repair

Base excision repair (BER) recognizes and repairs a wide variety of endogenous DNA damage incurred by oxidation, hydrolysis and simple alkylation, leading to abasic sites, deaminated nucleoside bases, and oxidative base damage (reviewed by Lindahl, 1993). Hydrolytic depurination has been estimated at 2,000–10,000 events/cell/day (Lindahl, 1993). DNA is also modified by naturally occurring reactive molecules in the nucleus, such as S-adenosylmethionine, which has been shown to methylate nucleophilic sites in DNA in a nonenzymatic reaction (Lindahl, 1993). BER corrects these types of endogenously derived DNA damage.

Although in vitro experiments demonstrate that only five proteins are required for efficient BER repair of a single DNA lesion (Kubota et al., 1996), BER involves a larger collection of proteins, due to a large number of DNA glycosylases, each of which recognizes a specific base damage or a subset of base damage and acts to remove the modified base, initiating the repair mechanism. BER often corrects a damaged base by single-base insertion, but repair patches of 7 or more nucleotides are also seen in a process that requires proliferating cell nuclear antigen (PCNA). Thus, at least two subpathways for BER exist in vivo, PCNA-independent short patch and PCNA-dependent long patch BER. No known human diseases have been shown to contain a defect in BER.

Ape/Ref-1 Knockout Mice

Mice deficient for *ref-1* (AP endonuclease, also known as *APE*), a pivotal enzyme in BER, die during early embryonic development (Xanthoudakis et al., 1996; and D.L. Ludwig, et al., unpublished observations). Morphologically, *ref-1* (–/–) embryos are severely abnormal by E5.5, and demonstrate elevated levels of pyknotic cells (Xanthoudakis et al., 1996). In addition to base excision repair, *ref-1* also catalyzes the reduction/oxidation modification of transcription factors such as Fos and Jun (Xanthoudakis et al., 1992). Thus, the lethal phenotype of *ref-1* knockout mice cannot be definitively attributed to lack of AP endonuclease activity.

polβ Mice

DNA polymerase β is the major polymerase utilized in BER, needed for excision of 5'-deoxyribose phosphate residues (created by *ref-1* excision at abasic sites) (Matsumoto and Kim, 1995), and for nucleotide insertion at the site of base damage (Sobol et al., 1996). Mice deficient for DNA polymerase β are lethal at midgestation (Gu et al., 1994). Although a detailed description of the deficient embryos has not been published, later reports indicate that knockout embryos were reduced in size and weight compared with control littermates (Betz et al., 1996). Mosaic mice were constructed by mixing cells that were wild type for *polβ* with cells deficient for *polβ* and subsequently examined to determine which tissues would contain few, if any, *polβ* (-/-) cells, suggesting that *polβ* was important for development or functionality in those tissue types (Betz et al., 1996). The overall percentage of *polβ* (-/-) cells in mosaic embryos markedly reduced from embryonic day 10.5 to birth (Betz et al., 1996). The reduction in *polβ* (-/-) cells was most pronounced in the thymus and least pronounced in the brain (Betz et al., 1996). Extracts of embryonic fibroblasts that were deficient for DNA polymerase β were defective for base excision repair (Sobol et al., 1996), indicating the importance of this function for BER.

Although DNA polymerase β is the predominant polymerase utilized in BER, a minor amount of activity has been detected using an alternative polymerase, indicating some redundancy in the system (Lindahl et al., 1995).

X-Ray Cross-Complementing-1 (Xrcc1) Knockout Mice

No enzymatic activity has been detected for XRcc1, however, the protein has been shown to increase the fidelity of BER and found to associate with several components of BER including DNA polymerase β and DNA ligase III, and thus possibly functions as a molecular scaffold for the BER complex (Kubota et al., 1996). Mice containing a null-mutation in XRcc1 have been constructed and, like *ref-1* (-/-) and *polβ* (-/-) mice, die during embryogenesis (Tebbs et al., 1996). Embryonic development of *Xrcc1* (-/-) embryos ceases just prior to gastrulation at embryonic day 6.5 (Tebbs et al., 1996). It appears that the embryonic tissues of the egg cylinder are uniquely affected by *Xrcc1* deficiency, demonstrating increased levels of apoptosis, whereas most of the extraembryonic tissues, including the ectoplacental cone, the parietal endoderm, and trophoblast cells, appeared relatively unaffected (R.S. Tebbs et al., unpublished observations). Embryos containing a double mutation in *Xrcc1* and *p53* survive approximately 24 hours longer than *Xrcc1* single mutants, indicating that at least some aspect(s) of the lethal phenotype are *p53*-independent (R.S. Tebbs, et al., unpublished observations).

Poly(ADP Ribose) Polymerase (PARP) Knockout Mice

PARP recognizes and binds to DNA breaks, whereupon it is induced to transfer ADP-ribose moieties to itself and other nuclear proteins (for reviews, see de Murcia and de Murcia, 1994; Satoh and Lindahl, 1994). As a consequence of PARP activation, NAD^+ levels are greatly diminished inside the cell (Heller et al., 1995). PARP has long been proposed to be an important component in BER, but its role in repair appears to be mainly passive, i.e., binding to breaks and transiently protecting them [possibly suppressing recombination (Lindahl et. al., 1995)] until they are repaired.

Mice deficient in poly(ADP-ribose) polymerase (PARP, or ADPRT) develop normally and are born healthy and fertile, however, there is a propensity toward developing severe skin disease in adult mice (Wang et al., 1995). The skin disease is characterized by extensive hair loss, enlarging erythemas, and occasional ulcerations (Wang et al., 1995). There is no evidence of defective DNA repair following exposure to UV or MNNG in *PARP* (–/–) fibroblasts, however, thymocytes from *PARP* (–/–) mice showed a reduced rate of proliferative recovery following exposure to sublethal doses of gamma-irradiation compared to wild-type animals (Wang et al., 1995). Pancreatic islet cells from *PARP* (–/–) mice did not demonstrate a significant reduction in NAD^+ levels following exposure to reactive oxygen intermediates, as was seen in cells from control mice (Heller et al., 1995).

Other BER Knockout Mice

Embryonic stem cells containing a null mutation at the 3-methyladenine DNA glycosylase gene (*Aag*) have been produced and, as expected, demonstrate hypersensitivity to simple alkylating agents (e.g., methyl methanesulfonate) (Engelward et al., 1996). Interestingly, *Aag* (–/–) cells are also sensitive to the bulky carcinogen mitomycin C and demonstrate an increased level of chromosome damage following exposure to methyl methanesulfonate (Engelward et al., 1996). Mice created from these cells are viable and develop normally (Friedberg et al., 1997).

Double-Strand Break Repair

Double-strand breaks are the major lethal lesion in DNA induced by ionizing radiation, and they also occur naturally as an intermediate in V(D)J recombination. Double-strand breaks can stimulate chromosomal deletions and rearrangements and are lethal if not repaired. In recent years several proteins have been shown to participate in rejoining these types of DNA breaks, in a reaction involving nonhomologous recombination (for recent reviews see Jeggo et al., 1995; Roth et al., 1995; Weaver, 1995). One of the main players in this repair mechanism is DNA-dependent protein kinase (DNA-PK), so named because DNA is required for its kinase activity. In addition to double-strand break repair and V(D)J recombination, DNA-PK has been implicated in transcription and DNA replication. DNA-PK consists of a catalytic subunit (DNA-PK$_{CS}$), and a DNA-binding protein complex named Ku. Ku is a heterodimer consisting of the Ku70 and Ku80 polypeptides. Although the details are not well understood, it appears that Ku recruits DNA-PK$_{CS}$ to altered forms of DNA containing double-stranded to single-stranded DNA junctions, leading to activation of DNA-PK kinase activity. In vitro, DNA-PK has been shown to phosphorylate many substrates including transcription factors (Sp1, Jun, Fos, Oct 1, Oct 2, SRF) and proteins associated with DNA repair (DNA-PK$_{CS}$, both subunits of Ku, RPA, and p53). Cell lines deficient for components of DNA-PK are sensitive to ionizing radiation and defective for V(D)J recombination.

Scid Mice

The *scid* (severe combined immunodeficiency) mouse was a serendipitous discovery by Bosma et al. in 1983 where, following quantification of immunoglobulin isotypes in mice maintained in a clean environment [specific pathogen-free (SPF) environment], a small number of mice with no detectable serum immunoglobulin were found in a litter of otherwise normal mice (Bosma et al., 1983). *Scid* mice are severely deficient for mature B and T lymphocytes and, consequently, unable to reject skin allografts (Bosma et al., 1983). With the exception of bone marrow, lymphoid organs (thymus, spleen, and lymph nodes) are diminished in size (1/10 or less normal size) and almost completely devoid of lymphoid cells in *scid* mice (Bosma et al., 1983; Custer et al., 1985). Fifteen percent of *scid* mice develop lymphoma, a neoplasm that is extremely rare in control mice of the same strain (Custer et al., 1985). If raised in an SPF environment, most *scid* mice live approximately three-fourths a normal life span, and they are fertile (Custer et al., 1985).

Some *scid* mice contain low levels of functional lymphocytes, demonstrating the so-called "leaky" phenotype (Bosma et al., 1988). The percentage of leaky *scid* mice increases with age, such that virtually all mice of the original CB17 strain are leaky for both B and T cells by 1 year of age (Carroll et al., 1989). The actual degree of leakiness is dependent upon the strain of the mouse (Nonoyama et al., 1993).

Early studies determined that the *scid* mutation results in a defect in V(D)J recombination, a process in which gene segments (V, D, and J) are joined to form variable regions for immunoglobulin and T-cell receptor genes during B- and T-cell maturation (Schuler et al., 1986). *Scid* cells correctly recognize and precisely cut recombination signal sequences, but demonstrate a defect in joining the cleaved ends (Malynn et al., 1988). It was further shown that the *scid* defect prevents correct formation of coding joints, whereas signal joints are recombined relatively normally (Lieber et al. 1988; Blackwell et al. 1989).

In addition to defective joining of antigen receptor genes during V(D)J recombination, *scid* cells also demonstrate a deficiency in the repair of DNA double-strand breaks induced by ionizing radiation, bleomycin, and restriction enzymes (Fulop and Phillips, 1990; Biedermann et al., 1991; Hendrickson et al., 1991; Chang et al., 1993). Double-strand breaks induced by ionizing radiation showed slower repair kinetics, as well as a reduced final level of repair after 24-hour incubation in *scid* fibroblast cells compared to control fibroblasts (Biedermann et al., 1991; Hendrickson et al., 1991). *Scid* mice show 2- to 3-fold enhanced radiosensitivity in vivo, evaluated in several tissue types, including spleen, bone marrow, gastrointestinal crypt cells, and skin, indicating that the *scid* defect is not specific to lymphoid cells (Fulop and Phillips, 1990; Biedermann et al., 1991). Curiously, defective repair of double-strand breaks appear to be unique to *scid* chromosomal DNA, as the joining of extrachromosomal plasmid DNA reflected wild-type frequencies in *scid* fibroblasts and lymphocytes (Harrington et al., 1992; Chang et al., 1993).

All of the evidence indicates that the gene encoding the catalytic subunit of DNA-PK (DNA-PK_{CS} or p350) is defective in *scid* mice. The DNA-PK_{CS} gene and the gene correcting the *scid* defect both map to the same location on human chromosome 8 (Blunt et al., 1995; Kirchgessner et al., 1995; Peterson et al., 1995). Only fragments of chromosome 8 which contain the DNA-PK_{CS} gene correct the radiosensitivity phenotype and the V(D)J defect of the Chinese hamster ovary mutant cell line V3, a mutant in the same complementation group as *scid* cells (Blunt et al., 1995; Kirchgessner et al., 1995). p350 is not recruited to double-stranded DNA oligonucleotides assayed by UV DNA-protein cross-linking (Blunt et al., 1995), and no DNA-PK activity can be detected in *scid* cell extracts, although addition of purified p350 corrects these deficiencies (Blunt et al., 1995; Kirchgessner et al., 1995; Peterson et al., 1995). p350 mRNA levels appear normal in *scid* cells, however, p350 protein levels are almost undetectable (Blunt et al.,

1996; Danska et al., 1996). Two laboratories have independently sequenced the carboxyl-terminal region of p350 in *scid* cells and identified a point mutation that creates a premature stop codon 83 amino acids upstream of the native stop codon, eliminating approximately 2% of the total protein (Blunt et al., 1996; Danska et al., 1996). The function of this region of DNA-PK$_{CS}$ is not known, but it is apparently distal to the conserved regions representing the kinase domain (Danska et al., 1996).

Ku80 Knockout Mice

Ku80 (–/–) mice, like *scid* mice, are severe combined immunodeficient, due to a block in lymphocyte development resulting from defective V(D)J recombination (Nussenzweig et al., 1996; Zhu et al., 1996). However, unlike *scid*, which are only defective in V(D)J coding joint formation, *Ku80* (–/–) mice show defects in both coding and signal joint formation (Nussenzweig et al., 1996; Zhu et al., 1996). Using ligation-mediated PCR, Zhu et al. have shown that coding ends accumulate in *Ku80* (–/–) mouse thymocytes with no detectable degradation of the hairpin-coding ends, and that the quantity of signal ends in *Ku80* (–/–) is similar to that found in wild-type control thymocytes and demonstrate no increased degradation of the blunted-signal ends (Zhu et al., 1996). However, this method does not detect large deletions that might be present due to extensive degradation of free ends. The stability of coding and signal ends found by Zhu et al. in *Ku80* (–/–) cells conflicts with other studies, which suggest that Ku protein is important for protecting double-stranded DNA ends before they are sealed (Taccioli et al., 1993; Liang and Jasin, 1996). Consistent with defective lymphocyte development, most lymphoid organs (spleen, thymus, and lymph nodes) in *Ku80* (–/–) mice are disproportionately small (Nussenzweig et al., 1996; Zhu et al., 1996).

Phenotypically, *Ku80* (–/–) mice are significantly smaller than their (–/+) and (+/+) siblings, first detected at embryonic day 15.5 and persisting to adulthood (with body weights 40–60% of sibling controls in live animals) (Nussenzweig et al., 1996). The viability of *Ku80* (–/–) pups is increased from 50 to 80% when larger siblings are removed (Nussenzweig et al., 1996). *Ku80* (–/–) females have small litters and are unable to nurture their newborn pups, and thus the pups must be nursed by a foster mother (Nussenzweig et al., 1996). Mouse embryonic fibroblasts isolated from *Ku80* (–/–) embryos grow at about half the rate of control embryonic fibroblasts (Nussenzweig et al., 1996).

Recombinational Repair

Homologous recombination ensures genetic variability by DNA shuffling during meiosis and plays an important function in the repair of DNA double-strand breaks and interstrand cross-links (Shinohara and Ogawa, 1995; Thompson, 1996). Although DNA double-strand breaks are repaired primarily by illegitimate recombination in mammalian cells (described above), double-strand breaks are also rejoined by a mechanism employing homologous recombination. It appears that homologous recombination, in collaboration with a form of nucleotide excision repair, corrects DNA interstrand cross-links formed by such agents as cisplatin and mitomycin C (Thompson, 1996). The molecular details of homologous recombination in mammalian cells are vague, and models are based largely on studies from better characterized systems such as in *E. coli* (for recent reviews see Camerini-Otero and Hsieh, 1995; Eggleston and West, 1996).

Rad51 Knockout Mice

S. cerevisiae RAD51 contains a high level of sequence homology to the *E. coli* recA protein (Shinohara et al., 1992), forms a recA-like filament on DNA (Ogawa et al., 1993), and, like recA, catalyzes stand exchange between homologous single- and double-stranded DNA in vitro (Sung, 1994). Homologs of RAD51 have been isolated from many eukaryotic organisms and were found to be one of the most highly conserved DNA repair proteins yet identified in humans (Thompson, 1996). A null-mutation in *Rad51* shows that the gene is required for embryonic development in the mouse and appears to be essential for cell viability (Tsuzuki et al., 1996a; Lim and Hasty, 1996). Tsuzuki et al. detect preimplantation *Rad51* (–/–) embryos of embryonic day 2-4 (E2-E4), but their occurrence was extremely rare (3 of 174 embryos), whereas Lim and Hasty detect *Rad51* (–/–) postimplantation embryos at the expected Mendelian ratio (~ 25%), which appeared relatively normal at E5.5, the stage at which *Rad51* (–/–) embryonic growth became retarded. Furthermore, when the *Rad51* mutation was introduced into a *p53* mutant background, the embryos appeared grossly normal to E7.5 (Lim and Hasty, 1996). *Rad51* (–/–) embryos demonstrate decreased cell proliferation, determined by BrdU incorporation as early as E5.5 and increased apoptosis at E7.5 (Lim and Hasty, 1996). In addition, E7.5 embryos showed multiple chromosome loss in mitotically dividing cells (Lim and Hasty, 1996). Attempts to create *Rad51* (–/–) cell lines in embryonic stem cells (ES), embryonal carcinoma cells (EC), or mouse embryo-derived fibroblasts (MEFs) derived from *Rad51-p53* double-mutant embryos were unsuccessful, suggesting Rad51 is essential for cell viability (Tsuzuki et al., 1996a; Lim and Hasty, 1996).

Brca1 Knockout Mice

Brca1 and *Brca2* are recently discovered genes which, when mutated, are responsible for familial breast cancer. *Brca1* and *Brca2* encode large proteins (1863 and 3418 aa, respectively) of unknown function. A recent discovery demonstrates *Brca1* and *Brca2* are potentially involved in DNA recombinational repair, an implication arising from experiments that show both proteins bind to *Rad51* (Scully et al., 1997; Sharan et al., 1997). Immunostaining demonstrates that *Brca1* and *Rad51* both form microscopic nuclear foci in mitotic cells during the S-phase of the cell cycle (Plug et al., 1996; Scully et al., 1997), and that the two proteins co-localize along synaptonemal complexes in the chromosomes of meiotic cells (Scully et al., 1997). As will be discussed below, both *Brca1* and *Brca2* (–/–) mice, like *Rad51* (–/–) mice, result in embryonic lethality at the same developmental stage. Thus, the evidence indicates a close connection between the three proteins.

   Brca1 knockout mice are developmentally retarded as early as E4.5 and completely arrest at the egg cylinder stage prior to gastrulation (Hakem et al., 1996; Liu et al., 1996). Reduced incorporation of bromodeoxyuridine into the DNA of E5.5-E6.5 *Brca1* (–/–) embryos demonstrates decreased cell proliferation (Hakem et al., 1996; Liu et al., 1996). Hakem et al. further demonstrated altered expression of a number of genes in *Brca1* (–/–) embryos associated with cell cycle regulation and cellular proliferation; cyclin E protein and *mdm-2* mRNA levels were decreased, while *p21* mRNA levels were increased. However, *p53* mRNA levels were normal in these embryos. There was no increase in apoptosis in *Brca1* (–/–) embryos (Hakem et al., 1996; Liu et al., 1996). Curiously, in situ hybridization using the early trophoblast cell marker *Mash-2* shows complete absence of diploid trophoblast in the extraembryonic region of E6.5 *Brca1* (–/–) embryos (Hakem et al., 1996). However, attempted rescue experiments indicate that the absence of diploid trophoblast is not the primary cause of early lethality (Hakem et al., 1996). Attempts to generate *Brca1* (–/–) ES cell lines with a second targeting vector were unsuccessful (Hakem et al., 1996), and blastocyst outgrowth experiments show impaired outgrowth of *Brca1* (–/–) inner cell mass (Liu et al., 1996). Mice heterozygous for the *Brca1* deletion are fertile and phenotypically normal, without any sign of cancer development within the first year (Hakem et al., 1996; Liu et al., 1996).

Gowen et al. deleted *Brca1* exon 11 in the mouse and found that embryos were phenotypically normal through E8.5 (Gowen et al., 1996). These embryos began to show developmental abnormalities at E9.5 to E10.5, with defects primarily localized in the neural tube (Gowen et al., 1996). It was later shown by others that mouse embryos contain at least two *Brca1* transcripts, likely arising through alternative splicing: an exon 11-positive and an exon 11-negative transcript (Hakem et al., 1996). Thus, the mouse constructed by Gowen et al. potentially contains the exon 11-negative transcript, which might explain the different phenotype. The difference in the two phenotypes does not appear to be genetic, as all three *Brca1* knockout mice contain the same genetic background (129/Ola).

Brca2 Knockout Mice

Mice nullizygous for *Brca2* are growth arrested at gastrulation around E6.5 (Sharan et al., 1997). *Brca2* (–/–) embryos produce some mesoderm, evidenced through gross observation and by in situ hybridization for *Brachyury* gene expression, and thus initiate gastrulation but cease further development (Sharan et al., 1997). In vitro cultures of *Brca2* (–/–) blastocysts are phenotypically equivalent to control embryos during one week of normal growth, but when exposed to γ-radiation during culture, (–/–) outgrowths are hypersensitive compared to controls (Sharan et al., 1997). Attempts to create *Brca2* (–/–) ES cells through two successive targeting experiments were unsuccessful (Sharan et al., 1997), thus *Rad51*, *Brca1*, and *Brca2* are all unable to produce (–/–) cell lines. Mice heterozygous for *Brca2* deficiency were healthy and fertile and showed no signs of tumor formation during the first 8 months of life (Sharan et al., 1997).

RAD54 Knockout Mice

Rad54 is a member of the SNF2 family of proteins (Eisen et al., 1995), and recent evidence indicates that Rad54 protein can interact directly with Rad51 (Jiang et al., 1996). However, the exact biochemical function of Rad54 has not been determined, and the protein has yet to be purified. *RAD54* (–/–) mice were viable and fertile and exhibited no gross abnormalities to at least 5 months of age (Essers et al., 1997). *RAD54* (–/–) ES cells were created by targeted disruption of both alleles and, compared to control cells, are hypersensitive to the lethal effects of γ-irradiation (2.3-fold), methyl methanesulfonate (3- to 4-fold), and mitomycin C (2- to 3-fold sensitization) (Essers et al., 1997). Furthermore, gene targeting at the *RB* and *CSB* locus were reduced 5- to 10-fold in *RAD54* (–/–) ES cells compared to heterozygous and wild-type control ES cells, suggesting that *RAD54* (–/–) cells have a reduced efficiency for homologous recombination (Essers et al., 1997).

Mismatch Repair

Mismatch repair corrects base-base mispairs and small insertions/deletions arising as errors during DNA replication (for reviews, see Umar and Kunkel, 1996; Modrich and Lahue, 1996; Kolodner, 1996). When mismatch repair is disabled, organisms demonstrate a mutator phenotype. Mismatch repair is also responsible for preventing recombination between divergent DNA homologs. In humans, a defect in mismatch repair predisposes to a hereditary form of colon cancer, known as hereditary nonpolyposis colon cancer (HNPCC).

MSH2 Knockout Mice

Human MSH2 protein binds to mismatch nucleotides and extrahelical sequences (insertion/deletion loop-type mismatches) (Fishel et al., 1994a,b), making it a functional homolog of the bacterial MutS mismatch repair protein, except that in human cells it functions as a heterodimer with MSH3 or MSH6 (Marsischky et al., 1996). *MSH2* (–/–) embryonic stem cells have an elevated frequency of spontaneous mutations, demonstrated by an *HPRT* forward mutation assay and by instability of simple repetitive sequences (microsatellite instability) (de Wind et al., 1995). Extracts from *MSH2* (–/–) cells do not display protein-binding activity to oligonucleotides that contain either a base-base mismatch or an extrahelical dinucleotide (de Wind et al., 1995). Furthermore, homologous recombination in *MSH2* (–/–) cells is just as efficient between nonisogenic DNA containing divergent sequences as with isogenic DNA derived from the same strain of mouse (de Wind et al., 1995), in contrast to the findings with wild-type cells (Deng and Capecchi, 1992; Riele et al., 1992; von Deursen and Wieringa, 1992).

MSH2 nullizygous mice are born healthy and fertile, however, young adult mice are predisposed to lymphoma (de Wind et al., 1995; Reitmair et al., 1995), such that by 12 months of age approximately 80% of *MSH2*-deficient mice develop lymphomas (Reitmair et al., 1996a). The increased incidence of lymphomas in *MSH2* (–/–) mice could be due in part from the genetic makeup of inbred mice, which show mild susceptibility to spontaneous lymphomas (as discussed by Reitmair et al., 1995). Lymphocyte development and T- and B-cell composition in lymphatic tissues of *MSH2* (–/–) mice appeared normal, indicating that lymphomas arise from otherwise normal hematopoietic cells (Reitmair et al., 1995). Lymphomas from *MSH2* (–/–) mice displayed microsatellite instability, demonstrating defective mismatch repair (Reitmair et al., 1995). Although no evidence of colon cancer was detected in young *MSH2* (–/–) mice, intestinal tumors were observed in 70% of *MSH2* (–/–) mice over 6 months of age (Reitmair et al., 1996a). All intestinal tumors were associated with a loss of the APC tumor suppressor protein (Reitmair et al., 1996a). Backcrossing *MSH2* (–/–) mice to *APC* (–/+) increased the multiplicity of intestinal adenomas and the rate of tumor progression (Reitmair et al., 1996b). In addition to lymphomas and intestinal neoplasms, 7% of older mice developed skin tumors (Reitmair et al., 1996a).

PMS2 Knockout Mice

The human proteins PMS2 and MSH1 have been shown to form a heterodimer, which together appear to be a functional homolog of the bacteria MutL protein (Li and Modrich, 1995). *PMS2* was disrupted in the mouse and found to be dispensable for normal embryonic development (Baker et al., 1995). DNA isolated from mouse tissue as well as tumors in *PMS2* (–/–) mice demonstrate elevated microsatellite instability, consistent with the functional knockout of mismatch repair (Baker et al., 1995). *PMS2* (–/–) mice display an elevated number of lymphomas and cervical sarcomas in the first year of life (Baker et al., 1995). Male *PMS2* (–/–) mice are infertile, producing grossly abnormal spermatozoa, whereas *PMS2* (–/–) female mice are completely fertile (Baker et al., 1995). Through the use of electron microscopy, a wide range of synaptic abnormalities associated with meiosis I were detected in *PMS2* (–/–) spermatocytes (Baker et al., 1995).

MLH1 Knockout Mice

Mice containing an *MLH1* knockout mutation have been constructed and are viable at birth (Edelmann et al., 1996; Baker et al., 1996). DNA isolated from *MLH1* (–/–) embryonic fibroblast cells and mouse tissue showed microsatellite instability, indicating deficiency in mismatch repair (Edelmann et al., 1996; Baker et al., 1996). Unlike *MSH2* (–/–) mice, which are fertile, or *PMS2* (–/–) mice, which only show male sterility, *MLH1* (–/–) mice are sterile in both sexes (Edelmann et al., 1996; Baker et al., 1996). Histological sectioning of *MLH1* (–/–) male gonads revealed an absence of late-stage spermatocytes and showed failure in spermatogenesis at the pachytene stage of meiosis (Edelmann et al., 1996; Baker et al., 1996). *MLH1* (–/–) females contain follicles at all stages of development as well as corpora lutea, indicating that ovulation can occur, although in vitro fertilized oocytes from mutant females did not progress beyond the two-cell stage (Edelmann et al., 1996; Baker et al., 1996). Meiotic nuclei in *MLH1* (–/–) male spermatocytes demonstrate normal formation of the synaptonemal complex, however, desynapsis occurred prematurely, leading to the hypothesis that crossing over of homologous chromosomes does not occur in MLH1-deficient gametes or, alternatively, that chiasmata formed at these sites are unstable (Baker et al., 1996). Interestingly, MLH1 protein localizes to the synapsed portion of the synaptonemal complex in normal meiotic nuclei (Baker et al., 1996). Long-term studies with *MLH1* (–/–) mice have not yet been reported, therefore no information is available on tumor susceptibility in these mice.

Other Mouse Repair Mutations

*Methyl Guanine Methyl Transferase (*MGMT*) Knockout Mice*

O^6-Methylguanine and O^4-methylthymine are two major premutagenic lesions formed in DNA upon exposure to alkylating agents, such as N-methyl-N'-nitro-N-nitrosoguanidine (MNNG) and N-methyl-N-nitrosourea (MNU). O^6-Methylguanine and O^4-methylthymine can mispair during DNA replication with thymine and guanine, causing G-to-A and T-to-C transition mutations, respectively. The enzyme O^6-methylguanine-DNA methyltransferase, product of the *MGMT* gene, catalyzes the transfer of a methyl group from O^6-methylguanine and O^4-methylthymine to a cysteine residue in its own molecule (Kawate et al., 1995). The reaction is a single-step reaction that inactivates the enzyme and thus requires a stoichiometric amount of protein for efficient repair.

   MGMT (–/–) mice are viable and appear normal, but show some growth retardation (Tsuzuki et al., 1996b). At 6 weeks after birth *MGMT* (–/–) mice were ~85%

the size of *MGMT* (+/+) littermates (Tsuzuki et al., 1996b). MNU injected into the interperitoneal cavity of *MGMT* (−/−) mice is lethal at levels that do not affect *MGMT* (+/+) mice (Tsuzuki et al., 1996b). MNU-induced toxicological effects include bone marrow and intestinal damage, depletion in the number of leukocytes and lymphocytes, and reduction in the size of the spleen and thymus (Tsuzuki et al., 1996b).

CONCLUSIONS

Inefficient DNA repair gives rise to several human diseases, including xeroderma pigmentosum, Cockayne syndrome, Fanconi's anemia, and the photosensitive form of trichothiodystrophy. Xeroderma pigmentosum and Fanconi's anemia patients are predisposed to develop cancer, predominantly skin cancer and leukemia, respectively. In addition, patients heterozygous for *BRCA1*, *BRCA2*, or the mismatch repair genes *hMSH2*, *hMLH1*, and *hPMS2* are predisposed to develop breast cancer and HNPCC, respectively, through a mechanism involving somatic mutation of the wild-type allele. Thus, there is a clear link between defective DNA repair and cancer predisposition, making DNA repair genes, in addition to oncogenes and tumor suppressor genes, an important category of gene endogenous to the cell that, when mutated, can accelerate the process of tumorigenesis. Mechanistically, defective DNA repair can promote cancer development by increasing the spontaneous mutation frequency, as most evidence indicates that cancer is a multistep process that requires an accumulation of genetic changes (Kinzler and Vogelstein, 1996, 1997).

   Most knockout mice deficient in DNA repair that survive to old age are cancer-prone, complementing the findings in human patients containing DNA repair gene defects. NER-deficient mice are predisposed to carcinogen-induced skin cancer. Mice deficient in double-strand break repair, as reported in the *Scid* mouse, and mice deficient for mismatch repair show a propensity toward developing lymphomas. Therefore, DNA repair deficiency in mice, as in humans, is also tightly linked to cancer predisposition.

   The particular tissue that is predisposed to malignancy depends upon which DNA repair mechanism has been rendered defective. Since most DNA repair pathways preferentially repair certain types of DNA damage, perhaps the tissue targeted for malignancy in any given DNA repair-deficient mouse reflects those tissues that are predominantly exposed to the corresponding class of carcinogen. For example, epidermis is exposed to UV radiation leading to DNA adducts preferentially repaired by NER, and therefore epidermis would be uniquely predisposed for malignancy in NER-deficient mice. Indeed, NER-deficient mice are predisposed to skin cancer. The predisposition of *Scid* mice to lymphomas might result from aberrant V(D)J recombination activity in these mice, as well as the high proliferative nature of T-lymphocytes. Furthermore, replication errors in the highly proliferative T-lymphocytes could also be responsible for the predisposition of lymphoma development in mismatch repair–deficient mice. Alternatively, or in addition, if each tissue possessed a different spectrum of DNA repair proteins, those tissues investing heaviest for a particular repair pathway would be more susceptible to knockout of that pathway.

   An intriguing observation that arose from knockout mouse studies was the finding that two repair systems, base excision repair (BER) and recombinational repair (REC), are both required for early embryonic development. This discovery was unexpected because of the relatively innocuous phenotypes of most nucleotide excision repair mutations, and it raises some intriguing questions. BER is the primary repair mechanism involved in correcting endogeneous forms of DNA damage (i.e., oxidative and hydrolytic, as well as simple alkylation-type DNA damage). Although deficiencies in specific DNA glycosylases do not result in embryonic lethality, disrupting proteins involved in the "core reaction," which repairs abasic sites, are all embryonic lethal. *Ref-1* and *Xrcc1* knockout embryo phenotypes are very similar, and both die just prior to gastrulation. Curiously, *polβ* knockout embryos also die prenatally, but at a later stage of gestation than *Ref-1* and *Xrcc1* knockout embryos. It has been noted that *ref-1* and *polβ* have multiple functions and that the exact biochemical function for *Xrcc1* is not known; however, the fact that they are collectively embryonic lethals suggests that the function they have in common, which is BER, is essential for embryonic development. In this regard it will be interesting to see whether DNA ligase III, which is also a component of the core reaction in BER, is embryonic lethal. Other questions include the following: Does the BER core complex have functions other than repair that result in embryonic lethality? If knockout of the repair capacity is causing the lethality, what is causing the damage in the embryo? Is there a markedly high level of damage in the pregastrulation-stage embryo that impedes further development (possibly arising from oxidative bursts associated with rapid cell cycle times characteristic of gastrulating embryos)? If there is a burst of DNA damage in peri-implantation stage embryos, is the damage cell type–specific or is the repair capacity cell type–specific? *Xrcc1* knockout embryos clearly demonstrate sensitivity to the embryonic portion of the epiblast, making this a compelling question.

Another intriguing issue about BER is the role it may play during meiotic recombination. Circumstantial evidence for such a role includes the high levels of DNA ligase III and *Xrcc1* expression at the pachytene stage of spermatogenesis, when repair of DNA damage involved in genetic recombination should be occurring (Chen et al., 1995; Walter et al., 1996). Moreover, DNA polymerase β is expressed at high levels during spermatogenesis (Alcivar et al., 1992). Because of the lethality of mutations in *Xrcc1*, apurinic endonuclease and DNA polβ, it will be necessary to approach the meiotic role of these genes using conditional lethal approaches. A clear meiotic role has already been demonstrated for the DNA mismatch repair genes, *PMS2* and *MLH1* (Baker et al., 1995, 1996; Edelmann et al., 1996). By contrast, mutations in the double-strand break repair gene, *Scid*, did not interfere with gametogenesis. Additional studies are needed to define the spectrum of DNA-repair activities needed for normal meiotic development.

In addition to BER, several recombinational repair-deficient mice are embryonic lethal. REC-deficient mice, like most BER-deficient mice, also die at the gastrulation stage of development. The details of recombinational repair in mammalian cells is very sketchy at best, most of the proteins involved are either not yet discovered or their functions have not been clearly defined. Many of the same questions mentioned above can be asked of REC-deficient mice: primarily, what is the substrate that REC-deficient mice cannot process that results in lethality? Since some of the NER proteins, including *ERCC1*, which is viable as a knockout, are thought to collaborate with REC repair proteins to correct DNA cross-links, cross-link damage is likely not the cause of lethality. If double-strand breaks are the offending substrate, then how could a substrate that is recognized and repaired by at least two separate repair pathways lead to lethality when only one pathway is rendered defective? To date, all DSB-deficient mice employing illegitimate recombinational repair develop normally (DNA-PK–mediated DSB repair), whereas most REC-deficient mice die during embryogenesis (homologous recombination-mediated DSB repair). Perhaps the answer lies in the fidelity of the two double-strand break repair mechanisms. The repair of double-strand breaks by homologous recombination is theoretically more accurate since the undamaged allele is used as a template. Accurate repair could be essential for stem cells, and therefore lethality in REC-deficient mice might reflect a heightened sensitivity of embryos to mutation-prone repair (discussed in Essers et al., 1997). However, the inability to create cell lines from several REC-deficient embryos raises the possibility that the REC system is essential for cell autonomous viability.

The effect of a *p53* (–/–) genotype in DNA repair-deficient mice is intriguing and highlights the direct involvement of *p53* in the response to DNA damage. In *XPC* (–/–) mice, *p53* deficiency exacerbates the NER-deficient phenotype, which is demonstrated physiologically by increased sensitivity to carcinogen-induced skin lesions and by a reduction in the latency period of carcinogen-induced skin cancer. In *Xrcc1* (–/–) BER-deficient mice and *Rad51* (–/–) REC-deficient mice, *p53* deficiency delays embryonic lethality 24–48 hours. Although *Xrcc1* (–/–) and *Rad51* (–/–) mice are destined to die early during embryonic development whether functional p53 is present or absent, the presence of p53 accelerates the process, thereby demonstrating the ability of p53 to recognize disturbed cells in the early embryo, which likely contain irreparable genomic abnormalities. p53 is activated by the presence of DNA damage, aparently leading to arrest in the G1 phase of the cell cycle (Kastan et al., 1991, 1992; Kuerbitz et al., 1992; Zhan et al., 1993), and if the damage is not repaired it is hypothesized that p53 activates an apoptotic program (for reviews, see Elledge and Lee, 1995; Enoch and Norbury, 1995). p53 mRNA expression has been detected in all cells of the embryo up to E10.5 (Schmidt et al., 1991), and it has been shown that most cells of early, but not late embryos, respond to DNA damage through the activation of p53 leading to apoptotic cell death (MacCallum et al., 1996; Gottlieb et al., 1997; Komarova et al., 1997). Furthermore, early embryos are much more sensitive to γ-radiation than late embryos (Hall, 1994). It is apparent that early embryos are closely monitored for DNA damage by p53 and that excessive damage can lead to embryonic lethality. Thus, analysis of DNA repair-deficient animals in a *p53* (–/–) background can reveal the physiological role of *p53* in response to DNA damage. Furthermore, evidence in *Xrcc1* (–/–) and *Rad51* (–/–) mice suggests that these mice are accumulating naturally derived p53-recognizable abnormalities at the pregastrulation stage of the embryo resulting in lethality, such that when p53 is absent the lethality is partially delayed.

Most knockout mice belonging to the same DNA-repair pathway show similar characteristics, although there are a few exceptions to this generalization. *ERCC1* (–/–) mice, unlike other NER-deficient mice, die shortly after birth of liver failure, and *Rad54* (–/–) mice, unlike other REC-deficient mice, are viable. These examples demonstrate the uniqueness of these two proteins. *ERCC1* might participate in other biochemical functions in additional to NER repair, and *Rad54* is apparently dispensible for the developmentally essential function of the REC-repair system. However, by seeking the similarities between mice of the same group, general conclusions can be drawn regarding the primary physiological function of a protien complex or multiprotein pathway; for example, in the case of DNA repair, NER-deficient mice are viable and prone to carcinogen-induced skin cancer, DSB-deficient mice are viable and immunodeficient due to defects in V(D)J recombination, BER- and REC-deficient mice die during embryogenesis, and MMR-deficient mice are viable and prone to lymphoma development. In the future, different knockout mice can be backcrossed in an effort to determine how multiple systems interact with each other.

The cumulative outcomes of studying such multiply deficient animals will provide unprecedented insights into the consequences and management of environmental stress in mammals.

ACKNOWLEDGMENTS

Work carried out by the authors was supported by NIEHS Grant ES08750 (R.S.T. and R.A.P.) and by an award from the University of California Energy Institute (R.S.T. and R.A.P.). R.S.T. was also supported by an Institutional National Research Service Award, 5T32 ES07106. The contents of this paper are solely the responsibility of the authors and do not represent the official views of the NIEHS/NIH or the University of California. We thank Dr. Mark MacInnes for helpful comments on the manuscript.

REFERENCES

Armstrong, J. F., Kaufman, M. H., Harrison, D. J., and Clarke, A. R. (1995). *Curr. Biol.*, 5: 931.

Avraham, K. B., Schickler, M. Sapoznikov, D., Yarom, R., and Groner, Y. (1988). *Cell, 54*: 823.

Baker, S. M., Bronner, C., E., Zhang, L., Plug, A. W., Robatzek, M., Warren, G., Elliott, E. A., Yu, J., Ashley, T., Arnheim, N., Flavell, R. A., and Liskay, R. M. (1995). *Cell, 82*: 309.

Baker, S. M., Plug, A. W., Prolla, T. A., Bronner, C. E., Harris, A. C., Yao, X., Christie, D-M., Monell, C., Arnheim, N., Bradley, A., Ashley, T., and Liskay, R. M. (1996). *Nat. Genet., 13*: 336.

Betz, U. A. K., Vobhenrich, C. A. J., Rajewsky, K., and Muller, W. (1996). *Current Biol., 6*: 1307.

Biedermann, K. A., Sun, J., Giaccia, A. J., Tosto, L. M., and Brown, J. M. (1991). *Proc. Natl. Acad. Sci. USA, 88*: 1394.

Bilinski, T., Krawiec, Z., Liczmanski, A., and Litwinska, J. (1985). *Biochem. Biophys. Res. Commun., 130*: 533.

Blackwell, T. K., Malynn, B. A., Pollack, R. R., Ferrier, P., Covey, L. R., Fulop, G. M., Phillips, R. A., Yancopoulos, G. D., and Alt, F. W. (1989). *EMBO J., 8*: 735.

Bloch, C. A., and Ausubel, F. M. (1986). *J. Bacteriol., 168*: 795.

Blunt, T., Finnie, N. J., Taccioli, G. E., Smith, G. C. M., Demengeot, J., Gottlieb, T. M., Mizuta, R., Varghese, A. J., Alt, F. W., Jeggo, P. A., and Jackson S. P. (1995). *Cell, 80*: 813.

Blunt, T., Bell, D., Fox, M., Taccioli, B. E., Lehmann, A. R., Jackson, S. P., and Jeggo, P. A. (1996). *Proc. Natl. Acad. Sci. USA, 93*: 10285.

Bonifer, C., Huber, M. C., Jagle, U., Faust, N., and Sippel, A. E. (1996). *J. Mol. Med., 74*: 663.

Bosma, G. C., Custer, R. P., and Bosma, M. J. (1983). *Nature, 301*: 527.

Bosma, G. C., Custer, F. M., Carroll, R. P., Gibson, D. M., and Bosma, M. J. (1988). *J. Exp. Med., 167*: 1016.

Bradley, A., Ramirez-Solis, R., Zheng, H., Hasty, P., and Davis, A. (1992). *CIBA Found. Symp., 165*: 256.

Brinster, R. L. (1993). *Int. J. Dev. Biol., 37*: 89.

Camerini-Otero, R. D., and Hsieh, P. (1995). *Annu. Rev. Genetics, 29*: 509.

Camper, S. A., Saunders, T. L., Kendall, S. K., Keri, R. A., Seasholtz, A. F., Gordon, D. F., Brikmeier, T. S., Keegan, C. E., Karolyi, I. J., Roller, M. L., et al. (1995). *Biol. Reprod., 52*: 246.

Carlioz, A., and Touati, D. (1986). *EMBO J., 5*: 623.

Carroll, A. M., Hardy, R. R., and Bosma M. J. (1989). *J. Immunol., 143*: 1087.

Ceballos-Picot, I., Nicole, A., Briand, P., Grimber, G., Delacourte, A., Defossez, A., Javoy-Agid, F., Lafon, M. Blouin, J. L., and Sinet, P. M. (1991). *Brain Res., 552*: 198.

Ceballos-Picot, I., Nicole, A., and Sinet, P-M. (1992a). *Free Radicals and Aging* (I. Emerit and B. Chance, eds.). Birkhauser Verlag, Basel, p. 89.

Ceballos-Picot, I., Nicole, Clement, M., Bourre, J.-M., and Sinet, P. M. (1992b). *Mutat. Res., 275*: 281.

Chan, P. H., Yang, G. Y., Chen, S. F., Carlson, E., and Epstein, C. J. (1991). *Ann. Neurol., 29*: 482.

Chang, C., Biedermann, K. A., Mezzina, M., and Brown, J. M. (1993). *Cancer Res., 53*: 1244.

Chen, J., Tomkinson, A. E., Ramos, W., Mackey, Z. B., Danehower, S., Walter, C. A., Schultz, R. A., Besterman, J. M., and Husain, I. (1995). *Mol. Cell. Biol., 15*: 5412.

Cheo, D. L., Meira, L. B., Hammer, R. E., Burns, D. K., Doughty, A. T. B., and Friedberg, E. C. (1996). *Curr. Biol., 6*: 1691.

Cheo, D. L., Ruben, H. J. T., Beira, L. B., Hammer, R. E., Burns, D. K., Tappe, N. J., van Zeeland, A. A., Mullenders, L. H. F., and Friedber, E. C. (1997). *Mutat. Res., 374*: 1.

Christofori, G., and Hanahan, D. (1994). *Semin. Cancer Biol., 5*: 3.

Cleaver, J. E., and Kraemer, K. H. (1995). *The Metabolic and Molecular Basis of Inherited Disease*, 7th ed., Vol. III (C. R. Scriver, A. L. Beaudet, W. S. Sly, and D. Valle, ed.). McGraw-Hill, New York, p. 4393.

Copp, A. J. (1995). *Trends Genet., 11*: 87.

Custer, R. P., Bosma, G. C., and Bosma, M. J. (1985). *Am. J. Pathol., 120*: 464.

Danska, J. S., Holland, D. P., Mariathasan, S., Williams, K. M., and Guidos, C. J. (1996). *Mol. Cell. Biol., 16*: 5507.

de Murcia, G., and de Murcia, J. M. (1994). *TIBS, 19*: 172.

Deng, C., and Capecchi, M. R. (1992). *Mol. Cell. Biol. 12*: 3365.

de Vries, A., van Oostrom, C. T. M., Hofhuls, F. M. A., Dortant, P. M., Berg, R. J. W., de Gruiji, F. R., Wester, P. W., van Kreijl, C. F., Capel, P. J. A., van Steeg, H., and Verbeek, S. J. (1995). *Nature, 377*: 169.

de Vries, A., van Oostrom, C. T. M., Dortant, P. M., Beems, R. B., van Kreijl, C. F., Capel, P. J. A., and van Steeg, H. (1997a). In Carcinogenesis in *XPA*-deficient mice. Doctoral thesis, University of Utrecht, The Netherlands. p. 78.

de Vries, A., Dolle, M. E. T., Broekhof, J. L. M., Muller, A. J. J. A., Kroese, E. D., van Kreijl, C. F., Capel, P. J. A., Vijg, J., and van Steeg, H. (1997b). In Carcinogenesis in *XPA*-deficient mice. Doctoral thesis, University of Utrecht, The Netherlands, p. 96.

de Wind, N., Dekker, M., Berns, A., Radman, M., and te Riele, H. (1995). *Cell, 82*: 321.

Donovan, P. J. (1994). *Curr. Topics Devel. Biol., 29*: 189.

Edelmann, W., Cohen, P. E., Kane, M., Lau, K., Morrow, B., Bennett, S., Umar, A., Kunkel, T., Cattoretti, G., Chaganti, R., Pollard, J. W., Kolodner, R. D., and Kucherlapati, R. (1996). *Cell, 85*: 1125.

Eggleston, A. K., and West, S. C. (1996). *TIG, 12* 20.

Eisen, J. A., Sweder, K. S., and Hanawalt, P. C. (1995). *Nucleic Acids Res., 23*: 2715.

Elledge, R. M., and Lee, W.-H. (1995). *BioEssays, 17*: 923.

Engelward, B. P., Dreslin, A., Christensen, J., Huszar, D., Kurahara, C., and Samson, L. (1996). *EMBO J., 15*: 945.

Enoch, T., and Norbury, C. (1995). Cellular responses to DNA damage: cell-cycle checkpoints, apoptosis and the roles of p53 and ATM. *TIBS, 20*: 426.

Epstein, C. J., Avraham, K. B., Lovett, M., Smith, S., Elroy-Stein, O., Rotman, G., Bry, C., and Groner, Y. (1987). *Proc. Natl. Acad. Sci. USA*, *84*: 8044.

Essers, J., Hendriks, R. W., Swagemakers, S. M. A., Troelstra, C., de Wit, J., Bootsma, D., Hoeijmakers, J. H. J., and Kanaar, R. (1997). *Cell*, *89*: 195.

Evans, M. J., and Kaufman, M. H. (1981). *Nature*, *292*: 154.

Fantel, A. G. (1996). *Teratology*, *53*: 196.

Fishel, R., Ewel, A., and Lescoe, M. K. (1994a). *Cancer Res.*, *54*: 5539.

Fishel, R., Ewel, A., Lee, S., Lescoe, M. K., and Griffith, J. (1994b). Binding of mismatched microsatellite DNA sequences by the human MSH2 protein. *Science*, *266*: 1403.

Fridovich, I. (1975). *Annu. Rev. Biochem.*, *44*: 6049.

Fridovich, I. (1978). *Science*, *201*: 875.

Friedberg, E. C., Walker, G. C., and Siede, W. (1995). *DNA Repair and Mutagenesis*. ASM Press, Washington, D.C.

Friedberg, E. C., Meira, L. B., and Cheo, D. L. (1997). *Mutat. Res.*, *383*: 183.

Fulop, G. M., and Phillips, R. A. (1990). *Nature*, *347*: 479.

Fung-Leung, W., and Mak, T. (1992). *Curr. Opin. Immunol.*, *4*: 189.

Gahlmann, R. (1993). *J. Exp. Animal Sci.*, *35*: 232.

Gottlieb, E., Haffner, R., King, A., Asher, G., Gruss, P., Lonai, R., and Oren, M. (1997). *EMBO J.*, *16*: 1381.

Gowen, L. C., Johnson, B. L., Latour, A. M., Sulik, K. K., and Koller, B. H. (1996). *Nat. Genet.*, *12*: 191.

Gu, H., Marth, J. D., Orban, P. C., Mossmann, H., and Rajewsky, K. (1994). *Science*, *265*: 103.

Hakem, R., de la Pompa, J. L., Sirard, C. Mo, R., Woo, M., Hakem, A., Wakeham, A., Potter, J., Reitmair, A., Billia, F., Firpo, E., Hui, C. C., Roberts, J., Rossant, J., and Mak, T. W. (1996). *Cell*, *85*: 1009.

Hall, E. J. (1994). *Radiology for the Radiologist*, 4th ed. Lippincott-Raven Publishers, Philadelphia.

Harrington, J., Hsieh, C-L, Gerton, J., Bosma, G., and Lieber, M. R. (1992). *Mol. Cell. Biol.*, *12*: 4758.

Heller, B., Wang, Z-Q., Wagner, E. F., Radons, J., Burkle, A., Fehsel, K., Burkart, V., and Kolb, H. (1995). *J. Biol. Chem.*, *270*: 11176.

Hendrickson, E. A., Qin, X-Q, Bump, E. A., Schatz, M. O., and Weaver, D. T. (1991). *Proc. Natl. Acad. Sci. USA*, *88*: 4061.

Hilliker, A. J., Duyf, B., Evans, D., and Phillips, J. P. (1992). *Proc. Natl. Acad. Sci. USA*, *89*: 4343.

Hogan, B. L. M., Beddington, R., Costantini, F., and Lacey, E. (1994). *Manipulating the Mouse Embryo*, 2nd ed., Cold Spring Harbor Laboratory Press, New York.

Huang, M. (1993). *Lab. Animal Sci.*, *43*: 156.

Jeggo, P. A., Taccioli, G. E., and Jackson, S. P. (1995). *Bioessays*, *17*: 949.

Jiang, H., Xie, Y., Houston, P., Stemke-Hale, K., Mortensen, U. H., Rothstein, R., and Kodadek, T. (1996). *J. Biol. Chem.*, *271*: 33181.

Joyner, A. L. (1991). *Bioessays*, *13*: 649.

Joyner, A. L., ed. (1993). *Gene Targeting—A Practical Approach*. Oxford University Press, Oxford.

Kastan, M. B., Onyekwere, O., Sidransky, D., Vogelstein, B., and Craig, R. W. (1991). *Cancer Res.*, *51*: 6304.

Kastan, M. B., Zhan, Q., El-Deiry, W. S., Carrier, F., Jacks, T., Walsh, W. V., Plunkett, B. S., Vogelstein, B., and Fornace, A. J. (1992). *Cell*, *71*: 587.

Kawate, J., Ihara, K., Kohda, K., Sakumi, K., and Sekiguchi, M. (1995). *Carcinogenesis*, *16*: 1595.

Kinouchi, H., Epstein, C. J., Mizui, T., Carlson, E., Chen, S. F., and Chan P. H. (1991). *Proc. Natl. Acad. Sci. USA*, *88*: 11158.

Kinzler, K. W., and Vogelstein, B. (1996). *Cell*, *87*: 159.

Kinzler, K. W., and Vogelstein, B. (1997). *Nature, 386*: 761.

Kirchgessner, C. U., Patil, C. K., Evans, J. W., Coumo, C. A., Fried, L. M., Carter, T., Oettinger, M. A., and Brown, J. M. (1995). *Science, 267*: 1178.

Kistner, A., Gossen, M., Zimmerman, E., Jerecic, J., Ullmer, C., Lubbert, H., and Bujard, H. (1996). *Proc. Natl. Acad. Sci. USA*, *93*: 10933.

Koller, B., and Smithies, O. (1992). *Ann. Rev. Immunol.*, *10*: 705.

Kolodner, R. (1996). *Genes Dev.*, *10*: 1433.

Komarova, E. A., Chernov, M. V., Franks, R., Wang, K., Armin, G., Zelnick, C. R., Chin, D. M., Bacus, S. S., Stark, G. R., and Gudkov, A. V. (1997). *EMBO J., 16*: 1391.

Kubota, Y., Nash, R. A., Klungland, A., Schar, P., Barnes, D. E., and Lindahl, T. (1996). *EMBO J., 15*: 6662.

Kuerbitz, S. J., Plunkett, B. S., Walsh, W. B., and Kastan, M. B. (1992). *Proc. Natl. Acad. Sci. USA*, *89*: 7491.

Kuhn, R., Schwenk, F., Aguet, M., and Rajewsky, K. (1995). *Science, 269*: 1427.

Lehmann, A. R., Bridges, B. A., Hanawalt, P. C., Johnson, R. T., Kanaar, R., Krokan, H. E., Kyrtopoulos, S., Lambert, B., Melton, D. W., Moustacchi, E., Natarajan, A. T., Radman, M., Sarasin, A., Seeberg, E., Smerdon, M. J., Smith, C. A., Smith, P. F., Thacker, J., Thomale, J., Waters, R., Weeda, G., West, S. C., van Zeeland, A. A., and Zdzienicka, M. Z. (1996). *Mutat. Res.*, *364*: 245.

Li, G.-M., and Modrich, P. (1995). *Proc. Natl. Acad. Sci. USA*, *92*: 1950.

Li, Y., Huang, T.-T., Carlson, E. J., Melov, S., Ursell, P. C., Olson, J. L., Noble, L. J., Yoshimura, M. P., Berger, C., Chan, P. H., Wallace, D. C., and Epstein, C. J. (1995). *Nat. Genet.*, *11*: 376.

Liang, F., and Jasin, M. (1996). *J. Biol. Chem.*, *271*: 14405.

Lieber, M. R., Hesse, J. E., Lewis, S., Bosma, G. C., Rosenberg, N., Mizuuchi, K., Bosma, M. J., and Gellert, M. (1988). The defect in murine severe combined immune deficiency: joining of signal sequences but not coding segments in V(D)J recombination. *Cell*, *55*: 7–16.

Lim, D.-S., and Hasty, P. (1996). *Mol. Cell. Biol.*, *16*: 7133.

Lindahl, T. (1993). *Nature, 362*: 709.

Lindahl, T. *J. Cell Sci.* (Suppl. *19*): 73.

Lindahl, T., Satoh, M. S., and Dianov, G. (1995). *Phil. Trans. R. Soc. Lond. B.*, *347*: 57.

Liochev, S. I., and Fridovich, I. (1991). *J. Biol. Chem.*, *266*: 8747.

Liu, C.-Y., Flesken-Nikitin, A., Li, S., Zeng, Y., and Lee, W.-H. (1996). *Genes Dev.*, *10*: 1835.

Longo, V. D., Gralla, E. B., and Valentine, J. S. (1996). *J. Biol. Chem.*, *271*: 12275.

MacCallum, D. E., Hupp, T. R., Midgley, C. A., Stuart, D., Campbell, S. J., Harper, A., Walsh, F. S., Wright, E. G., Balmain, A., Lane, D. P., and Hall, P. A. (1996). *Oncogene, 13*: 2575.

Malynn, B. A., Blackwell, T. K., Fulop, G. M., Rathbun, G. A., Furley, A. J. W., Ferrier, P., Heinke, L. B., Phillips, R. A., Yancopoulos, G. D., and Alt, F. W. (1988). *Cell, 54*: 453.

Marsischky, G. T., Filosi, N., Kane, M. F., and Kolodner, R. (1996). *Genes Dev.*, *10*: 407.

Martin, G. R. (1981). *Proc. Natl. Acad. Sci. USA*, *78*: 7634.

Matsumoto, Y., and Kim, K. (1995). *Science, 269*: 699.

McEntrye, J., ed. (1995). *TiBS, 20*: 381.

McWhir, J., Selfridge, J., Harrison, D. J., Squires, S., and Melton, D. W. (1993). *Nat. Genet.*, *5*: 217.

Miyauchi-Hashimoto, H., Tanaka, K., and Horio, T. (1996). *J. Invest. Dermatol.*, *107*: 343.

Modrich, P., and Lahue, R. (1996). *Annu. Rev. Biochem.*, *65*: 101.

Nakane, H., Takeuchi, S., Yuba, S., Saijo, M., Nakatsu, Y. Murai, H., Nakatsuru, Y., Ishikawa, T., Hirota, S., Kitamura, Y., Kato, Y., Tsunoda, Y., Miyauchi, H., Horio, T., Tokunaga,

T., Matsunaga, T., Nikaido, O., Nishimune, Y., Okada, Y., and Tanaka, K. (1995). *Nature*, *377*: 165.

Nizielski, S. E., Lechner, P. S., Croniger, C. M., Wang, N. D., Darlington, G. J., and Hanson, R. W. (1996). *J. Nutrition*, *126*: 2697.

Nonoyama, S., Smith, F. O., Bernstein, I. D., and Ochs, H. D. (1993). *J. Immunol.*, *150*: 3817.

Nussenzweig, A., Chen, C., Soares, V. d C., Sanchez, M., Sokol, K., Nussenzweig, M. C., and Li, G. C. (1996). *Nature*, *382*: 551.

Ogawa, T., Yu, X., Shinohara, A., and Egelman, E. H. (1993). *Science*, *259*: 1896.

Orr, W. C., and Sohal, R. S. (1992). *Arch. Biochem. Biophys.*, *297*: 35–41.

Orr, W. C., and Sohal, R. S. (1993). *Arch. Biochem. Biophys.*, *301*: 34.

Orr, W. C., and Sohal, R. S. (1994). *Science*, *263*: 1128.

Orr, W. C., and Sohal, R. S. (1995). *J. Biol. Chem.*, *270*: 15671.

Pedersen, R. A. (1994). *Reprod. Fertil. Devel.*, *6*: 543.

Pedersen, R. A., Papaioannou, V. E., Joyner, A., and Rossant, J. (1993). *Targeted Mutagenesis in Mice: A Vide Guide*. Cold Spring Harbor Laboratory Press, New York.

Peterson, S. R., Kurimasa, A., Oshimura, M., Dynan, W. S., Bradbury, E. M., and Chen, D. J. (1995). *Proc. Natl. Acad. Sci. USA*, *92*: 3171.

Phillips, J. P., Campbell, S. D., Michaud, D., Charbonneau, M., and Hilliker, A. J. (1989). *Proc. Natl. Acad. Sci. USA*, *86*: 2761.

Phillips, J. P., Trainer, J. A., Getzoff, E. D., Boulianne, G. L., Kirby, K., and Hilliker, A. J. (1995). *Proc. Natl. Acad. Sci. USA*, *92*: 8574.

Pinkert, C. A., ed., (1994). *Transgenic Animal Technology: A Laboratory Handbook*, Academic Press, Inc., San Diego, CA.

Plug, A. W., Xu, J., Reddy, G., Golub, E. I., and Ashley T. (1996). *Proc. Natl. Acad. Sci. USA*, *93*: 5920.

Przedborski, S., Jackson-Lewis, V., Kostic, V., Carlson, E., Epstein, C. J., and Cadet, J. L. (1992). *J. Neurochem.*, *58*: 1760.

Radford, I.R. (1986). *Int. J. Radiat. Biol.*, *49*: 611.

Ramirez-Solis, R., Davis, A., and Bradley, A. (1993). *Methods Enzymol.*, *255*: 855.

Reaume, A. G., Elliott, J. L., Hoffman, E. K., Kowall, N. W., Ferrante, R. J., Siwek, D. F., Wilcox, H. M., Flood, D. G., Beal, M. F., Brown, R. H., Scott, R. W., and Snider, W. D. (1996). *Nat. Genet.*, *13*: 43.

Reitmair, A. H., Schmits, R., Ewel, A., Bapat, B., Redston, M., Mitri, A., Waterhouse, P., Mittrucker, H. -W., Wakeham, A., Liu, B., Thomason, A., Griesser, H., Gallinger, S., Ballhausen, W. G., Fishel, R., and Mak, T. W. (1995). *Nat. Genet.*, *11*: 64.

Reitmair, A. H., Redston, M., Cai, J. C., Chuang, T. C. Y., Bjerknes, M., Cheng, H., Hay, K., Gallinger, S., Bapat, B., and Mak, T. W. (1996a). *Cancer Res.*, *56*: 3842.

Reitmair, A. H., Cai, J-C., Bjerknes, M., Redston, M., Cheng, H., Pind, M. T. L., Hay, Mitri, A., Bapat, B. V., Mak, T. W., and Gallinger, S. (1996b). *Cancer Res.*, *56*: 2922.

Reveillaud, I., Niedzwiecki, A., Bensch, K. G., and Fleming, J. E. (1991). *Mol. Cell. Biol.*, *11*: 632.

Reveillaud, I., Phillips, J., Duyf, B., Hilliker, A., Kongpachith, A., and Fleming, J. E. (1994). *Mol. Cell. Biol.*, *14*: 1302.

Rickert, R. C., Roes, J., and Rajewsky, K. (1997). *Nucleic Acids Res.*, *25*: 1317.

Riele, H. T., Maandag, E. R., and Berns, A. (1992). *Proc. Natl. Acad. Sci. USA*, *89*: 5128.

Rijkers, T., Peetz, A., and Ruther, U. (1994). *Transgenic Res.*, *3*: 203.

Rinkenberger, J. L., Cross, J. C., and Werb, Z. (1997). *Devel. Genet.*, *21*: 6.

Robbins, J. (1993). *Circ. Res.*, *73*: 3.

Robertson, E. J. (1987). *Teratocarcinomas and Embryonic Stem Cells—A Practical Approach*. Oxford University Press, Oxford.

Roth, D. B., Lindahl, T., and Gellert M. (1995). *Current Biol.*, *5*: 496.

Sah, V. P., Attardi, L. D., Mulligan, G. J., Williams, B. O., Bronson, R. T., and Jacks, T. (1995). *Nat. Genet.*, *10*: 175.

Sands, A. T., Abuin, A., Sanchez, A., Conti, C. J., and Bradley, A. (1995). *Nature*, *377*: 162.

Satoh, M. S., and Lindahl, T. (1994). *Cancer Res.*, *54*: 1899s.

Schaeffer, L., Roy, R., Humbert, S., Moncollin, V., Vermeulen, W., Hoeijmakers, J. H. J., Chambon, P., and Egly, J. M. (1993). *Science*, *260*: 58.

Schuler, W., Weiler, I. J., Schuler, A., Phillips, R. A., Rosenberg, N., Mak, T. W., Kearney, J. F., Perry, R. P., and Bosma, M. J. (1986). *Cell*, *46*: 963.

Scmid, P., Lorenz, A., Hameister, H., and Montenarh, M. (1991). *Development*, *113*: 857.

Scott, M. D., Meshnick, S. R., and Eaton, J. W. (1987). *J. Biol. Chem.*, *262*: 3640.

Scully, R., Chen, J., Plug, A., Xiao, Y., Weaver, D., Feunteun, J., Ashley, T., and Livingston, D. M. (1997). *Cell*, *88*: 265.

Seto, N. O. L., Hayashi, S., and Tener, G. M. (1990). *Proc. Natl. Acad. Sci. USA*, *87*: 4270.

Sharan, S. K., Morimatsu, M., Albrecht, U., Lim, D-S., Regel, E., Dinh, C., Sands, A., Eichele, G., Hasty, P., and Bradley, A. (1997). *Nature*, *386*: 804.

Shinohara, A., and Ogawa, T. (1995). *TIBS*, *20*: 387.

Shinohara, A., Ogawa, H., and Ogawa, T. (1992). *Cell*, *69*: 457.

Smithies, O. (1993) *Trends Genet.*, *9*: 112.

Sobol, R. W., Horton, J. K., Kuhn, R., Gu, H., Singhal, R. K., Prasad, R., Rajewsky, K., and Wilson, S. H. (1996). *Nature*, *379*: 183.

Staveley, B. E., Phillips, J. P., and Hilliker, A. J. (1990). *Genome*, *33*: 867.

Sung, P. (1994). *Science*, *265*: 1241.

Taccioli, G. E., Rathbun, G., Oltz, E., Stamato, T., Jeggo, P. A., and Alt, F. W. (1993). *Science*, *260*: 207.

Tebbs, R. S., Meneses, J. J., Pedersen, R. A., Thompson, L. H., and Cleaver, J. E. (1996). *Environ. Mol. Mutagen.*, *27*: 68.

Thompson, L. H. (1996). *Mutat. Res.*, *363*: 11.

Tsuzuki, T., Fujii, Y., Sakumi, K. Tominaga, Y., Nakao, K., Sekiguchi, M., Matsushiro, A., Yoshimura, Y., and Morita, T. (1996a). *Proc. Natl. Acad. Sci. USA*, *93*: 6236.

Tsuzuki, T., Sakumi, K., Shiraishi, A., Kawate, H., Igarashi, H., Iwakuma, T., Tominaga, Y., Zhang, S., Shimizu, S., Ishikawa, T., Nakamura, K., Nakao, K., Katsuki, M., and Sekiguchi, M. (1996b). *Carcinogenesis*, *17*: 1215.

Umar, A., and Kunkel, T. A. (1996). *Eur. J. Biochem.*, *238*: 297.

van der Horst, G. T. J., van Steeg, H., Berg, R. J. W., van Gool, A. J., de Wit, J., Weeda, G., Morreau, H., Beems, R. B., van Kreijl, C. F., de Gruijl, F. R., Bootsma, D., and Hoeijmakers, J. H. J. (1997). *Cell*, *89*: 425.

van Deursen, J., and Wieringa, B. (1992). *Nucleic Acids Res.*, *20*: 3815.

van Loon, A. P. G. M., Pesold-Hurt, B., and Schatz, G. (1986). *Proc. Natl. Acad. Sci. USA*, *83*: 3820.

Viney, J. L. (1995). *Cancer Metastasis Rev.*, *14*: 77.

Walter, C. A., Trolian, D. A., McFarland, M. B., Street, K. A., Gurram, G. R., and McCarrey, J. R. (1996). *Biol. Reprod.*, *55*: 630.

Wang, Z.-Q., Auer, B., Stingl, L., Berghammer, H., Haidacher, D., Schweiger, M., and Wagner, E. F. (1995). *Genes Dev.*, *9*: 509.

Wassarman, P. M., and DePamphilis, M., eds. (1993). *Guide to Techniques in Mouse Development*. Academic Press, Inc., San Diego, CA.

Weaver, D. T. (1995). *TIG*, *11*: 388.

White, C. W., Avraham, K. B., Shanley, P. F., and Groner, Y. (1991). *J. Clin. Invest.*, *87*: 2162.

Wispe, J. R., Warner, B. B., Clark, J. C., Dey, C. R., Neuman, J., Glasser, S. W., Crapo, J. D., Chang, L-Y., and Whitsett, J. A. (1992). *J. Biol. Chem.*, *267*: 23937.

Xanthoudakis, S., Miao, G., Wang, F., Pan, Y.-C. E., and Curran, T. (1992). *EMBO J.*, *11*: 3323.

Xanthoudakis, S., Smeyne, R., Wallace, J. D., and Curran, T. (1996). *Proc. Natl. Acad. Sci. USA*, *93*: 8919.

Yang, G., Chan, P. H., Chen, J., Carlson, E., Chen, S. F., Weinstein, P., Epstein, C. J., and Kamii, H. (1994). *Stroke*, *25*: 165.

Zhan, Q., Carrier, F., and Fornace, A. J. (1993). *Mol. Cell. Biol.*, *13*: 4242.

Zhu, C., Bogue, M. A., Lim, D.-S., Hasty, P., and Roth, D. B. (1996). *Cell*, *86*: 379.

7

Genes and Reproductive Behavior Resulting from Normal or Abnormal Development

Sonoko Ogawa and Donald W. Pfaff
The Rockefeller University, New York, New York

Work on the differentiation of reproductive behaviors during the last 30 years has proceeded under the guidance of a well-formed prevailing theory. This theory states that androgenic hormones produced in the developing testis circulate through the blood, enter the brain, and as a result alter the structure and function of neurons important for male-typical and female-typical reproductive behaviors. This theory does not explain all of the data available. First, there are peripheral effects of hormones (for example, in the adult, effects on somatosensory inputs) (Kow and Pfaff, 1973; Kow, et al., 1979), which could include alterations not only of sensory services but also of motor capacities. Frank Beach emphasized that effects of this sort during development could constitute essential mechanisms in the development of reproductive behavior. Second, circulating androgens cannot explain (Arnold, 1995) all of the data available from certain songbirds.

Nevertheless, the prevailing theory of the effects of androgens and their metabolites on brain remains the most useful way to organize data on the development of reproductive behavior. An entire scientific generation of experiments beginning with the work of Robert Goy and his collaborators and Roger Gorski and coworkers (Arnold et al., 1996; also Goy and McEwen, 1980) showed that administration of androgenic hormones just before or just after the birth of laboratory rodents could defeminize and correspondingly masculinize the behavior of genetic females. Conversely, castration of genetic males would permit the evocation of female reproductive behaviors while male-typical reproductive behaviors would decline.

To which neuroendocrine/behavioral systems have these thoughts been most productively applied? Masculine sex behaviors have some features to recommend them for analysis which are universal across the experimental vertebrate animals studied (Kelley and Pfaff, 1979). First, these behaviors are always reduced by castration and are always increased by testosterone administration. Second, androgens are always accumulated in a limbic-hypothalamic system of neurons with the preoptic area figuring prominently. Third, preoptic neurons are electrophysiologically sensitive to androgens (Pfaff and Pfaffmann, 1969), and preoptic neurons are essential for the normal performance of male-typical reproductive behaviors (reviewed by Kelley and Pfaff, 1979; Sachs and

Meisel, 1994). Putting these facts end-to-end, we can conclude that one important feature of the mechanisms for male-typical reproductive behavior is that circulating androgens are accumulated by androgen receptors in preoptic neurons, cause physiological alterations of those neurons, and as a result activate circuits important for mounting and intromission behaviors.

BACKGROUND ON FEMALE-TYPICAL BEHAVIOR

Despite these broadly stated findings about male reproductive behavior, mechanistic analyses of female-typical behaviors during development and in adulthood have proceeded more quickly. We see clearly that estrogenic treatments followed by progesterone are essential for the performance of lordosis behavior even to the point that the evocation of this behavior virtually comprises an "expression system" for estrogen-dependent gene expression in hypothalamic neurons. We know that the sequence of hormone delivery is crucial for successful lordosis behavior—estrogen priming must be given first followed by progesterone. Moreover, especially with respect to progesterone, the rate of change of hormonal levels is important: both for the stimulation of lordosis and for the ovulatory surge of leutinizing hormone from the pituitary, a sudden rise of progesterone levels will facilitate neuroendocrine and behavioral responses, while a long plateau of high progesterone levels actually will have the opposite effect (Attardi, et al. 1981; see Pfaff et al., 1994, for review). In the genetic female, estrogen followed by progesterone increases the probability that lordosis behavior will be exhibited in response to specific cutaneous stimuli (Kow et al., 1979,1980).

Most importantly, the neural circuitry for the interactions between specific environmental stimuli and circulating estrogen and progesterone has been spelled out (Pfaff, 1980; Pfaff et al., 1994). That is, somatosensory stimuli from the male activate neurons in the dorsal horn of the lumbar cord (Fig. 1, lower left). At the top of the obligatory supraspinal loop, estrogen-binding neurons emit an "enabling" signal, which triggers a hierarchical motor control system. Such a descending facilitation allows the excitation of the motor neurons for deep back muscles, which execute lordosis behavior (Fig. 1, lower right). Hypothalamic control regions for female and male mating behaviors and for parental behavior are strikingly different from each other (Fig. 2).

Hypothalamic mechanisms for lordosis behavior require estrogen-stimulated gene expression and protein synthesis. In fact, the estrogen effect on behavior is accompanied by the appearance of a novel protein spot on two-dimensional gels (Mobbs et al., 1988,1990). Figure 3 illustrates this spot in VMH and shows that this protein is transported to the midbrain central grey (MCG).

Thus, it is possible that fast progress with reproductive behavior in the female derived in part from the "existence proof" that discovery of the behavioral circuit was achievable. Moreover, proof of estrogen stimulation of gene expression for the progesterone receptor (Romano et al., 1989; Lauber et al., 1991) and the use of antisense DNA technology to show the importance of the messenger RNA for the progesterone receptor for behavior (Ogawa et al., 1994) comprised the first demonstration of a causal relationship between gene expression for a transcription factor and the performance of a specific behavior.

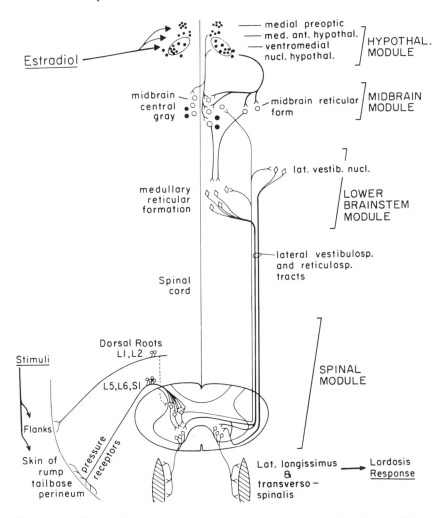

Figure 1 Circuitry for estrogen-dependent female-typical lordosis behavior. This was the first circuit determined for any vertebrate behavior. See text for elaboration. (From Pfaff et al., 1994.)

GENES IMPORTANT FOR THE DEVELOPMENT OF REPRODUCTIVE BEHAVIORS IN EXPERIMENTAL ANIMALS

Early on, it became apparent that some of the effects of circulating testosterone once it reached the brain might depend upon its aromatization to estradiol. Some of the masculinizing affects of testosterone could be prevented by administration of the aromatase inhibitor ATD (Parsons et al., 1984). We therefore used antisense DNA technology to test the hypothesis that gene expression for the estrogen receptor is important for the suppression of female reproductive behavior by neonatal androgen treatment of female rats (McCarthy et al., 1993). In these experiments, we chose a condition of neonatal masculinization mild enough that it might be blocked: namely, a 50-μg

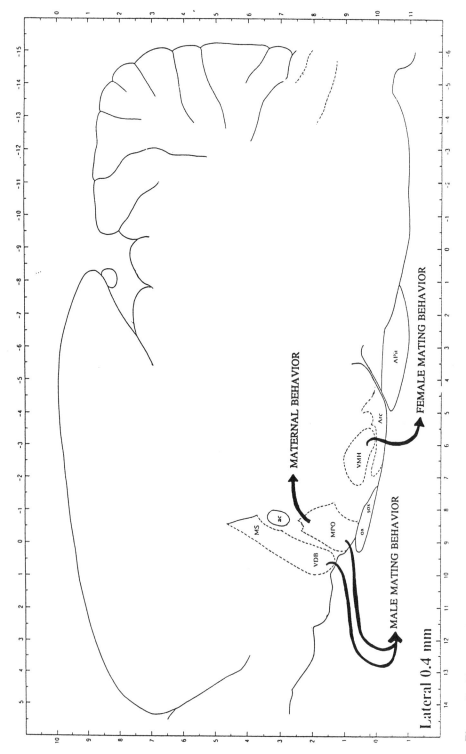

Figure 2 Ventromedial hypothalamic control regions for female rat mating behavior are different from the preoptic and basal forebrain control regions for male-typical and maternal behaviors. (From Pfaff and Lauber, 1994.)

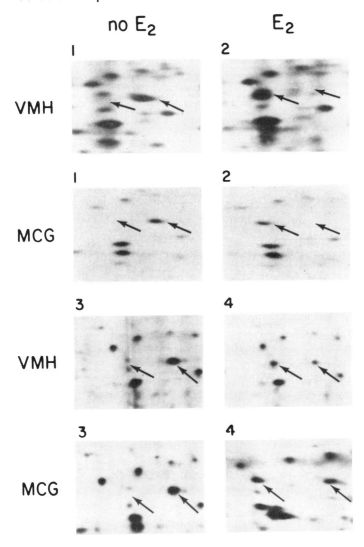

Figure 3 Systemic estradiol treatment (E2) caused the appearance of a novel spot (leftmost arrows) on two-dimensional gels of newly synthesized proteins in VMH. Two independent experiments are illustrated (panels 1 vs. 2 and 3 vs. 4). This protein was transported to midbrain central grey (MCG: panels 1 vs. 2 and 3 vs. 4). (From Mobbs et al., 1988.)

injection of testosterone propionate on day 3. Under these circumstances, some neonatal female rats received a 15mer which was antisense against the estrogen receptor mRNA, delivered directly to the hypothalamus by microinjection, whereas other females received a control treatment. We predicted that if behavioral masculinization induced in this way depended not only on aromatization but also upon binding of the estrogen (thus produced) to the estrogen receptor, then absence of adequate levels of messenger RNA for the estrogen receptor at exactly that time in development would block behavioral masculinization. Indeed, after the female rats had grown to adulthood and were tested for lordosis behavior, those that had received antisense DNA blocking estrogen receptor mRNA function had significantly higher levels of lordosis behavior

(McCarthy et al., 1993). Moreover, the masculinization of the sexually dimorphic pre-optic area was in part reversed. These results showed not only that defeminization and masculinization of behavior following neonatal androgen can depend upon estrogen receptor gene expression, but also demonstrate a neonatal gene expression effect manifested in behavior during adulthood.

Estrogen Receptor Knockout Mice

A different way of looking at those reproductive behaviors whose development would depend upon estrogen receptor function uses the knockout strategy. The striking demonstration of gene knockout technique applied to the estrogen receptor (Lubahn et al., 1993) made this possible. From the initial report it seems likely that lordosis behavior would be low because these animals were infertile and no lordosis behavior was observed. More detailed behavioral analysis (Ogawa et al., 1996a) reveals a somewhat more complicated picture indicating that the initial result was actually due to two separate phenomena. First, it may well be that hormone-dependent lordosis behavior is reduced, since the frequency of performance of an immobilization response coupled with dorsiflexion upon application of a strong cutaneous stimulus to the skin of the female mouse appears to be reduced (Ogawa et al., 1996a). The second phenomenon observed in the female reproductive behavior of these knockout animals involves an amazing change in the social behavior inherent in mating behavior tests. Upon introduction of the knockout mouse to the home cage of the stud male, the estrogen receptor knockout female is actually treated as a male intruder (Ogawa et al., 1996a). Thus, aggressive behavior rather than sexual behavior ensues. The absence of female typical reproductive behavior in these knockout female mice may therefore be a compound result of social behavior changes which discourage the initial application of adequate stimuli by the male, *plus* the lower sensitivity of the female to the adequate stimuli once applied.

Male estrogen receptor knockout mice also yielded unexpected results. From the infertility of these males and from the knowledge that a full panoply of male-typical reproductive behaviors may depend upon aromatization of testosterone to estradiol acting through the estrogen receptor in the preoptic neurons, we predicted that male-typical behavior would be absent in the knockout animals. This prediction was wrong. Mounting frequencies by the knockout males during standardized behavioral assays with female mice were similar to the wild-type control (Ogawa et al., 1996b,1997). Beyond simple mounting behaviors, however, something was defective about male mating reflexes such that intromissions rarely occurred and ejaculations were virtually absent.

Masculine Sexual Motivation Versus Penile Reflexes

The results of behavioral analysis to date appear to sort themselves into two differing major interpretations. First, the sexual motivation of these estrogen receptor knockout mice appears basically to be normal: mounting frequencies are high enough, mounting latencies are low enough, and the temporal profile of mounting behavior across long 3-hour tests is essentially the same in the homozygous knockout animals as in the wild-type controls (Ogawa et al., 1997). Now, we are furthering this interpretation by olfactory preference tests of the male toward odors from the female and by the application of formal quantitative behavioral tests of the motivation concept (Matthews et al., in press; Ogawa et al., 1996c). Dramatically different is the second type of data: the lack of ejaculation. By differential logic we are forced to conclude that the accuracy of the pelvic reflexes by the knockout male must be reduced such that penile stimulation necessary for ejaculation rarely reaches threshold.

GENES AND THE DEVELOPMENT OF SEXUAL BEHAVIOR IN HUMAN BEINGS

One type of disorder of sexual development in humans is the well-known androgen insensitivity syndrome. In such a case, a would-be male with a normal complement of XY chromosomes has the external features of a female, is raised as a female, and behaves and is treated as a female (Money et al., 1984). At a molecular level, clearly, two possibilities for mechanisms arise: mutations of the androgen receptor and mutations of enzymes responsible for metabolizing testosterone. The former presents us with a good animal model, the testicular feminizing mutation (TFM) mouse. TFM mice have androgen receptors in the brain with normal affinity for dihydrotestosterone, but these receptors are present in significantly lower concentrations than expected (Attardi et al., 1976). At the behavioral level, TFM mice do not mount females with normal frequencies, nor can they show intromissions or ejaculations (Olsen, 1992). Even if challenged with androgenic hormones, these TFM male mice show male-typical reproductive behaviors with even lower frequencies than control females (Olsen, 1992). Such results show the importance of normal androgen receptor gene expression for the development of masculine sex behaviors.

A much more complicated example derives from Kallmann's syndrome (hypogonadotropic, hypogonadism coupled with anosmia). With a sex ratio of 5 or 6 to 1 in favor of males, patients with this syndrome suffer from a complete lack of interest in the opposite sex—that is, a lack of sexual libido. They also retain an infantile reproductive tract and fail to develop appropriate secondary sex characteristics.

This human reproductive abnormality must be understood from the point of view of gonadotropin-releasing hormone (GnRH) neurons. GnRH not only controls the pituitary, causing it to release to release LH and FSH, but also influences reproductive behavior (Pfaff, 1973; Moss and McCann, 1973). In contrast to other neurons in the brain, GnRH neurons are not born in the ependymal linings with a short traverse to their final functional positions. Instead, GnRH neurons are born in the epithelium of the olfactory placode (Schwanzel-Fukuda and Pfaff, 1989). In all the species studied to date, from fish (Parhar et al., 1996) to human beings (Schwanzel-Fukuda et al., 1996a), these GnRH expressing cells migrate from the olfactory placode up the nasal septum along the base of the forebrain and then penetrate into the preoptic area and hypothalamus. One of the chemical cues for this migration is the neural cell adhesion molecule (NCAM), which beautifully adorns the migration route (Schwanzel-Fukuda et al., 1992) and whose perturbation will disrupt migration (Schwanzel-Fukuda et al., 1994). Nevertheless, we know from studies with NCAM knockout mice (Marlene abstract) that there must be other mechanical and chemical cues to guide migration. For example, an important part of the mechanism could be migration guidance by blood vessels, which are elaborated richly in the nasal mesenchyme (Schwanzel-Fukuda et al., 1996b). Not only could these specialized blood vessels provide guidance for migration through the olfactory apparatus, they could also aid GnRH neurons in penetrating the linings of the developing forebrain. Finally, because of the privileged association of

GnRH neuronal terminals with specialized blood vessels in the median eminence, it is even possible that GnRH neurons are "dragged forward" by the connection of their presumptive axon terminals with blood vessels, which move as they develop.

Our current theory of the GnRH neuronal migration is that these cells are born in the epithelium of the olfactory placode and leave, heading for the brain, along a predetermined route. Before they leave, "pioneer cells," which are not neurons, migrate toward the brain, laying down an NCAM trail. The vomeronasal nerve and terminal nerve follow that trail. In turn, GnRH neurons follow the mechanical and chemical cues along this trail thus established. Near the bottom of the brain, the pioneer cells form an aggregate with developing blood vessels. The GnRH-expressing cells migrate along the medial side of this aggregate into the basal forebrain. In some species, they stop in the preoptic area, while in others, they may get all the way to the infundibulum, the portion of the hypothalamus just above the median eminence.

Relating this GnRH neuronal migration during development to the subject of the present chapter, it is important to note that Kallmann's patients have no interest in the opposite sex, that is, they lack libido. We had the good fortune to be able to study tissue from an X-linked Kallmann's fetus. Many Kallmann's patients have damage in the neighborhood of Xp22.3. In the 19-week-old Kallmann fetus (Schwanzel-Fukuda et al., 1989) we observed that even though the GnRH gene was being expressed normally in this individual, the GnRH neurons did not migrate properly. Instead of making it into the basal forebrain, they were stuck in the olfactory apparatus. At this point, it becomes important to note that a gene at Xp22.3 absent or damaged in these Kallmann's individuals has been cloned (Legouis et al., 1991; Franco et al., 1991). The chemistry and anatomical distribution of this gene make it apparent that it could contribute to the establishment of a normal migration route.

The facts are thus in place to show the contribution of an individual gene by an indirect casual route to the development of a human reproductive behavior, male reproductive motivation. With experimental results well established at every step in this chain of reasoning, we can say that the reproductive behavior deficit is *due* to low testosterone, *due in turn* to low gonadotropin levels, *due in turn* to low concentrations of GnRH coming from the brain, *due in turn* to the absence of GnRH neurons in the brain, *due in turn* to the GnRH neuronal migration disorder, and *due ultimately* to genetic damage at Xp22.3.

In summary, the complexity of this mechanism serves to prevent oversimplified thinking about connections between genes and the development of reproductive behavior.

HORMONES AND MOOD

Finally, it is not the case that the developmental mechanisms important for reproductive behavior are limited to the generation of simple behaviors and reflex responses. Hormones also have massive effects on motivations and moods. For example, under the influence of estrogen and progesterone, even female rats will learn arbitrary responses to gain access to male rats (Matthews et al., 1996). Perhaps more to the point, in women, estrogens can lead to heightened mood (Lu et al., 1996; Schecter et al., 1996; Rubinow, 1996). In contrast, when progesterone is high enough that for a period of days estrogen:progesterone ratios are markedly lower, sadness, tears, and other aspects of poor feelings result (Halbreich et al., 1986; Schecter et al., 1996). A lesson from these findings is that in higher species where the repertoires of behaviors associated with reproduction are more complex, correspondingly, we should expect the range of developmental mechanisms to be more varied and subtle.

REFERENCES

Arnold, A., et al. (1996). *Hormones and Behavior*, Dec. issue (1996).

Attardi, B. (1981). *Endocrinology, 108*: 1487.

Attardi, B., Geller, L. N., and Ohno, S. (1976). *Endocrinology, 98*: 864.

Franco, B., Guioli, S., Pragliola, A., Incerti, B., Bardoni, B., Tonleorenzi, R., Carrozzo, R., Maestrini, E., Pieretti, M., Taillon-Miller, P., Brown, C. J., Willard, H. F., Lawrence, C., Persico, M. G., Camerino, G., and Ballabio, A. (1991). *Nature, 353*: 529.

Goy, R. W., and McEwen, B. S. (1980). *Sexual Differentiation of the Brain*. The MIT Press, Cambridge, MA.

Halbreich, U., Endicott, J., Goldstein, S., and Nee, J. (1986). *Acta Psychiatr. Scand., 74*: 576.

Kelley, D. B., and Pfaff, D. W. (1978). *Biological Determinants of Sexual Behavior* (J. Hutchison, ed.). Wiley, Chichester, England, p. 225.

Kow, L.-M., and Pfaff, D. W. (1973). *Neuroendocrinology, 13*: 299.

Kow, L.-M., Montgomery, M. O., and Pfaff, D. W. (1979). *J. Neurophysiol., 42*: 195.

Kow, L.-M., Zemlan, F. P., and Pfaff, D. W. (1980). *J. Neurophysiol., 43*: 27.

Lauber, A. H., Romano, G. J., and Pfaff, D. W. (1991). *Neuroendocrinology, 53*: 608.

Legouis, R., Hardelin, J. P., Levilliers, J., Claverie, J. M., Compain, S., Wunderle, V., Millasseau, P., Le Paslier, D., Cohen, D., Caterina, D., Bougueleret, L., Dellemarre-Van De Waal, H., Lutfalla, G., Weissenbach, J., and Petit, C. (1991). *Cell, 67*: 423.

Lu, R. B., Ko, H. C., Yao, B. L., Chang, F. M., Yeh, T. L., and Huang, K. E. (1996). *Biol. Psychiatry Abstract, 39*: 648.

Lubahn, D. B., Moyer, J. S., Golding, T. S., Couse, J. F., Korach, K. S., and Smithies, O. (1993). *Proc. Natl. Acad. Sci. USA, 90*: 11162.

Matthews, T. J., et al. (1996). Society for Neuroscience Abstract, *J. Exp. Anal. Behav.*, in press.

McCarthy, M. M., Schlenker, E., and Pfaff, D. W. (1993). *Endocrinology, 133*: 433.

Meisel, R. L., and Sachs, B. D. (1994). *The Physiology of Reproduction*, Vol. 2. p. 3.

Money, Schwartz and Lewis. (1984). *Psychoneuroendocrinology*.

Moss, R. L., and McCann, S. M. *Science, 181*: 177.

Ogawa, S., Olazabal, U. E., Parhar, I. S., and Pfaff, D. W. (1994). *J. Neurosci., 14*: 1766.

Ogawa, S., Taylor, J. A., Lubahn, D. B., Korach, K. S., and Pfaff, D. W. (1996a). *Neuroendocrinology, 64*: 467.

Ogawa, S., Lubahn, D. B., Korach, K. S., and Pfaff, D. W. (1996b). *Ann. NY Acad. Sci., 794*: 384.

Ogawa, S., Gordan, J. D., Taylor, J., Lubahn, D. B., Korach, K. S., and Pfaff, D. W. (1996c). *Hormones and Behavior*, in press.

Ogawa, S., Lubahn, D. B., Korach, K. S., and Pfaff, D. W. (1997). *Proc. Natl. Acad. Sci., USA*, in press.

Olsen, K. (1979). *Nature, 279*: 238.

Olsen, K. (1992). *Handbook of Behaviroal Neurobiology*, Vol. 11 (A. Gerall et al., eds.) Plenum Press, New York, p. 1.

Parhar, I. S., Iwata, M., Pfaff, D. W., and Schwanzel-Fukuda, M. (1996). *J. Comp. Neurol.*,

Parsons, B., Thomas, C. R., and McEwen, B. S. (1984). *Endocrinology, 115*(4): 1412.

Pfaff, D. W., and Pfaffmann, C. (1969). *Olfaction and Taste* (C. Pfaffmann, ed.). Rockefeller University Press, New York, p. 258.

Pfaff, D. W. (1973). *Science, 182*: 1148.

Pfaff, D. W. (1980). *Estrogens and Brain Function: Neural Analysis of a Hormone-Controlled Mammalian Reproductive Behavior.* Springer-Verlag, New York.

Pfaff, D. W., Schwartz-Giblin, S., McCarthy, M. M. and Kow, L.-M. (1994). *The Physiology of Reproduction*, 2nd ed. (E. Knobil and J. Neill, eds.). Raven, New York, p. 107.

Romano, G. J., Krust, A., and Pfaff, D. W. (1989). *Mol. Endocrinol., 3*: 1295.

Rubinow, D. R., and Schmidt, P. J. (1996). *Biol. Psychiatry Abstract, 39*: 613.

Schechter, D., Strasser, T. J., Endicott, J., Petkova, E., and Nee, J. (1996). *Biol. Psychiatry Abstracts, 39*: 646.

Schwanzel-Fukuda, M., and Pfaff, D. W. (1989). *Nature, 338*: 161.

Schwanzel-Fukuda, M., Bick, D., and Pfaff, D. W. (1989). *Mol. Brain Res., 6*: 311.

Schwanzel-Fukuda, M., Abraham, S., Crossing K. L., Edelman, G. M., and Pfaff, D. W. *J. Comp. Neurol., 321*: 1.

Schwanzel-Fukuda, M., Abraham, S., Reinhard, G. R., Crossin, K. L., Edelman, G. M., and Pfaff, D. W. (1994). *J. Comp. Neurol., 342*: 174.

Schwanzel-Fukuda, M., Crossin, K. L., Pfaff, D. W., Bouloux, P. M. G., Hardelin, J. P., and Petit, C. (1996a). *J. Comp. Neurol., 366*: 547.

Schwanzel-Fukuda, M., Pfaff, D. W., and Dellovade, T. (1996b). *Society for Neuroscience Abstract*, in press.

8

Reproductive Neuroendocrine Regulation in the Female: Toward a Neurochemical Mechanism for the Preovulatory Luteinizing Hormone Surge

William R. Crowley

University of Tennessee, College of Medicine, Memphis, Tennessee

Indeed what is there that does not appear marvelous when it comes to our knowledge for the first time? How many things, too, are looked upon as quite impossible until they actually have been effected? (Pliny the Elder)

OVERVIEW OF NEUROENDOCRINOLOGY

Neuroendocrinology can be viewed as an integrative discipline whose major concern deals with characterizing the interactions between nervous and endocrine systems. This field encompasses investigations that address both the means by which various hormones affect the structure and function of the nervous system and, in turn, how the nervous system regulates endocrine secretions. It is customary to differentiate two major neuroendocrine systems. The magnocellular hypothalamic neuroendocrine system consists of large neurosecretory cells contained within the hypothalamic supraoptic and paraventricular nuclei and several accessory clusters of cells, which secrete oxytocin and vasopressin into the systemic circulation in response to defined physiological stimuli (see Crowley and Armstrong, 1992, for review).

The parvicellular hypothalamic neuroendocrine division consists of numerous clusters of smaller neurosecretory cells that elaborate the hypothalamic hormones that stimulate or inhibit secretion of the hormones of the anterior pituitary (AP) gland; these include the gonadotropins [follicle-stimulating hormone (FSH) and luteinizing hormone (LH)], prolactin (PRL), thyrotropin (TSH), growth hormone (GH), and adrenocorticotropin (ACTH). Each of the above-mentioned "tropins" (i.e., FSH, LH, TSH, ACTH) is under the predominant influence of a single stimulatory hypothalamic hormone, while GH and PRL are controlled by separate excitatory and inhibitory hormones (Table 1).

Table 1 Hypothalamic Releasing Hormones and Their Cognate Anterior Pituitary Hormones

AP hormone	Hypothalamic hormone(s)
Luteinizing hormone, follicle-stimulating hormone	Gonadotropin-releasing hormone
Thyroid-stimulating hormone	Thyrotropin-releasing hormone
Adrenocorticotropic hormone	Corticotropin-releasing hormone
Prolactin	Dopamine (inhibitory)
	Thyrotropin-releasing hormone (stimulatory)
Growth hormone	Somatostatin (inhibitory)
	Growth hormone–releasing hormone (stimulatory)

Basic Principles of Anterior Pituitary Hormone Secretion

Several overarching principles govern the neuroendocrine control of AP hormone secretion (see Reichlin, 1992, for review). First, the regulation by the brain of AP secretion is mediated through the release of the hypothalamic hormones into the hypothalamohypophyseal portal vascular link between hypothalamus and anterior pituitary gland. Upon reaching the appropriate endocrine cells of the AP via the vasculature, the hypothalamic hormones act via their specific receptor and intracellular signaling mechanisms to inhibit or stimulate the secretion of their cognate AP hormone. Second, the secretion of both the hypothalamic and cognate pituitary hormones is episodic; such an intermittent secretory pattern is apparently essential for the expression of the respective biological activities of these hormones.

Third, feedback regulatory loops govern these systems (McEwen et al., 1979; Reichlin, 1992). In the case of the tropin AP hormones with defined endocrine gland targets (FSH, LH, TSH, ACTH), the target gland hormones exert the feedback control, while for AP hormones with more widespread somatic targets (GH, PRL), the feedback effects are exerted by the AP hormones themselves. Finally, whether long-loop or short loop, in general feedback control is predominantly inhibitory in nature, and is mediated by a centrally based suppression of releasing hormone secretion and/or action in the AP to inhibit the effect of the releasing hormone. The sole exception is the stimulatory feedback action of estradiol and progesterone on LH secretion, which is responsible for induction of the preovulatory LH surge, the mechanism for which constitutes the major emphasis of this chapter.

Objectives

The overall objective of this chapter is to review our current understanding of the neuroendocrine regulatory mechanisms governing reproductive cyclicity in the female. The first section of this chapter will review the basic phenomena of reproductive neuroendocrine regulation in the female, emphasizing studies in rats and primates, while the primary focus of the second section will be identifying potential neurochemical mechanisms by which the ovarian hormones exert their feedback actions, particularly those leading to the preovulatory LH surge.

The basic physiological process under regulation in this system is the maturation and development of the follicles within the ovary that contain the oocytes, followed by the rupture of the follicles to release the ova, in the process of ovulation (Freeman, 1988). These intraovarian events are governed by the integrated actions on the ovary exerted by the gonadotropins released from the AP, FSH, and LH. The secretion of these hormones is, in turn, under the influence of the decapeptide gonadotropin-releasing hormone (GnRH), which is the final neural signal that regulates reproduction (Kalra and Kalra, 1983). At a higher level of control, a number of neural systems, utilizing both classical as well as neuropeptide transmitters, exert excitatory and inhibitory influences over GnRH secretion (Kalra and Kalra, 1983; Crowley, 1987). The GnRH neurosecretory system must therefore integrate these neural inputs with the long-loop feedback actions of the ovarian hormones, estradiol and progesterone, in order to shape the physiologically appropriate pattern of gonadotropin secretion. Research over the past several decades has led to the concept that the ovarian hormones act primarily to govern the activity of these neuromessenger inputs (Kalra and Kalra, 1983; Crowley, 1987).

REPRODUCTIVE NEUROENDOCRINE REGULATION IN THE FEMALE

Patterns of Gonadotropin and GnRH Secretion

The pattern of LH secretion throughout the estrous or menstrual cycle is well known for various mammalian species (see Knobil, 1981; Kalra and Kalra, 1983; Karsch 1984; Freeman, 1988). One can conceptualize the cycle as having as its hallmark event the preovulatory LH surge, the massive discharge of the hormone that is primarily responsible for the rupture of the ovarian follicle. As noted above, it is now well established in a number of species that gonadotropins and GnRH are secreted in an episodic manner. In the female, this has been particularly well described for the primate and rat (see Knobil, 1980; Freeman, 1988 for reviews). In the female rhesus monkey, for example, relatively low amplitude pulses of LH occur with a periodicity of approximately 1.5–2 h during the follicular phase of the cycle. The preovulatory LH surge itself is characterized by high-amplitude, high-frequency pulses, followed during the luteal phase by a reduction in pulse frequency to one every 5–6 h, but with the individual amplitudes relatively high (Ferin et al., 1984). Similarly, in the female rat, LH release is pulsatile at each stage of the estrous cycle, with the pulse frequencies and amplitudes lowest on the day postovulation (estrus), more intermediate on the days of diestrus during follicular development, and highest during the preovulatory LH surge on the afternoon of proestrus (Gallo, 1981a,b; Fox and Smith, 1984). Frequent blood sampling during the surge reveals very rapid and high-amplitude LH pulses during the ascending limb and the plateau phase of the surge (Gallo, 1981a). For example, from a frequency of once per hour during the presurge period on proestrus, the frequency increases to 2–3 pulses/h during the initial period of the LH surge.

It is well established that physiological pattern of gonadotropin secretion is driven by the episodic release of GnRH from the hypothalamus (Kalra and Kalra, 1983). GnRH secretion has been monitored by techniques such as push-pull perifusion of the medial basal hypothalamus-median eminence region, push-pull perifusion or microdialysis of the AP, and by chronic cannulation or acute samplings of portal blood. In general, the major conclusions from these studies (1) confirm that GnRH secretion from the hypothalamus is indeed episodic (e.g., Ching, 1982; Levine and Ramirez, 1982; Levine et al., 1985; Park and Ramirez, 1989; Caraty et al., 1995), (2) demonstrate, when both can be measured concurrently, that there is a close temporal relationship between pulses of GnRH and of LH (e.g., Ching, 1982; Levine et al., 1985; Pau et al., 1993), and (3) reveal that the parameters of episodic GnRH secretion vary with the stage of the estrous cycle and with steroid hormone status of the animal (e.g., Levine and Ramirez, 1980, 1982; Dluzen and Ramirez, 1986).

It is now customary to refer to the neural mechanism that drives episodic GnRH and LH secretion as the "pulse generator" (Plant, 1986; Karsch, 1987; Evans and Karsch, 1995). However, we still have very little information as to what exactly constitutes this pulse generator at the anatomical or cellular level. Electrical recordings of multiple unit activity in the medial basal hypothalamus in the rhesus monkey reveal cyclic increases in activity that correspond closely to the periodicity of LH pulses (Wilson et al., 1984); at present, however, there is no evidence that these recordings directly reflect activity of GnRH neurons. Because the immortalized GT-1 hypothalamic cell lines that secrete GnRH show episodic release of the peptide in cell culture, it has been suggested that GnRH cells in vivo might be intrinsically pulsatile in their firing pattern (Wetsel et al., 1992). If so, however, this would require GnRH-GnRH interconnections in order to synchronize and coordinate the firing so that the peptide could be released into the portal capillaries as a bolus; the available evidence from a number of species indicates that GnRH-GnRH interconnections do exist, but are relatively uncommon (Silverman et al., 1994). Alternatively, perhaps the neurochemical inputs to the GnRH cells are the primary components of the pulse generator, as suggested some time ago by the observations that episodic LH secretion can interrupted by pharmacological agents acting centrally [e.g., noradrenergic receptor blockers (Drouva and Gallo, 1976; Pau et al., 1989)].

The GnRH System

GnRH Neuroanatomy

The anatomy of the GnRH neurosecretory system in various species, as assessed from immunocytochemical localization of perikarya, fibers, and terminals, and more recently by in situ hybridization to detect cell-specific expression of the mRNA encoding its precursor, has been the subject of numerous research reports and reviews (e.g., Sternberger and Hoffman, 1978; Silverman et al., 1979, 1994; Kawano and Daikoku, 1981; King et al., 1982; Shivers et al., 1983, 1986; Merchenthaler et al., 1984; Lehman et al., 1986; Goldsmith et al., 1990). As a basic generalization, GnRH cells are relatively few in number, and as stated succinctly by Silverman and coworkers (1994):

> The GnRH cells are not segregated into nuclear clusters but instead appear as a loose network spread through many classic cytoarchitectonic divisions. In most species, GnRH cells form a loose continuum from the telencephalic diagonal band of Broca and more dorsal septal areas (including the medial and triangular septal nuclei) to the bed nucleus of the stria terminalis, and diencephalic areas (including the periventricular area, medial and lateral preoptic areas, anterior hypothalamus and retrochiasmatic zone medial to the optic tract).

Major species differences do exist with respect to how far caudally the GnRH cells extend, i.e., whether they appear within the medial basal hypothalamus structures such as the arcuate nucleus. In the case of the rat, medial basal hypothalamic regions lack GnRH cells, whereas in guinea pig, sheep, and primates, including humans, GnRH cells are present in this area to varying degrees (Silverman et al., 1994).

GnRH-immunopositive nerve terminals are most abundant in the lateral portions of the external zone of the median eminence, which comprises one of the perivascular neurohemal contact zones; these terminals derive from anywhere between 50–75% of the GnRH perikarya (Silverman et al., 1994). Virtually every identified neurochemical system also projects fibers into the median eminence (Jacobowitz, 1988), some of them in close proximity to the GnRH neurosecretory endings. As discussed more fully below, the median eminence offers plentiful opportunities for nonclassical, terminal-to-terminal modes of neuronal communication between these neurochemical inputs and GnRH fibers and terminals.

Several other characteristics of GnRH anatomy are worth noting. One interesting discovery of the past several years is the demonstration that GnRH cells derive from the olfactory placode and migrate to their central neural locations during fetal and early neonatal life along defined olfactory pathways (Schwanzel-Fukuda et al., 1989; 1995; Wray et al., 1989). Second, many GnRH neurons innervating the median eminence express a second neuropeptide, galanin, with expression in females more prominent than in males (Merchenthaler et al., 1990; Coen et al., 1990; Liposits et al., 1995). As reviewed below, this neuropeptide may play an important role in modulating GnRH secretion and action during the LH surge. Third, it has been noted both in rat and primate that GnRH neurons in the medial preoptic area receive relatively few synaptic inputs, as compared to adjacent neurons, in part because of considerable glial ensheathment (Silverman et al., 1988; Goldsmith and Thind, 1995); subsequent sections of this chapter will identify those neurochemical inputs that have been demonstrated. However, one might view the GnRH system as relatively "privileged" in terms of the degree of afferent input.

GnRH Biochemistry

Discovery of the GnRH precursor and gene (Seeberg et al., 1987) has led to rapid advances in clarifying the steps involved synthesis of the mature peptide from its precursor. The proGnRH molecule varies in size across species from 59 to 60 amino acids (Wetsel, 1995). The GnRH decapeptide is processed from its precursor by enzymatic actions at a Gly-Lys-Arg locus (with Gly-11 providing the amidation), which separates GnRH from another peptide product, the 56-amino-acid residue gonadotropin-releasing hormone–associated peptide (GAP) (Wetsel et al., 1988; Wetsel, 1995); the biological function of this latter peptide has remained elusive. The gene encoding the precursor protein is structured into four exons and introns. The second exon contains the coding region for the signal peptide, GnRH, the enzymatic processing and amidation sites and the first (N-terminal) amino acid residues of the GAP peptide (Seeberg et al., 1987). As of this writing, a number of laboratories are conducting studies to characterize and map the regulatory elements of the GnRH gene (e.g., Weirman et al., 1995). As reviewed below, there have also been a number of attempts to test whether, in their feedback regulatory actions, steroid hormones might alter GnRH mRNA expression, but the results to date have been disparate.

GnRH Receptor and Signal Transduction

GnRH acts via a specific membrane receptor, the structure of which has recently been elucidated (Kaiser et al., 1992; Tsutsumi et al., 1992; Chi et al., 1993). As predicted from earlier ligand binding and signal transduction studies, the GnRH receptor belongs to the superfamily of membrane receptors, whose protein structures consist of seven regions of hydrophobic amino acids that span the cell membrane, interspersed with three extracellular and three intracellular domains. Such receptors, including the GnRH receptor, are well established to couple via G-proteins to second messenger systems; in the case of GnRH, coupling to Gq and/or G11 have been proposed (Hsieh and Martin, 1992).

As reviewed extensively (Huckle and Conn, 1988; Naor, 1990; Stojlkovic et al., 1990), the primary intracellular messenger transducing the GnRH signal in the gonadotrope is Ca^{2+}; stimulation of the GnRH receptor on AP gonadotropes produces an initial rapid rise of cytosolic Ca^{2+}, followed by a sustained plateau phase. Pharmacological and biochemical studies to dissect this primary cellular response indicate that the earliest increase occurs primarily from the mobilization of intracellular Ca^{2+} with some contribution from entry of extracellular Ca^{2+}, while the sustained phase depends primarily upon Ca^{2+} entry (Change et al., 1986, 1988; Hansen et al., 1987; Naor et al., 1988; Smith et al., 1987; Tasaka et al., 1988). Changes in Ca^{2+} occur consequent to coupling of the GnRH receptor to phospholipase C, which catalyzes the synthesis of the intracellular messengers inositol-1,4,5-trisphosphate, which acts primarily to mobilize Ca^{2+} from intracellular sources, and diacylglycerol, an activator of protein kinase C, which may act to facilitate Ca^{2+} entry (Andrews and Conn, 1986; Naor et al., 1986; Morgan et al., 1987). As with other systems, efforts are continuing to define the signaling cascade from the GnRH receptor, elevation in cytosolic calcium, and activation of protein kinases that ultimately lead to the exocytotic event associated with gonadotropin secretion. As reviewed below, changes in the responsiveness of the gonadotropes to GnRH are important mechanisms involved in ovarian hormone feedback regulation, and such influences could involve changes in GnRH receptor numbers, affinity, and/or second messenger couplings.

Feedback Regulation of GnRH/LH by Ovarian Hormones

The inhibitory and stimulatory feedback actions of ovarian hormones on LH secretion are easy to observe using ovariectomized (ovx) animals as the experimental model (Kalra and Kalra, 1983; Freeman, 1988; Brann and Mahesh, 1991). Removal of the ovaries invariably results in hypersecretion of LH, which is reflected in higher frequency and amplitude of the secretory pulses; this is usually interpreted as reflecting the removal of the restraining influence of ovarian secretions over the hypothalamic-pituitary axis. In ovx rats, administration of estradiol lowers mean plasma concentrations of LH and reduces pulse frequencies and amplitudes very rapidly, but not fully to the intact condition (Kalra et al., 1973). Progesterone alone typically does not produce such an inhibitory effect, but if given concurrently with or shortly after estradiol, enhances the suppression of circulating LH to that level seen in intact animals (Caligaris et al., 1971a; Goodman, 1978a).

However, both steroids also exert stimulatory actions, with both dosage and time of exposure the apparent critical factors in determining the switch from inhibition to stimulation of LH secretion. For example, in rats, beginning after approximately 24 h of estradiol exposure, the low levels of LH are interrupted by brief LH "minisurges" that occur daily in the late afternoon. These increases of LH do not reach the magnitude of the naturally occurring LH surge on proestrus, but occur each day near the time of the natural LH surge (Caligaris et al., 1971b; Legan and Karsch, 1975; Legan et al., 1975; Goodman, 1978a); this is though to reflect activity of a daily "neural clock" (Kalra and Kalra, 1983). LH subsequently decreases to low levels again until the appropriate time 24 h later, when another surge occurs. If a bolus injection of progesterone is made to ovx rats that have been estrogen-primed for at least 24–48 h, the afternoon LH surge occurs somewhat earlier and is magnified to a level approximating the physiological LH surge (Caligaris et al., 1971b; Simpkins et al., 1980). Interestingly, progesterone then induces a period of refractoriness to further LH stimulation by the estradiol treatment, which lasts for several days (Caligaris et al., 1971b). Similar positive and negative effects of the steroids are observed in rhesus monkey as well (Goodman and Knobil, 1981; Terasawa et al., 1984, 1987; Brann and Mahesh, 1991).

As reviewed extensively (Kalra and Kalra, 1983; Freeman, 1988; Brann and Mahesh, 1991; Kalra, 1993), studies in naturally cycling animals confirm that both estradiol and progesterone exert dual inhibitory and stimulatory actions on LH secretion at different phases of the cycle. In particular, it should be emphasized that stimulation of the preovulatory LH surge on proestrus in rats and at mid-cycle in rhesus monkeys involves an integrated action of both estradiol and progesterone for its expression (Kalra and Kalra, 1974; Schenken et al., 1976; Rao and Mahesh, 1986; DePaolo, 1988). Moreover, progesterone acts to both facilitate the preovulatory LH surge and prevent LH surges on subsequent days of the cycle (Freeman et al., 1976); this mechanism serves to limit the LH surge to the appropriate day, proestrus (Freeman, 1988).

That both steroids can inhibit as well as stimulate LH secretion in females presents a fascinating mechanistic problem to unravel. One aspect of this issue about which considerable disagreement still exists concerns the relative importance of central vs. AP actions of the steroids, i.e., whether modulation of GnRH secretion from the hypothalamus or of GnRH action in the AP, or both, is the primary mechanism underlying ovarian steroid action. Adding to the complexity are species differences and likelihood that negative and positive feedback are mediated by different mechanisms. One critical concept that can be stated with reasonable certainty is that in all species examined to date, the vast majority of GnRH neurons lack nuclear localized steroid receptors (Shivers et al., 1983; Herbison and Theodosis, 1992; Watson et al., 1992; Lehman and Karsch, 1993, Sullivan et al., 1995). While this does not rule out some nonclassical (e.g., membrane) action of the steroids directly on the GnRH neuron (e.g., Ke and Ramirez, 1987), it seems clear that the major target for either inhibitory or stimulatory action of ovarian hormones is not the GnRH neuron per se, but perhaps more likely, those neurochemical inputs to the GnRH neuron.

Estradiol and Negative Feedback

Arguing for a central site for the inhibitory actions of estradiol are early findings in rats that central implants of the steroid, particularly into the medial basal hypothalamic region, reduce circulating LH (Smith and Davidson, 1974; Blake, 1977). Also, the observation that estradiol can produce a long-lasting inhibition of LH secretion despite concurrently increasing the sensitivity of the pituitary gland to GnRH, i.e., producing an apparent stimulatory action (see below), would seem to argue that the steroid must be lowering the secretion of GnRH from the hypothalamus in order for LH to be reduced. Early studies measuring changes in the concentration of GnRH in portal blood of rats after ovx and/or steroid replacement appeared to support this view (Sarkar and Fink, 1979). However, as reviewed extensively (Kalra and Kalra, 1989), the preponderance of more recent evidence on GnRH release in vivo and in vitro, as well as on

GnRH immunoreactive peptide and mRNA levels, suggests the converse, i.e., that ovx lowers GnRH synthesis and release while estradiol actually increases these measures, even under conditions when LH secretion is suppressed. These findings are exactly the opposite of what one would predict if estradiol were to act centrally to decrease GnRH secretion.

Such findings therefore point to the gonadotropes of the AP as the primary site, and reduction of the response to GnRH as the primary mechanism, of estrogen negative feedback, particularly in the rat. Indeed, several lines of evidence from in vivo and in vitro studies on LH release support this view (Libertun et al., 1974; Vilchez-Martinez et al., 1980; Frawley and Neill, 1984). To some extent, the response to GnRH depends upon GnRH receptor numbers, which can vary with hormonal status (Clayton and Catt, 1981), and more recently Chin and coworkers (Kaiser et al., 1993) have shown that estradiol treatment of ovx rats can reduce the levels of the mRNA for the GnRH receptor. On the other hand, it is also the case that the inhibitory effects of estradiol on sensitivity to GnRH tend to be short-lived, lasting only a few hours, and convert ultimately to increased sensitivity to GnRH, apparently in preparation for the LH surge (Libertun et al., 1974; Vilchez-Martinez et al., 1980). At present, therefore, it is not possible to reconcile the findings in the rat that estradiol produces a relatively long-lasting suppression of LH secretion, while only briefly desensitizing the gonadotrope to GnRH.

In female rhesus monkeys, more consistent evidence exists for an AP site for estradiol negative feedback. In this species, estradiol can inhibit LH secretion without reducing GnRH release from the hypothalamus (Pau et al., 1990). Moreover, estradiol can inhibit LH secretion in the "hypophysiotropic clamp" preparation of Knobil and coworkers, in which female rhesus monkeys bearing lesions of the medial basal hypothalamus are administered an unvarying regimen of GnRH (see Wildt et al., 1981). Such observations can only be explained by an AP action.

Progesterone and Negative Feedback

It is important to consider an action of progesterone along with estradiol, as during the estrous or menstrual cycle, an action of one without the other actually may not be physiologically relevent. In rats, progesterone also produces biphasic effects on the LH response to GnRH, but the pattern is opposite to that of estradiol; i.e., progesterone transiently increases GnRH-induced LH release and then suppresses it (Lagace et al., 1980; Drouin and Labrie, 1981).

However, central sites may also be important for inhibitory progesterone action. For example, in rats, progesterone implants to the medial-basal hypothalamus decrease LH release (Blake, 1977). In female rhesus monkeys, a series of studies by Ferin and coworkers (Ferin et al., 1987) show that in the luteal phase progesterone stimulates the release of β-endorphin, which exerts an inhibition of GnRH secretion. As discussed more fully below, steroid-sensitive β-endorphin cells are found in the medial basal hypothalamus, and these may be the important central mediators for inhibitory progesterone feedback.

Estradiol and Positive Feedback

Antagonism of estrogen action by administration of pharmacological antagonists or immunoneutralization prevents the expression of the provulatory LH surge on proestrus

(Shirley et al., 1969; Ferin et al., 1969; Neil et al., 1971), and actions of the steroid in both brain and AP appear to be important. As noted above, estradiol increases the sensitivity of the gonadotrope to GnRH in vitro (Libertun et al., 1974; Drouin et al., 1976; Vilchez-Martinez et al., 1980) and in vivo studies in naturally cycling rats shows this effect to be important on proestrus, the day of the preovulatory LH surge (Aiyer et al, 1974; Baldwin and Downs, 1980). The sensitivity of the gonadotropes to GnRH appears to correlate well with GnRH numbers, which are also increased on the day of proestrus (Clayton et al., 1980; Savoy-Moore et al., 1980). Recent studies indicate that this is likely due to an effect of estradiol on GnRH receptor gene expression (Kaiser et al., 1993; Bauer-Dantoin et al., 1995).

While in the rat an action of estradiol in the AP undoubtedly contributes to positive feedback mechanisms through an influence on GnRH receptor synthesis, this alone does not account fully for positive feedback, as studies measuring GnRH output show that an increase in GnRH secretion accompanies the naturally occurring, as well as hormone-induced, LH surge (Sarkar et al., 1976; Ching, 1982; Levine and Ramirez, 1982). Indeed, the heightened episodic pattern of GnRH secretion on proestrus may contribute to the increase in GnRH receptors (Yasin et al., 1995). Estradiol can also induce LH release when implanted into discrete central sites and under conditions where the AP is unaffected by the steroid. Some of the most critical loci include the septum-diagonal band of Broca area, medial preoptic nucleus, bed nucleus of the stria terminalis, anterior hypothalamic nucleus, arcuate nucleus, and ventromedial nucleus (Terasawa and Kawakami, 1974; Kalra and McCann, 1975; Goodman, 1978; Kawakami et al., 1978). These structures also correspond to major locations of neurons containing estradiol receptors in their cell nuclei (Pfaff and Keiner, 1973; Sar et al., 1974; Rainbow et al., 1982). While some of these sites (particularly the ventromedial nucleus), undoubtedly play a primary role in reproductive behavior, based on lesion and stimulation studies, other areas in the medial preoptic and medial-basal hypothalamus regions are clearly involved in positive feedback mechanisms (Kalra and Kalra, 1983; Freeman, 1988). However, as noted above, the GnRH neuron per se lacks classical estradiol receptors; hence the primary targets for estradiol are other systems that influence GnRH secretion (see below).

Studies by Knobil and coworkers in rhesus monkeys garnered considerable evidence that the AP was the sole site of positive feedback in this species. For example, in their medial-basal hypothalamic-lesioned preparation in which an unvarying GnRH stimulus was provided to the AP, estradiol still exerted a positive feedback effect, seemingly arguing that the steroid acted to sensitize the gonadotrope to a constant central stimulus (Nakai et al., 1978; Wildt et al, 1981). However, arguing strongly for increased GnRH secretion as the major event in this species, as in others, are the definitive demonstrations that the LH surge in this species is associated with increased GnRH output (Xia et al., 1992; Pau et al., 1993). At which central location estradiol acts in this species to stimulate the GnRH/LH surge remains to be identified.

Progesterone and Positive Feedback

It is now clear that the stimulatory action of progesterone is also obligatory for the expression of the LH surge (Rao and Mahesh, 1986; DePaolo, 1988; Brann and Mahesh, 1991), and its actions also appear to be exerted primarily at central sites leading to activation of GnRH neurons (Lee et al., 1990), but to some extent also in the AP. Progesterone can acutely sensitize the gonadotrope to GnRH, but the effect is transitory and converts to inhibition at later times (Lagace et al., 1980). Progesterone also stimulates GnRH release in vivo in ovx, estrogen-primed rats (Levine and Ramirez, 1980) and rhesus monkeys (Woller and Terasawa, 1994). Progesterone stimulates LH release when applied to the septum-diagonal band, medial preoptic area and anterior hypothalamus, some of the same sites enumerated above for estrogen sensitivity (Kalra and McCann, 1975; Kawakami et al., 1978). Indeed, estrogen pretreatment is required for this action of progesterone. Neurons in these sites also possess progestin receptors (Parsons et al., 1982).

With the possible exception of a small population in the guinea pig, most GnRH cells in species examined to date lack nuclear progesterone receptors (King et al., 1995). Thus, as with estradiol (see above), the primary targets for progesterone in positive feedback are likely to be other systems that project to and regulate the GnRH system (see below).

Ovarian Hormones and GnRH mRNA Expression

When given in combination to induce an LH surge, estradiol plus progesterone, but not estradiol alone, also induce an important pattern of change in immunoreactive GnRH concentrations in the median eminence. Several hours prior to the onset of the LH surge, and before there is any increase in LH release from the AP, GnRH levels specifically in the median eminence undergo a dramatic rise, followed by a decline (Kalra et al., 1978; Kalra and Kalra, 1979; Wise et al., 1981a). This decline appears coincident with the initial increases in circulating LH and may therefore reflect the initial release into the portal blood. A similar pattern occurs naturally on proestrus during the hours prior to the surge (Kalra et al., 1973a; Kalra and Kalra, 1979; Wise et al., 1981b). Because pharmacological blockade of the LH surge (e.g., with noradrenergic antagonists, see below) also prevents the changes in GnRH accumulation, one can consider that this dynamic change in GnRH content of the median eminence antecedent to the LH surge is also an expression of ovarian hormone–positive feedback and moreover, an integral component of the LH surge mechanism. As discussed more fully below, there is some evidence for the neurochemical mechanisms that may mediate these changes.

It is still not quite clear, however, whether these changes primarily reflect pretranslational (i.e., de novo gene transcription in the rostral cell bodies, followed by transport to the median eminence) or posttranslational (e.g., processing of the mature peptide from the already preformed precursor in the nerve terminals) events in GnRH synthesis. Initial studies on GnRH mRNA expression using in situ hybridization indicated that the increase in GnRH mRNA levels at the perikarya, if it occurred at all, was manifest at the peak of the LH surge on proestrus (Zoeller and Young, 1988; Park et al., 1990) or after progesterone treatment to ovx, estrogen-primed rats (Kim et al., 1989). However, a more detailed study published recently (Porkka-Heiskanen et al., 1994) revealed that a population of GnRH cells in the medial preoptic area display increased levels of GnRH message several hours prior to the proestrus LH surge, while another subpopulation in more rostral basal forebrain areas shows an increase coincident with the surge. Similarly, using a biochemical approach to measure levels of the primary GnRH gene transcript as well as mature mRNA in the medial preoptic area, Gore and Roberts (1995) found that both measures were elevated in the medial preoptic area several hours prior to the LH surge on proestrus. Increased levels of GnRH mRNA have also been reported recently in the hours prior to the onset of the surge induced by estradiol alone or estradiol and progesterone (Cho et al., 1994; Petersen et al., 1995).

In summary, then, the preponderance of evidence suggests that ovarian steroid–negative feedback primarily involves an action at the AP to reduce the response of the gonadotrope to GnRH, while the positive feedback actions of the steroids are primarily mediated by stimulation of GnRH neurosecretion, in all likelihood also involving several steps in GnRH synthesis, possibly supplemented by effects on GnRH receptor synthesis that enhance the responsiveness of the gonadotrope to GnRH. Because the steroids do not act via nuclear receptors directly within the GnRH neuron to alter activity of these cells, the major central targets for ovarian steroid feedback actions are in all probability those neural systems that directly or indirectly regulate neurosecretion of GnRH.

A large number of specific neurotransmitter systems have been implicated in the control of LH secretion, and a number of authoritative reviews are available (Weiner and Ganong, 1978; Gallo, 1980; Barraclough and Wise, 1982; Kalra and Kalra, 1983; Kalra, 1986; Crowley, 1987; Kalra, 1993; Crowley et al., 1995). The next section provides a more focused discussion of the experimental evidence for involvement of a selected group of inhibitory and excitatory neurochemical systems in ovarian hormone positive feedback mechanisms. For each system, we will discuss its effects on LH/GnRH secretion, its mechanism of action, if known, its anatomical relationship to the GnRH system, and the evidence that it might participate in ovarian hormone–positive feedback. Where possible, studies in female primates will also be discussed.

NEUROCHEMICAL MECHANISMS IN POSITIVE FEEDBACK

The Inhibitory Neurochemical Systems

Endogenous Opioid Peptides

Effects on LH/GnRH. Perhaps the most extensively characterized of the inhibitory neurotransmitters regulating GnRH/LH secretion are the endogenous opioid peptides, of which β-endorphin, dynorphin A, and met- and leu-enkephalin have been the most studied with respect to neuroendocrine regulation. Opioid peptides, agonist analogs, and opioid agonist drugs such as morphine uniformly inhibit LH release in various experimental paradigms and in diverse species, and with a clearly central site of action (see Kalra and Kalra, 1983, 1984; Kalra, 1986, 1993; Crowley, 1988; van Wimersma Greidanus, 1991, for review). In particular, central administration of morphine or various endogenous opioid peptides (EOPs) in rats can inhibit the LH surge occurring on proestrus or induced in ovx rats by ovarian hormone treatment (Pang et al., 1977; Sylvester et al., 1980; Koves et al., 1981; Kalra and Simpkins, 1981; Hulse and Coleman, 1983; Kalra and Gallo, 1983; Leadem and Kalra, 1985a). Morphine treatment also prevents the rise in GnRH concentrations in the median eminence that precedes the LH surge (see above) (Kalra and Simpkins, 1981) and the increase in portal plasma GnRH during the surge (Ching, 1983). The three major classes of opioid receptor, μ, k, and δ, have been implicated in the inhibitory effects of these peptides (Leadem and Kalra, 1985b; Leadem and Yagenova, 1987; Crowley, 1988), but otherwise, relatively little is known concerning their cellular mechanisms.

Conversely, administration of the opioid receptor antagonist naloxone transiently increases LH release (Bruni et al., 1977; Blank et al., 1980; Gabriel and Simpkins,

1983, Adler and Crowley, 1984), and if administered under the appropriate conditions, can induce an LH surge very similar in character to that seen on proestrus (Allen and Kalra, 1986; Allen et al., 1988; Masotto et al., 1988), a finding with important implications for a prevalent hypothesis on positive feedback (see below). It is interesting to note the important similarity in this effect of naloxone and the action of progesterone (Masotto et al., 1988), and additional neurochemical parallels will be discussed below. Using an immature rat model, Schulz et al. (1981) showed that central immunoneutralization of β-endorphin or dynorphin, but not met-enkephalin, elevated serum LH in female rats. These findings, along with other evidence (see below), appear to more strongly implicate β-endorphin and perhaps dynorphin rather than the enkephalin peptides in LH suppression.

Anatomical Relationships. In vivo and in vitro studies point to the medial basal hypothalamus-median eminence complex and medial preoptic area as two key loci for inhibitory opioid effects on LH secretion (Rotsztejn et al., 1978; Drouva et al., 1980; 1981; Grandison et al., 1980; Weisner et al., 1984; Kalra, 1981; Wilkes and Yen, 1981; Leadem et al., 1985; Rasmussen et al., 1988). In particular, the evidence that opioids can reduce GnRH release from median eminence or medial basal hypothalamus preparations in vitro (Drouva et al., 1980, 1981; Wilkes and Yen, 1981; Rasmussen et al., 1988) suggest that EOP neurons may exert direct or indirect presynaptic control over GnRH release at the level of the nerve terminal.

Anatomical evidence, however, is not in agreement with this proposal. As reviewed by Hoffman et al. (1989), β-endorphin-, dynorphin-, and met-enkephalin-immunopositive fibers in the median eminence are not in proximity to GnRH fibers. However, interneuronal communication over a distance (volume transmission) may be characteristic of the median eminence, so a lack of close appositions may not be definitive.

On the other hand, β-endorphin–positive fibers are in apparent contact with GnRH perikarya in the medial preoptic area in rats (Leranth et al., 1988a; Hoffman et al., 1989). These fibers derive from perikarya in the arcuate nucleus of the medial basal hypothalamus (Akil et al., 1984). Such findings suggest that the hypothalamic β-endorphin system might exert a direct inhibitory regulation over GnRH secretion through actions at the GnRH cell body. In addition to this mechanism, another very likely possibility is that one or more EOPs inhibit GnRH secretion indirectly by interfering with the release or action of excitatory inputs to the GnRH network. Discussed more fully in subsequent sections of the chapter will be the substantial pharmacological and biochemical evidence that opioids exert a tonic inhibition over release of the adrenergic transmitters, norepinephrine and epinephrine, and neuropeptide Y, which are important stimulatory regulators of GnRH secretion.

Role in Positive Feedback. These observations have led to the hypothesis that one mechanism for ovarian hormone–positive feedback on proestrus involves removal of a tonic inhibitory tone over GnRH secretion exerted by one or more of the EOP, which then allows stimulation of the preovulatory GnRH/LH surge (Kalra and Kalra, 1983; Kalra, 1993). Evidence for a reduction in EOP influence at the time of the LH surge includes the observations that opioid antagonists do not elevate LH when given during the LH surge, suggesting loss of inhibitory tone at this time (Gabriel et al., 1983, 1986). More direct evidence for this hypothesis are the findings that immunoreactive β-endorphin concentrations in the medial basal hypothalamus and portal blood are reduced at the time of increased GnRH concentrations in portal blood during the LH surge

(Barden et al., 1981; Knuth et al., 1983; Sarkar and Yen, 1985; Kerdelhue et al., 1988). Consistent with this, the level of expression of the mRNA for proopiomelanocortin (POMC), the β-endorphin precursor, in the arcuate nucleus is reduced prior to the LH surge (Wise et al., 1990; Bohler et al., 1991; Petersen et al., 1993a); however, more recent studies suggest that reduction in POMC gene transcription is not specifically tied to the proestrus LH surge but may occur as part of the daily signal for LH release (Scarbrough et al., 1994). That EOP synthesis and release may be directly affected by the ovarian hormones is strongly suggested by the demonstrations that subsets of β-endorphin- and dynorphin-immunopositive neurons in the arcuate nucleus contain nuclear receptors for estradiol and progesterone (Morrell et al., 1985; Jirikowski et al., 1986; Fox et al., 1990).

In addition to a reduction in EOP synthesis and release, the LH surge is also associated with a decrease in opioid responsiveness as assessed pharmacologically (Berglund and Simpkins, 1988; Berglund et al., 1988). Consistent with this observation, there is a reduction in opioid receptor binding, primarily of the μ subtype, during the LH surge on proestrus or induced in ovx animals by ovarian steroids (Jacobson and Kalra, 1989; Weiland and Wise, 1990).

Two other peptidergic systems shown to exert inhibitory actions on LH release, possibly related to effects of stress, appear to act at least in part through the EOP inhibitory mechanism. Corticotropin-releasing hormone (CRH) is the releasing hormone for ACTH release from the AP, and thus is the final neural system integrating the influence of stress, which can inhibit LH release under some circumstances (Reichlin, 1992). Central administration of CRH inhibits LH release, and because this action can be blocked by opioid antagonists, such as naloxone, it appears to be the case that CRH neurons activate an inhibitory EOP, most likely β-endorphin (Petraglia et al., 1986; Almeida et al., 1988). Direct evidence that CRH releases β-endorphin has also been reported (Nikolarkis et al., 1986).

Neuropeptide Y is a major excitatory neuropeptide for LH release with a direct stimulatory action on GnRH release (see below), but can inhibit LH release under certain circumstances, most notably in ovx, hormonally untreated rats (Kalra and Crowley, 1984) and in intact rats after prolonged infusion (Catzeflis et al., 1993). It has been suggested that this inhibitory action may be associated with adverse metabolic conditions that may have a negative impact on reproductive function (Aubert et al., 1995). Recent studies indicate that the inhibition of LH release by NPY can be blocked with naloxone, and therefore may not be direct, but rather mediated by release of an EOP (Xu et al., 1993). Neuronal connections from hypothalamic NPY cells to nearby β-endorphin neurons provide anatomical support for this hypothesis (Horvath et al., 1992a).

Studies in Primates. In the female primate, a somewhat different role has been proposed for opioids in control of LH secretion during the menstrual cycle (see Ferin et al., 1984, for review). While the opioid antagonist naloxone increases LH release in female rhesus monkeys and humans, the effect is demonstrable only in the luteal phase of the cycle, not in the follicular phase (Van Vugt et al., 1983; 1984; Melis et al., 1984). These pharmacological results have been interpreted to mean that an EOP, most likely β-endorphin, mediates ovarian hormone–negative feedback during this period, but is not involved in the surge mechanism as hypothesized for rodents (Ferin et al., 1984). Recall that progesterone secreted from the corpus luteum, along with estradiol, reduces the frequency of LH pulses during this period. Also consistent with this view are findings showing that repeated morphine injections failed to block estrogen-induced LH surges in the female rhesus monkey (Ferin et al., 1982).

Studies in which the concentration of β-endorphin into the portal blood was used as an index of the activity of this system have shown that levels of this peptide tend to be higher in the luteal phase than in the follicular phase (Wardlaw et al., 1982). Moreover, ovariectomy, which increases LH release, is associated with reduced portal plasma β-endorphin (Wardlaw et al., 1982), and a negative feedback regimen of estradiol and progesterone elevated β-endorphin secretion (Wehrenberg et al., 1982). As with rodents, there is anatomical evidence in the monkey for direct synaptic connections between β-endorphin–positive cells and GnRH cell bodies (Thind and Goldsmith, 1988). Arcuate neurons in the monkey contain ovarian hormone receptors (Bethea et al., 1992), but whether these include β-endorphin–positive neurons as in the rat remains to be determined. One report in the green monkey (Leranth et al., 1992) indicated that progestin receptor–containing neurons of the arcuate nucleus in this species lacked β-endorphin.

γ-Aminobutyric Acid

Effects on LH/GnRH. A stimulatory effect of γ-Aminobutyric acid (GABA) on LH secretion has been observed in rats after central administration in vivo (Vijayan and McCann, 1978b; Negro-Vilar et al., 1980). Similarly, GABA and agonists at the $GABA_A$ receptor stimulate GnRH release from rat median eminence preparations (Nikolarkis et al., 1988; Masotto et al., 1989) and from the immortalized GT-1 GnRH-secreting cell lines in vitro (Favit et al., 1993; Hales et al., 1994; Martinez de la Escalera et al., 1994). The $GABA_A$ receptor is a classical ionotropic receptor, comprising a binding site–Cl^- ion channel complex, which when activated usually results in Cl^- influx and hyperpolarization of the postjunctional neuron. However, GnRH neurons are depolarized, and GnRH secretion increased, by GABA-A receptor stimulation, most likely by Cl^- efflux (Favit et al., 1993).

On the other hand, stimulation of a second GABA receptor subtype, the GABA-B receptor, inhibits LH secretion in vivo (Adler and Crowley, 1986; Masotto and Negro-Vilar, 1987; Akema and Kimura, 1991, 1992) and GnRH release in vitro (Masotto et al., 1989; Favit et al., 1993; Martinez de la Escalera et al., 1994). The $GABA_B$ stimulation with the agonist baclofen also decreases GnRH mRNA expression, as observed in experiments using in situ hybridization histochemistry (Bergen et al., 1991). $GABA_B$ receptor is coupled to inhibition of neurotransmitter release, most likely via a reduction in Ca^{2+} influx (Bowery et al., 1980); as suggested for EOP, GABAergic inhibition over GnRH release appears to be in part a direct effect at the GnRH neuron and in part indirect, by reducing excitatory noradrenergic inputs, as discussed in more detail below. A physiological role in control of LH secretion for the excitatory GABAergic effect has not emerged thus far, but substantial evidence exists that the inhibitory GABAergic action is important in controlling episodic LH release and particularly the LH surge; it is for this reason that GABA is grouped here with the inhibitory neuroregulators of LH secretion.

Anatomical Relationships. GABAergic neurons are widely distributed throughout the hypothalamus and medial preoptic area (Vincent et al., 1982). In particular, there are populations of GABAergic neurons in the medial preoptic area and medial basal hypothalamus that appear to play an important role in reproductive neuroendocrine regulation. A number of studies have demonstrated a close association of GABAergic neurons with GnRH cell bodies and nerve terminals. For example, Jennes et al. (1983) reported close appositions of fibers immunoreactive for glutamate decarboxylase (GAD), the GABA-synthesizing enzyme, with GnRH terminals in the median eminence, and perikarya in the medial preoptic area were apparent at the light microscopic level. As subsequently demonstrated with the electron microscope, GAD-positive terminals make synaptic contacts with GnRH cell bodies in the medial preoptic area (Leranth et al., 1985, 1988b). Consistent with important GABAergic inputs to GnRH cell bodies are the demonstrations that medial preoptic area GnRH neurons as well as the GT-1 cells in culture express the mRNA for the $GABA_A$ receptor (Favit et al., 1993; Petersen et al., 1993; Hales et al., 1994). Ligand binding studies also suggest the presence of $GABA_A$ and $_B$ receptor subtypes in GT-1 cells (Favit et al., 1993; Hales et al., 1994). These anatomical and biochemical findings support the proposal of direct actions of GABA at the level of the GnRH perikarya and nerve terminal region.

Role in Positive Feedback. A substantial amount of pharmacological and biochemical evidence has accumulated to support the hypothesis that a tonic inhibitory GABAergic tone over GnRH secretion is removed to allow expression of the preovulatory LH surge. First, pharmacological manipulations that elevate endogenous GABA or stimulate GABA receptors inhibit episodic LH release as well as the natural or ovarian hormone–induced LH surge in rats (Lamberts et al., 1983; Donoso and Banzan, 1984; Fuchs et al., 1984; Adler and Crowley, 1986; Akema et al., 1990; Akema and Kimura, 1991; Herbison and Dyer, 1991; Herbison et al., 1991; Jarry et al., 1991). Experiments in which the agents were applied centrally showed that the medial preoptic area was an effective site of action for GABAergic inhibition of LH release, and especially the LH surge (Lamberts et al., 1983; Herbison and Dyer, 1991; Jarry et al., 1991). Adler and Crowley (1986) were the first to specifically implicate the $GABA_B$ receptor in the blockade of the LH surge; this was subsequently confirmed by Akema and Kimura (1991). From studies performed to date, however, it does not appear possible to enhance LH release with a GABA receptor antagonist, as is the case with EOP antagonism (see above). It is possible that such a maneuver cannot override the influence of other important inhibitory systems, such as EOP.

There are now a large number of reports correlating the release of GABA, as measured by push-pull perifusion or microdialysis, in the medial preoptic area with alterations in LH secretion, particularly the LH surge. Jarry et al. (1988) reported that in ovx, hormonally untreated rats, LH secretory episodes were preceded by reductions in GABA release rates in the medial preoptic nucleus, as measured by push-pull perifusion. However, this relationship was not observed by Herbison et al. (1991). Wuttke and coworkers have reported in several communications the very important finding that GABA release rates in the medial preoptic nucleus are reduced prior to the estrogen-induced LH surge in ovx rats (Demling et al., 1985; Jarry et al., 1988, 1992, 1995). The most recent paper from this group (Jarry et al., 1995) noted no change in GABA release in the medial basal hypothalamus, highlighting the medial proptic area as the major locus. Taken together, these studies provide direct evidence in favor of the GABA disinhibition hypothesis mediating ovarian hormone positive feedback on GnRH secretion.

Moreover, it is particularly noteworthy that this research group has also demonstrated that a substantial proportion of medial preoptic GABAergic neurons contain estrogen receptors (Flugge et al., 1986). Hence, these investigators have suggested that in the rat, medial preoptic GnRH neurons are under the tonic inhibitory influence of

GABA; these GABAergic cells are directly estrogen-receptive, and the steroid acts to remove this GABAergic inhibition, allowing GnRH secretion for the surge (Jarry et al., 1988, 1992). How exactly estradiol accomplishes this has yet to be determined. However, Herbison et al. (1992) reported that the mRNA for GAD is reduced in the medial preoptic area during the hours on proestrus prior to the preovulatory LH surge, and Unda et al. (1995) found that progesterone treatment to ovx, estrogen-primed immature rats had the same effect; these studies suggest an effect of the steroids on GABA synthesis. Ovarian hormone also affect GABA receptors in complex ways (see Herbison and Fenelon, 1995), and effects on GABA receptive mechanisms may also contribute to the disinhibition mechanism.

It is obvious that there are many similarities in the two disinhibition hypotheses, involving an EOP and GABA, that have been proposed as mechanisms underlying ovarian hormone positive feedback. Both systems are proposed to exert a tonic inhibition over GnRH secretion that may be exerted both directly at the GnRH cell and indirectly, i.e., mediated by a decrease in delivery of excitatory inputs to the GnRH cell (see below). Estradiol and perhaps progesterone may act directly on the EOP and GABAergic neuron to decrease synthesis and release of their transmitters and may also impair postjunctional receptive mechanisms. As a result, the GnRH system is disinhibited to allow the LH surge. One may then ask whether the EOP and GABA system operate independent of each other or in series.

To date, the evidence on this question is inconclusive. Naloxone-stimulated LH release in vivo and GnRH release from median eminence in vitro can be blocked by elevations in GABA or stimulation of $GABA_B$ receptors with baclofen (Masotto and Negro-Vilar, 1987; Masotto et al., 1989; Brann et al., 1992). Although this is suggestive of an interaction, it is impossible to discern whether this means an EOP stimulates GABA release or that one can override the effect of EOP removal by elevating GABA, which acts independently. Horvath et al. (1992b) recently reported that a number of POMC-expressing arcuate neurons (β-endorphin) receive GABAergic innervation. However, other reports suggest that the GABA influence on β-endorphin synthesis is inhibitory (e.g., Blasquez et al., 1994).

Studies in Primates. Although there have been far fewer studies on the role of GABA in control of LH secretion in female primates, there is highly suggestive evidence for important GABAergic inhibition in such animals. In contrast to the rodent, ultrastructural studies in monkeys suggest that there are not direct GABAergic inputs to GnRH cells (Thind and Goldsmith, 1986). However, GABA neurons do innervate vasopressin neurons of the monkey hypothalamus, which do directly innervate GnRH cells (Thind et al., 1993). Also highly intriguing is the recent report that *all* of the progestin receptor-containing neurons of the medial basal hypothalamus in the green monkey are immunopositive for GAD (Leranth et al., 1992), suggesting they are GABAergic.

Studies by Terasawa and coworkers show that GABA exerts an inhibitory effect on LH secretion in monkeys. For example, infusion of bicuculline, a $GABA_A$ receptor antagonist, into the median eminence of prepubertal female monkeys increased GnRH release as measured by push-pull perifusion (Mitsushima et al., 1994). Similarly, infusion of an antisense oligonucleotide designed to impede GAD synthesis also enhanced GnRH release in the same model (Terasawa, 1995). Because GABA release rates in the medial basal hypothalamus fall coincident with the rise in GnRH release rates in female monkeys undergoing the transition to puberty, Terasawa has suggested that a removal of the inhibitory action of GABA may play a role in sexual maturation (Terasawa, 1995).

The Stimulatory Neurochemical Systems

The Adrenergic Transmitters: Norepinephrine and Epinephrine

Effects on LH/GnRH. The most extensively characterized of all the neuro-messengers affecting GnRH and LH secretion, the adrenergic transmitters also have the most clearly defined role in positive feedback mechanisms (see Barraclough and Wise, 1982; Kalra and Kalra, 1983; Crowley, 1986; Kalra, 1986; Ojeda et al., 1989; Crowley, 1995, for reviews). Central administration of norepinephrine (NE) can inhibit pulsatile LH release when given to ovx, hormonally untreated rats (Gallo, 1980), which suggests inhibitory influence of this transmitter. However, administration of α-adrenergic antagonists to such animals also reduces episodic LH release, suggesting stimulatory noradrenergic tone (Drouva and Gallo, 1976; Gnodde and Schuiling, 1976). It has been suggested that both influences of NE might be important in shaping episodic LH secretion and that different adrenergic receptor subtypes may mediate the differential effects (Kalra and Kalra, 1983).

When given centrally to ovarian intact or ovx plus estrogen- or estrogen-progesterone–treated rats, NE or α-adrenergic agonists stimulate LH release (Krieg and Sawyer, 1976; Vijayan and McCann, 1978a; Gallo and Drouva, 1979; Gallo, 1982; Kalra and Gallo, 1983). The requirement that animals be exposed to ovarian hormones in order for the stimulatory effect of neurotransmitters such as NE to be observed is well established, but the basis for the effect has never been clarified. Epinephrine (EPI) also stimulates LH release after central administration to ovarian hormone–pretreated rats and is even more potent in this regard than NE (Rubenstein and Sawyer, 1970; Vijayan and McCann, 1978; Gallo and Drouva, 1979; Kalra and Gallo, 1983).

Anatomical Relationships. The noradrenergic and adrenergic innervations of the hypothalamus derive from several clusters of cells in the lower brainstem and ascend via well-characterized routes (see Moore and Card, 1984; Hokfelt et al., 1984, for review). NE-containing fibers are abundant throughout the median eminence and medial preoptic area, i.e., in regions containing GnRH terminals and perikarya, respectively, while the distribution of EPI appears to be more limited to the medial basal hypothalamus. Electrical stimulation studies by Barraclough and coworkers have implicated NE systems arising from the ventro-lateral medulla (A1 cell group) and the pontine locus coeruleus (A6 cell group) in LH stimulation (Gitler and Barraclough, 1987, 1988). Interestingly, however, such stimulation is ineffective unless electrochemical stimulation is applied to the medial preoptic area; studies from this group reviewed below suggest that GnRH neurons are under a tonic inhibition that must be removed before stimuli may be effective.

A number of studies have addressed the question of whether adrenergic systems might be in direct or close contact with the GnRH neurons in the rat. When observed at the light microscopic level, presumptive NE- and EPI-containing fibers lie in close association with GnRH perikarya in the septal-preoptic areas (Jennes et al., 1982; Wray and Hoffman, 1986). However, according to one report at the EM level (Leranth et al., 1988a), catecholaminergic fibers do not make synaptic contact with GnRH cells; rather, they appear to contact GABAergic neurons, which are in direct contact. In the median eminence as well, the noradrenergic-adrenergic fibers tend to be most abundant in the interior zones, while the GnRH terminals have a periportal localization in the external zone (MacNeill et al., 1980; Jennes et al., 1982; Ciofi et al., 1991, 1993). Nonetheless, the median eminence has proven to be a major locus for the adrenegic stimulation of GnRH release. The effects of adrenergic transmitters and agonists at this level have been particularly well studied from a signal transduction standpoint, and the adrenergic stimulation of GnRH release may serve as model for nonclassical type of interneuronal communication involving a terminal-to-terminal mode of communication at a distance.

Signal Transduction. Results from studies using GnRH release from median eminence explants as the experimental approach have suggested that activation of an α_1-adrenergic receptor by NE or EPI is coupled to the synthesis of prostaglandin E_2 (PGE_2), which is the obligatory intracellular messenger for adrenergic-stimulated GnRH release (see Ojeda et al., 1989; Crowley, 1995, for review). PGE_2 appears to act primarily within the GnRH nerve terminal to release Ca^{2+} from an unidentified intracellular pool (Ojeda and Negro-Vilar, 1985; Ojeda et al., 1986a). It is well established that an elevation in intracellular Ca^{2+} ($[Ca^{2+}]i$) is a prerequisite for the release of neurotransmitters and neuropeptides by exocytosis, which is triggered by incompletely characterized, but clearly Ca^{2+}-dependent, events. This mechanism, then, is most applicable for a nerve terminal-to-nerve terminal type of interaction closely coupled to exocytosis; other mechanisms might be operational at the level of the GnRH perikarya.

The precise means by which α_1 receptor stimulation is coupled to PGE_2 synthesis and where this initial interaction takes place are not completely understood at present. A sequential mechanism whereby Ca^{2+}/calmodulin-dependent activation of phospholipase A_2, which releases arachidonic acid from membrane phospholipids, is followed by an action of cyclooxygenase, has been proposed, based on work showing that Ca^{2+} entry via voltage-sensitive Ca^{2+} channels and binding to calmodulin are essential for NE-induced PGE_2 production (Ojeda et al., 1986a) and that inhibitors of cyclooxygenase prevent NE-induced GnRH release (Ojeda et al., 1979). It is possible that this receptor-transduction pathway is located within the GnRH nerve terminals. The source of PGE_2 may not be the GnRH neuron, however, as another possibility is that NE activates α_1 receptors that are located on glial elements of the median eminence. Following activation of the PG synthetic pathway in these cells, PGE_2 may then be released to stimulate a membrane receptor on the GnRH nerve terminals, and thereby set in motion the cascade described above (Ojeda et al., 1986a).

Another consequence of noradrenergic-stimulated PGE_2 formation in the median eminence is the activation of another messenger pathway, involving cyclic 3′,5′-adenosine monophosphate (cAMP), which stimulates GnRH release by an independent mechanism involving Ca^{2+} entry via VSCC (Ojeda et al., 1986, 1989); one possibility involves phosphorylation of the L-type VSCC by the cAMP-activated protein kinase A to facilitate extracellular Ca^{2+} entry. The formation of cAMP probably occurs by a Ca^{2+} (from intracellular mobilization)/calmodulin-dependent activation of adenylyl cyclase (Ojeda et al., 1986). Seemingly, this pathway is not an obligatory mediator of NE-PGE_2 action (Ojeda et al., 1989), but may be a means of amplifying the noradrenergic signal by further increasing $[Ca^{2+}]i$.

Role in Positive Feedback. Several key lines of evidence suggest a role for the adrenergic systems in the underlying mechanism for ovarian hormone–positive feedback. First, LH surges on proestrus or induced in ovx rats by ovarian hormone treatment can be blocked by a variety of pharmacological maneuvers that disrupt adrenergic neurotransmission, strongly suggesting that the stimulatory adrenergic action on GnRH release is required for the LH surge (Kalra et al., 1972; Kalra and McCann, 1973, 1974). These initial pharmacological studies, which were subsequently confirmed by a number of laboratories using similar approaches, are a particular milestone in reproductive neuroendocrinology because they were the first to identify a potential neurochemical mechanism by which the ovarian hormones might induce an LH surge and opened the way for subsequent biochemical studies that have become standard experimental approaches to investigate feedback mechanisms.

It is important to emphasize that the use of selective synthesis inhibitors has shown that both NE and EPI are important in this regard. Crowley and coworkers were the first to implicate EPI in ovarian hormone–positive feedback using selective EPI synthesis inhibitors (Crowley and Terry, 1981; Crowley et al., 1982; Adler et al., 1983), and this was later confirmed by others (Coen and Coombs, 1983; Coen et al., 1985; Kalra, 1985). Both adrenergic transmitters also are critical for the accumulation of immunoreactive GnRH in the median eminence during the hours immediately preceding the surge (see above), based on studies with selective synthesis inhibitors and catecholamine neurotoxins (Simpkins and Kalra, 1979; Simpkins et al., 1980; Adler et al., 1983). Central administration of prazosin, an α_1-adrenergic antagonist, to female rats also reduces the level of GnRH mRNA that was increased at the time of the estrogen-induced surge (Weesner et al., 1993), consistent with the view that adrenergic transmitter inputs paly a role in regulation of GnRH biosynthesis.

One can interpret the pharmacological studies as showing that normal levels of adrenergic activity are necessarily permissive for the LH surge, or alternatively, that the ovarian hormones actually activate these systems and that an action of these transmitters is then necessary for release of GnRH. The development of techniques for assessing the functional state of adrenergic transmission by measuring synthesis or turnover rates in discrete regions of the brain after physiological manipulations in vivo allowed the definitive test of this hypothesis. Initial studies provided a good correlation between NE turnover increases and the LH surge in the medial preoptic area and medial basal hypothalamus (Crowley et al., 1978; Honma and Wuttke, 1980; Crowley, 1982), and more extensive time course studies showed that on proestrus and after estrogen alone or estrogen plus progesterone treatment there is increased NE turnover in the arcuate nucleus, median eminence, and medial preoptic nucleus in the hours preceding the surge, compared to earlier in the morning (Rance et al., 1982; Wise et al., 1981). Hence, there is an important interaction between hormonal condition and time of day. In particular, noradrenergic activity in the median eminence increases well before the onset of the LH surge (Rance et al., 1981; Wise et al., 1981b; Adler et al., 1983), consistent with involvement of this transmitter in stimulating GnRH synthesis occurring at this time.

After pharmacological studies had implicated EPI in surge mechanisms, turnover studies examining activity of this transmitter showed that similar to NE, EPI turnover is also increased prior to and during the LH surge on proestrus or after ovarian hormone priming, especially within the medial basal hypothalamus-median eminence (Adler et al., 1983; Sheaves et al., 1984, 1985). Finally, the release of both NE and EPI has been monitored by push-pull perifusion in the medial preoptic area by Demling et al. (1985), who observed marked increases in the release of both of these transmitters prior to and during the estrogen-induced LH surge in ovx rats.

Thus, the estradiol and progesterone treatment that induces an LH surge also increases the activity of NE and EPI innervation of the preoptic area and median eminence, and interference with NE and EPI transmission prevents the hormone-induced LH surge and antecedent increases in GnRH in the median eminence; similar events have been demonstrated on proestrus. Taken together, these findings suggest strongly that a critical component of the ovarian hormone–positive feedback mechanism involves activation of excitatory adrenergic inputs to GnRH neurons. Although this hypothesis has received wide support, it has not been firmly established how ovarian hormones exactly activate the adrenergic systems. There is no convincing evidence that ovarian hormones directly act on the adrenergic neurons, which as noted above are situated in the lower brainstem, quite distant from the well-known loci in the medial preoptic area and medial basal hypothalamus known to subserve positive feedback (see above).

Rather, as alluded to in the sections on opioids and GABA, the findings that ovarian hormone receptors are localized in hypothalamic and preoptic GABA neurons and in arcuate β-endorphin neurons, combined with studies showing that both of these systems exert an inhibitory presynaptic control over release of adrenergic transmitters, suggest that ovarian hormones activate NE and EPI release indirectly, i.e., by removing the influence of these intrahypothalamic inhibitory modulators. Such local control over the adrenergic inputs to the GnRH system was first suggested by Wuttke and co-workers (1982) for the GABAergic system, as they noted an inverse correlation between NE turnover and GABA turnover in the medial preoptic area during proestrus or after estrogen treatment. This relationship was strengthened by subsequent studies showing a similar inverse relationship between GABA release rates and NE and EPI release rates during an estrogen-induced LH surge (Demling et al., 1985). Moreover, central application of the $GABA_A$ agonist muscimol to the medial preoptic area decreased NE turnover rates in association with decreased LH secretion in ovx rats, suggestive of a presynaptic inhibitory GABAergic control over NE release (Lamberts et al., 1983; Fuchs et al., 1984).

Adler and Crowley (1986) further investigated this concept within the context of ovarian hormone–positive feedback. They showed that the blockade of the estrogen-progesterone–induced LH surge by muscimol and the $GABA_B$ agonist baclofen was associated with blockade of the ovarian hormone–induced increase in NE and EPI turnover, particularly in the medial basal hypothalamus. This concept is consistent with the well-established presynaptic action of the $GABA_B$ receptor to inhibit NE release (e.g., Bowery et al., 1980). It should be noted that GABA-NE anatomical relationships have not been reported in the medial-basal hypothalamus, especially median eminence, so a terminal-to-terminal mode of interaction for these two systems may well be operational in this structure.

Another important type of GABA-NE interaction may occur in the medial preoptic area, where anatomical studies have suggested direct GABAergic, but not noradrenergic inputs to GnRH cells (Leranth et al., 1988b). It has been commonly observed that rather high doses of NE are required to elicit LH release in vivo or GnRH release in vitro, and pharmacological studies suggest that this may be due in large part to a major inhibitory GABAergic influence that dampens the effect of excitatory messengers. For example, Hartman et al. (1990) and Akema and Kimura (1993) have found that prior blockade of either $GABA_A$ or $_B$ receptors markedly enhances the LH response to intraventricular NE.

Extensive pharmacological evidence also points to a similar action of EOP (see Kalra and Kalra, 1983, 1984; Crowley, 1988; Kalra, 1993, for reviews). The stimulation of LH release produced by the opioid antagonist naloxone can be blocked by inhibitors of NE and/or EPI synthesis or α-adrenegic receptor blockers (Kalra and Simpkins, 1981; Van Vugt et al., 1981; Kalra and Crowley, 1982), suggesting that removal of opioid tone might increase the release of the adrenergic transmitters for LH stimulation. Direct tests of this hypothesis showed that, indeed, naloxone increases NE and EPI turnover rates in the preoptic area and medial basal hypothalamus in association with LH release (Adler and Crowley, 1984; Akabori and Barraclough, 1986a), and increases NE and EPI release along with GnRH release from hypothalamic explants in vitro (Leadem et al., 1985). Note that naloxone shares all of these reactions with progesterone.

Conversely, morphine blocks the ovarian hormone–induced increase in NE and EPI turnover as well as the LH surge (Adler and Crowley, 1984; Akabori and Barraclough, 1986b). This notion of presynaptic inhibitory opioid receptors controlling release of the adrenergic transmitters receives support from pharmacological studies in other neural and peripheral systems (Taube et al., 1977).

Studies in Primates.   As in rodents, there is good evidence for an excitatory effect of NE in female monkeys, mediated by α_1 receptors, on GnRH release through an action in the median eminence that is associated with episodic hormone release. For example, Terasawa et al. (1988) have shown with push-pull perifusion of the median eminence in ovx female rhesus monkeys that GnRH pulses occur coincident with NE pulses. Moreover, the α_1 agonist methoxamine stimulates, while the α_1 antagonist prazosin reduces, GnRH release; suppression of episodic GnRH release and episodic LH by the α-blocker phentolamine was also reported by Pau et al. (1989).

Perhaps because of the uncertainty regarding central actions of the ovarian hormones in positive feedback in female monkeys (see above), there have been few studies directed at whether NE neurons play a role in this process as they undoubtedly do in rodents. However, Terasawa and coworkers (Gore and Terasawa, 1991; Terasawa, 1995) have implicated increased release of NE in the peripubertal period leading to first ovulation in this species.

Neuropeptide Y

Effects on GnRH/LH.   This 36-amino acid peptide is the most extensively characterized of the peptidergic co-transmitters that interact with classical neurotransmitters. Neuropeptide Y (NPY) was first co-localized with NE in the sympathetic nervous system (Lundberg et al., 1984); in several tissues innervated by this system, NPY acts chiefly to amplify the noradrenergic signal (Hakanson et al., 1986; Lundberg and Hokfelt, 1986; Lundberg et al., 1990). Neuroendocrinologists became interested in the possible role of NPY in control of LH shortly after the demonstration that this peptide is also synthesized in neurons of both adrenergic transmitters in the central nervous system (Everitt et al., 1984). A substantial literature now exists suggesting that NPY also exerts important amplifying actions in control of LH secretion (see Kalra and Crowley, 1992, for review).

The stimulatory action of NPY on LH release is also very well characterized, and this peptide shows a number of similarities, but also some important differences, with the action of NE/EPI. Like NE and EPI, central administration of NPY stimulates LH release, and ovarian hormone pretreatment is essential for this action (Kalra and Crowley, 1984). Also like NE and EPI, an important site of NPY action is the median eminence for stimulation of GnRH release (Crowley and Kalra, 1987). Because of these similarities in the action of NPY with the adrenergic transmitters and because of the possibility that NPY might be a co-transmitter with these amines, a series of collaborative studies by the author with the laboratory of Dr. Kalra has been directed at the question of whether NPY might act postjunctionally to amplify the excitatory signal provided by NE or EPI (see Kalra and Crowley, 1992; Crowley et al., 1995, for review).

Initial in vivo studies demonstrated that additive effects of NE and NPY on LH secretion can be seen under appropriate conditions (Allen et al., 1987). Subsequent experiments on the mode of NPY signal transduction showed that in contrast to the effects of NE, the stimulation of GnRH release by NPY cannot be blocked by cyclooxygenase inhibition, indicating that a prostaglandin does not mediate NPY's action (Kalra et al., 1992) Rather, the effects of NPY appear to be mediated largely through the Ca^{2+}/inositol phosphate messenger pathway, since NPY stimulation of GnRH release from median eminence persists in the absence of extracellular Ca^{2+}, but is abolished by agents that inhibit mobilization of intracellular Ca^{2+}, such as TMB-8 or ryanodine (Kalra et al., 1992). Hence, the amplifying or reinforcing action of NPY occurs through a signaling pathway that complements, rather than duplicates, that affected by NE.

In peripheral tissues such as blood vessels and in several brain areas, the Y1 subtype of NPY receptor is coupled via a G-protein to the calcium/inositol phosphate messenger system (Aakerlund et al., 1990), and evidence from our laboratories strongly supports this mechanism in the action of NPY (Kalra et al., 1992). For example, GnRH (and LH) release can be elicited only by Y1 NPY agonists, and not by Y2 agonists. The Y1 receptor requires nearly the entire NPY molecule for optimal binding and activation, probably achieved by the characteristic hairpin loop in the secondary structure of NPY that brings amino acid residues in both N- and C-terminals into juxtaposition (Grundemar and Hakanson, 1994). Our in vitro and in vivo studies with peptide fragments (Kalra et al., 1992) indicates that this concept applies to GnRH release as well. NPY's action on GnRH release can be blocked by the novel antagonist α-trinositol (Kalra et al., 1992), which is a noncompetitive but highly selective NPY antagonist, and which may act primarily by preventing the activation of phospholipase C consequent to Y1 receptor stimulation (Jansen et al., 1994).

The actions of NE and EPI on the LH surge are purely central and directed towards stimulating GnRH release. The stimulatory action of NPY, however, has a second component not shared by these amines. Our work indicates that analogous to the central action of NPY to augment the noradrenergic signals arriving at the GnRH neuron, NPY amplifies the signals traversing the GnRH-activated pathway in the gonadotrope. In the rat, NPY enhances GnRH-induced LH secretion, with no effect on basal secretion, in both in vivo and in vitro models (Crowley et al., 1987, 1990; Bauer-Dantoin et al., 1991; O'Connor et al., 1993; Woller et al., 1993). Our pharmacological studies in cultured rat anterior pituitary cells showed that the facilitatory effect of NPY is mimicked by an agonist at the L-type voltage-regulated calcium channel, Bay K 8644, and abolished by nitrendipine, an antagonist at such channels, suggesting that NPY might augment that portion of the GnRH signal involving Ca^{2+} entry via the L-type channel (Crowley et al., 1990). Support for this concept came from biochemical studies in which changes in $[Ca^{2+}]i$ in dispersed anterior pituitary cells were monitored by flow cytometry (Crowley et al., 1990). Brief pretreatment with NPY, which alone did not alter $[Ca^{2+}]i$, significantly enhanced the $[Ca^{2+}]i$ response occurring during the first minute of GnRH stimulation; blockade of this effect by EGTA implicated extracellular Ca^{2+} entry in this effect.

Direct NPY receptor–G-protein coupling to phospholipase C and or to the L-type calcium channel could account for this action, but our further studies have suggested a novel mechanism. Using several different binding paradigms, we found that NPY enhanced the binding of a radiolabeled GnRH analog to anterior pituitary membranes (Parker et al., 1991); this effect has recently been confirmed by another laboratory (Leblanc et al., 1994). Our kinetic studies indicated that NPY affected GnRH binding affinity, raising the possibility that NPY binding sites interact with the GnRH receptor in an allosteric regulatory manner.

In summary, the available evidence suggests a model in which adrenergic transmitters and NPY set in motion complementary and reinforcing transmembrane signaling cascades that act in concert to promote the exocytotic release of GnRH. The primary result of this interaction is the additive production of two intracellular messengers, inositol trisphosphate and PGE_2, that mobilize $[Ca^{2+}]i$ from intracellular pools. Indeed, NPY appears to be an excellent candidate to activate a non–PGE_2-mediated signaling pathway proposed by Ojeda and coworkers (1986b) based on pharmacological studies. Moreover, through the formation of cAMP via the NE-PGE_2 connection, and through the formation of diacylglycerol from NPY–phospholipase C coupling, protein kinases A (PKA) and C (PKC), respectively, also become activated. Along with the elevated cytosolic Ca^{2+}, these additional effectors are in position to promote exocytosis through their actions on multiple potential trigger mechanisms, such as ion channels, and proteins of synaptic vesicle membranes and/or the cytoskeleton. The anterior pituitary component of NPY action, which occurs as a result of the secretion of NPY released from the median eminence into the pituitary portal circulation, then enhances the GnRH action on the gonadotrope through unique mechanisms involving modulation of ligand binding, but with the same net result of signal amplification.

Anatomical Relationships.   The studies of Meister et al. (1989) and Ciofi and coworkers (1991, 1993) suggest strongly that the majority of NPY in the median eminence is co-localized with NE and/or EPI. NPY neurons in the brain are not confined to this system, however, and relevant to this discussion is a major cell group in the arcuate nucleus, not co-localized with adrenergic transmitters (Chronwall et al., 1985; Nakagawa et al., 1985; de Quidt and Emson, 1986). Lesion studies suggest that this system does not provide strong innervation, if any, to the median eminence, but probably does innervate other important regions such as the medial preoptic area (Sahu et al., 1988; Meister et al., 1989). Studies at the EM level have demonstrated NPY-positive nerve endings in the vicinity of GnRH neurons, with some evidence of synaptic inputs (Tsuruo et al., 1990).

Role in Positive Feedback.   As with NE/EPI, a combination of biochemical and pharmacological evidence strongly ties NPY to the mechanism of ovarian hormone–positive feedback. One of the first key observations was the demonstration by Crowley and coworkers (Crowley et al., 1985) that in the ovx, estrogen-plus-progesterone–treated rat, immunoreactive NPY concentrations specifically within the median eminence undergo a sequential accumulation, followed by decline, with exactly the same time course as shown by GnRH in the same tissue. Subsequently, Sahu et al. (1989) showed that the same phenomenon occurred on proestrus, and others showed a similar pattern in an immature rat model (Brann et al., 1991).

This observation has several important implications. First, it directly led to the subsequent discovery that NPY exerts a facilitatory modulation of GnRH action in the AP (see above) and consistent with this, to demonstrations that NPY is secreted into the hypophyseal portal blood (McDonald et al., 1987). Second, relevent to the central actions of NPY, the findings of parallel changes in NPY and GnRH concentrations in the median eminence suggested one or more of the NPY systems participating in neuroendocrine regulation become activated by the ovarian hormones. This would of course be expected if NPY were co-released with adrenergic transmitters, since as reviewed above it is known that these systems are activated in conjunction with the increases in median eminence GnRH.

Also consistent with this hypothesis is the demonstration that NPY release in the median eminence is increased prior to the LH surge, at the same time as its tissue levels are increased (Watanobe and Takebe, 1992). Implicating the arcuate NPY system in positive feedback as well are the findings that approximately 25% of the cells contain nuclear estrogen receptor (Sar et al., 1990). Moreover, the levels of preproNPY mRNA in the arcuate nucleus are increased prior to the LH surge on proestrus or induced by ovarian hormones (Sahu et al., 1994, 1995). An earlier study using in situ hybridization also suggested that NPY biosynthesis in arcuate neurons is increased during the LH surge (Bauer-Dantoin et al., 1992). Also of interest is the recent demonstration that GnRH responsiveness to NPY stimulation abruptly increases on proestrus, prior to the LH surge, suggestive of effects on NPY receptive mechanisms (Besecke and Levine (1994).

Interference with NPY action by immunoneutralization (Wehrenberg et al., 1989; Minami et al., 1990) blocks LH surges on proestrus or induced by ovarian hormones, indicating that the stimulatory actions of NPY are obligatory. Further, acute inhibition of NPY synthesis with a centrally administered antisense oligonucleotide also impairs the ovarian hormone–induced LH surge (Kalra et al., 1995). Hence, taken together, the findings that (1) NPY's stimulatory effects are obligatory for the expression of the LH surge, (2) NPY synthesis and release is increased, apparently in parallel with GnRH, by ovarian hormones in the hours immediately prior to the LH surge, and (3) at least some NPY neurons are directly hormone-sensitive, all implicate this peptide as a critical mediator of ovarian hormone–postive feedback.

Finally, several studies from our laboratories extend the concept of inhibitory presynaptic regulation by an EOP over the adrenergic excitatory inputs to GnRH to include control over NPY. For example, in vitro studies showed that the antagonist naloxone increased NPY release as well as GnRH release from medial basal hypothalamic explants; in vivo administration of naloxone also increased NPY concentrations in the median eminence. Further, naloxone increased NPY gene expression in the arcuate nucleus, and acute interference with NPY synthesis with antisense oligonucleotide prevented naloxone-induced LH release (Kalra et al., 1994). Thus, an EOP, most likely β-endorphin, inhibits NPY synthesis and release. These findings strengthen the view that disinhibition from opioid influences enhances the excitatory transmission to the GnRH neurosecretory cells. Additionally, it is possible to add the effects on NPY synthesis and release to the list of parallel actions of progesterone and naloxone.

Studies in Primates. Although the history of investigations on the neuroendocrine role of NPY is relatively recent, the action of NPY in the female primate has received more attention than any other neuro-messenger. Moreover, as of this writing, NPY stands as the sole neuro-messenger implicated in the central mechanisms of ovarian hormone–positive feedback in a primate. It is feasible to monitor the concurrent release patterns of NPY and GnRH in the median eminence in the primate and to draw inferences regarding NPY action in LH control. NPY release in this structure is pulsatile and shows close temporal correlation with GnRH release; the episodes of NPY release occur approximately 5 minutes prior to episodes of GnRH release, which occur approximately 5 minutes before LH pulses (Woller et al., 1992). That NPY stimulates GnRH release in this species was shown by the finding that immunoneutralization of NPY suppressed GnRH release (Woller et al., 1992), and studies showing that central infusion of NPY into the median eminence stimulates GnRH release; in the female monkey, estrogen exposure enhances, but is not required, for this effect as in the rat (Woller and Terasawa, 1991, 1992; Gore et al., 1993; Pau et al., 1995). NPY stimulates GnRH release from explants of primate median eminence in vitro; in cultured AP cells of the primate, NPY exerts a direct stimulatory action on LH secretion (Pau et al., 1991), in contrast to the rat, where its effect is strictly modulatory. Nevertheless, it is interesting that in the female primate as well as the rat, NPY does show the dual hypothalamic and AP sites of action.

Woller and Terasawa (1994) have also reported the highly significant finding that the progesterone-induced LH surge in ovx, estrogen-primed rhesus monkey is associated with an increase in release of both GnRH and NPY in the median eminence; the increase in NPY release was manifested as increased pulse frequency, and the NPY and GnRH pulses were highly correlated temporally as in their earlier study. They have proposed that enhancement of the frequency of NPY pulses may be an important central mechanism leading to GnRH discharge for the preovulatory LH surge in primates.

Galanin

Effects on LH/GnRH. Another peptidergic system that appears to play a key role in ovarian hormone–positive feedback is galanin (see Kalra, 1993; Merchenthaler et al., 1993, for review). Galanin's action on LH secretion is stimulatory and, like other systems, requires ovarian hormone exposure; unlike systems such as NE/EPI and NPY, however, no inhibitory actions of galanin are seen in ovx, untreated animals (Sahu et al., 1987). Similar to NPY, galanin exerts its stimulatory effects on LH secretion via a combination of central and pituitary actions. For example, galanin stimulates GnRH release from the median eminence, and in doing so appears to utilize the same prostaglandin E_2–mediated intracellular signal transduction pathway as does NE, since inhibition of PG synthesis blocks galanin-induced GnRH release (Lopez and Negro-Vilar, 1990). However, the action of galanin on GnRH release can also be blocked by α_1-adrenergic antagonists (Lopez and Negro-Vilar, 1990), suggesting the possibility that galanin might act primarily through an enhancement of the release of NE.

Galanin also is secreted into the pituitary portal blood, and its plasma profiles show a close temporal correlation with GnRH pulses. Galanin stimulates GnRH release directly and enhances GnRH-induced LH release from AP cells (Lopez et al., 1991; see also Coen et al., 1990). The latter effect may occur in part via modulation of GnRH binding (Parker et al., 1993), actions that are shared with NPY (see above).

Anatomical Relationships.   Galanin is a widely distributed neuropeptide, with a number of cell groupings present within various hypothalamic nuclei; particularly dense clusters of cells and terminal networks are seen in the medial preoptic area, and considerable, though more modest, innervation in the medial basal hypothalamus, including the median eminence (Merchenthaler et al., 1993). With regard to the important issue of direct galanin inputs to GnRH neurons, Merchenthaler et al. (1990) reported that galanin-positive nerve terminals were in direct synaptic contact with GnRH perikarya; perhaps more significantly, however, they and others (Coen et al., 1990) have demonstrated that a subset of GnRH cells in the medial preoptic area also express galanin; interestingly, this co-localization is more pronounced in the female (Merchenthaler et al., 1991) and can be influenced by hormones (see below). EM analysis of the median eminence of the rat confirms that GnRH and galanin are co-packaged in neurosecretory vesicles and that galanin occurs in approximately 90% of GnRH-positive fibers (Liposits et al., 1995). Give the established actions of galanin on GnRH release and GnRH-induced LH release cited above, galanin may serve a unique role in reproductive neuroendocrine regulation as a co-transmitter–modulator of GnRH secretion and action; additionally, independent galanin networks in the medial preoptic or medial basal hypothalamic regions might provide innervation to the GnRH system and reinforce these drives.

Role in Positive Feedback.   Strongly indicative of an important role for the stimulatory actions of galanin in the preovulatory LH surge in rats are the following observations. First, the ovarian hormone–induced LH surge can be blocked by galantide, a peptide antagonist of galanin receptors (Sahu et al, 1994b); the magnitude of the LH surge can also be reduced by galanin immunoneutralization (Lopez et al., 1993). These findings indicate that the stimulatory actions of galanin are required for the expression of the LH surge. Several studies have now provided evidence that ovarian hormones activate galanin synthesis within the GnRH neurons in conjunction with the LH surge (Marks et al., 1993, 1994; Brann et al., 1993). Marks et al. (1994) found that the ovarian hormone–induced increase in galanin mRNA expression in GnRH neurons could be blocked by the α-adrenergic antagonist phenoxybenzamine, which also blocks the LH surge. This important observation is consistent with the view that the influence of the gonadal steroids is transmitted to the GnRH cell through afferents such as the adrenergic system. Also relevent to this question are the findings of Bloch et al. (1992) that approximately 15% of the galanin-positive cells of the medial preoptic nucleus are estrogen receptor positive, suggesting direct steroid action on the non–GnRH-coexpressing galanin networks.

Glutamate

Effects on LH/GnRH.   The excitatory amino acid transmitter glutamate is a fourth major excitatory neuromessenger system for which significant evidence exists for involvement in ovarian hormone–positive feedback (see Brann and Mahesh, 1994; Brann, 1995, for review). Studies in the 1970s first implicated glutamate, acting through the *N*-methyl-D-aspartate (NMDA) receptor subtypes in the stimulation of LH release (Olney et al., 1976; Price et al., 1978), but there was little follow-up on these observations until the recent advances in excitatory amino acid pharmacology and molecular biology led to a "recrudescence" of interest in this area.

Glutamate and agonists at the NMDA as well as the DL-α-amino-3-hydroxy-5-methyl-4-isoxazole (AMPA)/kainate subtypes of glutamate receptor stimulate LH secretion after systemic or central administration in rats and monkeys (Brann and Mahesh, 1994; Brann, 1995, for review). In female rats, this effect is best seen on the day of proestrus or in ovx rats after ovarian hormone pretreatment; as for NE and NPY (see above), these agents have no effect or actually inhibit LH release in ovx, untreated rats (Arias et al., 1993; Luderer et al., 1993). This is suggestive of an effect of steroid hormones on excitatory amino acid receptor mechanisms, but studies to date have not demonstrated effects of estradiol or progesterone on NMDA receptor binding or expression of the receptor protein (Brann et al., 1993a).

The effectiveness of excitatory amino acid agonists after systemic administration points to an action outside the blood-brain barrier, e.g., in the median eminence. Consistent with this area as a prime site of action, numerous studies show that glutamate or glutamate agonists at AMPA and NMDA receptors stimulate GnRH release from median eminence preparations in vitro (Bourguignon et al., 1989a; Donoso et al., 1990, 1992; Lopez et al., 1992; Arias et al., 1993). Conversely, the NMDA channel antagonist MK-801 blunts episodic GnRH secretion from hypothalamic explants in vitro (Bourguignon et al., 1989b).

Glutamate, AMPA, kainate and NMDA also evoke GnRH release from GT-1 cells (Mahachoklertwattana et al., 1994a; Spergel et al., 1994), which also express a subunit of the NMDA receptor (Mahachoklertwattana et al., 1994a; Urbanski et al., 1994). Studies in these cells confirm the influx of Na^+ and Ca^{2+} in response to these agents, implicated in AMPA receptor- and NMDA receptor-mediated signaling in other systems.

Based on other experimental evidence (see below), the medial preoptic area may be an additional site of action for glutamate stimulation of the GnRH system. Several laboratories have now reported a very rapid increase in GnRH mRNA levels after systemic administration of NMDA or kainate (Petersen et al., 1991; Liaw and Barraclough, 1993; Gore and Roberts, 1994). However, studies by Gore and Roberts (1994) found no effects on the levels of GnRH primary transcript, suggesting a post-transcriptional action.

Nitric Oxide as a Mediator of Glutamate Action.   Considerable recent attention has been devoted to the diffusible gas nitric oxide (NO) as a novel signaling system in brain and other tissues that encompasses the features of both a neurotransmitter (i.e., first messenger) and a second messenger signal transduction mechanism (see Dawson and Snyder, 1994; Schuman and Madison, 1994, for review). For example, as noted by Schuman and Madison (1994), "whereas most neurotransmitters are packaged in synaptic vesicles and secreted in a Ca^{2+}-dependent manner from specialized nerve endings, NO is an unconventional transmitter which is not packaged in vesicles, but rather diffuses from its site of production in the absence of any specialized release machinery. The lack of a requirement for release apparatus raises the possibility that NO can be released from both pre- and postsynaptic neuronal elements." NO is also found in many hypothalamic neurons, including processes in the median eminence (Bhat et al., 1995), and now has been implicated in control of GnRH release in an interaction with several of the stimulatory systems discussed above, including glutamate.

For example, NO stimulates GnRH release from median eminence explants and GT-1 cells in vitro (Bonavera et al., 1993; Morreto et al., 1993; Rettori et al., 1993a; Mahachoklertwattana et al., 1994b), while inhibitors of the NO-synthesizing enzyme (NO synthase) impair episodic LH release (Rettori et al., 1993a) and prevent the ovarian hormone–induced LH surge in rats in vivo (Bonavera et al., 1993, 1994). As in other neural systems, NO appears to stimulate GnRH release by activating guanylyl cyclase, leading to cyclic GMP formation (Moretto et al., 1993). NO synthase inhibitors also reduce basal GnRH release in vitro, as does hemoglobin, a scavenger of NO (Mahachoklertwattana et al., 1994b); the implications from these findings are that NO is important in the GnRH release process and that NO may be released extracellularly after being formed.

Although studies in GT-1 cells indicate that NO is formed within this cell type (Mahachoklertwattana et al., 1994b), two reports indicate that GnRH neurons do not express the mRNA for NO synthase (Grossman et al., 1994; Bhat et al., 1995). Hence the source for NO is in all likelihood not the GnRH neuron itself. Immunohistochemical and in situ hybridization studies reveal the presence of numerous NO synthase-expressing cells near GnRH neurons, e.g., in the diagonal band of Broca and medial preoptic area (Grossman et al., 1994; Bhat et al., 1995). Also of importance in view of the effect of NO to stimulate GnRH release from median eminence preparations is the finding that there is considerable association of fibers containing NO synthase immunopositive staining and GnRH fibers in the medial eminence (Bhat et al., 1995).

The acute synthesis of NO in response to NMDA receptor activation as a result of Ca^{2+}-dependent activation of NO synthase has been demonstrated in a number of neural tissues (Schuman and Madison, 1994). The involvement of NO in the NMDA stimulation of GnRH release has also been suggested by the findings that in vivo stimulation of LH or in vitro stimulation of GnRH by NMDA or glutamate can be blocked by NO synthase inhibitors (Bonavera et al., 1993; Rettori et al., 1993b; Mahachoklertwattana et al., 1994b). Bhat et al. (1995) have reported that many NO synthase-positive neurons in the forebrain and medial preoptic area, in the vicinity of GnRH neurons, express the NMDA R1 receptor subunit. Taken together, the above findings suggest that NMDA receptor activation by glutamate may lead to the formation of NO, which may then act diffuse to GnRH neurons and act intracellularly, stimulating cGMP formation, to stimulate GnRH release. Because NO has also been implicated in the NE-induced formation of PGE_2 (Rettori et al., 1992), NO released from local neurons might activate this signaling pathway as well.

Anatomical Relationships.   Glutamate is considered the predominant excitatory transmitter in the mammalian brain, having a widespread distribution. However, much less is known than for the other systems discussed above concerning the detailed neuroanatomy of the glutamate networks that specifically influence GnRH secretion. Excitatory amino acid–containing nerve terminals are abundant in the arcuate nucleus (van den Pol et al., 1990; van den Pol, 1991), but little detailed information is available regarding innovations to other important sites such as the medial preoptic area or median eminence. Moreover, little is known regarding the specific question of direct glutamatergic inputs to GnRH neurons in the rat, and the most detailed information available has been obtained in the monkey (see below). In particular, whether glutamate-positive nerve endings in the median eminence have a close anatomical relationship with GnRH fibers is unknown.

Abbud and Smith (1995) and Bhat et al. (1995) have recently reported that GnRH neurons in the rat medial preoptic area do not contain the mRNA for the R1 subunit of the NMDA receptor, which is essential for a functional NMDA receptor–mediated signaling. However, many non-GnRH cells in this area showed labeling for this mRNA. Thus, despite the present of this receptor in GT-1 cells, these data suggest that GnRH neurons in situ may lack direct innervation by glutamate systems. Alternatively, glutamate action in the GnRH cell body region may be mediated by the AMPA/kainate receptor.

Role in Positive Feedback. Several studies have shown that agents blocking either the NMDA or AMPA/kainate receptor subtypes disrupt episodic LH secretion in ovx rats (Ping et al., 1994a) and suppress naturally occurring or hormone-induced LH surges in female rats (Lopez et al., 1990; Meijs-Roelofs et al., 1991; Urbanski and Ojeda, 1991; Brann and Mahesh, 1991; Brann et.al., 1993b, 1993c), again suggesting the obligatory participation of glutamate excitation. Interestingly, an NMDA antagonist also impaired the estrogen and progesterone-induced increase in GnRH message level (Seong et al., 1993).

Studies that have monitored the release of glutamate by push-pull perifusion or microdialysis have also shown that during estrogen-induced LH surge, the release of glutamate in the medial preoptic area is elevated (Demling et al., 1985; Jarry et al., 1992, 1995; Ping et al., 1994b). Studies on glutamate release in the medial basal hypothalamus showed no changes in response to estrogen (Jarry et al., 1995). Therefore, as with the other excitatory systems discussed above, this combination of pharmacological and biochemical evidence suggests strongly the importance of glutamate inputs to GnRH neurons in ovarian hormone–postive feedback mechanisms. NO released from as yet uncharacterized interneurons appears to be an obligatory mediator of the glutamate action at the NMDA receptor subtype.

Studies in Primates. The stimulatory actions of glutamate have also received considerable attention in primates, but mainly in the male and especially in the process of sexual maturation (e.g., Plant et al., 1989). However, as in female rats, systemic administration of NMDA to female monkeys acutely stimulates LH release (Wilson and Knobil, 1982), but only against a background of ovarian hormone exposure. NMDA inhibits LH release in ovx, hormonally untreated rhesus monkeys, but stimulates LH secretion if the animals are treated with ovarian hormones (Reyes et al., 1990, 1991).

Goldsmith and coworkers (1994) have provided an extensive analysis of the anatomical relationships between glutamate-immunopositive and GnRH-immunopositive elements in the monkey hypothalamus. Glutamate-positive neuronal cells were observed in the diagonal band, septal and preoptic regions of the basal forebrain, the magnocellular supraoptic and paraventricular nuclei of the hypothalamus, and in the arcuate-median eminence region. LM and EM analysis revealed some axo-dendritic (but few axo-somatic) glutamate-positive synapses with GnRH cells in the basal forebrain and arcuate nucleus. No indication of glutamate innervation of GnRH-positive fibers in the median eminence was observed. Interestingly, there were more examples of GnRH inputs to glutamate-positive cells and dendrites, particularly in the arcuate-median eminence region; the significance of this unexpected finding is unknown at present.

OVARIAN HORMONE–POSITIVE FEEDBACK: AN INTEGRATED NEUROCHEMICAL HYPOTHESIS

We can propose the fundamental elements and concepts of a neurochemical mechanism for ovarian hormone–positive feedback, based on work in the rat summarized above. Figure 1 shows a number of elements afferent to the GnRH system and interacting with each other; these systems appear to affect GnRH neurosecretion through actions exerted in regions containing the GnRH cell bodies, e.g., the medial preoptic area, and

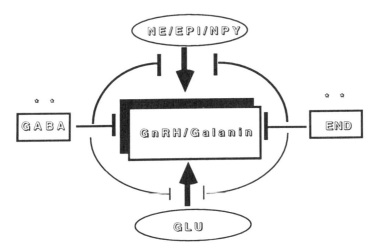

Figure 1 Interacting neurochemical systems involved in ovarian hormone-positive feedback on LH secretion in the rat. Arrows denote stimulatory influence; bars denote inhibitory influence. Asterisk denotes systems that are directly steroid-sensitive. END, β-Endorphin; EPI, epinephrine; GABA, γ-aminobutyric acid; GLU, glutamate; GnRH, gonadotropin-releasing hormone; NE, norepinephrine; NPY, neuropeptide Y.

nerve terminals in the median eminence, although the anatomical interrelationships may differ in the two sites.

The basic principles of this putative mechanism can be summarized as follows:

1. *Transmission in pathways mediated by the adrenergic transmitters, NPY, galanin, and glutamate is necessary for the unimpeded expression of the preovulatory LH surge.* Based on the available evidence, it appears that NPY acts to reinforce, via complementary second messenger pathways, the excitatory postjunctional signals provided to the GnRH neuron by the adrenergic transmitters. Galanin, released from the GnRH cell as well as from non-GnRH neurons, may presynaptically increase NE/EPI release, thereby reinforcing the presynaptic adrenergic signal to GnRH neurons. Glutamate provides additional excitatory drive, and one should envision important intracellular cross-talking between the ionotropic-NO mechanism activated by glutamate and the G-protein–coupled metabotropic mechanisms affected by NE/EPI/NPY. The excitatory inputs may not necessarily occur through classical synaptic connections with the GnRH neuron. At the level of the GnRH cells or nerve terminals, the signals may come from a distance and/or may involve as yet unidentified interneurons, e.g., those releasing NO.

2. *Estradiol and progesterone activate the excitatory inputs through disinhibition.* We can say with reasonable certainty that GABAergic cells in the medial preoptic and medial basal hypothalamic regions and β-endorphin–positive cells in the medial basal hypothalamus are directly steroid sensitive, and that one major effect of estradiol and progesterone is to decrease activity in these networks. These systems exert a tonic, inhibitory tone over GnRH secretion through direct inputs to the GnRH cell and by exerting presynaptic inhibition over release of the excitatory neuromessengers. Finally, these ex-

citatory systems are activated prior to the onset of the LH surge, as a result of the removal of the inhibitory GABAergic and EOP tone, and play an important role in the preparation of the GnRH system for hypersecretion into the portal blood, involving changes in gene expression and posttranslational events. These systems remain active during the surge to initiate and maintain GnRH hypersecretion into the portal blood.

3. *Multiple neurochemical signals reach the gonadotrope.* We can no longer consider the gonadotrope as responding only to GnRH. There is overwhelming evidence that other factors of central origin are secreted into the portal blood and target the gonadotrope to modulate GnRH action. Contributing to the activation of the LH surge is the facilitative effect of NPY, and possibly of galanin, co-secreted with GnRH.

4. *Similar mechanisms will operate in the female primate.* It seems highly likely that when investigated, all of the excitatory and inhibitory neuro-messengers affecting LH secretion in the rodent will also prove to exert similar effects in the primate. The investigation of central positive feedback mechanisms was perhaps impeded for a time by the prevailing view that ovarian hormones acted solely at the AP in these species. With the clear demonstrations that a surge of GnRH occurs in primates, as in all other animals studied, during the LH surge occurring either naturally or induced by ovarian hormones, this view is no longer tenable. Already, evidence that NPY is an important excitatory mediator of positive feedback has been presented.

REFERENCES

Abbud, R., and Smith, M. S. (1995). *Brain Res.*, *690*: 177.

Adler, B. A., and Crowley, W. R. (1984). *Neuroendocrinology*, *38*: 248.

Adler, B. A., and Crowley, W. R. (1986). *Endocrinology*, *118*: 91.

Adler, B. A., Johnson, M. D., Lynch, C. O., and Crowley, W. R. (1983). *Endocrinology*, *113*: 1431.

Aiyer, M. S., Fink, G., and Greig, F. (1974). *J. Endocrinol.*, *60*: 47.

Akabori, A., and Barraclough, C. A. (1986a). *Brain Res.*, *362*: 55.

Akabori, A., and Barraclough, C. A. (1986b). *Brain Res.*, *362*: 221.

Akema, T., and Kimura, F. (1991). *Brain Res.*, *562*: 169.

Akema, T., and Kimura, F. (1992). *Neuroendocrinology*, *56*: 141.

Akema, T., and Kimura, F. (1993). *Neuroendocrinology*, *57*: 28.

Akil, H., Watson, S. J., Young, E., Lewis, M. E., Khatchaturian, H., and Walker, J. M. (1984). *Annu. Rev. Neurosci.*, *7*: 223.

Allen, L. G., and Kalra, S. P. (1986). *Endocrinology*, 118: 2375.

Allen, L. G., Crowley, W. R., and Kalra, S. P. (1987). *Endocrinology*, *121*: 1953.

Allen, L. G., Hahn, E. Caton, D. and Kalra, S. P. (1988). *Endocrinology*, 122: 1004.

Almeida, O. F. X., Nikolarakis, K. E., and Herz, A. (1988). *Endocrinology*, *122*: 1034.

Andrews, W. V., and Conn, P. M. (1986). *Endocrinology*, *118*: 1148.

Arias, P., Jarry, H., Leonhardt, S., Moguilevsky, J. A., and Wuttke, W. (1993). *Neuroendocrinology*, *57*: 710.

Aubert, M. L., Gruaz, N. M., d'Alleves, V., Pierroz, D. D., Catzeflis, C., and Sizonenko, P. C. (1995). *The Neurobiology of Puberty* (T. M. Plant and P. A. Lee, eds). Journal of Endocrinology, Ltd, Bristol UK, p. 119.

Baldwin, D. M., and Downs, T. R. (1980). *Biol. Reprod.*, *24*: 581.

Barden, N., Merand, Y., Rouleau, D., Garaon, M., and Dupont, A. (1981). *Brain Res.*, *204*: 441.

Barraclough, C. A., and Wise, P. M. (1982). *Endocr. Rev.*, *3*: 91.

Bauer-Dantoin, A. C., Urban, J. H., and Levine, J. E. (1992). *Endocrinology*, *131*: 2953.

Bauer-Dantoin, A. C., Weiss, J., and Jameson, J. L. (1995). *Endocrinology*, *136*: 1014.

Bergen, H. T., Hejtmancik, J. F., and Pfaff, D. W. (1991). *Exp. Brain Res.*, *87*: 46.

Berglund, L. A. and Simpkins, J. W. (1988). *Neuroendocrinology*, *48*: 394.

Berglund, L. A., Derendorf, H., and Simpkins, J. W. (1988). *Endocrinology*, *122*: 2718.

Besecke, L. M., and Levine, J. E. (1994). *Endocrinology*, *135*: 63.

Bethea, C. L., Fahrenbach, W. H., Sprangers, S. A., and Freesh, F. (1992). *Endocrinology*, *130*: 895.

Bhat, G. K., Mahesh, V. B., Lamar, C. A., Ping, L., Aguan, K., and Brann, D. W. (1995). *Neuroendocrinology*, *62*: 187.

Blake, C. A. (1977). *Endocrinology*, *101*: 1130.

Blank, M. S., Panerai, A. E., and Friesen, H. (1980). *J. Endocrinol.*, *85*: 307.

Blasquez, C., Jegou, S., Feuilloley, M., Rosier, A., Vandesande, and Vaudry, H. (1994). *Endocrinology*, *135*: 2759.

Bloch, G. J., Kurth, S. M., Akesson, T. R., and Micevych, P. E. (1992). *Brain Res.*, *595*: 301.

Bohler, H. C. L., Tracer, H., Merriam, G. R., and Petersen, S. (1991). *Endocrinology*, *128*: 1265.

Bonavera, J. J., Sahu, A., Kalra, P. S., and Kalra, S. P. (1993). *Endocrinology*, *133*: 2481.

Bonavera, J. J., Sahu, A., Kalra, P. S., and Kalra, S. P. (1994). *Brain Res.*, *660*: 175.

Bourguignon, J.-P., Gerard, A., and Franchimont, P. (1989a). *Neuroendocrinology*, *49*: 402.

Bourguignon, J.-P., Gerard, A., Mathieu, J., Simons, J., and Franchimont, P. (1989b). *Endocrinology*, *125*: 1090.

Bowery, N. G., Hill, D. R., Hudson, A. L., Doble, A., Middlemiss, D. N., Shaw, J., and Turnbull, M. (1980). *Nature*, *283*: 92.

Brann, D. W. (1995). *Neuroendocrinology*, 61: 213.

Brann, D. W., and Mahesh, V. B. (1991). *Neuroendocrinology*, *53*: 107.

Brann, D. W., and Mahesh, V. B. (1994). *Front. Neuroendocrinol.*, *15*: 3.

Brann, D. W., McDonald, J. K., Putnam, C. D., and Mahesh, V. B. (1991). *Neuroendocrinology*, *54*: 425.

Brann, D. W., Zamorano, P. L., Putnam-Roberts, C. D., and Mahesh, V. B. (1992). *Neuroendocrinology*, *56*: 445.

Brann, D. W., Zamorano, P. L., Chorich, L. P., and Mahesh, V. B. (1993a). *Neuroendocrinology*, *58*: 666.

Brann, D. W., Chorich, L. P., and Mahesh, V. B. (1993b). *Neuroendocrinology*. *58*: 531.

Brann, D. W., Ping, L., and Mahesh, V. B. (1993c). *Mol. Cell. Neurosci.*, *4*: 292.

Bruni, J. F., Van Vugt, D. A., Marshall, S., and Meites, J. (1977). *Life Sci.*, *21*: 461.

Caligaris, L., Astrada, J. J., and Taleisnik, S. (1971a). *Endocrinology*, *89*: 331.

Caligaris, L., Astrada, J. J., and Taleisnik, S. (1971b). *Endocrinology*, *88*: 810.

Caraty, A., Evans, N. P., Fabre-Nys, C. J., and Karsch, F. J. (1995). *J. Reprod. Fert. Suppl.* *49*: 245.

Catzeflis, C., Pierroz, D. D., Rohner-Jeanrenaud, F., Rivier, J. E., Sizonenko, P. C., and Aubert, M. L. (1993). *Endocrinology*, *132*: 224–234.

Change, J. P., McCoy, E. E., Graeter, J. S., Tasaka, K., and Catt, K. J. (1986). *J. Biol. Chem.*, *261*: 9105.

Change, J. P., Stojlkovic, S. S., Graeter, J. S., and Catt, K. J. (1988). *Endocrinology*, *123*: 87.

Chi, L., Zhou, W., Prikhozhan, A., Flanagan, C., Davidson, J. S., Golembo, M., Illing, N., Millar, R. P., and Sealfon, S. C. (1993). *Mol. Cell. Endocrinol.*, *189*: R1.

Ching, M. (1982). *Neuroendocrinology*, *34*: 279.

Ching, M. (1983). *Endocrinology*, *113*: 2209.

Cho, B. N., Seong, J. Y., Cho, H., and Kim, K. (1994). *Brain Res.*, *652*: 177.

Chronwall, B. M., DiMaggio, D. A., Massari, V. J., Pickel, V. M., Ruggiero, D. A., and O'Donohue, T. L. (1985). *Neuroscience, 15*: 1159.

Ciofi, P., Fallon, J. H., Croix, D., Polak, J. M., and Tramu, G. (1991). *Endocrinology, 128*: 823.

Ciofi, P., Crowley, W. R., Pillez, A., Schmued, L. L., Tramu, G., and Mazzuca, M. (1993). *J. Neuroendocrinol., 5*: 399.

Clayton, R. N., and Catt, K. J. (1981). *Endocr. Rev., 2*: 186.

Clayton, R. N., Solano, A. R., Garcia-Vela, A., Dufau, M. L., and Catt, K. J. (1980). *Endocrinology, 107*: 699.

Coen, C. W. and Coombs, M. C. (1983). *Neuroscience, 10*: 187.

Coen, C. W., Simonyi, A., and Fekete, M. I. K. (1985). *Brain Res., 343*: 383.

Coen, C. W., Montagnese, C., and Opacka-Juffrey, J. (1990). J. *Neuroendocrinol., 2*: 107.

Crowley, W. R. (1982). *Neuroendocrinology, 34*: 381.

Crowley, W. R. (1987). *Integrative Neuroendocrinology*: *Molecular, Cellular and Clinical Aspects* (S. M. McCann and R. I. Weiner, eds.). Karger, Basel, p. 54.

Crowley, W. R. (1988). *Peptide Hormones*: *Effects and Mechanisms of Action* (A. Negro-Vilar and P. M. Conn, eds.). CRC Press, Inc., Boca Raton, FL, p. 79.

Crowley,, W. R., and Terry, L. C. (1981). *Brain Res., 204*: 231.

Crowley, W. R., and Kalra, S. P. (1987). *Neuroendocrinology, 46*: 97.

Crowley, W. R., and Armstrong, W. E. (1992). *Endocr. Rev., 13*: 33.

Crowley, W. R., O'Donohue, T. L., Wachslicht, H., and Jacobowitz, D. M. (1978). *Brain Res., 154*: 345.

Crowley, W. R., Terry, L. C., and Johnson, M. D. (1982). *Endocrinology, 110*: 1102.

Crowley, W. R., Parker, S. L., Sahu, A., and Crowley, W. R. (1995). *The Neurobiology of Puberty* (T. M. Plant and P. A. Lee, eds.). Journal of Endocrinology, Ltd, Bristol UK, p. 41.

Dawson, T. M., and Snyder, S. H. (1994). *J. Neurosci., 14*: 5147.

Demling, J., Fuchs, E., Baumert, M., and Wuttke, W. (1985). *Neuroendocrinology, 41*: 212.

DePaolo, L. V. (1988). *J. Endocrinol., 118*: 59.

de Quidt, M. E., and Emson, P. C. (1986). *Neuroscience, 8*: 545.

Dluzen, D. E., and Ramirez, V. D. (1986). *Neuroendocrinology, 43*: 459.

Donoso, A. O., and Banzan, A. M. (1984). *Acta Endocrinol., 106*: 298.

Donoso, A. O., Lopez, F. J., and Negro-Vilar, A. (1990). *Endocrinology, 126*: 414.

Donoso, A. O., Lopez, F. J., and Negro-Vilar, A. (1992). *Endocrinology, 131*: 1559.

Drouin, J., and Labrie, F. (1981). *Endocrinology, 108*: 52.

Drouin, J., Lagace, L., and Labrie, F. (1976). *Endocrinology, 99*: 1477.

Drouva, S. V., and Gallo, R. V. (1976). *Endocrinology, 99*: 651.

Drouva, S. V., and Gallo, R. V. (1977). *Endocrinology, 100*: 792.

Drouva, S. V., Epelbaum, J., Tapia-Arancibia, L., LaPlante, E., and Kordon, C. (1980). *Eur. J. Pharmacol., 61*: 411.

Drouva, S. V., Epelbaum, J., Tapia-Arancibia, L., LaPlante, E., and Kordon, C. (1981). *Neuroendocrinology, 32*: 163.

Everitt, B. J., Hokfelt, T., Terenius, L., Tatemoto, K., Mutt, V., and Goldstein, M. (1984). *Neuroscience, 11*: 443.

Favit, A., Wetsel, W. C., and Negro-Vilar, A. (1993). *Endocrinology, 133*: 1983.

Ferin, M., Tempone, A., Zimmerling, P. E., and Vande Wiele, R. L. (1969). *Endocrinology, 85*: 1070.

Ferin, M., Wehrenberg, W. B., Lam, N. Y., Alston, E. J., and Vande Wiele, R. L. (1982). *Endocrinology, 111*: 1652.

Ferin, M., Van Vugt, D., Wardlaw, S. (1984). *Rec. Progr. Horm. Res., 40*: 441.

Flugge, G., Oertel, W. H., and Wuttke, W. (1986). *Neuroendocrinology, 43*: 1.

Fox, S. R., and Smith, M. S. (1985). *Endocrinology, 116*: 1485.

Fox, S. R., Harlan, R. E., Shivers, B. D., and Pfaff, D. W. (1990). *Neuroendocrinology, 51*: 276.

Frawley, L. S. and Neill, J. D. (1984). *Endocrinology, 114*: 659.

Freeman, M. E. (1988). *The Physiology of Reproduction* (E. Knobil and J. Neill, eds.). Raven Press, Ltd, New York, p. 1893.

Freeman, M. E., Dupke, K. C., and Cruteau, C. M. (1976). *Endocrinology, 99*: 223.

Fuchs, E., Mansky, T., Stock, K.-W., Vijayan, E., and Wuttke, W. (1984). *Neuroendocrinology, 38*: 484.

Gabriel, S. M., and Simpkins, J. W. (1983). *Neuroendocrinology, 37*: 342.

Gabriel, S. M., Simpkins, J. W., and Kalra, S. P. (1986). *Endocrinology, 113*: 1806.

Gallo, R. V. (1980). *Neuroendocrinology, 30*: 122.

Gallo, R. V. (1981a). *Biol. Reprod., 24*: 100.

Gallo, R. V. (1981b). *Biol. Reprod., 24*: 771.

Gallo, R. V. (1982). *Neuroendocrinology, 35*: 380.

Gallo, R. V., and Drouva, S. V. (1979). *Neuroendocrinology, 29*: 149.

Gitler, M. S., and Barraclough, C. A. (1987). *Brain Res., 422*: 1.

Gitler, M. S., and Barraclough, C. A. (1988). *Neuroendocrinology, 48*: 351.

Gnodde, H. P., and Schuiling, G. A. (1976). *J. Endocrinol., 70*: 97.

Goldsmith, P. C., Thind, K. K., Song, T., Kim, E. J., and Boggan, J. E. (1990). *J. Neuroendocrinol, 2*: 157.

Goldsmith, P. C., Thind, K. K., Perera, A. D., and Plant, T. M. (1994). *Endocrinology, 134*: 858.

Goodman, R. L. (1978a). *Endocrinology, 102*: 142.

Goodman, R. L. (1978b). *Endocrinology, 102*: 151.

Goodman, R. L., and Knobil, E. (1981). *Neuroendocrinology, 32*: 57.

Gore, A. C., and Terasawa, E. (1991). *Endocrinology, 129*: 3009.

Gore, A. C., and Roberts, J. L. (1994). *Endocrinology, 134*: 2026.

Gore, A. C., and Roberts, J. L. (1995). *Endocrinology, 136*: 889.

Gore, A. C., Mitsushima, D., and Terasawa, E. (1993). *Neuroendocrinology, 58*: 23.

Grandison, L., Fratta, W., and Guidotti, A. (1980). *Life Sci., 26*: 1633.

Grossman, A. B., Rossmanith, W. G., Kabigtig, E. B., Cadd, G., Clifton, D., and Steiner, R. A. (1994). *J. Endocrinol., 140*: R5.

Grundemar, L., and Hakanson, R. (1994). *Trends Pharmacol. Sci., 15*: 153.

Hakanson, R., Wahlestedt, C., Ekblad, E., Edvinsson, L. and Sundler, F. (1986). *Progr. Brain Res., 68*: 279.

Hales, T. G., Sanderson, M. J., and Charles, A. C. (1994). *Neuroendocrinology, 59*: 297.

Hansen, J. R., McArdle, C. A., and Conn, P. M. (1987). *Mol. Endocrinol., 1*: 808.

Hartman, R. D., He, J.-R., and Barraclough, C. A. (1990). *Endocrinology, 127*: 1336.

Herbison, A. E., and Dyer, R. G. (1991). *Neuroendocrinology, 53*: 317.

Herbison, A. E., and Fenelon, V. S. (1995). *J. Neurosci., 15*: 2328.

Herbison, A. E., and Theodosis, D. T. (1992). *Neuroscience, 50*: 283.

Herbison, A. E., Chapman, C., and Dyer, R. G. (1992). *Exp. Brain Res., 87*: 2.

Hoffman, G. E., Fitzsimmons, M. D., and Watson, R. E. (1989). *Brain Opioid Systems and Reproduction* (R. G. Dyer and R. J. Bicknell, eds.). Oxford University Press, New York, p. 125.

Hokfelt, T., Johansson, O., and Goldstein, M. (1984). *Handbook of Chemical Neuroanatomy*, Vol. 2, Part 1 (A. Bjorklund and T. Hokfelt, eds.). Elsevier Science Publishers, New York, p. 157.

Honma, K., and Wuttke, W. (1980). *Endocrinology, 106*: 1848.

Horvath, T. L., Naftolin, F., Kalra, S. P. and Leranth, C. (1992a). *Endocrinology, 131*: 2461.

Horvath, T. L., Naftolin, F., and Leranth, C. (1992b). *Neuroscience, 51*: 391.

Hsieh, K.-P., and Martin, T. F. J. (1992). *Mol. Endocrinol., 6*: 1673.

Huckle, W. R., and Conn, P. M. (1988). *Endocr. Rev.*, *9*: 387.

Hulse, G. K., and Coleman, G. J. (1983). *Pharmacol. Biochem. Behav.*, *19*: 269.

Jacobowitz, D. M. (1988). *Synapse*, *2*: 186.

Jacobson, W., and Kalra, S. P. (1989). *Endocrinology*, *124*: 199.

Jarry, H., Perschl, A., and Wuttke, W. (1988). *Acta Endocrinol.*, *118*: 573.

Jarry, H., Leonhardt, S., and Wuttke, W. (1991). *Neuroendocrinology*, *53*: 261.

Jarry, H., Hirsch, B., Leonhardt, S., and Wuttke, W. (1992). *Neuroendocrinology*, *56*: 133.

Jarry H., Leonhardt, S., Schwarze, T., and Wuttke, W. (1995). *Neuroendocrinology*, *62*: 479.

Jennes, L., Beckman, W. C., Stumpf, W. E., and Grzanna, R. (1982). *Exp. Brain Res.*, *46*: 331.

Jennes, L., Stumpf, W. E., and Tappaz, M. L. (1983). *Exp. Brain Res.*, *50*: 91.

Jirikowski, G. F., Merchenthaler, I., Rieger, G. E., and Stumpf, W. E. (1986). *Neurosci. Lett.*, *65*: 121.

Kaiser, U. B., Zhao, D., Cardona, G. R. and Chin, W. W. (1992). *Biochem. Biophys. Res. Commun.*, *189*: 1163.

Kaiser, U. B., Jakubiak, A., Steinberger, A., and Chin, W. W. (1993). *Endocrinology*, *133*: 931.

Kalra, S. P., and McCann, S. M. (1973). *Progr. Brain Res.*, *39*: 185.

Kalra, S. P., and McCann, S. M. (1975). *Neuroendocrinology*, *19*: 289.

Kalra, S. P. Kalra, S. P., Krulich, L., Fawcett, C. P., and McCann, S. M. (1972). *Endocrinology*, *90*: 1168.

Kalra, S. P., Bonavera, J. J., Dube, M. G., Crowley, W. R., and Kalra, S. P. (1994). *Soc. Neurosci. Abstr.*, *20*: 1058.

Kalra, S. P. (1981). *Endocrinology*, *109*: 1805.

Kalra, S. P. (1985). *Neuroendocrinology*, *40*: 139.

Kalra, S. P. (1986). *Front. Neuroendocrinol.*, *9*: 31.

Kalra, S. P. (1993). *Endocr. Rev.*, *14*: 507.

Kalra, S. P., and Crowley, W. R. (1982). *Endocrinology*, *111*: 1403.

Kalra, S. P., and Crowley, W. R. (1984). *Life Sci.*, *35*: 1773.

Kalra, S. P., and Crowley, W. R. (1992). *Front. Neuroendocrinol.*, *13*: 1.

Kalra, S. P., and Gallo, R. V. (1983). *Endocrinology*, *113*: 23.

Kalra, S. P. and Kalra, P. S. (1974). *Endocrinology*, *95*: 1711.

Kalra, S. P., and Kalra, P. S. (1979). *Acta Endocrinol.*, *92*: 1.

Kalra, S. P., and Kalra, P. S. (1983). *Endocr. Rev.*, *4*: 311.

Kalra, S. P., and Kalra, P. S. (1989). *Biol. Reprod.*, *41*: 559.

Kalra, S. P., and McCann, S. M. (1974). *Neuroendocrinology*, *15*: 79.

Kalra, S. P., and Simpkins, J. W. (1981). *Endocrinology*, *109*: 776.

Kalra, S. P., Krulich, L., and McCann, S. M. (1973). *Neuroendocrinology*, *12*: 321.

Kalra, S. P., Kalra, P. S., Chen, C. L., and Clemens, J. S. (1978). *Acta Endocrinol.*, *89*: 1.

Kalra, S. P., Fuentes, M., Fournier, A., Parker, S. L., and Crowley, W. R. (1992). *Endocrinology*, *130*: 3323.

Karsch, F. J. (1987). *Ann. Rev. Physiol.*, *49*: 365.

Karsch, F. J., Bittman, E. L., Foster, D. L., Goodman, R. L., Legan, S. J., and Robinson, J. E. (1984). *Rec. Progr. Horm. Res.*, *40*: 185.

Kawakami, M., Yoshioka, N., Konda, J., Arita, J., and Vissesuvan, S. (1978). *Endocrinology*, *102*: 791.

Kawano, H., and Daikoku, S. (1981). *Neuroendocrinology*, *32*: 179.

Ke, F.-C., and Ramirez, V. D. (1987). *Neuroendocrinology*, *45*: 514.

Kerdelhue, B., Parnet, P., Lenoir, V., Schirar, A., Gaudoux, F., Levasseur, M.-C., Palkovits, M., Blacker, C., and Scholler, R. (1988). *J. Steroid Biochem.*, *30*: 1.

Kim, K., Lee, M. B. J., Park, Y., and Cho, W. K. (1989). *Mol. Brain Res.*, *6*: 151.

King, J. C., Tobet, S. A., Snavely, F. L., and Arimura, A. A. (1982). *J. Comp. Neurol.* *209*: 287.

King, J. C., Tai, D. W., Hanna, I. K., Pfeiffer, A., Haas, P., Ronsheim, P. M., Turcotte, J. C., and Blaustein, J. D. (1995). *Neuroendocrinology*, *61*: 265.

Knobil, E. (1980). *Rec. Progr. Horm. Res.*, *36*: 53.

Knuth, U. A., Sikand, G. S., Casanueva. F. F., Havlicek, F., and Friesen, H. G. (1983). *Life Sci.*, *33*: 1443.

Koves, K., Martin, J., Molnar, J., and Halasz, B. (1981). *Neuroendocrinology*, *32*: 82.

Krieg, R. J., and Sawyer, C. H. (1976). *Endocrinology*, *99*: 411.

Lagace, L., Massicotte, J., and Labrie, F. (1980). *Endocrinology*, *106*: 684.

Lamberts, R., Vijayan, E., Graf, M., Mansky, T., and Wuttke, W. (1983). *Exp. Brain Res.*, *52*: 356.

Leadem, C. A., and Kalra, S. P. (1985a). *Endocrinology*, *117*: 684.

Leadem, C. A., and Kalra, S. P. (1985b). *Neuroendocrinology*, *41*: 342.

Leadem, C. A., and Yagcnova, S. V. (1987). *Neuroendocrinology*, *45*: 109.

Leadem, C. A., Crowley, W. R., Simpkins, J. W., and Kalra, S. P. (1985). *Neuroendocrinology*, *40*: 497.

Lee, W.-S., Smith, M. S., and Hoffman, G. E. (1990). *Endocrinology*, *127*: 2604.

Legan, S. J., and Karsch, F. J. (1975). *Endocrinology*, *96*: 57.

Legan, S. J., and Coon, G. A., and Karsch, F. J. (1975). *Endocrinology*, *96*: 50.

Lehman, M. N., and Karsch, F. J. (1993). *Endocrinology*, *133*: 887.

Lehman, M. N., Robinson, J., Karsch, F. J., Silverman, A. J. (1986). *J. Comp. Neurol.* *244*: 19.

Leranth, C., MacLusky, N. J., Sakamoto, H., Shanabrough, M., and Nafolin, F. (1985). *Neuroendocrinology*, *40*: 536.

Leranth, C., MacLusky, N. J., Shanabrough, M., and Naftolin, F. (1988a). *Brain Res.*, *449*: 167.

Leranth, C., MacLusky, N. J., Shanabrough, M., and Naftolin, F. (1988b). *Neuroendocrinology*, *48*: 591.

Leranth, C., MacLusky, N. J., Brown, T. J., Chen, C., Redmond, D. E., and Naftolin, F. (1992). *Neuroendocrinology*, *55*: 667.

Levine, J. E., and Ramirez, V. D. (1980). *Endocrinology*, *107*: 1782.

Levine, J. E., and Ramirez, V. D. (1982). *Endocrinology*, *111*: 1439.

Levine, J. E., Norman, R. L., Gliessman, P. L., Oyama, T. T., Bangsberg, D. R., and Spies, H.G. (1985). *Endocrinology*, *117*: 711.

Liaw, J.-J., and Barraclough, C. A. (1993). *Mol. Brain Res.*, *17*: 112.

Libertun, C., Orias, R., and McCann, S. M. (1974). *Endocrinology*, *94*: 1094.

Liposits, Z., Reid, J. J., Negro-Vilar, A., and Merchenthaler, I. (1995). *Endocrinology*, *136*: 1987.

Lopez, F. J., and Negro-Vilar, A. (1990). *Endocrinology*, *127*: 2431.

Lopez, F. J., Donoso, A. O., and Negro-Vilar, A. (1990). *Endocrinology*, *126*: 1771.

Lopez, F. J., Merchenthaler, I., Ching, M., Wisniewski, M. G., and Negro-Vilar, A. (1991). *Proc. Natl. Acad. Sci. USA*, *88*: 4508.

Lopez, F. J., Donoso, A. O., and Negro-Vilar, A. (1992). *Endocrinology*, *130*: 1986.

Lopez, F. J., Meade, E. H., and Negro-Vilar, A. (1993). *Endocrinology*, *132*: 795.

Luderer, U., Storbl, F., Levine, J. E., Schawartz, N. B. (1993). *Biol. Reprod.*, *48*: 857.

Lundberg, J. M., and Hokfelt, T. (1986). *Progr. Brain Res.*, *68*: 241.

Lundberg, J. M., Terenius, L., Hokfelt, T., and Tatemoto, K. (1984). *J. Neurosci.*, *4*: 2376.

Mahachoklertwattana, P., Sanchez, J., Kaplan, S. L., and Grumback, M. M. (1994a). *Endocrinology*, *134*: 1023.

Mahachoklertwattana, P., Black, S. M., Kaplan, S. L., Bristow, J. D., and Grumbach, M. M. (1994b). *Endocrinology*, *135*: 1709.

Mansky, T., Mestres-Ventura, P. and Wuttke, W. (1982). *Brain Res.*, *231*: 353.

Marks, D. L., Smith, M. S., Vrontakis, M., Clifton, D. K., and Steiner, R. A. (1993). *Endocrinology*, *132*: 1836.

Marks, D. L., Lent, K. L., Rossmanith, W. G., Clifton, D. K., and Steiner, R. A. (1994). *Endocrinology*, *134*: 1991.

Martinez de la Escalera, G., Choi, A. L. H., and Weiner, R. I. (1994). *Neuroendocrinology*, *59*: 420.

Masotto, C., and Negro-Vilar, A. (1987). *Endocrinology*, *121*: 2251.

Masotto, C., Wisniewski, G., and Negro-Vilar, A. (1989). *Endocrinology*, *125*: 548.

Masotto, C., Sahu, A., Dube, M. G., and Kalra, S. P. (1990). *Endocrinology*, *126*: 1990.

McDonald J. K., Koenig, J. I., Gibbs, D. M., Collins, P., and Noe, B. D. (1987). *Neuroendocrinology*, *46*: 538.

McEwen, B. S., Davis, P. G., Parsons, B., Pfaff, D. W. (1979). *Ann. Rev. Neurosci.*, *2*: 65.

Meijs-Roelofs, H. M., Kramer, P., and van Leeuwen, E. C. (1991). *J. Endocrinol.*, *131*: 435.

Meister, B., Ceccatelli, S., Hokfelt, T., Anden, N.-E., Anden, M., and Theodorsson, E. (1989). *Exp. Brain Res.*, *76*: 343.

Melis, G. B., Paoletti, A. M., Gammbiaccini, M., Mais, V., and Fioreti, P. (1984). *Neuroendocrinology*, *39*: 60.

Merchenthaler, I., Gores, T., Setalo, G., Petrusz, P., and Flerko, B. (1984). *Cell Tiss. Res.*, *237*: 15.

Merchenthaler, I., Lopez, F. J., and Negro-Vilar, A. (1990). *Proc. Natl. Acad. Sci. USA*, *87*: 6326.

Merchenthaler, I., Lopez, F. J., Lennard, D. E., and Negro-Vilar, A. (1991). *Endocrinology*, *129*: 1977.

Minami, S., Frautschy, S. A., Plotsky, P. A., Sutton, S. W., and Sarkar, D. K. (1990). *Neuroendocrinology*, *52*: 112.

Mitsushima, D., Hei, D. L., and Terasawa, E. (1994). *Proc. Natl. Acad. Sci. USA*, *91*: 395.

Moore, R. Y., and Card, J. P. (1984). *Handbook of Chemical Neuroanatomy*, Vol. 2, Part 1 (A. Bjorklund and T. Hokfelt, eds.). Elsevier Science Publishers, New York, p. 123.

Moretto, M., Lopez, F. J., and Negro-Vilar, A. (1993). *Endocrinology*, *133*: 2399.

Morgan, R. O., Chang, J. P., and Catt, K. J. (1987). *J. Biol. Chem.*, *262*: 1166.

Morrell, J. I., McGinty, J. F., and Pfaff, D. W. (1985). *Neuroendocrinology*, *41*: 417.

Nakagawa, Y., Shisaka, S., Emson, P. C., and Tohyama, M. (1985). *Brain Res.*, *361*: 52.

Nakai, Y., Plant, T. M., Hess, D. L., Keogh, E. J., and Knobil, E. (1978). *Endocrinology*, *102*: 1008.

Naor, Z. (1990). *Endocr. Rev.*, *11*: 326.

Naor, Z., Ayrad, A. Limor, R., Zahut, H., and Lotan, M. (1986). *J. Biol. Chem.*, *261*: 12506.

Naor, Z., Capponi, A. M., Rossier, M. F., Ayalon, D., and Limor, R. (1988). *Mol. Endocrinol.*, *2*: 512.

Negro-Vilar, A., Vijayan, E., and McCann, S. M. (1980). *Brain Res. Bull.*, *5*: 239.

Neill, J. D., Freeman, M. E., and Tillson, S. A. (1971). *Endocrinology*, *89*: 1448.

Nikolarkis, K. E., Almeida, O. F. X., and Herz, A. (1986). *Brain Res.*, *399*: 152.

Nikolarkis, K. E., Loeffler, J.-Ph., Almeida, O. F. X., and Herz, A. (1988). *Brain Res. Bull.*, *21*: 677.

Ojeda, S. R., and Negro-Vilar, A. (1985). *Endocrinology*, *116*: 1763.

Ojeda, S. R., Negro-Vilar, A., and McCann, S. M. (1979). *Endocrinology*, *104*: 617.

Ojeda, S. R., Urbanski, H. F., Katz, K. H., and Costa, M. E. (1986a). *Neuroendocrinology*, *43*: 259.

Ojeda, S. R., Urbanski, H. F., Katz, K. H., Costa, M. E., and Conn, P. M. (1986b). *Proc. Natl. Acad. Sci.*, *83*: 4932.

Ojeda, S. R., Urbanski, H. F., Junier, M.-P., and Capdevila, J. (1989). *Ann. NY Acad. Sci.*, *559*: 192.

Olney, J. W., Cicero, T. J., Meyer, E. R., and de Gubareff, T. (1976). *Brain Res.*, *112*: 420.

Pang, C. N., Zimmermann, E., and Sawyer, C. H. (1977). *Endocrinology*, *101*: 1726.

Park, O.-K., and Ramirez, V. D. (1989). *Neuroendocrinology*, *50*: 66.

Park, O-K., Gugneja, S., and Mayo, K. E. (1990). *Endocrinology*, *127*: 365.

Parsons, B., Rainbow, T. C., MacLusky, N. J., and McEwen, B. S. (1982). *J. Neurosci.*, *2*: 1446.

Pau, K.-Y. F., Hess, D. L., Kaynard, A. H., Ju, W.-Z., Gliessman, P. L., and Spies, H. G. (1989). *Endocrinology*, *124*: 891.

Pau, K.-Y.F., Gliessman, P. L., Hess, D. L., Ronnekleiv, O. K., Levine, J. E., and Spies, H. G. (1990). *Brain Res.*, *517*: 229.

Pau. K.-Y. F., Kaynard, A. H., Hess, D. L., and Spies, H. G. (1991). *Neuroendocrinology*, *53*: 396.

Pau, K.-Y., Berria, M., Hess, D. L., and Spies, H. G. (1993). *Endocrinology*, *133*: 1650.

Pau, K.-Y.F., Berria, M., Hess, D. L., and Spies, H. G. (1995). *J. Neuroendocrinol.*, *7*: 63.

Petersen, S. L., McCrone, S., Keller, M., and Gardner, E. (1991). *Endocrinology*, *129*: 1679.

Petersen, S. L., Keller, M. L., Carder, S. A., and McCrone, S. (1993a). *J. Neuroendocrinol.*, *5*: 643.

Petersen, S. L., McCrone, S., Coy, D., Adelman, J. P., and Mahan, L. C. (1993b). *Endocr. J.*, *1*: 29.

Petersen, S. L., McCrone, S., Keller, M., and Shores, S. (1995). *Endocrinology*, *136*: 3604.

Petraglia, F., Vale, W., and Rivier, C. (1986). *Endocrinology*, *119*: 2445.

Pfaff, D., and Keiner, M. (1973). *J. Comp. Neurol.*, *151*: 121.

Pfaff, D. W., and Schwanzel-Fukuda, M. (1995). *The Neurobiology of Puberty* (T. M. Plant and P.A. Lee, eds.). Journal of Endocrinology, Ltd, Bristol, U.K., p. 3.

Ping, L., Mahesh, V. B., and Brann, D. W. (1994a). *Endocrinology*, *135*: 113.

Ping, L., Mahesh, V. B., Wiedmeier, V. T., and Brann, D. W. (1994b). *Neuroendocrinology*, *59*: 318.

Plant, T. M. (1986). *Endocr. Rev.*, 7: 75.

Porkka-Heiskanen, T., Urban, J. H., Turek, F. W., and Levine, J. E. (1994). *J. Neurosci.* *14*: 5548.

Price, M., Olney, J. W., and Cicero, T. W. (1978). *Neuroendocrinology*, *26*: 352.

Rainbow, T. C., Parsons, B., MacLusky, N. J., and McEwen, B. S. (1982). *J. Neurosci.*, *2*: 1439.

Rance, N., Wise, P. M., Selmanoff, M. K., and Barraclough, C. A. (1981). *Endocrinology*, *108*: 1795.

Rao, I. M., and Mahesh, V. B. (1986). *Biol. Reprod.*, *35*: 1154.

Rasmussen, D. D., Kennedy, B. P., Ziegler, M. G., and Nett, T. M. (1988). *Endocrinology*, *123*: 2916.

Reichlin, S. (1992). *William's Textbook of Endocrinology*, 8th ed. (J. D. Wilson and D. W. Foster, eds.). W. B. Saunders, Philadelphia. p. 135.

Rettori, V., Gimeno, M., Lyson, K., and McCann, S. M. (1992). *Proc. Natl. Acad. Sci.*, *89*: 11543.

Rettori, V., Belova, N., Dees, W. L., Nyberg, C. L., Gimeno, M. and McCann, S. M. (1993a). *Proc. Natl. Acad. Sci. USA*, *90*: 10130.

Rettori, V., Kamat, A., and McCann,, S. M. (1993b). *Brain Res. Bull.*, *33*: 501.

Reyes, A., Luckhaus, J., and Ferin, M. (1990). *Endocrinology*, 127: 724.

Reyes, A., Xia, L. and Ferin, M. (1991). *Neuroendocrinology*, *54*: 405.

Rotsztejn, W. H., Drouva, S. V., Patou, E., and Kordon, C. (1978). *Eur. J. Pharmacol.*, *50*: 285.

Rubinstein, L., and Sawyer, C. H. (1970). *Endocrinology*, *86*: 988.

Sahu, A., Crowley, W. R., Tatemoto, K., Balasubramanian, A., and Kalra, S. P. (1987). *Peptides*, *8*: 291.

Sahu, A., Kalra, S. P., Crowley, W. R., and Kalra, P. S. (1988). *Brain Res.*, *457*: 376.

Sahu, A., Jacobson, W., Crowley, W. R., and Kalra, S. P. (1989). *J. Neuroendocrinol.*, *1*: 83.

Sahu, A., Crowley, W. R., and Kalra, S. P. (1990). *Endocrinology*, *126*: 876.

Sahu, A., Crowley, W. R., and Kalra, S. P. (1994a). *Endocrinology, 134*: 1018.

Sahu, S., Xu, B., and Kalra, S.P. (1994b). *Endocrinology, 134*: 529.

Sahu, A., Crowley, W. R., and Kalra, S. P. (1995). *J. Neuroendocrinol., 7*: 291.

Sar, M., Sahu, A., Crowley, W. R., and Kalra, S. P. (1990). *Endocrinology, 127*: 2752.

Sarkar, D. K., and Fink, G. (1979). *J. Endocrinol., 80*: 303.

Sarkar, D. K., and Yen, S. S. C. (1985). *Endocrinology, 116*: 2075.

Sarkar, D. K., Chiappa, S. A., Fink, G., and Sherwood, N. M. (1976). *Nature, 264*: 461.

Savoy-Moore, R. T., Schwartz, N. B., Duncan, J. A., and Marshall, J. C. (1980). *Science, 209*: 942.

Scarbrough, K., Jakubowski, M., Levin, N., Wise, P. M., and Roberts, J. L. (1994). *Endocrinology, 134*: 555.

Schenken, R. S., Werlin, L. B., Williams, R. F., Prihoda, T. J., and Hodgen, G. D. (1985). *J. Clin. Endocrinol. Metab., 60*: 886.

Schulz, R. Wilhelm, A., Pirki, K. M., Gramsch, C., and Herz, A. (1981). *Nature, 294*: 757.

Schuman, E. M., and Madison, D. V. (1994). *Annu. Rev. Neurosci., 17*: 153.

Schwanzel-Fukuda, M., and Pfaff, D. W. (1989). *Nature, 338*: 161.

Seeberg, P H., Mason, A. J., Stewart, T. A., and Nikolics, K. (1987). *Rec. Progr. Horm. Res., 43*: 69.

Seong, J.-Y., Lee, Y.-K., Lee, C. C., and Kim, K. (1993). *Neuroendocrinology. 58*: 234.

Sheaves, R., Warburton, E., Laynes, R., and MacKinnon, P. (1984). *Brain Res., 323*: 326.

Sheaves, R., Laynes, R., and MacKinnon, P. C. B. (1985). *Endocrinology, 116*: 542.

Shirley, B., Wolinsky, J., and Schwartz, N. B. (1968). *Endocrinology, 82*: 959.

Shivers, B. D., Harlan, R. E., Hejmancik, J. F., Conn, P. M., and Pfaff, D. W. (1986). *Endocrinology, 118*: 883.

Shivers, B. D., Harlan, R. E., Morrell, J. I., and Pfaff, D. W. (1983). *Nature, 304*: 345.

Silverman, A. J., Krey, L. C., and Zimmerman, E. A. (1979). *Biol. Reprod., 20*: 98.

Silverman, A.-J., Livne, I., and Witkin, J. W. (1988). *The Physiology of Reproduction* (E. Knobil and J. Neill, eds.). Raven Press, Ltd, New York, p. 1683.

Simpkins, J. W. and Kalra, S. P. (1979). *Brain Res. 170*: 475.

Simpkins, J. W., Kalra, P. S., and Kalra, S. P. (1980). *Endocrinology, 107*: 573.

Smith, E. R., and Davidson, J. M. (1974). *Endocrinology, 95*: 1566.

Smith, C. E., Wakefield, I., King, J. A., Naor, Z., Millar, R. P., and Davidson, J. P. (1987). *FEBS Lett., 225*: 247.

Spergel, D. J., Krsmanovic, L. Z., Stojlkovic, S. S., and Catt. K. J. (1994). *Neuroendocrinology, 59*: 309.

Sternberger, L. A., and Hoffman, G. E. (1978). *Neuroendocrinology, 25*: 111.

Stojlkovic, S. S., Stutzin, A., Izumi, S., Dufour, S., Torsello, A., Virmani, M. A., Rojas, E., Catt, K. J. (1990). *New Biol., 2*: 272.

Stumpf, W. E., Sar, M., and Keefer, D. A. (1975). *Anatomical Neuroendocrinology* (W. E. Stumpf and L. D. Grant, eds.), Karger, Basel, p. 104.

Sullivan, K. A., Witkin, J. W., Ferin, M., and Silverman, A.-J. (1995). *Brain Res., 685*: 198.

Sylvester, P. W., Chen, H. T., and Meites, J. (1980). *Proc. Soc. Exp. Biol. Med., 164*: 207.

Tasaka, K., Stojlkovic, S. S., Izumi, S.-I., and Catt, K. J. (1988). *Biochem. Biophys. Res. Commun., 154*: 398.

Taube, H. D., Starke, K., and Borowski, E. (1977). *Naunyn-Schmiedeberg's Arch. Pharmacol., 299*: 123.

Terasawa, E. (1995). *The Neurobiology of Puberty* (T. M. Plant and P. A. Lee, eds.). Journal of Endocrinology, Ltd, Bristol, UK, p. 139.

Terasawa, E., and Kawakami, M. (1974). *Endocrinol. Japon., 21*: 51.

Terasawa, E., Yeoman, R. R., and Schultz, N. J. (1984). *Biol. Reprod., 31*: 732.

Terasawa, E., Krook, C., Eman, S. Watanabe, G., Bridson, W. E., Sholl, S. A., and Hei, D. L. (1987). *Endocrinology, 120*: 2265.

Terasawa, E., Krook, C., Hei, D. L., Gearing, M., Schultz, N. J., and Davis, G. A. (1988). *Endocrinology*, *123*: 18008.

Thind, K. K., and Goldsmith, P. C. (1986). *Brain Res.*, *383*: 215.

Thind, K. K., and Goldsmith, P. C. (1988). *Neuroendocrinology*, *47*: 203.

Thind, K. K., Boggan, J. E., and Goldsmith, P. C. (1993). *Neuroendocrinology*, *57*: 289.

Tsuruo, Y., Kawano, H., Kagotani, Y., Hisano, S., Daikoku, S., Chichara, K., Zhang, T., and Yanihara, N. (1990). *Neurosci. Lett.*, *110*: 261.

Tsutsumi, M., Zhou, W., Millar, R. P., Mellon, P. L., Roberts, J. L., Flanagan, C. A., Dong, K., Gillo, B., and Sealfon, S. C. (1992). *Mol. Endocrinol.*, *6*: 1163.

Unda, R., Brann, D. W., and Mahesh, V. B. (1995). *Neuroendocrinology*, *62*: 562.

Urbanski, H. F., and Ojeda, S. R. (1990). *Endocrinology*, *126*: 1774.

Urbanski, H. F., Fahy, M., Dashel, M., and Mesul, C. (1994). *J. Reprod. Fert.*, *100*: 5.

Van Vugt, D. A., Aylsworth, C. A., Sylvester, P. W., Leung, F. C., and Meites, J. (1981). *Neuroendocrinology*, *33*: 261.

Van Vugt, D. A., Bakst, G., Dyrenfurth, I. and Ferin, M. (1983). *Endocrinology*, *113*: 1858.

Van Vugt, D. A., Lam, N. Y., and Ferin, M. (1984). *Endocrinology*, *115*: 1095.

van Wimersma Greidanus, T. B., and Grossman, A. B. (1991). *Progr. Sens. Physiol. 12*: 2.

Vijayan, E. E., and McCann, S. M. (1978a). *Neuroendocrinology*, *25*: 150.

Vijayan, E. E., and McCann, S. M. (1978b). *Brain Res.*, *155*: 35.

Vilchez-Martinez, J. A., Arimura, A., Debeljuk, L., and Schally, A. V. (1974). *Endocrinology*, *94*: 1300.

Vincent, S. R., Hokfelt, T., and Wu, J.-Y. (1982). *Neuroendocrinology*, *34*: 117.

Wardlaw, S. L., Wehrenberg, W. B., Ferin, M., Antunes, J. L., and Frantz, A. G. (1982). *J. Clin. Endocrinol. Metab.*, *55*: 877.

Watanobe, H., and Takebe, K. (1992). *Neurosci. Lett. 146*: 57.

Watson, R. E., Langub, M. C., and Landis, J. W. (1992). *J. Neuroendocrinol. 4*: 311.

Weesner, G. D., Krey, L. C., and Pfaff, D. W. (1993). *Mol. Brain Res.*, *17*: 77.

Wehrenberg, W. B., Wardlaw, S. L., Frantz, A. G., and Ferin, M. (1982). *Endocrinology*, *111*: 879.

Weiland, N. G., and Wise, P. M. (1990). *Endocrinology*, *126*: 84.

Weiner, R. I., and Ganong, W. F. (1978). *Physiol. Rev.*, *58*: 905.

Weirman, M. E., Bruder, J. M., Jacobsen, B. M., Neeley, C. I., and Kepa, J. K. (1995). *The Neurobiology of Puberty* (T. M. Plant and P. A. Lee, eds.). Journal of Endocrinology, Ltd, Bristol, UK, p. 15.

Weisner, J. B., Koenig, J. I., Krulich, L., and Moss, R. L. (1984). *Life Sci. 34*: 1463.

Wetsel, W. C. (1995). *The Neurobiology of Puberty* (T. M. Plant and P. A. Lee, eds.). Journal of Endocrinology, Ltd, Bristol, UK, p. 25.

Wetsel, W. C., Culler, M. D., Johnston, C. A., and Negro-Vilar, A. (1988). *Mol. Endocrinol. 2*: 22.

Wetsel, W. C., Valenca, M. M., Merchenthaler, I., Liposits, Z., Lopez, F. J., Weiner, R. I., and Negro-Vilar, A. (1992). *Proc. Natl. Acad. Sci. USA.*, *91*: 1423.

Wildt, L., Hausler, A., Hutchinson, J. S., Marshall, G., and Knobil, E. (1981). *Endocrinology*, *108*: 2011.

Wilkes, M. M., and Yen, S. S. C. (1981). *Life Sci.*, *28*: 2355.

Wilson, R. C., and Knobil, E. (1982). *Brain Res.*, *248*: 177.

Wilson, R. C., Kesner, J. S., Kaufman, J.-M., Uemura, T., Akema, T., and Knobil, E. (1984). *Neuroendocrinology*, *39*: 256.

Wise, P. M., Camp-Grossman, P. and Barraclough, C. A. (1981a). *Biol. Reprod.*, *24*: 820.

Wise, P. M., Rance, N., Selmanoff, M. K., and Barraclough, C. A. (1981b). *Endocrinology*, *108*: 2179.

Wise, P. M., Rance, N., and Barraclough, C.A. (1981c). *Endocrinology*, *108*: 2186.

Wise, P. M., Scarbrough, K., Weiland, N. G., and Larson, G. H. (1990). *Mol. Endocrinol. 4*: 886.

Woller, M. J., and Terasawa, E. (1991). *Endocrinology, 128*: 1144.

Woller, M. J., and Terasawa, E. (1992). *Neuroendocrinology, 56*: 921.

Woller, M. J., and Terasawa, E. (1994). *Endocrinology, 135*: 1679.

Woller, M. J., McDonald, J. K., Reboussin, D. M., and Terasawa, E. (1992). *Endocrinology, 130*: 2333.

Woller, M. J., Campbell, G. T., Liu, L., Steigerwalt, R. W., and Blake, C. A. (1993). *Endocrinology, 133*: 2675.

Wray, S. W., and Hoffman, G. E. (1986). *Brain Res., 399*: 327.

Wray, S., Grant, P., Gainer, H. (1989). *Proc. Natl. Acad. Sci. USA, 86*: 8132.

Xia, L., Van Vugt, D., Alston, E. J., Luckhaus, J., and Ferin, M. (1992). *Endocrinology, 131*: 2812.

Xu, B., Sahu, A., Crowley, W. R., Leranth, C., Horvath, T., and Kalra, S. P. (1993). *Endocrinology, 133*: 7474.

Yasin, M., Dalkin, A. C., Haisenleder, D. J., Kerrigan, J. R., and Marshall, J. C. (1995). *Endocrinology, 136*: 1559.

Zoeller, R. T., and Young, W. S. (1988). *Endocrinology, 123*: 1688.

9

The Hypothalamus and Pituitary as Targets for Reproductive Toxicants

Ralph L. Cooper and Jerome M. Goldman
U.S. Environmental Protection Agency, Research Triangle Park, North Carolina

Lee Tyrey
Duke University Medical Center, Durham, North Carolina

INTRODUCTION

In contrast to the numerous studies identifying gonadal cells as primary target sites for toxicant effects in adult animals, there are relatively few studies examining the extent to which exposure to environmental agents may modify reproductive function through alterations of the hypothalamus and pituitary. In this chapter, we review the evidence that the central nervous system (CNS) of the developing animal represents a sensitive target site for reproductive toxicants and that exposure during this period of life can produce irreversible effects on the neuroendocrine system. We also explore the extent to which toxicant exposure in adulthood can modify CNS control of gonadal function and address the influence of gender on the impact of such CNS toxicants on reproductive function.

EFFECT OF ENVIRONMENTAL TOXICANTS ON SEXUAL DIFFERENTIATION OF THE BRAIN

The developing nervous system appears particularly sensitive to endocrine influences. Brief exposure to certain hormones can lead to developmental effects that will irreversibly alter the early organization of the hypothalamic-pituitary-gonadal axis and modify normal neuroendocrine control during adulthood. For example, several studies have shown that early exposure to certain drugs and environmental chemicals will alter the

The research described in this article has been reviewed by the National Health and Environmental Effects Research Laboratory, U.S. Environmental Protection Agency, and approved for publication. Approval does not signify that the contents necessarily reflect the views and policies of the Agency, nor does mention of trade names or commercial products constitute endorsement or recommendation for use.

processes of sexual differentiation of the brain. Prior to differentiation, the brain is inherently female or at least bipotential (Gorski, 1986), meaning that functional and structural sex differences in the CNS are not due directly to disparities in neuronal genomic expression, but rather are imposed by the gonadal steroid environment during development. Toward the end of pregnancy, the fetal testes become active for a brief period and release appreciable amounts of testosterone, which in turn leads to masculinization of the brain. Sexual differentiation of the CNS can be modified by experimental treatments administered shortly before or shortly after birth, unlike gonadal and reproductive tract differentiation, which can be modified from early gestation onward.

In the CNS, testosterone is metabolized to both estradiol and dihydrotestosterone (DHT). In the rat, mouse, and hamster, aromatization of testosterone to estradiol is responsible for CNS sex differentiation, whereas in certain other mammals (e.g., rhesus monkey) the androgenic (DHT) pathway appears to be a critical factor (McEwen, 1980). If one administers exogenous steroids, such as testosterone propionate, to the genotypic female rodent within the first week of postnatal life, her neuroendocrine system will differentiate phenotypically male (i.e., her brain is masculinized). Such a masculinization of the female brain by the aromatization of testosterone to estrogen explains similar masculinizing effects paradoxically observed with low doses of estrogen or diethylstilbestrol (DES), treatments that are without effect on the genotypic male's brain. In adulthood, these "masculinized" females (1) do not ovulate, (2) have polyfollicular ovaries, (3) display persistent vaginal estrus, (4) do not show positive feedback to gonadal hormones [i.e., an ovulatory surge of luteinizing hormone (LH) cannot be stimulated], and (5) exhibit sexual behavior more typical of that observed in the genetic male. In contrast, different effects are seen following castration in early postnatal life. Removal of the ovaries from the neonate is without major effect on neuroendocrine status in the female. However, if the testes are removed prior to the third postnatal day of life, male rats exhibit neuroendocrine characteristics in adulthood that are typical of the female, including both the ability to release a surge of LH and the elicitation of feminine levels of lordosis behavior. The timing of the important developmental endocrine events responsible for sexual differentiation of the human brain remains poorly defined, but the critical events appear to occur earlier in pregnancy.

A number of xenoestrogens, including Kepone (chlordecone) (Gellert, 1978), methoxychlor (Gray et al., 1989), and zearalenone (Kumagai and Shimizu, 1982), have been shown to "masculinize" female rats. Investigations in the neonatal rat also indicate that analogs of DDT other than methoxychlor, i.e., 1-(o-chlorophenyl)-10(p-chlorophenyl)-2,2,2-trichloroethane (o,p'-DDT), may also have estrogenic activity at the neuroendocrine level. Heinrichs et al. (1971) found that rats given o,p'-DDT as neonates exhibited advanced puberty (vaginal opening), persistent vaginal estrus after a period of normal cycling, follicular cysts, and a reduction in the number of corpora lutea (anovulation) (Bulger and Kupfer, 1985). In contrast, proported antiestrogens, such as tamoxifen (Dohler et al., 1984), demasculinize the male's brain, including an effect on the sexually dimorphic nucleus of the preoptic area (SDN-POA), so that its size resembles that observed in the female. Moreover, Faber and Hughes (1993) reported that this effect on the SDN-POA occurred in response to treatment with the phytoestrogen genistein in a dose-related fashion.

In the male rat, perinatal treatment with aromatase inhibitors such as fenarimol has been hypothesized to induce infertility by inhibiting normal masculinization of the brain (Hirsch et al., 1986). The antiandrogen vinclozolin, which acts as an androgen (DHT) receptor blocker, can disrupt normal reproductive tract development in the male rat (Gray et al., 1994). However, vinclozolin has no effect on the aromatization of testosterone to estrogen and thus does not alter male sexual behavior when the animals are evaluated in adulthood. Interestingly, another DDT metabolite, p,p'-DDE, has been found to bind to the androgen receptor and interfere with normal testosterone-mediated cellular changes. Thus, this "antiandrogenic" metabolite may alter sexual development of peripheral reproductive tissues in a manner similar to vinclozolin (Kelce et al, 1995). Although the hormonal influence on sexual differentiation of the CNS may vary somewhat among different species, some role for gonadal hormone modulation of CNS development is indicated for most animals studied.

Interestingly, exposure to agents other than those with steroidogenic activity has also been shown to influence sexual differentiation of the brain, although these have been studied less systematically. Serotonergic and catecholaminergic agents, when given alone or in combination with testosterone or estrogen, have been reported to influence the volume of the SDN-POA (see Dohler, 1991, for review). The masculinizing effects of androgens on the female brain can be partially blocked by neuroactive drugs such as reserpine and chlorpromazine, while pentobarbital and phenobarbital provide more complete protection against the influence of testosterone (Arai and Gorski, 1968). The mechanisms by which such interactions occur remain to be elucidated, but these observations suggest that other mechanisms involved in sexual differentiation of the CNS may render this process susceptible to disruption by environmental compounds that do not necessarily possess steroidogenic activity.

Serotonin is known to exert trophic effects within the CNS during development (Lauder, 1983). Interestingly, both stimulation and inhibition of serotonin synthesis have profound effects on development and differentiation of the sexually dimorphic nucleus (Dohler, 1991). Dohler and colleagues also reported that NE receptor activation augmented the masculinizing effects of testosterone in male and female neonates. At the very least, these data imply that serotonergic and adrenergic input may play an important role during differentiation of the SDN-POA.

Certain aspects of neuroendocrine function appear to be modified via nonsteroidogenic mechanisms after birth. For example, it appears that early postnatal exposure to prolactin (PRL) may be important for normal hypothalamic regulation of prolactin at puberty and in the adult. Prolactin is present in the milk of lactating rats and, when ingested by the pups, can pass through the gut and enter the systemic circulation. Shyr et al. (1986) found that suppression of PRL levels in the milk by administration of a dopamine receptor agonist (bromocriptine) on postnatal days 2–5 (but not days 9–12) resulted in very low PRL levels in the blood of neonates during treatment. Subsequently, when measured between days 30 and 35, the concentration and turnover of dopamine (DA) within the tuberoinfundibular tract of the pups was depressed, thereby elevating serum PRL. Importantly, regular postpubertal ovarian cycles were absent in the hyperprolactinemic females until day 60. That laboratory also found that, when tested in vitro, the lactotrophs from the pituitaries of hyperprolactinemic animals were unresponsive to DA regulation and tended to secrete more PRL in response to thyrotropin-releasing hormone (TRH) stimulation (Shah et al., 1988). That this effect was dependent on PRL exposure was supported by the observation that these alterations could be reversed if the mother was treated with PRL at the time bromocriptine was administered. Experimentally induced hyperprolactinemia in the weanling rat can advance the onset of puberty. Advis and Ojeda (1978) showed that induction of hyperprolactinemia by exposing pups to a dopamine receptor blocker beginning on postnatal day 22 will significantly advance puberty, as indicated by early vaginal opening. Moreover, after puberty there were disturbances noted in the ovarian cycles of these animals. These results raise the possibility that one function of milk-derived PRL in the neonate involves maturation of the inhibitory dopamine control over PRL secretion and that disruption of this process may have long-lasting consequences.

The tuberoinfundibular dopaminergic pathway is known to be involved in the regulation of PRL secretion under a variety of conditions. Since PRL plays a key role in the initiation and maintenance of pregnancy in the rat, it is not surprising that inhibition of PRL release following treatment with dopamine agonists (i.e., bromocriptine) can disrupt pregnancy (Cummings et al., 1991). Although, there are several pharmaceutical agents that mimic endogenous dopamine, we know of no such environmental agents with similar effects. However, it is possible that the loss of pregnancy following carbon tetrachloride exposure (Narotsky and Kavlock, 1995) may reflect a disruption of CNS regulation of the pituitary and that this compromises the hormonal support of pregnancy.

Exposing the weanling rat to certain pesticides has been shown to delay the onset of puberty in the rat. For example, we found that lindane (γ-hexachlorohexane) caused a dose-dependent delay in vaginal opening when treatment was initiated on day 21 (Cooper et al., 1989). Lindane is a well-known neurotoxicant that will induce seizure activity as a result of its effect on the chloride ion channels associated with the GABAa receptor. Lindane inhibits the influx of the chloride ion into the cell and thus leads to depolarization. Initially, it appeared that these effects were a result of an "antiestrogenic" action of lindane because in related experiments with prepubertal rats this compound inhibited the estrogen-induced increase in uterine weight. A more recent series of binding studies showed that lindane did not bind to the estrogen receptor in vitro, nor did it alter the induction of progesterone receptors in response to estrogen stimulation (Laws et al., 1994). This lack of effect on the estrogen receptor suggests that lindane's influence on puberty in the female rat is more likely related to its effect on GABAa receptor function and that interference with the action of the chloride ion channel may influence hypothalamic-pituitary activity.

Delayed sexual maturation in female rats following prepubertal exposure to delta-9-tetrahydrocannabinol (delta-9-THC), the primary psychoactive agent in marijuana, illustrates reproductive toxicity that may result from naturally-occurring compounds (Field and Tyrey, 1984; Wenger et al., 1988). Unlike many such compounds, THC is not estrogenic (Okey and Truant, 1975). Postweaning rats subjected to continuing twice-daily treatment with THC exhibited a delay of 4–5 days in the onset of vaginal cornification and first ovulation (Field and Tyrey, 1984, 1990). The apparent delay in prepubertal follicular development could not be attributed to general toxic or nutritional effects, since the rates of body weight gain over the course of treatment were equivalent for treated and control animals. It is noteworthy that the delay was critically linked to the timing of treatment, insofar as treatment initiated at 27 days of age was effective, but that initiated at 24 or 30 days of age was not (Field and Tyrey, 1990).

The existence of a period of particular vulnerability to the delaying effects of THC emphasizes the crucial importance of the timing of exposure in detecting developmental toxicities, even at the relatively advanced age of puberty. Moreover, the fact that the earliest treatment was ineffective in delaying puberty, despite its continuation through the time of the later effective treatment, clearly points to the need to consider the possibility of tolerance or refractoriness to the effects of toxic agents with continued exposure. In the case of THC, only a few days of treatment will induce tolerance in pubertal rats (Field and Tyrey, 1986). Thus, if tolerance is induced, chronic treatment might fail to unmask developmental effects that depend on toxicant action during limited periods of vulnerability.

The mechanism through which THC retards sexual maturation is unknown. Reductions in serum LH and PRL levels during the initial phase of THC treatment were observed, but neither event was definitively linked to the delayed onset of first estrus (Field and Tyrey, 1990). Nonetheless, those hormones are known factors in pubertal development, and their suppression has been associated with delayed puberty under other circumstances (Advis et al., 1981; Bronson, 1986). If the THC-induced delay results from an inhibition of LH or PRL release, its action is almost certainly mediated by altered hypothalamic regulation of pituitary function. THC neither interferes with the LH-releasing action of GnRH nor directly inhibits PRL release from pituitary lactotrophs (Ayalon et al., 1977; Tyrey, 1978; Hughes et al., 1981).

In summary, sexual differentiation of the CNS is affected by a variety of environmental compounds, and although much of the research has focused on those compounds reported to have steroidogenic activity, it may be premature to assume that other nonsteroidal compounds are without effect on sexual differentiation of the brain.

EFFECT OF ENVIRONMENTAL TOXICANTS ON HYPOTHALAMIC-PITUITARY CONTROL OF GONADAL FUNCTION IN THE ADULT

The synthesis and release of pituitary hormones are under the feedback control of steroidal and other hormonal agents circulating in the blood, as well as of the actions of releasing and inhibiting factors manufactured within specialized neurons located in the hypothalamus. The releasing hormones, in turn, are regulated by several types of feedback signals and multiple neurochemicals, which include the classical neurotransmitters (acetylcholine, catecholamines, serotonin, etc.) and several neuropeptides [opioids, galanin, neuropeptide-Y, etc. (Kalra and Kalra, 1983)]. As a result, many diverse pharmaceutical agents can modify pituitary hormone secretion. This may be brought about by direct action on the pituitary by synthetic steroids (e.g., DES-induced increase in PRL synthesis), by agents that act on pituitary membrane receptors (e.g., bromocriptine inhibition of PRL release), or by compounds that affect neurotransmitter synthesis or neuropeptide regulation of hypothalamic-releasing factors. The effects of various therapeutic agents on reproductive function are well established. These drugs may either depress CNS activity (i.e., anesthetics, analgesics, and tranquilizers) or stimulate it (i.e., antidepressants and hallucinogens). In fact, a variety of such agents are often used to probe the central control of neuroendocrine function. Drugs of abuse also have been shown to alter the hormonal control of reproduction through a CNS mechanism. For example, delta-9-THC significantly reduces LH, follicle-stimulating hormone (FSH), PRL, and testosterone concentrations in the blood and causes decrements in sexual organ weights (Dalterio et al., 1978). In the female rat, delta-9-THC suppresses serum gonadotropin secretion, disrupts estrous cyclicity, and delays sexual development (Tyrey; 1978; Chakravarty et al., 1979; Field and Tyrey, 1984). There is general consensus that the influence of delta-9-THC on reproductive function is mediated through changes in the hypothalamic control of pituitary function. Similarly, opiates also appear to exert their primary effect on the hypothalamic-pituitary axis. Such changes in central regulation of the neuroendocrine axis result in dysfunction of the gonads and sex accessory organs in both humans and animals.

A number of recent studies have examined the effect of environmental toxicant exposure on the regulation of the ovulatory surge of LH in the rat. The timing of this endocrine event is critical for normal fertilization and pregnancy. Although there are differences in the ovarian cycle length, considerable homology exists between the rat and human in the CNS-pituitary mechanisms controlling LH secretion. The generation of the LH surge is under control of the pulsatile release of hypothalamic gonadotropin-releasing hormone (GnRH). This releasing factor is in turn regulated by hypothalamic neurotransmitters (especially norepinephrine) and opioid peptides (enkephalins).

In the rat, the LH surge occurs during the afternoon of vaginal proestrus in response to changes in the neuronal regulation of pulsatile GnRH release. These changes occur during the so-called sensitive or critical period, which occurs roughly between 1400 and 1600 hours on proestrus (see Everett, 1989, for review). Early studies showed that if administered just before this critical period, general anesthetics [e.g., sodium pentobarbital and phenobarbital (Everett and Sawyer, 1950)], agents that disrupt the synthesis of norepinephrine [e.g., fusaric acid, α-methyl-para-tyrosine (Kalra and Kalra, 1983)], or agents that interfere with α-noradrenergic receptor stimulation [e.g., phenoxybenzamine and phentolamine (Ratner, 1971; Plant et al., 1978; Weick 1978)] will disrupt the pattern of GnRH secretion and consequently impair the LH surge. Similarly, morphine exerts an inhibitory effect on LH secretion in male and female of several species (Cooper et al., 1986, for review). On the other hand, injections too far outside this time period were shown to have limited effectiveness (Everett, 1989). Importantly, if the surge is blocked by drugs administered before the critical period on proestrus, it will occur approximately the same time on the following day, resulting in a corresponding 24-hour delay in ovulation (Everett and Sawyer, 1950).

The formamidine pesticides chlordimeform (CDF) and amitraz (AMI) have been reported to block norepinephrine binding to the α_2 receptor (Costa et al., 1988). We found that a single exposure to chlordimeform administered immediately prior to the critical period inhibited the ovulatory surge of LH (Goldman et al., 1991). The site of action was most likely central in nature since systemically administered CDF does reach the brain (Johnson and Knowles, 1983; Costa et al., 1988) and has been found to inhibit the stimulated release of GnRH in vitro without having a parallel direct effect on pituitary LH secretion (Goldman et al., 1990). As mentioned above, if the ovulatory surge of LH is blocked in intact rats on the afternoon of vaginal proestrus, it will typically occur the following day. That is, the CNS mechanisms involved in the control of this circadian event become functional again 24 hours later (and are still influenced by the light-dark cycle). This is true for the formamidine pesticides, in that a single administration of amitraz would induce a one-day delay in the afternoon LH surge (Goldman and Cooper, 1993), thus causing a corresponding one-day delay in ovulation. When ovulation did occur, no effects were seen on the number of ova shed.

The technique of delaying ovulation in the female rat with anesthetics has been used extensively to identify time-dependent alterations in follicular function and oocyte viability in the rat (Peluso and Butcher, 1974; Braw and Tsafriri, 1980; Uilenbroek et al., 1980) and hamster (Terranova, 1980). However, relatively few studies have evaluated the effect of ovulatory delays on pregnancy outcome. Butcher and colleagues (Fugo and Butcher, 1966; Butcher and Fugo, 1967; Butcher et al., 1969) did report that delaying the LH surge for 2 days with sodium pentobarbital could lead to decreased fertility and an increase in the number of abnormal fetuses. However, the possible effects on pregnancy outcome caused by CNS-toxicant–induced ovulatory delays has received virtually no attention in reproductive toxicology. In a series of recent studies, we evaluated whether or not nongenotoxic environmental contaminants, such as chlordimeform, could also delay ovulation by blocking the LH surge and, if so, whether this delay would have an adverse impact on the ensuing pregnancy. The results showed that a CDF-induced delay did not prevent the females from mating on the evening of the delayed surge (i.e., 24 hours later than normal) (Cooper et al., 1994b), allowing a substantial percentage of the CDF-delayed females to become pregnant. The assessment of ovaries after the first week of pregnancy revealed a normal number of corpora lutea (CL). However, when dams were evaluated on gestation day 20, there was a substantial reduction in litter size of the treated animals. Whether this reduction was attributable to the presence of luteinized unruptured follicles, impaired fertilization, preimplantation loss, or early resorptions remains unresolved. However, in a subsequent study described below, we were able to focus more specifically on such issues using ovulatory delays in response to the fungicide thiram.

Thiram (tetramethylthiuram disulfide) is a dithiocarbamate known to inhibit dopamine β-hydroxylase activity and thus decrease hypothalamic norepinephrine concentrations (Stoker et al., 1995) by blocking its conversion from dopamine. We found that exposure to thiram could block ovulation at doses as low as 12 mg/kg if injected immediately before the critical period for the neural trigger of the LH surge (Stoker, et al., 1993). A single dose of thiram administered on the afternoon of vaginal proestrus delayed the LH surge and ovulation without affecting the number of ova shed when ovulation occurred 24 hours later. Importantly, we also found that the delay in ovulation did not alter the number of embryos implanting when the pregnant dams were examined on gestation day 8. By gestation day 11, the number of implantation sites was still unaffected, although the overall development of the embryos was impaired (Stoker et al.,1996). As was seen following a CDF-induced delay in ovulation, the thiram-induced delay led to a significant reduction in the number of fetuses present on gestation day 20. Thus, a very brief exposure to a toxicant affecting the noradrenergic mechanisms involved in pituitary LH release can delay ovulation and as a result cause a decrease in the implantation rate and a decrease in the number of live fetuses present late in gestation.

Since a number of neurotransmitters and neuropeptides are known to regulate the ovulatory surge of LH (Everett and Tyrey, 1982), it is not surprising that environmental compounds targeting different hypothalamic mechanisms can interfere with generation of the surge, delay ovulation, and exert adverse effects on fertility (see Table 1). For example, sodium valproate (Cooper et al., 1994a) interferes with GABA receptor–mediated mechanisms (Jones, 1991). The significance of the neuroendocrine-disrupting effects of valproic acid is important to the reproductive toxicology of environmental compounds, since the structurally related industrial byproduct, hexanoic acid, has been shown to be teratogenic in rodents. By virtue of its GABA-like activity, it may also have effects on the control of LH that are similar to valproic acid. Thus, a toxicant-induced delay in the LH surge may result from a disruption of different neurotransmitter and neuropeptide systems. But, at the same time, a common effect of the delayed LH surge on pregnancy outcome is that litter size is reduced, regardless of the agent used to disrupt the surge.

Because steroid hormones have a significant role in the regulation of anterior pituitary function, it is not surprising that environmental estrogens may modify this normal feedback influence on the hypothalamus and pituitary. In the male, many of the adverse effects of exposure to xenoestrogens on testicular function have been attributed to a direct action on the testes (e.g., Cooper et al., 1986, for review). However, adverse effects of estrogens on male reproduction could conceivably be mediated by an influence on the hypothalamus and pituitary, tissues rich in estrogen receptors (Pfaff and Keiner, 1973). Furthermore, changes in pituitary hormone secretion have been noted sooner and at lower doses of DES than those required to alter any testicular measures (Cooper et al., 1985a,b; Rehnberg et al., 1986). Doses of methoxychlor that had no detectable effects on testicular function or reproductive behavior in the male rat (i.e., 25 and 50 mg/kg/day) elevated pituitary prolactin levels (Goldman et al., 1986). However, the extent to which such modifications in the endocrine control of testicular function contribute to altered testicular physiology or fertility remains to be determined.

The effect of chlordecone on the neuroendocrine control of the female reproductive system has been studied in detail by Hong and coworkers (1985) and Uphouse (1985). Chlordecone can lead to a persistent estrus in adult females and, as indicated above, masculinize the brain if administered during early development. However, this compound can alter CNS activity. Hong et al. (1985) argued that the changes in hypothalamic and pituitary neuropeptide concentrations were similar to those induced by

Table 1 Effect of Delayed Ovulation on Pregnancy and Litter Size

	% Pregnant	Number pups	Number CL
Control	100	13.3	15.0
Pentobarbital	63*	9.3*	14.1
Chlordimeform	77*	8.3*	15.0
Thiram	42*	8.5*	15.1
Methanol	30*	6.3*	12.5
Valproic Acid	43*	8.0*	12.3

Four-day-cyclic females were given a single injection of the compound on the afternoon of vaginal proestrus. Each treatment was known to block the LH surge and delay ovulation for 24 hours. On the following day, the successfully delayed females were still sexually receptive. The percent pregnant is based on the number of females that were either sexually receptive or had a sperm-positive smear after mating. Pup and CL (means ± SEM) are from confirmed pregnant females only. Pup and CL measures were obtained on gestation day 20. *$p < 0.05$ as compared to controls.

estrogen, suggesting that the estrogenic action of this peptide is the mechanism responsible for the central effects. Likewise, Uphouse argued that since chlordecone can decrease serum LH, increase serum PRL, and induce vaginal cornification in the ovariectomized female, it is reasonable to assume that this compound disrupts reproductive function because it is a weak estrogen. However, a precautionary note must be added here. As is often the case, compounds that have been identified as possessing a steroidlike action can also cause CNS changes unrelated to such activity. Chlordecone, for example, can induce tremors (Uphouse and Eckols, 1986) at doses employed for their estrogenic activity (Hong et al., 1985). Moreover, the rapid onset of various alterations, such as a reduction in sexual behavior (Brown et al., 1991), make it unlikely that steroidal mechanisms underlie their observed behavioral changes. Alternative possibilities could involve chlordecone-induced effects on CNS neurotransmitter action (Gandolfi et al., 1984; Chen et al., 1985; Williams et al., 1988), which by themselves may be influenced by alterations in calcium transport (Hoskins and Ho, 1982).

As with compounds that affect the CNS control of LH secretion by interfering with neurotransmitter or neuropeptide regulation of GnRH, steroidogenic compounds can influence LH release. This involvement of endogenous steroids in regulation of the LH surge represents another CNS-pituitary mechanism through which environmental compounds with steroidogenic activity may alter reproductive outcome in the female. Everett (1948) showed that by experimentally increasing serum estradiol or progesterone concentrations during periods of the estrous cycle when they would normally be low could alter the timing of ovulation. Injection of estradiol at 1200 hours on diestrus 1 advanced the LH surge and ovulation by 24 hours, although fewer ova were released (Everett, 1989). The presumption that this advancement in the timing of the surge involves activation of the CNS mechanisms regulating LH secretion was confirmed by Krey and Everett (1973), who demonstrated that the advanced surge could be blocked by nembutal. In contrast, an injection of progesterone on diestrus 1 can delay the LH surge and ovulation (i.e., extend the cycle of the 4-day rat to 5 days) (Everett, 1948).

Ying and Greep (1972) reported that advancing the surge by 24 hours did not alter the fertilizability of the eggs released. Nonetheless, they reported that these eggs failed to implant and that females in whom the surge was advanced by estradiol mated, ovulated, and became pseudopregnant. This conclusion was based on an examination of the vaginal smears after mating (predominantly leukocytic) and the absence of implantations sites on gestation days 9 and 20. Recently, we evaluated the effect of an estradiol-induced advance of ovulation and found minimal evidence that any of the females so treated were pregnant when examined on gestation day 20. However, when the uterine contents of the females were examined on gestation day 8, we found a number of implantation sites. Thus, in our study, the eggs from the female in whom the surge was advanced did fertilize and implantation did occur (Fig. 2). Because of a number of procedural differences, we cannot speculate as to why Ying and Greep (1972) failed to observe implantation sites in their study of the estrogen-advanced female, but these data demonstrate that the eggs from these females can fertilize and implant. But, as is the case with delayed ovulation, the viability of the embryos from the advanced female is markedly impaired.

We also evaluated the effect of delaying the LH surge for 24 hours with injections of progesterone on diestrus 1. In contrast to the effects seen following treatment on vaginal proestrus (when the antral follicles contains ripened oocytes), this progest-

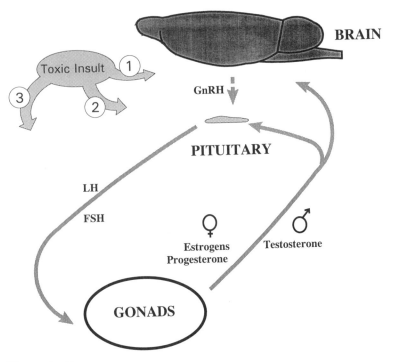

Figure 1   The neuroendocrine control of gonadal function presents a number of target sites for toxicant insult. Various compounds may act directly on the (1) brain, (2) pituitary, or (3) gonads, thereby impairing hormonal communication among these structures. Alternatively, a compound may also affect any of these sites in combination.

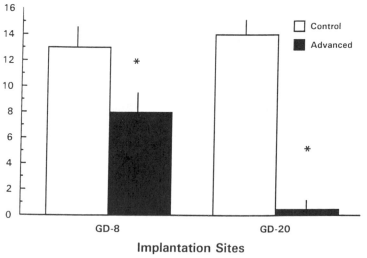

Figure 2   Number of implantation sites present on gestation days 8 and 20 observed in females in whom the LH surge was advanced by a single injection of estradiol on diestrus 1. Although fewer than control, there were an number of implantation sites present on gestation day 8. On gestation day 20 there was no indication of any implantation sites in 19 of 21 females. In two of the gestation day 20 females a single resorption was observed. It should be noted that the number of ova present in the advanced female was also reduced.

erone-induced delay (administered when the follicle and oocyte are still maturing) was without effect on the ensuing pregnancy or litter size. These data again stress the selective vulnerability of the female's reproductive success based on the timing of the CNS insult. Although we have not used this protocol for advancing ovulation with any of the xenoestrogens or antiestrogens discussed above, it is plausible that they could mimic the effects seen with estradiol and progesterone, if the timing and dose of the single injection were selected properly.

Other compounds that have been reported to interfere with reproductive cycling in the adult may exert their effects directly on the hypothalamus and pituitary. For example, rats fed the chlorotriazine herbicide atrazine will show an increase in the rate of pseudopregnancy (characterized by prolonged periods of diestrus, and large corpora lutea present in the ovaries, and elevated serum progesterone) (Cooper et al., 1995). At high doses, however, this compound will lead to an anestrous condition accompanied by unstimulated ovaries, small uteri, and low to undetectable concentrations of serum estrogen and progesterone. Interestingly, the animals fed the higher doses of atrazine also have very low serum LH, FSH, and prolactin concentrations, indicating that atrazine either blocked the compensatory rise in gonadotropin secretion after a severe reduction in ovarian hormone levels or that atrazine had a direct CNS action that reduced pituitary hormone secretion and subsequently ovarian function. Using a very brief exposure to atrazine, we found that this chlorotriazine can block the secretion of both LH and FSH, most likely at the hypothalamic level (Cooper et al., 1995). However, the mechanism of this presumed CNS site of action is not currently known.

The above studies demonstrate that environmental toxicants can disrupt ovarian cycling, ovulation, and pregnancy outcome by altering the hypothalamic-pituitary control or reproduction. Although such clear effects can be seen in the female, similar studies in the adult male indicate that few, if any, environmental compounds will lead to infertility through a disruption of hypothalamic and pituitary control of the gonads. In the male, most, if not all, reproductive toxicants studied to date appear to disrupt fertility by direct testicular or epididymal alterations. For example, in contrast to the marked effect of a single exposure to CDF on the female's reproductive capacity, we found that single or multiple doses of CDF in the male had minimal effects on gonadal function. Following two injections of CDF, serum LH and testosterone were significantly lower than in vehicle-treated controls (Goldman et al., 1990). However, as shown in Figure 3, after 4 days of CDF exposure, both hormones returned to control values (Laws et al., 1991). Similarly, neither pituitary LH levels nor serum LH and testosterone concentrations were different in male rats treated with CDF for 7, 14, or 21 days. (R. L. Cooper et al., unpublished). We also examined the effect of CDF on sexual behavior and fertility in males exposed to CDF for 1, 7, 14, or 21 days. At each test interval, sexual activity was impaired (i.e., longer latency to first mount, first mount with intromission, and first ejaculation) when measured within one hour of treatment. However, this effect on behavior was transient, since the CDF-treated males all produced sperm-positive smears when housed with a proestrous female overnight. The cohabitated females all became pregnant, and there were no differences in litter size following any of the CDF exposure periods. Finally, evaluations of sperm number, morphology, and motility revealed no noticeable effect of CDF on any of these parameters.

Sex differences in fertility were also noted in rats treated with methanol. In the female, we found that a single dose (1.6 g/kg) disrupted the LH surge, delayed ovula-

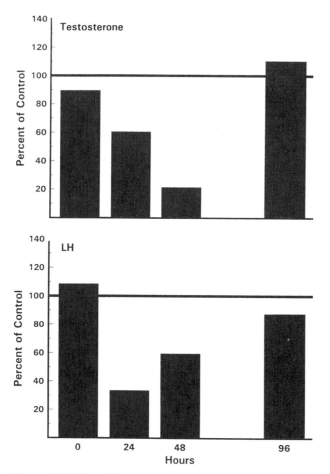

Figure 3 Effect of two chlordimeform injections (50 mg/kg 12 hours apart) on serum testosterone and LH in male rats presented as a percent of baseline measures taken at time zero. By 96 hours both hormones returned to pretreatment values. (Adapted from Goldman et al., 1990; Laws et al., 1991.)

tion, and reduced litter size. A single dose of methanol to the male lowered serum LH and testosterone (Cooper et al., 1992), but these acute neuroendocrine changes were without long-range effects on any testicular measure. On the other hand, repeated exposure to methanol caused significant changes in testes weight, sperm number, and sperm morphology (Cooper et al., 1992). However, the serum hormone levels in these males did not reflect impairment of the hypothalamic-pituitary control of the testes even at very high doses of this alcohol. These data led us to conclude that the disruption of testicular function in these animals was the result of a direct action of methanol on the testes. Since acute methanol exposure did not suppress serum LH, they also suggest that the central gonadotropic axis develops a tolerance with repeated exposures.

There likely are multiple reasons for such discrepancies in the response of the male and female reproductive systems to acute toxic insult. Disparities between the sexes in the metabolism of an administered compound (Gustafsson et al., 1983) may result in differences in the concentration of that compound delivered to a target site. Furthermore, there may also be differences in the activity of these target site cells. While such factors may contribute to the discrepant responsiveness of the two reproductive systems, the fundamental differences in the patterns of hormonal regulatory control may be more important in accounting for differential susceptibility. In the male, hypothalamic-pituitary-gonadal interrelationships are characterized by static endocrine signaling, in which blood hormonal concentrations are maintained at basal levels that exist within a fairly broad functional range. For example, levels of testosterone within the seminiferous tubules can fall by as much as 80% before effects on spermatogenesis will be seen (Zirkin, 1989). But sustaining the impact of a toxic insult to the hypothalamus enough to affect adversely the maintenance of gonadal function may be precluded by the emergence of adjustive CNS changes, which in many cases may represent a developing tolerance to xenobiotic exposure. For example, as mentioned previously, continued exposure in the male to CDF caused an initial depression in serum levels of LH and testosterone. However, over a period of a few days that values returned to control levels (Laws et al., 1991). Comparable adjustive effects on the hypothalamic control of pituitary function also have been seen following methanol exposure (Cooper, 1992). In the male, we have not identified an environmental toxicant that can affect fertility through alterations in neuroendocrine control of the testes, although acute exposure may cause transient disruptions in gonadotropin and gonadal hormone secretion. In contrast, if the testes are also a target site for the toxicant (as is the case with methanol), then continued exposure may indeed cause a direct gonadal toxicity.

In contrast to the male, normal reproductive function in the cycling female requires the periodic presence of the midcycle surges of LH and FSH. These functional endocrine events induce a sequelae of ovarian alterations necessary to drive the processes participating in oocyte release. As described above, a perturbation in the LH surge will affect ovulation, causing either a premature release of oocytes (as in response to an early estradiol exposure) or a delay in follicular rupture (seen following a toxicant-induced blockade in the surge). More extended xenobiotic exposures may also modify the persistence of such effects on the hypothalamic control of pituitary function. We have examined the effect of a 3-week exposure of CDF, nembutal, and methanol on the female's ovarian cycle and fertility. In each case, initial exposure resulted in some perturbation of regular cycles. However, after 7–10 days of treatment, regular cycles reemerged in the majority of females receiving a dose of CDF, nembutal, or methanol known to block the LH surge in response to a single exposure. Furthermore, if the females were mated at the end of such prolonged exposures, litter size was not altered. Thus, in the female acute exposure to a neuroendocrine toxicant can alter fertility as a consequence of delayed ovulation. But extended treatment with the same dose can frequently be without effect. We feel that these observations are important to the assessment of potential reproductive toxicants. If tolerance to such compounds develops, then it is unlikely that most of the standard reproductive risk-assessment protocols (which evaluate fertility only and emphasize prolonged treatment) would detect any effect on fertility unless the gonads are direct targets. This may help explain why compounds such as CDF were found to be without effect on reproduction when tested in studies involving extended exposures given prior to mating (Blackmore, 1969).

In summary, the hypothalamus and pituitary are important target sites for reproductive toxicants. Numerous studies demonstrate that in the developing animal, brief exposure to a variety of agents may result in permanent alterations in reproductive function in adulthood. Although there are fewer studies that have focused on the neuroendocrine regulation of the gonads in the adult, there are also clear examples of neurotoxicants that may alter fertility. However, there appear to be fundamental sex differences in the sensitivity of neuroendocrine control to such toxicants. These differences, together with the possible development of tolerance following extended xenobiotic exposure, raise serious questions about the suitability of current testing protocols used to characterize adverse reproductive effects, an issue that is particularly relevant for evaluations in the female.

REFERENCES

Advis, J. P., and Ojeda, S. R. (1978). *Endocrinology, 108*: 924.

Advis, J. P., White, S. S., and Ojeda, S. R. (1981). *Endocrinology, 109*: 1321.

Arai, Y., and Gorski, R. A. (1968). *Endocrinology, 82*: 1005.

Ayalon, D., Nir, I., Cordova, T., Bauminger, S., Puder, M., Naor, Z., Kashi, R., Zor, V., Harell, A., and Lindner, H. R. (1977). *Neuroendocrinology, 23*: 31.

Blackmore, R. H. (1969). Unpublished report cited in *WHO Pesticide Residue Series No. 1, Evaluation of Some Pesticide Residues in Food*, World Health Organization, Geneva, p. 9.

Braw, R. H., and Tsafriri, A. (1980). *J. Reprod. Fertil., 59*: 259.

Bronson, F. H. (1986). *Endocrinology, 118*: 2483.

Brown, H. E., Salamanca, S., Stewart, G., and Uphouse, L. (1991). *Toxicol. Appl. Pharmacol., 110*: 97.

Bulger, W. H., and Kupfer, D. (1985). *Endocrine Toxicology* (J. A. Thomas, K. S. Korach, and J. A., McLachlan, eds.), Raven Press, New York, p. 1.

Butcher, R. L., and Fugo, N. W. (1967). *Fertil. Steril., 18*: 297.

Butcher, R. L., Blue, J. D., and Fugo, N. W. (1960). *Fertil. Steril., 20*: 223.

Chakravarty I., Shah, P. G., Sheth, A. R., and Ghosh, J. J. (1979). *J. Reprod. Fertil., 57*: 113.

Chen, P. H., Tilson, H. A., Marbury, G. D., Caroum, F., and Hong, J. S. (1985). *Toxicol. Appl. Pharmacol., 77*: 158.

Cooper, R. L., Goldman, J. M., Rehnberg, G. L., Booth, K. C., McElroy, W. K., Hein, J. (1985a). *EPA/ARMY Workshop to Evaluate a Protocol for Reproductive Assessment*, Charleston, SC, p. 27.

Cooper, R. L., Goldman, J. M., Gray, L. E., Jr., Lyles, K. W., and Ellison, D. L. (1985b). *Toxicologist, 5*: 182.

Cooper, R. L., Goldman, J. M., and Rehnberg, G. L. (1986). *Environ. Health Perspect., 70*: 177.

Cooper, R. L., Mole, M. L., Rehnberg, G. L., Goldman, J. M., McElroy, W. K., Hein, J. F., and Stoker, T. E. (1992). *Toxicology, 71*: 69.

Cooper, R. L., Barrett, M. A., McElroy, W. K., Goldman, J. M., and Stoker, T. E. (1994a). *Toxicologist, 14*; 77.

Cooper, R. L., Barrett, M. A., Goldman, J. M., Rehnberg, G. L., McElroy, W. K., and Stoker, T. E. (1994b). *Fund. Appl. Toxicol., 22*: 474.

Cooper, R. L., Parrish, M. B., McElroy, W. K., Rehnberg, G. L., Hein, J. F., Goldman, J. M., Stoker, T. E., and Tyrey, T. E. (1995). *Toxicologist, 15*: 294.

Cooper, R. L., Stoker, T. E., Goldman, J. M., Parrish, M. B., and Tyrey, L. (1996). *Reprod. Toxicol., 10*: 257.

Costa, L. G., Olibet, G., and Murphy, S. D. (1988). *Toxicol. Appl. Pharmacol., 93*: 319.

Cummings, A. M., Perreault, S. D., and Harris, S. T. (1991). *Fund. Appl. Toxicol., 17*: 563.

Dalterio, S. Bartke, A., Roberson, C., Watson, D., and Burstein, S. (1978). *Pharmacol. Biochem. Behav.*, *8*: 673.

Dohler, K. D. (1991). *Int. Rev. Cytol.*, *131*: 1.

Dohler, K. D., Srivastava, S. S., Shryne, J. E., Jarzab, B., Sipos, A., and Gorski, R. A. (1984). *Neuroendocrinology*, *38*: 297.

Everett, J. W. (1948). *Endocrinology*, *43*: 389.

Everett, J. W. (1989). *Neurobiology of Reproduction of the Female Rat*. Springer-Verlag, New York.

Everett, J. W., and Sawyer, C. H. (1950). *Endocrinology*, *47*: 198.

Everett, J. W., and Tyrey, L. (1982). *Endocrinology*, *111*: 1979.

Faber, K. A. and Hughes, C. L. Jr. (1993). *Reprod. Toxicol.*, *7*: 35.

Field, E., and Tyrey, L. (1984). *Life Sci.*, *35*: 1725.

Field, E., and Tyrey, L. (1986). *J. Pharmacol. Exp. Ther.*, *238*: 1034.

Field, E., and Tyrey, L. (1990). *J. Pharmacol. Exp. Ther.*, *254*: 171.

Fugo N. W., and Butcher, R. L. (1966). *Fertil. Steril.*, *17*: 804.

Gandolfi, O., Cheney, D. L., Hong, J. S., and Costa, E. (1984). *Br. Res.*, *303*: 117.

Gellert, R. J. (1978). *Environ. Res.*, *16*: 123.

Goldman, J. M., and Cooper, R. L. (1993). *Female Reproductive Toxicology* (J. J. Heindel and R. E. Chapin, eds.). Academic Press, San Diego, p. 79.

Goldman, J. M., Cooper, R. L., Rehnberg, G. L., Hein, J. F., McElroy, W. K., and Gray, L. E., Jr. (1986). *Toxicol. Appl. Pharmacol.*, *86*: 474.

Goldman, J. M., Cooper, R. L., Laws, S. C., Rehnberg, G. L., Edwards, T. L., McElroy, W. K., and Hein, J. F. (1990). *Toxicol. Appl. Pharmacol.*, *104*: 25.

Goldman, J. M., Cooper, R. L., Edwards, T. L., Rehnberg, G. L., McElroy, W. K., and Hein, J. F. (1991) *Pharmacol. Toxicol.*, *68*: 131.

Gorski, R. A. (1986). *Environ. Health Perspect.*, *70*: 163.

Gray, L. E. Jr., Ferrell, J., Ostby, J., Rehnberg, G., Linder, R., Cooper, R., Goldman, J., Slott, V., and Laskey, J. (1989). *Fundam. Appl. Toxicol.*, *12*: 92.

Gray, Jr. L. E., Ostby, J. S. and Kelce, W. R. (1994). *Toxicol. Appl. Pharmacol.*, *129*: 46.

Gustafsson, J.-A, Mode, A., Norstedt, G., et al. (1983). *Ann. Rev. Physiol.*, *45*: 51

Heinrichs, W. L., Gellert, R. J., Bakke, J. L., and Lawrence, N. L. (1971). *Science*, *173*: 642.

Hirsch, K. S., Adams, E. R., Hoffman, D. G., Markham, J. K., and Owen, N. V. (1986). *Toxicol. Appl. Pharmacol.*, *86*: 391.

Hong, J. S., Hudson, P. M., Yoshikawa, K., Ali, S. F., and Mason, G. A. (1985). *Neurotoxicology*, *6*: 167.

Hoskins, B., and Ho, I. K. (1982). *J. Toxicol. Environ. Health*, *9*: 535.

Hughes, C. L., Jr., Everett, J. W., and Tyrey, L. (1981). *Endocrinology*, *109*: 876.

Johnson, T. L., and Knowles, C. O. (1983). *Gen. Pharmacol.*, *14*: 591.

Jones, T. H. (1991). *Horm. Res.*, *35*: 8285.

Kalra, S. P., and Kalra, P. S. (1983). *Endo. Rev.*, *4*, 311.

Kelce, W. R., Stone, C. R., Laws, S. C., Gray, L. E., Kemppainen, J. A., and Wilson, E. M. (1995). *Nature*, *375*: 581.

Krey, L. C., and Everett, J. W. (1973). *Endocrinology*, *93*: 377.

Kumagai, S., and Shimizu, T. (1982). *Arch. Toxicol.*, *50*: 279.

Lauder, J. M. (1983). *Psychoneuroendocrinology*, *8*: 121.

Laws, S. C., Rehnberg, G. L., Hart, D. W., Goldman, J. M., and Cooper, R. L. (1991). *Toxicologist*, *11*: 76.

Laws, S. C., Carey, S. A., Hart, D. W., and Cooper, R. L. (1994). *Toxicology*, *92*: 127.

McEwen, B. S. (1980). *Biol. Reprod.*, *22*: 43.

Narotsky, M. G., and Kavlock, R. J. (1995). *J Toxicol. Environ. Health 45*: 145.

Okey, A. B., and Truant, G. S. (1975). *Life Sci.*, *17*: 1113.

Peluso, J. J., and Butcher, R. L. (1974). *Fertil. Steril.*, *25*: 494.

Pfaff, D., and Keiner, M. (1973). *J. Comp. Neurol.*, *151*: 121.

Plant, T. M., Nakai, Y., Belchetz, P., Keogh, E., and Knobil, E. (1978). *Endocrinology*, *102*: 1015.

Ratner, A. (1971). *Proc Soc. Exp. Biol. Med.*, *138*: 995.

Rehnberg, G.L., Hein, J.F., McElroy, K., Goldman, J.M., Cooper, R.L. (1986). *Toxicologist*, *6*: 287.

Shah, G. V., Shyr, S. W., Grosvenor, C. E., and Crowley, W. R. (1988). *Endocrinology*, *122*: 1883.

Shyr, S. W., Crowley, W. R., and Grosvenor, C. E. (1986). *Endocrinology*, *119*: 1212.

Stoker, T. E., Goldman, J. M., and Cooper, R. L. (1993). *Reprod. Toxicol.*, *7*: 211.

Stoker, T. E., Cooper, R. L., Goldman, J. M., Rehnberg, G. L., McElroy, W. K., Hein, J. F., and Parrish, M. B. (1995). *Toxicologist*, *15*: 294.

Stoker, T. E., Cooper, R. L., Goldman, J. M., and Andrews, J. E. (1996). *Neurotoxicol. Teratol.*, *18*: 277.

Terranova, P. F. (1980). *Biol. Reprod.*, *23*: 92.

Tyrey, L. (1978)., *Endocrinology*, *102*: 1808.

Uilenbroek, J. T. J., Woutersen, P. J. A., and van der Schoot, P. (1980). *Biol. Reprod.*, *23*: 219.

Uphouse, L. (1985). *Neurotoxicology 6*: 191.

Uphouse, L., and Eckols, K. (1986). *Neurotoxicology*, *7*: 25.

Weick, R. F. (1978). *Neuroendocrinology*, *26*: 108.

Wenger, T., Croix, D., and Tramu, G. (1988). *Biol. Reprod.*, *39*: 540.

Williams, J., Eckols, K., Stewart, G., and Uphouse, L. (1988). *Neurotoxicology, 9*: 597.

Ying, S.-Y., and Greep, R. O. (1972). *Proc. Soc. Exp. Biol. Med.*, *139*: 741.

Zirkin, B. R., Santulli, R., Awoniyi, C. A., and Ewing, L. L. (1989). *Endocrinology*, *124*: 3043.

10

Developmental Neurotoxicology

Gaylia Jean Harry
National Institute of Environmental Health Sciences, Research Triangle Park, North Carolina

INTRODUCTION

Neurotoxicity is defined as a structural change or a functionally adverse response of the nervous system to a chemical, biological, or physical agent. Characterization of toxicity in the nervous system encompasses multiple levels of organization and complexity, including structural, biochemical, physiological, and behavioral. This is due to the complexity of the nervous system and its activities necessary to maintain the balance of all of the various organ systems of the body. The smooth coordination of body motion and functioning as well as the integration of complex cognitive functions such as speech, emotion, learning, and memory are dependent upon the normal functioning of the nervous system. Interaction with the external environment is dictated by the nervous system and dependent upon the normal complex arrangement and network of anatomical structures and biochemical and physiological integrative functioning. The temporal and spatial organization of nervous system development is a precise and complex process, with the basic framework of the nervous system laid down in a stepwise fashion in which each step is dependent upon the proper completion of the previous one. Thus, a relatively minor disturbance resulting in a perturbation of the developmental interactions between selective cells for a limited time period may result in a major deleterious outcome. This chapter will present information on critical processes in the normal development of the nervous system and evidence that alterations in the environment of the system can result in long-term damage.

BASIC PRINCIPLES OF NERVOUS SYSTEM DEVELOPMENT

The development of the nervous system starts in the fetus and for humans is not complete until approximately the time of puberty. Organogenesis for the nervous system occurs during the period from implantation through midgestation with synaptogenesis and myelination predominent during late gestation and early neonatal period. The complex architecture of the brain requires that different cell types develop in a precise spatial relationship to one another. To accomplish this, the embryo must not only establish a coordinate system for itself, but must also ensure that the appropriate cell types are correctly generated within this coordinate system (for review, see Jacobson, 1991). The problems of pattern formation and cell type determination are intimately related. The principal cell types of the central nervous system (CNS) neurons and macroglia (astrocytes and oligodendrocytes), arise from a pseudostratified epithelium of ectodermal sheet origins. Neural ectoderm is induced by an interaction of the mesoderm with the overlying ectoderm of the trilaminar embryo. This neural plate is committed to develop into neural tissue, and local areas of the plate along its anterior-posterior axis are predestined to develop into specific brain regions. Neural plate cells migrate toward the midline and begin to acquire different cell shapes. Differential changes in cell morphology result in the edges of the neural plate folding in to form the neural groove and adhere to form the neural tube. Cells on the margins of the plate migrate into the region between the surface ectoderm and the dorsal aspect of the neural tube and become the cells of the neural crest. Cells emerge from the neural crest and migrate to specific sties in the periphery according to characteristic migration patterns determined by the local environment. Cell adhesion and extracellular matrix molecules have a regulatory effect on neural crest cell motility and morphology. Some neural crest derivatives retain developmental plasticity and the commitment of terminal differentiation is delayed. These cells can undergo transdifferentiation without cell divisions in response to environmental conditions.

The pseudostratified epithelial cells continue to proliferate and increasing fluid pressure within the central canal leads to "ballooning" of the rostral end of the tube to form the three brain vesicles that define the major divisions of the brain. The early neural tube demonstrates structural signs of specialization between regions, and Homeobox gene expression corresponds to different brain segments and appears to regulate the expression of segment-specific traits (Holland and Hogan, 1988; Bally-Cuif et al., 1992; McMahon et al., 1992). Eventually, this tube becomes grossly subdivided into regions, corresponding roughly to the future forebrain, midbrain, hindbrain, and spinal cord. Different types of neurons develop in each of these regions. Initially the neuroepithelium contains only one undifferentiated population of stem cells, which give rise to two main lineages, neuronal and glial stem cells, and subsequently the many sublineages of the stem cell type.

BASIC STRUCTURAL COMPONENTS OF THE NERVOUS SYSTEM

The nervous system is comprised of two components: the CNS, which is composed of the brain and the spinal cord, and the peripheral nervous system (PNS), which is composed of ganglia and the peripheral nerves which lie outside of the brain and spinal cord. In the nervous system, the neurons are responsible for reception, intergration, transmission, and storage of information. They are similar in structure to other cells with the additional feature of axons and dendrites, which allow for nerve impulses and communication with other neural cells. The afferent or sensory neurons carry information into the nervous system, while motor neurons carry commands to the muscles and glands. Interneurons with the system transfer information within the system and process local information.

The various glial cells in the nervous system, the astrocytes and oligodendrocytes of ectodermal origin, the microglia of mesodermal origin, and the ependymal cells of ectodermal origin retain the ability to proliferate, unlike the neurons. Glial cells have a dynamic nature and provide critical processes for the development (growth factors and cell adhesion factors) and functioning (local pH levels and ionic balance) of the nervous system. In some parts of the nervous system, particular kinds of cells are generated from committed progenitors or "blast" cells. These cells proliferate symmetrically and then differentiate. Examples of such progenitors are the sympathoadrenal progenitor and the O2A progenitor from optic nerve. The O2A progenitors first give rise to oligodendrocytes around the time of birth and begin to generate type 2 astrocytes during the second postnatal week. An important mitogen for the O2A progenitor is platelet-derived growth factor (PDGF), which stimulates the progenitors to divide and thereby increases the number of oligodendrocytes that develop. In contrast to the intrinsically timed differentiation of oligodendrocytes, the differential of type 2 astrocytes is timed by cell-extrinsic factors. Ciliary neuronotrophic factor (CNTF) is a diffusible signal that, in association with other signals of the extracellular matrix, is required to induce O2A progenitors to develop into type 2 astrocytes.

The cerebral hemispheres develop from the region of the rostral dorsal part of the neural tube named the telencephalic pallium. The continued proliferation of the clonogenic neuroepithelial germinal cells causes the outward bulging of the pallial walls to form the cerebral vesicles. This primary radial organization is distorted by differentiation and growth of neurons and glial cells and by nerve fibers. Neurogenesis begins during fetal development with temporal differences in cell production maintained with the first-generated neurons reaching their final position before subsequent generations of neurons (for review, see Jacobson, 1991). The radial glial cell processes serve to guide neurons from the zone of neuronal generation to the zones for final settlement (Astrom, 1967; Mugnaini and Forstronen, 1967; Sidman and Rakic, 1973; O'Rourke et al., 1992). The laminar features of the cortex are generated over time by differential movement of groups of neurons born at different times (Hicks and D'Amato, 1968; Rakic, 1974; Smart and Smart, 1982; Luskin and Shatz, 1985; Bayer and Altman, 1991). Given that different populations of neurons in each structure are produced at different stages of development, the timing of any given environmental stimuli can produce very different effects on the final formation of each brain region.

BLOOD-BRAIN BARRIER

The blood-brain barrier (BBB) is a regulatory interface between the blood and the nervous system. It serves to regulate the environment of the brain to ensure that it remains independent of the fluctuations of normal blood substances. The selectivity of the BBB to circulating substances is dependent upon both lipid solubility and mechanisms of transendothelial cell passage. A series of specific carrier-mediated transport systems or facilitated diffusion mechanisms exist through which required nutrients, hormones, amino acids, proteins, peptides, and fatty acids are able to reach the brain. The BBB also serves as an enzymatic barrier. For example, the presence of degrading enzymes in the brain endothelial cells restrict both endogenous neuroactive substances like the monoamines and potentially neurotoxic exogenous compounds. While the barrier controls the entry of normal physiological substances into the nervous system, it cannot prevent the exchange of substances from the blood to the brain when the transport is dictated by the same physiological properties of lipid solubility or active transport carrier mechanisms.

Certain regions of the nervous system are devoid of a BBB; these circumventricular organs are highly vascularized structures that are in contact with the brain median ventricular system and the cerebral spinal fluid (CSF). The dense and permeable capillaries allow direct contact with plasma filtrates, and the cells produce hormones or act as chemoreceptors or CSF in the case of the choroid plexus (Arvidson, 1986; Valois and Webster, 1987). The BBB is considered to be less tight in the immature brain. Evidence for this comes from the studies of dye exclusion and the observation that adult brains stain significantly less than immature brains when a circulating dye is administered. In addition, the concentration of serum proteins is higher in the fetal cerebral spinal fluid relative to the adult (Cavanagh et al., 1983). The tight junctions between cerebral endothelial cells develop very early, while brain barrier transport mechanisms involving carriers or enzymes appear later in the brain development (Saunders, 1992). As brain capillaries develop tight junctions they begin to express the glucose transporter (GLUT-1) (Pardridge et al., 1990). When there are few perivascular glial, vascular invasion at approximately embryonic day 12–14 is by vessels permeable to horseradish peroxidase (HRP) (Dermietzel and Krause, 1991). As the perivascular glia increase in number by E15, the vessels become more impermeable to HRP. Drugs that are amphiphilic and induce lipidosis, for example, triparanol, AY-9944, and perhexiline, produce lipidosis in the CNS of young rats, while they produce few if any changes in the adult brain (Drenckhahn and Lullmann-Rauch, 1979). When exposed to the herbicide paraquat, 2-week-old rats contained a higher concentration in the brain compared to 3-month-old animals (Corasaniti et al., 1991). The BBB of the developing animal has been reported to be "less tight" with regard to cadmium than that of the mature animal (Arvidson, 1986; Valois and Webster, 1987). The pattern of susceptibility of neurons, petechial hemorrhages, and endothelial vacuolation following postnatal exposure to cadmium is thought to parallel the maturation of capillaries in the brain parenchyma (Webster and Valois, 1981.)

The nerves are also protected by a blood-nerve barrier, which is similar in structure and function to the blood-brain barrier (for review, see Rechthand and Rapoport, 1987). This barrier is formed by the perineurium and the endoneurial blood vessels separated by tight junctions joining lamellae of the perineurial sheath. The blood vessels are permeable, while those in the endoneurium are relatively impermeable. With regard to plasma proteins, the endoneurium is permeable to varying degrees and there is usually increased protein leakage in pathological conditions. The perineurial permeability is incomplete in newborn rats and mice and does not fully develop until the end of the third postnatal week (Kristensson and Olsson, 1971). Areas existing in the absence of a blood-nerve barrier are the optic nerve (Flage, 1977), the olfactory bulbs (Balin et al., 1986), the brain arterioles (Westergaard and Brightman, 1973), the dorsal root, and autonomic ganglia (Brightman et al., 1970; Jacobs, 1977). In addition the nerve networks lying between muscle layers of the gastrointestinal tract from the smooth muscle part of the esophagus to the internal anal sphincter known as the myenteric plexuses are devoid of a blood-nerve barrier (Jacobs, 1977). The terminal end of the axon of a motor nerve, the neuromuscular junction, is separated from the muscle by only the basement membrane of the Schwann cell and not a barrier (Broadwell and Brightman, 1976).

CELL MIGRATION AND AXON GUIDANCE

The function of the mature nervous system depends on the actions of distinct neuronal circuits, which function correctly because all are connected appropriately to each other.

The diversity of connections formed by a single nerve cell is one of the key features that distinguishes neurons from cells in other tissues of the body. In brain morphogenesis, migration of cells plays a significant role. Most neurons are required to travel long distances through complex areas to reach final position. The migration of neuronal precursors plays a role in establishing the identity of some neurons and defining the functional properties and connections of the neuron (Sidman and Rakic, 1973). Cell translocation is achieved by a combination of the extension of cell process, its attachment to the substratum, and subsequent pulling of the entire cell by means of contractile proteins associated with an intracellular network of microfilaments. Directional control is exercised by cells moving along "guide" cells or guided by a concentration gradient (Sidman and Rakic, 1973). The migration of neural crest cells into the spaces surrounding the neural tube depends on the nature of the extracellular matrix. For successful migration a number of signals are required to be regulated in a coordinated manner. Components of the extracellular matrix, collagen, fibronectin, laminin, proteoglycans, and hyaluronic acid regulate neural crest cell migration. Early in neural development the extracellular matrix in both the central and peripheral nervous systems contains glycoproteins that are effective in promoting axon growth. Two such proteins, laminin and fibronectin, persist in the periphery but are virtually absent from the brain and spinal cord of adult mammals. Ion regulation of cell adhesion and process extension is an important factor in neuronal differentiation (for review, see Kater and Mills, 1991).

NEUROTROPHIC FACTORS

Interactions between cells in direct contact and between cells and components of the intercellular matrix are mediated by cell surface adhesion molecules and cell surface recognition molecules. Additional interactions are by means of diffusible molecules, such as growth factors and trophic agents or by agents that require transport over long distances. Studies on neural development have identified several neurotrophic factors released by the targets of neurons that trigger biochemical changes in the neuron and are important for its survival and growth. Nerve growth factor (NGF) is the best-characterized neurotrophic factor (Levi-Montalcini and Angeletti, 1968). NGF is a neurotrophic factor required for survival and neurite growth of cholinergic neurons of the basal forbrain, sympathetic postganglionic neurons, and sensory ganglion cells derived from the neural crest. It is produced by cells in peripheral targets of sympathetic and sensory axons, Schwann cells, and macrophages in developing and regenerating peripheral nerves. The highest level of NGF occurs in the hippocampus and the cerebral cortex, which are targets for cholinergic megnocellular neurons of the basal forebrain. The critical need for NGF is demonstrated in experiments showing that antibodies to NGF cause death of neurons during their time of contact formation, while exogenous NGF will rescue neurons that would otherwise die.

Brain-derived neurotrophic factor (BDNF) is a 12.3 kDa basic protein belonging to the same family as NGF and found in very low abundance in the CNS (Leibrock et al., 1989). The maximal effect of BDNF occurs during the time when embryonic neurons contact targets in the CNS. BDNF supports the survival of the proprioceptive neurons in the dorsal root ganglion that are not responsive to NGF (Davies et al., 1986a). It stimulates neurite outgrowth and supports the survival of sensory neurons derived from ectodermal placodes—nodose ganglion neurons (Hofer and Barde, 1988).

Neurotrophin 3 (NT-3) also belongs to the NGF family of proteins and promotes neurite outgrowth and survival of neurons from both the nodose ganglion and dorsal root ganglion (Maisonpierre et al., 1990).

Ciliary neurotrophic factor (CNTF) is an acidic protein which is present in peripheral nervous tissues and promotes the survival of sympathetic, parasympathetic, motor, and sensory neurons during development and in the adult (Barbin et al., 1984; Manthorp et al., 1986; Blottner et al., 1989; Sendtner et al., 1990). CNTF may be responsible for sympathetic ganglion cell proliferation and differentiation and type 2 astrocyte differentiation (Hughes et al., 1988; Lillien et al., 1988). Basic fibroblast growth factor (FGF) is produced by neurons in normal CNS (Pettmenn et al., 1986; Finkelstein et al., 1988) and by astrocytes in culture (Ferrara et al., 1988; Hatten et al., 1988). Both the basic and acidic forms of FGF stimulate outgrowth of neurites and reduce the effects of injury and enhance regeneration. This can be shown in peripheral nerve regeneration, cholinergic neuronal survival after septohippocampal pathway lesion, and delayed death of retinal ganglion cells following sectioning of the optic nerve.

Migration is also influenced by adhesion properties of the cells with a direct interaction between a cell and the extracellular matrix. These adhesion molecules are cell adhesion molecules (CAMs), intracellular adhesion molecules (I-CAMs), integrins, and cadherins. CAMs are a family of high molecular weight cell surface glycoproteins that possess morphoregulatory properties during neural development. The members of this family include neural CAM (N-CAM), neuronal-glial CAM (Ng-CAM-NILE or L1), Tenascin (TAG-1), and adhesion molecule on glia (AMOG/β_2 isoform of the membrane Na, K-ATPase pump). N-CAM is found widespread early in embryogenesis and throughout the development of the nervous system in both glia and neuronal cells (Rutishauser et al., 1976). It mediates Ca^{2+}-independent homophilic binding and aggregation of neuronal cells. It may contribute to processes such as glial guidance of axonal processes, neurite fasciculation, axon–target cell interactions, and cell-positioning relationships (Rutishauser and Jessel, 1988). The glycosylation state of N-CAM is developmentally regulated. Early in development, N-CAM is in a highly glycosylated form with a sialic acid content of approximately 30%. In this form, cells are less adhesive, it is not until the more mature less sialic acid form appears that N-CAM–mediated cell adhesion occurs (Rutishauser and Jessel, 1988). N-cadherin (A-CAM) is a member of a family of non-Ig glycoproteins that are important in cell-cell interaction functioning in a Ca^{2+}-dependent manner (Grunwald et al., 1980; Hatta and Takeichi, 1986; Crittenden et al., 1987; Lagunowich et al., 1991). The ectoderm shows an early expression of N-cadherin and may serve to induce the development of the neural plate (Detrick et al., 1990) and mediate closure of the neural tube. During development, N-cadherin plays a role in the guidance of growth cones during neurite extension (Hatta and Takeichi, 1986; Matsunaga et al., 1988a,b). Altered expression of N-cadherin during specific times of development results in a disruption of the normal pattern of nervous system development (Edelman and Thiery, 1985; Takeichi, 1988). L1 is found only on neuronal cells and displays Ca^{2+}-independent binding. It is involved in heterotypic binding between neuronal and neuroglial cells. L1 expressed on astrocytes in culture will increase neurite outgrowth of co-plated dorsal root ganglion cells. TAG-1 is transiently expressed on the surface of a subset of neurons in the CNS and is considered to play a role in the initial stage of axonal growth and guidance over neuroepithelial cells. A temporal pattern of expression of these adhesion factors exists with the loss of TAG-1 with the onset of expression of L1. Tenascin is a large extracellular matrix glycoprotein implicated in cell proliferation and neural cell attachment and possibly playing a role in neurite outgrowth and cell migration. It is one of the factors controlling neural crest migration and differentiation.

Integrins are membrane receptors with ligands consisting of I-CAMs and other matrix components such as collagen, laminin, and fibronectin. Integrin activation can lead to rapid changes in cell adhesion properties in the local environment and can signal intracellular events. These receptors provide the developing neural cells a system for linking adhesion/migration information with other developmental signals controlling proliferation and differentiation.

NEURITE OUTGROWTH

Elongation of neurites occurs exclusively at their growth cones, the actively growing tips of all branches of axons and dendrites. The activities of the growth cone involve making and breaking adhesive contacts with components of the extracellular matrix. For this purpose migrating neurons and growth cones secrete proteases and plasminogen activators, which result in activation of proteolytic enzymes, which break down components of the extracellular matrix (Pittman and Buettner, 1989). Growth cones are structurally and functionally different from the rest of the neuron. The growth cone is the site of most of the interactions of the developing neuron with its environment. As the dendrite or axon elongate the cell body and proximal segments of the axon or dendrite are attached and stationary, only the growth cone moves through the environment. Ultrastructure of the growth cone is correlated with its functional activities; elongate and growth; motility and active exploration; uptake of materials; secretion of enzymes, neurotransmitters, and trophic agents. The morphology of the growth cone can be modified by the type of neuron, stage of development, functional state, contact with different molecules in the substratum, and local environment (NGF, laminin, and fibronectin), which promote axon elongation. While the rate of axonal elongation and regeneration is set by the slow axonal transport delivery of membrane components from the cell body, other activities are regulated by the local conditions. For example, the direction of elongation of the growth cone can be influenced by mechanical guidance by the substratum, extracellular matrix molecules, diffusible extracellular growth factors, and electrical fields (Kater and Letourneau, 1985; Lander, 1987; Bray and Hollenbeck, 1988; Van Hooff et al., 1989). Specific neurotransmitters such as serotonin and dopamine can also alter neurite elongation and growth cone movement. Serotonin inhibits neuronal outgrowth of specific subsets of neurons. Adhesion molecules on the cell or on the extracellular matrix mediate neurite-neurite and neurite-glial interactions and influence neurite outgrowth. Intracellular pathways are important in neurite outgrowth with cAMP and inositol phospholipids acting as intracellular regulators. Ca^{2+} influx can regulate both neurite elongation and growth cone movements (Katter and Mills, 1991). This may occur via (1) the Ca^{2+}-dependent kinases that can phosphorylate cytoskeletal proteins, (2) a direct effect on tubulin, or (3) the calcium-binding protein calmodulin, which interacts with tau proteins and MAP2, which stabilize microtubules by forming cross-bridges or actin microfilament polymerization.

AXON

During the period of active growth, intricate cellular networks are formed. The axon is the first process to develop from the neuron, and it grow from the cell body along a specific pathway to form a synaptic connection with a specific target. At the growing tip of each axon is the growth cone, which responds to environmental cues and "guides" the axon to a final terminal connection. The proper functioning of the nervous system depends on the outgrowth of axons to make connections with the correct postsynaptic targets. The axon is the first process to develop from the young neuron, and the axon always grows from the cell body in a specific direction, along a specific pathway, to form synaptic connections with specific targets. The final pattern of neuronal connections may be regarded as the result of a series of distinct processes: outgrowth of axons and selection of pathways to their destinations, dendritic outgrowth and formation of specific morphology, selection of specific targets, elimination of incorrect and redundant synapses, and the establishment of a final pattern of synaptic connections.

The initial outgrowth of the axon normally occurs in the direction it must take to reach its correct destination. In each region of the CNS, the axons grow out in characteristic and consistent direction with only slight variations of their directions and course of growth. Growth in the nervous system is controlled by both stimulatory and inhibitory substances (for review see Schwab et al., 1993). Other factors that can modulate growth of the developing axon include matrix proteoglycans (Reichardt and Tomaselli, 1991), heparin proteoglycans (Unsicker et al., 1993), and chondroitin sulfate (Snow et al., 1991). It has also been postulated that there exist specific signals to stop axon movement and initiate synapse formation (Raper and Grunewald, 1990). The limbic system–associated membrane protein (LAMP) is a cell surface protein expressed in cortical and subcortical areas of the limbic system, including the frontal cortex, basal forebrain, hippocampus, limbic thalamus, hypothalamus, amygdala, and other areas. It is required for the formation of the septohippocampal circuit in vitro (Keller and Levitt, 1989; Keller et al., 1989) and has been shown to modulate the growth of mossy fibers in the neonatal hippocampus in situ (Barbe and Levitt, 1992). Alterations in LAMP result in subtle changes in circuit formation rather than gross structural alterations.

DENDRITE FORMATION

After the neurons have migrated to their final positions, there is often a long delay before full differentiation of the dendrites occurs. Branching results in a great increase in the surface area of the dendrites, which form the major postsynaptic surface of the neuron. Each type of neuron has a distinct size and pattern shape of dendritic branching. During normal development, initiation of axonal and dendritic growth is influenced by extrinsic conditions. Environmental factors that can influence the final pattern include:

1. Passive guidance by oriented structures, such as, collagen, cartilage, blood vessels
2. Local conditions including electrical fields, extracellular matrix molecules, cell adhesion molecules, and integrins
3. Active modification of the extracellular matrix by enzymes released by the growth cone

4. Growth factors released by cells in the axonal pathways and by axonal targets
5. Inhibition of axonal growth and stabilization by conditions at the targets
6. Modification of axons and dendrites in response to nerve activity

The dendrites of local circuit neurons with short axons differentiate later than dendrites of the principal neurons, which have long axons. Maturation of dendrites tends to occur in ventrodorsal sequence, thus, the dendrites of the deep layers of the cortex mature before the superficial layers. The timing between dendrites of different neurons follows the sequence of dendritic maturation on neurons with long axons prior to those of short axons (Marin-Padilla, 1970; Pupura, 1975). In general, the motor neurons develop before the sensory neurons in the same region of the CNS. In the human brain, dendritic growth is maximal from the late fetal periods to about 1 year of age, and it continues at a slower pace for several years thereafter (Schade and Groenigen, 1961). During this period of dendritic growth and maturation, the configuration of the dendritic spine changes from a long thin filopodium to a stubby, mushroom-shaped structure. In the brain of infants with mental retardation or chromosomal abnormalities, immature, thin filopodium like spines have been well documented. It is unclear, however, whether such change results from the transformation of preexisting spines or from lack of spine maturation. The proper functioning of the dendrites depends on the proper relationship with axons that are destined to form synapses on them. Dendritic spines are the main postsynaptic sites on most neurons and because inhibitory and excitatory inputs are often spatially segregated on different regions of the dendrites, spines are important for integration of synaptic inputs (Feldman, 1984).

SYNAPTOGENESIS

The onset of synaptogenesis occurs according to a specific timetable and occurs during the same period as dendritic outgrowth and proceeds very rapidly. Excessive production of synapses, followed by elimination of redundant synapses, occurs in many regions. The temporal sequence of formation of synaptic connections between different neurons is correlated with the times of neuronal origin and differentiation. Synaptogenesis generally follows the same ventrodorsal pattern as described for neurons, with synaptogenesis occurring earlier in the deeper layers of the cerebral cortex and cerebellum (Jones, 1983). Synaptic differentiation is measured by the number of synaptic vesicles per terminal, with the mean number of synaptic vesicles per terminal increasing about two- or threefold during the first month of postnatal development in rodents (Dyson and Jones, 1980; Blue and Parnavelas, 1983). From the prenatal period to young adulthood there is an increase in the number of synapses in many regions of the central nervous system (Aghajanian and Bloom, 1967; Brand and Rakic, 1984), and this pattern can be altered by such things as undernutrition (Dyson and Jones, 1976). The timing of an increase may significantly differ among different species and also vary in the different regions of the brain even in the same species.

Through transsynaptic or transneuronal stimulation, neurons may exercise an influence on others, either directly across a single synapse or indirectly via interneurons in a chain or circuit, and is required by many types of neurons to complete their development. Usually it is the final maturation and continued vitality of neurons that depends on transsynaptic stimulation. During development, neurons are much more sensitive to direct injury as well as to deafferentation and to removal of their targets. Cerebral cortical dendrites and synapses develop postnatally and form excitatory synapses. There are many reports of changes in dendritic spines in the cerebral cortex following either sensorimotor deprivation or increased functional activity.

NEURONAL DEATH AND DEVELOPMENT

Morphogenic cell death is an important natural aspect of development occurring in most regions of the CNS (Cowan et al., 1984). It occurs in many parts of the nervous system during cavitation, fusion, folding, and bending of the neural plate and neural tube. A large number of cells die during reduction in the thickness of the dorsal part of the neural tube. This accompanies the formation of the dorsal raphe in the mesencephalon and the choroid plexuses of the 3rd and 4th ventricles. Histogenetic cell death occurs during histogenesis and remodeling of various cell groups in the CNS and ganglia of the PNS. It is highly dependent upon the extracellular environment in such things as cell-cell interactions and on nutritional, hormonal, and trophic influences. Among the mechanisms known to cause histogenesis is the increase in intracellular calcium ions caused by excitatory amino acids.

Dying cells endocytose extracellular proteins and transport them to lysosomes for destruction, thus, much of the documentation of cell death has come from uptake studies of HRP. Dying neurons can autophagosize (Ericcson, 1969) or undergo apoptosis with a condensation and dissolution of chromatin and transfer of chromatin into cytoplasmic autophagic vacuoles (Clarke and Hornung, 1989). Neuronal death may also be caused by failure to obtain critical amounts of neurotrophic factors and the proper retrograde transport of such factors. Although neurons originate, migrate, and differentiate in the absence of their normal targets, such as absence of synaptic connections will result in a neuron's death. When a neuron makes multiple connections and therefore has multiple sources of trophic factors, the cell can adjust for the loss of one synaptic target site by the retraction or elimination of the axon or of the collaterals without neuronal death. There are many such cases during development. For example, in the mature cerebellum there is one single climbing fiber innervating each Purkinje cell; however, during development multiple climbing fibers innervate each Purkinje cell (Crepel et al., 1976). This retraction is possibly due to a competition between fibers for a limited amount of trophic factors. Other examples are the thalamocortical and callosal innervation of the visual cortex (Lund et al., 1984; Innocenti et al., 1986), the ipsilateral corticocortical projections, callosal projections to the somatosensory cortex (Caminiti and Innocenti, 1981; Ivy and Killacky, 1982), and corticospinal projections (Bates and Killacky, 1984).

NEURONAL SIGNALING

Neuronal signaling depends on rapid changes in the electrical potential differences across nerve cell membranes. These rapid changes in potential are made possible by ion channels, a class of integral proteins that traverse the cell membrane. These channels conduct ions, recognize and select among specific ions, and open and close in response to specific electrical, mechanical, or chemical signals. The common synaptic transmitters are low molecular weight molecules, however, a wide variety of peptides can serve as synaptic messengers. The nerve cell uses both electrical synapses and chemical synapses. In the brain the chemical synapse is more common than electrical synapses and can endure changes in effectiveness. This plasticity is critical for memory and other higher functions of the brain. Chemical synapses can amplify neuronal signals, allowing a small presynaptic nerve terminal to alter the potential of a large postsynaptic cell. A single action potential releases thousands of neurotransmitter molecules allowing amplification of the synaptic response. There are two major types of chemical transmission that differ according to the postsynaptic receptor that is activated. In one type, the postsynaptic receptor gates an ion channel directly; in the other it does so indirectly by means of a second messenger. However, the chemical synaptic response is more easily modified than electrical transmission. Thus, many synaptic receptors are selective targets for disease and for drug therapy.

A central nerve cell receives both excitatory and inhibitory inputs from hundreds of neurons that use different chemical transmitters. The different inputs can reinforce as well as cancel one another. No one presynaptic neuron in the CNS is capable of exciting a postsynaptic cell sufficiently to reach the threshold for an action potential. This complex converging input to a single cell is mediated by different kinds of synaptic receptors sensitive to the different transmitters, and these receptors control distinct ion channels, some directly gated and some gated indirectly by second messengers. Thus, a central neuron must integrate a diverse set of inputs into a coordinated response. Neuronal integration depends on the summation of synaptic potentials that spread passively to the trigger zone. This process is affected by two passive membrane properties of the neuron: the temporal summation and the spatial summation.

DEVELOPMENT OF NEUROTRANSMITTER SYSTEMS

In rodents, the development of the neurotransmitter systems begins during the late fetal period. Morphological development generally follows a caudal-rostral pattern, with maturation of cell bodies in the brain stem and basal forebrain first, followed by the development of axonal projections. Neurotransmitter systems exhibit a variety of developmental profiles, beginning with the very early monoamine systems in the brain stem (for review, see Levitt, 1982), which innervate a large number of areas in the brain (for review, see Moore and Bloom, 1978, 1979; Moore and Card, 1984). This continues during development to the late-expressing neuropeptides and amino acid transmitters in cerebral cortical neurons.

The noradrenergic, dopaminergic, serotonergic, and cholinergic pathways originate from cell bodies in the brain stem and basal forebrain, which send projections into the cortex and hippocampus. The cells forming these pathways all differentiate at about the same time, but the cholinergic pathway develops relatively late in comparison to the dopaminergic, noradrenergic and serotonergic pathways, which mature early in the postnatal period. The noradrenergic pathway is one of the earliest to develop and may be important in regulating the developmental process of other systems. Noradrenergic neurons in the locus ceruleus appear in the mid-embryonic period (E10–E13) in rodents (Lauder and Bloom, 1974) and between E27 and E36 in primates (Levitt and Rakic, 1982). Immunocytochemical studies have shown that in rats, projections from these cells reach the cerebral cortex by the late embryonic period (E17), although the rate of synthesis of noradrenaline does not reach adult levels until about 2 months of age. Dopaminergic neurons develop in the ventral tegmental area and the substantia nigra around embryonic day 13 (Lauder and Bloom, 1974) in the rodent and on fetal days 36–43 in monkeys (Johnson, 1985). By birth, dopamine fibers penetrate the cortical plate and extend caudally into the cingulate cortex, reaching a mature innervation pattern by postnatal days (PND) 7–14. The concentration of dopamine reaches 15–29% of the adult levels by birth in rodents and then rises slowly between birth and PND 10 (Coyle and Henry, 1973). At that point, there is a rapid acceleration in the production and accumulation of dopamine, so that dopamine reaches 50% of adult levels by PND 30 and 75% of adult levels by PND 40. The dopaminergic system approaches maturity at around PND 55–60.

Serotonergic neurons also appear between E11 and E15 in the medial and dorsal raphe in the brain stem of rodents (Lauder and Bloom, 1974). Projections innervate the cortical regions within one week, and adult patterns of innervation are seen by PND 12 (Wallace and Lauder, 1983). Serotonin (5-hydroxytryptamine; 5-HT) concentrations in the neonatal cortex are approximately 33% of adult values reaching adult levels around PND 60. The serotonergic system also develops relatively early in primates, with adult patterns of innervation becoming evident by 6 weeks of age.

The cell bodies of cholinergic neurons appear in the basal forebrain between E12 and E15 and projections innervate the cortex (Eckenstein and Baughman, 1984; Bayer, 1985). The birthdate of cholinergic neurons in the basal forebrain is approximately E12–E15, and development occurs slowly until the end of the first postnatal week. At birth, the concentration of acetylcholine is approximately 40% of adult levels (Coyle and Yamamura, 1976). The development of cortical cholinergic innervation in humans parallels that of rodents, with acetylcholine esterase activity developing between 12 and 22 weeks in fetuses (Candy et al., 1985).

γ-Aminobutyric acid (GABA), the excitatory amino acids (glutamate, aspartate, and glycine), and the N-peptides (e.g., encephalins, somatostatin, cholecystokinin, substance P) develop late but reach adult levels relatively quickly. GABA neurons are primarily intrinsic interneurons and appear in the cerebral cortex of rats between E14 and E29 (Miller, 1985). Concentrations of GABA are low in the newborn, but reach adult levels by 3 weeks of age (Johnson and Coyle, 1980). Somatostatin neurons develop in the cortex between PND 1 and 14, and glutamate reaches adult concentrations around PND 15 (Kvale et al., 1983).

Although the morphological development of the neurotransmitter systems is complete 1–2 weeks after birth, the rate of synthesis and the concentration of some neurotransmitters do not reach adult levels until, in some cases, up to 1 or 2 months after birth in rodents. In humans, it may take 3 or 4 years after birth to reach the adult level. Most of the neurotransmitter systems that have been studied show an "overshoot" in innervation and/or synthesis of neurotransmitter at some point during development, which is then "pruned" back to the adult level. The early developmental pattern of central monoamine and acetylcholine systems suggested a role in controlling cell proliferation and dendritic growth (Blue and Parnavelas, 1982; Coyle et al., 1986; Hohmann et al., 1988; Lauder, 1988; Mattson, 1988; Leslie, 1993; Lauder, 1993). Catecholamines affect cell survival (Rosenberg, 1988). Dopamine and serotonin modify axon and dendritic outgrowth (for review, see Mattson, 1988; Lipton and Kater, 1989). Both the cholinergic and the monoaminergic systems are critical in regulating the response of the visual cortex to visual deprivation during a critical period of development (Bear and

Singer, 1986; Cline et al., 1987; Costantine-Paton et al., 1990). The dopaminergic system has been suggested to play a major role in neuronal differentiation. Following prenatal cocaine exposure, the anterior cingulate cortex, a cortical region receiving dense dopaminergic innervation, exhibits decreases in and altered patterns of apical dendrite growth (Jones et al., 1992). Cocaine exposure in rodents has also been reported to change cell-proliferative behavior (Gressens et al., 1992), temporal development of monoamines (Akbari et al., 1992), and dopamine receptor binding (Scalzo et al., 1990). The significance of the dopamine system in developmental cocaine exposure manifests itself in the studies of Wang and Murphy (1993) where it was shown that intravenous administration during the latter two thirds of gestation in the rabbit produced pyramidal neurons with stunted and twisted dendrites in certain cerebral cortical regions with high dopamine innervation. The mediation of specific patterns of glutamate-induced activity could impact survival of specific neuronal populations, cell motility, fiber outgrowth, and final connectivity patterns (for review, see Lipton and Kater, 1989; Leslie, 1993).

FUNCTION OF SYNAPTIC RECEPTORS

Synaptic receptors have two major functions: recognition of specific transmission and activation of effectors. The receptor recognizes and binds a transmitter in the external environment, which then alters the cell's biochemical state. Receptors for neurotransmitters can be grouped based upon how the receptor and effector functions are coupled. The nicotinic acetylcholine (Ach), the GABA, the glycine, the AMPA (kainate-quisqualate), and the N-methyl-D-aspartate (NMDA) class of glutamate receptors are receptors that gate ion channels directly. The other group of receptors that gate channels indirectly, G-protein coupled receptors, include the α- and β-adrenergic, serotonin, dopamine, and muscarinic Ach receptors and receptors for neuropeptides and rhodopsin. The receptor molecule is coupled to its effector molecule by a guanosine nucleotide-binding protein (G-protein), which produces a diffusible second messenger [cyclic adenosine monophosphate (cAMP), diacylglycerol, or an inositol polyphosphate] that initiates a biochemical cascade (activation of specific protein kinases for protein phosphorylation or mobilize Ca^{2+} ions from intracellular stores).

During brain development, the basal level of adenylate cyclase activity remains constant while agonist-stimulated adenylate cyclase increased with age (Keshles and Levitzki, 1984; Ma et al., 1991). The concentration of cyclic AMP increases with age, and cyclic GMP remains relatively stable in the rat forebrain. In the postnatally developing cerebellum, the basal level of cyclic GMP increases with age and glutamate-stimulated cyclic GMP accumulation reaches a peak at postnatal day 8 in cells undergoing differentiation (Spano et al., 1975; Garthwaite and Balazs, 1978). Cyclic AMP–dependent protein kinases remain relatively stable during postnatal development (Schmidt et al., 1980). The growth inhibitory properties of ethanol, which include alterations in cell proliferation and migration (Miller, 1986; Miller and Potempa, 1990) and morphological differentiation of neurons (West et al., 1986), have been proposed to be linked to interactions with levels of cyclic AMP and cyclic AMP–dependent protein kinases (Pennington, 1992).

During postnatal development, acetylcholine and glutamate are present in high concentrations in the neonatal rat brain. Most rat receptors—muscarinic, glutaminergic,

α_1-adrenergic, or histamine receptors—develop postnatally and reach adult levels at about 30–45 days of age. The development profile of phosphoinositide metabolism elicited by activation of these receptors varies for each neurotransmitter and in each brain region (Palmer et al., 1990; Balduini et al., 1991). The ability of norepinephrine and histamine to stimulate inositol metabolism increases with age (Schoeppe and Rutledge, 1985) in parallel to their receptors. The ontogenic profile associated with the presence of neurotransmitters and respective receptors suggests a critical role in the maturation of the nervous system (Lauder, 1988; Mattson and Hauser, 1991). In the rat, a majority of dendritic aborization, establishment of synaptic contacts, and astroglial cell proliferation occurs during the "brain growth spurt" between postnatal days 1 and 15 (Dobbing, 1974; Rodier, 1980). This is similar to the time when certain neurotransmitters—acetylcholine and glutamate—are present in high concentrations, suggesting a role in brain development. For example, acetylcholine stimulates DNA synthesis through the activation of phospholipase C–coupled muscarinic receptors in an age-dependent fashion in primary neonatal astroglial cells (Ashkenazi et al., 1989). Excitatory amino acid stimulation of inositol phospholipid metabolism reduced proliferation of cultured astrocytes (Nicoletti et al., 1990). In developing neuronal cultures, glutamate promotes growth and differentiation at low concentrations (Aruffo et al., 1987) and reduction in dendritic length and suppression of axonal outgrowth at higher concentrations (Farooqui and Horrocks, 1991).

Glutamate or related excitotoxins have been suggested to be responsible for neuronal degeneration in a variety of neurological disorders, including ischemia, seizure, and neurodegenerative diseases (Choi, 1988; Guidotti, 1990). While it has been suggested that excitatory amino acids may be particularly toxic to the developing brain, there is also a role for them in various processes of brain development. EAAs play multiple roles during development, e.g., regulating neuronal survival, growth and differentiation, neuronal circuitry, and cytoarchitecture (McDonald and Johnston, 1990). The sensitivity of neurons to EAA neurotoxicity changes during development. The susceptibility to various EAA agonists is different between brain regions and neuronal populations. For example, kainic acid is extremely toxic in the adult brain yet, relatively nontoxic in the immature rat brain (Coyle, 1983). The neurotoxicity of NMDA is augmented in the developing brain, and for some reason the effect is most severe on postnatal day 7 (McDonald et al., 1988). At this same time (postnatal days 7–14) quisqualate and AMPA peak in neurotoxicity (Silverstein et al., 1986; McDonald et al., 1992). The toxicity of EAAs may be linked to stimulation of phosphoinositide metabolism. Both quisqualate and glutamate stimulate a more pronounced breakdown of phosphoinositide in the neonatal hippocampal slice (Nicoletti et al., 1986).

MYELINATION

The myelinating cells in the nervous system are responsible for forming a sheath around the axon, which requires the small soma to produce and support many more times its own volume of membrane and cytoplasm. The myelin sheath allows for efficient conductance of signaling. The oligodendroglia serves in this capacity in the CNS and the Schwann cell in the PNS. Signaling interactions between the axonal membrane and the myelinating cells plays a critical role in maintaining both the functional and structural integrity of the system. Myelination is a critical process in the maturation of the nervous system. During defined periods of development, myelination occurs within a relatively brief interval, thus making it susceptible to perturbation during this time such as genetic disorders, viral infections, substances of abuse, environmental factors, toxicants, and nutrition (for reviews, see Wiggins, 1986). This time period corresponds to the final prenatal months and the first few years of postnatal life in humans and between days 15 and 30 of postnatal life in rats. Myelination starts after a burst of proliferation of the myelin-forming cells—Schwann cells in the PNS and oligodendroglia in the CNS. During the rapid deposition of myelin, a large portion of the brain's metabolic activity and protein and lipid synthesis are involved. Thus any metabolic insult during this "vulnerable period" could result in decreased myelin formation (Morell et al., 1993). If a disruption occurs during the proliferation of myelinating cells, there exists an irreversible deficit of myelin-forming cells and hypomyelination. Perturbation at a later time can result in a deficit in myelin amount that can be reversed.

The initial condition of myelination is the proliferation of myelin-forming cells, namely, Schwann cells in the PNS and oligodendroglia in the CNS (myelination gliosis). In the peripheral nerves, the Schwann cells surround bundles of axons initially, but gradually as the Schwann cells proliferate, they associate with fewer numbers of axons. The axons destined to be myelinated eventually establish one-to-one relationships with myelinating Schwann cells (Webster, 1971; Webster et al., 1973). A Schwann cell myelinates an internodal segment of a single axons, while an oligodendroglia myelinates an internodal segment of many axons. Loosely packed myelin sheaths become compact and the thickness of myelin sheaths around axons increase with time, proportional to the size of axons (Raine, 1984). The pattern of myelination is similar in mammalian species, and in general, myelination in the CNS advances in the descending order of the spinal cord, brain stem, cerebellum, and cerebrum (Jacobson, 1963). However, each fiber tract has its own spatio-temporal pattern in myelination, and thus the degree of myelination differs in different fiber tracts at a given stage of development. For example, in the spinal cord, myelination proceeds in a rostral to caudal gradient (Schwab and Schnell, 1989), and in optic nerve, myelination progresses with a retinal to chiasmal gradient (Skoff et al., 1980). With the aid of molecular biological techniques, complicated patterns of CNS myelination have been identified. In parallel with these morphological changes, significant biochemical changes, in particular in lipid metabolism, take place during the period of progressive myelination and stabilization.

LOCALIZATION OF FUNCTION

The CNS consists of six main parts: the spinal cord, the medulla oblongata, the pons (and cerebellum), the midbrain, the diencephalon, and the cerebral hemispheres. Through a variety of experimental methods, specific functions have been assigned to each of these brain regions. Although different regions are specialized for different functions, many sensory, motor, and other mental functions are subserved by more than one neural pathway. When one region or pathway is damaged, others are able to compensate partially for the loss. The compensatory process is known as parallel processing and can obscure a behavioral evidence for localization. The critical signaling functions of the brain—the processing of sensory information, the programming of motor and emotional responses, learning memory—are carried out by interconnected sets of neurons. Sensory pathways include neurons that link the receptor at the periphery with the spinal cord, brain stem, thalamus, and the cerebral cortex. Sensory systems receive information from the environment through receptors at the periphery of the body and transmit this information to the CNS.

To produce a behavior, each participating sensory and motor nerve cell generates, in sequence, four types of signals at four different sites within the neuron: an input signal (receptor potential in the sensory neuron and a synaptic potential in the interneuron or motor neuron), an integration signal, a conducting signal, and an output signal. Each component is located at a particular region in the neuron and carries out a special function in signaling. All of these signals depend on the electrical properties of the cell membrane. Almost all pathways transmitting sensory information to the cerebral cortex make connections in the thalamus, and the thalamic neurons of each sensory system project to a specific primary sensory area of the cerebral cortex. The somatic sensory system (the skin senses) uses the neurons of the dorsal root ganglion as its sensory receptor, and it detects input distinguished as pain, temperature, touch, and limb proprioception. This information is conveyed to the brain relayed through the spinal cord, where it is conveyed along two major ascending systems: one that carries information about tactile sensation and another that carries information about pain and temperature. The information continues through the thalamus, and it is not until it reaches the somatic sensory areas in the cortex that the inputs from the various submodalities interact.

In the human, brain functions relating to language are located primarily in the cerebral cortex, which is divided into four anatomically distinct lobes: frontal, parietal, occipital, and temporal. Each lobe has been associated with specific functions. The frontal lobe is responsible for planning for future action and control of movement, the parietal lobe for somatic sensation and body image, the occipital lobe for vision, and the temporal lobe for hearing and aspects of learning, memory, and emotion. The functions localized to discrete regions in the brain are not complex faculties of mind, but elementary operations. Elaborate faculties are constructed from the serial and parallel interconnections of several brain regions. The interruption of a single link within a pathway disrupts only the one pathway and need not interfere permanently with the performance of the system as a whole. To a limited extent, the remaining system can reorganize to perform the lost function.

MOTOR SYSTEM

The rodent is born with limited behavioral ability and undergoes extensive and rapid development of motor ability prior to the time of weaning. Early assessment of motor function, therefore, usually relies on reflexive tests of responding. The techniques used are observational and typically sensitive to only the most overt deviations in ontogeny or the ability of the subject. With maturation, most motor responses become stronger and show superior behavioral integration. Although limb muscles can be innervated functionally by motorneurons at any level of the spinal cord, the mechanisms for coordinated limb movements are restricted to the brachial and lumbosacral regions of the spinal cord; thoracic segments of the spinal cord or cranial motor nuclei are unable to participate in the control of limb movements. The character of limb movements is determined by the region of spinal cord from which it is innervated (Szekely, 1963). Early in development, the spinal cord segments at limb levels develop functional specificity for fore- or hindlimbs. There is evidence to suggest that the motor functions of the spinal neurons are specified in the late neurula and that specification increases progressively during development (Szekely, 1963).

A few responses are present only during early stages of development, such as "rooting," a necessary event that terminates in nursing, and pivoting. When these responses do not diminish with maturity, it is considered indicative of cerebral cortex damage. Thus, delays in the sequence of motor system development can be as important as a specific age-related deficit in determining the impact of the alteration. In these early stages of motor system development, there is no evidence to suggest a gender difference prior to weaning.

In addition to observational measurements of reflex responses, motor behavior can be assessed by monitoring movement from one location to another. One accepted approach for measuring locomotion is the use of photocell chambers, which monitor both vertical and horizontal movement. This automated measurement procedure limits the confounding variable of sensory feedback to the test animal and can be accompanied by visual observation to determine characteristics of the response. Mazes have also been used to assess motor behavior in young rodents. Both the figure-eight maze and the radial arm maze can be used for testing rats as young as postnatal day 12 and can continue through adulthood (Norton et al., 1975; Reiter et al., 1975; Olton, 1977). With animals tested as a litter, the developmental pattern of activity in a figure-eight maze shows that activity increases across the preweaning period. Initially, young pups will display a burst in activity during the first 5 minutes, which subsides, while by postnatal day 20 the activity becomes distributed across the test session. The response of the young pups is likely analogous to "huddling," since when tested individually in a locomotor activity test environment, young pups display high levels of activity at an earlier age.

Swimming ability has also been used to assess development of the motor system. A distinct developmental sequence for swimming ability has been described (Schapiro et al., 1970). In general, prior to PND 6 the young rat pup is unable to remain above water and would not survive. Between PND 6 and 11, pups show a progressive development of the required coordinated movements of both front and forelimbs to stay afloat. It is not until PND 12 that rat pups develop the ability to swim in a straight line.

Gait can be assessed by observational techniques or quantitated as foot splay at surface contact. Development of limb strength is dependent upon muscle development as well as neural innervation and can be quantitatively assessed by rotarod, forelimb/hindlinb grip strength, and inverted screen apparatus (Dunham and Miya, 1957; Coughenour et al., 1977; Meyer et al., 1979).

Wall climbing, a behavior specific to PND 7–17 (Barrett et al., 1982; Scalzo and Burge, 1992), has been used to assess the effects of drugs and chemicals. This technique emphasizes a specific topography of motor behavior rather than amount of movement and may be mediated initially by catecholaminergic activation and with maturation shifts to cholinergic activity (Spear et al., 1985; Linville and Spear, 1988). Exposure to environmental factors can dramatically increase the frequency of wall climbing. For example, incidence can be enhanced by elevations in temperature or by exposure to shock and the drugs PCP, cocaine, amphetamine, apomorphine, and clonidine.

Evaluation of motor behavior in various settings can offer information concerning other aspects of nervous system functioning. For example, motor behavior can be normal in a familiar environment, while subtle changes in motor functioning are noted when animals are placed in a stressful or novel environment (Ferguson et al., 1993). Therefore, depending on the experimental design and the test environment, one can either assess the integrity of the motor system or use motor behavior to detect shifts in an underlying neural process such as emotionality. For example, a variety of motor systems–associated endpoints were assessed in rhesus monkeys developmentally exposed to lead. Normal primates show different motor topographies in an open-field situation and a home-cage environment. No difference in behavior between test environments was seen in 6-year-old rhesus monkeys exposed to lead during their first year (Ferguson et al., 1993). In additional studies, lead has been shown to produce increased latencies to enter an open field, a decrease in locomotion while in the open field, as well as an increase in environmental exploration (Ferguson and Bowman, 1990, 1992).

In assessing human development, reflex tests, many of which emphasize various motor abilities, are incorporated. Many of these reflexes, such as crawling, grasping, Moro, rooting, stepping, and the tonic neck reflex, are behaviorally stereotyped at birth and disappear with maturation. This loss of stylized motor responses is dependent on the postnatal development of the cerebral cortex, and any persistence of the behavior is indicative of incomplete neurological development. One obtains greater voluntary control over motor responses as cortical development ensues. Since the use of fine motor skills is highly developed in humans and an important aspect of normal functioning, sequential testing of motor capabilities leading to walking and fine motor skills associated with writing are conducted in the developing human infant.

NEURAL SUBSTRATES OF BEHAVIORAL DEVELOPMENT

Extensive reviews on the learning capacities of developing animals and methods of assessing these functions can be found elsewhere (Spear and Rudy, 1991; Mactutus, 1994). A number of developmental neurotoxicants are known to produce impairments in cognitive functioning, with impaired intellectual development in humans and impaired performance of tasks of learning and memory in animals (Stanton and Spears, 1990). These neurotoxicants include ethanol, lead, methylmercury, polychlorinated biphenyls, antiproliferative agents, anticonvulsants, opioids, amphetamines, cocaine, and nicotine. The majority of the assessments have been conducted after developmental exposure, during either the juvenile or adult period, with limited research conducted during the early stages of development. Areas of exception include the animal model of fetal alcohol syndrome (Goodlett et al., 1987; Driscoll et al., 1990; Wigal and Amsel, 1990) and limited work with cocaine (Spear et al., 1989; Heyser et al., 1990) and heavy metals (Noland et al., 1982; Stanton, 1991). Types of tests used include olfactory conditioning (Barron et al., 1988), taste aversion learning (Riley et al., 1984), passive avoidance (Riley et al. 1979), appetite learning and extinction (Wigal and Ammel, 1990), and water maze (Goodlett et al., 1987).

In a working definition form, learning is usually defined as associative and nonassociative. Within these categories are additional subdivisions. Nonassociative learning refers to changes in behavior due to a single stimulus. These include habituation (a decrease in response with repeated presentations not due to fatigue or sensory adaptation) and sensitization (increasing response due to previous exposure to a stimulus). In the rat, habituation is present at birth; however, the ontogeny is influenced by the maturation of the response capabilities of the animal. For example, at birth simple reflex habituation is present, whereas habituation of more complex behaviors presents in line with the ontogeny of that behavior.

In associative learning, i.e., Pavlovian conditioning and instrumental learning, behavior changes as a function of the relationship between two or more events. Both of these procedures of associative learning have been used successfully in the developing animal (Spear and Rudy, 1991). The choice of sensory system, motivational state, and response system and the complexity of the learning phenomenon critically influence the developmental stage at which learning can be observed. Appetitive behavior in rats undergoes major changes during postnatal development. Sensitivity to shock reinforcement does not appear to change during the postnatal period. Developmental of the sensory system determines the type of stimuli that can be detected. In the rat, the tactile and chemical senses (olfaction and taste) guide behavior during the first 2 weeks of life, while auditory and visual senses being to function at the third week. Usually the development of a sensory system precedes by a couple of days the ability to learn a conditioned response based on the system. The aversive Pavlovian conditioning of an olfactory stimulus with footshock can be used in the neonatal period and has been used successfully in developmental neurotoxicology (Spear et al., 1989). In the rat, appetitive choice behavior (learning of a position of reinforcement) can be learned as early as postnatal day 7, whereas a more complicated task of delayed alternation or conditioned spatial discrimination requires use of recent memory and does not appear until between the second and third week postnatal (Green and Stanton, 1989). The distinctions between the ontogeny of tasks could be used to assess alterations in cognitive development in neurotoxicology. A comparison between simple learning tasks and tests of higher-order learning and memory may help characterize the developmental and neurobiological nature of a neurotoxicant's effects.

The understanding of the neurobiological basis of different forms of behavior has advanced dramatically over the last two decades. This has made it possible to identify and use a test paradigm that will detect alterations in an underlying mechanism or, in the reverse, to make alterations in a specific component of a test paradigm having a greater likelihood of being able to identify a region of injury in the nervous system. The emphasis in neurobiology and learning has been on the identification of critical brain regions and neural circuits. Following the identification of neural circuits, attempts are made to analyze the cellular and molecular mechanisms associated with neural plasticity (for review, see Cotman and Lynch, 1988). Once a circuit for a behavior has been identified, effort is made to identify the interactions between the various brain regions that contribute to the behavior in question. For example, research efforts on the neural basis for memory have identified the frontal and temporal regions of the brain as significant structures. The memory system impaired in amnesia consists of a circuit of sensory cortical areas. These systems include the limbic structures (hippocampus and amygdala) and associated cortical areas in the temporal lobe, specific thalamocortical systems, and projection systems from the basal forebrain (Mishkin and Appenzeller, 1987; Bachevalier, 1990). The maturation of these regions is a contributory factor in the development of cognitive functioning of rodents, monkeys, and humans (Goldman-Rakic et al., 1983; Nonneman, 1984; Diamond, 1990; Freeman and Stanton, 1991). While the prefrontal cortex has been identified as a potential target site of chemicals that alter memory development (Diamond et al., 1992), the hippocampus has been identified to be susceptible to a range of insults, e.g., antiproliferative agents (Altman and Bayer, 1975; Rodier, 1986), heavy metals (Stanton, 1992), and ethanol (Wigal and

Amsel, 1990; Goodlett and West, 1992). In the developing animal, there are a number of learning and memory tasks that can be used to assess early hippocampal damage (Altman and Bayer, 1975).

The learning of a specific response can involve different nervous system regions depending on the reflex or behavioral task required as the response. Eyeblink conditioning is a learning paradigm that has recently been applied in neurotoxicology and involves a neural circuit of the brain stem and cerebellum (Thompson, 1988). In this model the conditioned stimulus of a tone is conveyed to the cerebellum via the mossy fiber projections of the pontine nuclei. In the cerebellum the signal is conveyed by the axons of the granule cells to the Purkinje cells along with collateral input to the deep nuclei of the cerebellum. When the unconditioned stimulus input is an airpuff, the deep nuclei of the cerebellum are stimulated similarly by the Purkinje cells and collateral input, however, the stimulus is conveyed by the climbing fibers from the dorsal assessory olive. Thus, in an associative learning paradigm pairing a tone with an airpuff, the cerebellar cortex and deep nuclei are converged on by both the mossy fiber and climbing fiber input, respectively. While the cerebellum is essential for the learned conditioned response, it is not for a general eyeblink response. The neural linkage of the conditioned eyeblink response is an input stimulated by the conditioned response to the contralateral red nucleus back to the ipsilateral motor nuclei controlling the eyeblink response, while the linkage for the airpuff unconditioned stimulus is through a reflex pathway located in the brain stem. Thus, lesions of the cerebellar deep nuclei can abolish the conditioned responding of an eyeblink without affecting the reflexive eyeblink (Skelton, 1988). If the dorsal assessory olive is damaged, conditioning of the response cannot be achieved or maintained in previously trained animals (McCormick et al., 1985). Additional support for the involvement of these neural circuits in the conditioned eyeblink response comes from the observation that direct electrical stimulation of either region can serve to replace the associated stimulus. The maturation of the cerebellum, which occurs over a protracted period, is critical for the occurrence of eyeblink conditioning. Thus, this learning paradigm can allow for the assessment of alterations in structural maturation following an early insult. In the rat, the cerebellum matures postnatally, extending past the time of weaning. Linked to this is a late postnatal ontogeny of eyeblink conditioning (Stanton et al., 1992).

In assessing the developing organism, tests that can be used very early in the developmental process are limited but are also critical for the detection of an early lesion prior to any compensatory changes made by the nervous system that could mask the effect. One procedure that has been used successfully in the rat is olfactory learning since this is an early-developing sensory modality. There are also efforts to determine the neural basis of such learning, which should lead to an understanding of the neural circuitry similar to that known for the eyeblink conditioning previously mentioned.

The ability to assess the developing organism for behavioral functioning is dependent upon the maturation of both the sensory and motor systems. Four processes are engaged when a learned behavior is performed in response to a stimulus: sensory, motivational, learning, and motor. Any one of these can be affected by a neurotoxicant, and a careful distinction must be made between learning and performance deficits. For example, the previously mentioned paradigm of eyeblink conditioning is not only dependent upon the development of the cerebellum in the neural circuitry, it is also dependent upon the time of eye opening and the development of the auditory system to attend to the tone stimulus. If an assessment is made of the functioning of a sensory system without regard to the state of the motor system required for the recorded response, an incorrect evaluation can be made. This is also true in the mature organism. For example, if one is assessing learning and memory by a task requiring a motor response, e.g., bar press or running, caution must be taken to ensure that the paradigm used is not compromised by a decrease in performance ability.

ALTERATIONS IN NERVOUS SYSTEM DEVELOPMENT: STRUCTURE AND FUNCTION

Undernutrition and Brain Development

In the human infant, several studies have provided evidence supporting the concept of a critical period from birth to about 2 years of age during which the nervous system is most vulnerable to malnutrition and not responsive to nutritional therapy. The production of neurons is virtually completed by about the midpoint of gestation, after which glial cell production continues to the end of gestation and into the second postnatal year. In the rat, the production of neurons continues throughout gestation. The vulnerability of the developing nervous system to various factors is determined by the developmental stage of the cellular activities targeted by a specific insult. Therefore, the effects of an agent or condition will be different at different times during development. No one vulnerable period can be identified; it will vary depending on the agent and the timetable of developmental events in different species (Dobbing et al., 1971). Such general factors like undernutrition can have maximal effects on processes that are most active during what has been called the "brain growth spurt" (Dobbing and Sands, 1979). Thus, depending on the developmental process ongoing at the time of exposure, an alteration can be produced in either number of neurons, glial cells, or myelination. The effects of malnutrition on the cerebral cortex of the rat are correlated with the times of development of different neurons: layer V pyramidal cells are affected most in the early postnatal age, layer III pyramidal cells at early and later postnatal ages, and interneurons during the late postnatal age. The most striking effects of developmental undernutrition are the reductions in number of dendritic spines and number of synapses per neuron (Bedi et al., 1980, 1989; Warren and Dedi, 1982). In rats, malnutrition during the first 2–3 weeks after birth results in a 30% deficit in the synapse-to-neuron ratio in the cerebral cortex, which is augmented by rearing in isolation (Warren and Bedi, 1982). Malnutrition also delays the development of the cortical barrels and reduces the number of neurons (Vongdokmai, 1980). Such an effect could be due to a direct effect on the cerebral cortex or could be a secondary effect as a result of alterations in the sensory relay nuclei or the vibrissal follicles. In the mouse, severe neonatal malnutrition can result in the loss of neurons and axon terminals (Cragg, 1972).

The cerebellar cortex develops rapidly in the neonatal period and is particularly vulnerable to the effects of malnutrition (Persson and Sima, 1975). In the rat this is manifest as a stunting of growth of Purkinje cell dendrites. Neonatal underfeeding results in a reduction of Purkinje cell dendritic length and branching and density of dendritic spines. These effects on Purkinje cells are probably the result of a complex etiology. The effects could be direct on synthesis and assembly of membrane and cytoskeletal components of dendrites or indirect as an effect of loss of granule cells or glial cells. Neonatal undernutrition affects more severely the cerebellum of females than of males with regard to synaptogenesis, myelination, and total volume.

Myelination occurs mainly postnatally in both humans and rat and continues for the same time relative to lifespan (see Wiggins, 1986, for review). The process of myelination in both the central and peripheral nervous systems appears to be sensitive to undernutritional factors. Reduced myelination in the corpus callosum is associated with a reduction in number oligodendrocytes. There is a reduction in total myelin proteins and gangliosides. In the PNS, the Schwann cell ultrastructure is abnormal and the number of myelin lamellae is reduced.

Thyroid Hormone

The thyroid gland produces two hormones: L-thyroxine (T4) and L-triiodothyronine (T3). Thyroxine is the main circulating form, but it is converted to T3, which is the active hormone. T3 receptors are expressed on neurons in the mammalian brain during the fetal period while neuron production is active. The receptors continue to be expressed on both neurons and glia throughout the postnatal period. Thyroid hormone stimulates neural development in several ways: increasing cell proliferation, synthesis of microtubule-associated proteins (MAPs) and tubulin, increasing microtubule assembly, inducing axonal and dendritic outgrowth, synaptogenesis, and myelination.

T3 receptors are detectable in the rat brain on gestational day 20 (Schwartz and Oppenheimer, 1978), and the density reaches a maximum at birth in the cerebellum and at PND 9 in the cerebrum and falls to adult levels by PND 30. In humans, the neuron production in the cerebral cortex continues until 25 weeks of gestation, and the T3 receptors are detected as early as 10 weeks gestation and increase substantially during the next 6 weeks (Bernal and Pekonen, 1984). In the adult mammalian nervous system, there is a high level of expression of the T3 receptor in the hippocampus, amygdala, and cerebral neocortex (Ruel et al., 1985). T4 has a general action of increasing metabolic rate and enhancing tubulin synthesis, thus promoting neurite outgrowth. There are several reports of reduced axonal and dendritic growth in hypothyroid rats and of the converse in hyperthyroid rats. In hypothyroid rats, there is a reduction in length of parallel fibers and diminished branching of Purkinje cell dendrites (Nicholson and Altman, 1972). Thyroid hormone treatment of neonatal rats increases the size of cell body and dendrites of pyramidal cells in area CA3 of the hippocampus (Gould et al., 1990).

Neonatal thyroidectomy of rats decreases myelination, reduces neuronal volume with no change in number, and decreases dendritic branching (Eayrs, 1955) and number of axodendritic synapses in the cerebral cortex (Balazs et al., 1971), resulting in a decrease in brain growth after day 14. In the cerebellum of hypothyroid rats there is a reduction in the number of synapses in the molecular layer (Nicholson and Altman, 1972). However, the cerebellar axodendritic synapses appear normal and the effect is on dendritic growth. The rate of migration of granule cells in the cerebellum and outgrowth of parallel fibers is slowed, resulting in a decrease in length of fibers and possible inability to make adequate connections with Purkinje cell dendritic spines. This would lead to subsequent granule cell death. This retardation of neuronal migration may be associated with the concurrent retardation of differentiation of cerebellar astrocytes (Clos et al., 1982). This would include effects on the Bergmann glial cells, which are necessary for granule cell migration. Postnatal treatment of rats with T4 produces a decrease in brain weight, body weight, acceleration in the histogenesis and morphogenesis of the cerebellar cortex (Nicholson and Altman, 1972), increase in the number of dendritic spines in layer IV of the visual cortex (Schapiro et al., 1972), and an increase in S100 astrocyte protein in the cerebellum (Clos et al., 1982). The decrease in brain weight has been found to be due to a 30–40% reduction in the number of brain cells formed after birth.

Thyroid hormone receptors are present in neurons of the hippocampal formation, thus this brain region is very sensitive to alterations in the levels of thyroid hormones during development. Hyperthyroidism in newborn rats results in overgrowth of the hippocampal granule cell axons, the mossy fibers, which synapse on the hippocampal pyramidal cell dendrites (Lauder and Mugnaini, 1980). In hypothyroidism the mossy fibers appear normal while the volume of the hippocampus is decreased as the result of less granule cells. The primary effect of hypothyroidism on the developing cerebellum is the stunting of dendritic development, reduced outgrowth of parallel fibers leading to a failure in synaptogenesis, death of some granule cells, basket cells, and stellate cells. Similar effects are seen in the development of the cerebrum with retardation of differentiation and growth resulting in a reduction of dendritic growth and synaptogenesis.

Effects of Adrenal Glucocorticoids on Brain Development

There are two types of glucocorticoid receptors in the mammalian CNS: type 1 and type 2 (Funder and Sheppard, 1987). Type 1 receptors have a high affinity for corticosterone, and they are localized in the dentate gyrus granule cells and CA1 and CA2 pyramidal cells of the hippocampus. Type 2 receptors have the highest affinity for synthetic glucocorticoids and are localized on glial cells and neurons throughout the nervous system. Glucocorticoid receptors can be seen in the rat brain on gestational day 17 (Kitraki et al., 1984) and increase gradually to adult levels by postnatal day 20 (Clayton et al., 1977). In the cerebellar cortex, the expression of glucocorticoid receptors is correlated with the growth and disappearance of the external granular layer (Pavlik and Buresova, 1984). When administered during the neonatal period, glucocorticoids have diverse effects on the nervous system. They inhibit proliferation of neurons and neuroglial cells, delay cell differentiation (neurite outgrowth, synaptogenesis, myelination), and induce the catecholamine synthetic enzymes—tyrosine hydroxylase, dopamine β-hydroxylase, and phenylethanolamine N-methyltransferase—in neurons of neural crest origin. They induce glutamine synthetase in astrocytes and glycerol-3-phosphate dehydrogenese in oligodendrocytes (Meyer, 1985).

Sex Hormones in Brain Development

Initially the nervous system is indifferent to gender, and intracellular estradiol receptors develop in both males and females. Female brain organization takes place in the absence of testosterone during the critical period of birth $\pm$ 5 days, while male sexual behavior is under the influence of estrogen derived from neural aromatization of testosterone. Sexual dimorphism in the CNS is the result of sex hormones and can alter such things as number of neurons, reducing cell death, increasing cell growth, dendritic branching, synaptogenesis, and regulating patterns of synaptic function. It appears as if differential cell death is the major mechanism for the development of sex differences in the brain. The development of sexual behavior in mammals is determined by sex hormones that bind to specific receptors in the medial preoptic area, limbic forebrain structures, and ventromedial hypothalamic nuclei. A single injection of testosterone given to a female or castrated male rat within the first few days following birth results in mature male sexual behavior (Levine and Mullins, 1966). In both sexes, estradiol is the hormone that alters sexual behavior, and the effect of estradiol on the developing brain is to produce male sexual behavior as mediated by estrogen receptors. Estrogen binds to a protein in the blood and is not free to enter the brain, whereas testosterone or a synthetic estrogen is not bound by this protein and can freely enter the brain (McEwen et al., 1975). In the adult, testosterone is bound by the hypothalamus in the male and estradiol by the female (Plapinger and McEwen, 1973). This preference is not present in the developing animal. The binding of estradiol can be eliminated by experimental androgenization of female rats and normal development of the male (Vertes et al., 1973).

In mammals, the level of testosterone in the blood is high during the perinatal period and is required for the organization of neural mechanisms associated with male sexual behavior. The levels then decrease, only to rise again at puberty where testosterone is required for activation of some of the neural mechanisms for masculine behavior. The anatomical distribution of sex hormone–binding sites in the brain has been determined by radiolabeling techniques. The neurons in the preoptic region, tuberal region of the hypothalamus, rostral regions of the limbic system and hypothalamus, and the mesencephalon underneath the tectum are sites of sex steroid binding in all vertebrates.

Inorganic Lead

The developing nervous system has long been recognized as a primary target site for lead-induced toxicity. Subtle neuronal changes following developmental exposure to lead include reduction in hippocampal axonal and dendritic arborization (Petit et al., 1983) and synaptic elaboration (Petit and LeBoutillier, 1979). Similar reductions have been reported for cortical ontogenesis (Krigman and Hogan, 1974) and synaptic elaboration (Averill and Needleman, 1980; McCauley et al., 1982; Bull et al., 1983). Developmental alterations are not limited to the neuronal population. The astroglial cell population has been shown to be altered following lead exposure. Krigman and coworkers (1974) reported an increase in the number of astroglia in the cortex, while others report both structural and functional changes of astroglia in response to lead exposure (Goyer and Rhyne, 1973; Holtzman et al., 1987; Cookman, 1988; Tiffany-Castiglioni et al., 1989; Legare et al., 1993; Dave et al., 1993; Buchheim et al., 1994). Recently it was reported that early developmental exposure to lead acetate results in a disruption of the normal developmental profile of expression for both GAP-43 mRNA, a protein associated with the axonal growth cone (Schmitt et al., 1996), and GFAP mRNA (glial fibrillary acidic protein), and astrocyte specific protein (Harry et al., 1996). In addition, the cell adhesion molecule, N-cadherin has been shown to be altered in function following lead exposure accompanied by a delay in maturation and neural fold fusion (Lagunowich et al., 1993). Postnatal lead exposure has been reported to cause abnormalities in the development of cerebellar Purkinje neuron dendrites which may be related to differences in the development of synaptic contacts with these neurons. During development, the accumulation of brain myelin is decreased with a continued lower level until adulthood (Toews et al., 1980; 1983). Low doses of lead which can produce microscopic hemorrhagic encephalopathy in the cerebellum do not depress myelination (Sundstrom and Karlsson, 1987).

Organic Lead

The active trialkyl metabolites trimethyl lead (TML) and triethyl lead (TEL) are potent neurotoxicants. Morphologically, the lesions produced following exposure are similar to those seen following TMT exposure, although the neuronal damage may be somewhat more diffuse (Seawright, 1984). Neuronal damage is seen in the hippocampus, frontal cortex, and cerebellar Purkinje cells (Ferris and Cragg, 1984). TEL administered during development severely inhibited myelination (Konat and Clausen, 1984). The impairment appeared to more specific for myelination in animals exposed between 20 and 24 days of age compared to an earlier time of PND 14–18. In the forebrain, synthesis of myelin proteins was more severely suppressed than total protein, offering evidence for myelin specificity (Konat et al., 1984). In rats 2–4 weeks of age, the formation of myelin is the major biosynthetic process ongoing in the brain, and it is critically dependent on posttranslational processing and transport of certain proteins. Although the protein composition of myelin isolated from TEL-intoxicated young rats was identical to controls, some aspects of posttranslational processing in the synthesis of integral membrane proteins were inhibited (Konat and Clausen, 1980; Konat and Offner, 1982). Behavioral alterations seen following developmental exposure to tetraethyl lead are similar to those found in the adult. Body weight is marginally decreased, and brain growth is decreased comparatively more (Booze et al., 1983; Konat, 1984). This is accompanied by tremor, hyperreactivity, social aggression, and seizures/convulsions.

Cadmium

The acute effect of cadmium toxicity is on the vascular endothelium. However, not all vascular endothelia in the body are affected. Mice, rats, and rabbits that received intraperitoneal or subcutaneous injections of cadmium during the immediate postnatal period developed encephalopathy characterized by extensive petechial hemorrhages and endothelial vacuolation (Webster and Valois, 1981). Mice that received a subcutaneous injection on postnatal day 1 and were killed 24 hours later showed early degenerative changes in the cortical neurons in addition to widespread petechial hemorrhages in the cerebrum, cerebellum, and brain stem. The mice that survived after treatment on day 1 were smaller than controls, hypoactive, and showed deficits in coordination. The cerebrum and cerebellum of these mice examined at 8 weeks of age were small with dilated ventricles and thin cortex. Susceptibility of the brain to cadmium toxicity diminishes with age. Cadmium exposure at PND 15 showed hemorrhages limited to the cerebellum, although neuronal degeneration was still noted. No visible changes were noted in the brain of mice exposed to cadmium at PND 22. This decrease in susceptibility was thought to parallel the maturation of capillaries in the brain parenchyma (Webster and Valois, 1981).

Mercury

Congenital MeHg poisoning is also called fetal Minamata disease due to the massive outbreak of methylmercury poisoning in Minamata Bay and Niigata, Japan, during the

1950s. A similar poisoning occurred as late as 1971–1972 in Iraq. The brains of a young child and an infant with Minamata disease showed widespread neuronal damage with associated gliosis and cytoarchitectural abnormalities such as ectopic cell masses and disorganization of the cortical layers (Takeuchi and Eto, 1972). The major neuropathological findings included disturbed neuronal migration, and laminar cortical organization in the cerebrum and cerebellum were reported in the infants born to mothers who had ingested methylmercury-contaminated bread during pregnancy (Choi et al., 1978). Deficits in myelination and abnormalities in the cerebellar development also have been reported in experimental animals (Burbacher et al., 1990). The cases of Minamata disease are similar in disturbances in neuronal migration, defective lamination of the cerebral cortex, changes in the pattern of the cerebral gyri, and either gliosis or neuroglial heterotopias. It has been proposed that organomercurials induce their effects by inhibiting neuronal migration. While this has been difficult to confirm in an experimental model, a transient inhibition of neuronal migration has been reported in exposed rodent pups (Reuhl and Change, 1979; Choi, 1983; Change, 1984b), however, by weaning the cytoarchitecture of the brain appears normal (Inouye et al., 1985). Since it has been difficult to confirm permanent neuronal misplacement, research efforts have been directed toward processes known to be intricately involved in neuronal migration. Interest has focused on components of the cytoskeleton, in particular the microtubular system. As previously mentioned, microtubules are cytoplasmic organelles involved in mitosis, cell movement, process formation, and axonal transport (for review, see Dustin, 1984). MeHg has antimitotic activities (Ramel, 1967), and exposure during mitosis results in arrest during the late prometaphase of metaphase stages of division (Ramel, 1967; Rodier et al., 1984). Cytoplasmic microtubules rapidly depolymerize in the presence of micromolar concentrations of MeHg (Miura et al., 1984; Sager and Syversen, 1986). In culture, neuroblastoma cells and neurons in brain explants retract neurites and growth cones with exposure to MeHg. The major constitutive protein of the growth cone is growth associated protein-43 (GAP-43). A loss of microtubules could interfere with neuronal migration in several ways, including the disruption of microtubule-dependent cell-cell communication required for oriented neuronal movement. A delay in neuronal mitosis could result in neurons migrating asynchronously with other parts of the brain. Such a process has been proposed to occur with ethanol exposure (Miller, 1986). In postmigratory neurons, exposure to MeHg could result in disassembled microtubules in neurites altering dendritic arborization, synaptic organization, or cell activities. The antimicrotubule effects of MeHg have been shown to be reversible in the absence of MeHg (Sager and Syversen, 1986). Cell adhesion molecules such as N-CAM are altered following methylmercury poisoning (Reuhl and Change, 1979). In the mouse, neonatal exposure rapidly converts the immature form of N-CAM to the more mature one and alters the posttranslational modification (Lagunowich et al., 1991; Graff et al., 1992).

Methylmercuury has also been shown to inhibit the incorporation of sulfate into myelin-characteristic lipid sulfatide (Grundt et al., 1974). Methylmercury also inhibited the specific enzyme activity of UDP-galactose (ceramide galactosyltransferase) and of phosphoadenosine-5′-phosphosulfate (cerebroside sulfotransferase) (Grundt and Neskovic, 1980). The effects of methylmercury on the neurotransmitter systems of the developing brain include alterations in the catacholaminergic and serotonergic systems. Neonatal exposure to 2.5 or 5 mg/kg methylmercury daily for 21 days increased dopamine turnover and decreased synaptosomal [^{3}H]-dopamine uptake (Bartolome et al., 1982).

Norepinephrine turnover and steady-state levels were also increased. Administration of 5 mg/kg MeHg on PND 5–7 resulted in a decrease in whole-brain concentrations of tryptophan, serotonin, 5-hydroxyindoleacetic acid, dopamine, and norepinephrine, tryptophan hydroxylase activity, and serotonin turnover at PND 8 (Taylor and DiStefano, 1976). This was followed by an increase in serotonin, 5-hydroxyindolacetic acid, dopamine, and norepinephrine concentrations at PND 15 with the increases in serotonin persisting until PND 60. Inhibition of uptake, along with stimulation of uptake and release of dopamine, serotonin, and norepinephrine, have also been observed (Komulainen and Tuomisto, 1981; Komulainen et al., 1983). The activity of glutamic acid decarboxylase, a marker for GABA neurons, was decreased in the striatum and cortical areas by postnatal methylmercury exposure (O'Kusky and McGeer, 1985). Since the GABA systems are known to modulate monoaminergic neurotransmission, the monoamine changes noted may be related to effects on the GABA systems.

Tellurium

Toxicant-induced demyelination in the developing animal is rapidly generated in the PNS following early postweaning exposure to tellurium. In this model, the primary effect is on the Schwann cells, resulting in a preferential loss of the large myelin intermodes. The demyelination is associated with a specific metabolic block in the biosynthesis of cholesterol with the inhibition of squalene epoxidase (Harry et al., 1989). This effect is seen primarily in the young animal and may be associated with an absence of sufficient cholesterol for the structural formation of myelin. The affected Schwann cells differentiate, proliferate, and autophagocytize their myelin sheath. The highly synchronized pattern of demyelination is rapidly followed by remyelination. Prenatal administration results in hydrocephalus (Duckett, 1971).

6-Aminonicotinamide

6-Aminonicotinamide (6-AN) is an antimetabolite of nicotinamide and an inhibitor of NADP-dependent dehydrogenases. Gestational exposure to 6-AN produces congenital hydrocephalus and excencephaly in rats (Chamberlain and Nelson, 1963; Turbow and Chamberlain, 1968). In these hydrocephalic brains, various degrees of necrosis and hemorrhages were noted, but there was no evidence of obstruction of the ventricular system. Postnatal exposure causes vacuolation of glial cells and degeneration of ependymal cells, causing an age-dependent obstructive hydrocephalus due to aqueductal stenosis (Aikawa and Suzuki, 1987). Vulnerability appeared to be higher between PND 1 and 5, with a decrease by PND 10. When the injection was carried out 1 or 5 days postnatally, incidence of hydrocephalus was 100%, but dropped to 65% in 10-day-old mice. The glial cells appear to be more susceptible to 6-AN toxicity than neurons, and due to the disruption of neuroglial interactions during development, the cerebellum is malformed in rats receiving early repeated postnatal injections of 6-AN (Soteol and Rio, 1980).

Trimethyltin

Trimethyltin (TMT) produces a distinct pattern of neuronal degeneration in the hippocampus. In the adult rat this pattern is characterized by necrosis of pyramidal cells in the Ammon's horn. In the neonatal rat, the hippocampus is most vulnerable at post-natal days 13–15, with stunted development of both the body and brain and total destruction of the Ammon's horn observed (Chang, 1984a). A selective and specific neuronal destruction can be observed in various subfields or segments of the Ammon's horn when the animals are exposed at various neonatal ages. The age-dependent pattern of pathology was correlated with the development and functional maturity of the hippocampal neurons. The functional manifestation of such changes include agression, hyperirritability, tremor, spontaneous seizures, hyperreactivity, and changes in schedule-controlled behavior (Brown et al., 1979; Dyer et al., 1982).

In the mouse, the sensitivity of the fascia dentata to TMT toxicity increases rapidly, while that of the Ammon's horn decreases, until by the end of the second week of postnatal life, the mouse pup displays the adult response to acute treatment (Reuhl et al., 1983; Reuhl and Cranmer, 1984). The correlation between the growing functional maturity of the fascia dentata and its vulnerability to TMT may be associated with TMT acting as a excitotoxin requiring synaptic maturity to express its neurotoxicity. The seizure activity seen with TMT in animals older than 2 weeks might contribute to the neurotoxicity by the release of endogenous excitotoxins such as glutamate. Neonatal rats dosed with TMT (1 mg/kg/every other day/PND 3–30) showed decreased GABA concentrations in the hippocampus by PND 55. Dopamine levels in the striatum were also decreased, whereas concentrations of DOPAC were unchanged, suggesting compensation by an increase in release (Mailman et al., 1983).

Triethyltin

The basic CNS change induced by exposure to triethyltin (TET) is massive cerebral edema, restricted primarily to the white matter or myelin (Magee et al., 1957; Torak et al., 1960, 1970) with the formation of myelin vacuoles (Jacobs et al., 1977). The pathological effect varies with the age of the animal (Suzuki, 1971). When newborn rats were exposed to TET, the brains became swollen and petechial hemorrhages were observed, particularly in the cerebellum. Necrotic cells were found diffusely throughout the brain (Watanabe, 1977). When older animals (postnatal day 8) were exposed, both the hemorrhagic and necrotic changes occurred, however, damage was seen in the myelinated fibers of the brain stem and cerebellum. The morphological alteration in myelin dissipates over time, however, biochemical evidence suggests that the amount of myelin produced is decreased and myelin protein deficits persist until adulthood (Blaker et al., 1981; Toews et al., 1983). In all cases, the protein composition of myelin remained intact (Blaker et al., 1981). Exposure of PND 5 rat pups with 6–9 mg/kg TET induced an anterior-posterior atrophy of the brains with cortical thinning and degenerative neuronal changes, infiltration of microglia, and increased cellular density in the cortex (Veronesi et al., 1982; Veronesi and Bondy, 1986). Developmental exposure to TET results in persistent behavioral changes including alterations in learning and memory and dopaminergic functioning (Harry and Tilson, 1981; Reiter et al., 1981). Exposure to neonates results in permanent behavioral changes, which become particularly evident as the animal matures (Miller, 1984). With high levels of exposure there is a high degree of developmental delay followed by hyperactivity with maturation and diminished male sexual performance. Brain weight shows a permanent dose-related decrease. In 5-day-old rat pups, the in vivo brain concentrations of tin following a neurotoxic dose of TET are sufficient to inhibit mitochondrial adenosine triphosphatase

(ATPase) and thus may lead to cell death, while in the adult the tin binds to myelin thus decreasing the free amount of tin available to inhibit mitochondrial ATPase.

Hexachlorophene

2,2g162-Methylene bis(3,4,6-trichlorophenol) is an antimicrobial agent that has been previously used in soaps and detergents and was used in the bathing of newborn babies to prevent bacterial infections (Herter, 1959; Powell, et al., 1973; for review, see Towfighi, 1980). In young rats, edema of the myelin sheath is evident after postnatal day 15, allowing for the accumulation of myelin membrane wrapping and the appearance of interperiod lines, thus offering a site for fluid accumulation and a hydrophobic reservoir for the compound (Nieminen et al., 1973). Developmental exposure results in a decrease of the normal accumulation of myelin during development (Matthieu et al., 1974).

Cuprizone

Bis-cyclohexanone oxalydihydrazone is a copper chelator that results in CNS demyelination following dietary exposure to weanling mice. The loss can reach as much as 70% in white matter regions of the cerebrum (Carey and Freeman, 1983). In this model, the mRNA for myelin-associated glycoprotein, a protein located at the myelin-axonal interface, is downregulated during demyelination and returns to normal levels following cessation of exposure (Fujita et al., 1990).

Excitotoxicants

In the developing organism, the immaturity of the blood-brain barrier may enable passage of chemicals into the nervous system that would be excluded in the adult brain. This raises concerns with regard to such substances as glutamate, which can be consumed at high dietary levels in the form of monosodium glutamate (MSG). Excessive amounts of glutamate are highly toxic to neurons, especially immature neurons (Murphy and Baraban, 1990; Murphy et al., 1990). Since glutamate is the major excitatory transmitter in the brain, almost all cells in the brain have receptors that respond to it. In tissue culture even a brief exposure to high concentrations of glutamate will kill many neurons, an action called glutamate toxicity. This is thought to be primarily due to an excessive inflow of Ca^{2+} through NMDA-activated channels. High concentrations of intracellular Ca^{2+} may activate Ca^{2+}-dependent proteases and may produce free radicals that are toxic to the cell. Glutamate toxicity may contribute to cell damage after stroke, to the cell death that occurs with persistent seizures in status epilepticus, and to degenerative diseases such as Huntington's chorea.

It has been proposed that early ingestion of glutamate by young children may cause disruption of normal growth and development by an interaction with the neuroendocrine regulatory systems (see Olney, 1995, for reviews). In 1976, Olney and coworkers demonstrated that a subtoxic dose of glutamate to young adult rats induced a rapid increase in blood levels of luteinizing hormone (LH). Glutamate and the excitatory amino acid analogs of glutamate act upon the LH axis via secretion by hypothalamic neurons of gonadotrophin-releasing hormone (Olney and Price, 1980; Cicero et al., 1988). This secretion into the portal blood delivers GnRH to the pituitary and stimulates the release of LH. The administration of NMDA to prepubescent male rhesus monkeys can cause premature onset of puberty (Plant et al., 1989, 1990). In addition, glutamate or aspartame will elevate plasma LH, prolactin, and growth hormone in the same age monkey (Medhamurthy et al., 1990). A systemic injection of L-cysteine (cys) results in a generalized pattern of damage throughout an immature brain (Olney et al., 1972). When injected directly into the brain of an immature rat, both cysteine and NMDA produce damage in the frontoparietal neocortex, hippocampus, septum, caudate nucleus, and thalamus (Ikonomidou et al., 1989). MK-801, a NMDA antagonist, can protect against both chemical patterns of damage (Olney et al., 1990).

The sensitivity of the nervous system to NMDA is developmentally regulated. The sensitivity of the 7-day-old rat brain to NMDA is approximately 60 times greater than that of an adult rat brain (McDonald et al., 1992). Each subpopulation of neurons may express different timetables for receptor maturity, thus varying onset and duration of peak sensitivity. This variability in maturity may account for the various patterns of brain damage as a function of developmental stage in which the neurotoxic process is triggered. By comparison, the potent neurotoxicant kainic acid is much less potent in the developing brain as compared to the adult. The quisqualate receptor is also developmentally regulated, reaching peak sensitivity slightly later than the NMDA receptor.

SUMMARY

In many cases, the underlying mechanisms responsible for neurotoxicity have yet to be determined. However, when a substance alters the sensory or motor functions, disrupts the processes of learning and memory, or produces detrimental behavioral effects, it is characterized as neurotoxic. The vulnerability of the developing nervous system to toxic substances is due to a number of characteristics inherent in the biological system. Minor changes in the structure or function of the nervous system can lead to profound consequences for neurological, behavioral, and related body functions. Components of the nervous system display an ontological profile and demand that the normal developmental profile be examined in addition to any evaluation of a shift in development due to toxicant exposure. A combination of basic science research into the process of normal development and evaluation of the system following perturbation may allow for an understanding of the critical processes involved in the normal functioning of the nervous system.

REFERENCES

Aghajanian, G. K., and Bloom, F. E. (1967). *Brain Res.* 6: 716.
Aikawa, H., and Suzuki, K. (1987). In *Stroke and Microcirculation* (J. Cervos-Navarro and R. Ferszt, eds.). Raven Press, New York, p. 351.
Akbari, H. M., Kramer, H. K., Whitaker-Azmitia, P. M., Spear, L. P., and Azmitia, E. C. (1992). *Brain Res.* 572: 57.
Altman, J., and Bayer, S. (1975). In: *The Hippocampus: Structure and Development.* Vol. 1 (R. L. Isaacson and K. H. Pribram, eds.). Plenum Press, p. 95.
Aruffo, C., Ferszt, R., Hildebrandt, A. G., and Cervos-Navarro, J. (1987). *Dev. Neurosci, 9:* 228.

Arvidson, B. (1986). *Neurotoxicology*, 7: 89.

Ashkenazi, A., Ramachandran, J., and Capon, D. J. (1989). *Nature* 340.

Astrom, K. E. (1967). *Prog. Brain Res.*, 26: 1.

Bachevalier, J. (1990). In: *The Development and Neural Basis of Higher Cognitive Functions* (A. Diamond, ed.). New York Academy of Sciences Press, New York, p. 457.

Baird, D. H., Baptista, C. A., Wang, L.-C., and Mason, C. A. (1992). *J. Neurobiol.*, 23: 579.

Balazs, R., Kovacs, S., Cocks, W. A., Johnson, A. L., and Eayrs, J. T. (1971). *Brain Res.*, 25: 555.

Balduini, W., Candura, S. M., and Costa, L. G. (1991). *Dev. Brain Res.* 62: 115.

Balin, B. J., Broadwell, R. D., Salcman, M., and El-Kalliny, M. (1986). *J. Comp Neurol.*, 251: 260.

Bally-Cuif, L., Alvarado-Mallart, R. M., Darnell, D. K., and Wassef, M. (1992). *Development*, 115: 999.

Barbin, G. Manthorpe, M., and Varon, S. (1984). *J. Neurochem.*, 43: 1468.

Barrett, B. A., Caza, P., Spear, N. E., and Spear, L. P. (1982). *Physiol. Behav.*, 29: 501.

Bartolome, J., Trepanier, P., Chait, E. A., Seidler, F. J., Deskin, R., and Slotkin, T. A. (1982). *Toxicol. Appl. Pharmacol.*, 65: 92.

Bates, C. A., and Killackey, H. (1984). *Dev. Brain Res.*, 13: 265.

Bayer, S. (1985). *Int. J. Dev. Neurosci.*, 3: 229

Bayer, S., and Altman, J. (1991). *Neocortical Development*. Raven Press, New York.

Bear, M. F., and Singer, W. (1986). *Nature*, 320: 172.

Bedi, K. A., Hall, R., Davies, C. A., and Dobbing, J. (1980). *J. Comp. Neurol.*, 193: 863.

Bernal, J., and Pekonen, F. (1984). *Endocrinology*, 114: 677.

Blaker, W. D., Krigman, M. R., Thomas, D. J., Mushak, P., and Morell, P. (1981). *J. Neurochem.*, 36: 44.

Blottner, D., Briggemann, W., and Unsicker, K. (1989). *Neurosci. Lett.*, 105: 316.

Blue, M. E., and Parnavelas, J. G. (1982). *J. Comp. Neurol.*, 205: 199.

Blue, M. E., and Parnavelas, J. G. (1983). *J. Neurocytol.*, 12: 697.

Booze, R. M., Mactutus, C. F., Annau, Z., and Tilson, H. A. (1983). *Neurobehav. Toxicol. Teratol*, 5: 367.

Brand, S., and Rakic, P. (1984). *Anat. Embryol.*, 169: 21.

Bray, D., and Hollenbeck, P. J. (1988). *Ann. Rev. Cell Biol.*, 4: 43.

Brightman, M., Klatzo, I., Olsson, Y., and Reese, T. (1970). *J. Neurol Sci.*, 10: 215.

Broadwell, R. D., and Brightman, M. W. (1976). *J. Comp. Neurol.*, 166: 257.

Brown, A. W., Aldridge, W. N., Street, B. W., and Verschoyle, R. D. (1979). *Am. J. Pathol.*, 97: 59.

Buchheim, K., Moack, S., Stoltenburg, G., Lilienthal, H., and Winneke, G. (1994). *Neurotoxicology*, 15: 665.

Bully, R., McCauley, P., Taylor, D., and Croften, K. (1983). *Neurotoxicology*, 4: 1.

Caminiti, R., and Innocenti, G. M. (1981). *Exp. Brain Res.*, 42: 53.

Campbell, B. A., Lytle, L. D., and Fibiger, H. C. (1969). *Science*, 166: 635.

Candy, J., Perry, E., Bloxham, C., Thompson, J., Johnson, M., Oakely, A., and Edwardson, J. (1985). *Neurosci. Lett.*, 61: 91.

Carey, E. M., and Freeman, N. M. (1983). *Neurochem. Res.*, 8: 1029.

Cavanagh, M. E., Corelis, M. E., Dziegielewska, K. M., Evans, C. A. N., Lorscheider, F. L., Mollgard, K., Reynolds, M. L., and Saunders, N. R. (1983). *Dev. Brain Res.*, 11: 159.

Chamberlain, J. G., and Nelson, M. M. (1963). *Proc. Soc. Exp. Biol. Med. 112*: 836.

Chang, L. W. (1984a). *Bull. Environ. Contam. Toxicol.*, 33: 295.

Chang, L. W. (1984b). *Toxicology and the Newborn* (S. Kacew and M. J. Reasor, eds.). Elsevier, Amsterdam, p. 175.

Choi, B. H. (1983). In: *Reproductive and Developmental Toxicity of Metals* (T. W. Clarkson, G. F. Nordberg, and P. R. Sager, eds.). Plenum, New York, p. 473.

Choi, B.H., Lapham, L.W., Amin-Zaki, L. and Saleem, T. (1978). *J. Neuropathol. Exp. Neurol.*, *37*: 719.

Choi, D. W. (1988). *Neuron*, *1*: 623.

Clarke, P. G. H., and Hornung, J. P. (1989). *J. Comp. Neurol.*, *283*: 438.

Clayton, C. J., Grosser, R. I., and Stevens, W. (1977). *Brain Res.*, *134*: 445.

Cline, H. T., Debski, E. A., and Constantine-Paton, M. (1987). *Proc. Natl. Acad. Sci. USA*, *84*: 4342.

Clos, J., Legrand, C., Legrand, J., Ghandour, M. S., Labourdette, G., Vincendon, G., and Gombos, G. (1982). *Dev. Neurosci.*, *5*: 285.

Constantine-Paton, M., Cline, H. T., and Debski, E. (1990). *Annu. Rev. Neurosci.*, *13*: 129.

Cookman, G., Hemmes, S., Keane, G., King, W., and Regan, C. (1988). *Neurosci. Lett*, *86*: 33.

Corasaniti, M. T., Defilippo, R., Rodino, P., Nappi, G., and Nistico, G. (1991). *Funct. Neurol.*, *6*: 385.

Cotman, C. W., and Lynch, G. S. (1988). In: *Learning Disabilities: Proceedings of the National Conference* (J. F. Kavanagh and T. J. Truss, Jr., eds.). York Press, Parkton, MD, p. 1.

Coughenour, L. L, McClean, J. R., and Parker, R. B. (1977). *Pharmacol. Biochem. Behav.*, *6*: 351.

Cowan, W. M., Fawcett, J. W., O'Leary, D. M., and Stanfield, B. B. (1984). *Science*, *225*: 1258.

Coyle, J. T. (1983). *J. Neurochem*, *4*: 1.

Coyle, J., and Henry, J. (1973). *J. Neurochem.*, *21*: 61.

Coyle, J., and Yamamura, H. (1976). *Brain Res.*, *118*: 429.

Coyle, J. T., Oster-Granite, M. L., and Gearhart, J. D. (1986). *Brain Res. Bull.*, *16*: 773.

Cragg, B. G. (1972). *Brain*, *95*: 143.

Crepel, F., Mariani, J., and Delhaye-Bouchard, N. (1976). *J. Neurobiol.*, *7*: 567.

Crittenden, S. L., Pratt, R. S., Cook, J. S., Balsamo, J., and Lilien, J. (1987). *Development*, *101*: 729.

Cunningham, T. J. (1982). *Intl., Rev. Cytol.*, *74*: 163.

Dave, V., Vitarella, D., Aschner, J. L., Fletcher, P., Kimelberg, H. K., and Aschner, M. (1993). *Brain Res.*, *618*: 9.

Davis, M. R., and Constantine-Paton, M. (1983). *J. Comp. Neurol.*, *221*: 444.

Dermietzel, R., and Krause, D. (1991). *Int. Rev. Cytol.*, *127*: 57.

Detrick, R. J., Dickey, D., and Kintner, C. (1990). *Neuron*, *4*: 493.

Diamond, A. (1990). *The Development and Neural Basis of Higher Cognitive Functions*. New York Academy of Sciences Press, New York.

Dobbing, J. (1974). In: *Scientific Foundations of Pediatrics* (J. A. Davis and J. Dobbing, eds.). W. B. Saunders Co., Philadelphia, p. 565.

Dobbing, J., and Sands, J. (1979). *Early Hum. Dev.*, *3*: 79.

Dobbing, J., Hopewell, J. W., and Lynch, A. (1971). *Exp. Neurol.*, *32*: 439.

Drenckhahn, D., and Lullman-Rauch, R. (1979). *Neuroscience*, *4*: 697.

Duckett, S. (1971). *Exp. Neurol.*, *31*: 1.

Dunham, N. W., and Miya, T. S. (1957). *J. Am. Pharm. Assoc.*, *46*: 208.

Dustin, P. (1984). *Microtubules*. Springer-Verlag, Berlin.

Dyer, R. S., Walsh, T. J., Wonderlin, W. F., and Bercegeay, M. (1982). *Neurobehav. Toxicol. Teratol.*, *4*: 127.

Dyson, S. E, and Jones D. G. (1976). *Prog. Neurobiol.*, *7*: 171.

Dyson, S. E., and Jones, D. G. (1980). *Brain Res.*, *183*: 34.

Eayers, J. T. (1955). *Acta Anat.*, *25*: 160.

Eckenstein, F., and Baughman, R. (1984). *Nature*, *309*: 153.

Edelman, G. M., and Thiery, J. P., eds (1985). *The Cell in Contact: Adhesions and Junctions as Morphogenetic Determinents*. John Wiley & Sons, New York.

Farooqui, A., and Horrocks, L. A. (1991). *Brain Res. Rev.*, *16*: 171.

Feldman, M. L. (1984). In: *The Cerebral Cortex*, Vol. 1 (A. Peters and E. G. Jones, eds.). Plenum Press, New York, p. 123.

Ferguson, S. A., and Bowman, R. E. (1990). *Neurotoxicol. Teratol.*, *12*: 92.

Ferguson, S. A., and Bowman, R. E. (1992). *Neurotoxicol. Teratol.*, *14*: 73.

Ferguson, S. A., Medina, R. O., and Bowman, R. E. (1993). *Neurotoxicol. Teratol.*, *154*: 145.

Ferrara, N., Ousley, F., and Gospodarowicz, D. (1988). *Brain Res.*, *462*: 223.

Ferris, N. J., and Cragg, B. G. (1984). *Acta Neuropathol.*, *63*: 306.

Finkelstein, S. P., Apostolides, P. J., Caday, C. G., Prosser, J., Phillips, M. F., and Klagsbrun, M. (1988). *Brain Res.*, *460*, 253.

Flage, T. (1977). *Acta Ophthalmol.*, *55*: 652.

Freeman, J. H., Jr., and Stanton, M. E. (1992). *Behav. Neurosci.*, *106*: 924.

Fujita, N., Ishiguro, H., Sato, S., Kurihara, T., Kuwano, R., Sakimura, K, Takahashi, Y., and Miyatake, T. (1990). *Brain Res.*, *513*: 152.

Funder, J. W., and Skeppard, K. (1987). *Ann. Rev. Physiol.*, *49*: 397.

Garthwaite, J., and Balazs, R. (1978). *Nature*, *275*: 328.

Goldman-Rakic, P. S., Isseroff, A., Schwartz, M. L., and Bugbee, N. M. (1983). In: *Handbook of Child Psychology*: *Infancy and Developmental Psychobiology* (P. Mussen, ed.). Wiley, New York.

Goodlett, C. R., and West, J. R. (1992). In: *Maternal Substance Abuse and the Developing Nervous System* (I. S. Sagon and T. Slotkin, eds.). Academic Press, New York, p. 45.

Goodlett, C. R., Kelly, S. J., and West, J. R. (1987). *Psychobiology*, *15*: 64.

Gould, E., Westlind-Danielsson, A., Frankfurt, M., and McEwen, B. S. (1990). *J. Neurosci.*, *9*: 3347.

Goyer, R. A., and Rhyne, B. C. (1973). *Int. Rev. Exp. Pathol.*, *12*: 1.

Graff, R. D., Lagunowich, L. A., and Reuhl, K. R. (1992). *Toxicologist*, *12*: 312.

Green, R. J., and Stanton, M. E. (1989). Behav. *Neurosci.*, *103*: 98.

Gressens, P., Kosofsky, B. E., and Evrard, P. (1992). *Neurosci. Lett*, *140*: 113.

Grundt, I., Offner, H., Konat, G., and Clausen, J. (1974). *Environ. Physiol. Biochem.*, *4*: 166.

Grundt, I. K., and Neskovic, N. M. (1980). *Environ. Res.*, *23*: 282.

Grunwald, G. B., Geller, R. L., and Lilien, J. (1980). *J. Cell Biol.*, *85*: 766.

Guidotti, A. (1990). *Neurotoxicity of Excitatory Amino Acids*. Raven Press, New York, p. 340.

Harry, G. J., and Tilson, H. A. (1981). *Neurotoxicology*, 2: 283.

Harry. G. J., Toews, A. D., Krigman, M. R., Morell, P. (1985). *Toxicol. Appl. Pharmacol.*, *77*: 458.

Harry, G. J., Goodrum, J. F., Bouldin, T. W., Wagner-Recio, M., Toews, A. D., and Morell, P. (1989). *J. Neurochem*, *52*: 938.

Harry, G. J., Schmitt, T. J., Gong, Z., Brown, H., Zawia, N., and Evans, H. L. (1996). *Toxicol. Appl. Pharmacol.*, *139*: 84.

Hatta, K., and Takeichi, M. (1986). *Nature*, *320*: 447.

Hatten, M. E., Lynch, M., Rydel, R. E., Sanchez, J., Joseph-Silverstein, J., Moscatelli, D., and Rifkin, D. B. (1988). *Dev. Biol.*, *125*: 280.

Herter, W. B. (1959). *Kaiser Found. Med. Bull*, 7: 228.

Heyser, C. J., Chen, W. J., Miller, J. Spear, N. E., and Spear, L. P. (1990). *Behav. Neurosci.*, *104*: 955.

Hicks, S. P., and D'Amato, C. J. (1968). *Anat. Rec.*, *160*: 619.

Hofer, M. M., and Barde, Y. A. (1988). *Nature*, *331*: 261.

Hohmann, C. F., Brooks, A. R., and Coyle, J. T. (1988). *Dev. Brain Res.*, *42*: 253.

Holland, P. W. H., and Hogan, B. L. M. (1988). *Development*, *102*: 159.

Holtzman, D., Olson, J., DeVries, C., and Bensch, K. (1987). *Toxicol. Appl. Pharmacol.*, *89*: 211.

Hughes, S. M., Lillien, L., Raff, M. C., Rohrer, H., and Sendtner, M. (1988). *Nature*, *335*: 70.

Ikonomidou, C., Price, M. T., Mosinger, J. L., Frierdich, G., Labruyere, J., Salles, K. S., and Olney, J. W. (1989). *J. Neurosci.*, *9*: 1693.

Innocenti, G. M., Clarkem S., and Kraftsik, R. (1986). *J. Neurosci.*, *6*: 1384.

Inouye, M., Murao, K., and Kajiwara, Y. (1985). *Neurobehav. Toxicol. Teratol.*, *7*: 227.

Ivy, G. O., and Killackey, H. P. (1982). *J. Neurosci.*, *2*: 735.

Jacobs, J. M. (1977). *J. Neurocytol.*, *6*: 607.

Jacobs, J. M., Cremer, J. E., and Cavanagh, J. B. (1977). *Neuropathol. Appl. Neurobiol.*, *3*: 169.

Jacobson, M. (1991). *Developmental Neurobiology*, 3rd ed. Plenum Press, New York.

Jacobson, S. (1963). *J. Comp. Neurol.*, *121*: 5.

Johnson, I. B., and Hall, W. G. (1980). *J. Comp. Physiol. Psychol.*, *94*: 977.

Johnson, M. (1985). In: *Developmental Neurochemistry* (R. Wiggins, D. McCardless, and E. Euna, eds.). University of Texas Press, Austin, p. 193.

Johnson, M., and Coyle, J. (1980). *J. Neurochem.*, *34*: 1429.

Jones, D. G. (1983). In: *Advances in Cellular Neurobiology*, Vol. 4 (S. Federoff and L. Hertz, eds.). Academic Press, New York, p. 163.

Jones, L., Fischer, I., and Levitt, P. (1992). *Neurosci. Abstr.*, *18*: 421.

Kater, S., and Letourneau, P. C. (1985). *J. Neurosci Res.*, *13*: 1.

Kater, S. B., and Mills, L. R. (1991). *J. Neurosci.*, *11*: 891.

Keller, F., and Levitt, P. (1989). *Neuroscience*, *28*: 455.

Keller, F., Rimvall, K., Barbe, M. F., and Levitt, P. (1989). *Neuron*, *3*: 551.

Keshles, O., and Levitzki, A. (1984). *Biochem. Pharmacol.*, *33*: 3231.

Kitraki, E., Alexis, M. N., and Stylianopoulou, F. (1984). *J. Steroid Biochem*, *20*: 263.

Komulainen, H., and Tuomisto, J. (1981). *Acta Pharmacol. Toxicol.*, *48*: 214.

Komulainen, H., Pietarinen, R., and Tuomisto, J. (1983). *Acta Pharmacol. Toxicol.*, *52*: 381.

Konat, G. (1984). *Neurotoxicology*, *5*: 87.

Konat, G., and Clausen, J. (1980). *J. Neurochem.*, *35*: 382.

Konat, G., and Offner, H. (1982). *Exp. Neurol.*, *75*: 89.

Krigman, M. R., and Hogan, E. L. (1974). *Environ. Health Perspect.* 7: 187.

Kristensson, K., and Olsson, Y. (1971). *Acta Neuropathol.*, *17*: 127.

Kvale, I., Fosse, V., and Fonnum, F. (1983). *Dev. Brain Res.*, *7*: 137.

Lagunowich, L. A., and Grunwald, G. B. (1991). *Differentiation*, *47*: 19.

Lagunowich, L. A., Bhambhani, S., Graff, R. D., and Reuhl, K. R. (1991). *Soc. Neurosci. Abstr.*, *17*: 715.

Lagunowich, L. A., Stein, A. P., and Reuhl, K. R. (1993). *Toxicologist*, *13*: 168.

Lander, A. D. (1987). *Mol. Neurobiol.*, *1*: 213.

Lauder, J. M. (1988). *Prog. Brain Res.*, *73*: 365.

Lauder, J. M. (1993). *TINS*, *16*: 233.

Lauder, J., and Bloom, F. (1974). *J. Comp. Neurol.*, *155*: 469.

Lauder, J. M., and Mugnaini, E. (1980). *Dev. Neurosci.*, *3*: 248.

Legare, M. E., Castiglioni, A. J., Jr., Rowles, T. K., Calvin, J. A., Snyder-Armstead, C., Tiffany-Castiglioni, E. (1993). *Neurotoxicology*, *14*: 77.

Leibrock, J., Lottspeich, F., Hohn, A., Hofer, M., Hengerer, B., Masiakowski, P., Thoenen, H., and Barde, Y.-A. (1989). *Nature*, *341*: 149.

Leslie, F. M. (1993). In: *Neurotrophic Factors* (S. E. Laughlin and J. H. Fallon, eds.). Academic Press, New York, p. 565.

Levi-Montalcini, R., and Angeletti, P. U. (1968). *Physiol. Rev.*, *48*: 534.

Levine, S., and Mullins, R. F. (1966). *Science*, *152*: 1585.

Levitt, P. (1982). In: *Gilles de la Tourette Syndrome* (A. J. Friedhoff and T. N. Chase, eds.). Raven Press, New York, p. 49.

Levitt, P., and Rakic, P., (1982). *Dev. Brain Res.*, *4*: 4630.

Lillien, L. E., Sendtner, M., Rohrer, H., Hughes, S. M., and Raff, M. C. (1988). *Neuron*, *1*: 485.

Linville, D. G., and Spear, L. P. (1988). *Psychopharmacology*, *95*: 200.

Lipton, S. A., and Kater, S. B. (1989). *TINS*, *12*: 265.

Lund, R. D., Chang, R. L. F., and Land, P. W. (1984). *Dev. Brain Res.*, *14*: 139.

Luskin, M. B., and Shatz, C. J. (1985). *J. Comp. Neurol.*, *242*: 611.

Ma, F. H., Okhuma, S., Kishi, M., and Kuriyama, K. (1991). *Int. J. Dev. Neurosci.*, *9*: 347.

Mactutus, C. F. (1994). In: *Principles of Neurotoxicology* (L. W. Chang, ed.). Marcel Dekker, New York, p. 397.

Magee, P. N., Stoner, H. B., and Barnes, J. M. (1957). *J. Pathol. Bact.*, *73*: 102.

Mailman, R. B., Krigman, M. R., Frye, G. D., and Hanin, I. (1983). *J. Neurochem.*, *40*: 1423.

Maisonpierre, P. C., Belluscio, L., Squinto, S., Ip, N. Y., Further, M. E., Lindsay, R. M., and Yancopoulos, G. D. (1990). *Science*, *247*: 1446.

Manthorpe, M., Skaper, S., Williams, L. R., and Varon, S. (1986). *Brain Res.*, *367*: 282.

Marin-Padilla, M. (1970). *Brain Res.*, *23*: 167.

Marin-Padilla, M. (1974). *Brain Res.*, *66*: 373.

Marin-Padilla, M. (1976). *J. Comp. Neurol.*, *167*: 63.

Matsunaga, M., Hatta, K., Nagafuchi, A., and Takeichi, M. (1988a). *Nature*, *334*: 62.

Matsunaga, M., Hatta, K., and Takeichi, M. (1988b). *Neuron*, *1*: 289.

Matthieu, M.-M., Zimmerman, A. W., Webster, H. deF., Ulsamer, A. G., Brady, R. O., and Quarles, R. H. (1974). *Exp. Neurol.*, *45*: 558.

Mattson, M. P. (1988). *Brain Res. Rev.*, *13*: 179.

Mauk, M. D., Steinmetz, J. E., and Thompson, R. F. (1986). *PNAS*, *83*: 5349.

McCauley, P., Bull, R., Tonti, P., Lutkenhoff, S., Meister, M., and Doerger-Stober, J. (1982). *J. Toxicol. Environ. Health*, *10*: 639.

McCormick, A. A., Steirmetz, J. E., and Thompson, R. F. (1985). *Brain Res.*, *359*: 120.

McDonald, J. W., and Johnston, M. V. (1990). *Brain Res. Rev.*, *15*: 41.

McDonald, J. W., Silverstein, F. S., and Johnston, M. V. (1988). *Brain Res.*, *459*: 200.

McDonald, J. W., Trescher, W. H., and Johnston, M. V. (1992). *Brain Res.*, *583*: 54.

McEwen, B. S., Plapinger, L., Chaptal, C., Gerlach, J., and Wallach, G. (1975). *Brain Res.* *96*: 400.

McMahon, A. P., Joyner, A. L., Bradley, A., and McMahon, J. A. (1992). *Cell*, *69*: 581.

Meyer, J. S. (1985). *Physiol. Rev.* *65*: 946.

Meyer, O. A., Tilson, H. A., Byrd, W. C., and Riley, M. T. (1979). *Neurobehav. Toxicol. 1*: 223.

Miller, D. B. (1984). *Toxicol. Appl. Pharmacol.*, *72*: 557.

Miller, M. (1985). *Dev. Brain Res.*, *23*: 187.

Miller, M. W. (1986). *Science*, *233*: 1308.

Miller, M. W., and Potempa, G. (1990). *J. Comp. Neurol.*, *293*: 92.

Mishkin, M., and Appenzeller, T. (1978). *Sci. Am.*, *256*: 80.

Miura, K., Inokawa, M., and Imura, N. (1984). *Toxicol. Appl. Pharmacol.*, *73*: 218.

Moore, R. Y., and Bloom, F. E. (1978). *Annu. Rev. Neurosci.*, *1*: 129.

Moore, R. Y., and Bloom, F. E. (1979). *Annu. Rev. Neurosci.*, *2*: 113.

Moore, R. Y., and Card, J. P. (1984). In: *Handbook of Chemical Neuroanatomy*, Vol. 2. *Classical Transmitters in the CNS*, Part I (A. Bjorklund and T. Hokfelt, eds.). Elsevier, Amsterdam, p. 123.

Morell, P., Quarles, R. H., and Norton, W. T. (1993). In: *Basic Neurochemistry*, 5th ed. (R. W. Albers, G. W. Siegel, P. Molinoff, and E. Agranoff, eds.). Raven Press, New York, p. 117.

Mugnaini, E., and Forstronen, P. F. (1967). *Z. Zellforsch.*, *77*: 115.

Murphy, T. H., and Baraban, J. M. (1990). *Dev. Brain Res.*, *57*: 146.

Murphy, T. H., Schaar, R. L., and Coyle, J. T. (1990). *FASEB J.*, *4*: 1624.

Nadel, L., and Zola-Morgan, S. (1984). In: *Infant Memory* (M. Moscovitch, ed.). Plenum Press, New York.

Nicholson, J. L., and Altman, J. (1972). *Science*, *176*: 530.

Nicoletti, F., Jadarola, M. J., Wrobleski, J. T., and Costa, E. (1986). *Proc. Natl. Acad. Sci. USA*, *83*: 1931.

Nicoletti, F., Magri, G., Ingras, F., Bruno, V., Cantani, M. V., Dell'Albani, P., Condorelli, D. F., and Avola, R. (1990). *J. Neurochem.*, *54*: 771.

Nieminen, L., Bjondahl, K., Mottonen, M. (1973). *Food Cosmet. Toxicol.*, *11*: 635.

Noland, E. A., Taylor, D. H., and Bull, R. J. (1982). *Neurobehav. Toxicol. Teratol.*, *4*: 539.

Nonneman, A. J., Corwin, J. W., Sahley, C. L., and Vicedomini, J. P. (1984). In: *Early Brain Damage*, Vol. 2 (S. Finger and R. Almli, eds.). Academic Press, New York, p. 139.

Norton, S., Culver, B., and Mullenix, P. (1975). *Behav. Biol.*, *15*: 317.

O'Kusky, J. R., and McGeer, G. (1985). *Dev. Brain Res.*, *21*: 299.

Olney, J. W. (1980). In: *Experimental and Clinical Neurotoxicity* (P. S. Spencer and H. H. Schaumberg, eds.). Williams and Wilkins, Baltimore, p. 490.

Olney, J. W., and Price, M. T. (1980). *Brain Res. Bull.*, *5*(Suppl.2): 361.

Olney, J. W., Ho., O. L., Rhee, V., and Schainker, B. (1972). *Brain Res.*, *45*: 309.

Olney, J. W., Misra, C. H., and Rhee, V. (1976). *Nature*, *264*: 659.

Olney, J. W., Zorumski, C., Price, M. T., and Labruyere, J. (1990). *Science*, *248*: 596.

Olton, D. (1977). *Sci. Am.*, *236*: 82.

O'Rourke, N. A., Dailey, M. E., Smith, S. J., and McConnel S. K. (1992). *Science*, *255*: 373.

Palmer, E., Nagel-Taylor, K., Krause, J. D., Roxas, A., and Cotman, C. W. (1990). *Dev. Brain Res.*, *51*: 132.

Pardridge, W. M., Boado, R. J., and Farrell, C. R. (1990). *J. Biol.* Chem., *265*: 18035.

Pavlik, A., and Buresova, M. (1984). *Dev. Brain Res.*, *12*: 13.

Pennington, S. N. (1992). In: *Development of the Central Nervous System*: *Effects of Alcohol and Opiates* (M. W. Miller, ed.). Wiley-Liss, Inc., New York, p. 189.

Persson, L., and Sima, A. (1975). *Neurobiology*, *5*: 151.

Petit, T., and LeBoutillier, J. (1979). *Exp. Neurol.*, *64*: 482.

Petit, T., Alfano, D., and LeBoutillier, J. (1983). *Neurotoxicology*, *4*: 79.

Pettmann, B., Weibel, M., Sensenbrenner, M., and Labourdette, G. (1985). *FEBS*, *189*: 102.

Pittman, R. N., and Buettner, H. M. (1989). *Dev. Neurosci.*, *11*: 361.

Plant, T. M., Gay, V. L., Marshall, G. R., and Arslan, M. (1989). *Proc. Natl. Acad. Sci. USA*, *86*: 2506.

Plapinger, L., and McEwen, B. S. (1973). *Endocrinology*, *93*: 1119.

Powell, H., Searner, O., Cluck, L., and Lampert, P. (1973). *J. Pediatr.*, *82*: 976.

Pupura, D. P. (1975). In: *Advances in Neurology*, Vol. 12, *Physiology and Pathology of Dendrite* (G. W. Kreutzberg, ed.). Raven Press, New York, p. 91.

Raine, C. S. (1984). In: *Myelin*, 2nd ed. (P. Morell, ed.). Plenum Press, New York, p. 1.

Rakic, P. (1974). *Science*, *183*: 425.

Raper, J. A., and Grunewald, E. B. (1990). *Exp. Neurol.*, *109*: 70.

Rechthand, E., and Rapoport S. I. (1987). *Prog. Neurobiol.*, *28*: 303.

Reichardt, L. F., and Tomaselli, K. J. (1991). *Annu. Rev. Neurosci.*, *14*: 531.

Reinstein, D. K., and Issacson, R. L. (1981). *Neurosci. Lett.*, *26*: 251.

Reiter, L. W., Anderson, G. E., Laskey, J. W., and Cahill, D. F. (1975). *Environ. Health Perspect.*, 12: 119.

Reiter, L. W., Heavner, G. B., Dean, K. R., and Ruppert, R. H. (1981). *Neurobehav. Toxicol. Teratol.*, *3*: 285.

Reuhl, K. R., and Chang, L. W. (1979). *Neurotoxicology*, *1*: 21.

Reuhl, K. R., and Cranmer, J. M. (1984). *Neurotoxicology*, *5*: 1987.

Reuhl, K. R., Smallridge, E. A., Chang, L. W., and Mackenzie, B. A. (1983). *Neurotoxicology*, *5*: 187.

Riley, E. P., Lochry, E. A., and Shapiro, N. R. (1979). *Psychopharmacology*, *62*: 47.

Riley, E. P., Barron, S., Driscoll, C. D., and Chen, J. S. (1984). *Teratology*, *29*: 325.

Rodier, P. M. (1976). *Neurotoxicology*, 7: 69.

Rodier, P. M. (1980). *Dev. Med. Child Neurol.*, 22: 525.

Rodier, P. M. (1986). In: *Handbook of Behavioral Teratology* (E. P. Riley and C. V. Vorhees, eds.). Plenum Press, New York, p. 185.

Rodier, P. M., Aschner, M., and Sager, P. R. (1984). *Neurobehav. Toxicol. Teratol.*, 6: 379.

Rosenberg, P. A. (1988). *J Neurosci.*, 8: 2887.

Ruehl, J., Faure, R., and Dussault, J. H. (1985). *J. Endocrinol. Invest.*, 8: 343.

Rutishauser, U., and Jessel, T. M. (1988). *Physiol. Rev*, 68: 819.

Rutishauser, U., Thiery, J. P., Brackenbury, R., Sela, B., and Edelman, G. M. (1976). *Proc. Natl. Acad. Sci. USA*, 73: 577.

Sager, P. R., and Syversen, T. L. (1986). In: *The Cytoskeleton. A Target for Toxic Agents*. Plenum, New York, p. 97.

Saunders, N. R. (1992). In: *Physiology and Pharmacology of the Blood-Brain Barrier* (M. W. B. Bradbury, ed.). Springer-Verlag, Heidelberg, p. 327.

Scalzo, F. M., and Burge, L. J. (1992). *Dev. Psychobiol.*, 25: 597.

Scalzo, F. M., Ali, S. F., Framdes, N. A., and Spear, L. P. (1990). *Pharmacol. Biochem. Behav.*, 37: 371.

Schade, J. P., and Groenigen, W. B. (1961). *Acta Anat.*, 47: 74.

Schapiro, S., Salas, M., and Vukorick, K. (1970). *Science*, 193: 146.

Schapiro, S., Vukovich, K., and Globus, A. (1973). *Exp. Neurol.*, 40: 286.

Schmidt, M. J., Palmer, G. C., and Robinson, G. A. (1980). In: *Psychopharmacology of Aging* (C. Eisdorfer and W. E. Rann, eds.). Spectrum, New York, p. 213.

Schmitt T. J., Zawia N., and Harry G. J. (1996). *Neurotoxicology*, 17: 407.

Schoeppe, D. D., and Rutledge, C. D. (1985). *Biochem. Pharmacol.*, 34: 2705.

Schwab, M. E., and Schnell, L., (1989). *J Neuorcytol.*, 18: 161.

Schwab, M. D., Kapfhammer, J. P., and Bandtlow, C. E. (1993). *Annu Rev. Neurosci.*, 16: 565.

Schwartz, H. L., and Oppenheimer, J. H. (1978). *Endocrinology*, 103: 267.

Seawright, A. A., Brown, A. W., Ng, J. C., and Hardlicka, J. (1984). In: *Biological Effects of Organolead Compounds* (P. Grandjean and E. C. Grandjean, eds.). CRC Press, Boca Raton, FL, p. 178.

Sendtner, M., Kreutzberg, G. W., and Thoenen, H. (199). *Nature*, 345: 440.

Sidman, R. L., and Rakic, P. (1973). *Brain Res.*, 62: 1.

Silverstein, F. S., Chen, R., and Johnston, M. V. (1986). *Neurosci. Lett*, 71: 13.

Skelton, R. W. (1988). *Behav. Neurosci.*, 102: 586.

Skoff, R. P., Toland, D., and Nast, E. (1980). *J. Comp. Neurol.*, 191: 237.

Smart, I. H. M., and Smart, M. (1982). *J. Anat.*, 134: 273.

Snow, D. M., Watanabe, M., Letourneau, P. C., and Silver, J. (1991). *Development*, 113: 1473.

Sotelo, C., and Rio., J. P. (1980). *Neuroscience*, 5: 1737.

Spano, P. F., Kumakura, K., Govoni, S., and Trabucchi, M. (1975). *Pharmacol. Res. Comm.*, 7: 223.

Spear, N. E., and Rudy, J. W. (1991). In: *Developmental Psychobiology: New Methods and Changing Concepts* (H. N. Shair, G. A. Barr, and M. A. Hofer, eds.). Oxford University Press, New York, p. 84.

Spear, L. P., Enter, E. K., and Linville, D. G. (1985). *Neurobehav. Toxicol. Teratol.*, 7: 691.

Spear, L. P., Kristen, C. L., Bell, J., Yoottanasupun, V., Greenbaum, R., O'Shea, J. Hoffman, H., and Spear, N. E. (1989). *Neurotoxicol. Teratol*, 11: 57.

Stanton, M. E., Freeman, J. H., Jr., and Skelton, R. W. (1992). *Behav. Neurosci.*, 106: 657.

Sundstrom, R., and Karlsson, B. (1987). *Arch. Toxicol.*, 59: 3412.

Suzuki, K. (1971). *Exp. Neural.*, 31: 207.

Szekely, G. (1963). *J. Embryol. Exp. Morphol.*, 11: 431.

Takeichi, M. (1988). *Development*, 102: 639.

Takeuchi, T. (1977). *Pediatrician*, 6: 69.

Takeuchi, T., and Eto, K. (1972). In: *Minamata Disease: Methylmercury Poisoning in Minamata an Niigata* (T. Tsubaki and K. Irukayama, eds.). Kodansha, Tokyo, p. 103.

Taylor, L. L., and DiStefano, V. (1976). *Toxicol. App. Pharmacol.*, *38*: 489.

Thompson, R. F. (1988). *Trends Neurosci.*, *11*: 152.

Tiffany-Castiglioni, E., Sierra, E., Wu, J., and Rowles, T. (1989). *Neurotoxicology*, *10*: 417.

Toews A. D., Krigman M. R., Thomas, D. J., and Morell, P. (1980). *Neurochem Res.*, *5*: 605.

Toews, A. D., Blaker, W. D., Thomas, D. J., Gaynor, J. J., Krigman, M. R., Mushak, P., and Morell, P. (1983). *J. Neurochem.*, *41*: 816.

Torak, R. M., Terry, R. D., and Zimmerman, H. M. (1960). *Am. J. Pathol.*, *36*: 273.

Torak, R. M., Gordon, J., and Prokop, J. (1970). *Int. Rev. Neurobiol.*, *12*: 45.

Towfighi, J. (1980). In: *Experimental and Clinical Neurotoxicology* (P. S. Spencer and H. H Schaumburg, eds.). Williams & Williams, Baltimore, p. 440.

Turbow, M. M., and Chamberlain, J. G. (1968). *Teratology*, *1*: 103.

Unsicker, K., Grothe, G., Ludecke, G., Otto, D., and Westermann, R. (1993). In: *Neurotrophic Factors* (S. E. Loughlin and S. H. Fallon, eds.). Academic Press, New York, p. 313.

Valois, A. A., and Webster, W. S. (1987). *Neurotoxicology*, *8*: 463.

Van Hooff, C. O. M., Oestreicher, A. B., DeGraan, P. N. E., and Gispen, W. H. (1989). *Mol. Neurobiol.*, *3*: 101.

Veronesi, B., and Bondy, S. (1986). *Neurotoxicology*, *7*: 69.

Veronesi, B., Brady, A., and Reiter, L. W. (1982). *Neurotoxicology*, *3*: 136.

Vertex, M., Barnea, A., Lindner, H. R., and King, R. J. B. (1973). *Adv. Exp. Med. Biol.*, *36*: 137.

Vongdokmai, R. (1980). *J. Comp. Neurol.*, *191*: 283.

Wallace, J., and Lauder, J. (1983). *Brain Res. Bull*, *10*: 459.

Wang, X. H., and Murphy, E. H. (1993). *Soc. Neurosci. Abstr.*, *19*: 50.

Warren, M. A., and Bedi, K. S. (1982). *J. Comp. Neurol.*, *210*: 59.

Watanabe, I. (1977). In: *Neurotoxicology* (L. Roizin, H. Shiraki, and N. Grcevic, eds.). Raven Press, New York, p. 317.

Webster, H. deF. (1971). *J. Cell Biol.*, *48*: 348.

Webster, H. deF., Martin, J. R., and Connell, M. F. (1973). *Dev. Biol.*, *32*: 401

Webster, W. S. and Valois, A. A. (1981). *J. Neuropathol. Exp. Neurol.*, *40*: 247.

West, J. R., and Pierce, D. R. (1986). In: *Alcohol and Brain Development* (J. R. West, ed.). Oxford University Press, New York, p. 120.

Westergaard, E., and Brightman, M. W. (1973). *Comp. Neurol.*, *152*: 17.

Wigal, T., and Amsel, A. (1990). *Behav. Neurosci.*, *104*: 11.

Wiggins, R. C. (1986). *Neurotoxicology*, *7*: 103.

11

Lead Neuroendocrine Toxicity

Rebecca Z. Sokol
University of Southern California, Los Angeles, California

INTRODUCTION

The male reproductive axis consists of the hypothalamus (H), pituitary (P), testicles (T), and sex accessory glands. The HPT axis is physiologically a closely regulated system (1). Control and coordination of reproductive function occurs via feedback signals, both positive and negative, exerted by the hormones secreted at each level of the HPT axis (Fig. 1). These signals include (1) stimulation and inhibition of gonadotropin-releasing hormone (GnRH) by neurotransmitters, (2) stimulation of luteininzing hormone (LH) and follicle-stimulating hormone (FSH) by GnRH, (3) stimulation of testosterone by LH, (4) inhibition of hypothalamic GnRH secretion and pituitary (LH) responsiveness to GnRH by testicular steroids, and (5) inhibition of pituitary FSH release by testicular inhibin and, possibly, circulating estrogens (2,3).

The reproductive hormones ultimately regulate spermatogenesis. During puberty, changes in the secretion of the pituitary hormones precede the maturation of the testes (4). LH plays an important role in the initiation and maintenance of sperm production, primarily by stimulating testosterone secretion by a direct action on the interstitial cells (5). FSH is also necessary for the initiation of spermatogenesis at the time of puberty (6). Although inhibin is also involved in the feedback regulation of FSH secretion, the precise role of gonadotropins and possibly gonadal steroids in the regulation of inhibin secretion remains to be elucidated (7).

Testicular testosterone clearly exerts negative feedback on the hypothalamic-pituitary unit. For example, orchiectomy leads to an immediate increase in the level of gonadotropin secretion. The administration of exogenous testosterone suppresses the gonadotropin response (7).

Any disruption by toxicants of the delicately coordinated interactions among the components of the axis may lead to reproductive abnormalities. Reproductive toxicants may disrupt the reproductive axis directly or indirectly (8). Direct disruption occurs if the chemical structure of the toxicant is similar to that of one of the hormones. As a result, the chemical substitutes for the hormone at its target receptor site possibly interfering with hormone binding, release, or synthesis. Alternatively, a toxicant may disrupt the axis by altering the structure of the hormone during synthesis or glycosylation

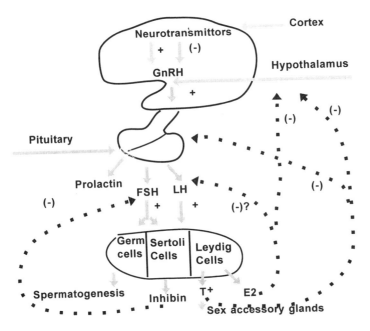

Figure 1 Hypothalamic-Pituitary Testicular Axes: Positive and negative feedback signals.

or by changing the metabolism of the hormone. The end result in all cases is the ultimate decrease in fertility potential.

A testicular toxicant will cause a decline in testosterone levels and/or spermatogenesis with a corresponding elevation in gonadotrophins. A H/P toxicant will suppress gonadotrophins, which results in a corresponding decline in circulating testosterone levels and some suppression of spermatogenesis. Sufficient disruption of the normal androgenic control of the testes by a toxicant will result in a decrease in the spermatid count (9,10). A decline in testosterone levels will also diminish the fertilizing potential of spermatozoa (11).

LEAD

Lead's reproductive toxicity was documented by the Greeks and Romans centuries ago. In spite of extensive studies evaluating the pathophysiology of lead toxicity on the reproductive axis, controversy still persists. Whereas the majority of studies suggest that the primary site of lead's toxicity is the central nervous system (CNS), some studies report direct testicular toxicity, and a small number of others question if lead is truly toxic to the reproductive system (Table 1).

Animal Studies

Early studies evaluating lead as a reproductive toxicant reported that oral exposure induced changes in spermatogenesis and fertility (12). More recently, in a series of experiments designed to determine the pathophysiology of the induced infertility, male

Table 1 Results of Animal Studies of Lead Toxicity

Species	Positive studies	Negative studies
Rat	20	3
Mouse	5	1
Rabbit	1	1
Monkey	1	0

rats were dosed with increasing concentrations of lead acetate containing water. Parameters evaluated included sperm concentration, serum LH and FSH, and serum and intratesticular testosterone levels. Lead exposure resulting in blood lead levels greater than 30 μg/dl depressed serum and intratesticular testosterone levels and spermatogenesis. Gonadotropin levels in these animals were not elevated in response to the inhibition of gonadal function indicating that lead's primary site of toxic action was the hypothalamus and/or pituitary (13–15).

In studies designed to identify more specifically the site of lead's toxic actions on the HPT axis, the response of lead-treated male rats as compared to control animals to stimulation testing was evaluated (16). Stimulation testing determines the level of the axis that is altered. The animals underwent stimulation testing with GnRH, naloxone, and hCG (human chorionic gonadotropin). The GnRH stimulation test evaluates the functional capacity of the gonadotrophs to release LH. Naloxone, on the other hand, is reported to produce a dose-dependent rise in serum LH by a mechanism involving increased endogenous hypothalamic GnRH output. This increase in GnRH output is postulated to occur via naloxone's inhibition of β endorphin actions, which in turn increases dopamine secretion and, consequently, LH responsiveness. The drug has no effect on GnRH-stimulated release of LH by the anterior pituitary (17). The ability of the testes to secrete testosterone is conventionally tested with the administration of hCG, which mimics the stimulatory actions of LH. Lead-treated animals responded to both GnRH stimulation and LH stimulation in a hyperresponsive manner, indicating an intact pituitary-testicular axis. However, lead-treated animals manifested a blunted LH response to naloxone stimulation, suggesting the lead interfered with GnRH release (16) (Figs. 2–4). A significant dose-dependent increase in mRNA levels of GnRH and LH and a significant increase in pituitary stores of β-LH was also noted in lead-exposed male rats (18).

Age at the time of exposure affects the severity of toxicity. Male rats who are exposed to lead prepubertally at a time preceding the onset of rapid maturation of the male hypothalamic-pituitary-testicular axis (HPT) are the least vulnerable to the toxic effects of lead (19). At this stage of rat development, HPT function is relatively inactive. Therefore, perturbations to the axis might yield limited changes. On the other hand, the HPT axis of the pubertal is most sensitive to disruption.

Interaction between lead, neurotransmitters, and gonadotropins may explain these age-related effects. Neurotransmitters appear to play an active role in inhibiting gonadotrophin release before the onset of puberty. This inhibitory action on GnRH release decreases with the onset of puberty (20–22). Changes in neurotransmitter release proteins are associated with alterations in the secretory pattern of hypothalamic GnRH. In response, the secretory pattern of the gonadotropins released from the pituitary is also

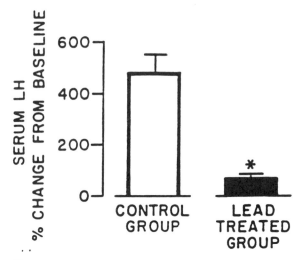

Figure 2 Effect of lead acetate on the responsiveness of LH to naloxone stimulation in Group
I. *p < 0.05.

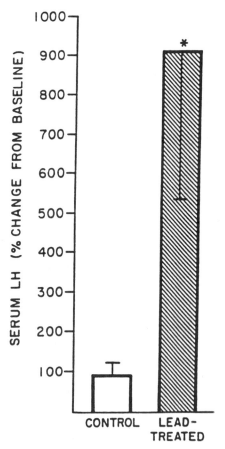

Figure 3 Effect of lead acetate on the responsiveness of LH to GnRH stimulation in Group
II. *p < 0.001.

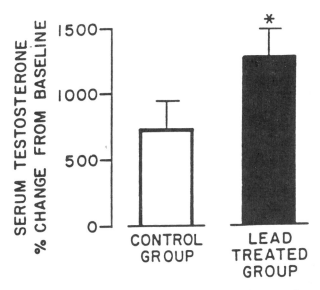

Figure 4   Effect of lead acetate on the responsiveness of testosterone LH stimulation in Group III. *$p < 0.05$.

altered. As the animal matures, the testes become more sensitive to the stimulatory actions of the gonadotropins and testosterone levels and sperm numbers increase (23).

Lead is postulated to affect the release of neurotransmitters, including norepinephrine and dopamine (24). It is theoretically possible that those animals who are prepubertal at the onset of lead treatment have a built-in resistance to lead's toxic actions on neurotransmitters. In other words, a greater dose of lead would be required to disrupt the inhibitory control of opiates on GnRH release in the prepubertal animal than the dose that would induce a disruption of the signals in the animal who has entered puberty. Thus, the pubertal or young adult animal would manifest a lead-induced decline in testosterone production and sperm production more readily than would the prepubertal animal. Autoradiographic studies have localized lead to the median eminence of the hypothalamus further implicating the hypothalamus as a primary site of the neurotoxic actions of lead (25).

Although these data are consistent with the hypothesis that lead's toxicity is directed at the hypothalamic-pituitary axis, direct testicular toxicity may also exist. A number of investigators have reported on the toxicity of lead on the testis and spermatogenesis. Histological evidence of testicular damage and inhibited spermatogenesis is reported in mice and rats (26,27). Whereas one investigator observed an increased frequency of abnormal sperm morphology in mouse semen samples collected at 30 days after acute exposure to lead acetate by peritoneal injection, others reported no morphological changes in sperm after prolonged lead exposure (28,29).

A recent study evaluated whether lead exposure in vivo altered the morphology of the spermatozoa, the proportion of different cell types in the testes as evaluated by DNA flow cytometry, and/or the ability of sperm to fertilize ova in vitro (30). The lead-treated groups fertilized significantly fewer eggs than did sperm from the control group. No ultrastructural changes were noted in the spermatozoa of animals treated with lead compared to control animals. There were no differences in weights or in the histogram patterns of testicular cells collected from lead-treated animals. A critical level of test-osterone is necessary for sperm to achieve full fertilizing potential (31). These data suggest that lead alters sperm function by altering the hormonal control of spermato-genesis rather than by direct toxic action on spermatozoa.

These findings confirmed those reported earlier (32). Using a mouse model, in-vestigators reported that long-term exposure to inorganic lead reduced the ability of sperm harvested from lead-exposed males to fertilize mouse eggs in vitro (32). The in vitro fertilization assay is a sensitive method for detecting subtle changes in the fertili-zation capacity of spermatozoa (33). The relative insensitivity of breeding studies could explain the conflicting data published regarding the effects of lead exposure on fertil-ity outcome (33).

Studies conduced on male monkeys and male rabbits, in general, confirm the conclusion that lead is primarily a CNS toxicant, but the data also suggest the testicle as a site of toxicity. Sexually mature male rabbits treated with subcutaneous injections of lead acetate for 15 weeks to achieve blood lead levels between 50 ng/dl and 110 μg/dl were compared to control animals given injections of a 5% dextrose solution. Blood and semen were collected weekly. As in earlier rat studies, there was an inverse relationship between blood lead and sperm count. Serum testosterone levels were lower in the lead-treated groups than the controls. The FSH levels were marginally increased, but the pattern of LH response was variable. No weight differences in the sex acces-sory organs were found and no changes in sperm morphology were seen with the elec-tron microscope (34). They did find abnormalities of rabbit sperm morphology using light microscopy. The rabbit data suggest a dual site of toxicity, both at the hypotha-lamic-pituitary axis and at the testicle.

Studies conducted on male monkeys also suggest a dual level of toxicity (35). Nine-year-old male monkeys were dosed with lead acetate (1500 μg/kg/day) either (1) from birth onward, (2) beginning at postnatal day 300, or (3) from postnatal days 0 to 400. Control animals were treated with vehicle only. At the end of each treatment pe-riod, baseline and GnRH-stimulated plasma levels of LH, FSH, and T and inhibin were measured. No differences among the treatment groups were found in baseline LH, FSH, and testosterone levels. Mean plasma levels of LH were significantly lower in the life-time exposure group compared to controls at 5 and 10 minutes post–GnRH stimula-tion. Subtle decreases in the inhibin:FSH ratio were noted in both the lifetime and postinfancy groups compared to the control groups. Based on these results, the authors concluded that chronic low level lead exposure may be associated with an impaired pituitary LH response to GnRH stimulation and subtle changes in Leydig (testosterone) and Sertoli cell (inhibin) function. They hypothesized that chronically exposed animals may develop compensatory mechanisms (36).

Rat studies also suggest an adaptation to toxicity. All parameters measured, in-cluding hormone levels, sperm parameters, and measurements of intracellular stores of gonadotropin, point to an acclimation of the reproductive axis during lead exposure (37). The reversibility of the toxic effects also suggests a system that reaches a new steady state during exposure and returns to the original state when exposure ends (38).

Although the majority of animal studies support the conclusion that lead is a re-productive toxicant, some studies do not (12,29,39–41). Discrepancies in results may be explained by differences in experimental design.

CLINCIAL STUDIES

Clinical studies lend support to the conclusion that lead is toxic to the reproductive system. However the exact effects on the reproductive hormones are inconsistent. In an early study, Lancranjan and coworkers evaluated the semen analysis, serum testosterone levels, and urinary gonadotropins of men who worked in a battery plant and were exposed to lead on the production line and compared the results to those collected from men who worked in a low-lead office environment (42). They found a dose-related suppression of spermatogenesis, normal or decreased serum testosterone, and normal urinary gonadotropins. Although they concluded that lead was primarily a testicular toxicant, the fact that gonadotrophins were not elevated in response to the low testosterone also supports the conclusion that lead acts on both the testis and the hypothalamic-pituitary axis.

Subsequent studies reporting lead's effects on the reproductive axis of men suggest CNS and/or testicular sites of toxicity (43–47). Because of differences in hormone and sperm function assay methods and, most importantly, the small numbers of men evaluated in each of these studies, the data are not conclusive.

To overcome the issue of the small number of subjects studied, investigators extracted data from the relevant literature to develop a regression model relating sperm counts to blood lead levels (48). Data from 271 workers were collated. The ranges for means of blood lead and sperm concentration were 13–90 µg/dl and 32–101 million sperm/ml, respectively. A linear least squares regression model provided evidence for an inverse relationship between blood lead and sperm count. This model relates a 10 µg/dl increase in blood lead to a decrease of 4.7 million sperm/ml. Data generated in the rat model are in agreement with this clinical data.

In summary, lead's interference with spermatogenesis is most likely secondary to a disruption of the signals between the hypothalamus, pituitary, and testes. By disrupting the signals between the hypothalamus and pituitary, the appropriate gonadotropin stimulation of the testes is diminished. Intratesticular and circulating testosterone levels and spermatogenesis are decreased. In turn, sperm function and fertility are diminished. Duration of exposure increases toxicity to a certain point, with the animal adapting to the toxicity of the hypothalamic-pituitary level possibly through a change in receptor number or sensitivity. Finally, the toxic effects on the HPT axis are reversible.

REFERENCES

1. Sokol, R. Z. (1992). In *Infertility Reproductive Medicine Clinics of North America* (J. W. Overstreet, ed.). WB Saunders Co., Philadelphia, p. 389.
2. Winters, S. J., and Troen, P. (1975). *J. Clin. Endocrinol. Metab.*, *65*: 842.
3. Desmoulin, A., Hustin, J., Lambotte, R. and Franchimont, P. (1981). *Intragonadal Regulation of Reproduction* (P. Franchimont and C. P. Channing, eds.). Academic Press, New York, p. 327.
4. Kalra, S. P., and Kalra, P. S. (1983). *Endo. Rev.*, *4*: 311.
5. Barraclough, C. A. (1982). *Endo. Rev.*, *3*: 1991.
6. Plant, T. M. (1986). *Endo. Rev. 7*: 75.
7. Matsumoto, A. M., and Bremner, W. J. (1987). *J. Clin. Endocrinol. Metab.*, *1*: 71.
8. Mattison, D. R. (1983). *Am. J. Indust. Med.*, *4*: 65.

9. Russell, L. D., Malone, J. P., and Karpas, S. L. (1981). *Tissue Cell, 13*: 369.

10. Chapin, R. E., and Williams, J. (1989). *Toxicol. Pathol., 17*: 446.

11. Orgebin-Crist, M. C., Danzo, B. J., and Davies, J. (1975). In: *Handbook of Physiology* (D. W. Hamilton and R. O. Greepo, eds.). American Physiological Society, Washington, DC, p. 319.

12. Winder, C. *Reproduct. Toxicol., 3*: 221.

13. Sokol, R. Z., Madding, C., Swerdloff, R. S. (1985). *Biol. Reprod. 33*: 722.

14. Thoreaux-Manley, A., Velez de la Calle, F. J., Oliver, M. F., et al. *Toxicology, 100*: 101.

15. Ronis, M. J. J., Badger, T. M., Shema, S. J., et al. (1996). *Toxicol. Appl. Pharmacol., 136*: 361.

16. Sokol, R. Z. (1987). *Biol. Reprod., 37*: 1135.

17. Cicero, T. J., Schandber, B. A., and Meyer, E. R. (1979). *Endocrinology, 104*: 1286.

18. Klein, D., Wan, J. J. Y., Kaokyab, S., Okuda, H., and Sokol, R.Z. (1994). *Biol. Reprod., 50*: 802.

19. Sokol, R. Z., and Berman, N. (1992). *Toxicology, 69*: 269.

20. Ramirez, D. V., and McCann, S. M. (1963). *Endocrinology, 72*: 452.

21. Blank, M. S., Panerai, A. E., and Friesen, H. G. *Science, 203*: 1129.

22. Wilkinson, M., and Bhanot, R. (1982). *Endocrinology, 110*: 1046.

23. Raum, W. S., and Swerdloff, R. S. (1986). *Endocrinology, 119*: 168.

24. Silbergeld, E. K. (1987). In: *Lead Versus Health* (M. Rutter and R. E. Jones, eds.). Wiley, London, p. 191.

25. Stumpf, W. E., Sar, M., and Grant, L. (1980). *Neurotoxicology, 1*: 593.

26. Hilderbrand, D. C., Der, R., Griffin, W. T., and Fahim, M. S. (1973). *Am. J. Obstet. Gynecol., 115*: 1058.

27. Eyden, B. P., Maisen, J. R., and Mattelin, G. (1978). *Bull. Environ. Contemp. Toxicol., 19*: 266.

28. Varma, M. M., Joshi, S. R., and Adeyemi, A. O. *Experientia, 30*: 486.

29. Valor, B. A., Kimmel, C. A., Woods, J. S., McConnel, E. P., et al. *Toxicol. Appl. Pharmacol., 56*: 59.

30. Sokol, R. Z., Okuda, H., Nagler, H., and Berman N. (1994). *Toxicol. Appl. Pharmacol., 124*: 310.

31. Robaire, B., and Hermo, L. (1988). In: *The Physiology of Reproduction* (E. Knobil, J. Neall, et al., eds.). Raven Press, New York, p. 999.

32. Johansson, L., Sjoblom, P., and Wide, M. (1987). *Environ. Res., 42*: 140.

33. Amann, R. P. (1982). *Fundam. Appl. Toxicol., 2*: 13.

34. Moorman, W. J., Clark, J. C., Skaggs, S. R., Turner, T. W., et al. (1995). Validation of the rabbit model for assessing reproductive toxicants. High Dose Phase Report, NIOSH Report.

35. Halloway, A. J., Moore, H. D., and Foster, P. D. M. (1990). *Reprod. Toxicol., 4*: 21.

36. Foster, W. P., McMahon, A., Yeung-Lai, E. V., Hughs, E. G., and Rice, D. C. (1993). *Reprod. Toxicol., 7*: 203.

37. Sokol, R. (1990). *J. Androl., 11*: 521.

38. Sokol, R. Z. (1989). *Reprod. Toxicol., 3*: 175.

39. Coffigny, H., Thopreau-Manlay, A., Pino-Lataillade, G., et al. (1994). *Hum. Exp. Toxicol., 13*: 241.

40. Nathan, E., Huang, H. F., Pogach, L., et al. (1992). *Arch. Environ. Health, 47*: 370.

41. Willems, M. I., deSchepper, G. G., Wihowo, A. A., et al. (1982). *Arch. Toxicol., 50*: 149.

42. Lancranjan, I., Popescuh, H. I., Javanescu, O., and Klepsch, I. (1975). *Arch. Environ. Health*, 30: 396.

43. Assenato, G., Paci, C., Baser, M. E., Molinini, I., e al. (1987). *Arch. Environ. Health, 41*: 124.

44. Telisman, S., Cvitkovic, P., Gavella, M., Pongracic, J. et al. (1990). International Symposium on Lead and Cadmium Toxicity, Leidaihe, China, August, p. 18.

45. Lerda, D. (1992). *Am. J. Indust. Med.*, *22*: 567.
46. Gustafson, A., Hedner, P., Schutz, A., and Skerfving, S. (1989). *Int. Arch. Occup. Environ. Health*, *61*: 277.
47. Ng, T. P., Goh, N. H., Ng, Y. P., Ong, N. Y., et al. (1991). *Br. J. Indust. Med.*, *48*: 485.
48. Moorman, W. J., Shaw, P. B., and Schrader, S. M. (1993). *J. Androl.*, *14*:1 (Ab).

12

Female Reproduction and Toxicology of Estrogen

James H. Clark
Baylor College of Medicine, Houston, Texas

INTRODUCTION

The normal maturation of the female animal depends on the secretion of steroid hormones by the ovary. Yet inappropriate exposure to steroid hormones can disrupt the normal flow of maturational events and cause reproductive dysfunction. Thus, steroid hormones can be thought of as normal physiological agents as well as toxins. This chapter will describe what is generally known about both the normal and toxic influences of steroid hormones as they relate to sexual maturation in the rat and human female.

One of the most difficult tasks in reproductive toxicology and endocrinology is extrapolating from one species to another. There are tremendous differences between species, and failure to recognize these differences has led to mistakes in judgment on the part of toxicologists and governmental regulators. This chapter will point out the problems incurred when attempting to use rodents to predict what will happen in humans, and no attempt will be made to discuss other species. Even comparisons between rats and mice are sometimes difficult and can be very misleading.

BASIC REPRODUCTIVE DIFFERENCES BETWEEN RATS AND HUMANS

Perhaps one of the biggest and most important differences between rats and humans is their exposure to endogenous estrogen during pregnancy. The fetal rat is exposed to very small amounts of endogenous estrogen during most of pregnancy—in the low picomolar range. This is very important because estrogens cause masculinization of female rat fetuses. In contrast, the human fetus is exposed to very high levels of estrogen, in the micromolar range, and the human female is not masculinized by estrogen exposure (MacLusky and Naftolin, 1981; Albrecht and Pepe, 1988).

Thus, in the human the fetus is exposed to very high levels of estrogen, which would masculinize female fetuses in the rat but have no effect on sexual centers in the brain that control reproductive cyclicity in the human. Either these centers in the human brain are insensitive to estrogens or estrogens are not available for receptor binding. It is known that steroid hormone–binding globulins are very high during pregnancy and that these proteins bind estrogens with relatively high affinity (Hammond, 1990). However, it is not known whether these blood-binding proteins are protective.

REPRODUCTIVE CYCLICITY AND ESTROGEN EXPOSURE

As discussed above, exposure to exogenous estrogen during pregnancy or the neonatal period in the rat will cause the hypothalamic-pituitary-gonadal axis to be acyclic or masculinized. This occurs normally in the male rat as a result of the conversion of fetal testicular testosterone to estradiol by the enzyme armoatase (MacLusky and Naftolin, 1981). In female rats there is no testosterone and very low levels of estrogen, therefore they develop as females. In all mammals studied thus far the female developmental pathway is autonomous and does not appear to require a hormonal signal.

Reproductive cyclicity in the female and the control of ovulation depends on the cyclic release of gonadotropin-releasing hormones (GNRH) from the hypothalamus. This cyclic secretion of GNRH causes cyclic release of follicle-stimulating hormone (FSH) and luteinizing hormone (LH) from the pituitary. These hormones stimulate the synthesis and secretion of estrogen and progesterone in a cyclic fashion. In addition, the action of FSH and LH on the ovary causes the cyclic release of eggs. Estrogen and progesterone interact with the hypothalamus and pituitary via negative and positive feedback mechanisms and influence the cyclic secretion of GNRH, FSH, and LH.

Exposure of female rats to androgens or estrogens during fetal or neonatal life results in the disruption of the mechanisms that control cyclic secretion of the various hormones. This leads to the development of the persistent estrus syndrome, characterized by anovulatory ovaries which secrete a continuous moderate amount of estrogen and no progesterone. Therefore, these animals do not exhibit an estrous cycle (Gorski, 1963).

This effect of estrogen or androgen takes place only if the animal is exposed during the critical period of development that extends from day 18 of pregnancy to days 8–10 of life in the rat. During this period, the hypothalamic structures believed to be involved in control of normal cyclic hormone secretion are undergoing neuronal maturation. Disruption or modification of hypothalamic maturation has a permanent effect that results in a malelike or acyclic pattern of hormone release and infertility (MacLusky and Naftolin, 1981; McEwen, 1981).

In the rat and other rodent species, the persistent estrus syndrome appears to result from the action of estrogens on hypothalamic development. Even the effects of androgens are thought to occur by their metabolic conversion to estrogens (MacLusky and Naftolin, 1981). Physiological estrogens, such as estradiol, estriol, and estrone, nonphysiological estrogens such as diethylstilbestrol (DES), kepone, o,p'-DDT, and estrogen agonist/antagonists such as clomiphene are known to cause this acyclic syndrome in the rat (Gellert et al., 1974; Clark and McCormack, 1977; Eroschenko and Palmiter, 1980). Thus, it appears that estrogenicity of an agent is correlated with the toxic effect.

In contrast to the rat, human females fetuses are not masculinized by fetal exposure to the high blood concentrations of estrogens that occur during pregnancy. If human females were masculinized, there would be no reproductively active females and there would be no human race.

The mechanism involved in masculinizing the hypothalamus in human males is not known, but it does not appear to be via estrogen and is probably occurring as a result of fetal testicular androgens. Some investigators believe that 5-α-dehydro-testosterone, an active metabolite of testosterone, is involved in masculinizing the male hypothalamus. However, males who suffer from 5-α reductase deficiency are behaviorally males at puberty. Therefore, other mechanisms must be involved.

These big differences in reproductive endocrinology lead to big differences in reproductive outcome that result from inappropriate estrogen exposure during fetal or neonatal life. As explained previously, the persistent estrus syndrome in rats is characterized by ovarian follicles that do not ovulate (because the hypothalamic-pituitary axis is not producing cyclic GnRH and LH) and consequently do not produce a corpus luteum, which produces progesterone. So we have sterile female animals that are exposed to significant levels of endogenous estrogen in a continuous manner in the absence of progesterone. Such a condition will lead to reproductive tract abnormalities in some species and neoplastic changes in others.

Since female humans are not masculinized by estrogen exposure, they do not suffer from the persistent estrus syndrome and are not sterile or acyclic as adults. Consequently they are not exposed to continuous ovarian estrogen and are not likely to suffer from the same reproductive tract abnormalities as rats. Women who were exposed to diethylstilbestrol (DES), a synthetic potent estrogen, during fetal life generally have normal cycles and are not sterile (Barnes, 1979). Therefore, exposure to estrogenic toxins is not likely to lead to defeminized or masculinized patterns of secretion of gonadotropins or infertility in the human female. This also appears to be the case for exposure to androgenic substances, which are thought to be converted to estrogens. Goy and Resko (1972) produced pseudo-hermaphroditic monkeys by injecting the mothers with testosterone proprionate during gestation; however, no changes were observed in the adult menstrual cycle of the offspring. This insensitivity of the hypothalamic control centers to androgens is also exemplified by the congenital adrenal hyperplasia syndrome. This disease is characterized by large amounts of adrenal androgens that are produced during development and masculinize the external genitalia of females; however, the reproductive cycle is not affected.

Therefore, any extrapolation concerning reproductive cyclicity from the rat to human in the case of inappropriate estrogen exposure during fetal development is not valid and bespeaks a lack of basic knowledge of reproductive biology.

REPRODUCTIVE TRACT ABNORMALITIES AND ESTROGEN EXPOSURE

Fetal exposure to estrogens during the first trimester of pregnancy will cause reproductive tract abnormalities in the human, but they differ form those in the rat both in cause and outcome and therefore cannot be directly compared. During the early latter part of the first trimester of pregnancy in the human, the Müllerian ducts are differentiating into the uterus and the upper third of the vagina (Herbst et al., 1975). If the ducts are exposed to excessive estrogen at this time, the epithelium of the Müllerian duct differentiates prematurely in both the uterus and upper vagina. As development continues, the uterinelike epithelium remains in the upper vagina and is not replaced by squamous epithelium as is the rest of the vagina. This uterine epithelium develops glands similar to the endometrial lining of the uterus. At puberty when reproductive cyclicity begins and the ovary produces estrogen, this left over uterine lining in the vagina begins to grow as it does in the uterus. This growth produces vaginal adenosis, a disorder characterized by the presence of uterine glands where none should be.

In contrast to the effects of fetal exposure to estrogens in humans, fetal or neonatal exposure of mice, rats, and hamsters to estrogens results in preneoplastic and neoplastic changes in the vagina, uterus, pituitary, and mammary gland (Cramer and Horning, 1936; Allen and Gardner, 1941; Gardner et al., 1959; Dunn and Green, 1963; Takasugi and Bern, 1964; Forsberg, 1972, 1975; Furth et al., 1973; Leavitt et al., 1981). All of these abnormal changes are probably the result of the persistent estrus syndrome. That is, they result from the continous exposure to endogenous estrogen being produced by an acyclic ovary, and not as a primary result of fetal or neonatal exposure to estrogen.

It is apparent that perinatal estrogen exposure is far more detrimental to the rat that it is to the human and that extrapolations made from studies done in the rat to potential outcomes in humans are invalid and nonscientific.

DES AND REPRODUCTIVE TOXICOLOGY

The differences between estrogens and their potential toxicity will be discussed later. This section will be devoted to DES because of its former use in medicine and agriculture. DES is a synthetic estrogen to which many people assign special significance, and it is often considered to act via different mechanisms than to physiological estrogens. This attitude was first promulgated by the popular press and later by serious scientists, who believed that DES has special "cancer-inducing" properties. These reports originated from the observations of Herbst et al. (1971), who described in increased incidence of vaginal adenosis and clear-cell adenocarcinomas in young women whose mothers had taken large quantities of DES during pregnancy. This report and many of the studies of DES that followed failed to consider the well-known differences between rat and human reproduction and have led to much confusion. They have also ignored the subsequent papers by Herbst et al. (1975), Lanier et al. (1973), and Bibbo et al. (1977), which have failed to substantiate the claim that DES causes vaginal cancer.

Vaginal cancer is a very rare disease, and of the thousands of human females who were exposed as fetuses to DES only a very few have adenocarcinoma of the vagina. Therefore, it is not valid to say that DES causes this cancer. It is clear, however, that DES exposure during early pregnancy does cause vaginal adenosis. The reasons for this have been discussed above and are straightforward. DES is simply acting like any other estrogen—it is just present at the wrong time during development of the human internal genitalia. Nothing is mysterious about its mechanism of action. Another estrogen would have done the same thing if it had been given during the first trimester of pregnancy. Not only does this time coincide with the period of Müllerian duct development, but it is the time when endogenous estrogens are low. Therefore, any exogenous estrogen given at this time will result in inappropriate estrogen exposure.

After the first trimester blood estrogens in the human become very high and the addition of another estrogen would have little consequence. This explains why all of the cases of vaginal adenosis occur in women whose mothers were given DES during the first trimester, not later (Herbst et al., 1975). However, this fact is often ignored, and considerable confusion has arisen concerning the relative ability of blood-binding proteins to sequester estrogens, as a protective mechanism against estrogen action. This confusion once again arose because investigators ignored the differences between rats and humans. It was well known that the blood of pregnant rats and humans contains large quantities of α-fetoprotein (AFP) during fetal and neonatal development. AFP from the rat binds physiological estrogens with high affinity and thus reduces the level of free hormone in the blood. Conversely, AFP binds DES weakly and would not prevent receptor binding and estrogenic stimulation. Estrogens that do not bind well to AFP are very effective estrogenic agents and would be capable of disrupting normal reproduction in the rat (Clark and Peck, 1979; Sheehan et al., 1980; McEwen, 1981). Although these concepts appear to be valid for the rodent model, they are not applicable to the human. In the human, AFP does not bind physiological estrogens very tightly and cannot readily be called a protective protein (Nunez et al., 1976). Therefore, once again, extrapolations from rat to human break down.

DO ALL ESTROGENS ACT IN THE SAME WAY?

As discussed above, DES is considered by some investigators to act in a fashion different from other estrogens. This is also the case for other estrogenic compounds and adds to the confusion and controversy concerning the toxicity of estrogens. In order to examine this topic further, we must first discuss the types of estrogenic substances. Estrogenic compounds can be divided into three major groups, which will be discussed below.

Physiological Estrogens

These include the steroidal estrogens such as 17β-estradiol, estrone, estriol, and their natural metabolites. All physiological estrogens appear to act in the same way. All of them bind to the estrogen receptor and cause tight binding in the nucleus (Clark and Peck, 1979). They share the ability to activate RNA polymerase (Hardin et al., 1976) and induce the expression of creatine kinase (Pentecost et al., 1990), epidermal growth factor (EGF), and its receptor (Lingham et al., 1988), insulinlike growth factor I (Murphy et al., 1987), tissue factor (a factor involved in the coagulation of blood) (Henrikson et al., 1992), and alterations in the structure of collagen (Pastore et al., 1992), and the proto-oncogenes c-Fos, c-jun, and jun-B and c-myc (Weisz and Bresciani, 1988; Loose-Mitchell et al., 1988; Papa et al., 1991; Chiappetta et al., 1992). Therefore, it appears that physiological estrogens bind to the estrogen receptor in a similar way and stimulate the transcription of identical genes.

Nonphysiological Estrogens

These are nonsteroidal estrogens such as kepone, DDT, and DES. Even estrogens of the nonphysiological class appear to act in a fashion similar to the physiological estrogens. Clomiphene, DES, kepone and, o,p'-DDT have been shown to bind to cause high-affinity nuclear binding of the estrogen receptor and to stimulate many of the same biosynthetic events and growth responses as physiological estrogens (Clark et al., 1980; Eroschenko and Palmiter, 1980; Kupfer and Bulger, 1980; Sheehan, 1981). Since these responses are the result of gene transcription, one can assume that nonphysiological estrogens are activating the same genes as physiological estrogens.

Thus, the question arises: Is the reproductive toxicity of an agent simply a function of its estrogenicity? The answer to this question is still forthcoming; however, it is clear that all estrogenic agents, whether physiological or nonphysiological, act as reproductive toxins if administered at critical periods of development or in a continuous fashion in the adult. Therefore, it is likely that estrogenicity and toxicity are closely related.

As mentioned before, DES is considered by some to be an exception to the above statements. It is also possible that DES is active by mechanisms that do not involve estrogenicity. DES is metabolized to various active intermediate metabolites that could interact with cellular constituents and disrupt normal function (Metzler et al., 1980). This probably is not a general mechanism of action for DES, however, since both physiological estrogens and DES cause the same reproductive-tract abnormalities in rodent species. Since the metabolic intermediates of both compounds are quite different, it seems improbable that such metabolic intermediates are involved.

A principal contribution to the pharmacokinetic properties of any drug or hormone is made by the metabolic alterations it may undergo. It is well known that many substances are not toxic unless they are metabolized to active metabolites. This concept has led some investigators to suggest that estrogens act as toxins by such activation mechanisms. It has been proposed that the carcinogenic potential of DES results from its metabolic conversion to active intermediates, such as E-DES-3,4-oxide (Metzler et al., 1980). This compound is very active in the sister chromatid exchange assay and is estrogenic. Since E-DES-3,4-oxide also binds to the estrogen receptor, it has been proposed that the receptor transports E-DES-3,4-oxide to the nucleus where it may become covalently bound to DNA. The presumed covalent interaction would lead to altered genomic function and the development of abnormal cell growth. It has also been proposed that the DES metabolite DES-4',4"-quinone forms DNA adducts, and these are instrumental in the cause of kidney cancer in hamsters (Gladek and Liehr, 1989). All of these studies use heroic doses of estrogen, which are far above the levels to which organisms would be exposed. For example, Gladek and Liehr (1989) aministered DES at a dose of 20 mg/kg to 10-day-old hamsters. This is approximately 50,000 times the physiological dose and thousands of times any dose that might be present in the environment.

Some DES metabolites manifest the interesting property of very similar receptor-binding characteristics, but they do not stimulate all of the uterotropic responses stimulated by estradiol (Korach et al., 1979, 1985). Whether these observations are related to the toxicity of DES is unknown; however, one fact is clear: DES binds to the estrogen receptor and stimulates the same estrogenic response stimulated by estradiol. Therefore, it is likely that any metabolite formed is an insignificant factor in the overall estrogenic response.

It has been suggested that estrogens, such as estradiol-17β, manifest their deleterious effects because they are metabolized to epoxide or quinone intermediates (Purdy et al., 1982). These authors have measured the ability of various steroids to cause the transformation of Balb/c 3T3 cells in culture. They concluded that DES, estrone, ethynyl estradiol, and estradiol will cause transformation while moxestrol and estriol will not. Since the first four compounds are good substrates for catechol estrogen formation and the latter two are not, these authors conclude that transformation is linked to the formation of active intermediates and not to the estrogenicity of these hormones. As in the case of DES metabolism discussed in the previous paragraph, this is an interesting and potentially important concept; however, it will be necessary to show that transformation and metabolism occur at concentrations of hormone expected under in vivo exposure conditions. The observed cell transformations were seen only when the concentration of estrogen was in the μM range, and no effects were seen below this level of exposure. Such high concentrations are 100–1000 times those that would be expected for hormone exposure in vivo. In addition, it has been demonstrated that estriol will cause adverse effects on reproduction, including mammary cancer, and yet estriol is a poor substrate for conversion to catechol estrogens (Noble et al., 1975). Such observations will require further examination before the metabolic conversion concept can be accepted as valid.

The hydroxylation of estradiol at the 4 position has been suggested as a mechanism of tumorigenesis (Liehr et al., 1995). Unfortunately, this study, like those cited above, suffers from the use of very high doses of estrogen in in vitro preparations. Once again, these concentrations are thousands of times higher than would ever occur in the human. Whether any of these reactions actually take place under conditions that cause cancer is debatable.

For the most part, estrogens of these two classes appear to be acting in the same fashion, and any toxicity is probably due to their estrogenicity.

Triphenylethylene Estrogen Agonist/Antagonists

These include antiestrogens such as tamoxifen, nafoxidene, and clomiphene. Because these drugs are mixed agonist/antagonist, it would appear that they have a different mechanism of action. But even in this case we have to be careful to keep the differences between rat and human in mind and to consider the response as a function of time after exposure. Also, data obtained from in vitro studies do not always agree with those derived from in vivo experiments, and they often add more confusion than clarification to the matter.

From the very beginning it was apparent the clomiphene acted differently in the rat than in the human. It blocked ovulation in the rat and caused ovulation in the human. In fact the triphenylethylene antiestrogens appear to have many species-specific actions that are difficult to explain by standard hormone receptor concepts. As an example, nafoxidine is a mixed estrogen agonist/antagonist in the rat and an agonist in the mouse. (For many more examples of species differences, see Clark and Markaverich, 1982.)

In the rat these compounds stimulate a full range of genetic transcription that cannot be differentiated from estradiol (Clark and Mani, 1994). These responses occur during the first hours after administration of the compounds and extend through the first 24 hours. If only this time period were observed, these compounds would be called estrogen agonists. It is only after 24 hours that antagonism begins to appear. Therefore, these drugs are acting as agonists for several hours and then as antagonists.

Obviously the mechanism of action of these drugs is very complex and has yet to be determined.

Most of the work in humans has been done with clomiphene, which is a mixture of *cis* and *trans* forms. The *cis* form is considered to be estrogenic and the *trans* to be antiestrogenic. Although this complicates the interpretation of most of these data, it is clear that triphenylethylene drugs are much less estrogenic in humans than in the rat. In the rat both forms of clomiphene will stimulate uterine epithelial cell hypertrophy with equal potency (Clark and Guthrie, 1981). Yet when the whole uterus is considered, *trans* clomiphene is clearly an estrogen antagonist. Endometrial hyperplasia is known to be a side effect of long-term treatment with tamoxifen. Yet, tamoxifen is considered to be primarily an antiestrogen in the human and is used to block the effects of estrogen on the proliferation of estrogen receptor–containing breast cancer cells.

It should be noted, as it almost never is, that the rationale for treating estrogen receptor positive breast cancer with tamoxifen is somewhat suspect. Although it appears to make sense on the surface it ignores the fact that treatment with other compounds, such as adrenal steroids, are just as effective. Hence the actual mechanism is up for question. This is also apparent from the fact that estrogen receptor negative patients also respond to tamoxifen treatment.

IN VITRO AND IN VIVO COMPARISONS OF ESTROGEN AGONIST/ANTAGONISTS

Any examination of the mechanism of action of type 3 compounds must consider the time-dependent differences in their action discussed in the previous section. Even the traditional method for testing these drugs is misleading because it fails to observe the early period of estrogenic activity and only examines the result after 3–4 days of injections. More recent attempts to examine their mechanism of action are totally dependent on tissue culture cells in vitro and require validation in vivo. The confusion that is generated by these in vitro experiments and their general lack of correlation with in vivo observations are typified by the following. Tamoxifen and 4-hydroxytamoxifen block the estrogen-stimulated increases in several specific proteins in MCF-7 cells and have no agonistic activity (Westley and Rochefort, 1979, 1980), whereas, as pointed out above, this drug acts in vivo as an agonist during the first 24 hours after administration. Tamoxifen and nafoxidine inhibit [^{3}H]-thymidine incorporation, DNA polymerase activity, and reduce cell number in MCF-7 cell cultures (Lippman et al., 1976; Edwards et al., 1980). In contrast, tamoxifen has been reported to stimulate cell proliferation in pituitary cells in culture (Amara and Dannies, 1986), while others have found no effect on proliferation (Prysor-Jones et al., 1983; Shull et al., 1992). Estrogen-stimulated prolactin synthesis is inhibited by tamoxifen, and no agonistic effect is seen with the drug alone (Lieberman et al., 1983a,b), yet others report a stimulation of prolactin expression by tamoxifen in pituitary cell culture (Prysor-Jones et al., 1983; Martinez-Campos et al., 1986; Shull et al., 1992).

The most recent experiments are performed in transfected cells in vitro (McDonnell et al., 1995), and the results obtained by this method also contradict those found in vivo. It is suggested by McDonnell et al. (1995) that antiestrogens, like tamoxifen, cause a different conformation in the estrogen receptor than does estradiol, and this alters the ability of the complex to stimulate transcriptional activity. This idea has been suggested many times and certainly is logical; however, it is not substantiated by work done in vivo. See previous discussions on this point.

The use of transfected cells in vitro involves cells which carry genes in the form of naked DNA which has been inserted into them. These pieces of naked DNA express the estrogen receptor and contain a response gene. Such naked genes do not resemble cellular genes which are covered with various interacting proteins and are in a chromatin comformation (Spelsberg et al., 1983). Therefore, any observation made in such an in vitro system must be very carefully validated in the animal (in vivo).

ESTROGENICITY AND TOXICITY

From the previous discussions of type 1 and 2 estrogen, it is clear that estrogenicity and toxicity are closely correlated. The conclusion seems to be that the estrogenic nature of these compounds is the cause of their toxicity. This does not rule out the possibility of action of these compounds in systems other than reproductive. The toxicity of type 3 compounds is more complex because of their dual action. Estrogenicity of type 3 compounds appears to be responsible for the toxic effects of perinatal exposure. Clomiphene administration in neonatal rats is clearly estrogenic and results in the persistent estrus syndrome in the adult female rat (Clark and McCormack, 1977). As discussed previously, the reproductive abnormalities result from the continuous exposure to ovarian estrogen and are only indirectly caused by inappropriate clomiphene exposure. Thus there is no great mystery concerning the mechanism of action of clomiphene in causing this toxic outcome. In these experiments clomiphene was acting as an estrogen agonist, not as an antagonist. Hence, had such an assay been the sole source of knowledge concerning this compound, it would have been classified as an estrogen. Such findings point to the need to consider not only the species but the age or developmental stage of the organism.

Certainly the antagonistic effects of these drugs could have toxic effects. As noted previously, administration of clomiphene during the critical period of sexual differentiation could block the action of estrogen at the hypothalamic level in the rat and cause feminization of male neonates (reviewed in Clark and Markaverich, 1982). Clomiphene exposure in adult rats will block ovulation; whereas, in women it induces ovulation. These effects of clomiphene are generally thought to be due to its antiestrogenic properties, although it can be argued that the estrogenic properties are involved.

USE OF HIGH-DOSE ESTROGENS IN TOXICOLOGY STUDIES

Very often high doses of estrogens and other putative toxins are administered to animals and the results are extrapolated to lower doses. This is an unfortunate practice, which as led to much controversy and confusion. It is not only an incorrect way to assess the activity of any compound, but it is likely to be counterproductive. The reason for this is simple: high doses of almost any compound will inhibit the very response being studied. It is well known that high doses of estrogens will inhibit uterine growth (Terenius, 1971). This principle has been used for many years in the treatment of breast cancer with high doses of DES. Therefore, it is possible that many high-dose studies have not shown any effects due to inhibition of the very effects of interest. If lower doses had been used, the effects would have be evident. Also, if lower doses had been used and a valid dose-response curve obtained, a valid potency estimate could have been made.

Toxic outcomes of treatment with very high doses of hormones or toxins may be the result of mechanisms that are entirely different from the normal physiological ones. Therefore, they do not represent a valid outcome on which to anticipate actions at lower doses.

PLANT ESTROGENS AND CANCER

Several investigators have suggested that bioflavonoids act as inhibitors and/or contributors to the estrogen load of human females and may be involved in the inhibition or facilitation of breast neoplasia. In the first place, it would be very surprising if blood levels of bioflavonoids reached significant levels. These compounds are extensively metabolized in the gut by bacteria and less than 0.1% appear to be absorbed (Harbone, 1975). This is very fortunate because most of these compounds are highly mutagenic. Therefore, if these substances were absorbed to any significant degree, humans would be exposed continuously to very potent mutagens. Carcinogenic effects would be seen to increase in vegetarians and other groups whose main source of nutrition is plants. Since this not the case—indeed just the opposite is the case—it seems very unlikely that bioflavonoids make any contribution to the estrogenic load of the body.

Even if concentrations of phytoestrogens in the blood were high enough to cause estrogen receptor binding, they would probably act as weak estrogens and make little contribution to the overall estrogenic load of the individual. These compounds would be analogous to other weak physiological estrogens which are agonists at high concentrations or if their concentrations are maintained over long periods of time. (For a complete discussion of weak or short-acting estrogens, see Clark and Mani, 1994.)

Some investigators have suggested that phytoestrogens might act as estrogen antagonists and thereby protect against endogenous estrogens (Adlercreutz et al., 1992, 1993). However, this hypothesis appears to be unlikely since weak estrogens are not antagonistic when they are present for long periods of time, as they would be if they were eaten (reviewed in Clark and Peck, 1979; Clark and Mani, 1994). Weak or short-acting estrogens act on estrogen antagonists only when they are injected in saline and are cleared rapidly from the blood. If they are present for several hours, as they would be from intestinal absorption, they simply act as weak estrogens and have no antagonistic effect. Since they do not bind very tightly to the estrogen receptor (Markaverich et al., 1995), their contribution to the estrogenic capacity of the endogenous estrogens would be very little.

ESTROGENS AS PERMISSIVE FACTORS IN NEOPLASIA

As discussed above, humans do not suffer from persistent estrus syndrome and consequently are not exposed to continuous ovarian estrogen. However, there are circumstances in which humans are exposed to continuous estrogen, which does have detrimental effects. Endometrial hyperplasia and cancer occur in women who have been exposed to either endogenous or exogenous estrogens for prolonged periods of time

(Gusberg and Kardon, 1971; Fechner and Kaufman, 1974; Lucas, 1974; Dewhurst et al., 1975). These cases include women with ovarian tumors that produce estrogens, women who fail to ovulate and as a result are exposed to estrogen without the normal intervention of the luteal phase of the cycle, and women who have taken estrogens for many years because they lack functional ovaries. These adverse effects caused by continuous exposure to estrogens probably occur due to the lack of normal cyclic elevations in progesterone. Progesterone is known to antagonize and modify the ability of estrogen to act on estrogen target tissues (Clark and Peck, 1979).

So, in the case of continuous estrogen exposure, is estrogen a carcinogen? The answer may be yes, but only indirectly. There is no question that cells must divide in order for neoplasia to manifest itself. Since estrogens are mitogens, it is obvious that they should be considered at least indirect carcinogens. However, whether they are direct carcinogens as discussed above remains an open question. It is my opinion that estrogens should be considered permissive agents and not direct carcinogens.

ESTROGENS, EVOLUTION, AND TOXICITY

An understanding of how estrogens control and/or interfere with reproductive function requires some understanding of our evolutionary past and how it relates to our current physiology. Most of the problems we presently exhibit are linked to our evolutionary past and are the result of modern culture interacting with a physiology and anatomy that was adapted to cope with paleolithic conditions. Humans are physiologically no different now than they were in paleolithic times. Physiological evolution is much too slow to adapt in such a short time. At the beginning of civilization, approximately 12,000 years ago, humans were adapted and designed by natural selection to be hunter-gatherers and had existed as such for 90% of their existence. It is only in the last few thousand years that our physiology and anatomy have been transported by agriculture and civilization to our current predicament.

Loss of Estrus Behavior in Humans

Most mammals, including nonhuman primates, have a behavioral estrus period or sexual heat period during the time of the reproductive cycle when ovulation occurs (Short, 1986). Females in estrus will permit and encourage copulation by males and by doing so increase the likelihood that copulation will result in fertilization. Estrus behavior is an estrogen-regulated desire to copulate—it may even be painful (as in cats and other animals with penile spines)—and is most certainly not necessarily enjoyed by the female (Zarrow and Clark, 1968). Such a system of synchronized ovulation and mating is clearly of selective advantage and increases the probability of fertile mating.

A momentous event in the evolution of humans was the loss of this behavioral estrus. Such a loss was a must for the development of civilization as we know it because it eliminated the disruptive effect of females in heat. Most primate species, although they do menstruate somewhat like humans, retain a midcycle behavioral estrus.

Several authors have speculated about the loss of estrus and its importance to the development of society and civilization; however, to my knowledge none have proposed reasons for it. My opinion is that the loss came as a result of a decreased sensitivity to estrogen in the brain during fetal development. As discussed previously, humans are unique in that they have very high blood levels of estrogen during pregnancy. Such levels in other species would masculinize the brain of the fetus and produce all male-like offspring, i.e., acyclic with no female behavior. Therefore, the human fetus must have evolved an insensitivity to maternal estrogens, otherwise women would not have menstrual cycles and would be behavioral males. The desensitization of brain regions to estrogen may well have desensitized the behavioral centers of the brain that control estrus behavior.

The physiological reason for such a high level of estrogen during pregnancy is not known. As mentioned above, most other animals do not have high levels of estrogen during pregnancy. Only some of the great apes, considered to be our closest relatives, have intermediate levels of estrogen during pregnancy. The tendency toward loss of midcycle estrus occurs in the chimpanzee, which has levels of estrogen in the blood during pregnancy that are intermediate between the human and other primates (Albrecht and Pepe, 1988). Thus, the chimpanzee may represent a transitional stage in the evolutionary pathway similar to that which led to humans.

The evolution of high levels of estrogen in the blood during pregnancy, possibly associated with brain development in early hominids, may have cause a decline in reproductive capacity in hominid types that did not develop an insensitivity to estrogen. Thus, as estrogen blood levels rose and selection pressures for increased brain size were maintained, only those early hominids that coordinately developed an insensitivity to estrogens would have survived. Others would have been masculinized and infertile.

The loss of behavioral estrus set the stage of the formation of civilization as we know it. Other forms of civilized culture are possible, but their social order and family life would be very different. Unfortunately, the endocrine changes associated with these alterations also set the stage for problems with estrogen-induced cancers.

Breast Cancer and Endogenous Estrogens

This human retention of a paleolithic physiology in the modern world presents several health hazards. Humans were adapted and designed by natural selection to be hunter-gatherers and have existed as such for 90% of our historical existence. The reproductive strategy in hunter-gatherers is quite different from that of the modern "civilized" human. In such groups women spend most of their adult life either pregnant or lactating (Short, 1984). Thus, menstrual cycles are the unnatural state and pregnancy followed by lactation is the natural state. Natural, as used here, refers to the natural wild state, which I contend reflects our present physiology and has remained unchanged since paleolithic times. Pregnancy is characterized by high levels of progesterone and estrogen over many months and lactation by acyclicity and low levels of ovarian steroid hormones. Thus, the wild (natural) state for women is to go from months of exposure to high levels of estrogen and progesterone during pregnancy to months, perhaps years, of exposure to relatively low levels of estrogen and progesterone.

Therefore, it seems that menstrual cycles occurring over the greater part of a woman's lifespan may be potentially dangerous. Ovarian activity has been shown to be a risk factor in breast cancer. Premenopausal ovariectomy is protective against breast cancer in proportion to the reduction in years of menstrual cycles (MacMahon and Feinleib, 1960). This protective effect of early ovariectomy is eliminated by the administration of estrogen (Hoover et al., 1976). In addition, a higher risk of breast cancer exists for women who have an early menarche and late menopause (Yuasa and MacMahon, 1971). It has been suggested by Henderson et al. (1985) that increased risk for breast cancer is related to the cumulative number of regular ovulatory cycles and that reducing the frequency of ovulatory cycles is associated with reduced risks. In contrast, Korenman (1980) proposed that decreased exposure to luteal progesterone in the presence of follicular phase estrogen increases the risk for breast cancer. These opposite views relate to each author's interpretation of the actions of progesterone. Korenman (1980) considers progesterone to be protective because it is an estrogen antagonist, which decreases the effectiveness of estrogen as a mitogen. Such anti-estrogenic actions of progesterone are well known for organs such as the uterus and vagina; however, Henderson et al. (1985) point out that mitotic activity in breast tissue is enhanced during the luteal phase of the cycle when progesterone is elevated.

These conflicting views concerning the action of progesterone are based on considerable supposition and on epidemiological data which are argumentative. Sherman et al. (1982) reported that no relationship between menstrual cycle length and breast cancer exists. However, considering our evolutionary past, it seems more likely that exposure to continuous menstrual cycles of any type (regular or irregular) are unnatural and may expose breast tissue to stimulatory events, which predispose the tissue to neoplastic transformation. This unnatural exposure coupled with the equally unnatural longevity, which we all enjoy, establishes conditions that permit the expression of breast cancer. Therefore, it can be argued that a reproductive lifetime consisting of pregnancy followed by lactation would be protective against breast cancer. Some data suggest that this may be the case: early pregnancy (before age 20) decreases the relative risk of breast cancer from 1.0 in nulliparous women to 0.5 (MacMahon et al., 1970), and pregnancy at all ages appears to have a protective effect (Kelsey, 1979). However, there is no agreement on the protective effect of lactation (reviewed by Vorherr, 1980). Many investigators report a protective effect, while others report none. In certain select groups, such as the Canadian Eskimos, there is a clear-cut protective effect associated with lactation. In Eskimos breast cancer is extremely rare: 1.3 per 100,000 per year as compared to 72.5 per 100,000 per year in the United States. Eskimos breast-fed their infants 3 years longer and many of them are pregnant or lactating continuously from 17 to 50 years of age.

The latter half of pregnancy and the entire period of lactation is associated with a much reduced mitotic rate in the human breast (Kaiser, 1969). If reduced mitotic rate is related to reduced predisposition to breast cancer, then pregnancy and lactation are likely to be protective.

The protective effect of pregnancy in humans has also been observed in rats (Sinha et al., 1988). In addition, treatment with estrogen and progesterone protected against carcinogen-induced mammary cancer (Grubbs et al., 1985). Therefore, in this case, it appears at first glance that rats treated with estrogen plus progesterone might be a good model for the study of the protective effect of pregnancy. However, caution is warranted because during pregnancy rats have very low levels of estrogen in the blood, while humans have very high levels. Therefore, the effects of estrogen during pregnancy in the rat are insignificant when compared to the human. So the mechanisms may be very different. The fact that estrogen/progesterone treatment does protect from mammary cancer in the rat may be unrelated to the true mechanism and may simply be a pharmacological artifact.

Breast cancer occurs most frequently during the postmenopausal years when ovarian steroid hormones are not secreted. If these hormones are associated with carcinogenesis, why do cancers develop in their absence? In my view, the answer is simple: the neoplastic event occurred during the premenopausal years and the cancer becomes evident years later during menopause. It is well known that the latency period between the carcinogenic event and the onset of neoplasia can last many years.

Some investigators have attempted to explain menopausal cancer by invoking specific endocrine alterations that occur during menopause. Siiteri et al. (1974) proposed that endometrial cancer is associated with estrone formation from the aromatization of adrenal androstenedione in adipose tissue. Since progesterone is very low in postmenopausal women, estrone would act as an unopposed estrogen and stimulate endometrial proliferation. Although this is an interesting hypothesis, it should be noted that postmenopausal women do not display any evidence of the presence of physiological active estrogens. The vagina in postmenopausal women is atrophic, and since the vagina is a very sensitive indicator of the presence of estrogen, it is apparent that there is insufficient estrone present to cause endometrial proliferation.

Therefore, the reason why most reproductive tract and breast cancers appear during menopause is that women live long enough for them to develop. The carcinogenic event(s) occurred during the premenopausal period when the cells of these tissues were actively growing and regressing under the influence of estrogen and progesterone. Since these cycles of cell growth and regression as they occur during the menstrual cycle represent the unnatural state, not the wild state of continuous pregnancy and lactation, they increase the probability of carcinogenic insult. Our paleolithic ancestors, whose reproductive physiology persists to this day, did not suffer from breast cancer because they never reached menopausal age.

Estrogen Replacement Therapy and Menopause

It makes a great deal of sense to replace the hormones that a person loses either by disease or by menopause. If someone has diabetes, the physician prescribes insulin. By the same reasoning, if a women's ovaries stop producing estrogen and progesterone at 50 years of age, she should receive replacement of these two hormones. It is clear that such replacement decreases the incidence of osteoporosis, atherosclerotic heart disease, vaginal atrophy, and hot flashes (Carr, 1994). These important benefits greatly outweigh the slight risks of uterine cancer (Smith et al., 1975). Women who receive estrogen therapy and have endometrial cancer outlive those with the cancer who are not taking estrogen (Collins et al., 1980).

The real questions is how estrogen and progesterone should be replaced. It is generally thought that these hormones should be given so as to simulate those found during the menstrual cycle. However, as discussed above, menstrual cycles may themselves be hazardous. So, if this is the case, should we try to replace them as they would have been during pregnancy and lactation in order to simulate our ancestral physiology? Unfortunately, this would probably be too difficult. So the best solution is to give estrogen and progesterone daily in doses that will maintain their levels in the blood so as to block menopausal symptoms and disease. Such a regimen will benefit the older female while adding few risks.

REFERENCES

Adlercreutz, H., Mousavi, Y., and Hockerstedt, K. (1992). *Acta Oncol.*, *31*: 175.

Albrecht, E. D., and Pepe, G. J. (1988). *Nonhuman Primates in Perinatal Research* (Y. W. Brans, T. J. Kuehl, eds.). John Wiley and Sons, New York.

Allen, E., and Gardner, V. (1941). *Cancer Res.*, *1*: 359.

Amara, J. F., and Dannies, P. S. (1986). *Mol. Cell Endocrinol.*, *47*: 183.

Barnes, A. B. (1979). *Fertil. Steril.*, *32*: 148.

Bibbo, M., Gill, W. B., Azizzi, F., Blough, R., Fang, W. S., Rosenfeld, R. L., Schumacher, G. F., Sleeper, K., Sonek, M. G., and Wied, G. L. (1977). *Obstet. Gynecol.*, *49*: 1.

Carr, B. R. (1992). *Williams Textbook of Endocrinology* (J. D. Wilson and N.W. Foster, eds.). W. B. Saunders, Philadelphia.

Chiappetta, C., Kirkland, J. L., Loose-Mithcell, D. S., Murthy, L., and Stancel, G. M. (1992). *Steroid Biochem. Mol. Biol.*, *41*: 113.

Clark, J. H., and McCormack, S. A. (1977). *Science*, *197*: 164.

Clark, J. H., and Peck, E. J., Jr. (1979). *Female Sex Steroids*: *Receptors and Functions*. Springer-Verlag, Berlin.

Clark, J. H., and Markaverich, B. M. (1982). *Pharmacol. Ther. 15*: 468.

Clark, J. H., and Mani, S. K. (1994). *The Physiology of Reproduction* (E. Knobil and J. D. Neill, eds.). Raven Press, New York, p. 1011.

Clark, J. H., Paszko, Z., and Peck, E. J., Jr. (1977). *Endocrinology*, *100*: 91.

Clark, J. H., McCormack, S. A., Kling, R., Hodges, D., and Hardin, J. W. (1980). In: *Hormones and Cancer* (S. Iacobelli, ed.). Raven Press, New York, p. 295.

Cole, P., and MacMahon, B. (1969). *Lancet*, *I*: 604.

Collins, J., Allen, L. H., and Donner, A. (1980) *Lancet*, *2*: 961.

Cramer, W., and Horning, E. S. (1936). *Lancet*, *I*: 247.

Dewhurst, C. J., DeKoos, E. B., and Haines, R. M. (1976). *Am. J. Obstet. Gynecol.*, *82*: 412.

Dunn, T. B., and Green, A. W. (1963). *J. Natl. Cancer Inst.*, *31*: 425.

Edwards, D. P., Murphy, S. R., and McGuire, W. L. (1980). *Cancer Res.*, *40*: 1722.

Eroschenko, V., and Palmitter, R. (1980). In: *Estrogens in the Environment* (J. McLachlan, ed.). Elsevier/North Holland, New York, p. 305.

Fechner, R. E., and Kaufman, R. H. (1974). *Cancer*, *34*: 444.

Forsberg, J.-G. (1972). *Am. J. Obstet. Gynecol.*, *113*: 83.

Forsberg, J.-G. (1975). *Am. J. Obstet. Gynecol.*, *121*: 101.

Furth, J., Ueda, G., and Clifton, K. H. (1973). In: *Methods in Cancer Research* (H. Busch, ed.). Academic Press, New York, p. 201.

Gardner, W. U., Pfeiffer, C. A., and Trentin, J. J. (1959). In: *The Physiopathology of Cancer* (F. Humburger, ed.). Harper & Row, New York, p. 152.

Gellert, R. J., Heinrichs, W. L., and Sverdloff, R. (1974). *Neuroendocrinology*, *16*: 84.

Gladek, A., and Lehr, J. G. (1989). *J. Biol. Chem.*, *264*: 16847.

Gorski, R. (1963). *Am. J. Physiol.*, *205*: 842.

Goy, R. W., and Resko, J. A. (1972). *Rec. Prog. Horm. Res.*, *18*: 707.

Grubbs, C. J., Farnell, D. R., Hill, D. L., and McDonough, K. C. (1985). *J. Natl. Cancer Inst.*, *74*: 927

Gusberg, S. B., and Kardon, P. (1971). *Am. J. Obstet. Gynecol.*, *11*: 633.

Hammond, G. L. (1990). *Endocrine Rev.*, *11*: 65.

Haney, A. F., Hammond, C. B., Soules, M. R., and Creasman, W. T. (1979). *Fertil. Steril.*, *31*: 142.

Harbone, J. B., Mabry, T. J., and Mabry, H. (1975). *The Flavonoids*. Acedemic Press, New York.

Henderson, B. E., Ross, R. K., Judd, H. L., Krailo, M. D., and Pike, M.C. (1985). *Cancer*, *56*: 1206.

Henrikson, K. P., Greenwood, J. A., Penticost, B. T., Jazin, E. E., and Dickerman, H. W. (1992). 130: 2669.

Herbst, A.L., H. Ulfelder, and D.C. Poskanzer. (1971). *N. Engl. J. Med. 284*: 878.

Herbst et al. (1975). *N. Engl. J. Med., 292*: 344.

Hobson, W. C., Fuller, G. B., Mueller, W., Korte, F., and Coulston, F. (1978). *Ecotox. Environ. Safety, 2*: 257.

Hoover, R., Gray, L. A., and Cole, P. (1976). *N. Engl. J. Med., 295*: 401.

Horwitz, K. B., and McGuire, W. L. (1978). *J. Biol. Chem., 253*: 2223.

Jackson, R. L., and Dockerty, M. B. (1957). *Am. J. Obstet. Gynecol., 67*: 161.

Kaiser, R. (1969). *Geburtshilfe Frauenheilkd., 29*: 420.

Kaufman, R. H., Binder, G. L., Gray, P. M., and Adam, E. (1977). *Am. J. Obstet. Gynecol., 128*: 51.

Kelsey, J. L. (1979). *Epidemiol. Rev., 1*: 74.

Kincl, F. A., Folch, Pi, A., Maqueo, M., Lasso, L. H., Oriol, A., and Dorfman, R. I. (1965). *Acta Endocrinol. 49*: 193.

Korach, K. S., Metzler, M., and McLachlan, J. A. (1979). *J. Biol. Chem., 254*: 8963.

Korach, K. S., Fox-Davies, C., Quarmby, V. E., and Swaisgood, M. H. (1985). *J. Biol. Chem., 260*: 15420.

Korenman, S. G. (1980). *Cancer, 46*: 874.

Kupfer, D., and Bulger, W. H. (1980). In: *Estrogens in the Environment* (J. McLachlan, ed.). Elsevier/North Holland, New York, p. 239.

Lanier, A. P., Noller, K. L., Decker, D. G., Elveback, L. R., and Kurland, L. (1973). *Mayo Clin. Proc., 48*: 793.

Larson, J. A. (1954). *Am. J. Obstet. Gynecol., 3*: 551.

Leavitt, W. W., Evans, R. W., and Hendry, W. J., III. (1981). In: *Hormones and Cancer* (W. W. Leavitt, ed.). Plenum Publishing Co., New York.

Lemon, H. M., Miller, D. M., and Foley, J. F. (1971). *Natl. Cancer Inst. Monogr., 34*: 77.

Lieberman, M. E., Gorski, J., and Jordan, V. C. (1983a). *J. Biol. Chem. 258*: 4741.

Lieberman, M. E., Jordan, V. C., Fritsch, M., Santos, M. A., and Gorski, J. (1983b). *J. Biol. Chem., 258*: 4734.

Liehr, J. G., Ricci, M. J., Jefcoate, C. R., Hannigan, E. V., Hokanson, J. A., and Zhu, B. T. (1995). *Proc. Natl. Acad. Sci. USA, 92*: 9220.

Lingham, R. B., Stancel, G. M., and Loose-Mithell, D. S. (1988). *Mol. Endocrinol., 2*: 230.

Lippman, M., Bolan, G., and Huff, K. (1976) *Cancer Res., 36*: 4595.

Lobl, R. T., and Maenza, R. M. (1975). *Biol. Reprod., 13*: 255.

Loose-Mitchell, D. S., Chiappetta, C., and Stancel, G. M. (1988). *Mol. Endocrinol., 2*: 946.

Lucas, W. E. (1974). *Obstet. Gynecol. Surv., 29*: 507.

MacLusky, N. J., and Naftolin, F. (1981). *Science, 211*: 1294.

MacMahon, B., and Feinleib, M. (1960). *J. Natl. Cancer Inst., 24*: 733.

MacMahon, B., Cole, P., and Lin, T. M. (1970) *Bull. WHO, 43*: 209.

Markaverich, B.M., Upchurch, S., McCormack, S. A., Glasser, S. R., Clark, J. H. (1981). *Biol. Reprod., 24*: 171.

Markaverich, B. M., Webb, B., Densmore, C. L., and Gregory, R. R. (1995). *Environ. Health Perspect., 103*: 574.

Martinez-Campos, A., Amara, J. F., and Dannies, P. S. (1986). *Mol. Cell Endocrinol., 48*: 172.

McEwen, B. S. (1981). *Science, 211*: 1301.

Metzler, M., Gottschlich, R., and McLachlan, J. A. (1980). In: *Estrogens in the Environment* (J. McLachlan, ed.). Elsevier/North Holland, New York, p. 293.

Moore, P. H., Jr., and Rhim, J. S. (1982). In: *International Workshop on Catechol Estrogens* (M. Lipsett and G. R. Merriam, eds.). Raven Press, New York, p. 123.

Murphy, L. J., Murphy, L. C., and Friesen, H. G. (1987). *Mol. Endocrinol., 1*: 445.

Noble, R. L., Hochacka, B. C., and King, D. (1975). *Cancer Res., 35*: 766.

Nunez, E. A., Benassayag, C., Savu, L., Vallette, G. V., and Jayle, M. F. (1976). In: *Onco-developmental Gene Expression* (W. H. Fishman and S. Sell, eds.). Academic Press, New York, p. 365.

Papa, M., Mezzogiorno, V., Bresciani, F., and Weisz, A. (1991). 175: 480.

Pastore, G. N., Dicola, L. P., Doooahon, N. R., and Gardner, R. M. (1992). *Biol. Reprod.*, 47: 83.

Pentecost, B. T., Mattheiss, L., Dickerman, H. W., and Kumar, S. A. (1990) *Mol. Endocrinol.*, 4: 1000.

Pomerance, W. (1973). *Obstet. Gynecol.*, 42: 12.

Prysor-Jones, R. A., Silverlight, J. J., and Jenkins, J. S. (1983). *J. Endocrinol.*, 97: 261.

Purdy, R. H., Goldzieher, J. W., LeQuesne, P. W., Abdel-Baky, S., Durocher, C. K., Moore, P. H., Jr., and Rhim, J. S. (1982). In: *International Workshop on Catechol Estrogens* (M. Lipsett and G. R. Merriam, eds.). Raven Press, New York, p. 123.

Rudali, G., Apiou, F., and Muel, B. (1975). *Eur. J. Cancer*, 11: 39.

Sheehan, D. M., Branham, W. S., Medlock, K. L., Olson, M. E., and Zehr, D. (1980). *Teratology*, 21: 68A.

Sherman, B. M., Wallace, R. B., and Bean, J. A. (1982). *Cancer Res.*, 42(Suppl.): 3286.

Short, R. V. (1984) *Sci. Am.*, 250: 35.

Short, R. V. (1986). *Reproduction in Mammals. Book 3* (C. R. Austin and R. V. Short, ed.). Cambridge University Press, London.

Shull, J. D., Beams, F. E., Baldwin, T. M., Gilcrist, C. A., and Hrbek, M. J. (1992). *Mol. Endocrinol.*, 6: 529.

Siiteri, P. K., Schwarz, B. E., and MacDonald, P. C. (1974). *Gynecol. Oncol.*, 2: 226.

Sinha, D. K., Pazid, J. E., and Dao, T. L. (1988). *Br. J. Cancer*, 57: 390.

Smith, D. C., Prentice, R., and Thompson, D. J. (1975). *N. Engl. J. Med.*, 293: 1164.

Spelsberg, T. C., Littlefield, B. A., and Seelke, R. (1983). *Recent Prog. Horm. Res.*, 39: 463.

Takasugi, N. (1976). *Int. Rev. Cytol.*, 44: 193.

Takasugi, N., and Bern, H. A. (1964). *J. Natl. Cancer Inst.*, 33: 855.

Terenius, L. (1971). *Acta Endocrinol.*, 66: 431.

Vorhevrr, H. (1980). *Breast Cancer*. Urban and Schwarzenberg Inc., Baltimore.

Weisz, A., and Bresciani, F. (1988). *Mol. Endocrinol.*, 2: 816.

Westley, B. R., and Rochefort, H. (1979). *Biochem. Biophys. Res. Commun.*, 90: 410.

Westley, B. R., and Rochefort, H. (1980). *Cell*, 20: 353.

Wotiz, H. H., Sjhane, J. A., Vigersky, R., and Brecher, P. I. (1968). In: *Prognostic Factors in Breast Cancer* (A. P. M. Forrest and P. B. Kunkler, eds.). Livingston Press, Edinburgh, p. 368.

Yuasa, S., and MacMahon, B. (1971). *Bull. WHO*, 42: 195.

Zarrow, M. X., and Clark, J. H. (1968). *J. Endocrinol.*, 40: 343.

13
Phytoestrogens and Reproductive Medicine

Claude L. Hughes, Jr.
Duke University Medical Center, Durham, North Carolina

Ginger Tansey
Wake Forest University School of Medicine, Winston-Salem, North Carolina

INTRODUCTION

Terminology and Initial Concepts

If we are to attempt consideration of the notion that compounds of plant origin can influence mammalian reproduction by mimicry of reproductive hormones, then clarity of the meaning of terms is necessary even if there is no consensus. The term "phyto-estrogen" implies the existence of a chemical or class of chemicals produced by plants that acts like an "estrogen" (in animals). Standard definitions should help. Estrus is "heat; that portion or phase of the sexual cycle of female animals characterized by willingness to accept the male" (*Stedman's Medical Dictionary*, 22nd ed., 1972). Thus, estrus is a specific pattern of behavior. Estrogen is a "generic term for any substance, whether naturally occurring or synthetic, that exerts biological effects characteristic of estrogenic hormones, such as estradiol, named for its ability to induce estrus in lower mammals" (*Stedman's Medical Dictionary*, 22nd ed., 1972). Thus, as an absolute minimum, any putative estrogen should be capable of inducing estrus in "lower mammals." By this proper, simple, and explicit terminology there is hardly any occurrence of truly "estrogenic" compounds in plants, because the only phytochemicals that have been specifically shown to elicit estrus are the classical estrogenic steroids, estrone, estradiol, and estriol. The comprehensive review of Farnsworth et al. (1975) described only 13 plants that contained scant quantities of steroidal estrogens with content on the order of 0.1–0.3 mg/kg of plant material (Jacobsohn et al., 1965).

It is noteworthy that the vast preponderance of research, literature, and scientific interest in compounds called phytoestrogens over the last half-century has been directed toward understanding the mammalian actions of plant-derived isoflavonoids and coumestans. These compounds are quantitatively prominent in many plants, including several used for feed for livestock and food for human beings, and indeed have fascinating biological properties. However, since there has never been even a single report indicating that these isoflavonoids and coumestans can induce estrus in lower animals, then we must conclude that these compounds are not estrogens at all and we have been in error in characterizing them as such! We are thus left with the established usage of

277

the term "phytoestrogen" to mean something less than or other than a "real" estrogen. Whereas estradiol and other steroidal estrogens are known to elicit certain specific effects in several target tissues other than the portions of the brain regulating sexual behavior (and probably other complicated integrative nervous system functions), the extent to which the isoflavonoids and coumestans mimic estradiol in these other targets does suggest these compounds play important biological roles in animals that ingest plants that contain these classes of "phytoestrogens." The potential importance of phytoestrogens in human health and disease is certain to derive both from their mimicry of steroidal estrogens and their failure to closely mimic the actions of steroidal estrogens. While the role of these compounds in chronic diseases is considered elsewhere in this book (Chap. 14), we will focus on the actions of phytoestrogens on reproductive physiology per se and some of the concepts that may be relevant for appreciating the consequences of exposure of mammals to exogenous naturally occurring or xenobiotic environmental putative estrogens.

There is currently great scientific and public concern about the potential ecological and human health impact of environmental chemicals that mimic steroidal mammalian estrogenic reproductive hormones. Legitimate controversy often arises when the biological consequences of exposure to different classes of estrogenic chemicals are compared and contrasted. Unprejudiced consideration of the effects (whether adverse, beneficial, or nil) of environmental estrogens of all types and from all sources is required if scientific reason is to prevail in the discourse in society at large.

Human diets contain both archiestrogens (ancient, naturally occurring) and xenoestrogens (novel, man-made). Apparent contradictions regarding the potential for risk or benefit from exposure to these classes of agents imply that some clarifying concepts should be invoked. We contend that use of the biological concepts of co-evolution of plants and animals and natural selection of animal populations by dietary phytochemicals leads to sets of distinct and testable expectations for the mammalian effects of both ancient and novel estrogens.

Structure and Occurrence of Phytoestrogens

Dietary phytoestrogens are naturally occurring constituents of plants that elicit estradiol-like effects in one or more target tissue in animals. Nearly 70 years ago, it was noted that certain plants could induce estrus in animals (reviewed by Bradbury and White, 1954). Subsequently, after adoption of bioassay methods that assessed vaginal or uterine effects of putative estrogens, over 300 plants were found to possess some degree of estrogen-like activity (Bradbury and White, 1954; Farnsworth et al., 1975). These phytoestrogens are predominantly from two chemical classes (coumestans and isoflavones) and their metabolites, such as equol. The isoflavones and the coumestans have 15-carbon structures similar to the 17-carbon structure of estradiol-17β (Fig. 1). Other naturally occurring estrogens are fungal in origin and, taken together, the mycoestrogens and phytoestrogens can be grouped together as archiestrogens for consideration of co-evolutionary hypotheses.

Some plants that contain consequential amounts of phytoestrogens are shown in Table 1. Quantitatively the isoflavones are quite important. The richest sources among foodstuffs are legumes and grains, with soy content of the isoflavones genistein, daidzein, and their conjugates on the order of 0.5–3 mg/g of soy protein. For human health considerations, a focus on the isoflavones from soybeans is justifiable due to the

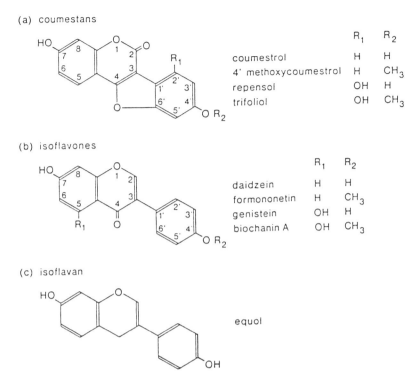

(a) coumestans

	R$_1$	R$_2$
coumestrol	H	H
4' methoxycoumestrol	H	CH$_3$
repensol	OH	H
trifoliol	OH	CH$_3$

(b) isoflavones

	R$_1$	R$_2$
daidzein	H	H
formononetin	H	CH$_3$
genistein	OH	H
biochanin A	OH	CH$_3$

(c) isoflavan

equol

Figure 1 Structures of the common plant estrogens. (From Adams, 1989.)

facts that consumption of soybeans and soy food products is increasing (Messina and Messina, 1991), and soybeans are the major source of genistein and daidzein in human diets (Franke et al., 1994). Content is influenced by cultivar and locale, thus implying effects of cultural conditions such as temperature, rainfall, timing of harvest, soil fertility, pest and disease damage, etc. As an illustration, the range of dietary content has been shown for soybeans. Several hundred varieties or cultivars of soybeans exist, and their phytoestrogen content can vary from 50 to 300 mg/100 g (Eldridge and Kwolek, 1983). Thus, the effects of dietary soy alone could vary widely and would obviously depend upon amount ingested, pattern of intake, and the relative content of phytoestrogen in the soy product.

In human beings dietary phytoestrogens are readily absorbed, circulate in the plasma, and are excreted in the urine (Adlercreutz et al., 1993a; Morton et al., 1994; Xu et al, 1995). The plasma levels range up to hundreds of nmol/liter in persons consuming diets that are rich in these compounds (Adlercreutz et al., 1993b). The relative estrogenic biological potency of dietary phytoestrogens in mammals in influenced by gut metabolism (Adlercreutz et al., 1986; Xu et al., 1995). Estrogenic effects can be decreased due to degradation or increased by conversion of less potent compounds to more potent compounds by action of gut microorganisms. Some differences in the gut metabolism of dietary phytoestrogens can be readily explained by general features such as ruminant versus monogastric digestive systems. On the other hand, production of equol from less potent isoflavones is, at least in primates, more individually unique

Table 1 Occurrence and Content of Phytoestrogens in Common Plants

Family	Examples	Phytoestrogens	Estimates of content
Chenopodiaceae	Beet	Unidentified	+
Compositae	Sunflower	Coumestans[a]	+
Cruciferae	Rape	Unidentified	+
Euphorbiaceae	Castor oil plant	Unidentified	+
Gramineae	Barley Bluegrass Oats Orchardgrass Rye Ryegrass Rice Wheat	Isoflavonoids[b] coumestans, and resorcyclic acid lactones	+ to +++
Labiatae	Sage	Unidentified	++
Leguminosae	Alfalfa Clovers (red, strawberry, subterranean, and others) Soybean	Isoflavonoids and coumestans	++
Liliaceae	Garlic	Unidentified	+++
Malvaceae	Hollyhock	Unidentified	+
Palmaceae	Date palm	Unidentified	+
Polygonaceae	Rhubarb	Unidentified	+
Rosaceae	Apple Cherry Plum	Unidentified	+
Rubiaceae	Coffee	Unidentified	+
Saficaceae	Willow	Unidentified	++
Solanaceae	Potato	Unidentified	+
Umbelliferae	Parsley	Unidentified	+

+ = weakly estrogenic; ++ = moderately estrogenic; +++ = strongly estrogenic.
[a]Coumestans include coumestrol and related compounds.
[b]Isoflavonoids include genistein, daidzein, formononetin, Biochanin A, their conjugates, and related compounds.

such that some persons are "equol formers" and others are not. Evidence that these differences relate to populations of microbial flora include alteration in status due to use of oral antibiotics. The relative estrogenicity of a given diet can therefore differ among individuals due to use of antimicrobial medications or variations in immune system regulation of the composition of gut flora.

In summary, accurate characterization of the exposure of any individual to dietary phytoestrogens requires a multifaceted assessment of diet composition, sources of foodstuffs, patterns of dietary intake, common therapies for intercurrent illnesses, and individual differences in gut metabolism.

MECHANISMS OF ACTION

Beyond acknowledgment that these phytochemicals fail to elicit estrus, other estradiol-like effects can be produced by these compounds. The impact of these other estradiol-like actions of phytoestrogens on the reproductive physiology of mammals can be quite prominent (Cheng et al., 1953, 1955; Biggers and Curnow, 1954) and came to scientific attention as the cause of devastating fertility problems for sheep producers in Australia about 50 years ago (Bennetts et al., 1946). The nature of these estradiol-like actions of phytoestrogens is partially but incompletely understood. Although these phytochemicals might affect mammalian reproductive physiology solely by mimicry of estradiol via estrogen receptors, there is no a priori justification for assuming that all mammalian reproductive effects of these compounds (that have some degree of estrogen-like action) will be estrogen receptor–mediated or that observed effects will necessarily be limited to mechanisms that are known to be affected by steroidal estrogens. Thus, in addition to potential pharmacokinetic and metabolic effects of phytoestrogens on production and patterns of secretion of endogenous sex hormones, a phytoestrogen may:

1. Act as an estrogen receptor agonist (the simplest case)
2. Act as an estrogen receptor antagonist (or mixed agonist/antagonist)
3. Have estrogen receptor–independent estradiol-like actions (as an agonist or antagonist)
4. Have no action on particular target tissues or processes
5. Have effects of particular target tissues or processes that involve nonestrogenic mechanisms. In this case, several possibilities are known to exist for isoflavones.

Estrogen Receptor Binding

Many different phytoestrogens have been shown to bind the estrogen receptors (ER) and effect nuclear translocation (Martin et al., 1978). Phytoestrogens appear to be mixed estrogen agonists/antagonists due to low-affinity binding to the ER (Verdeal et al., 1980). However, not all effects of phytoestrogens can be identified as ER-induced actions; there are effects that are clearly not related to ER binding. Additionally, it must be noted, as emphasized in the Introduction, that the estrogenic potency of these phytoestrogens is highly tissue dependent. Generalizations about relative estrogenic potencies (Bickoff et al., 1962) cannot be simply derived such as from ER-binding affinities and presumptions about pharmacokinetics of these agents. An example of the tissue-dependent relative potency can be seen when comparing the affinity of genistein for uterine cytosol receptors, which is 1/100–1/1000 the affinity of estradiol, to the potency in the hypothalamus and/or pituitary, where genistein is approximately 1/10 the potency of estradiol (Hughes, 1987/88; Faber and Hughes, 1991). In addition, the binary pattern of response to GnRH-induced luteinizing hormone (LH) secretion for both estradiol and genistein implies a differential response at a single site or more than one site or mechanism of action (Hughes, 1987/88; Faber and Hughes, 1991).

Compounds that competitively bind to the estrogen receptor (ER) can mimic estradiol via this classical mode of action. Phytoestrogens are known to bind to the ER with low affinity. Phytoestrogens clearly show estrogen agonist activity, especially in animals that have relatively low levels of endogenous estrogens. Some other studies have shown blunting of the actions of more potent estrogens by concurrent treatment with phytoestrogens suggesting but not proving receptor-dependent antagonism of estrogen action.

Other Receptor Binding

Several possibilities other than the ER exist which might function as alternative receptor-mediated mechanisms for manifestation of phytoestrogen action. A specific mechanistic alternative to the ER offers the possibility of rectifying some of the phenomenological differences between observed effects of phytoestrogens and more traditional ER agonists. Furthermore, since phytoestrogens are not endogenous ligands per se within animals, these compounds may not act exclusively through a single mechanism.

Nonclassical mechanisms of estrogen or phytoestrogen action are known or supposed: membrane actions; microsomal antiestrogen binding sites; antioxidant properties; inhibition of several classes of enzymes including tyrosine kinases, DNA topoisomerases and aromatase; and ER Type II receptors.

Membrane Binding to erb-B2 Protein

Extremely rapid effects (response in seconds) of estrogens on target tissues have been described in some studies. Since membrane binding of estradiol has been shown for the erb-B2 protein, which is closely related to the epidermal growth factor (EGF) receptor (Matsuda et al., 1993), other putative estrogens might act in a similar fashion to influence target tissue growth or differentiation.

"Antiestrogen Binding Site"

Extensive studies over more than a decade characterized a microsomal (Watts and Sutherland, 1986) high-affinity specific binding site for tamoxifen and related compounds that was distinct from the nuclear ER (Sutherland et al., 1980; Murphy and Sutherland, 1981, 1983; Gross et al., 1993). Once the evidence began to suggest that the AEBS was not consistently associated with inhibition of proliferation of target cells by tamoxifen (Sheen et al., 1985), questions arose regarding the role of these sites and the consequences of ligand binding and cellular processes. Other recent studies, however, seem to indicate that AEBS and ER may both be involved in the development of antiestrogen-resistant target cells (Pavlik et al., 1992). In MCF-7 mammary cancer cell lines (ER+), which are sensitive to estrogens and antiestrogens, the ratio of AEBS:ER was lower than in LY-2 cells, which are ER+ but are insensitive to antiestrogens (Pavlik et al., 1992). When excess AEBS activity was found, the antiestrogen was sequestered at the AEBS and was not available to bind the ER. Since it is known that sterols such as the oxygenated derivatives of cholesterol can bind the AEBS (Lin and Hwang, 1991), if the actual ligands are classes of untested compounds like isoflavones, then some of the apparent failures of tamoxifen–like agents to elicit biological effects through this mechanism would be less surprising.

Type II Estrogen Binding Sites

Investigations over more than a decade strengthened the notion that Type II estrogen binding site (ER Type II) is linked to biologically important processes such as proliferation of certain estrogen-responsive target tissues (Markaverich and Clark, 1979; Markaverich et al., 1987; Markaverich and Gregory, 1991). Other work regarding estrogen dependency of uterine eosinophils (Perez et al., 1996) and peroxidase activity associated with eosinophils seems to confound a simple concept of the role(s) of these intracellular entities. From the perspective of dietary exposure to phytochemicals, it is noteworthy that an "endogenous" ligand for the ER Type II that was laboriously and carefully identified in tissues of animal origin turned out to be a derivative of a dietary phytochemical (Markaverich et al., 1988a). In fact, the known ligands with the greatest affinity for ER Type II are the flavonoids luteolin and quercetin (Markaverich et al., 1988b). These phytochemicals are common dietary constituents and are structurally related to but distinct from the major phytoestrogens. While some data are contradictory, it seems that phytoestrogens do not show substantial binding to ER Type II (Markaverich et al., 1988b). In turn, the flavonoids are not active at the classical ER (Markaverich et al., 1988b).

Isoflavone "Orphan" Receptor

A testable speculation is that a specific "isoflavone receptor" will be demonstrated which will account for many (but probably not all) of the effects of these and related diphenolic dietary compounds. After the fashion of the retinoic acid receptor family, DNA sequences have been identified that are distinct from the ER but have a high degree of sequence homology (Giguere et al., 1988). These two estrogen-related receptors (ERR1 and ERR2) belong to a subfamily of the nuclear receptor superfamily, the steroid hormone/thyroid hormone superfamily. Members of this family have been implicated in regulation of expression of estrogen-responsive genes such as lactoferrin (Liu et al., 1993). Recent studies have shown that ERR1 binds upstream from the ERE, and that mutations at the ERR1 binding site result in a reduction in ER-mediated response (Shi et al., 1996). While it is of interest to note that ERR1 and ERR2 appear to exhibit constitutive activity in the absence of an exogenous ligand (Lydon et al., 1992), these two "orphan" receptors have not yet been well characterized and specific putative ligands have simply not been identified. Estradiol-like compounds such as phytoestrogens could be ligands for these or other such "orphan" receptors.

Some "orphan" receptors may be an example of a general interspecific cuing mechanism by which animals specifically detect phytochemical signals that would be of important biological consequence for survival and or reproduction. Phytochemicals involved in such interspecific cuing relationships can be called "interacoids." Such nonnutrient, nonvitamin phytochemicals would constitute environmental cues to which animals could specifically respond but for which there are not (strictly speaking) endogenous ligands of animal origin (hormones). This is a logical alternative to our usual expectation that phytochemicals showing potent effects on animals *must* be mimics of endogenous ligands (in the strict sense of synthesized by the animal for physiological signaling) as was proven to be the case with opioid narcotics/endogenous opiates. It is easy to suppose that detection of status of dietary plants could be of immense utility for animals in need of optimally matching growth or reproductive effort to seasonality, availability and quality of food, or deferring growth or reproduction until later in the life span when conditions could be less hazardous (Whitten, 1992). If this general hypothesis is true, then phytoestrogens (especially the isoflavones) may be interacoids, which act primarily through specific although as yet unidentified ("orphan") receptor-mediated mechanisms.

Aryl Hydrocarbon Receptor

There is growing evidence of important actions of isoflavones upon aryl hydrocarbon receptor (AhR)–mediated events. The most complete studies suggest that isoflavones inhibit phosphorylation of heat shock protein 90 (HSP90) (Gradin et al., 1994; Poellinger et al., 1995). HSP90 appears to modulate AhR activity by one of several mechanisms: (1) it allows correct folding of the ligand-binding domain (i.e., acts as a chaperone), (2) it interferes with Arnt hetero-dimerization, and (3) it assists in folding of the receptor into a DNA-binding domain, in this way acting as a regulator of transcription (Antonsson et al., 1995). Release of HSP90 is a necessary prerequisite for receptor-mediated transcription of AhR responsive genes. HSP90 has also been shown to be an integral component of steroid receptor signaling pathways (Nathan and Lindquist, 1995). Thus, isoflavones may not directly affect HSP90 function, but certainly could affect indirectly the function of HSP90 by interfering with the kinase second messenger pathway. It is quite possible that some of the known antiproliferative or developmental effects of isoflavones will be due to these actions that involve antagonism of the AhR. Perhaps the biological role of the AhR will ultimately be defined in terms of inhibitory or stimulatory (Poellinger et al., 1995) actions of dietary phytochemical cues.

Receptor-Independent Actions

Tyrosine Kinases

The isoflavone genistein is a well-known inhibitor of tyrosine kinases (Akiyama et al., 1987) that are central components in signal transduction processes. Most tissues could be influenced by agents that act through such universal mechanisms if the target tissue–specific doses are sufficient. Since there is evidence of antagonism of steroidal estrogen actions by isoflavones (Folman and Pope, 1966), the report (Migliaccio et al., 1993) of extremely rapid induction (within 10 seconds) of protein tyrosine phosphorylation by estradiol in human breast cancer cells (MCF-7) suggests that the antagonistic actions of isoflavones could be independent of the classical ER-mediated transcriptional mechanism. It is still not clear whether the target tissue dose of isoflavones following dietary intake is or is not sufficient to effect inhibition of this important class of enzymes.

Other Enzymes

Several other actions of isoflavones on enzymatic activity have been reported. Genistein has been shown to:

> Inhibit DNA topoisomerases I and II (Okura et al., 1988)
> Inhibit aromatase (Gangrade et al., 1991)
> Stimulate prostaglandin H synthase (Degen, 1990)
> Inhibit 17β-hydroxysteroid oxidoreductase (Makela et al., 1995a)
> Inhibit ribosomal S6 kinase (Linassier et al., 1990)

The physiological and target-specific impact of inhibition of this remarkable spectrum of critically important enzymes could produce an astonishing array of differential responses.

EFFECTS ON TARGET TISSUES DURING DIFFERENT PHASES OF LIFE

Effects of exposure to exogenous estrogens will depend upon the endocrine status of the exposed individual. Levels and pattern of release of sex hormones differ according to sex; age; phase of life (in terms of reproduction); prior developmental, nutritional, toxicological or reproductive history; and genetics (including uniquely sensitive or resistant individuals or subpopulations). As regards dietary phytoestrogens, evidence does suggest that different responses can be associated with one or more of these variables. Other than accounting for the sex of exposed individuals or groups, few of these factors have been studied. A logical way to consider whether actions of these agents are more likely to be adverse versus nil versus beneficial is by phase of life. Several reports taken together allow a preliminary comparison of the similarities and differences in the actions of phytoestrogens during different phases of life. In anthropocentric terms the simplest divisions of the reproductive phases of life for this purpose are developmental, reproductive-age, and postreproductive intervals. Potential risks or benefits of phytoestrogens will thus depend not only on dose, potency, duration, and pattern of exposure to the phytochemicals. Risk-benefit assessments must be considered relative to the developmental phase or life stage of the individual at the time of exposure. Furthermore, since phytoestrogens may have estrogenic, antiestrogenic, or nonestrogenic effects (as reviewed above), each phytochemical (or subclass) will have to be carefully studied for effects at all pertinent mammalian organ/target tissue sties throughout the life span.

Our review of selected studies regarding the effects of phytoestrogens on mammalian reproductive physiology will illustrate some of the potential risks and benefits of dietary phytoestrogens with particular emphasis on the different concerns or opportunities in the different phases of life. Obviously, some effects will not be limited to a particular age or phase of life (Fig. 2).

Developmental Interval

Fetal/Neonatal Period

In addition to sheep (Adams, 1978, 1995), reproductive tract and ovulatory alterations have been observed in rodents following developmental exposure to phytoestrogens. Neonatal exposure of mice to phytoestrogens results in persistent vaginal cornification, hyperplastic lesions of the uterus, and cervico-vaginal abnormalities (Burroughs, 1995). In the pregnant mouse, Burroughs (1995) found that coumestrol caused uterine enlargement, decreased the ovulation rate, and increased embryo degeneration.

Many studies have shown adverse effects of exogenous estrogens on sexual differentiation following exposure of the fetus or the neonate (Mori and Nagasawa, 1988; Kincl, 1990). Given the widespread occurrence of naturally occurring and man-made environmental estrogens, exposures to these compounds are inevitable, and a comprehensive understanding of their toxicity must include neuroendocrine and neurological characterizations.

The neonatal hormone environment profoundly affects the sexually differentiated pattern of behaviors, reproductive physiology, and central nervous system (CNS) anatomy and neurochemistry. Sexual differentiation of the CNS depends upon exposure to gonadal steroid hormones (Gorski, 1986; Beyer and Feder, 1987; McEwen,

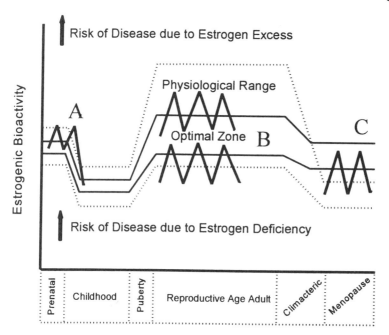

Figure 2 Schematic of the possible relationship between estrogenic bioactivity and risk of disease over a woman's life span. The optimal zone includes the age-specific levels of total (endogenous plus exogenous) estrogenic bioactivity that produces the lowest risk of disease due to relative excess or deficiency. Without exogenous exposure or overt endogenous endocrine disease, the optimal zone will fall within the physiological range, except after menopause. Levels outside of the age-specific physiological range constitute frank endocrine disorders with high risk of other disease states. Levels within the physiological range but outside the optimal zone are associated with increased risk of estrogen-dependent diseases, but this risk is more modest than that posed by nonphysiological states. If the simplest assumption is made that phytoestrogens are solely weak nonaccumulative estrogen agonists, then episodic dietary exposures would be represented as shown (A, B, and C). A: Exposures during the fetal-neonatal interval could easily episodically reach or exceed the upper physiological limits. B: Exposures during the reproductive-age adult interval would be strongly modulated by the status of ovarian function. C: Exposures during the postreproductive interval would consistently lift the level of estrogenic bioactivity in a "favorable" direction. (With permission and modified from Clarkson et al., 1995.)

1987). Morphological differences between the sexes in several nuclear regions of the CNS have been shown at the ultrastructural as well as the light microscopic levels (Walsh et al., 1982).

Neonatal exposure of females to aromatizable androgens or estrogens significantly increases the volume of the sexually dimorphic nucleus of the preoptic area (SDN-POA) (Walsh et al., 1982; Tarttelin and Gorski, 1988). Since the enzyme aromatase is present in the brain and testosterone is readily converted locally to estradiol, the local conversion of testosterone to estradiol seems to be the primary mechanism (Gorski, 1986). Neonatal androgenization of female rats produces diminished LH secretory pattern characteristic of males (Mennin et al., 1974). As adults, such androgenized females display generally delayed pubertal onset, decreased regularity of ovarian cyclicity, and cessation of cyclicity at an earlier age (Gellert 1978a,b; Gellert and Wilson, 1979).

Phytoestrogen exposure can similarly alter the neuroendocrine development of progeny. In rats, treatment of dams during pregnancy by injection with coumestrol produces female progeny that exhibit premature cessation of ovarian cyclicity (Whitten et al., 1993). In comparison, when dams are injected with genistein during late gestation, there is reduced anogenital distance in both male and female progeny and delayed vaginal opening (Levy et al., 1995). However, neonatal exposure of pups to coumestrol results in altered uterine growth, gland genesis, and estrogen receptor expression.

Mice treated neonatally with coumestrol (100 µg on postnatal days 1–5) demonstrate long estrous cycles and persistent vaginal cornification at 60 days of age (Burroughs et al., 1990). Various doses of coumestrol can elicit ovarian changes such as hemorrhage, polyovular follicles, and lack of corpora lutea (Burroughs et al., 1990). Levels of uterine progestin receptors were increased by treatment with coumestrol (Whitten and Naftolin, 1992). The coumestrol treatment also resulted in earlier vaginal opening and onset of estrous cycles, while later in life the treated females had earlier onset of acyclicity (reproductive senescence). This effect was strongest in females lactationally exposed to coumestrol (Whitten et al., 1995). Our previous studies (Faber and Hughes, 1991, 1993) have shown that neonatal exposure to the phytoestrogen genistein can alter pituitary secretion of LH and enlarge ("masculinize") the SDN-POA in female rats. In contrast, neonatally exposed males show increased responsiveness and no change in SDN volume (Register et al., 1995). Coumestrol and β-sitosterol also decreased basal LH secretion in treated females in a fashion similar to the decrease seen following DES treatment (Register et al., 1995).

The well-known teratogenic effects of diethylstilbestrol (DES) on the developing genital tract of both male and female human fetuses (Kincl, 1990) serve as a warning to be cognizant of potential adverse effects of other estrogenic agents. While there is no epidemiological evidence of potential developmental effects of phytoestrogens on the genital tract in populations who consume large quantities of phytoestrogen-rich soyfoods, the undoubtedly adverse cervical and uterine effects observed in sheep grazed on estrogenic clovers apparently is cumulative (Adams, 1989).

Certain points deserve emphasis regarding the known or potential developmental effects of phytoestrogens.

1. In the developmental interval phytoestrogens have been shown to have effects. Investigations have been limited primarily to rats, and the data suggest a mixture of estrogenic and either antiestrogenic or nonestrogenic actions.

2. Dietary exposure of dams to coumestrol during gestation and/or lactation produced female progeny that had premature cessation of ovarian cyclicity. The implication is that developmental exposure to this phytoestrogen can alter (advance) neuroendocrine senescence.

3. Depending upon agent and dose, neonatal injection of pups with genistein or coumestrol altered basal or induced LH secretion in males and females or CNS anatomy (enlarge the sexually dimorphic nucleus of the medial preoptic area) in the postpubertal interval. The implication is that neonatal treatment of pups can influence pituitary physiology and sexual development of the CNS.

4. Injection of dams late in gestation with some doses of genistein reduced anogenital distances in male and female progeny and delayed vaginal opening. No effects were found regarding postpubertal SDN-POA volumes or induced secretion of LH. The implication is that prenatal exposure to this

phytoestrogen may influence development of some genital tract sex hormone target tissues and pubertal onset.

5. Neonatal injection of pups with coumestrol or equol altered uterine growth, gland genesis, and expression of uterine estrogen receptor with the direction of effects depending upon timing of treatment. The implication is that neonatal treatment of pups can affect female genital tract development.

6. Neonatal injection of pups with genistein increased latency and reduced the incidence and multiplicity of mammary tumors due to DMBA treatment on day 50 of life (Lamartinere et al., 1995). The implication is that neonatal treatment of pups with this phytoestrogen can protect against chemically induced mammary cancer.

7. Regarding these developmental effects, it is difficult to consider comparisons between species since most data are from rodents. In rodents the findings with phytoestrogens are generally comparable to the results found in with other steroidal and stilbene estrogens in the past. Nevertheless, some of the dose-response relationships diverge in ways that suggest that simple mimicry of estradiol does not readily explain all of the observations with phytoestrogens. Although it is difficult to strictly correlate developmental stages of rodents and primates in the prenatal to neonatal phases, since the rat studies with coumestrol (Whitten et al., 1993, 1995) found that the more important exposure interval for these developmental effects was the lactational phase, the observation that infants ingesting soy-based formulas excrete large quantities of phytoestrogens and metabolites in the urine (Cruz et al., 1994) raises concern regarding possible influences of these agents on neuroendocrine developmental processes in human neonates.

Peripubertal Period

Whether consuming high amounts of phytoestrogens during the peripubertal period will affect onset of puberty is an open question. Whitten et al. (1993, 1995) have conducted detailed studies of the effect of coumestrol (a phytoestrogen found in red clover, alfalfa sprouts, and other plants, excluding soybeans) on the onset of puberty in female rats. They have provided strong evidence that in coumestrol-treated rats, vaginal opening occurs at a lower body weight, and first and second estrus occur sooner. We are unaware of any data indicating that high consumption of isoflavones (such as in soy foods) results in earlier puberty in females or later puberty in males. Preliminary data from our center (Hughes et al., 1996) show that the maturation index of vaginal cytologies is modestly increased (estrogenized) in peripubertal nonhuman primates maintained on a diet that contains soy phytoestrogens; however, the experimental design did not permit detection of any change in the onset of puberty per se. Further studies are needed in models such as the juvenile monkey in which the peripubertal hormonal changes are similar to those of human beings.

Reproductive-Age Interval

Females

The classic case illustrating reproductive actions of phytoestrogens is clover disease in sheep grazed on subterranean clover in western Australia (Bennetts, 1946; Schinckel,

1948). In this disorder, prolonged exposure to the estrogenic pasture caused permanent infertility in ewes. Morphological changes of several reproductive tissues are observed in the exposed ewes (reviewed by Adams, 1989). Mammary gland enlargement and hypertrophy of the teats is common in both ewes (and wethers). Changes in cervical morphology are prominent, as are uterine effects. In sheep there appears to be a range of exposure that does not profoundly impair ovulation but does cause cumulative compromise of other genital tract components (Adams, 1989). Estrogenic effects can also be demonstrated in other tissues in the ewe, including the pituitary, adrenal, and thyroid glands, suggesting that effects on hormonal control of fertility may also occur in this species (Adams, 1989). Studies in several other species have also shown effects of phytoestrogens on fertility or patterns of secretion of reproductive hormones. Evidence from several animal species demonstrates that certain dietary phytoestrogens can disrupt reproductive processes (Moule et al., 1963; Hughes, 1988; Kaldas and Hughes, 1989; Adams, 1989). Whereas effects on males are slight, alterations in females include changes in structure and function of the genital tract and disruption of central endocrine control of ovulation. Although human populations who traditionally consume large quantities of phytoestrogen-enriched soyfoods appear reproductively successful, one initial study has shown effects of soy ingestion on menstrual cycle length and the LH surge in women (Cassidy et al., 1994). In reproductive-age adults, the pattern of effects is reasonably consistent, with depression of fertility observed in studies of cattle (Adler and Trainin, 1960), cheetahs (Setchell et al., 1987), guinea pigs (East, 1952), mice (Leavitt and Wright, 1963; Fredricks et al., 1981), quail (Leopold et al., 1976), rabbits (Wright, 1960), and sheep (Bennetts, 1946; Schinckel, 1948) due to dietary intake of phytoestrogens (reviewed in Hughes, 1988). Generally, effects seem much more distinct in females than in males, but studies in more nonruminant species would be useful. The inevitable conclusion is that phytoestrogens are antifertility agents. Potential sites of antifertility actions of phytoestrogens in females include the genital tract, the ovary, pituitary, and the brain.

Genital Tract. Multiple studies in rats and mice show uterotrophic effects, which may be estrogenic or antiestrogenic in nature depending upon the experimental design. Extensive studies in sheep demonstrate either transient or permanent alterations in the female genital tract including morphological and biochemical changes of the cervix (Adams, 1995) and the uterus (Tang and Adams, 1981, 1982), which can credibly explain reversible and irreversible loss of fertility. Limited trials in women (Wilcox et al., 1990; Baird et al., 1995) and monkeys (Hughes et al., 1995; Clinc et al., 1996) suggest that phytoestrogens elicit minimal estrogenic effects on the maturation of vaginal epithelium and do not antagonize the actions of steroidal estrogens on the vagina. On the other hand, stimulation of some estrogen-dependent histochemical and histomorphometric markers in the uterus by dietary steroidal estrogens is diminished by concurrent inclusion of soy phytoestrogens in the diet (Tansey et al., 1996). The implication is that the female genital tract is an important group of distinct targets of phytoestrogen action. The patterns of effect may not be simply that of a weak estrogen agonist.

Ovary. Limited evidence suggests that genistein can influence maturation of oocytes and the function of granulosa cells. One report has shown that genistein reversibly inhibits the in vitro maturation of porcine oocytes (Jung et al., 1993). The effect seemed to be related to changes in both protein phosphorylation and protein synthesis.

While the study did not address all of the potential alternative mechanisms, the data was certainly suggestive of protein kinase inhibitory effects as the most likely action.

Kaplanski et al. (1981) showed that phytoestrogens could inhibit progesterone production by bovine granulosa cells in vitro. In terms of possible mechanisms of action, the results of one study regarding the inhibition of TGF-α stimulated aromatase activity in rat granulosa cells by genistein in vitro (Tilly et al., 1992) and inhibition of human chorionic gonadotropin-induced progesterone secretion from luteinized human granulosa cells in vitro (Gassman et al., 1995) were both interpreted by the investigators as indicative of protein kinase inhibitory effects of genistein. Whether other phytoestrogen mechanisms of action are pertinent or not, these preliminary studies strongly suggest that phytoestrogens may act directly on the ovary to alter reproductive processes.

Pituitary and Brain. Dietary exposure of sheep (Moule et al., 1963; Chamley et al., 1985; Montgomery et al., 1985), acute injection of rats (Hughes, 1988), and dietary exposure of women (Cruz et al., 1994; Nicholls et al., 1995; Cassidy et al., 1996) alter pituitary secretion of LH. The effect seems to revolve around diminished responsivity of the pituitary to the stimulatory actions of gonadotropin-releasing hormone rather than suppression of "basal" secretion of gonadotropins. The most recent study in young (presumably Caucasian) adult women (Cassidy et al., 1996) suggests that a dose-response relationship exists with daily intake of 45 mg conjugated isoflavones significantly suppressing the midcycle surge of LH, whereas daily intake of 25 mg unconjugated isoflavones did not significantly alter the midcycle surge of LH. The implication is that phytoestrogen exposure appears to reversibly diminish responsivity of the pituitary to gonadotropin hormone–releasing hormone stimulation. Since there are suggestive data indicating diminished symptoms of vasomotor instability in menopausal women consuming phytoestrogen-rich diets (Haines et al., 1994; Murkies et al., 1995), at least some of the observed effect on LH secretion may be attributable to hypothalamic action. On the other hand, as reviewed in the Introduction, the inability of phytoestrogens to elicit estrus behavior strongly implies that these compounds are not effective estrogens within the central nervous system of adult females. In fact, the most extensive sexual behavior studies have been conducted in clover-affected ewes, which suggest antiestrogenic effects. Clover-diseased ewes are slower than normal ewes to exhibit estrus behavior and to allow mounting by the ram in response to either a single or several daily doses of estradiol (Adams, 1978; Adams, 1981; Adams, 1983). In comparison to normal ewes, clover-affected ewes also show increased aggressive behavior, such as challenging and head bunting of rams and other ewes following administration of testosterone (Adams, 1981). This indicates that the clover-affected ewes are relatively defeminized compared to the normals.

Males

Evidence of reproductive effects of phytoestrogens in the male seem limited but appear to be consistent with expectations for exogenous administration of a weak bioactive estrogen. For example, rams grazed on estrogenic clover have reduced sperm counts (reviewed in Setchell et al., 1984), but it is not clear whether fertility is affected. Additional observational and experimental evidence suggests a protective action of phytoestrogens as regards prostatic carcinoma which is consistent with hypothetical mechanisms of action, including (1) direct antiproliferative effects on the prostate, or

(2) indirect effects that would favorably alter androgenic status (e.g., diminish androgenic bioactivity) by suppressing testosterone stimulation (by brain or pituitary action), or by enhanced hepatic metabolic clearance of androgens (Adlercreutz, 1990). Experimental studies in animals and in vitro systems show that these soy phytoestrogens have antiestrogenic (Makela et al., 1995b) actions in a murine model of DES-induced prostatic dysplasia and antiproliferative properties even in cell lines that lack estrogen receptors (Messina et al., 1994).

Postreproductive Interval (Females)

As mentioned above regarding vasomotor symptoms in menopausal women, limited information indicates that phytoestrogens do have effects in the postreproductive phase of life. Some effects suggested by demographic data may prove extremely important regarding reduction of chronic disease risk in western countries. This prospect derives from the prominent differences in the incidence of breast cancer, endometrial cancer, prostate cancer, colon cancer, and coronary artery atherosclerosis between Asian populations consuming traditional Oriental diets and humans consuming a contemporary western diet. For all of these diseases "protection" is associated with Asian diets, which prominently feature phytoestrogen-containing soyfoods. Beyond these population-based comparisons, only a few small-scale studies have been reported in postreproductive, menopausal, or aged humans, although studies are underway in some sites regarding a potential role for phytoestrogens in the prevention of cancer and coronary artery atherosclerosis. These extremely important potential healthful benefits of dietary phytoestrogens are extensively analyzed in Chapter 14 of this book and will not be elaborated upon here. In brief, results of studies to date show only beneficial effects of dietary soy phytoestrogens in the postreproductive interval. These benefits appear to include reduction of risk for several endocrine-dependent cancers and coronary artery atherosclerosis. Given the prolonged time span over which these chronic diseases develop, it is likely that inclusion of soy phytoestrogens in the diets of younger adults and perhaps even older children would reduce the risk of these diseases later in life.

PHYTOESTROGENS AS REPRODUCTIVE TOXICANTS

Plant-Animal Relationships

Many phytochemicals are known to be involved in plant-animal interactions (Table 2). Much of the "payoff" to the plant for investing in extravagant biosynthetic secondary metabolism is defense from predators. Subtle and precise coexistence relationships between some plant-insect pairs are clear examples of co-evolution of plants and animals. Only slightly less stringent evidentiary demands make this reasoning applicable to vertebrates (Fig. 3). Such co-evolution of plants and animals can be described as a chemical "arms race" between the "creativity" of plant secondary metabolism versus mammalian adaptation by natural selection.

Application of this concept to the particular case of phytoestrogens suggests that phytochemical mimicry of mammalian reproductive hormones is an antifertility defensive "strategy" of plants. Phytoestrogenic compounds, such as the isoflavonoids, appear to function in plants as defensive substances (Hughes, 1988). In times of stress (drought, grazing, etc.) plant levels of phytoestrogens increase. In some animals, these

Table 2 Phytochemicals Involved in Plant-Animal Interactions

Class	Number of compounds	Types of Effects
Alkaloids	5500	Toxic Bitter-tasting
Other nitrogen compounds	600	Toxic Bitter-tasting Foul-smelling Poisonous Hallucinogenic
Terpenoids	4000	Toxic Bitter-tasting Pleasant-smelling Hemolytic Allergenic
Phenolics	1700	Colors Antifeedants Metabolic reproductive disturbances
Polyacetylenes	650	Toxic

Source: Chapin et al., 1996.

indicators of plant stress could serve as early warning signals that plant growth will be poor, and the animal population will be at risk as well (Whitten, 1992). One example of this appears to be the concentration of isoflavonoids in forage plants of California quail. In this instance drought stress leads to increased phytoestrogen content of the plants and seemingly, in turn, the reproduction of the quail is suppressed (Leopold et al., 1976). This environmental signal may help the quail avoid the burden of attempting reproduction when the prospective harvest will be poor. This signaling would be of more use in animals that reproduce within a short period of time such as on the order of weeks. In contrast, larger animals with longer gestation periods that are removed in time from the harvest might not "benefit" from such signaling to temporarily delay reproduction and would instead face a more all-or-none outcome of reproductive success or failure (at least for the season). These species could be expected to have a different pattern of response to phytoestrogens (Whitten, 1992).

The response of mammalian populations to these dietary phytochemical hormonal mimics is selection for traits, including (1) more effective detoxification/metabolism/disposition mechanisms, (2) redundancies in physiological processes that are critical for successful (if not maximal) reproduction, and (3) behavioral adaptations. Behavioral adaptations to cope with dietary phytochemical hormonal mimics could include (1) mating behavior per se, but more likely (2) foraging behavior leading to dietary diversity with food preferences driven by cues for selection, satiety, and tolerance that would be reflected in spacial/temporal patterns of food intake, (3) group composition and distribution relative to "patchy" versus homogeneous food sources, and (4) interaction with seasonal events (weather and migrations) in pursuit of food "quantity." This implies that assessments of the interaction of food quality with food quantity/availability in the foraging behavior of animals is incomplete without factoring in the content of minor constituents (nonnutritive phytochemicals). Furthermore, at least for some species,

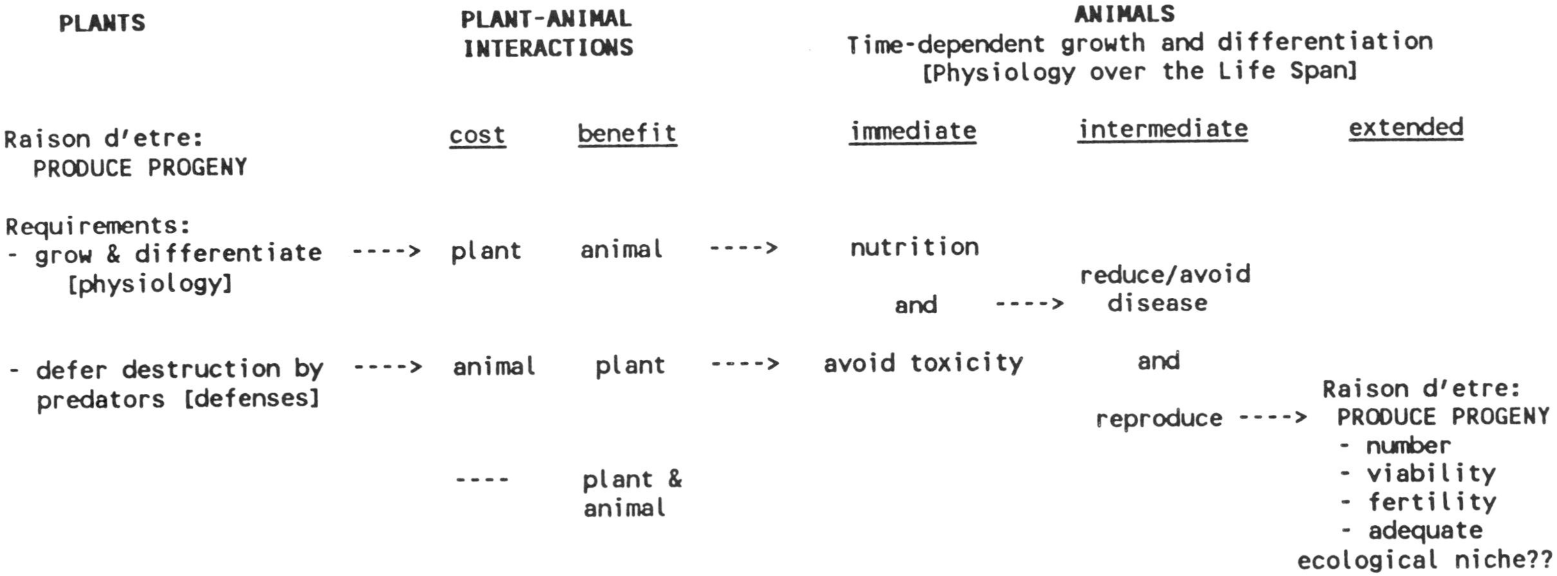

Figure 3 Schematic of the interdependent variables underlying phytochemical toxicity and cueing that characterize plant-animal interactions.

behavioral adaptations would be geographically contingent vis-à-vis relative seasonality or aseasonality of reproduction.

Severity of Effect of Phytoestrogens on Animals

The concept of co-evolution of plants and animals implies that the demonstrable effects of these phytochemicals on mammalian reproduction should be constrained rather than flagrant. Ancestral animals that suffered profound disruption of reproduction would not have progeny in subsequent generations and the susceptible genotype(s) would diminish or disappear. Present animal populations must consist of the progeny of ancestors that successfully reproduced in spite of some level of exposure to phytochemical hormonal mimics.

Differential Selection

If certain populations of animal species have been differentially selected by exposure to dietary phytoestrogens, then an interesting anthropocentric question is: Have certain subpopulations of humans been subjected to different patterns of natural selection due to traditional (ancient and persistent) differences in dietary compositions? A co-evolutionary perspective would predict that groups of humans with distinctively different dietary histories should have different degrees of resistance or tolerance to adverse effects of dietary phytoestrogens during developmental and reproductive-age intervals. Precisely this sort of experimental "natural selection" effect was observed in Merino sheep maintained on either estrogenic or nonestrogenic pasture (Croker et al., 1979). Fertility of the third generation of those that did manage to reproduce on estrogenic pasture was the same as the original population that had not been subjected to this dietary selective pressure. These findings may be relevant to humans because phytoestrogens are found in specific foods that are prominent in certain diets. Large quantities of phytoestrogens and their metabolites can be identified in human urine samples (Barnes et al., 1994) from persons consuming such food items, which indicates that substantial exposures do occur. What effect this level of exposure has on human development and reproductive function remains to be determined. Studies of dietary effects of phytoestrogens on fertility and reproductive endocrine processes in nonhuman primates and different ethnic groups of humans are needed for comparison to the extensive ovine data.

If such selection due to dietary composition does or has occurred within or among human populations, then the next important question is: Would such selection be relevant to all intervals of human life? The simplest answer is a reasonably confident, No! Since effects in the postreproductive interval of life are virtually irrelevant in evolutionary terms, there is no reason to predict substantial differences between different groups of humans regarding effects in aging humans or as regards diseases or conditions that manifest predominantly in the postreproductive phase of life.

Phytoestrogens as Environmental Estrogens

An additional critical concern regards expectations about comparisons of the reproductive and developmental effects of phytoestrogens versus xenoestrogens. In contrast to the plant-animal relationships mediated by "archiestrogens" (and other phytochemicals), a somewhat opposite expectation should apply to chemicals that are, in evolutionary terms, novel. There is no reason to suppose that present animal populations could have undergone specific prior selection for resistance to the adverse reproductive effects of xenobiotic hormonal mimics. Thus, substantial portions of such unselected populations would be expected to show sensitivity to novel agents and display reproductive or developmental alterations as a result.

CONCLUSION

Dietary phytoestrogens affect many physiological processes in mammals. The estrogenic isoflavones are particularly relevant for human health concerns. At present, the emerging impression is a balance of concern regarding possible risk of adverse effects to fetal/neonatal nervous and reproductive system development and adult female reproductive endocrine function, as well as enthusiasm regarding potential benefit on risk of several chronic "western" diseases (e.g., breast and prostate cancers and cardiovascular disease).

We have shared an inhomogeneous but intimate and ancient coevolutionary history with phytochemical reproductive hormonal mimics. The degree of potential risks and possible benefits of phytoestrogens are best understood as derivatives of that biological history. Additionally, a cogent perspective regarding likely differences between archiestrogens and xenoestrogens seems logically apparent as well. The balance of the relationship of potential risks and possible benefits of dietary estrogens is best understood as a result of our biological history.

REFERENCES

Adams, N. (1978). *J. Reprod. Fertil.*, *53*: 203.

Adams, N. R. (1981). *J. Reprod. Fertil. 30* (Suppl.): 223.

Adams, N. R. (1983). *J. Reprod. Fertil.*, *68*: 113.

Adams, N. R. (1989). In: *Toxicants of Plant Origin*, Vol., IV, *Phenolics* (P. R. Cheeke, ed.). CRC Press, Boca Raton, FL, p. 23.

Adams, N. R. (1995). *Proc. Soc. Exp. Biol. Med.*, *29*: 87.

Adler, J. H., and Trainin, D. (1960). *Refuah Vet*, *17*: 115.

Adlercreutz, H., Fotsis, T., Bannwart, C., Wahala, K., Makela, T., Brunow, G., and Hase, T. (1986). *Steroid Biochem.*, *25*: 791.

Adlercreutz, H. (1990). *Scand. J. Clin. Lab. Invest.*, *50* (Suppl.201): 3.

Adlercreutz, H., Fotsis, T., Lampe, J., Wahala, K., Makela, T., Bronow, G., and Hase, T. (1993a). *Scand. J. Clin. Lab. Invest.*, *53*(Suppl.215): 5.

Adlercreutz, H., Markkanen, H., Watanabe, S. (1993b). *Lancet, 342*: 1209.

Akiyama, T., Ishida, J., Nakagawa, S. Ogawara, H., Watanabe, S., Itoh, N., Shibuya, M., and Fukami, Y. (1987). *J. Biol. Chem.*, *262*: 5592.

Antonsson, C., Whitelaw, M. L., McGuire, J., Gustafsson, J. A., and Poellinger, L. (1995). *Mol. Cell Biol.*, *15*(2): 756.

Baird, D. D., Umbach, D. M., Lansdell, L., Hughes, C. L., Setchell, K. D. R., Weinberg, C. R., Haney, A. F., Wilcox, A. J., and McLachlan, J. A. (1995). *J. Clin. Endocrinol. Metab.*, *80*: 1685.

Barnes, S., Peterson, G., Grubbs, C., and Setchell, K. D. R. (1994). In: *Diet and Cancer: Markers, Prevention, and Treatment* (M. M. Jacobs, ed.). Plenum, New York, p. 135.

Bennetts, H. W., Underwood, E. J., and Shier, F. L. (1946). *Aust. Vet. J.*, *22*: 2.

Beyer, C., and Feder, H. H. (1987). *Annu. Rev. Physiol.*, *49*: 349.

Bickoff, E. M., Livingston, A. L., Hendrickson, A. P., and Booth, A. N. (1962). *Agr. Food Chem.*, *10*: 410.

Biggers, J. D., and Curnow, D. H. (1954). *Biochem. J.*, *58*: 278.

Bradbury, R. B., and White, D. C. (1954). *Vitamin Horm.*, *12*: 207.

Burroughs, C. D., Mills, K. T., and Bern, H. A. (1990). *Reprod. Toxicol.*, *4*: 127.

Burroughs, C. D. (1995). *Proc. Soc. Exp. Biol. Med.* *208*: 78.

Cassidy, A. Bingham, S., and Setchell, K. D. R. (1994). *Am. J. Clin. Nutr.*, *60*: 333.

Cassidy, A. Bingham, S., and Setchell, K. D. R. (1996). *Br. J. Nutr.*, *74*: 587.

Chamley, W. A., Clarke, I. J., and Moran, A. R. (1985). *Austr. J. Biol. Sci.*, *38*: 109.

Chapin, R. E., Stevens. J. T., Hughes, C. L., Kelce, W. R., Hess, R. A. and Daston, G. P. (1996). *Fundam. Appl. Toxicol.*, *29*: 1.

Cheng, E., Story, C. D., Yoder, L., Hale, W. H., and Burroughs, W. (1953). *Science*, *118*: 164.

Cheng, E. W., Yoder, L., Story, C. D., and Burroughs, W. (1955). *Ann. NY Acad.* Sci., *61*: 652.

Clarkson, T. B., Anthony, M. S., and Hughes, C. L. (1995). *Trends Endocrinol. Metab.* 6: 11.

Cline, J. M., Obasanjo, I. O., Paschold, J. C., Adams, M. R., and Anthony, M. S. (1996). *Fertil. Steril.*, *65*: 1031..

Croker, K. P., Lightfoot, R. J., Johnson, T. J., Adams, N. R., and Carrick, M. J. (1989). *Austr. J. Agr. Res.*, *40*: 165.

Cruz, M. L. A., Wong, W. W., Mimouni, F., Hachey, D. L., Setchell, K. D. R., Klein, P. D., and Tsang, R. C. (1994). *Pediatr. Res.*, *35*: 135.

Degen, G. H. (1990). *J. Steroid Biochem.* *35*: 473.

East, J. (1952). *Aust. J. Sci. Res. B*, *5*: 472.

Eldridge, A. C., and Kwolek, W. F. (1983). *J. Agric. Food Chem.*, *31*: 394.

Faber, K. A., and Hughes, C. L. (1991). *Biol. Reprod.* 45: 649.

Faber, K. A., and Hughes, C. L. (1993). *Reprod. Toxicol.*, *7*: 35.

Farnsworth, N. R., Bingel, A. S., Cordell, G. A., Crane, F. A., and Fong, H. H. S. (1975). *J. Pharm. Sci.*, *64*: 717.

Folman, Y., and Pope, G. S. (1966). *J. Endocrinol.*, *34*: 215.

Franke, A. A., Custer, L. J., Cerna, C. M., and Narala, K. (1995). *Proc. Soc. Exp. Biol. Med.*, *208*: 18.

Fredricks, G. R., Kincaid, R. L., Bondioli, K. R., and Wright R. W. (1981). *Proc. Soc. Exp. Biol. Med.*, *167*: 237.

Gangrade, B. K., Davis, J. S., and May, J. V. (1991). *Endocrinology*, *129*: 2790.

Gassman, A., Gagliardi, C. L., Weiss, G., Von Hagen, S., and Goldsmith, L. T. (1995). Program and Abstracts of the 77th Annual Meeting of the Endocrine Society, Abstract #P1-325, p. 194.

Gellert, R. J. (1978a). *Environ. Res.* 16: 123.

Gellert, R. J. (1978b). *Environ. Res.*, *16*: 131.

Gellert, R. J., and Wilson, C. (1979). *Environ. Res.*, *18*: 437.

Giguere, V., Yang, N., Segui, P., and Evans, R. M. (1988). *Nature*, *331*: 91.

Gorski, R. A. (1986). *Environ. Health Perspect.*, *70*: 163.

Gradin, K. Whitelaw, M. L., Toftgard, R., Poellinger, L., and Berghard, A. (1994). *J. Biol. Chem.*, *269*: 23800.

Gross, C., Yu, M., van Herle, A. J., Giuliano, A. E., and Juillard, G. J. (1993). *J. Clin. Endocrin. Metab.*, *77*(5): 1361.

Haines, C. J., Chung, T. K. H., and Leung, D. H. Y. (1994). *Maturitas*, *18*: 175.

Hughes, C. L. (1987/88). *Reprod. Toxicol.*, *1*: 179.

Hughes, C. L. (1988). *Environ. Health Perspect.*, *78*: 171.

Hughes, C. L., Tansey, G., Cline, J. M., and Lessey, B. (1995). Estrogenic, anti-estrogenic and non-estrogenic effects of phytoestrogens in reproductive tissues from female primates [abstr]. Third International Conference on Phytoestrogens, Little Rock, AR, Dec. 3-6.

Jacobsohn, G. M., Frey, M. J., and Hochberg, R. B. (1965). *Steroids*, *6*: 93.

Jung, T., Fulka, J., Lee, C., and Moor, R. M. (1993). *J. Reprod. Fertil.*, *8*: 529.

Kaldas, R. S., and Hughes, C. L. (1989). *Reprod. Toxicol.*, *3*: 81.

Kaplanski, O., Shemesh, M., and Berman, A. (1981). *J. Endocrinol.*, *89*: 343.

Kincl, F. A. (1990). *Hormone Toxicity in the Newborn*. Springer-Verlag, New York.

Lamartiniere, C. A., Moore, J., Holland, M., and Barnes, S. (1995). *Proc. Soc. Exp. Biol. Med.*, *208*: 120.

Leavitt, W. W., and Wright, P. A. (1963). *J. Reprod. Fertil.*, *6*: 115.

Leopold, A. S., Erwin, M., Oh, J., and Browning, B. (1976). *Science*, *191*: 98.

Levy, J. R., Faber, K. A., Ayyash, L., and Hughes, C. L. (1995). *Proc. Soc. Exp. Biol. Med.*, *208*: 60.

Lin, L., and Hwang, P. L. (1991). *Biochim. Biophys. Acta*, *1082*(2): 177.

Linassier, C., Pierre, M., LePeco, J.-B., et al. (1990). *Biochem. Pharmacol.*, *39*: 187.

Liu, Y., Yang, N., and Teng, C. T. (1993). *Mol. Cell Biol.*, *13*(3): 1836.

Lydon, J. P., Power, R. F., and Conneely, O. M. (1992). *Gene Expr.*, *2*(3): 273.

Makela, S. I., Poutanen, M., Lehtimaki, J., Kostian, M.-L., Santii, R., and Vihko, R. (1995a). *Proc. Soc. Exp. Biol. Med.*, *208*: 51.

Makela, S. I., Pylkkanen, L. H., Santti, R. S. S., and Adlercreutz, H. (1995b) *J. Nutr.*, *125*: 437.

Markaverich, B. M., and Clark, J. H. (1979). *Endocrinology*, *105*: 1458.

Markaverich, B. M., Adams, N. R., Roberts, R. R., Alejandro, M., and Clark, J. H. (1987). *J. Steroid Biochem.*, *28*: 599.

Markaverich, B. M., Gregory, R. R., Alejandro, M. A., Clark, J. H., Johnson, G. A., and Middleditch, B. S. (1988a). *J. Biol. Chem.*, *263*: 7203.

Markaverich, B. M., Roberts, R. R., Alejandro, M. A., Johnson, G. A., Middleditch, B. S., and Clark, J. H. (1988b). *J. Steroid Biochem*, *30*: 71.

Markaverich, B. M., and Gregory, R. R. (1991). *Biochem. Biophys. Res. Commun.*, *177*: 1283.

Martin, P. M., Horwitz, K. B., Ryan, D. S., and McGuire, W. (1978). *Endocrinology*, *35*: 1860.

Matsuda, S., Kadowaki, Y., Ichino, M., Akiyama, T., Toyoshima, K., and Yamamoto, T. (1993). *Proc. Natl. Acad. Sci. USA*, *90*: 10803.

McEwen. B. S. (1987). *Environ. Health Perspect.*, *74*: 177.

Mennin, S. P., Kubo, K., and Gorski, R. A. (1974). *Endocrinology*, *95*: 412.

Messina, M., and Messina, V. (1991). *J. Am. Diet. Assoc.*, *91*: 836.

Messina, M. J., Persky, V., Setchell, K. D. R., and Barnes, S. (1994). *Nutr. Cancer*, *21*: 113.

Migliaccio, A., Pagano, M., and Auricchio, F. (1993). *Oncogene*, *9*: 2183.

Montgomery, G. W., Martin, G. B., Le Bars, J., and Pelletier, J. (1985). *J. Reprod. Fertil.*, *73*: 457.

Mori, T., and Nagasawa, H. (1988). *Toxicity of Hormones in Perinatal Life*. CRC Press, Boca Raton, FL.

Morton, M. S., Wilcox, G., Wahlqvist, M. L., and Griffiths, K. (1994). *J. Endocrinol.*, *142*: 251.

Moule, G. R., Braden, A. W. H., and Lamond, D. (1963). *Anim. Breed Abstr.*, *31*: 139.

Murkies, A. L., Lombard, C., Strauss, B. J. G., Wilcox, G., Burger, H. G., and Morton, M. S. (1995). *Maturitas*, *21*: 189.

Murphy, L. C., and Sutherland, R. L. (1981). *J. Endocrinol.*, *91*: 155.

Murphy, L. C., and Sutherland, R. L. (1983). *J. Clin. Endocrinol. Metab.*, *57*: 373.

Nathan, D. F., and Lindquist, S. (1995). *Mol. Cell Biol.*, *15*(7): 3917.

Nicholls, J., Lasley, B. L., Gold, E. B., Nakajima, S. T., and Schneeman, B. O. (1995). *J. Nutr.*, *125*(3S): 803S.

Okura, A., Arakawa, H., Oka, H., Yoshinari, T., and Monden, Y. (1988). *Biochem. Biophys. Res. Commun.*, *157*: 183.

Pavlik, E. J., Nelson, K., Srinivasan, S., Powell, D. E., Kenady, D. E., DePriest, P. D., Gallion, H. H., and van Nagell, J. R., Jr. (1992). *Cancer Res.*, *52*(15): 4106.

Perez, M. C., Furth, E. F., Matzumura, P. D., and Lyttle, C. R. (1996). *Biol. Reprod.*, *54*: 249.

Poellinger, L., Whitelaw, M. L., McGuire, J., Anotsson, C., Pongratz, I., Lindebro, M., Kleman, M., and Gradin, K. (1995). Regulation of dioxin receptor function by genistein and phytoantiestrogens [abstr]. Third International Conference on Phytoestrogens, Little Rock, AR, Dec. 3-6.

Register, B., Bethel, M. A., Thompson, N., Walmer, D., Blohm, P., Ayyash, L., and Hughes, C. L. (1995). *Proc. Soc. Exp. Biol. Med.*, *208*: 72.

Schinckel, P. G. (1948). *Austr. Vet. J.*, *24*: 289.

Setchell, K. D. R., Borriello, S. P., Hulme, P., Kirk, D. N., and Axelson, M. (1984). *Am. J. Clin. Nutr.*, *40*: 569.

Setchell, K. D. R., Gosselin, S. J., Welsh, M. B., Johnston, J. O., Balistreri, W. F., Kramer, L. W., Dresser, B. L., and Tarr, M. J. (1987). *Gastroenterology*, *93*: 225.

Setchell, K. D. R. (1993). Dietary estrogens - methods for detection and measurement [abstr]. Second International Conference on Phytoestrogens, Little Rock, AR, Oct. 17-20.

Sheen, Y. Y., Simpson, D. M., Katzenellenbogen, B. (1985). *Endocrinology*, *117*: 561.

Shi, H., Shigeta, H., Yang, N., Fu, K., and Teng, C. (1996). Human estrogen-related receptor 1 gene: identification of multiple transcripts and promoter usage [abstr]. 1996 Triangle Conference on Reproductive Biology, Chapel Hill, NC, Jan. 20.

Stedman's Medical Dictionary, 22nd ed. The Williams and Wilkins Co., Baltimore, 1972.

Sutherland, R. L., Murphy, L. C., Foo, M. S., Green, M. D., and Whybourne, A. M. (1980). *Nature*, *288*: 273.

Tang, B. Y., and Adams, N. R. (1981). *J. Endocrinol.*, *89*: 365.

Tang, B. Y., and Adams, N. R. (1982). *Austr. J. Biol. Sci.*, *35*: 527.

Tansey, G. Hughes, C. L., Cline, J. M., Krummer, S., Walmer, D. K. and Schmotzer, S. (1996). *Proc. Soc. Exp. Biol. Med.* (in press).

Tarttelin, M. F., and Gorski, R. A. (1988). *Brain Res.*, *456*: 271.

Tilley, J. L., Billig, H., Kowalski, K. I., and Hsueh, A. J. W. (1994). *Mol. Endocrinol.*, *6*: 1942.

Verdeal, K., Brown, R. R., Richardson, T., and Ryan, D. S. (1980). *J. Natl. Cancer Inst.* *64*: 285.

Walsh, R. J., Brawer, J. R., and Naftolin, F. (1982). *J. Anat.*, *135*: 733.

Watts, C. K. W., and Sutherland, R. L. (1986). *Biochem. J.*, *236*: 903.

Whitten, P. L. (1992). *Am. J. Phys. Anthropol. Suppl.*, *14*: 172.

Whitten, P. L., and Naftolin, F. (1992). *Steroids*, *57*: 56.

Whitten, P. L., Lewis, C., and Naftolin, F. (1993). *Biol. Reprod.*, *49*: 1117.

Whitten, P. L., Lewis, C., Russell, E., and Naftolin, F. (1995). *Proc. Soc. Exp. Biol. Med.*, *208*: 82.

Wilcox, G., Wahlqvist, M. L., Burger, H. G., and Medley, G. (1990). *BMJ*, *301*: 905.

Wright, P. A. (1960). *Proc. Soc. Exp. Biol. Med.*, *105*: 428.

Xu, X., Harris, K. S., Wang, H.-J., Murphy, P. A., and Hendrich, S. (1995). *J. Nutr.*, *125*: 2307.

14

Human Health and Phytoestrogens

Herman Adlercreutz
University of Helsinki and Folkhälsan Center, Helsinki, Finland

INTRODUCTION

An abundant literature during the last 20–30 years dealing with both epidemiological and migrant studies, supports the view that the western diet is mainly responsible for the high incidence of chronic and degenerative diseases in Western Europe, the United States, and Canada and now also in many of the East European countries (Dunn, 1975; Trowell and Burkitt, 1981; Reddy and Cohen, 1986a, 1986b; Rose et al., 1986; Stanford et al., 1995; Griffiths et al., 1996). These diseases include the major hormone-dependent cancers, colon cancer, and coronary heart disease. In the Mediterranean region as well as in Finland, particularly in the east and north, incidence of breast, colon, and prostate cancer is lower than in Western Europe and the United States but higher than in some Asian countries like Japan and China, and this may be at least partly explained by diet (Teppo et al., 1980; Reddy et al., 1985, 1987; Rose et al., 1986; Pukkala et al., 1987; Adlercreutz, 1990b; Rose, 1992; Minami et al., 1993; Kliewer and Smith, 1995; Tominaga and Kuroishi, 1995; Willett et al., 1995; Griffiths et al., 1996). On the basis of epidemiological and animal experimental studies, it has been concluded that high fat and low fiber intake are important contributors to the increased incidence of the above-mentioned diseases. Recently the conclusions have, however, been questioned with regard to breast cancer (Willett et al., 1992). Independent of which position one takes in this debate, it seems that causes other than an abundance of fat and lack of fiber should receive consideration for the high incidence of cancer and coronary heart disease.

Due to the fact that all these diseases are to a various extent related to sex hormones or sex hormone metabolism, we have postulated that the western diet compared to the vegetarian or semivegetarian diet in some developing and Asian countries may alter hormone production, metabolism, or action at the cellular level by some biochemical mechanisms (Adlercreutz et al., 1982a, 1984, 1990b, 1991d, 1995c; Griffiths et al., 1996). The detection and identification in human body fluids of compounds, lignans, and isoflavonoids of plant origin with molecular weights and structures similar to those of steroids gave us the idea that they could be important modulators of steroid hormone metabolism and action (Adlercreutz et al., 1982a, 1986c; Setchell et al., 1984; Adlercreutz, 1984, 1988a, 1990b; Bannwart et al., 1984a; Setchell and Adlercreutz, 1988). The mammalian lignans (Setchell and Adlercreutz, 1979; Setchell et al., 1980a,c;

Adlercreutz et al., 1982a) were previously unknown compounds, but the isoflavonoids, the most important group of the so-called phytoestrogens, have been well known in the veterinary field (Price and Fenwick, 1985).

However, despite their importance in veterinary medicine, there was no evidence indicating a pathogenetic (Lindner, 1976) or beneficial role of the isoflavonoids in humans. When the mammalian lignans were discovered, it was soon realized that both groups of diphenols may be of importance and perhaps act in concert due to their similar structures. In fact, they have now been shown to influence not only sex hormone metabolism and biological activity but also intracellular and steroid metabolic enzymes, protein synthesis, growth factor action, malignant cell proliferation, angiogenesis, calcium transport, Na^+/K^+ ATPase, vascular smooth muscle cells, lipid oxidation, and cell differentiation, making them strong candidates not only for a role as cancer-protective compounds but also in the protection against other degenerative diseases like coronary heart disease.

There is an increasing literature on lignans and isoflavonoids and an increasing production and use of foods containing isoflavonoids and lignans for the prevention of western diseases, including cancer. This calls for a critical review of the role of these compounds in human health. To understand the significance of phytoestrogens for human health, it is, however, necessary to understand human physiology and the metabolism of these compounds. Therefore, these aspects will be dealt with in this chapter. (The reader is also referred to Setchell and Adlercreutz, 1988; Adlercreutz, 1990, 1995c; Messina et al., 1994a,b; Clarkson, et al., 1995a; Adlercreutz et al., 1995a; Knight and Eden, 1995; Griffiths et al., 1996.)

DEFINITION OF PHYTOESTROGENS

By definition, phytoestrogens are estrogenic compounds of plant origin. More than 300 plants have estrogenic activity (Bradbury and White, 1954; Farnsworth et al., 1975), but some contain steroidal estrogens (Labov, 1977), which in my opinion should not be included among the phytoestrogens. The estrogenic steroids in plants are identical with those found in human urine (estrone, estradiol, estriol). Less than half of all the plants having estrogenic activity are consumed by animals or humans. The number of nonsteroidal compounds with estrogenic activity in plants is constantly increasing, and to the best of my knowledge the number now exceeds 40. For practical reasons, because many are not consumed by humans, we include in the phytoestrogen group in this discussion the isoflavonoids formononetin, daidzein, biochanin A, and genistein, their glycosides and other conjugates, their mammalian metabolites, and the coumestane coumestrol. Previously some estrogens (e.g., zearalenone, a resorcyclic acid lactone) that are associated with molding grain (Price and Fenwick, 1985) have been included but should be defined as fungal estrogens. The fungal estrogens will not be considered here because they should not normally be present in food. Also, the xenobiotic environmental estrogens are not included in this discussion. Phytoestrogens occurring in food for human consumption should not be compared with these estrogens, as has been done numerous times in the lay press. Two other isoflavones, pratensin and prunetin, are of limited occurrence, and nothing is known about the effect on humans; therefore, they will not be discussed. The biologically very active miroestrol (Labov, 1977) will also not be included.

Because the mammalian lignans are weakly estrogenic (Jordan et al., 1985; Welshons et al., 1987) and bind weakly to rat uterine cytosol (J. Clark and H. Adlercreutz, unpublished), they should be included among the phytoestrogens. Recently it has been found that some chalcones, flavonones, and flavonols bind to the estrogen receptor and compete with estradiol (Miksicek, 1993, 1995) and cause estrogen responses. Consequently, it seems now that these compounds should be included. However, because practically nothing is known about their occurrence in the mammalian organism, these latter compounds will not be dealt with in this connection. ß-Sitosterol is also weakly estrogenic but is normally absorbed only in negligible amounts (Salen et al., 1970), and therefore it will not be considered here.

LIGNANS AND ISOFLAVONOIDS IN FOODS AND IN HUMANS: OCCURRENCE AND METABOLISM

Phytoestrogen Concentrations in Various Foods

Lignans

Flaxseed (linseed) is the most abundant source of lignans in foods. Lignans have been determined by high-performance liquid chromatography (HPLC) in flaxseed and chapparal after mild acid hydrolysis of the glycosides. Flaxseed was found to contain 80 mg secoisolariciresinol per 100 g (Obermeyer et al., 1993, 1995) (for structures, see Fig. 1) corresponding to values found for urinary excretion of lignans after feeding flaxseed meal to rats (Axelson et al., 1982a). Thompson et al. (1991) using an indirect method assaying by gas chromatography (GC) the production of enterolactone and enterodiol from foods incubated with fecal bacteria found a secoisolariciresinol concentration in flaxseed meal of 67.5 mg/100 g, and 52.7 mg/100 g in flaxseed flour. Later they reported that flaxseed varieties grown in the same location in Canada had

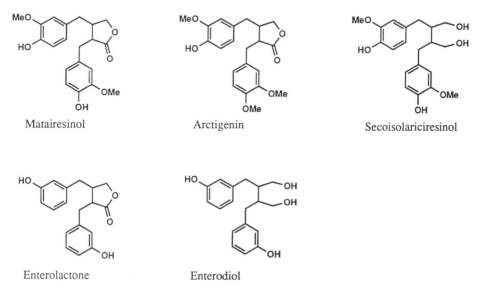

Matairesinol   Arctigenin   Secoisolariciresinol

Enterolactone   Enterodiol

Figure 1 Structures of lignans.

lignan values ranging from 25.8 to 72.9 mg/100 g (Thompson, 1995a). They also reported values for a large number of food samples (some are shown for comparison purposes in Table 1). In our hands the method did not give reliable results because we found that the fecal bacteria used for the conversion of lignan precursors to enterolactone and enterodiol contained high amounts of these mammalian lignans. Furthermore, the conversion was not quantitative. In the study of Thompson et al. (1991), the background values from the bacteria were not reduced from the values obtained after incubation of various food samples. A further problem with the method is that the GC peak of enterodiol in fecal extracts frequently contains some unknown compound, making the GC determination nonspecific for this lignan. This unknown compound may interfere with enterodiol assays in methods involving blind detection systems also in other biological samples. In our laboratory attempts to eliminate these problems were unsuccessful. However, it seems that the method of Thompson et al. (1991) has some advantages over our own method. We have found that cereals contain unknown precursors, which are converted to enterolactone by intestinal bacteria, and such compounds can at present only be measured by the bacterial indirect procedure converting them to enterolactone as in the method of Thompson et al. (1991) (see also below).

We have recently developed new methodologies for the simultaneous assay of all important phytoestrogens in meal, flour, and soy products (Mazur et al., 1996) and found by isotope dilution gas chromatography–mass spectrometry in the selective ion monitoring mode (ID-GC-MS-SIM) the lignan secoisolariciresinol in soy meal and matairesinol in flaxseed (Table 1). The concentrations are low compared to the content of the main phytoestrogens, and mass spectra of these compounds have not been obtained. We found about 4.5–5 times more secoisolariciresinol in flaxseed (Table 1) compared to the results of other investigators using indirect or direct methods (Axelson et al., 1982a; Thompson et al., 1991; Obermeyer et al., 1993, 1995). Different brands of flaxseed may give different results, but more likely our hydrolysis method is more effective in liberating the lignan aglycones from their glycosidic form. Another reason is that we can measure the dehydrated anhydrosecoisolariciresinol product of secoisolariciresinol, which has not been done with other methods. The detection of matairesinol is probably a result of the more sensitive detection system (GC-MS) rather than the GC and HPLC methods. Table 1 shows some values in different food specimens obtained with this methodology compared with values described in the literature.

The values obtained in Dr. Thompson's laboratory by the indirect method are for the most part much higher than ours. We have found out that the intake of a certain amount of bread with a measured content of secoisolariciresinol and matairesinol leads to a much higher amount of mammalian lignan metabolites in urine than expected (see below). It seems that the bacteria are able to use other substrates than we can measure in the grain for production of enterolactone and enterodiol, which explains to a great extent the differences in the results between the laboratories. Much work is probably needed before we know all precursors for the mammalian lignans.

The mammalian lignans seem to be mainly derived from grains, seeds, fruits, berries, and some vegetables (Adlercreutz et al., 1987, and unpublished data). We have now established by GC-MS techniques that the plant lignans in grain are localized in the outer fiber-containing layers, with the highest concentration in the aleurone layer (Adlercreutz, 1990b; Nilsson et al., 1996) containing phytin, polyphenols, enzyme inhibitors, and other compounds generally regarded as antinutritional factors. This one- to three-cell-thick layer is tightly bound to the fiber layer; the liberation of the lignan precursors from these very resistant cells is difficult and requires a relatively complicated three-step hydrolytic procedure developed by us (Mazur et al., 1996). Small amounts of secoisolariciresinol seem to occur in other parts of the rye grain (Nilsson et al., 1996). Modern milling techniques usually eliminate the aleurone layer because of its close association with the outer fiber layer, which therefore does not, with some exceptions, occur in the products supplied to the market for consumption. This is true particularly in western societies.

Recent analyses in our laboratory have revealed that the lignan content of berries is relatively high (up to 1500 µg/100 g dry weight) (Adlercreutz et al., 1987, and unpublished). Some vegetables like garlic (about 400 µg/100 g dry weight) contain relatively high amounts of lignans.

In a well-controlled study (Kirkman et al., 1995) in which phytoestrogens were measured in urine, men and women consumed three different diets: soy, carotenoid vegetable, and cruciferous vegetable. The carotenoid and cruciferous vegetable diets led to higher lignan excretion than when the subjects were on their basal diet. Women excreted more enterodiol on the cruciferous diet and less enterolactone than men on all the test diets. In another similar study it was shown that enterodiol excretion increased on a vegetable/fruit diet compared to the basal and legume/allium diet (Hutchins et al., 1995).

In Japanese subjects urinary lignan excretion correlates to the intake of whole soybeans (Adlercreutz et al., 1991e), which is probably due to the presence of the small amounts of secoisolariciresinol (Table 2).

Within a year, more data with regard to lignan contents of various foods will be available because we are continuously analyzing all kinds of foods to establish a data base.

Isoflavonoids

Since 1931 it has been well known that soybeans contain very high amounts (up to 1000-3000 µg/g) of the glycosides of the two isoflavones, daidzein and genistein (Walz, 1931; Walter, 1941; Ahluwalia et al., 1953; Ohta et al., 1979; Ohta et al., 1980; Eldridge and Kwolek, 1983). Much later, a third major compound, glycitein (Fig. 2), was found, mainly as a glycoside (glycitin) (Naim et al., 1973), and confirmed by other (Pratt and Birac, 1979; Kudou et al., 1991) (Fig. 2). Small amounts of these three compounds occur in the free form. Fermented soy may contain a catecholic conversion product of glycitein, 6,7,4'-trihydroxyisoflavone (György et al., 1964; Klus et al., 1993). Furthermore, small amounts (about 5µg/100 g) of the isoflavone coumestrol have been found in soybeans (Lookhart et al., 1978). However, we have not been able to detect any coumestrol in soybean meal using our very sensitive ID-GC-MS-SIM method. Coumestrol, however, is present in soybean sprouts (Murphy, 1982) and in relatively high amounts in alfalfa and clover sprouts (Franke et al., 1994a), which we have confirmed by GC-MS. Using GC a peak with the same retention time as coumestrol is seen in many types of beans including soybeans. The compound is, however, not coumestrol. Soy sauce does not contain any isoflavones (Murphy, 1982) but does contain the lignan precursor coniferyl alcohol (Yokotsuka, 1986) and perhaps a minimal amount of secoisolariciresinol. It has been shown that extracting the oil from soybeans with hexane does not remove the isoflavones or the isoflavone glucosides (Eldridge and Kwolek, 1983). However, Franke et al. (1994a) reported a loss of 30–40% after defatting powdered soy foods with hexane. This is difficult to explain because the isoflavones and their glycosides are not soluble in hexane.

Table 1 Mean lignan content (μg/100 g dry weight)[a] of various meal products, seeds, and soybeans analyzed by gas chromatography (GC) or combined gas chromatography-mass spectrometry (GC-MS).

Product analyzed	SECO	Matairesionol	Total	Enterolactone after incubation with fecal bacteria	Enterodiol after incubation with fecal bacteria	Total	Ref. (Method)
Flaxseed	369,900	1,087	371,000				Mazur et al., 1996 (GC-MS)
Flaxseed	81,700	0	81,700				Obermeyer et al., 1995 (HPLC-MS)
Flaxmeal				9.841	68,204	78,045	Thompson et al., 1991b (GC)
Flaxmeal	226,000	0	226,000				Obermeyer et al., 1995 (HPLC-MS)
Flaxseed, crushed and defatted	546,000	1,300	547,300				Mazur et al., 1996 (GC-MS)
Flaxseed flour				12,980	44,877	57,857	Thompson et al., 1991 (GC)
Soybean				767	188	955	Thompson et al., 1991 (GC)
Clover seed	13.2	0	13.2				Mazur et al. 1996 (GC-MS)
Sunflower seed				216	209	425	Thompson et al., 1991 (GC)
Sunflower seed	609.5	0	609.5				Mazur et al., 1996 (GC-MS)
Rapeseed				1,111	177	1,288	Thompson et al., 1991 (GC)
Poppy seed	14.0	12.1	26.1				Mazur et al. 1996 (GC-MS)
Caraway seed	220.7	5.7	226.4				Mazur et al. 1996 (GC-MS)
Wheat (whole-grain)				476	91	567	Thompson et al., 1991 (GC)
Wheat (whole-grain)	32.9	tr	32.9				Mazur et al. 1996 (GC-MS)
Wheat white meal	8.1	0	8.3				Mazur. 1996 (GC-MS)
Wheat bran				296	327	623	Thompson et al., 1991 (GC)

Wheat bran	98.3	0	98.3	Mazur et al. 1996 (GC-MS)
Oat (whole grain)		279	99	378	Thompson et al., 1991 (GC)
Oat bran		303	440	743	Thompson et al., 1991 (GC)
Oat meal	13.4	0.3	13.7	Mazur et al. 1996 (GC-MS)
Oat bran	23.8	155	178.8	Mazur et al. 1996 (GC-MS)
Barley (whole grain)		46	83	129	Thompson et al., 1991 (GC)
Barley (whole grain)	58.0	0	58.0	Mazur et al. 1996 (GC-MS)
Barley bran		276	158	434	Thompson et al., 1991 (GC)
Barley bran	62.6	0	62.6	Mazur et al. 1996 (GC-MS)
Rye meal (whole grain)		78	103	181	Thompson et al., 1991 (GC)
Rye meal (Amando, whole grain)	47.1	65.0	112.1	Nilsson et al., 1996 (GC-MS)
Rye bran (Amando)	132	167	299	Nilsson et al., 1996 (GC-MS)
Triticale (whole grain)	38.8	9.4	48.2	Mazur et al. 1996 (GC-MS)
Triticale meal (mean of four different brands)	21.4	10.7	32.1	Mazur et al. 1996 (GC-MS)

[a]Dry weight in all publications from L. Thompson's and H. Adlercreutz's laboratories.
SECO = Secoisolariciresinol; tr = traces.

Table 2 Mean (μg/g) Isoflavonoid Content of Various Soybean Products and Components Analyzed by HPLC or GC-MS[a]

Product analyzed[b]	Genistin	Genistein	Total genistein	Daidzin	Daidzein	Total daidzein	Glycitin	Glycitein	Ref. (Method)
Soybeans, Weber	1024 ± 55	24 ± 0	1048	0	22 ± 11	22			Murphy, 1982 (HPLC)
Soybeans Amsoy	747 ± 18	40 ± 11		117 ± 15	1 ± 3	118			Murphy, 1982 (HPLC)
Soybean full-fat flakes	2041	44	2085	1185	20	1205	9[c]	10	Eldridge and Kwolek, 1983 (HPLC)
Soybean meal (whole)		187	1000		145	706			Petersson and Kiessling, 1984 (HPLC)
Soybean defatted flakes	1601 ± 196	51 ± 11	1652	200 ± 13	1 ± 1	201			Murphy, 1982 (HPLC)
Soybean defatted flakes	1885	44	1929	1140	25	1165	8[c]	12	Eldridge and Kwolek, 1983 (HPLC)
Soybean flour (defatted)	1198	267	1465	617	328	945	129[c]	10	Eldridge, 1982 (HPLC)
range	580-1540	40-460	770-1800	480-770	80-480	650-1140	60-220	0-30	
Soybean flour, (Soyolk)[c]			969		674				Mazur et al., 1996 (GC-MS)
Soybean meal (defatted)		97 ± 11	753 ± 186		49 ± 2	616 ± 173			Petersson and Kiessling, 1984 (HPLC)
Soy protein concentrate	688	96	784	302	112	414	78[c]	12	Eldridge, 1982 (HPLC)
range	40-1910	10-220	50-2130	30-760	20-200	60-870	10-220	0-40	
Soy flakes (defatted)	2150	67	2217	596	56	652			Seo and Morr, 1984 (HPLC)
	88	8		81	7				
Soy protein isolate, acid	300 ± 13	29 ± 8	329	10 ± 1	0	10	40[c]	18	Murphy, 1982 (HPLC)
Soy protein isolate	654	150	804	206	138	344			Eldridge, 1982 (HPLC)
range	550-800	50-220	710-990	140-300	80-210	240-510			
Soy protein isolate	672 ± 42	103 ± 13	775	134 ± 19	56 ± 15	190			Seo & Morr, 1984 (HPLC)
Tofu brand F[c]	51 ± 3	54 ± 6	105	35 ± 4	47 ± 11	82			Murphy, 1982 (HPLC)
Tofu brand H[c]	75 ± 16	78 ± 16	153	35 ± 2	117 ± 13	152			Murphy, 1982 (HPLC)
Tofu, "Weber"	104 ± 4	29 ± 8	133	0	0	0			Murphy, 1982 (HPLC)
Kikkoman Firm tofu			213 ± 81			76 ± 27			Dwyer et al., 1994 (GC-MS)
Nasoya Soft tofu			187 ± 7			73 ± 22			Dwyer et al., 1994 (GC-MS)
Hatcho Miso		145	145		137	137			Mazur et al., 1996 (GC-MS)

Food							Reference
Soy-milk formula, (ProSobee)		21.8			17.1	0	Setchell and Welsh, 1987b (HPLC)
Soy-milk formula (Isomil)		22.6			19.1	15	Setchell and Walsh, 1987b (HPLC)
Soy drink, (First Alternative)		21 ± 1.1			7 ± 0.3	0	Dwyer et al., 1994 (GC-MS)
Soy-milk formula (Jevity Isotonic)	0	3.1 ± 2.5	0		0.3 ± 0.3		Dwyer et al., 1994 (GC-MS)
Soy sauce	0	0	0		0		Murphy, 1982 (HPLC)
Soybean hulls[d]							
Amsoy variety	5	33	66	28	94	0[c]	Eldridge and Kwolek, 1983 (HPLC)
Tiger variety	15	89	86	74	160	0[c]	
Suzuyutaka strain	10	30	20	0	20	0	Kudou et al., 1991 (HPLC)
Soybean hypocotyl[d]							
Amsoy variety	247	300	10315	190	10505	6641[c]	Eldridge and Kwolek, 1983 (HPLC)
Tiger variety	242	333	7599	140	7739	5888[c]	
Suzuyutaka strain[f]	160	2620	8380	350	8730	10040[c]	Kudou et al., 1991 (HPLC)
Soybean cotyledon[d]							
Amsoy variety	28	1167	375	14	389	17[c]	Eldridge and Kwolek, 1983 (HPLC)
Tiger variety	59	2117	1028	28	1056	16[c]	
Suzuyutaka strain[g]	140	2240	1450	110	1650	0[c]	Kodou et al., 1991 (HPLC)

[a]In addition to the compounds shown, we also measured formononetin, biochanin A, and coumestrol in all samples. Coumestrol was not found in any of the foods.

[b]Authors' soy product name used.

[c]Glycitin-7β-glucoside.

[d]Soybeans contain about 6.3–8% hull, 2–2.2% hypocotyl, and 90–91.5% cotyledon.

[e]Also 0.3 μg/g of formononetin, 0.7 μg/g of biochanin A, and 1.3 μg/g of secoisolariciresinol were found in this flour.

[f]In addition, 570 μg/g of 6″-O-acetyldaidzin and 390 μg/g of 6″-O-acetylgenistin.

[g]In addition, 80 μg/g of 6″-O-acetyldaidzin and 10 μg/g of 6″-O-acetylgenistin.

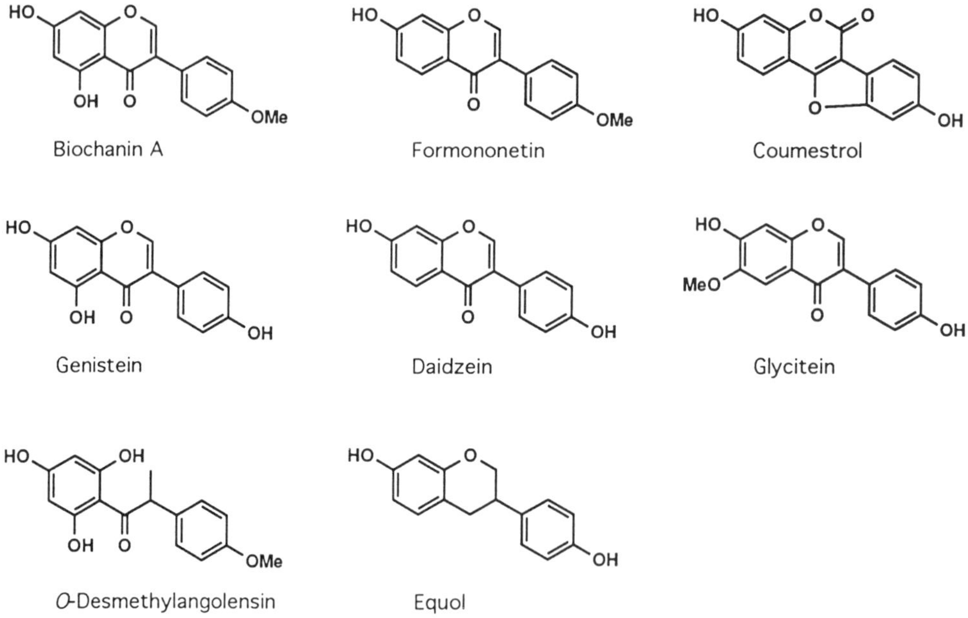

Figure 2 Structures of isoflavonoids and coumestrol.

Because the isoflavonoids occur mainly as glycosides in soybeans, methods based on HPLC were developed for the separation and assay of these conjugates, and the contents of daidzein, genistein, and glycitein glycosides in various soybean products were determined (Ohta et al., 1979; Farmakalidis and Murphy, 1985; Kudou et al., 1991; Coward et al., 1993; Wang and Murphy, 1994b). Nine different glucosides were quantified: daidzin, genistin, glycitein, 6"-O-acetyldaidzin, 6"-O-acetylgenistin, 6"-O-acetylglycitin, 6"-O-malonyldaidzin, 6"-O-malonylgenistin, and 6"-O-malonylglycitin (Wang and Murphy, 1994b). Because all these glycosides are readily hydrolyzed and the aglycones absorbed in the gut, their individual measurement is not necessary for the evaluation of the usefulness of foods for human consumption. Whether they can be absorbed as such is not known, but it seems unlikely based on an earlier experience with conjugated estrogens (Adlercreutz and Martin, 1980). On the other hand, it was recently suggested that quercetin glycoside is more readily absorbed when coupled to glucose (Hollman et al., 1995). This observation needs confirmation. Because to the best of my knowledge no glycosidases seem to occur in human organs, absorbed glycosides should appear in plasma and urine.

A summary of HPLC assay results for the main isoflavonoids in some selected soybean products and components is shown in Table 2. The isoflavones are concentrated in the hypocotyl (germ), and their content in soybean hulls is low (Eldridge and Kwolek, 1983; Kodou et al., 1991). (More information as to the content of isoflavonoids in soybean products may be obtained in the following publications: Coward et al., 1993; Dwyer et al., 1994; Franke et al., 1994a; Wang and Murphy, 1994b.) The isoflavonoid content of soybean seeds varies depending on growth conditions; temperature during seed development has a great influence (Tsukamoto et al., 1995), and there are clear effects on isoflavonoid glycoside pattern of seed genetics, crop year, and growth location (Wang and Murphy, 1994a). Total isoflavonoid content shows a threefold variation. The crop year had a greater effect than the location.

Another rich source of isoflavonoids is clover, clover seeds, and clover sprouts. Some beans other than soybeans contain isoflavonoids, but the concentrations are lower (Table 2).

Biochanin A and ß-sitosterol have been identified in bourbon (Rosenblum et al., 1987) and genistein and daidzein in beer (Rosenblum et al., 1992). The identification of these four compounds was based on GC-MS using selected ion monitoring of only four ions, but the relative abundances of the ions for standards and samples were not documented. Full spectra were not obtained, probably due to the low amounts present. The identification of these compounds in bourbon and beer must therefore be regarded as tentative. Quantitative data revealed a concentration between 7.1 and 20.6 µg/100 ml of ß-sitosterol in bourbon whiskey (Rosenblum et al., 1991).

Phytoestrogens Identified in Humans

Lignans

About 18 (Setchell and Adlercreutz, 1979; Setchell et al., 1980a,b,c, 1981; Stitch et al., 1980a,b; Adlercreutz et al., 1981) years ago two cyclically occurring unknown compounds, now called enterolactone and enterodiol, were detected in the urine of the female vervet monkeys and women, subsequently identified separately and independently by two groups (Setchell et al., 1980a; Stitch et al., 1980b). Furthermore, small amounts of four plant lignans—matairesinol, lariciresinol, isolariciresinol, and secoisolariciresinol—were identified (Bannwart et al., 1984b, 1989), and 7'-hydroxymatairesinol and 7'-hydroxyenterolactone were detected in human urine (Bannwart et al., 1988a). Enterolactone was measured in human and bovine plasma and semen by GC-MS (Dehennin et al., 1982). Enterolactone, enterodiol, matairesinol, and secoisolariciresinol have been measured in human plasma by ID-GC-MS-SIM (Setchell et al., 1983; Adlercreutz et al., 1993b,c,1994a). Recently the two main mammalian lignans, enterolactone and enterodiol, were identified in human feces by comparing their mass spectra with authentic synthesized standards (Adlercreutz et al., 1995e). This confirmed our early observation on the presence of enterolactone in feces (Setchell et al., 1981). Also the two known precursors of the mammalian lignans, matairesinol and secoisolariciresinol, have been detected by ID-GC-MS-SIM in feces (Adlercreutz et al., 1995e). Enterolactone alone or together with enterodiol has been measured in urine in rats and chimpanzees (Axelson and Setchell, 1981; Adlercreutz et al., 1986c; Musey et al., 1995; Mäkelä et al., 1995a), and in cow milk (Adlercreutz et al., 1986b), human breast cyst fluid, saliva, and prostatic fluid (Finlay et al., 1991; Griffiths et al., 1996). An unknown enterolactone-like compound with an extra nonphenolic hydroxyl group and probably not identical with 7-hydroxyenterolactone has been detected and its mass spectrum described (Joannou et al., 1995).

Isoflavonoids and Coumestrol

The isoflavonoid phytoestrogens are heterocyclic phenols with a close similarity in structure to estrogens. Their diphenolic (better bisphenolic) character makes them similar to lignans as well. They occur mainly as glycosides in numerous plants, and many studies have shown that after modification of their original structures by intestinal bacteria they have hormonal effects in animals (Price and Fenwick, 1985; Setchell and

Adlercreutz, 1988; Müller et al., 1989). The most well-known condition resulting from ingestion of subterranean clover is the "clover disease" in sheep (Bennets et al., 1946; Shutt, 1976).

The following isoflavonoid phytoestrogens have been identified or detected in human urine in this laboratory: formononetin, methylequol, daidzein, dihydrodaidzein, O-desmethylangolensin, genistein, and 3',7-dihydroxyisoflavan (Bannwart et al., 1984a,b,1986, 1987, 1988b; Adlercreutz et al., 1991c) (Fig. 2). Now, after synthesis of all compounds, their structures have been confirmed. Equol was first identified in the urine of many animals (review in Setchell and Adlercreutz, 1988) and identified in human urine independently in two laboratories (Axelson et al., 1982b; Adlercreutz et al., 1982a). Recently glycitein was also identified in human urine and five isoflavonoid metabolites (6'-hydroxy-O-desmethylangolensin, dihydrogenistein, dehydro-O-desmethylangolensin, and two isomers of tetrahydrodaidzein) were tentatively identified (Joannou et al., 1995; Kelly et al., 1993). In the latter study the identification of dihydrodaidzein, after a challenge of volunteers with soy, was confirmed, but no 3',7-isomer of equol could be detected in these subjects. Our previous observation on the presence of this compound in human urine may be explained by its presence in cow milk, from which it may have originated (Bannwart et al., 1986; Adlercreutz et al., 1986b). Only once have we observed minimal amounts of coumestrol in human urine in a vegetarian subject.

Daidzein, genistein, equol, and O-desmethylangolensin have been measured in human plasma by ID-GC-MS-SIM, but due to the relatively low amounts, mass spectra were not obtained (Adlercreutz et al., 1993b,c, 1994b). Recently we determined by the same method all these isoflavonoids in feces and identified definitely daidzein, genistein, and equol by comparing with authentic standards and deuterated compounds. The identity of O-desmethylangolensin was well established for ions with a mass higher than m/z 180 (Adlercreutz et al., 1995b). In the same study, lignans and isoflavonoids were determined in the feces of 10 omnivorous and 10 vegetarian women.

Origin, Formation, and Metabolism of Phytoestrogens in Humans

Studies on the origin, formation, and metabolism of the phytoestrogens in animals (Price and Fenwick, 1985; Müller et al., 1989) and in humans (Adlercreutz, 1988a, 1991d; Setchell and Adlercreutz, 1988) have been reviewed, and in this connection mainly recent studies in human subjects will be described.

Lignans

The mammalian lignans enterolactone and enterodiol are formed from plant precursors by the action of intestinal bacteria (Setchell et al., 1981, 1982; Axelson and Setchell, 1981; Borriello et al., 1985). Flaxseed secoisolariciresinol has been found to occur in the form of a diglycoside (Axelson et al., 1982a; Obermeyer et al., 1995), and it is likely that matairesinol also occurs in plants mainly in the glycosidic form. These glycosides are hydrolyzed in the proximal colon: it was shown that feeding of rye bran containing both secoisolariciresinol and matairesinol to ileostomy patients did not change the excretion of enterolactone and enterodiol in urine but slightly increased the matairesinol excretion (Hallmans et al. 1996). The intestinal microflora converts matairesinol to enterolactone and secoisolariciresinol to enterodiol, and the latter may be oxidized to enterolactone (Setchell et al., 1982; Borriello et al., 1985; Setchell and Adlercreutz, 1988). In rats the plant lignans undergo an enterohepatic circulation, and it is likely that this is also the case in humans as shown for phenolic estrogens (Adlercreutz et al., 1979; Adlercreutz and Martin, 1980). Lignans are excreted both in urine and feces, the fecal excretion pathway in human subjects seems to be more important than for estrogens (Adlercreutz and Järvenpää, 1982b; Adlercreutz et al., 1979), since the amounts are similar or only slightly lower in feces than in urine (Adlercreutz et al., 1995b). It could be that some of the formed mammalian lignans escape absorption because their formation also occurs in the distal colon (Bach Knudsen and H. Adlercreutz, unpublished observations).

As mentioned, the richest source of lignans in human foods is flaxseed. Early studies indicated that omnivorous, lactovegetarian, and macrobiotic American women living in Boston excreted proportionally more enterodiol of total lignans in urine than did Finnish women (Adlercreutz et al., 1986a). This was thought to be due to the high intake of antibiotics in the United States both as drugs and in the form of residual antibiotics in food. Administration of antibiotics almost completely eliminates the formation of enterolactone and enterodiol from plant precursors in the gut (Setchell et al., 1981; Adlercreutz et al., 1986a) and leads, after initial rapid lowering of the lignan levels in urine, to a relative increase in the enterodiol: enterolactone ratio (Adlercreutz et al., 1986a). Both the Boston and Finnish subjects studied had been free from antibiotics for 3 months, but frequent earlier intake of antibiotics by the Boston subjects or antibiotics in food may still have changed their gut flora. We do not, in fact, know how long it takes to normalize the gut flora with regard to lignan metabolism after intake of antibiotics because the subjects in our study on the effect of oxytetracyclin on lignan metabolism were followed for only 40 days (Adlercreutz et al., 1986a), at which time the enterodiol: enterolactone ratio was still very high. However, because flaxseed is a more common ingredient in American than Finnish food and this seed contains enormous amounts of secoisolariciresinol, the immediate precursor of enterodiol, another explanation for the relatively high enterodiol excretion in American women is a higher intake of flaxseed and lower consumption of whole-grain bread.

Urinary lignan excretion in Finnish women correlates with the intake of total vegetable fiber, and fiber from berries and fruit, and with legume fiber intake (Adlercreutz et al., 1981, 1986a, 1987, 1988b). The highest correlation coefficients were found if fiber intake was calculated per kg body weight.

Following intake of 10 g of flaxseed/day in young women, total lignan excretion in urine increased 13-fold and in feces 18-fold (Lampe et al., 1994; Kurzer et al., 1995). There was no change in the excretion of isoflavonoids. Enterolactone excretion increased 8.8-fold in urine and 16.1-fold in feces, and enterodiol excretion increased 17.9-fold in urine and 32-fold in feces. This shows that fecal excretion is a relatively important pathway of elimination. The relative amount of enterodiol increased considerably probably due to the very high secoisolariciresinol content of flaxseed. Obviously the amount of enterodiol formed by intestinal bacteria was so high that the conversion to enterolactone did not take place to a normal extent. After intake of 10 mg of flaxseed per day; total lignan excretion increased by 56.1 μmol/day, corresponding to an intake of at least 20.3 mg secoisolariciresinol per day, which is much higher than the suggested content of secoisolariciresinol in flaxseed (about 8 mg/10 g) (Axelson et al., 1982a; Obermeyer et al., 1995) except for the recent value of 36.9 mg/10 g obtained in this laboratory (Mazur et al., 1996) (Table 1). Because matairesinol and secoisolariciresinol were not measured in urine (Lampe et al., 1994; Kurzer et al., 1995) and secoisolari-ciresinol as well as conjugated metabolites were not measured in feces, the total urinary + fecal lignan excretion value after flaxseed intake in these studies is lower than the true one. The conclusion is that the lignan excretion in this experiment is more in agreement with the higher value found by us in flaxseed. This suggests that the lignan content of flaxseed may be much higher than previously thought.

Recently we found that the content of lignans in rye bread measured with our method is much lower than the urinary excretion values. Ten women were first kept 2 weeks on a diet containing no whole grain bread. Their urinary excretion of entero-lactone and enterodiol was measured by GC-MS. After a mean total intake of 671 nmol/day of rye bread lignans (matairesinol and secoisolariciresinol) (about 280 g of rye bread/day), the urinary excretion of mammalian lignans was 7.3-fold higher than the intake of precursors. Considering that substantial amounts must have been excreted by the fecal route, too, a rough estimate would be that about 10–12 times more was excreted as enterolactone and enterodiol than was measured in the rye in the form of matairesinol and secoisolariciresinol. This means that rye most likely contains other precursors for the mammalian lignans than those measured. At present we are involved in the identification of these precursors. It has been shown that the lignan arctiin is converted to arctigenin in the rat intestine (Nose et al., 1992), and plasma levels of metabolites were measured (Nose et al., 1993). Other similar compounds like dimeric (Han et al., 1994) or trimeric butyrolactone lignans could perhaps be found in this cereal because intestinal bacteria could convert all these lignans to enterolactone.

In the above-described experiment in 12 women, the urinary excretion of lignans varied from 6.9 to 26.2 nmol/g of rye bread (calculated from the difference between values during white wheat bread consumption and rye bread consumption). Because the fecal lignan values were not measured, it is not known whether differences in intestinal metabolism between individuals were responsible for this variation.

Isoflavonoids and Coumestrol

The metabolism of formononetin, daidzein, and biochanin A and genistein has been studied, particularly in sheep (Batterman et al., 1965; Braden et al., 1967; Nilsson et al., 1967; Shutt et al., 1970). Good reviews of these studies have been published (Shutt et al., 1970; Lindner, 1976; Price and Fenwick, 1985; Müller et al., 1989). Biochanin A is converted to genistein, which is further metabolized to p-ethylphenol. Formononetin is converted to daidzein and daidzein to O-desmethylangolensin and equol. The metabolism is different in various animals, and metabolism in humans may not, therefore, be identical to that found in sheep or cattle (Braden et al., 1971). As suggested for lignans (Borriello et al. 1985), the hydrolysis of the glycosides of flavonoids probably takes place in the proximal colon, and several bacteria have been found that produce the necessary enzymes (Hackett, 1986; Bokkenheuser et al., 1987; Bokkenheuser and Winter, 1988). Recently we found that in pigs the formation of enterolactone from its food precursors proceeds throughout the colon (Bach Knudsen et al., unpublished).

In humans equol (Axelson et al., 1984; Setchell et al., 1984) and O-desmethyl-angolensin are most likely formed, as in sheep, by intestinal bacterial action from formononetin and daidzein present in foodstuffs like soy products (Setchell and Adlercreutz, 1988). Some people are unable to produce equol or excrete this isoflavan in very low amounts (Axelson et al., 1984; Setchell et al., 1984; Setchell and Adlercreutz, 1988; Adlercreutz et al., 1991e).

Recently it was found (Kelly et al., 1993; Joannou et al., 1995) that subjects with low equol excretion after a soy load have high excretion of O-desmethylangolensin and vice versa. The metabolism of genistein has been further studied and a new metabolic pathway proposed leading from genistein to dihydrogenistein (saturation of C-ring) and further to 6'-hydroxy-O-desmethylangolensin (Kelly et al., 1993; Joannou et al., 1995). This pathway is very likely to occur. Another pathway from daidzein to equol was also proposed differing from that published by us (Adlercreutz et al., 1987). Joannou et al. (1995) suggest that the formation of equol proceeds via dihydrodaidzein and tetrahydrodaidzein instead of via dehydroequol. This is a possible pathway. The pathway via dehydroequol has not yet been established, because the synthesis of dehydroequol has been found to be very difficult. However, we have found a compound that could be dehydroequol in human urine. Joannou et al. (1995) also propose that dihydrodaidzein is converted first to 2-dehydro-O-desmethylangolensin and then to O-desmethylangolensin. In our hands the silylation procedure used by Joannou et al. (1995) converts dihydrodaidzein partly to 2-dehydro-O-desmethylangolensin, and, therefore, we cannot yet accept this new pathway as being valid.

Soybeans and soy products seem to contain less daidzein than genistein (Table 2), with the exception of some soy milk products (Xu et al., 1994a). In plasma of subjects consuming soy, the mean concentrations of genistein compared to those of daidzein are usually higher, but in urine the mean daidzein excretion tends to be similar or is higher than the mean genistein excretion (Adlercreutz et al. 1991c,e, 1993b, 1995b). This would suggest that daidzein is more bioavailable than genistein, as was also found in a study with soy milk powder in 12 women (Xu et al., 1994a). In this latter study the total recovery of daidzein was approximately 21% and that of genistein 9% when subjects consumed soy milk. However, only 1–2% was found in feces. This seems very low because we found in subjects consuming their habitual diet that the fecal isoflavonoid excretion per day is about 15–70% of the corresponding urine values depending on compound and diet (Adlercreutz et al., 1986a, 1995a,b). After intake of soy milk powder, no equol was found in the 12 women investigated by Xu et al. (1994a), despite a recovery of 62–74% in all spiked samples. In the study of Xu et al. (1994a), plasma and urine samples were analyzed by HPLC according to Lundh et al. (1988), but no precision or recovery values were presented. Fecal samples were determined with a simple procedure involving extraction with acetonitril in HCL, purification on a Sep-Pak C_{18} column, and HPLC. The reliability of this method was not documented.

In another recent study using the same methodology, Xu et al. (1995a) found that after intake of three different doses of soy milk powder in five subjects with low excretion of fecal isoflavonoids, the recovery of soy isoflavonoids in urine was low, but in two subjects with high fecal excretion the recovery was much higher. In the subjects excreting high amounts of isoflavonoids in feces, fecal recovery was about 5–8% for the two compounds and urinary recovery 10–17%; in the other, the fecal recovery was about 0.4–0.8% and the urinary recovery about 30–38%. Xu et al. conclude that the relative ability of gut microflora to degrade isoflavonoids determines the bioavailability of genistein and daidzein. They also showed that the half-life of genistein and daidzein, when incubated with gut bacteria in anaerobic conditions, was short (genistein-3.3 h; daidzein-7.5 h). Perhaps the subjects in the first study belonged to the group with high degradation of phytoestrogens in the gut.

Recently it was found that the parent compounds genistein and daidzein have much shorter in vivo half-lives than equol and O-desmethylangolensin (Kelly et al., 1995). These authors also found that the individual variability with regard to the metabolic response to a challenge with isoflavonoids as measured by the excretion of different metabolites was extensive. The most stable response was obtained for daidzein (variability 4x) followed by genistein (7x). The two metabolites of daidzein, equol (1527x) and O-desmethylangolensin (17x), showed the highest response variability. Subjects who excreted high amounts of equol excreted much lower amounts of O-desmethylangolensin and vice versa.

In a study of six male subjects consuming soy milk for 26 days, the recovery of ingested daidzin + daidzein was $46.9 \pm 15.2\%$, and for genistin + genistein it was $14.6 \pm 9.2\%$ (Lu et al., 1995a). The value for daidzein is considerably higher than reported by Xu et al. (1994a). It was suggested that the metabolic pathways did not change during chronic treatment. However, the absorption $t_{1/2}$ of daidzein and the formation and absorption of equol increased during chronic treatment. Chronic soy ingestion affected the relative absorption $t_{1/2}$ of daidzein and genistein. the absorption rate of daidzein was initially faster than that of genistein, but after chronic soy milk ingestion genistein was absorbed more quickly than daidzein. Chronic ingestion of soy milk also led to a slower excretion rate of all three compounds studied. The absorption of isoflavonoids seems to be slower in humans compared to sheep. Peak excretion occurred for daidzein at 7.1 ± 1.9 hours, for genistein at 6.7 ± 2.1 hours, and for equol at 13.5 hours (one subject), and the major part was excreted during the first 24 hours (Lu et al., 1995a). Only one of the six men produced equol from daidzein. Later the same group found that men excrete less daidzein after soy milk consumption than women and have lower daidzein/genistein ratios in urine (Lu et al., 1995b).

Plasma pharmacokinetics of genistein in mice after intravenous injection of genistein has been studied. From their results the authors concluded that it seems impossible to achieve or at least sustain a plasma level corresponding to those used for inhibition of growth of malignant cells in cell culture studies (Supko and Malspeis, 1995).

Because equol and other isoflavone metabolites occur in cow milk (Adlercreutz et al., 1986b), their presence in human urine is not necessarily the consequence of intestinal metabolism of their precursors. In countries with low soy consumption like Finland, the low basal levels of isoflavonoids found are, therefore, probably a result of intake of dairy products and meat. Even fish given soy-containing food, including some ready-formed isoflavonoid metabolites, may be responsible for the small amounts excreted in urine of Finns, or it may result from food (e.g., bread) additives containing minor amounts of soy protein with isoflavonoids.

As shown for estrogens (Adlercreutz et al., 1979; Adlercreutz and Martin, 1980) and flavonoids (Hackett, 1986), the isoflavonoids seem to undergo an enterohepatic circulation, at least in the rat (Axelson and Setchell, 1981). There are no human data on isoflavonoid metabolites in bile.

In a study in which the phytoestrogen excretion was measured for a basal diet (typical American food) and then for a diet containing canned garbanzo beans, onions, garlic, humus, and rice (legume/allium diet), the subjects excreted more isoflavonoids compared to the basal diet (Hutchins et al., 1995).

In 60 subjects consuming a soy protein beverage containing 34 g of soy powder (Altima, Protein Technologies International, St. Louis, MO), it was found that 35%

of the subjects excreted more than 2000 nmol/24 h of equol (Lampe et al., 1995). The "nonexcreters" excreted 21–233 nmol/24 h. There was no difference in incidence between men and women. Surprisingly they found that urinary excretion of daidzein, genistein and O-desmethylangolensin was similar between excreters and nonexcreters and between men and women. This contradicts the findings by Kelly et al. (1993), who found higher O-desmethylangolensin excretion in subjects with low equol excretion. The equol excreters consumed a significantly higher percentage of energy as carbohydrates and significantly greater amounts of plant protein and dietary fiber, both as soluble and insoluble fiber, than nonexcretors (Lampe et al., 1995). On the other hand, we found in Japanese subjects consuming their habitual diet a positive correlation between intake of fat and meat and equol excretion in urine (Adlercreutz et al., 1991e).

Chimpanzees in captivity consuming their regular food excrete large amounts of phytoestrogens in urine, particularly equol (Musey et al., 1995). When shifted to a high-fat diet the urinary excretion decreased by more than 90%, reaching levels found in omnivorous subjects or women with breast cancer. It seems that equol formation in chimpanzees is much higher than in human subjects, which must depend on the intestinal microflora.

After tofu consumption, relatively high values of coumestrol were found in urine (Franke and Custer, 1994b). The fact that the mean excretion of other phytoestrogens was very different for the subjects consuming tofu less than once a week compared to those consuming it more than once a week and was similar for both groups with regard to coumestrol excretion suggests that the HPLC analysis was not specific and an unknown compound measured which was not affected by soy intake. We have never found coumestrol in tofu products (Table 2) and only once in human urine in a vegetarian. Using thin-layer chromatography, coumestrol has been found both in the alfalfa hay fed to ewes and in conjugated form in the plasma of the ewes (Newsome and Kitts, 1977).

BIOLOGICAL EFFECTS AND MECHANISM OF ACTION OF PHYTOESTROGENS

In this connection we will mainly discuss phytoestrogens that have been detected or measured in the human organism.

Estrogenic and Antiestrogenic Effects

Many studies have shown that phytoestrogens bind to estrogen receptors and show significant estrogenic effects in animals, in humans, and in cell cultures (Noteboom and Gorski, 1963; Shutt and Cox, 1972; Martin et al., 1978; Tang and Adams, 1980; Newsome and Kitts, 1980; Jordan et al., 1985; Price and Fenwick, 1985; Setchell et al., 1987a; Müller et al., 1989; Gavaler et al., 1991; van Thiel et al., 1991; Whitten and Naftolin, 1991; Mäkelä et al., 1991, 1995c; Nwannenna et al., 1995). The most well-known estrogenic effect of phytoestrogens is the "clover disease" in Australian sheep leading to infertility (Price and Fenwick, 1985). This disease is caused by the high intake of red clover and high production of equol by intestinal bacteria in the sheep. Equol is antagonistic to estradiol by competing with estradiol-receptor complex for nuclear binding and yet fails to initiate the replenishment of estrogen receptors effectively (Tang and Adams, 1980). In most animal experiments causing great estrogenic effects, the intake or administration of phytoestrogens has been very high. In the studies by Nwannenna et al., 6.5 g of phytoestrogens were given to ewes weighing about 64 kg, corresponding to the consumption of more than 3 kg of soymeal per day for a human subject of the same weight.

In other studies coumestrol and zearalanol have been used to study the estrogenic effects of phytoestrogens. Coumestrol is a relatively rare component of the human diet, as it occurs practically only in various sprouts. Zearalanol is a fungal estrogen, which should not occur in our diet. Coumestrol is structurally more closely related to estradiol than any other of the phytoestrogens occurring in human diet. The dose tested has usually been 100 µg/day for mice, but some estrogenic effects have been seen with a dosage of $\geq 5\mu g$/day (Burroughs, 1995) corresponding to a dose of 15–17.5 mg coumestrol/day for a human subject weighing 60–70 kg. That would equal the amount of coumestrol contained in about 30 kg/day of alfalfa sprouts (dry weight!), but in that case the subject would also consume about 1.2 g of formononetin and 30 mg of biochanin A per day. The richest source of coumestrol in the human diet is mung-bean sprouts, containing 20 times the amount of coumestrol (about 1 mg/100 g) that alfalfa sprouts contain. Mung sprouts contain no formononetin but high amounts of daidzein (~ 700 µg/100/g dry weight) and genistein (~ 2000 µg/100 g dry weight).

Definite antiestrogenic effects have been observed in vivo because synthetic or natural estrogens seem to be counteracted by administered isoflavonoids or their presence in the diet (Folman and Pope, 1966, 1969; Tang and Adams, 1980; Kitts et al., 1983; Mäkelä et al., 1991). On the other hand, it has been found that equol, genistein, and coumestrol act through estrogen-receptor–mediated processes and do not show any antiestrogenic effects in human breast cancer cells in culture (Welshons et al., 1987; Mäkelä et al., 1994). The effects could be blocked by antiestrogens. In fact, it seems that antiestrogenic effects have been mainly observed in intact animals, with some exceptions (Mäkelä et al., 1995c), and the antiestrogenic mechanism is largely unknown. It is in fact doubtful whether there exists a true antiestrogenic effect of phytoestrogens at the cellular level. Recently, Salo et al. (1995) found that genistein transiently increased the expression of the estrogen-sensitive c-*fos* proto-oncogene in the prostatic urethra of the adult castrated neoDES mouse, but the response lasted longer after genistein than after estradiol treatment. However, genistein failed to induce the lactotransferrin gene in the same experimental model. In addition, when given neonatally, genistein, unlike DES, did not permanently increase the expression of the lactotransferrin gene.

The effect of phytoestrogens on estrogen-responsive cells may not be a complete estrogen response (Markaverich et al., 1995; Mäkelä et al., 1995c) and may be dependent on age and the prevailing endogenous estrogen level as well as on diet (Adams, 1995). An important estradiol effect like the implantation-inducing one cannot be achieved by any concentration of a phytoestrogen such as genistein or coumestrol (Perel and Lindner, 1970). Additive effects on estradiol stimulation of uterine weight and reduction of cytosolic estrogen receptor binding were observed following oral, but not parenteral, administration of coumestrol (Whitten et al., 1994). The authors also found antiestrogenic effects of coumestrol in the rat uterus. The problem of phytoestrogen effects is complex, and the authors urge specific identification of the action of each chemical. It is well known that oral estrogens stimulate the production of sex hormone–binding globulin (SHBG) resulting in higher plasma SHBG concentrations. However, with regard to their effect on production of SHBG in HepG2 cells in culture, the effects of both estradiol and phytoestrogens do not even seem to be mediated via the estrogen receptor (Loukovaara et al., 1995b,d).

The lignans enterolactone and enterodiol bind weakly to rat uterine cytosol (J. H. Clark and H. Adlercreutz, unpublished) but did not show any detectable estrogenic activity in vivo in mice (Setchell et al., 1981). However, in vitro in four sensitive assays in tissue culture, including breast cancer cell lines, the lignans were stimulatory and the effect was blocked by the antiestrogen tamoxifen. No antiestrogenic properties were observed (Jordan et al., 1985). In another study enterolactone in vivo inhibited estrogen-stimulated RNA synthesis in rat uterine tissue when administered 22 hours before estradiol (Waters and Knowler, 1982). The concentrations of enterolactone used were very low, and it is doubtful whether this result can be repeated. We observed stimulatory effect of enterolactone on MCF-7 breast cancer cells in the absence of estradiol, but a slightly stimulatory or nonstimulatory concentration of estradiol combined with a slightly stimulatory concentration of enterolactone did not cause any stimulation or tendency to inhibition (Mousavi and Adlercreutz, 1992). The enterolactone concentration was 1 μmol/liter, which can be regarded as physiological. Enterolactone, but not enterodiol, was shown to stimulate pS2 expression in MCF-7 cells (Mäkelä et al., 1994; Sathyamoorthy et al., 1994). These diverging results are difficult to explain, but it has been suggested (Adlercreutz, 1990b; Whitten and Naftolin, 1991) that the effect of exogenous weak estrogens may be either agonistic or antagonistic depending on the level of endogenous estrogens, and this has been experimentally confirmed with regard to coumestrol (Whitten and Naftolin, 1991).

Phytoestrogens in bourbon seem to have biological effects both in ovariectomized rats and perhaps in women (Gavaler et al., 1987, 1991). The principal compounds are suggested to be ß-sitosterol and biochanin A.

Coumestrol was recently shown to cause different responses in immature ovariectomized and immature intact rats, and it was concluded that ovarian estrogens may be a part of the phytoestrogen response. It was shown in immature intact rats that coumestrol enhances the activity of subsequent estradiol administration in spite of the fact that coumestrol itself did not show a complete estrogen response in the immature ovariectomized rat (Markaverich et al., 1995).

It is concluded that all phytoestrogens occurring in higher amounts in human food have relatively low estrogenic activity and that they may have antiestrogenic effects in certain situations. What will be the net result in human subjects is an unsettled question. The effect will probably not be the same in different organs.

Effects on Enzymes

Inhibition of 17ß-hydroxysteroid dehydrogenase type I by coumestrol and genistein has been observed (Mäkelä et al., 1995b). However, this inhibition did not correlate to inhibition of proliferation of breast cancer cells (Mäkelä et al., 1994). This enzyme is expressed in certain target tissues of estrogen action, such as normal and malignant human breast and endometrium (Poutanen et al. 1995), keeping the tissue:plasma ratio of estradiol high and stimulating proliferative activity of cancers. Another very recent observation is a strong inhibition of ß-hydroxysteroid dehydrogenases of *Pseudomonas testosteroni* by daidzein, genistein, formononetin, and biochanin A (Keung, 1995). Daidzein was shown to inhibit the oxidation of testosterone as well as that of pregnenolone, pregnanolone, estradiol, epiandrosterone, and dehydroepiandrosterone, which could have significant immplications for steroid actions in the brain (Keung, 1995) and for steroid biosynthesis in the ovaries. These observations were the consequence of the finding the genistein, daidzein, biochanin A, and formononetin strongly inhibit the human mitochondrial alcohol dehydrogenase I (ADH) (Keung, 1993b), which is the only alcohol dehydrogenase isozyme capable of catalyzing 3ß-hydroxysteroid oxidation (McEvily et al., 1988). The 7-O-glucosyl derivatives of these isoflavonoids did not inhibit ADH but they are potent aldehyde dehydrogenase (ALDH) inhibitors (Keung, 1993b; Keung and Vallee, 1993a). These effects have been utilized in the treatment of alcoholism in China.

Genistein, equol, biochanin A, formonetin, daidzein, and enterolactone and to a lesser degree coumestrol and enterodiol were found to inhibit 5α-reductase. Enterolactone, genistein, biochanin A, and enterodiol and to a lesser degree equol, formononetin, daidzein, and coumestrol inhibited 17ß-hydroxysteroid dehydrogenase in genital skin fibroblasts. A concentration of 100 μmol/liter resulted in an almost 100% inhibition of both enzymes by genistein and enterolactone, respectively. These phytoestrogens also inhibited 5α-reductase in prostate tissue homogenates (Evans et al., 1995). A cocktail of seven compounds, each at 10 μM concentration, inhibited 5α-reductase by 77% and 17ß-hydroxysteroid dehydrogenase by 94% in human genital skin monolayers. The inhibition of 5α-reductase by the phytoestrogen cocktail was of the same magnitude as that caused by a 10 μM concentration of Finasteride, a potent drug used for the inhibition of 5α-reductase in benign prostatic hyperplasia.

Enterolactone, the most abundant mammalian lignan, is a moderate inhibitor of placental aromatase and competes with the natural substrate androstenedione for the enzyme (Adlercreutz et al., 1993a). A theoretical intermediate between matairesinol and enterolactone, 4,4'-dihydroxyenterolactone, showed the strongest inhibition. Other experiments with a choriocarcinoma cell line (JEG-3) showed that enterolactone is very readily transferred from cell culture media into the cells and inhibits the aromatase (Adlercreutz et al., 1993a). Secoisolariciresinol, a precursor of enterodiol, showed no activity (Gansser and Spiteller, 1995). Flavonoids, occurring in very high amounts in the diet, are also inhibitors of the aromatase enzyme (Kellis and Vickery, 1984a; Ibrahim and Abul-Hajj, 1990; Campbell and Kurzer, 1993). Studies in human preadipocytes show inhibition of the aromatase enzyme to various degrees by lignans, the most effective being didemethoxymatairesinol (also 4,4'-enterolactone) (Campbell and Kurzer, 1993; Wang et al., 1994). Most lignans are only weak inhibitors. However, a diet rich in vegetables, fruits, and berries may, due to the abundance of these compounds and flavonoids, lead to sufficient concentrations, e.g., in fat or cancer cells, to reduce conversion of androstenedione to estrone, lowering risk for estrogen-dependent cancer (Henderson et al., 1988).

At the time of the identification of genistein in human biological materials (Adlercreutz et al., 1991c, 1993b, 1995b), the inhibitory effect of this phytoestrogen on a great number of tyrosine-specific protein kinases was well known (Akiyama et al. 1987; Ogawara et al., 1989; Akiyama and Ogawara, 1991). Since then more than 150 publications have appeared using genistein as a tyrosine kinase inhibitor in numerous different types of experiments. The compound is specific for tyrosine protein kinases and is a competitive inhibitor with respect to ATP. It seems to inhibit most known members of this group of protein kinases, except the p40 protein-tyrosine kinase, including those belonging to the src family. Using a targeting antibody genistein has recently been effectively used in the treatment of B-cell precursor leukemia in mice (Uckun et al., 1995). Genistein also inhibits topoisomerase I and II (Okura et al., 1988; Markovits et al., 1989; Kondo et al., 1991; Constantinou et al., 1995a) and protein histidine kinase (Huang et al., 1992). Protein-tyrosine kinase activity is associated with cellular receptors for epidermal growth factor (EGF), insulin, insulinlike growth factor I (IGF-I), platelet-derived growth factor (PDGF), and mononuclear phagocyte growth factor (colony stimulating factor 1, CSF-1). Both tyrosine kinases and topoisomerases play an important role in cell proliferation and transformation. Tyrosine kinases have been associated with oncogene products of the retroviral src gene family and are correlated with the ability of retrovirus to transform cells (Akiyama et al., 1987; Akiyama and Ogawara, 1991; Markovits et al., 1989; Ogawara et al., 1989; Teraoka et al., 1989). Tyrosine kinase activity is also associated with breast cancer oncogene expression (Le Cam, 1991; Lehtola et al., 1992).

It is concluded that phytoestrogens inhibit to a greater or lesser extent the most important steroid biosynthetic enzymes as well as many important enzymes involved in cell transformation and proliferation. It is likely that these activities form some of the most essential parts of their disease-preventive effects.

Phytoestrogens and Nuclear Estrogen Type II Binding Sites

Several isoflavonoids and lignans compete with estradiol for the rat uterine so-called nuclear type II estrogen-binding sites (Adlercreutz et al., 1992a; Markaverich et al., 1995). The highest affinity with regard to type II site binding of the diphenolic compounds found and measured by us in human urine is shown by the isoflavones daidzein and equol, but some lignans, such as matairesinol, isolariciresinol, and enterolactone, also show relatively good competition. Previously it was thought that genistein does not bind to these sites, but this was obviously due to its low solubility, and a recent publication demonstrates that genistein and daidzein have about equally good affinity for these sites (Markaverich et al., 1995).

The nuclear type II sites have been suggested to constitute a component of the genome that regulates estrogen-stimulated growth (Markaverich and Clark, 1979; Markaverich et al., 1981). These authors also demonstrated that p-hydroxyphenyllactate (MeHPLA) is an endogenous ligand for the nuclear type II sites and is derived from bioflavonoid or tyrosine metabolism (Markaverich et al., 1984; Markaverich et al., 1988a). They suggest that this ligand and some bioflavonoids through binding to these sites may be important cell growth–regulating agents (Markaverich et al., 1988b; Markaverich et al., 1992). If this is true, an antiproliferative and antiestrogenic effect of phytoestrogens could be mediated via these binding sites.

However, despite the abundant presence of nuclear type II sites in numerous tissues and tumors (Markaverich et al., 1995), the protein has not, despite great efforts, been isolated and identified. It has been suggested that eosinophils are the source of uterine type II estrogen-binding sites and that the type II binding site is peroxidase (Lyttle et al., 1984; Lee et al., 1989). These authors also suggested that there is an estrogen-stimulated production of a chemotactic factor, which causes the influx of eosinophils, and this was blocked by treatment with the antiestrogen tamoxifen and by inhibitors of RNA and protein synthesis (Lee et al., 1989). That eosinophils are a source of the type II sites has been disputed by the group that has been working mainly with these sites (Markaverich et al., 1986). Interestingly, it has been shown that diethylstilbestrol (DES), genistein, zearalenone, and zearalanol are competitive inhibitors of rat uterine peroxidase. Coumestrol was a noncompetitive inhibitor (Shore and Lyttle, 1986). Recently it was shown that dietary flavonoids inhibit thyroid peroxidase (Divi and Doerge, 1996), suggesting that the type II estrogen nuclear binding sites could be a mixture of different peroxidases. The type II binding sites have been suggested to be associated with tyrosinaselike activity (Garai et al., 1992a; Garai and Clark, 1992b, 1994) but the type II sites were later separated from this enzyme activity (Densmore et al., 1994). Type II binding sites have also been found in MCF-7 human breast cancer cells (Markaverich et al., 1994).

It seems that the problem of bioflavonoid interaction with nuclear type II estrogen-binding sites and possible growth regulation must await the isolation of these sites. It is likely that several proteins (enzymes) are binding the bioflavonoids with less affinity than that of the estrogen receptor. One of the main candidates seems at present to be the peroxidase enzymes. Before identification of these sites, their role in estrogen-mediated cell growth and proliferation remains unsolved.

Phytoestrogens and SHBG

Lignans and isoflavonoids seem to stimulate SHBG synthesis in the liver and in this way most likely reduce the biological effects of sex hormones (Adlercreutz et al., 1987, 1988b, 1990a, 1992a) (Fig. 3). An increase in SHBG results in a lowering of the percentage of free testosterone (%FT) and the percentage of free estradiol (%FE2) and reduction in both the albumin-bound and free fraction of the sex hormones. This reduces the metabolic clearance rate (MCR) of the steroids and reduces in this way their biological activity. The biological role of SHBG has now been found to be even more

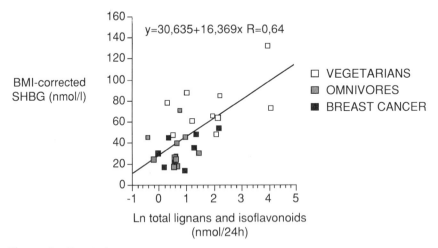

Figure 3 Correlation between total lignans and isflavonoids in urine and plasma sex hormone binding globulin (SHBG) in three groups of young women (omnivores, vegetarians, and breast cancer patients) (Finlandia study). Every SHBG assay result is the mean value obtained by analysis of two samples of three pooled separate blood collections (winter and summer) in duplicate. The urine phytoestrogen values (sum of enterolactone, enterodiol, daidzein, equol and *O*-desmethylangolensin) represent the mean values of two 72-h urine collections (winter and summer).

important than previously believed because it has been found to bind to a receptor on the cell surface and regulate the bioavailability or action of the hormones in a highly selective or targeted fashion (Hammond, 1995). Very recently it was found that SHBG exerts a negative control on estradiol action in MCF-7 human breast cancer cells (Fortunati et al., 1996). A stimulation of SHBG production with resulting higher levels would, therefore, theoretically lower risk for sex hormone–dependent cancer.

In Finnish women total fiber intake, total fiber intake/kg body weight, and grain fiber intake/kg body weight correlate positively with urinary excretion of total lignans and isoflavonoids (Adlercreutz et al., 1987, 1988b). The urinary excretion of the two groups of compounds and enterolactone alone in both pre- and postmenopausal Finnish women correlate positively with plasma SHBG and negatively with plasma %FE2 and %FT (Adlercreutz et al., 1987, 1988b, and unpublished results). This correlation occurred also if corrections were made for the body mass index of the subjects.

In vitro studies using HepG2 liver cancer cells showed that enterolactone (Adlercreutz et al., 1992a), genistein (Mousavi and Adlercreutz, 1993; Loukovaara et al., 1995d), and daidzein and equol (Loukovaara et al., 1995d) stimulate SHBG synthesis. Daidzein and equol increased SHBG levels in parallel intra- and extracellularly, whereas genistein in the studies with HepG2 cells by Loukovaara et al. (1995d) increased SHBG levels only within the cells, resembling in that sense the in vitro effect of estradiol. The effect on SHBG production appeared to occur at the posttranscriptional level and is perhaps caused by stabilizing the SHBG mRNA and seems not to be mediated via the estrogen receptor as seems to be the case also for estradiol (Loukovaara et al., 1995b,c,d).

These observations seem to explain the higher SHBG values seen in vegetarians with normal weight (Armstrong et al., 1981; Adlercreutz et al., 1989a, 1988b). However, in several studies no SHBG increase was observed after intake of flaxseed with high concentration of secoisolariciresinol or soy containing isoflavonoids (Shultz et al., 1991; Phipps et al., 1993; Cassidy et al., 1994; Cassidy et al., 1995; Baird et al., 1995). This may be partly due to the relatively low number of subjects in some studies and the great between- and within-individual variations in SHBG levels in others, making it almost impossible to detect a small change. Another problem is that the mean SHBG basal values were already relatively high. Practically all our omnivorous subjects and subjects with breast cancer showed values below the basal mean values of the women in these other studies. It is possible that there is an upper limit for the effect of phytoestrogens. On the other hand, it is likely that vegetarians have lower plasma insulin levels, which may increase plasma SHBG (Loukovaara et al., 1995a) independent of intake of phytoestrogens. There may, in addition, be an effect of the menstrual cycle on the values. In an unpublished study, we found that feeding 200–300 g rye bread/day to 12 women resulted in an almost significant increase in SHBG after 2 weeks ($p < 0.07$) (Adlercreutz, 1990b). Recently a similar result was obtained in a collaborative study with Dr. A. Brzezinski at Hadassah Medical Center in Jerusalem. Preliminary results after a 3-month treatment suggested that soy and flaxseed intake in postmenopausal women results in a significant increase in SHBG ($p < 0.01$). However, these increases could have been due to some other components of soy or flaxseed. It is important that this question be settled because low SHBG and high estrone (Lipworth et al., 1996) and probably high androstenedione and testosterone (Adlercreutz et al., 1989a) are associated with breast cancer risk.

Soy or flaxseed feeding experiments need to be carried out in subjects with low initial SHBG levels. Only then can the possible in vivo effect of these compounds be adequately judged. More than one sample is needed for the basal values, and the samples should not be too old because SHBG concentration changes with time in stored samples (Adlercreutz et al., 1989a). The effect of phytoestrogens on plasma SHBG may be related to the disease-preventive effects of these compounds.

Phytoestrogens and the Menstrual Cycle

As described above, in sheep the consumption of subterranean clover leads to an infertility syndrome called clover disease characterized by anovulatory estrus and infertility. The large formation of relatively estrogenic equol from formononetin in clover with high concentrations in plasma of the sheep leads to this complication (Nilsson et al., 1967; Lindner, 1976; Shutt, 1976; Price and Fenwick, 1985). Genistein and coumestrol bind to hypothalamic and pituitary receptors (Mathieson and Kitts, 1980), and the high concentrations seem to cause alterations in these receptors (Tang and Adams, 1978). This may result in decreased sensitivity of the receptors to feedback regulation of estradiol at physiological concentrations. The condition resembles the persistent estrus syndrome occurring in rodents after neonatal estrogenization.

In human subjects the plasma concentrations of equol are relatively low compared to the concentrations of daidzein and genistein, even after soy consumption (Adlercreutz et al., 1994a), and in 91 postmenopausal women 165 mg of isoflavones per day showed at most a small estrogenic effect on vaginal cytology (Baird et al., 1995). In sheep, consumption of phytoestrogenic alfalfa or subterranean clover leads to higher plasma luteinizing hormone (LH) concentrations (Bindon et al., 1982; Hettle and Kitts, 1983/1984). In postmenopausal subjects after intake of 165 mg of isoflavones per day, there was no effect on gonadotropins (Baird et al., 1995). However, in premenopausal women intake of 45 mg of conjugated isoflavones per day LH and follicle-stimulating hormone (FSH) caused a significant decrease (Cassidy et al., 1995), and there was a slight significant prolongation of the follicular phase and therefore the progesterone peak came slightly later. In this study plasma LH tended also to be lower on the isoflavonoid-free soy diet. Recently in another study in premenopausal women similar results with regard to plasma LH were obtained, but in addition plasma estradiol and progesterone decreased and the menstrual cycle was prolonged (Lu et al., 1996). A decrease in plasma dehydroepiandosterone sulfate (DHEAS) was also found, which is in disagreement with the observation that Adventist women who are vegetarians with high soy intake and a lower risk of breast cancer have higher levels of DHEAS (Persky and van Horn, 1995). The results of the two soy load studies (Cassidy et al., 1995; Lu et al., 1996) resemble with regard to the effect on plasma LH and reduction of the progesterone peak those obtained during continuous administration of 6 mg estriol/day to normal premenopausal women (Vähäpassi and Adlercreutz, 1975), but in the estriol study the cycle length was unaffected. The identification of ovulation and synchronization of the cycles for calculating mean values are always a great problem in this kind of study, and exactly how this was done was not clearly described (Cassidy et al., 1995).

When we calculated approximative phytoestrogen intake levels from the dietary study in a village outside Kyoto, Japan (Adlercreutz et al., 1991e), we arrived at intakes of about 19 mg of genistein and 5 mg of daidzein in the women and 10 mg of genistein and 6 mg of daidzein in the men. In these calculations we used our own quantitative results for soy products obtained by GC-MS. The values for women are considerably less than those used in the above-described human studies. In a letter to the editor, Messina (1995b) calculated that the intake for our subjects was higher, about 50 mg isoflavonoids per day. We do not known whether glycitein was included. Because the exact manner of calculation is not known, it seems that the only way is to calculate the total intake of phytoestrogens by analyzing Japanese homogenized diets. This we are doing at present.

Flaxseed rich in lignans also changes the menstrual cycle by prolonging the luteal phase and increasing significantly the luteal phase progesterone:estradiol ratio without significantly affecting luteal phase progesterone or estradiol levels (Phipps et al., 1993). There were no anovulatory cycles during flaxseed intake (10 g/day) but 3 of 36 cycles were anovulatory during the control period. Plasma estrone, estradiol, DHEAS, prolactin, and SHBG in early follicular phase were unaffected. Mean follicular phase testosterone decreased but early follicular phase testosterone did not change. In rats 1.5 mg or 3.0 mg of secoisolariciresinol diglucoside per day or 10% flaxseed in the diet or tamoxifen treatment cause significant changes in the cycling pattern by lengthening cycles and leading to acyclic periods (Orcheson et al., 1993).

More detailed discussions on the physiological and pathological effects of phytoestrogens in animals, particularly sheep, can be found in Müller et al. (1989), Price and Fenwick (1985), and Setchell and Adlercreutz (1988). The effect of soy and whole grain products and seeds on the menstrual cycle need to be further studied. However, the results already obtained suggest that these effects are relatively small and would not probably affect fertility.

Phytoestrogen Effects on Vascular Endothelial Cells, Angiogenesis, and Tumor Invasion

Urinary extracts containing genistein and synthesized genistein inhibit bFGF-stimulated endothelial cell (bovine brain-derived capillary endothelial cells) proliferation. Cenistein also inhibits the proliferation of vascular endothelial cells derived from adrenal cortex (ACE) or aorta (BAE) (Fotsis et al., 1993). In cells stimulated by bFGF, > 25 μM genistein causes cell death, but at lower concentrations the inhibitory effect is reversible (Fotsis et al., 1995a). The results were the same on uncoated and gelatin-coated substrata. Genistein does not affect quiescent cells up to concentrations of 200 μM (Fotsis et al., 1993). These data clearly suggest that genistein targets only proliferating cells, leaving quiescent nondividing cells unaffected.

Having established the inhibitory effect of genistein on the proliferation of endothelial cells, we investigated its effect on angiogenesis. A prerequisite for the formation of new capillaries is the breakdown of the basement membrane of the parent vessels through the finely regulated production of proteolytic enzymes and their inhibitors (Pepper and Montesano, 1990) followed by migration of endothelial cells and invasion. Proteolytic degradation of the extracellular matrix by endothelial cells is controlled by angiogenic factors, which induce the production of urokinase type plasminogen activator and its physiological inhibitor plasminogen activator inhibitor-1 (Pepper and Montesano, 1990). It has been shown that genistein reduces the production of plasminogen activator and plasminogen activator inhibitor-1 (Fotsis et al., 1993) in cloned bovine microvascular endothelial (BME) cells. Moreover, genistein inhibited the bFGF-induced migration of endothelial cells in wounded confluent monolayers of endothelial cells. Thus, genistein also interferes with some early events of angiogenesis.

The effect of genistein was studied in an experimental in vitro system that mimics angiogenesis in vivo. BME cells, seeded on the surface of collagen gels, invade the gels when exposed to bFGF and form capillarylike tubes beneath the gel surface (Montesano and Orci, 1985). Genistein alone had no effect on confluent BME cultures. However, when added together with bFGF, it inhibited their ability to invade the gels and generate capillary-like structures (Fotsis et al., 1993). Quantitative results revealed that the half-maximal concentration was high (150 µM), due, it was later discovered, to poor solubility of genistein with inappropriate diluents (M. S. Pepper, unpublished), and later a half-maximal concentration of 8 µM of genistein was obtained (Fotsis et al., 1995b) corresponding to the half-maximal concentration of 6 µM for the in vitro antiproliferative effect on endothelial cells (Fotsis et al., 1993). In fact we believe that in many other studies the poor solubility of genistein may have affected the results obtained.

Using the rabbit cornea neovascularization model (Rastinejad et al., 1989), the in vivo inhibition of angiogenesis by genistein has now been established (Fotsis et al., 1995b). The mechanism of action of the antiangiogenetic effect of genistein may be mediated by inhibition of tyrosine kinases, e.g., FGF receptors (Ullrich and Schlesinger, 1990), or by inhibiting S6 kinase (Linassier et al., 1990), an enzyme activated by bFGF. Genistein could also affect angiogenesis by inhibiting topoisomerase I and II (Okura et al., 1988; Yamashita et al., 1990).

In conclusion, our studies indicate that genistein is a nontoxic compound with regard to cells not actively dividing but inhibits proliferation of tumor and FGF-stimulated endothelial cells and inhibits their invasion and the proteolytic enzyme involved. It thus inhibits angiogenesis both in early and later phases of the events. Our recent studies, with a large number of similar compounds revealing the structural requirements for the antiangiogenetic and antiproliferative action, pave the way for the possible development of new therapeutic drugs (Fotsis et al., 1997).

Phytoestrogens and Insulin

The insulin receptor tyrosine kinase belongs to the tyrosine kinases that could be inhibited by genistein (Ogawara et al., 1989). Using rat adipocytes it was found that genistein did not influence insulin receptor ß-subunit autophosphorylation or tyrosine kinase substrate phosphorylation either in vivo or in vitro (Abler et al., 1992). Genistein inhibited insulin-stimulated glucose oxidation in a concentration-dependent manner, however, at very high concentrations (90–370 µmol/liter) not occurring in the human organism. It is likely that the poor solubility of genistein has affected these results, because it is not possible to dissolve genistein in such high concentrations. Thus, genistein can at high concentrations inhibit certain responses to insulin, but this cannot necessarily be attributed to inhibition of tyrosine receptor kinase activity. This was confirmed in insulin-stimulated CSV3-1 cells (Wang and Scott, 1994).

Subsequent studies with adipocytes suggested that genistein inhibited glucose transport by decreasing intrinsic activity, rather than number, of the plasma membrane–associated glucose transporters GLUT4 (Smith et al., 1993). Insulin also increases epithelial Na$^+$ reabsorption, and many of its actions involve tyrosine kinases. Genistein was found to reduce the number of active Na^+ channels in cell-attached patches of A6 cells (Matsumoto et al., 1993).

Genistein, on the other hand, stimulates the release of insulin from MIN6 cells, a glucose-sensitive insulinoma cell line, but only at stimulatory concentrations of glucose (Ohno et al., 1993). Genistein caused an accumulation of cAMP in the cells, which was inhibited by calcium antagonists. The results showed that the insulin release and the cAMP accumulation was modulated by calcium and depended on extracellular calcium. This effect cannot be explained by inhibition of tyrosine kinase (Akiyama et al., 1987), phosphitidylinositol turnover (Imoto et al., 1988; Dean et al., 1989; Dhar et al., 1990; Gaudette and Holub, 1990; Hill et al., 1990; Kondo et al., 1991; Makishima et al., 1991), or topoisomerase (Okura et al., 1988; Markovits et al., 1989), all of which are reported to be sensitive to genistein, because these inhibitory effects are not likely to cause elevation of intracellular calcium concentration. Paradoxically, genistein increases insulin release from the ß-cell islet of normal mice with simultaneous reduction of Ca^{2+} influx in these cells. The concentrations needed were high (40–370 µmol/liter) and again may suggest a solubility problem for the genistein used in these studies. Daidzein, which is not a tyrosine kinase inhibitor, was slightly less potent than genistein on K^+ and Ca^{2+} channels, but increased insulin secretion in a similar way (Jonas et al., 1995). In insulin-secreting INS-1 cells, genistein abolished the inhibitory effect of insulin or insulinlike growth factor I (IGF-1), but daidzein had no effect. The authors arrived at the conclusion that tyrosine kinase is involved in the inhibitory effects of insulin and IGF-I on insulin release in INS-1 cells (Verspohl et al., 1995).

The high concentrations needed to affect insulin release and inhibition of glucose oxidation suggest that the studied phytoestrogens can hardly have any beneficial or deleterious effects on glucose turnover or insulin action in humans. The possible problem with solubilization of the genistein used must be considered in the future.

Phytoestrogens and the Immune System

Many plant lignans have been shown to have antiviral, bactericidic, and fungistatic activities (Rao, 1978; Markkanen et al., 1981; Ayres and Loike, 1990). Whether these activities play a role outside the plant is not known, but lignans could play a significant role as protectors of the body against pathogenic organisms, at least in the gastrointestinal tract, because their concentration may be high (Adlercreutz et al., 1995b). Recently arctigenin, a lignan with a structure closely related to matairesinol (plus one methyl group) (Fig. 1), was found to have immunomodulatory activity (Eich et al., 1996). It was shown that this compound is an inhibitor of human immunodeficiency virus Type-1 integrase. It should be mentioned that both matairesinol and arctigenin showed strong suppressive effect on the growth of human leukemic HL-60 cells. Because about 10–12 times more of the mammalian lignans enterolactone and enterodiol have been found in body fluids after consumption of a measured amount of rye lignan (unpublished results), arctigenin could be another precursor (Nose et al., 1992). We are now searching for this compound in various human foods.

Genistein inhibits nitric oxide (NO·) production in murine macrophages, suggesting a modulatory effect on immune response (Krol et al., 1995). It has also been shown that genistein is a powerful immunosuppressive agent because it inhibits CD_{28} monoclonal antibody–stimulated human T-cell proliferation (Atluru and Gudapaty, 1993a;

Atluru et al., 1993b). Furthermore, it was shown that genistein inhibited interleukin 2 synthesis and interleukin receptor expression as well as production of leukotriene B(4).

The evidence available suggests that phytoestrogens may be active in the human organism because of their immunomodulating effects as well as a possible direct toxic action on bacteria, fungi, and virus. This could occur particularly in the gastrointestinal tract and should definitely be the subject of further study.

Other Effects of Phytoestrogens

Because numerous signaling events in the cells involve tyrosine phosphorylation and genistein is a specific tyrosine kinase inhibitor (albeit with low specificity with regard to the numerous different tyrosine kinases), it follows that there is an abundant literature on the effect of genistein on a high number of biochemical events. All these actions could be the target for another review. In most of these experiments the genistein concentrations have been rather high, in fact sometimes higher than the actual solubility of the compound. Some observations of no effects have later been revised when the compound has been dissolved in dimethylsulfoxide (DMSO). The relevance of studies with concentrations of genistein exceeding 100 μmol/liter for human health is doubtful, and even those with concentrations below 100 μM may have been hampered by solubility problems. Because of the very abundant literature we will include some of these studies on other effects of phytoestrogens, not mentioned previously, when discussing the relation of phytoestrogens to human disease.

EPIDEMIOLOGY OF LIGNANS AND ISOFLAVONOIDS

A summary of all of our results regarding urinary lignan and isoflavonoid excretion in various dietary groups of women and men, including two groups of breast cancer patients, has recently been presented (Adlercreutz et al., 1995a). None of the subjects investigated had been treated with antibiotics during the previous 3 months. This is important, because antibiotics affect particularly lignan metabolism.

With regard to lignans, macrobiotics subjects living in Boston had the highest values followed by other vegetarian groups living in Boston and Helsinki. The lowest lignan values among the subjects living in western societies were found in the breast cancer groups and Boston omnivorous women. However, even lower lignan values were found in recent immigrant Asian women in Hawaii, and low values were found in the Japanese men and women, who, however, had very high isoflavonoid values. The recent immigrants to Hawaii had similar isoflavonoid values as those consuming an omnivorous western diet, showing that they very rapidly omit soy products from their diet. Because they did not consume whole grain bread, the result was very low lignan and isoflavonoid values in urine. They still consumed a very low-fat diet, which may be the reason why they have some protection against breast cancer. They also have low plasma and urinary estrogens but high fecal excretion of estrogens (Goldin et al., 1986; Adlercreutz et al., 1994b), obviously due to interruption of the enterohepatic circulation (Adlercreutz et al., 1994b).

Recent measurements of urinary and plasma phytoestrogens in Japanese and Finnish men revealed that despite low urinary lignan values, the values for the free plus sulfate-conjugated lignans in plasma was as high in the Japanese men as in the Finnish men. However, the glucuronide values for enterolactone were significantly lower (p < 0.001) (unpublished results in 30 Japanese and 30 Finnish men). Thus, the Japanese men have more diphenolic compounds in the biologically active fraction than the Finnish men. Like estrone sulfate, being a precursor and "storage form" of biologically active estrone and activated by sulfatases, we believe that the phytoestrogen sulfates occurring in high concentrations in plasma (Adlercreutz et al., 1993b,c, 1994a) are potentially biologically active. Unfortunately no assays have been carried out in plasma of American men.

In an earlier study we found that lignan excretion in Japanese subjects was related to the intake of whole soybeans (Adlercreutz et al., 1991e), agreeing well with our recent detection of secoisolariciresinol in soybean meal (Mazur et al., 1996). In other soy products no lignans were detected.

In Finland, higher lignan excretion is seen in subjects living in North Karelia, with lower breast, colon, and prostate cancer risk compared to those living in southwest Finland (Adlercreutz et al., 1986a). In Japan the risks of breast, prostate, and colon cancer are low—Japanese have the highest plasma values we have measured for isoflavonoids as well as high values for the biologically active lignans. The Finnish values are somewhat less, and the American population has the lowest values. Exceptions are vegetarians, who always have higher values but whose risk of cancer is also lower. Interestingly, the urinary phytoestrogen excretion in Australian women is relatively high (Murkies et al., 1995), and their breast cancer incidence is lower than in Canada (Kliewer and Smith, 1995).

A recent study shows that the phytoestrogen level in the prostatic fluid of men is the lowest in England and much higher in Portugal, Beijing, and Hong Kong (Griffiths et al., 1996). The prostate cancer incidence is higher in England compared to the other countries, and we have suggested that consumption of isoflavonoid-rich soy products (Adlercreutz et al., 1991e) or lignan-rich whole grain rye bread (Adlercreutz, 1990b) may prevent prostate cancer (see below).

The relatively scanty evidence we have suggests that populations with higher phytoestrogen consumption have lower cancer mortality and/or incidence than western populations with low intake of these compounds. Further discussion on the measurements of phytoestrogens in various subjects will be included in the next section.

PHYTOESTROGENS AND CANCER

There was much evidence indicating that soybean products prevent cancer long before it was suggested that this could be due to their content of phytoestrogens. In these studies it was proposed that protease inhibitors, phytic acid or ß-sitosterol, could be the active component(s) (Kennedy, 1995). The suggestion that lignans and isoflavonoids may prevent breast (Adlercreutz et al., 1982a, 1984, 1986b; Bannwart et al., 1984a; Setchell et al., 1984), prostate (Adlercreutz et al. 1991e), or colon cancer (Setchell et al., 1981; Adlercreutz 1984) has led to numerous studies to evaluate this hypothesis (see below). The relationship between phytoestrogens and cancer is far from being clear, but the view is emerging that the anticancer effects are more likely to be due to a concerted action of many types of phenolic compounds including flavonoids, isoflavonoids and lignans than to single compounds. Furthermore, it seems that the beneficial anticancer effects may perhaps not be a result of an antiestrogenic effect via the estrogen receptor by blocking the action of more potent estrogens as originally believed, but to a number of other effects at many different levels. Phytoestrogens influence hormone production, gonadotropin release, many steroid biosynthetic enzymes, and protein binding. They affect the mechanism of action of hormones and their metabolism and influence various cellular events both at the pre- and posttranscriptional level without involving binding to the estrogen receptor. Like steroids (Wehling, 1994), these compounds seem to have many nongenomic effects. Recent studies show that several single compounds, like the isoflavonoids, genistein, daidzein, perhaps equol, and also the lignan enterolactone, reach plasma levels, which would be compatible with biological activity in humans. On the other hand, practically nothing is known about the tissue concentrations of any of the compounds found in humans. This lack of information significantly affects our ability to judge the role of phytoestrogens in human health.

Phytoestrogens and Breast Cancer

Phytoestrogens and breast cancer have been discussed in many reviews (Adlercreutz et al., 1986a, 1987, 1990b, 1991d, 1992c, 1993d,e, 1995a; Setchell and Adlercreutz, 1988; Messina and Barnes, 1991; Rose, 1992, 1993; Messina et al., 1994a,b, 1995a; Hawrylewicz et al., 1995; Clarkson et al., 1995a; Griffiths et al., 1996). Table 3 shows a list of studies involving breast cancer. Some new aspects and the most recent relevant studies will be discussed here.

Two large studies show that a high intake of various soy products protects against breast cancer (Hirayma, 1986; Lee et al., 1991). In a third study (Nomura et al., 1978) the diet of the spouses of breast cancer patients among Japanese in Hawaii was studied and it was assumed that the diet of the wives was similar to that of their husbands. In one group the husbands of the breast cancer patients consumed less green tea (containing high amounts of flavonoids) and seaweeds (which may contain daidzein) (Adlercreutz et al., 1991e) than the husbands of the control women, but with regard to soybean products the consumption tended to be less in the breast cancer families, although there was no statistically significant difference. The husbands of the breast cancer patients consumed significantly more butter and red meat. In the second study group (Nomura et al., 1978) the tendency was similar; the husbands of the breast cancer patients showed significantly less consumption of Japanese soup, usually containing some soy in addition to green and yellow vegetables. There was also a tendency to lower intake of tofu in the breast cancer group. This study is, due to its nature and results, not convincing with regard to any effect of phytoestrogens. Recently an epidemiological study in China did not support the hypothesis that high intake of soy protein protects against breast cancer (Yuan et al., 1995).

Subjects with breast cancer or at high risk of breast cancer excrete low amounts of lignans and isoflavonoids (Adlercreutz et al., 1982a, 1988b), but subjects living in areas with low risk of hormone-dependent cancers and colon cancer have higher levels (Adlercreutz et al., 1986a, 1987, 1991e, 1992b, 1993c, 1995a). Higher lignan mean values in women were observed in North Karelia compared to women in the Helsinki area, which also correlates with breast cancer risk. Intake of fruits and berries in Finnish women has a strong correlation with lignan excretion (Adlercreutz et al., 1987). Berries contain the seeds of the plant, which are rich in lignan precursors; in addition the berries contain a lot of flavonoids (unpublished observation). In Finland a common tradition is consumption of rye or oat porridge with lingonberries. This traditional meal contains plenty of lignans and flavonoids.

Table 3 Studies Involving or Related to Phytoestrogens and Breast Cancer[a]

Compound or study design	Species, cell type, or population	Effect or result	Ref.
Coumestrol, genistein	Mice	Antiestrogenic effects in uterus	Folman and Pope, 1966
Genistein, coumestrol	MCF-7 cells	Competition with estradiol	Martin et al., 1978
Genistein, coumestrol	MCF-7 cells	Stimulation of proliferation	Martin et al., 1978
Coumestrol orally	rat mammary tumor model ovariectomized	No stimulation of growth	Verdeal et al., 1980
Diet and phytoestrogen excretion	American women	Low urinary excretion in women at higher risk	Adlercreutz et al., 1982a
Diet and phytoestrogen excretion	Finnish and American women	Low urinary excretion in women at higher risk	Adlercreutz et al., 1986b
Diet and phytoestrogen excretion	Chimpanzees in captivity	High urinary excretion	Adlercreutz et al., 1986c
Diet and phytoestrogen excretion	Finnish, American and Japanese women	Low urinary excretion in women at higher risk Correlation with plasma SHBG	Adlercreutz et al., 1987
Equol, enterolactone	MCF-7, T47D cells	Stimulation of growth	Welshons et al., 1987
Enterolactone (high concentrations)	MCF-7, T47D cells	Inhibition of growth	Welshons et al., 1987
Biochanin A	Hamster embryo cells	Inhibition of of benzo(a) pyrene metabolism and binding to DNA	Cassady et al., 1988
Diet and phytoestrogen excretion	Finnish women	Lowest urinary excretion in breast cancer	Adlercreutz et al., 1988b
Daidzein	ZR-75-1 cells	Inhibition	Hirano et al., 1989b
Soybean chips (also heated)	Rat mammary tumor model	Inhibition of tumor growth	Barnes et al., 1990

(continued)

Table 3 Continued

Compound or study design	Species, cell type, or population	Effect or result	Ref.
Enterolactone, enterodiol, and methylated derivatives	ZR-75-1 cells	Inhibition	Hirano et al., 1990
Diet and phytoestrogen excretion	Japanese women	High excretion associated with low risk	Adlercreutz et al., 1991e
Flaxseed	Rat mammary tumor model	Protective	Serraino and Thompson, 1991
Soy food	Women in Singapore	High intake associated with low risk	Lee et al., 1991
Genistein, biochanin A, daidzein	MCF-7, MDA-468, MCF-7-D40	Inhibition of proliferation	Peterson & Barnes, 1991
Genistein (low dose)	MCF-7 cells	Stimulation of proliferation	Peterson & Barnes, 1991
Heated soybean protein isolate	Rat mammary tumor model	Inhibition of tumor progression	Hawrylewicz et al., 1991
Enterolactone	MCF-7 cells	Inhibition of proliferation at physiological concentrations in the presence of low amounts of estradiol	Mousavi and Adlercreutz, 1992
Enterolactone	MCF-7 cells	Stimulation of proliferation in the absence of estradiol or at high concentrations in the presence of estradiol	Mousavi and Adlercreutz, 1992
Flaxseed	Rat mammary tumor model	Inhibits promotional phase	Serraino and Thompson, 1992a
Flaxseed	American women	Prolongation of luteal phase No effect on plasma estrone, estradiol, DHEAS, prolactin or SHBG	Phipps et al., 1993

Daidzein, equol, enterolactone	MCF-7 cells	Stimulation of pS2 expression; enterodiol inactive	Sathyamoorthy et al., 1994
Genistein	MCF-7 cells	Inhibition of proliferation; G(29/M cell cycle arrest; apoptosis	Pagliacci et al., 1994
Soy milk intake	American women	Decease in urinary estrogens and plasma LH	Goldin, 1994
Coumestrol, genistein	HeLa cell estrogen receptor gene construct with CAT reporter gene	Increased activity of the reporter gene	Mäkelä et al., 1994
Coumestrol, genistein	MCF-7 cells	pS2 gene expression increased	Mäkelä et al., 1994
Coumestrol, biochanin A, genistein	MCF-7 cells	Stimulation of proliferation	Mäkelä et al., 1994
Coumestrol, genistein	Purified placental 17β-hydroxy-steroid dehydrogenase type I	Inhibition	Mäkelä et al., 1994, 1995b
Coumestrol, genistein	MCF-7 cells	No inhibition of conversion of estrone to estradiol	Mäkelä et al., 1994, 1995b
Coumestrol, genistein	T47D cells	Inhibition of conversion of estrone to estradiol; stimulation of proliferation	Mäkelä et al., 1994, 1995b
Genistein	MCF-7/WT, MCF-7/ADR(R) MDA-231	Inhibition of proliferation	Monti and Sinha, 1994
Genistein	BALB/c mammary cancer (410.4), highly metastatic	Inhibition of invasion	Scholar and Toews, 1994
Soy intake	American postmeno-pausal women	No effect on plasma LH, FSH, or SHBG	Baird et al., 1995
Soy intake	British women	Plasma LH and FSH decrease; prolongation of follicular phase	Cassidy et al., 1995
Genistein	Rat mammary tumor model	Reduced marginally tumor incidence and multiplicity	Constantinou et al., 1995b

(continued)

Table 3 Continued

Compound or study design	Species, cell type, or population	Effect or result	Ref.
Genistein, daidzein or their combination	ICR mice	Inhibited sister chromatid exchanges in bone marrow cells and perhaps DNA adduct formation in liver and mammary glands	Giri & Lu, 1995
Daidzein	Rat mammary tumor model	Daidzein reduced multiplicity without affecting incidence	Constantinou et al., 1995b
Genistein (and Kievitone)	MCF-7, T47D, SKBR3	Inhibition of proliferation	Hoffman, 1995
Genistein s.c. postpartum	Rat mammary tumor model	Reduced incidence and multiplicity; lower progesterone; enhancement of maturation	Lamartiniere et al., 1995b
Diet and phytoestrogen	Male chimpanzee	"Western diet" reduces phyto-estrogen excretion to a minimum	Musey et al., 1995
Coumestrol	Immature, ovariectomized mice	Sensitizes uterus to subsequent estrogen stimulation; does not stimulate uterine cellular hyperplasia	Markaverich et al., 1995
Secoisolariciresinol diglucosidein diet	Rat mammary tumor model	Smaller tumor volume and decrease in total tumor number	Thompson et al., 1995b

Flaxseed oil with no lignans in diet	Rat mammary tumor model	Smaller tumor volume, but no decrease in total number	Thompson et al., 1995b
Flaxseed 2.5 and 5% in diet	Rat mammary tumor model	Smaller tumor volume and decrease in total tumor number	Thompson et al., 1995b
Genistein or soy protein in diet	Female CD rat	Reduced DNA binding of 7,12-dimethylbenz(a)anthracene in mammary glands	Upadhyaya and El-Bayoumy, 1995
Formononetin s.c. (high dose)	Castrated female BALB/c mice	Mammary gland proliferation	Wang et al., 1995
Soy intake	Chinese women	No effect of soy intake on breast cancer risk	Yuan et al., 1995
Soy in diet	Rat mammary cancer model	Growth inhibition, partly due to low methionine content	Hawrylewicz et al., 1995
Soy intake	American women	Reduction in plasma LH, FSH and progesterone, prolongation of cycle	Lu et al., 1996

[a]Mainly studies discussing the role of isoflavonoids in breast cancer or using pure isoflavonoids have been included, with some exceptions of special relevance to the problem.

In addition to possible inhibitory effects on cancer cell proliferation, on production of estrogens from androgens by inhibition of the aromatase enzyme, and on biological activity of sex hormones due to an effect on SHBG plasma concentration, there is evidence suggesting an effect of both lignans and isoflavonoids on the secretion of gonadotrophins, plasma levels of estradiol and progesterone, and the length of the menstrual cycle (Adlercreutz, 1990b; Orchesone et al., 1993; Phipps et al., 1993; Cassidy et al., 1995; Goldin, 1994; Lu et al., 1996). However, in a study of Baird et al. (1995) no effects on plasma gonadotropins or SHBG were observed in postmenopausal women. In the first study by Cassidy et al. (1994), the follicular phase estradiol level tended to increase during soy consumption, but when three additional subjects were included this effect disappeared (Setchell et al., 1995).

Lowering of estrogen levels could be due to inhibition of the aromatase enzyme or indirectly via decrease in plasma gonadotropin concentration. In granulosa cells TGFα is more potent than FSH in stimulating estrogen production, and genistein attenuates this effect, suggesting that the action of TGFα is mediated through protein tyrosine kinase (Gangrade et al., 1991). Another direct inhibition of estradiol synthesis in the ovaries is also possible because of the combined effect of lignans and isoflavonoids on two other enzymes involved in estradiol biosynthesis: 3ß-hydroxysteroid dehydrogenase and 17ß-hydroxysteroid oxidoreductase type I. This would reduce the risk for breast cancer and may be one explanation for the lower estrogen plasma concentrations in Asian women (Bernstein et al., 1990; Key et al., 1990; Shimizu et al., 1990). However, there seem to be some other responsible factors like short status and high fecal estrogen excretion (Adlercreutz et al., 1994b).

It is somewhat surprising that recent prospective epidemiological studies do not show any protective effect of fiber with regard to breast cancer risk (Willett et al., 1992) despite many studies suggesting that fiber or low-fat, high-fiber diets protect against the disease (e.g., Lubin et al., 1986; van't Veer et al., 1990; Rose, 1992; Baghurst and Rohan, 1994; Yuan et al., 1995). This would imply that lignans play no role in breast cancer development because of the correlation between fiber intake and lignan excretion (Adlercreutz et al., 1981, 1986a, 1987, 1988b). However, in the American study only the amount of total dietary fiber was determined. Total fiber intake in a subject seems to be unrelated to lignan content of the U.S. diet because of the low intake of grain and particularly whole-grain bread. Furthermore, total fiber intake tells us rather little about the effect on the enterohepatic circulation of estrogens affecting plasma estrogen levels (Adlercreutz, 1990b, 1991a) because it is the insoluble fiber, without or combined with a low-fat diet, that causes this effect. In agreement with the results of the study of Pryor et al. (1989) showing that fiber intake from grain during adolescence reduced breast cancer risk, it has been observed that there is a positive correlation between fiber intake and menarcheal age in girls (Hughes, 1990; de Ridder et al., 1991).

The favorable effect of grain fiber intake may be due to an effect on the enterohepatic circulation of estrogens reducing plasma estrogen levels, to a loss of energy by increased fecal excretion of fat, or to hormonal effects of phytoestrogens associated with the fiber. Further evidence is obtained from the results on the administration of bran to rats, which postponed menarche (Arts et al., 1992b). It should be mentioned that in contradiction to earlier suggestions that wheat bran does not contain any lignans (Setchell et al., 1983), we have now found that the lignan content is relatively high. Setchell et al. (1983) did not observe any increase in urinary lignan excretion following consumption of wheat bran. It must be kept in mind that it takes some time for the gut flora to adapt to bran intake and produce mammalian lignans. Early menarche is associated with increased breast cancer risk (Henderson and Bernstein, 1991). Results of studies in postmenopausal women in Boston (Adlercreutz et al., 1989a) and in premenopausal women in Helsinki (Adlercreutz et al., 1989b) showed that the main and, in fact, the only significant difference between the diets of the breast cancer patients, the omnivorous and the vegetarian control women was a lower intake of grain products in the breast cancer subjects. Comparing the diets of the Boston and Finnish women studied by us, the main difference is in the grain and grain fiber intake, being much higher in the Helsinki women with a lower risk of breast cancer than in the Boston women. This disparity results in differences in lignan excretion, but also in significant differences in fecal bulk important for the effect on enterohepatic circulation (Adlercreutz, 1991a). Further, the fat:grain fiber ratio (g/g) seems to be an important additional determinant of the enterohepatic circulation of estrogens (Adlercreutz, 1990b, 1991a, 1994b,c).

In the Finnish women the significance of the positive correlation between the excretion of lignans and isoflavonoids in urine and plasma SHBG and the negative correlations with %FE2 and %FT are stronger than the separate correlations for each group of compounds (Adlercreutz et al., 1987). In two studies we found the lowest SHBG values in breast cancer patients compared to control omnivorous and vegetarian subjects (Adlercreutz et al., 1988b, 1989a). In the second part of the Finlandia project dealing with groups of postmenopausal women studied for one year, we again found the lowest SHBG values in the breast cancer and omnivorous women and higher values in the vegetarians. In the same research there was a significant positive correlation between urinary total diphenol excretion and plasma SHBG ($r = 0.64$; $p < 0.001$) (Adlercreutz et al., 1992a), which was independent of body mass index. Plasma SHBG levels are inversely correlated to plasma insulin (Hautanen et al., 1993) and androgens, the latter hormones tending to be high in breast cancer (Adlercreutz et al., 1989a), and these androgenic hormones are likely to play a role in the pathogenesis of both breast and prostate cancer.

Our hypothesis with regard to the protective role of phytoestrogens in breast cancer (Adlercreutz et al., 1982a; Adlercreutz, 1984; Bannwart et al., 1984a; Setchell and Adlercreutz, 1988) has received support from many studies (Table 3). It was shown that powdered soybean chips, both before and after denaturation of protease inhibitors, decrease mammary tumor formation in a rat breast cancer model (Barnes et al., 1990). Chimpanzees in captivity are very resistant to breast cancer in various experimental systems and excrete very high amounts of phytoestrogens in urine (Bannwart et al., 1984a; Musey et al., 1995). In a recent study on MNU-induced mammary tumors in rats, genistein reduced marginally both tumor incidence and multiplicity but daidzein reduced only multiplicity without affecting incidence (Constantinou et al., 1995b). Peterson and Barnes (1991), Monti and Sinha (1994), Pagliacci et al. (1994), and Hoffman (1995) reported that genistein inhibits the growth of various breast cancer cells, both receptor positive and receptor negative, and leads to cell cycle arrest and apoptosis. A very interesting finding is that genistein inhibits the growth of primary human mammary epithelial (HME) cells stimulated by epidermal growth factor (EGF) or insulin much more effectively than it inhibits MCF-7 cells. It was suggested that this could be due to the lower degree of metabolism of genistein in HME cells compared to that in MCF-7 cells (Peterson and Barnes, 1994). The isoflavonoid kievitone was found to be more potent than genistein (Hoffman, 1995). Genistein inhibits invasion in Nucleopor invasion chambers in vitro of a highly metastatic mammary cancer (Scholar and Toews, 1994). It was reported that DNA binding of DMBA in mammary glands of CD rats was inhibited by genistein or soy protein intake (Upadhyaya and El-Bayoumy, 1995).

Linseed containing high amounts of lignans inhibits mammary carcinogenesis in rats (Serraino and Thompson, 1991, 1992a). Some lignans, including mammalian lignans, have been shown to be antiproliferative with regard to breast cancer cells (Hirano et al., 1989b). Genistein, found in high amounts by us in human, chimpanzee, and cow urine and in human plasma and feces, is anticarcinogenic, probably due to its inhibitory effect on protein tyrosine kinases (Akiyama et al., 1987; Markovits et al., 1989; Ogawara et al., 1989; Teraoka et al., 1989; Reddy et al., 1992) and on angiogenesis (Fotsis et al., 1993) and perhaps due to its antioxidative properties (Wei et al., 1993) (Table 3). When fed to rats, soy protein isolates, which in our experience always contain isoflavonoids, inhibit mammary tumor progression (Hawrylewicz et al., 1991). Furthermore, epidemiological evidence obtained in Singapore indicates that soy intake protects premenopausal women against breast cancer (Lee et al., 1991). However, postmenopausal women were not protected. Enterolactone alone (0.5–10 μM) stimulates the growth of MCF-7 cells, but in the presence of estradiol, added in concentrations slightly stimulating growth and in lower amounts, growth did not differ from the control or tended to be less (Mousavi and Adlercreutz, 1992). The mechanism of this phenomenon is unknown.

Some highly interesting studies show that genistein administered on days 2,4, and 6 or 16,18, and 20 postpartum in rats receiving dimethylbenz(a)anthracene (DMBA) on day 50 resulted in longer latency to first mammary tumor developed and lower incidence and multiplicities of tumors (Lamartiniere et al., 1995a,b,c). A positive correlation was found between chemoprevention and reduced cellular proliferation and differentiation of terminal ductal structures. These results are supported by the study showing that genistein and daidzein inhibit DMBA-induced sister chromatid exchanges in bone marrow cells and possibly DNA adduct formation in liver and mammary glands of female ICR mice (Giri and Lu, 1995).

Despite all of these positive effects of phytoestrogens or intake of soy or linseed on breast cancer risk, it must be kept in mind that phytoestrogens are weak estrogens and that during certain experimental conditions they are always stimulatory on cell proliferation or estrogen-dependent gene expression (Mäkelä et al., 1994). This effect seems to be mediated via the estrogen receptor. However, most, if not all the positive effects of phytoestrogens on breast cancer risk, do not seem to be mediated through a signal pathway involving the estrogen receptor. In in vivo studies no effect of the most estrogenic natural phytoestrogen coumestrol on a rat mammary tumor model could be seen (Verdeal et al., 1980). To my knowledge in only one study in castrated female BALB/c mice did very high subcutaneous doses of formononetin, which in itself is only weakly estrogenic, result in mammary gland proliferation (Wang et al., 1995). In this case it would have been of interest to measure the concentrations of the more estrogenic metabolites daidzein and equol in these animals. Formononetin does not occur normally in the human diet (only in certain clover seeds), and the amounts given were enormous.

A summary of the possible mechanisms by which phytoestrogens could protect against breast cancer is shown in Figure 4. It must be kept in mind, however, that the cancer-preventive effect of soy intake may not be only due to their isoflavonoid con-

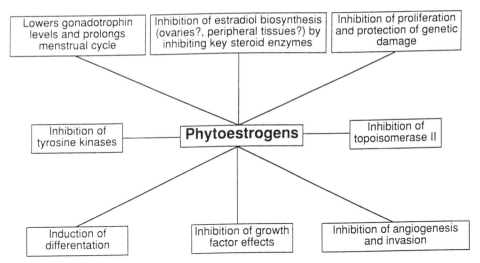

Figure 4 Suggested mechanisms by which phytoestrogens may protect against breast cancer.

tent. Many other compounds and mechanisms probably contribute to this effect (Hawrylewicz et al., 1995; Kennedy, 1995). The same is true for the phytoestrogens in whole grain products.

Phytoestrogens and Prostatic Disease

In Japan and some other Asian countries, despite of the same incidence of latent and small or noninfiltrative prostatic carcinomas as in the Western countries, prostate cancer mortality rates are low (Ota and Misu, 1958; Breslow et al., Yatani et al., 1982). In 1985, after having found very high urinary excretion of isoflavonoids in Japanese men (Adlercreutz et al., 1988c, 1991e), we suggested that this may be due to the effect of phytoestrogens, particularly isoflavonoids, inhibiting the growth of the latent cancer (Adlercreutz, 1990b; Adlercreutz et al., 1993c). In epidemiological studies fat and meat show a positive and cereals a negative association with prostate cancer mortality (Rose et al., 1986). Decreased prostate cancer risk has been found in Adventist men (Mills et al., 1989) consuming large amounts of beans, lentils, peas and some dried fruits (all dietary sources of flavonoids) and in men of Japanese ancestry in Hawaii (Severson et al., 1989) who continue to consume rice and tofu, a soybean product containing isoflavonoids (Lindner, 1967; Setchell and Welsh, 1987b).

When we measured isoflavonoids in several lots of four different tofu products by GC/MS (Dwyer et al., 1994), the daidzein and genistein contents varied from 57.3 to 117.0 and from 159 to 306 µg/g, respectively (Table 2). Soybeans contain larger amounts of phytoestrogens than tofu (Table 2). Mean consumption of soy products (except soy sauce) in Japanese men was found to be 39.2 ± 36.4 g/day, and the intake of various soy products in Japanese men and women showed a strong positive association with urinary excretion of isoflavonoids (Adlercreutz et al., 1991e). Calculation of the approximate intake of isoflavonoids in these men using mean values we have obtained for several soy products and various soybeans gave a value of about 6 mg of daidzein and 10 mg of genistein per day. Japanese men have very high levels of isoflavonoids in plasma (Adlercreutz et al., 1993c, 1994a), the highest value exceeding 2 μmol/liter. The free + sulfate fraction of the lignans is relatively high (Adlercreutz et al., 1994a). Lignan excretion in Japanese subjects showed only a positive association with the intake of pulses and beans and boiled unprocessed soybeans. Preparation of tofu products seems, therefore, to eliminate the lignan precursors from the beans. There are, however, small amounts of secoisolariciresinol in soy milk and soy protein drinks and in soy meal and flakes.

Soy has been found to have a protective effect with regard to prostatitis in rats (Sharma et al., 1992). Recently it was found that soy is protective with regard to prostatic dysplasia in a neonatally estrogenized mouse model (Mäkelä et al., 1991, 1995a), and it was reported that genistein and biochanin A, the precursor of genistein, inhibit in cell cultures the growth of both androgen-dependent and androgen-independent prostate cancer cells (Peterson and Barnes, 1993; Rokhlin and Cohen, 1995) (confirmed by us). The well-known therapeutic effect of estrogens in prostate cancer would suggest that phytoestrogens may inhibit prostate cancer cell growth during the promotional phase of the disease or may influence differentiation as shown for genistein with regard to numerous different types of leukemia and other malignant cells (see Table 5). Genistein may inhibit prostate cancer growth due to interference with growth factor tyrosine kinases (Akiyama et al., 1987; Peterson and Barnes, 1993; Rokhlin and Cohen, 1995), or topoisomerase II (Markovits et al., 1989) and by inhibiting angiogenesis (Fotsis et al., 1993, 1995a) (Table 4). Angiogenesis is probably of substantial importance when a latent cancer is converted to a more aggressive one (Furusato et al., 1994). The lignan enterolactone inhibits 5α-reductase isoenzyme 1 and 2 and genistein the isoenzyme 2 (Evans et al., 1995). A lowering of the conversion of testosterone to 5α-dihydrotestosterone (5α-DHT) would theoretically reduce the intracellular androgen effect in the prostate and slow down the growth of prostate cancer cells. In fact, it has been shown that plasma 3α,17ß-androstanediol and androsterone glucuronide, suggested to be indicators of peripheral 5α-reductase activity, are lower in young Japanese men compared to young adult white and black men in the United States (Ross et al., 1992). A reduction in LH secretion caused by intake of phytoestrogens as seen in women (Adlercreutz, 199b; Cassidy et al., 1995; Lu et al., 1996) could lead to decreased testosterone synthesis and lower prostatic cancer risk. However, plasma testosterone does not seem to be lower in Japanese men than in U.S. white males (Ross et al., 1992).

A very interesting new mechanism by which genistein might inhibit cancer initiation involves the inhibition of the release of the molecular chaperone Hsp90 from the dioxin receptor before its activation (Coumailleau et al., 1995). This effect is probably due to the inhibition of tyrosine phosphorylation (Poellinger et al., 1995). Dioxin and related compounds are very potent tumor promoters and induce a number of enzymes converting procarcinogens to carcinogens (particularly cytochrome P4501A1). The inhibition of the activation of the ligand-receptor complex and its binding to DNA would protect the cell from the effect of these very toxic compounds.

An epidemiological study showed that environmental factors such as diet of Japanese subjects can substantially impact the likelihood of developing clinically detectable prostate cancer (Shimizu et al., 1991). Abandoning the soy-rich, low-fat, low–animal protein diet for a western diet increases the risk of prostate cancer. An abundance of animal fat and beef in the diet appears to be associated with increased risk for prostate cancer (Le Marchand et al., 1994). Despite high animal fat intake the prostate cancer incidence in Finland, particularly in northeast Finland (Teppo et al., 1980), has been much lower than in the United States, but higher than in Japan (Griffiths et al. 1996).

The higher production of lignans in the gut due to a relatively high intake of whole grain products, particularly rye bread and berries in the low-incidence rural areas, may perhaps explain this phenomenon. The lignans are weaker estrogens than the isoflavonoids, but a possible protective effect may well be independent from the estrogenic effect (see above).

Recently we obtained results in rats with transplanted prostate cancer cells suggesting that rye bran or soy in the diet delays the growth of these transplants (Landström et al., 1997). These results support the view that both soy and rye bran contain compounds that inhibit prostate cancer growth.

The role of estrogens in prostate cancer development is still a controversial matter. Therefore, the role of exogenous phytoestrogens in reducing prostatic cancer risk is far from being clear. Treatment of prostate cancer with estrogens results in inhibition of cancer growth, but estrogens have also been shown to be associated with growth of both benign prostatic hyperplasia (Winter et al., 1995) and prostatic cancer (Shirai et al., 1994). Rats receiving a combination of testosterone and ethinylestradiol (EE) in the postinitiation stage of prostatic cancer stimulated progression of the cancer, but EE in combination with 5α-DHT did not. Animal experiments support the estrogen hypothesis (Leav et al., 1989; Pylkkänen et al., 1991; Mäkelä et al., 1995c). It has been shown that estrogens regulate the androgen receptor in stromal cells of the human hyperplastic prostate (Collins et al., 1994) and LNCaP human prostatic cancer cells are stimulated by estradiol (Veldscholte et al., 1994; Castagnetta et al., 1995). On the other hand, we have found that plasma estradiol concentrations are significantly lower in prostatic cancer patients compared with patients with benign prostatic hyperplasia (Rannikko and Adlercreutz, 1983). Because lignans and some weakly estrogenic flavonoids inhibit the aromatase enzyme and some isoflavonoids the 17ß-hydroxysteroid dehydrogenase type I, a possible deleterious effect of estrogens could be reduced by intake of phytoestrogens in foods. An increase in plasma SHBG concentration caused by high intake of phytoestrogens would also reduce uptake of both androgens and estrogens in the cells of target organs.

Thus, epidemiological studies as well as cell culture and animal experiments provide evidence suggesting that isoflavonoids and lignans are protective and lower the risk of prostate cancer. A summary of the possible protective mechanisms is shown in Figure 5. The protective effects seem to occur mainly during the promotional phase of the disease, because of the similar incidence of latent cancers in Japan and the western world. However, as long as the role of estrogens in prostate cancer as well as the mechanism of action of phytoestrogens in the prostate are unclear, the potential beneficial role of phytoestrogens in lowering prostate cancer risk remains a hypothesis.

Phytoestrogens and Gastrointestinal Cancer

Lignans may be protective with regard to both breast and colon cancer (Adlercreutz, 1984, 1990b). We have observed a higher urinary lignan excretion in subjects consuming a diet lowering the risk of colon cancer (Korpela et al., 1992, and unpublished) or living in areas with low colon cancer risk (Teppo et al., 1980; Adlercreutz et al., 1986a). Epidemiological evidence obtained in Japan (Watanabe and Koessel, 1993) points to lower colon cancer incidence in areas with high tofu consumption. This is now being further investigated.

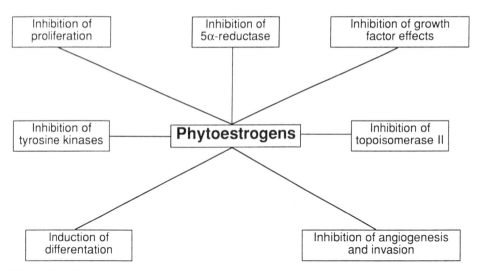

Figure 5 Suggested mechanisms by which phytoestrogens may protect against prostate cancer.

Reproductive factors seem to play a role in the development of colon cancer (Potter and McMichael, 1983; Conteas et al., 1988; Potter, 1995), and estrogen receptors have been found in normal mucosa, adenomatous polyps, and cancer cells in colon (Xu and Thomas, 1995b). Their concentration is low, but corresponds to that in normal breast tissue (Singh and Langman, 1995). Many other factors suggest some role of estrogens in colon cancer (Adlercreutz, 1984, 1990b; Xu and Thomas, 1994b, 1995b; Singh and Langman, 1995). Estrogen-replacement therapy in postmenopausal women diminishes mortality from bowel cancer (Calle et al., 1995), and antiestrogen treatment with tamoxifen may enhance risk of colon cancer in women (Rutqvist et al., 1995). These results would suggest that phytoestrogens may play some role in colon carcinogenesis. It is likely that the concentration of phytoestrogens and their precursors is high in colon ascendens, but their concentration in feces is also considerable, particularly in vegetarians (Adlercreutz et al., 1995b). In such subjects we found an excretion for single compounds of up to 2.5 μmol/24 h for equol and 6.6 μmol/24 h for enterolactone. The highest concentrations were 9.2 nmol/g for equol and 46.5 nmol/g for enterolactone. Because all these phytoestrogens are unconjugated in the colon, they most likely have biological effects.

A summary of colon cancer studies and phytoestrogens is shown in Table 4. Experimental studies in rodents support the view that phytoestrogens may reduce colon cancer risk. Five percent linseed, rich in lignans, in the diet of rats seems to protect against colon cancer (Thompson and Serraino, 1990; Serraino and Thompson, 1992b), and genistein and biochanin A inhibit the proliferation of gastric, esophagus, and colon cancer (Matsukawa et al., 1993; Yanagihara et al., 1993). Some recent studies in rat colon cancer models indicate that genistein, or in one study soy intake, decreases colon cancer risk (Helms and Gallaher, 1994; Steele et al., 1995). Cell culture studies using many different colon cancer cell lines indicate that genistein inhibits proliferation (Heruth et al., 1995; Kuo and Summers, 1995). Our own studies indicate that matairesinol and enterolactone at higher concentrations inhibit colon cancer cell proliferation, but secoisolariciresinol has a weak stimulary effect on cell growth.

Because cereal fiber–rich food is protective with regard to colon cancer and contains large amounts of lignans, it is difficult to separate the positive effect of fiber on metabolism of e.g., bile acids (Korpela et al., 1992) or production of short-chain fatty acids and the possible beneficial effects of lignans. This I have discussed to some extent relatively recently (Adlercreutz, 1990b) and will not further treat it in this connection.

Thus the evidence for a beneficial role of isoflavonoids and lignans in colon cancer is at present inconclusive and further studies are necessary. It seems, however, that cereal fiber in many different ways may decrease colon cancer risk, one of the possible mechanisms being an effect of lignans on the mucosal cells.

Other Anticancer Effects of Phytoestrogens

In Table 5 a number of other studies on anticancer effects of phytoestrogens are described, but will be only briefly discussed here.

Genistein inhibits the cell cycle and is a potent inducer of differentiation in many types of leukemia and melanoma cells and effectively inhibits their proliferation. Recently Uckun et al. (1995) in a very interesting study took advantage of this property in treating mice with B-cell precursor leukemia by targeting of genistein to the B-cell–specific receptor CD19 with a monoclonal antibody. At less than one-tenth the maximum tolerated dose, more than 99.999% of human BCP leukemia cells were killed, which led to 100% long-term event-free survival from an otherwise invariably fatal leukemia. Another interesting aspect is the reversal of non-P glycoprotein multidrug resistance caused by genistein (Takeda et al., 1992, 1994; Versantvoort et al., 1993, 1994).

Extracts from human urine containing genistein and synthetic genistein have been shown to inhibit the growth of cells from solid pediatric tumors like neuroblastomas (with both normal and enhanced MYCN expression), rhabdomyosarcomas, and Ewing's sarcomas (Schweigerer et al., 1992). A very important property of genistein is the inhibition of angiogenesis (Fotsis et al., 1993, 1995a), which has been discussed above. If genistein could be targeted to the capillaries invading solid tumors by coupling it to a suitable antibody, a general cancer drug might be developed.

PHYTOESTROGENS AND OTHER DISEASES

Phytoestrogens and Heart Disease

The effect of soy intake of lowering plasma lipids (Anderson et al., 1995b) has been well known for a long time, but until recently it was thought (Setchell, 1985; Adlercreutz, 1990b) that the phytoestrogens in soy caused or at least contributed to this effect. Several studies have demonstrated a lowered risk of cardiovascular disease in women treated postmenopausally with estrogens (Stampfer et al., 1991a; Stampfer and Colditz, 1991b; Gura, 1995) showing favorable effects on plasma lipids and risk of myocardial infarction. A logical consequence of this is that it may be the phytoestrogens in soy that have favorable effects. However, it must be kept in mind that the soy protein is a plant protein and may in itself affect lipid metabolism (Terpstra et al., 1984).

In addition to the flavonoids, genistein has been shown to be a powerful antioxidant in vitro (Pratt and Birac, 1979; Wei et al., 1993, 1995; Ferrer and Barcelo, 1994;

Table 4 Studies Involving or Related to Phytoestrogens and Prostate and Colon Cancer[a]

Compound or food	Species, cell type or population	Effect or result	Ref.
Prostate cancer (disease)			
Soybean paste soup intake	Japanese men	No association with cancer	Hirayama, 1979
Soy products in diet	Men of Japanese ancestry	Less cancer risk	Severson et al., 1989
Diet high in beans, lentils, peas, and dried fruits	Adventist American men	Less cancer risk	Mills et al., 1989
Soy in diet	Male mice	Inhibition of dysplasia	Mäkelä et al., 1991
Flaxseed	American men	No effect on sex hormone metabolism	Shultz et al., 1991
Soy in diet	Rat	Preventive effect on prostatitis	Sharma et al., 1992
Genistein	LNCaP, DU-145 cells	Inhibition of proliferation	Peterson and Barnes, 1993
Genistein	MAT-LyLu rat prostate cancer cells; PC-3 cells	Inhibition of proliferation	Naik et al., 1994
Genistein	Natural killer-rich cells	Decreased cytotoxicity with regard to K562 cells	Nishio et al., 1994
Genistein	Rats implanted with MAT-LyLu rat prostate cancer cells	No effect	Naik et al., 1994
Biochanin A, genistein formononetin, daidzein, equol	LNCaP cells	Increase in testosterone glucuronidation	Sun et al., 1995
Genistein	PC3M and PC3 cells	Inhibition of proliferation	Kyle et al., 1995
Genistein	DU145, PC3, ND1, LNCaP ALV31 and JCA1 cells	Inhibition of proliferation	Rokhlin and Cohen, 1995
Rye bran or soy	Rats with implanted prostate cancer	Inhibition of progression	Landström et al., 1997

Colon cancer

Flaxseed	Rat colon cancer model	Protective	Thompson and Serraino, 1990; Serraino and Thompson, 1992b
Soy intake	Japanese men, women	Reduced risk	Watanabe and Koessel, 1993
Genistein, biochanin A	Colon cancer cells; stomach cancer cells	Inhibition of proliferation	Yanagihara et al., 1993
Soy, genistein	Rat colon cancer model	Less aberrant crypts	Helms and Gallaher, 1994
Genistein	Colon cancer cells HCT8, SW837	Inhibition of proliferation; decrease in c-myc RNA	Heruth et al., 1995
Genistein	Multidrug-resistant colon cancer cells (S1-MI)	Reversal of resistance	Rabindran et al., 1995
Genistein	Human colon cancer cells Caco 2 and HT-29	Growth inhibition and apoptosis	Kuo and Summers, 1995
Genistein	F344 rat colon cancer model	Inhibition of aberrant colon cancer crypts	Steele et al., 1995
Matairesinol, secoisolariciresinol, enterolactone	Colon cancer cells	Inhibition of proliferation	Elomaa et al., unpublished

[a]Mainly studies discussing the role of isoflavonoids in cancer or using pure isoflavonoids have been included.

Table 5 Other Studies Related to Possible Anticancer Effects of Isoflavonoids and Lignans Detected in Humans[a]

Type of cancer	Compound	Species or cell type	Effect or result	Ref.
(VAL 12)Ha-*ras*-transformed cells	Genistein	NIH 3T3 cells	Inhibition of proliferation and of topoisomerase I and II	Okura et al., 1988
Mitogen-induced proliferation	Daidzein	Human lymphocytes	Inhibition of proliferation	Hirano et al., 1989c
Erythroleukemia	Genistein	Mouse MEL cells	Induction of differentiation	Watanabe et al., 1989
Melanoma	Genistein	Five cell lines	Induction of differentiation	Kiguchi et al., 1990
Myeloid leukemia	Genistein	Human K562 cells	Induction of differentiation	Honma et al., 1990
Leukemia	Genistein	Human HL-60, K-562 cells	Induction of differentiation	Constantinou et al., 1990
Myeloid leukemia	Genistein	ML-1, HL-60 cells	Induction of differentiation	Makishima et al., 1991
Myeloid leukemia	Genistein	MO7E cells	Inhibition of proliferation	Kuriu et al., 1991
Embryonal carcinoma	Genistein	Mouse F9 cells	Induction of differentiation	Knodo et al., 1991
Mitogen-induced proliferation	Plant lignans	Human lymphocytes	Inhibition of proliferation	Hirano et al., 1991
Leukemia	Genistein human cells	MOLT-4, HL-60	Inhibits cell cycle, progression, and growth	Traganos et al., 1992
Solid pediatric tumors	Genistein	Neuroblastoma, sarcoma rhabdomyosarcoma, 7 cell types	Inhibition of proliferation	Schweigerer et al., 1992
Liver cancer	Enterolactone	HepG2 cells	Stimulation of SHBG synthesis	Adlercreutz et al., 1992a
Non-P-glycoprotein Mediated multidrug-resistant cells	Genistein	K562/TPA	Reversal of resistance	Takeda et al., 1992, 1994
Leukemia and TPA-stimulated PMN cells	Genistein	HL-60 and TPA-stimulated PMN cells	Inhibition of hydrogen peroxide formation	Wei et al., 1993
Endothelial cells	Genistein	Many different endothelial cells	Inhibition of angiogenesis	Fotsis et al., 1993
Myeloid leukemia	Daidzein	HL-60 cells	Induction of differentiation	Jing et al., 1993

Disease/condition	Compound	Cell type/model	Effect	Reference
Non-P-glycoprotein mediated multidrug-resistant cells	Genistein	Many different cell types	Modulation of decreased drug accumulation	Versantvoort et al., 1993, 1994
Gastric cancer	Genistein	HGC-27 cells	Growth inhibition; arrest of cell cycle at G(2)M	Matsukawa et al., 1993
Monoblastic leukemia	Genistein	U937 cells	Induction of differentiation	Makishima et al., 1993
Liver cancer	Genistein	HepG2 cells	Inhibition of proliferation	Mousavi and Adlercreutz, 1993
TPA-mediated skin tumor	Genistein	Mouse	Inhibition	Bowen et al., 1993
Monocytic leukemia	Genistein	Mouse cell line	Cytotoxicity	Kanatani et al., 1993
Prolactin-simulated lymphoma	Genistein	Rat Nb2 lymphoma cells	Growth inhibition	Buckley et al., 1993
v-H-*ras*-transformed cells	Genistein	NIH 3T3 cells	Induction of differentiation	Kuo et al., 1993
Gastric cancer	Gensitein	AGS human gastric cancer cells	Growth inhibition; inhibition of EGF-receptor tyrosine kinase	Piontek et al., 1993
Neutron-induced liver cancer	Soy	Mice	Decreased the frequency and multiplicity of tumors	Ito et al., 1993
Leukemia	Genistein	Jurkat and P388 leukemia sublines with "atypical" multidrug resistance	No effect	Finlay et al., 1994
Myeloid leukemia	Genistein, matairesinol arctigenin[b]	HL-60	Growth inhibition	Hirano et al., 1994
Myeloid leukemia	Daidzein, enterodiol	HL-60	No effect	Hirano et al., 1994
T-lymphocytic leukemia	Arctigenin	MOLT-4	Growth inhibition	Hirano et al., 1994
T-cell leukemia	Genistein	Jurkat cells	Growth inhibition, apoptosis; arrest of cell cycle at G(2)M	Spinozzi et al., 1994
EGF-stimulated osteoblasts	Genistein	G292 cells	Inhibition of proliferation	Stephan and Dziak, 1994

(continued)

Table 5 Continued

Type of cancer	Compound	Species or cell type	Effect or result	Ref.
Leukemia, melanoma	Genistein	HL-60, K-562, SK-MEL-131	Inhibition of proliferation; induction of differentiation	Constantinou and Huberman, 1995
Follicular thyroid cancer, TGF-α–stimulated	Genistein	FTC133 cells	Inhibits growth and invasion	Holting et al., 1995
Erythroleukemia	Genistein	Mouse (MEL)	Induction of differentiation; growth inhibition	Jing and Waxman, 1995
Erythroleukemia	Biochanin A	Mouse (MEL)	No effect on differentiation Growth inhibition	Jing and Waxman, 1995
Melanoma with varying p 53 content	Genistein	UISO-MEL-4, -6, -7, and -8 cells	p53-negative cells more sensitive to growth inhibition	(Rauth et al., 1995
Neuroblastoma	Genistein	TS12 and SJNKP cells	Growth inhibition; induction of differentiation	Rocchi et al., 1995
Various experimental systems	Genistein		Effective scavenger of hydrogen peroxide	Record et al., 1995b

[a]Mainly studies discussing the role of isflavonoids in cancer or using pure isoflavonoids have been included.
[b]Arctigenin is included because of its very close structural similarity with matairesinol.

Record et al., 1995), and soy intake inhibits oxidation of low-density lipoprotein (LDL) isolated from the subjects and studied in vitro (Tikkanen et al., 1995). Another interesting biological effect of lignans is the increase in vitamin E activity caused by sesame seed lignans (Yamashita et al., 1992). It is not known whether mammalian lignans in the human organism have the same effect. Other mechanisms by which phytoestrogens could affect the formation of atherosclerotic plaques, including inhibition of growth factors, angiogenesis, and cell adhesion inhibition, has been discussed by Raines and Ross (1995). Genistein has been found to potentiate the induction of LDL receptor gene expression by hepatocyte growth factor (Kanuck and Ellsworth, 1995).

Phytoestrogens inhibit chemotaxis and multiple steps of the cell cycle of vascular smooth muscle cells (Shimokado et al., 1994, 1995) and inhibit endothelial cell proliferation (Fotsis, 1993; Vargas et al., 1993). These effects are probably responsible for the beneficial effect of isoflavonoid-containing soy fed to atherosclerotic monkeys after iliac artery balloon injury (Honoré et al., 1995). There was a clear inhibition of the neointimal growth. A vasodilatation effect was also observed. Genistein increases whole cell Ca^{2+}-activated K^+ channel currents in vascular endothelial cells (Xiong et al., 1995). Clarkson et al. (1995a) recently discussed this problem in connection with their own very interesting studies in rhesus monkeys showing that soy protein diet enriched in phytoestrogens, when compared with soy protein devoid of phytoestrogens, resulted in significantly improved cardiovascular risk factors in females but not in males (Anthony et al., 1995). Soy containing isoflavonoids clearly decreased LDL cholesterol in surgically postmenopausal monkeys, as well as conjugated equine estrogens together with isoflavonoid-free soy. After treatment the high-density lipoprotein (HDL) cholesterol was significantly higher and triglycerides significantly lower compared to the group fed equine estrogens plus isoflavonoid-free soy (Clarkson et al., 1995b). Isoflavonoids may also prevent or minimize plaque enlargement in atherosclerotic arteries following angioplasty surgery. The continuation of these studies is awaited with interest, and similar ones should be carried out in human subjects. Recently a study in 20 men (Gooderham et al., 1996) receiving 60 mg soy protein per day for one month containing 131 mg of isoflavonoids increased the mean plasma levels of genistein and daidzein to 907 and 498 nmol/liter, respectively, but did not affect the normal cholesterol levels.

Recently it was found that intake of dietary fiber, particularly cereal fiber, decreased substantially the risk of coronary death in smoking middle-aged Finnish men. An increase of 10g in dietary fiber intake reduced the risk by 18% (Pietinen et al., 1995). Because fiber-rich food, particularly cereal fiber, contains lignans, it is also possible that the lignans have protective effects against cardiovascular disease.

It has been shown that genistein inhibits platelet aggregation in vitro (Dhar et al., 1990; Gaudette and Holub, 1990; Asahi et al., 1992; Rendu et al., 1992), which could be due to inhibition of tyrosine kinases. This could not be confirmed after administration of soy to human male subjects (Gooderham et al., 1996). It has also been suggested that genistein and daidzein could inhibit platelet aggregation by acting as a thromboxane A_2 receptor antagonist (Nakashima et al., 1991). Tyrosine kinases are also involved in fibrinogen receptor activation, and it was shown that the aggregation inhibition was due to suppression of GPIIb-IIIa activation (Furman et al., 1994). Genistein also inhibits thrombin-evoked Ca^{2+} entry into the cell but not the Ca^{2+} release from internal stores (Asahi et al., 1992; Ozaki et al., 1993; Sargeant et al., 1993; Filipeanu et al., 1995). Ca^{2+} concentration affects smooth muscle contraction and genistein in this way relaxes smooth muscles in many different organs (Herrera et al., 1992; Yang et al., 1992, 1993; DiSalvo et al., 1994; Gokita et al., 1994; Abebe and Agrawal, 1995; Gould et al., 1995; Hatakeyama et al., 1995; Steusloff et al., 1995). Furthermore, genistein affects the regulation of Na^+ and pH in the platelets inhibiting the Na^+/H^+ exchange induced by vasopressin (Aharonovits et al., 1992).

In this connection it should be mentioned that the mammalian lignan enterolactone, in concentrations between 10^{-3} and 10^{-4} M, has been reported to display ouabain from its binding sites on the cardiac digitalis receptor and inhibits Na^+/K^+ ATPase activity, thus having digitalislike effects (Braquet et al., 1986; Fagoo et al., 1986). The doses needed were 2–3 orders of magnitude higher than those required for ouabain. Flavonoids do not seem to have any effect in this regard (Hirano et al., 1989a).

It is concluded that phytoestrogens, particularly in the form of soy intake, seem to have great potentialities for the prevention of cardiovascular diseases. Because this area is really in its very beginning of development, we must await further studies to prove the beneficial effects of soy and isoflavonoids in this respect. Recent studies suggest that lignans may also have some protective effects.

Phytoestrogens, Menopause, and Osteoporosis

An editorial by Lock (1991) and a subsequent letter from our group (Adlercreutz et al., 1992b) seem to have resulted in several studies on the effect of soy on menopausal symptoms. At present, only one seems to have been completed (Murkies et al., 1995). In this study soy flour or unbleached wheat flour was administered in a random double-blind fashion to postmenopausal women. The basal urinary mean values for daidzein and enterolactone were high, mean values being 2.13 and 6.13 µmol/24 h, respectively, for the soy group and 1.71 and 7.78 µmol/24 h for the wheat group, respectively, much higher than we have seen in any western society. Both daidzein and enterolactone values were much higher than in Finnish women. The mean equol value was low (62 nmol/24 h) in the soy group and 9.4 times higher (583 nmol/24 h) in the wheat group. After 12 weeks of treatment with soy flour the mean daidzein and equol values had increased 21- and 58-fold, respectively, and the enterolactone values 1.8-fold. In the wheat flour group the daidzein values increased 1.8-fold, and the equol values remained the same. The enterolactone mean values decreased slightly. Wheat flour in Australia is frequently fortified with soy flour, and this seems to be the only good explanation for the results. Consequently it is no surprise that hot flushes decreased significantly in both groups—in the soy group by 40% and in the wheat group by 25%—as did the symptom scores. The vaginal maturation indexes did not change. However, there was no significant difference between the groups at the end of the study with regard to symptoms, which may be explained by the soy flour content of both treatments, in spite of quantitative differences. White wheat flour contains very little phytoestrogens and only minimal amounts of lignans (Table 1). The results with regard to the effect on vaginal smear are in agreement with the results obtained in another study in postmenopausal women showing only marginal effects after intake of soy products for one month, increasing the urinary phytoestrogen concentrations 105-fold (Baird et al., 1995).

In an ongoing collaborative study with Dr. A. Brzezinski at Hadassah Medical Center in Jerusalem, it appears that a 3-month treatment with soy and whole grain in the food in postmenopausal women has resulted in an improvement of the menopausal symptoms.

Some very interesting results with regard to osteoporosis have been obtained both in rats and in women. In rats it was shown that a low dose (1 mg/day) but not a high dose (3.2 or 10 mg/day) of genistein was equivalent to conjugated equine estrogens (5 µg/day) in maintaining bone mass in ovariectomized rats (Anderson et al., 1995a). Another study in this field showed that 10 g of soy protein containing 10 mg of genistein twice daily reduced hypoestrogenic symptoms ($p < 0.004$ vs. placebo) in post- and premenopausal women and had a borderline significant effect on general symptoms ($p < 0.07$) and sleep quality ($p < 0.1$). LDL cholesterol decreased ($p < 0.01$), but HDL cholesterol and triglycerides remained unchanged with this regimen. Serum alkaline phosphatase decreased ($p < 0.03$), suggesting an effect on the bones. Recently it was found that coumestrol in an experimental system both inhibits bone resorption and stimulates bone mineralization (Tsutsumi, 1995).

In conclusion, there is great interest among women in treating menopausal symptoms by natural food products instead of using synthetic drugs containing estradiol or estriol or even equine estrogens extracted from horse urine. Subjects with treated breast cancer are a particular problem. In addition to alleviating symptoms, the food should have a protective effect with regard to coronary heart disease and osteoporosis. Climacteric symptoms are relatively rare in China and Japan (Lock, 1991; Tang, 1994), and the incidence of hip fractures is also lower than in the United States and Europe (WHO Study Group, 1994). Japanese women have lower femoral neck bone mass but still have substantially lower incidence of hip fractures. This seems to be due to shorter femoral necks and, to a lesser degree, to a smaller femoral angle (Nakamura et al., 1994). Osteoporosis is today becoming a serious problem in Japanese society and begins already in the early postmenopausal period (Tsunenari et al., 1995). This may be due to a shift from traditional diet to a more westernized one with a reduction of isoflavonoid-rich foods.

POSSIBLE TOXIC EFFECTS OF PHYTOESTROGENS

Because other chapters in this book will deal with toxicological problems involving phytoestrogens, this discussion will only deal with those aspects relevant to human health and compounds that occur in sufficient amounts in the human diet to have biological activity. Table 6 shows the relative estrogenic effects of various phytoestrogens in vitro in human endometrial cancer cells (Ishikawa cells) (Markiewicz et al., 1993). Enterolactone and enterodiol have not been studied, but the estrogenic activity of enterolactone in three other test systems involving various estrogen-sensitive cells is very low (Welshons et al., 1987), being 100,000–1,000,000 times less than that of estradiol and about 1000 times less than that of equol. Enterodiol is an even weaker estrogen. Therefore, it seems that possible toxic or negative effects of lignans definitely cannot be due to their estrogenicity.

As pointed out above, some studies on toxic effects of phytoestrogens have been carried out using fungal estrogens, which should not normally be present in foods and are therefore not discussed here. Other studies used coumestrol in high amounts. Coumestrol, first isolated by Bickoff et al. (1957), is the phytoestrogen most resembling estradiol in its action, but is 500 times less estrogenic in the Ishikawa cell test (Markiewicz et al., 1993) than this endogenous estrogen. In the mouse uterus weight test (Bickoff et al., 1962), coumestrol was 200 times less active than estrone, which

Table 6 Relative Estrogenicity of Isoflavonoids Using the Ishikawa Cell Alkaline Phosphatase Assay

Estradiol	100
Estrone	10
Coumestrol	0.2
Genistein	0.08
Equol	0.06
Daidzein	0.013
Biochanin A	<0.006
Formononetin	<0.0006

Source: Markiewicz et al., 1993.

means about 2000 times less active than estradiol. However, in the sheep uterine cystosol estrogen receptor binding test, coumestrol had only 20 times less affinity than estradiol (Shutt and Cox, 1972). Daidzein has a 1000 times lower affinity for the estrogen receptor than estradiol (Shutt and Cox, 1972), and in the Ishikawa cell test the estrogenic activity is 7700 times less than that of estradiol. Genistein binds with 110 times lower affinity to the estrogen receptor (Shutt and Cox, 1972), but in the Ishikawa cell test it is 1250 times less estrogenic than estradiol. Equol is 1700 times less estrogenic than estradiol in the Ishikawa cell test but in the receptor binding test the affinity is only 250 times lower than that of estradiol (Shutt and Cox, 1972). The binding affinity for the estrogen receptor does not seem to correlate with the other tests, suggesting that at least some phytoestrogens are not complete estrogens, as suggested recently for coumestrol (Markaverich et al., 1995). It should be emphasized that biochanin A and formononetin are metabolized in the gut to genistein and daidzein, respectively (daidzein is further metabolized to equol and to other weakly estrogenic metabolites), and therefore the binding test or Ishikawa cell test (Markiewicz et al., 1993) does not give any information as to the real biological activity of these compounds if they are not metabolized in the cells.

In the review of Price and Fenwock (1985) animal studies are described in which various factors were evaluated after administration of pure phytoestrogens. The doses of phytoestrogens, if adjusted to the weight of human subjects, were much higher than human subjects would consume in soy products, various sprouts, or other phytoestrogen-containing foods. The lowest dose used to cause uterine hypertrophy in mice corresponds to 1.5 g of genistein or 300 g of genistein/day for a woman weighing 60 kg. In mice infertility is caused in both sexes by giving 15 mg/day of genistein, corresponding to a dose in a 60-kg human subject of 45 g genistein/day. To cause testes atrophy in a male rat, 0.5% of the diet must be genistein (Price and Fenwick, 1985). Thus, the amounts used in these experiments seem not to be relevant for human health.

Three areas have been investigated for possible negative effects of phytoestrogens: (1) infertility in adults, (2) reproductive abnormalities in newborn children or development disturbance with later reproductive dysfunction, or (3) stimulation of cancer growth, particularly breast and endometrial cancer.

Infertility in Adults

Phytoestrogens in mammalian reproduction as been discussed in several reviews (Labov, 1977; Price and Fenwich, 1985; Müller et al., 1989). The first (?) review in this field was written by Bickoff (1963). Phytoestrogens may cause infertility in cattle and sheep ingesting clover or other plants. However, desert rodents and red kangaroos (*Megaleia rufa*) need phytoestrogenic plants to increase their levels of fertility (Labov,1977). The infertility syndrome in sheep called clover disease (Bennets et al., 1946; Shutt, 1976) is due to ingestion of subterranean clover with very high concentrations of phytoestrogens, particularly formononetin. The active compound is the metabolite equol formed via daidzein from formononetin. The formation of equol seems to be much higher in sheep compared to human subjects. After ingestion of clover plasma, levels of equol could reach concentrations of 300–400 μg/100 ml or 12.4–16.5 μmol/liter. The highest total plasma equol value found by us in a Japanese male subject was 0.38 μmol/liter, but most subjects had values below 0.01 μmol/liter. In the human soy feeding experiments described above, relatively large doses of soy products have been given, but the effects on gonadotropins and menstrual cycle length as well as on plasma hormones have been relatively small and obviously not sufficient to cause infertility in women. Bickoff (1963) describes one known effect in women. Near the end of World War II, people in Holland consumed large quantities of tulip bulbs due to a severe food shortage. Tulip bulbs have high estrogen activity and many women who ate them showed manifestations of estrogen imbalance including uterine bleeding and abnormalities of the menstrual cycle. The nature of the estrogens in tulip bulbs was not described.

To my knowledge there are no published long-term studies in men including determination of plasma phytoestrogens after high intake of soy products or any other phytoestrogen occurring in common foods. The highest plasma total phytoestrogen value found by us in a Japanese male subject was >2 μmol/liter of genistein. This subject was consuming his habitual diet and the plasma sample was obtained in the morning. To the best of my knowledge there are no particular fertility problems among Japanese men. In one study (Gooderham et al., 1996) soy protein was administered in the form of a beverage powder (Altima HP-20, Protein Technologies International, St. Louis) providing 80.3 mg of genistein, 35.6 mg of daidzein, and 15.1 mg of glycitein per day to 10 healthy men. The mean plasma concentrations measured by GC-MS after one month were 907 ± 245 nmol/liter for genistein and 498 ± 102 nmol/liter for daidzein, which exceed the mean values for Japanese men. A longer study using similar doses needs to be carried out measuring sperm counts, plasma hormones, and other parameters related to male fertility.

According to present knowledge the only foods containing significant amounts of coumestrol in addition to other phytoestrogens and consumed in substantial amounts by human vegetarian subjects are alfalfa and other sprouts. Our analyses by GC-MS have revealed that alfalfa sprouts contain about 4000 μg of formononetin, 110 μg of biochanin A, 60 μg of daidzein, 32 μg of secoisolariciresinol, and 45 μg of coumestrol per 100 g dry weight, total amount being about 4250 μg/100 g dry weight. Alfalfa produces a hyperestrogenic syndrome in cattle, but antiestrogenic effects in vivo have also been observed (Adler, 1962). This antiestrogenic effect could be separated from coumestrol by extracting it by chloroform, which would not remove the estrogenic activity. The author mentions that this antiestrogenic effect may be one reason for the extreme rarity of primary mammary cancer in cows and sheep. To my knowledge this antiestrogenic compound has never been identified. We have now decided to investigate this matter. Studies in vegetarian women in this laboratory revealed the presence of coumestrol in the urine of one subject. We do not think that coumestrol in the diet could cause infertility problems in human subjects even after daily intake of sprouts. The metabolism of coumestrol in humans is to my knowledge not understood.

The mode of action of phytoestrogens in causing infertility was first suggested to be due to a gonadotropin effect directly on the ovaries; later experiments suggested that they acted as releasing factors on the hypophysis. Other studies showed that coumestrol blocks the release of gonadotropins (Labov, 1977). The studies in human subjects cited above indicate that if any effect occurs on gonadotropin secretion after high intake of soy phytoestrogens, it is inhibitory. In this connection I will not discuss the possible mechanisms by which phytoestrogens could increase fertility (Labov, 1977).

In summary, there seems to be no evidence that intake of phytoestrogens from soy or other leguminosae or from sprouts may cause infertility in adult human subjects. To cause such effects enormous amounts of the particular foods must be ingested. As long as the plasma levels of phytoestrogens do not exceed the maximal ones measured in Japanese subjects consuming their habitual diets, infertility problems should not occur.

Effects in Pregnancy or in Newborn Babies Consuming Soy Milk

There has been an intense discussion in the medical (Irvine and Fitzpatrick, 1995) and lay press as well as during the Third International Conference on Phytoestrogens in Little Rock in December 1995 regarding the possible deleterious effects of soy milk formulas used for infant feeding. There are at least four important factors to be considered before an objective evaluation of this problem can be made:

1. Cow milk, which is the other alternative to mother's milk, contains phytoestrogens depending on the time of year the cow is grazing or the nature of the feed. We have analyzed samples of cow milk from Finnish cows during both winter and summer and found 0.97–335 nmol/liter of isoflavonoids (daidzein, equol, and genistein). The highest values converted to mass units are about 85 µg/liter. It is likely that the values are higher in other countries, e.g., in Australia, where the phytoestrogen content in clover is high.

2. Soy milk formulas do not as far as we know contain any steroidal estrogens. However, whole cow milk contains according to our experience 800–1700 pmol/liter of estrone + estradiol + estriol. The highest values in milk are found in the spring when the cattle start grazing outdoors. About 45% of the steroidal estrogens is estradiol with a biological activity about 1250 times higher than that of genistein (Markiewicz et al., 1993). We have calculated that the mean content of steroidal estrogens in cow milk bought from the shop corresponds with regard to estrogenic activity to about 200 µg/liter of genistein. The samples with the highest estrogen values correspond with regard to biological activity to a concentration of genistein of about 300 µg/liter. To this we have to add the content of phytoestrogens—95 µg/liter—giving maximally 385 µg/liter "genistein equivalents." But we have to remember that higher clover intake by cows in countries other than Finland will

considerably increase this value. Soy milk formula contains 400–4000 µg/liter of genistein and less daidzein as measured by GC-MS (Dwyer et al., 1994). Thus the estrogenic activity of ordinary cow milk seems to be in the same range as that of soy milk formula with the lowest values. In the study of Cruz et al. (1994) the mean urinary excretion of genistein in infants on soy milk formula was about 300 µg/liter, but on cow milk it was much less, about 30 µg/liter (approximated from a figure), but the concentrations of estrogens and phytoestrogens in the milk were not reported. Infants consume during the first 4 months of life approximately 60–130 g/day of the formula, which corresponds to a genistein intake of about 240–640 µg/day or about 150 µg/kg body weight per day. The amount of available biologically active genistein is much less (most likely less than 20%) due to destruction in the gut or lack of absorption (see below). This is the dose that should be used for calculations when genistein intake is compared to that administered *orally* to experimental animals. In an adult rat maximally 45 µg of genistein per day taken orally corresponds to the dose a human infant is obtaining from soy formula with high amounts, and in a newborn rat the corresponding amount is less than 1 µg/day. If such low doses of genistein are administered to rats in the form of soy containing the other isoflavonoids, it is unlikely that any estrogenic or antiestrogenic effects would be observed. Daidzein occurs in lower amounts and is biologically about six times less active than genistein (Markiewicz et al., 1993).

3. The bacteria of the infant gut may not hydrolyze isoflavonoid glycosides and steroidal glucuronides. However, in a 4-month study on male infants considerable absorption of soy isoflavonoids was found, which were excreted in the urine (Cruz et al., 1994). The exact urine collection dates were not indicated, and therefore it is impossible to judge at which times during these months the gut is able to absorb the phytoestrogens.

4. The newborn's liver, particularly the prematurely born infant's liver, has less glucuronyl transferase enzyme activity. Consequently the biological activity of steroids and phytoestrogens may be relatively higher in the newborn than in an older child because of less activation. Premature infants have even less capacity for glucuronidation. The captive cheetah with reduced liver glucuronyl transferase activity is sensitive to soy isoflavonoids and develops a venoocclusive disease (Setchell et al., 1987a). It is also interesting to note in this connection that there may be racial differences in phytoestrogen metabolism (Adlercreutz et al., 1994a).

The discussion on phytoestrogens in infant formulas is at first a little confusing because antiestrogenic (Irvine and Fitzpatrick, 1995) effects are considered to be the problem, with an effect on estrogen imprinting. At the very beginning of human studies we suggested that the beneficial effects of phytoestrogens could be due to an antiestrogenic effect (Adlercreutz et al., 1982a; Bannwart et al., 1984a; Setchell and Adlercreutz, 1988). It has appeared that many if not most of the effects may be related to other mechanisms (see above). However, at low doses phytoestrogens have different effects than at higher doses. Genistein at low doses in early life has nonandrogenizing, pituitary-sensitizing effects in postpubertal castrated female rats, while at higher doses genistein mimics more the typical effects of estrogens (Faber and Hughes,

1993). Only at pharmaceutical doses of 500–1000 μg/day is the volume of the sexually dimorphic nucleus (SDN-POA) increased. The low dose was 10 μg/day subcutaneously, 10 times higher than it should be to be compared to the dose calculated above for a newborn rat and corresponding to the amount ingested by an infant in early life using soy milk formula. Furthermore, the activity is much higher when given parenterally, probably at least 5 times higher. The dose used in the experiments by Faber and Hughes (1993) is consequently at least 50 times higher than that ingested orally by an infant. Earlier the same authors did a similar study in castrated male rats (Faber and Hughes, 1991) and no effect on SDN-POA was observed, but males exposed neonatally to 100 μg/day of genistein (subcutaneously) showed increased responsiveness to GnRH-induced LH release and higher doses attenuated or decreased LH secretion in response to GnRH. However, this dose is also very high and not comparable to that ingested with soy milk formula if calculated per kg body weight.

Hughes et al. (1995) summarized their views on the effects of soy protein phytoestrogens in reproductive tissues from rats, monkeys, and human subjects. Because these studies are underway, the results cannot yet be evaluated.

Studies on the effect of coumestrol and equol on the developing rat uterus were carried out using doses between 10 and 100 μg subcutaneously per newborn rat given for 5–10 days early in life (Medlock et al., 1995a,b). In these experiments the effect of equol was inconsistent with either an estrogenic or antiestrogenic action in the uterus, but that of coumestrol resembled that of DES, although 1000 times more compound was needed. Similar studies were carried out earlier giving neonatally coumestrol to mice for 5 days, resulting in reproductive abnormalities. Even very low doses of coumestrol demonstrated effects.

A big problem has been and still is that in most studies the rodent diet probably already contains plenty of phytoestrogens. A basal phytoestrogen level in the diet, not sufficient to cause effects, may more rapidly result in a positive response when a small dose of a relatively strong phytoestrogens is added. We were recently informed that this problem has not received sufficient consideration in U.S. laboratories and particularly from rodent food producers. In most studies the exact composition of the diet was described and no analyses were carried out with regard to phytoestrogen concentrations in diet or in urine or plasma. In my opinion, all such studies should be repeated giving the rodents a soy-free diet. Until then no toxicity studies can be regarded as reliable, not even those carried out with DES, as the phytoestrogens are antiestrogenic in the presence of DES.

We have, in collaboration with Dr. Shaw Watanabe at the Epidemiological Division, National Cancer Center Research Institute, Tokyo, studied the concentration of phytoestrogens in maternal and cord plasma and amniotic fluid at birth in Japanese women consuming a traditional Japanese diet. We found that the phytoestrogen levels are about as high in cord blood as in maternal plasma, and relatively high concentrations are also found in amniotic fluid. Therefore, Japanese fetuses are most likely exposed to considerable concentrations of phytoestrogens. I have not been able to find any indications of an increased rate of reproductive abnormalities in Japanese subjects that could be related to the phytoestrogen-rich diet. The amount of phytoestrogens in meconium is probably very high in the Japanese newborn, and this will be analyzed in the near future. High phytoestrogen levels during fetal life may protect females against future breast cancer.

In conclusion, there is no evidence in the literature that soy protein with iso-flavonoids could cause reproductive or other abnormalities in human subjects even if administered during the first months of life. The doses used in rodent experiments have been higher than those ingested by human infants, and, in addition, we know that most rodent feeds have high levels of phytoestrogens, which usually has not been taken into account. The milk of the dams could also contain phytoestrogens because of the high amounts in rodent feed.

Earlier calculations as to the amount obtained by newborns based on methods with blind detection systems may be erroneous. Therefore, new calculations should be made as to the amount of isoflavonoids consumed by newborns using specific measurements of the isoflavonoid concentrations in soy milk formulas. The methods are constantly developing and have recently been considerably improved. It should be emphasized that most studies in rodents showing effects on reproductive organs were carried out by subcutaneous administration of the pure compounds. Because at present our knowledge is rather limited, it seems that during the first 4 months of life soy milk formulas with high phytoestrogen concentration should be avoided, but formulas with low isoflavonoid contents may in my opinion be used. Because cow and human milk always contains both steroidal estrogens and phytoestrogens, it seems that small amounts of estrogens or phytoestrogens are not deleterious to the human fetus and may even be beneficial for growth.

Possible Cancer-Stimulating Effects

There is no evidence in the literature suggesting that phytoestrogens, in the amounts present in human foods, could stimulate already existing cancer or that such phyto-estrogens could initiate cancer. The high plasma levels in Japanese subjects having low breast, prostate, and colon cancer risk would also suggest that soy consumption is not associated with any risk. However, the Japanese have traditionally consumed a very low-fat diet and their endogenous estrogen levels tend to be lower than in western populations (see above). The combination of high phytoestrogen intake with a western diet and high plasma estrogen levels may not be beneficial. On the other hand, it seems that phytoestrogens tend to act as antiestrogens when estrogen levels are high and as estrogens when the levels are low.

Whether xenoestrogens can cause breast cancer remains an open question (Davis et al., 1993; Safe, 1995). Safe (1985) states that the "results would suggest that the linkage between dietary or environmental estrogenic compounds and breast cancer has not been made."

Measurements of plasma levels with methods more simple than GC-MS, GC, or HPLC will soon be available, making it possible to adjust the levels after intake of soy, flaxseed, or other foods with very high phytoestrogen concentrations for maximal benefit and for avoidance of negative effects. Such methods will soon be available in this laboratory.

GENERAL CONCLUSIONS

The original detection of lignans in the estrogen fraction of human urine in 1979 soon led to the finding that many diphenolic compounds are present in human body fluids and that their concentration is high in populations or groups of subjects, like vegetarians, with low cancer risk. The soy isoflavonoid genistein, in particular, has been in the center of interest because of its numerous biological activities mainly related to its inhibitory effect on tyrosine kinases. Soy has become an increasingly interesting food for the prevention of cancer, but the evidence is still not sufficient to state that soy prevents cancer or coronary heart disease. Many other factors of the diet and other environmental factors, social status, and reproductive history may be involved. However, there is no evidence suggesting that soy intake is harmful—large populations, including children, ingest soy products daily and have a low cancer risk. Recent experiments suggest that the isoflavones may prevent breast cancer when consumed before puberty. Animal experiments carried out to study the toxicity of phytoestrogens have used very high doses of the phytoestrogens or phytoestrogens not occurring normally in human food and/or the diets were not controlled and proven phytoestrogen-free before the experiments were initiated.

With regard to the lignans, it seems that we have only scratched the surface of this interesting group of compounds. Recent findings indicate that the levels of lignans in the human organism are much higher than estimated from analysis of the two identified precursors, matairesinol and secoisolariciresinol, in the diet. Their low estrogenicity combined with their presence in fiber-rich foods having other disease-preventing properties and their recently discovered inhibitory effects on steroid biosynthetic enzymes make the lignans interesting candidates for a role in prevention of cancer. However, much work is still needed to verify this hypothesis.

There is no doubt that this field is today expanding exponentially. The direct effects of phytoestrogens on many events related to cell transformation and proliferation and their modulatory role with regard to our hormonal system have put these compounds in the frontline of research in the field of nutritional effects on cancer, coronary heart disease, and other chronic diseases.

ACKNOWLEDGMENTS

The research carried out in this laboratory since 1979 has been supported by the Medical Research Council of the Academy of Finland, the Sigrid Jusélius Foundation, Helsinki, and by the Finnish Cancer Foundations, Helsinki, later by NIH grant 1 R01 CA56289-01 and a grant from Nordic Industrial Foundation, and recently by NIH grant 2 R01 CA56289-04 and EC contract FAIR-CT95-0894 and grants from the King Gustav Vth and Queen Victoria's Foundation, Sweden.

REFERENCES

Abebe, W., and Agrawal, D. K. (1995). *J. Cardiovasc. Pharmacol., 26*: 153.

Abler, A., Smith, J. A., Randazzo,, P. A., Rothenberg, P. O., and Jarrett, L. (1992). *J. Biol. Chem., 267*: 3946

Adams, N. R. (1995). *Proc. Soc. Exp. Biol. Med., 208*: 87.

Adler, J. H. (1962). *Vet. Record, 74*: 1148.

Adlercreutz, H., Martin, F., and Järvenpää, P. (1979). *Contraception, 20*: 201.

Adlercreutz, H., and Martin, F. (1980). *J. Steroid Biochem., 13*: 231.

Adlercreutz, H., Setchell, K. D. R., Lawson, A. M., Mitchell, F. L., Kirk, D. N., and Axelson, M. (1981a). *Rec. Prog. Horm. Res., 37*: 295.

Adlercreutz, H., Fotsis, T., Heikkinen, R., Dwyer, J. T., Goldin, B. R., Gorbach, S. L., A. M. L., and Setchell, K. D. R. (1981b). *Med. Biol., 59*: 259.

Adlercreutz, H., Fotsis, T., Heikkinen, R., Dwyer, J. T., Woods, M., Goldin, B. R. and Gorbach, S. L. (1982a). *Lancet, 2*, 1295.

Adlercreutz, H. and Järvenpää, P. (1982b). *J. Steroid Biochem., 17*, 639.

Adlercreutz, H. (1984). *Gastroenterology, 86*, 761.

Adlercreutz, H., Fotsis, T., Bannwart, C., Wähälä, K., Mäkelä, T., Brunow, G., and Hase, T. (1986a). *J. Steroid Biochem., 25*: 791.

Adlercreutz, H., Fotsis, T., Bannwart, C., Mäkelä, T., Wähälä, K., Brunow, G., and Hase, T. (1986b). *Advances in Mass Spectrometry-85, Proceedings of the 10th International Mass Spectrometry Conference* (J. F. J. Todd, ed.). John Wiley, Chichester, Sussex, p. 661.

Adlercreutz, H., Musey, P. I., Fotsis, T., Bannwart, C., Wähälä, K., Mäkelä, T., Brunow, G., and Hase, T. (1986c). *Clin. Chim. Acta, 158*: 147.

Adlercreutz, H., Höckerstedt, K., Bannwart, C., Bloigu, S., Hämäläinen, E., Fotsis, T., and Ollus, A. (1987). *J. Steroid Biochem., 27*: 1135.

Adlercreutz, H. (1988a). *Progress in Diet and Nutritin, Frontiers of Gastrointestinal Research 14* (C. Horwitz and P. Rozen, eds.). S. Karger, Basel, p. 165.

Adlercreutz, H., Höckerstedt, K., Bannwart, C., Hämäläinen, E., Fotsis, T., and Bloigu, S. (1988b). *Progress in Cancer Research and Therapy*, Vol. 35: *Hormones and Cancer 3* (F. Bresciani, R. J. B. King, M. E. Lippman, and J.-P. Raynaud, eds.). Raven Press, New York, p. 409.

Adlercreutz, H., Honjo, H., Higashi, A., Fotsis, T., Hämäläinen, E., Hasegawa, T. and Okada, H. (1988c). *Scand. J. Clin. Lab. Invest., 48* (suppl 190): 190.

Adlercreutz, H., Hämäläinen, E., Gorbach, S. L., Goldin, B. R., Woods, M. N., and Dwyer, J. T. (1989a). *Am. J. Clin. Nutr., 49*: 433.

Adlercreutz, H., Fotsis, T., Höckerstedt, K., Hämäläinen, E., Bannwart, C., Bloigu, S., Valtonen, A., and Ollus, A. (1989b). *J. Steroid Biochem., 34*: 527.

Adlercreutz, H. (1990a). *Ann. NY Acad. Sci. 595*: 281.

Adlercreutz, H. (1990b). *Scand. J. Clin. Lab. Invest., 50* (suppl 201): 3.

Adlercreutz, H. (1991a). *Nutrition, Toxicity, and Cancer* (I. R. Rowland, ed.). CRC Press, Boca Raton, FL, p. 137.

Adlercreutz, H. (1991b). *Clin. J. Sports Med., 1*: 149.

Adlercreutz, H., Fotsis, T., Bannwart, C., Wähälä, K., Brunow, G., and Hase, T. (1991c). *Clin. Chim. Acta, 199*: 2638.

Adlercreutz, H., Mousavi, Y., Loukovaara, M., and Hämäläinen, E. (1991d). *The New Biology of Steroid Hormones* (R. Hochberg and F. Naftolin, eds.). Serono Symposia Publications, Raven Press, New York, p. 145.

Adlercreutz, H., Honjo, H., Higashi, A., Fotsis, T., Hämäläinen, E., Hasegawa, T., and Okada, H. (1991e). *Am. J. Clin. Nutr., 54*: 1093.

Adlercreutz, H., Mousavi, Y., Clark, J., Höckerstedt, K., Hämäläinen, E., Wähälä, K., Mäkelä, T., and Hase, T. (1992a). *J. Steroid Biochem. Mol. Biol., 41*: 331.

Adlercreutz, H., Hämäläinen, E., Gorbach, S., and Goldin, B. (1992b). *Lancet, 339*: 1233.

Adlercreutz, H., Mousavi, Y., and Höckerstedt, K. (1992c). *Acta Oncol., 31*: 175.

Adlercreutz, H., Bannwart, C., Wähälä, K., Mäkelä, T., Brunow, G., Hase, T., Arosemena, P. J., Kellis, T. J., Jr., and Vickery, L. E. (1993a). *J. Steroid Biochem. Molec. Biol., 44*: 147.

Adlercreutz, H., Fotsis, T., Lampe, J., Wähälä, K., Mäkelä, T., Brunow, G., and Hase, T. (1993b). *Scand. J. Clin. Lab. Invest., 53* (suppl 215): 5.

Adlercreutz, H., Markkanen, H., and Watanabe, S. (1993c). *Lancet, 342*: 1209.

Adlercreutz, H., Carson, M., Mousawi, Y., Palotie, A., Booms, S., Loukovaara, M., Mäkelä,

T., Wähälä, K., Brunow, G., and Hase, T. (1993d). *Food and Cancer Prevention: Chemical and Biological Aspects* (K. W. Waldron, I. T. Johnson, and G. R. Fenwick, eds.). The Royal Society of Chemistry, Cambridge, p. 349.

Adlercreutz, H. (1993e). *Advances in Steroid Analysis '93* (S. Görög, ed.). Akadémiai Kiadó, Budapest, p. 511.

Adlercreutz, H., Fotsis, T. Watanabe, S., Lampe, J., Wähälä, K., Mäkelä, T., and Hase, T. (1994a). *Cancer Detect. Prev., 18*: 259.

Adlercreutz, H., Gorbach, S. L., Goldin, B. R., Woods, M. N., Dwyer, J. T., and Hämäläinen, E. (1994b). *J. Natl. Cancer Inst., 86*: 1076.

Adlercreutz, H., Gorbach, S. L., Goldin, B. R., Woods, M. N., Dwyer, J. T., Höckerstedt, K., Wähälä, K., Hase, T., Hämäläinen, E., and Fotsis, T. (1994c). *Polyc. Aromatic Comp., 6*: 261.

Adlercreutz, C. H. T., Goldin, B. R., Gorbach, S. L., Höckerstedt, K. A. V., Watanabe, S., Hämäläinen, E. K., Markkanen, M. H., Mäkelä, T. H., Wähälä, K. T., Hase, T. A., and Fotsis, T. (1995a). *J. Nutr., 125*: 757S.

Adlercreutz, H., Fotsis, T., Kurzer, M. S., Wähälä, K., Mäkelä, T. and Hase, T. (1995b). *Anal. Biochem., 225*: 101.

Adlercreutz, H. (1995c). *Environ. Health Perspect., 103* (suppl 7) : 103.

Adlercreutz, H., Fotsis, T., Kurzer, M. S., Wähälä, K., Mäkelä, T., and Hase, T. (1995e). *Anal. Biochem., 225*: 101.

Aharonovits, O., Zik, M., Livne, A. A., and Granot, Y. (1992). *Biochim. Biophys. Acta, 1112*: 181.

Ahluwalia, V. K., Bhasin, M. M., and Seshadri, T. R. (1953). *Current Sci.* (India), *22*: 363.

Akiyama, T., Ishida, J., Nakagawa, S., Ogawara, H., Watanabe, S.-I., Itoh, N., Shibuya, M., and Fukami, Y. (1987). *J. Biol. Chem., 262*: 5592.

Akiyama, T., and Ogawara, H. (1991). *Protein Phosphorylation, Pt B*. Academic Press, San Diego, p. 362.

Anderson, J. J., Ambrose, W. W., and Garner, S. C. (1995a). *J. Nutr., 125*: 799S.

Anderson, J. W., Johnstone, B. M., and Cooknewell, M. E. (1995b). *N. Engl. J. Med., 333*: 2762.

Anthony, M. S., Clarkson, T. B., Weddle, D. L., and Wolfe, M. S. (1995). *J. Nutr., 125*: 803S.

Armstrong, B. K., Brown, J. B., Clarke, H. T., Crooke, D. K., Hähnel, R., Maserei, J. R., and Ratajzak, T. (1981). *J. Natl. Cancer Inst., 67*: 761.

Arts, C. J. M., Grovers, C. A. R. L., Vandenberg, H., and Thijssen, J. H. H. (1992b). *Acta Endocrinol., 126*: 451.

Asahi, M., Yanagi, S., Ohta, S., Inazu, T., Sakai, K., Takeuchi, F., Taniguchi, T., and Yamamura, H. (1992). *FEBS Lett., 309*: 10.

Atluru, D., and Gudapaty, S. (1993a). *Vet. Immunol. Immunopathol., 38*: 113.

Atluru, D., Gudapaty, S., Odonnell, M. P., and Woloschak, G. E. (1993b). *J. Leukocyte Biol., 54*: 269.

Axelson, M., and Setchell, K. D. R. (1981). *FEBS Lett., 123*: 337.

Axelson, M., Sjövall, J., Gustafsson, B. E., and Setchell, K. D. R. (1982a). *Nature, 298*: 659

Axelson, M., Kirk, D. N., Farrant, R. D., Cooley, G., Lawson, A. M., and Setchell, K. D. R. (1982b). *Biochem. J., 201*: 353.

Axelson, M., Sjövall, J., Gustafsson, B. E., and Setchell, K. D. R. (1984). *J. Endocrinol., 102*: 49.

Ayers, D. C., and Loike, J. D. (1990). *Lignans, Chemical. Biological and Clinical Properties*. Cambridge University Press, Cambridge.

Baghurst, P. A., and Rohan, T. E. (1994). *Int. J. Cancer, 56*: 173.

Baird, D. D., Umbach, D. M., Lansdell, L., Hughes, C. L., Setchell, K. D. R., Weinberg, C. R., Haney, A. F., Wilcox, A. J., and Mclachlan, J. A. (1995). *J. Clin. Endocrinol. Metab., 80*: 1685.

Bannwart, C., Fotsis, T., Heikkinen, R., and Adlercreutz, H. (1984a). *Clin. Chim. Acta, 136*: 165.

Bannwart, C., Adlercreutz, H., Fotsis, T., Wähälä, K., Hase, T., and Brunow, G. (1984b). *Finn. Chem. Lett.*: 120.

Bannwart, C., Adlercreutz, H., Fotsis, T., Wähälä, K., Hase, T., and Brunow, G. (1986). *Advances in Mass Spectrometry—85, Proceedings of the 10th International Mass Spectrometry Conference* (J. F. J. Todd, ed.). John Wiley, Chichester, p. 661.

Bannwart, C., Adlercreutz, H., Wähälä, K., Brunow, G., and Hase, T. (1987). *International Symposium on Applied Mass Spectrometry in the Health Sciences*. Fira de Barcelona, Palau de Congressos, Barcelona, P. 169.

Bannwart, C., Adlercreutz, H., Wähälä, K., Brunow, G., and Hase, T. (1988a). *11th International Congress on Mass Spectrometry*, Bordeaux August 29–September 2, 1988.

Bannwart, C., Adlercreutz, H., Wähälä, K., Kotiaho, T., Hesso, A., Brunow, G., and Hase, T. (1988b). *Biomed. Environ. Mass Spectrom., 17*: 1.

Bannwart, C., Adlercreutz, H., Wähälä, K., Brunow, G., and Hase, T. (1989). *Clin. Chim. Acta, 180*: 293.

Barnes, S., Grubbs, C., Setchell, K. D. R., and Carlson, J. (1990). *Mutagens and Carcinogens in the Diet* (M. W. Pariza, H. U. Aeschbacher, J. S. Eton, and S. Sato, eds.). Wiley-Liss, Inc. New York, p. 239.

Barnes, S., Peterson, G., Grubbs, C., and Setchell, K. (1994). *Diet and Cancer: Markers, Prevention, and Treatment* (M. M. Jacobs, ed.). Plenum Press, New York, p. 135.

Barnes, S., Peterson, T. G., and Coward, L. (1995). *J. Cell Biochem.* (Suppl. 22): 181.

Batterman, T. J., Hart, N. K., and Lamberton, J. A. (1965). *Nature* 4980: 509.

Bennets, H. W., Underwood, E. J. and Shier, F. L. (1946). *Aust. Vet. J., 22*: 2.

Bernstein, L., Yuan, J.-M., Ross, R. K., Pike, M. C., Hanisach, R., Lobo, R., Stanczyk, F., Gao, Y.-T., and Henderson, B. E. (1990). *Cancer Causes Control, 1*: 51.

Bickoff, E. M., Booth, A. N., Lyman, R. I., Livingstone, A. L., Thompson, C. R., and DeEds, F. (1957). *Science, 126*: 969.

Bickoff, E. M., Livingston, A. L., Hendrickson, A. P., and Booth, A. N. (1962). *Agric. Food Chem., 10*: 410.

Bickoff, E. M. (1963) *Proc. 22nd Biology Colloquium, Physiology of Reproduction*. Oregon State University, p. 93.

Bindon, B. M., Adams, N. R., and Piper, L. R. (1982). *Anim. Reprod. Sci., 5*: 7.

Bokkenheuser, V. D., Shackleton, C. H. L., and Winter, J. (1987). *Biochem. J., 248*: 933.

Bokkenheuser, V. D., and Winter, J. (1988). *Plant Flavonoids in Biology and Medicine II: Biochemical, Cellular, and Medicinal Properties*. Alan R. Liss, Inc., New York, p. 143.

Borriello, S. P., Setchell, K. D. R., Axelson, M., and Lawson, A. M. (1985). *J. Appl. Bacteriol., 58*: 37.

Bowen, R., Barnes, S., and Wei, H. (1993). *Proc. Am. Assoc. Cancer Res., 34*: 555.

Bradbury, R. B., and White, D. E. (1954). Vitamins and Hormones. *Advances in Research and Applications* (R. S. Harris, G. F. Marrian, and K. V. Thimann, eds.). Academic Press, Inc., New York, *XII*: 207.

Braden, A. W. H., Hart, N. K., and Lamberton, J. A. (1967). *Aust. J. Agric. Res., 18*: 335.

Braden, A. W. H., Thaun, R. I., and Shutt, D. A. (1971). *Aust. J. Agric. Res., 22*: 663.

Braquet, P., Senn, N., Robin, J.-P., Esanu, A., Godfraind, T., and Garay, R. (1986). *Pharmacol. Res. Commun., 18*: 227.

Breslow, N. E., Chan, C. W., Dhom, G., Drury, R. A. B., Frnaks, L. M., Gellei, B., Lee, Y. S., Lundberg, S., Sparke, B., Sternby, N. H., and Tulinius, M. (1977). *Int. J. Cancer, 20*: 680.

Buckley, A. R., Buckley, D. J., Gout, P. W., Liang, H. Q., Rao, Y. P. and Blake, M. J. (1993). *Mol. Cell. Endocrinol., 98*: 17.

Burroughs, C. D. (1995). *Proc. Soc. Exp. Biol. Med., 208*: 78.

Calle, E. E., Miracle-McMahill, H. L., Thun, M. J., and Heath, C. W. (1995). *J. Natl. Cancer Inst., 87*: 517.

Campbell, D. R., and Kurzer, M. S. (1993). *J. Steroid Biochem. Mol. Biol., 46*: 381.

Cassady, J. M., Zennie, T. M., Chae, Y.-H., Ferin, M. A., Portuondo, N. E., and Baird, W. M. (1988). *Cancer Res., 48*: 6257.

Cassidy, A., Bingham, S., and Setchell, K. D. R. (1994). *Am. J. Clin. Nutr., 60*: 333.

Cassidy, A., Bingham, S., and Setchell, K. (1995). *Br. J. Nutr., 74*: 587.

Castagnetta, L. A., Miceli, M. D., Sorci, C. M. G., Pfeffer, U., Farruggio, R., Oliveri, G., Calabro, M., and Carruba, G. (1995). *Endocrinology, 136*: 2309.

Clarkson, T. B., Anthony, M. S., and Hughes, C. L. (1995a). *Trends Endocrinol. Metab., 6*: 11.

Clarkson, T. B., Anthony, M. S., and Hughes, C. L., Jr. (1995b). *Third International Conference on Phytoestrogens*, National Center for Toxicological Research, Food and Drug Administration, Little Rock, AR.

Collins, A. T., Zhiming, B., Gilmore, K., and Neal, D. E. (1994). *J. Endocrinol., 143*: 269.

Constantinou, A., Kiguchi, K., and Huberman, E. (1990). *Cancer Res., 50*: 2618.

Constantinou, A., Mehta, R., Runyan, C., Rao, K., Vaughan, A., and Moon, R. (1995a). *J. Nat. Prod.-Lloydia, 58*: 217.

Constantinou, A., Thomas, C., Mehta, R., Runyan, C., and Moon, R. (1995b). *Proc. Am. Assoc. Cancer Res., 36*: 115.

Constantinou, A., and Huberman, E. (1995c). *Proc. Soc. Exp. Biol. Med., 208*: 109.

Conteas, C. N., Desai, T. K., and Arlow, F. A. (1988). *Gastroenterol. Clin. North Am., 17*: 761.

Coumailleau, P., Poellinger, L., Gustafsson, J.-Å., and Whitelaw, M. L. (1995). *J. Biol. Chem., 270*: 25291.

Coward, L., Barnes, N. C., Setchell, K. D. R., and Barnes, S. (1993). *J. Agric. Food Chem., 41*: 1961.

Cruz, M. L. A., Wong, W. W., Mimouni, F., Hachey, D. L., Setchell, K. D. R., Klein, P. D., and Tsang, R. C. (1994). *Pediatr. Res., 35*: 135.

Davis, D. L., Bradlow, H. L., Wolff, M., Woodruff, T., Hoel, D. G., and Anton-Culver, H. (1993). *Environ. Health. Perspect., 101*: 372.

de Ridder, C. M., Thijssen, J. H. H., Van't Veer, P., Van Duuren, R., Bruning, P. F., Zonderland, M. L., and Erich, W. B. M. (1991). *Am. J. Clin. Nutr., 54*: 805.

Dean, N. M., Kanemitsu, M., and Boynton, A. L. (1989). *Biochem. Biophys. Res. Commun., 165*: 795.

Dehennin, L., Reiffsteck, A., Joudet, M., and Thibier, M. (1982). *J. Reprod. Fertil., 66*: 305.

Densmore, C. L., Schauweker, T. H., Gregory, R. R., Webb, B., Garcia, E., and Markaverich, B. M. (1994). *Steroids, 59*: 282.

Dhar, A., Paul, A. K., and Shukla, S. D. (1990). *Mol. Pharmacol., 37*: 519.

DiSalvo, J., Pfitzer, G., and Semenchuk, L. A. (1994). *Can. J. Physiol. Pharmacol., 72*: 1434.

Dunn, J. E. (1975). *Cancer Res., 35*: 3240.

Dwyer, J. T., Goldin, B. R., Saul, N., Gualtieri, L., Barakat, S., and Adlercreutz, H. (1994). *J. Am. Diet Assoc., 94*: 739.

Eich, E., Pertz, H., Kaloga, M., Schulz, J., Fesen, M. R., Mazumder, A., and Pommier, Y. (1996). *J. Med. Chem., 39*: 86.

Eldridge, A. (1982). *J. Agric. Food Chem., 30*: 353.

Eldridge, A., and Kwolek, W. F. (1983). *J. Agric. Food Chem., 31*: 394.

Evans, B. A. J., Griffiths, K., and Morton, M. S. (1995). *J. Endocrinol., 147*: 295.

Faber, K. A., and Hughes, C. L. (1991). *Biol. Reprod., 45*: 649.

Faber, K. A., and Hughes, C. L. (1993). *Reprod. Toxicol., 7*: 35.

Fagoo, M., Braquet, P., Roibin, J. P., Esanu, A., and Godfraind, T. (1986). *Biochem. Biophys. Res. Commun., 134*: 1064.

Farmakalidis, E., and Murphy, P. A. (1985). *J. Agric. Food Chem., 33*: 385.

Farnsworth, N. R., Bingel, A. S., Cordell, G. A., and Crane, F. A. (1975). *J. Pharm. Sci., 64*: 717.

Ferrer, M. A., and Barcelo, A. R. (1994). *J. Plant Physiol., 144*: 64.

Filipeanu, C. M., Brailoiu, E., Huhurez, G., Slatineanu, S., Baltatu, O., and Branisteanu, D. D. (1995). *Eur. J. Pharmacol., 281*: 29.

Finlay, E. M. H., Wilson, D. W., Adlercreutz, H., and Griffiths, K. (1991). *J. Endocrinol., 129*: (abstract 49).

Finlay, G. J., Holdaway, K. M., and Baguley, B. C. (1994). *Oncol. Res., 6*: 33.

Folman, Y., and Pope, G. S. (1966). *J. Endocrinol., 34*: 215.

Folman, Y., and Pope, G. S. (1969). *J. Endocrinol., 44*: 213.

Fortunati, N., Fissore, F., Fazzari, A., Becchis, M., Comba, A., Catalano, G. M., Berta, L., and Frairia, R. (1996). *Endocrinol., 137*: 686.

Fotsis, T., Pepper, M., Adlercreutz, H., Fleischmann, G., Hase, T., Montesano, R., and Schweigerer, L. (1993). *Proc. Natl. Acad. Sci. USA, 90*: 2690.

Fotsis, T., Pepper, M., Adlercreutz, H., Hase, T., Montesano, R., and Schweigerer, L. (1995a). *J. Nutr., 125*: S790.

Fotsis, T., Pepper, M. S., Aktas, E., Breit, S., Rasku, S., Adlercreutz, H., Wähälä, K., Montesano, R., and Schweigerer, L. (1997). *Cancer Res. 57*: 2916.

Franke, A. A., Custer, L. J., Cerna, C. M., and Narala, K. K. (1994a). *J. Agric. Food Chem., 42*: 1905.

Franke, A. A., and Custer, L. J. (1994b). *J. Chromatogr. B-Bio. Med. Appl., 662*: 47.

Furman, M. I., Grigoryev, D., Bray, P. F., Dise, K. R., and Goldschmidt-Clermont, P. J. (1994). *Circ. Res., 75*: 172.

Furusato, M., Wakui, S., Sasaki, H., Ito, K., and Ushigome, S. (1994). *Br. J. Cancer, 70*: 1244.

Gangrade, B. K., Davis, J. S., and May, J. V. (1991) *Endocrinology, 129*: 2790.

Gansser, D., and Spiteller, G. (1995). *Planta Med., 61*: 138.

Garai, J., Tiller, A. A., and Clark, J. H. (1992a). *Steroids, 57*: 183.

Garai, J., and Clark, J. H. (1992b). *Steroids, 57*: 248.

Garai, J., and Clark, J. H. (1994). *J. Steroid Biochem. Mol. Biol., 49*: 161.

Gaudette, D. C., and Holub, B. J. (1990). *Biochem. Biophys. Res. Commun., 170*: 238.

Gavaler, J. S., Galvao-Teles, A., Monteiro, E., Van Thiel, D. H., and Rosenblum, E. (1991). *Hepatology, 14*: 87A.

Gavaler, J. S., Rosenblum, E. R., Van Thiel, D. H., Eagon, P. K., Pohl, C. R., Campbell, I. M., and Gavaler, J. (1987). *Alcohol. Clin. Exp. Res., 11*: 399.

Giri, A. K., and Lu, L. J. W. (1995). *Cancer Lett., 95*: 125.

Gokita, T., Miyauchi, Y., and Uchida, M. K. (1994). *Gen. Pharmacol., 25*: 1673.

Goldin, B. R. (1994). *Dietary Phytoestrogens: Cancer Cause or Prevention?* National Cancer Institute, Herndon, VA.

Goldin, B. R., Adlercreutz, H., Gorbach, S. L., Woods, M. N., Dwyer, J. T., Conlon, T., Bohn, E., and Gershoff, S. N. (1986). *Am. J. Clin. Nutr., 44*: 945.

Gooderham, M. J., Adlercreutz, H., Ojala, S. T., Wähälä, K., and Holub, B. J. (1996). *J. Nutr., 126*: 2000.

Gould, E. M., Rembold, C. M., and Murphy, R. A. (1995). *Am. J. Physiol.-Cell Physiol., 37*: C1425.

Griffiths, K., Adlercreutz, H., Boyle, P., Denis, L., Nicholson, R. I., and Morton, M. S. (1996). *Nutrition and Cancer*. ISIS Medical Media, Oxford.

Gura, T. (1995). *Science, 269*: 771.

György, P., Murata, K., and Ikehata, H. (1964). *Nature, 203*: 870.

Hackett, A. M. (1986). *Plant Flavonoids in Biology and Medicine: Biochemical, Pharmacological, and Strcuture-Activity Relationships* (V. Cody, C. V. Middleton Jr., and J. B. Harborne, eds.). Alan R. Liss, Inc., New York, p. 125.

Hallmans, G., Zhang, J.-X., Lundin, E., Landström, M., Sylvan, A., Åman, P., Adlercreutz, H., Härkönen, H., and Bach Knudsen, K. E. (1995) *International Rye Symposium: Technology and Products*. VTT Symposium 161, Technical Research Centre of Finland, Espoo, p. 61.

Hammond, G. L. (1995). *Trends Endocrinol. Metab., 6*: 298.

Han, B. H., Kang, Y. H., Yang, H. O., and Park, M. K. (1994). *Phytochemistry, 37*: 1161.

Hatakeyama, N., Wang, Qu., Goyal, R. K., and Akbarali, H. I. (1995). *Am. J. Physiol.-Cell Physiol., 37*: C877.

Hautanen, A., Sarna, S., Pelkonen, R., and Adlercreutz, H. (1993). *Metabolism, 42*: 870.

Hawrylewicz, E. J., Huang, H. H., and Blair, W. H. (1991). *J. Nutr., 121*: 1693.

Hawrylewicz, E. J., Zapata, J. J., and Blair, W. H. (1995). *J. Nutr., 125*: S698.

Helms, J. R., and Gallaher, D. D. (1994). *J. Nutr., 125* (suppl): 802S.

Henderson, B. E., Ross, R., and Bernstein, L. (1988). *Cancer Res., 48*: 246.

Henderson, B. E., and Bernstein, L. (1991). *Breast Cancer Res. Treat., 18*: S11.

Herrera, M. D., Marheunda, E., and Gibson, A. (1992). *Planta Med., 58*: 314.

Heruth, D. P., Wetmore, L. A., Leyva, A., and Rothberg, P. G. (1995). *J. Cell Biochem., 58*: 83.

Hettle, J. A., and Kitts, W. D. (1983/1984). *Anim. Reprod. Sci., 6*: 233.

Hill, T. D., Dean, N. M., Mordan, L. J., Lau, A. F., Kanemitsu, M. Y., and Boynton, A. L. (1990). *Science, 248*: 1660.

Hirano, T., Oka, K., and Akiba, M. (1989a). *Life Sci., 45*: 1111.

Hirano, T., Oka, K., and Akiba, M. (1989b). *Res. Commun. Chem. Pathol. Pharmacol., 64*: 69.

Hirano, T., Oka, K., Kawashima, E., and Akiba, M. (1989c). *Life Sci., 45*: 1407.

Hirano, T., Fukuoka, K., Oka, K., Naito, T., Hosaka, K., Mitsuhashi, H., and Matsumoto, Y. (1990). *Cancer Invest., 8*: 595.

Hirano, T., Wakasugi, A., Oohara, M., Oka, K., and Sashida, Y. (1991). *Planta Med., 57*: 331.

Hirano, T., Gotoh, M., and Oka, K. (1994). *Life Sci., 55*: 1061.

Hirayama, T. (1979). *Natl. Inst. Monogr., 53*: 149.

Hirayam, T. (1986). *Diet. Nutrition and Cancer* (Y. Hayashi, M. Nagao, T. Sugimura, S. Takayama, L. Tomatis, L. W. Wattenberg, and G. N. Wogan, eds.). *Jpn Sci. Soc. Press* and VNU Sci. Press, Tokyo, Utrecht, p. 41.

Hoffman, R. (1995). *Biochem. Biophys. Res. Commun., 211*: 600.

Hollman, P. C. H., Devries, J. H. M., Vanleeuwen, S. D., Mengelers, M. J. B., and Katan, M. B. (1995). *Am. J. Clin. Nutr., 62*: 1276.

Holting, T., Siperstein, A. E., Clark, O. H., and Duh, Q. Y. (1995). *Eur. J. Endocrinol., 132*: 229.

Honma, Y., Okabe-Kado, J., Kasukabe, T., Hozumi, M., and Umezawa, K. (1990). *Jpn. J. Cancer Res., 81*: 1132.

Honoré, E. K., Williams, J. K., Anthony, M. S., and Clarkson, T. B. (1995). *Third International Conference on Phytoestrogens*, National Center for Toxicological Research, Food and Drug Administration, Little Rock, AR.

Huang, J. M., Nasr, M., Kim, Y. H., and Matthews, H. R. (1992). *J. Biol. Chem., 267*: 15511.

Hughes, C., Tansey, G., Cline, J. M., and Lessey, B. (1995). *Third International Conference on Phytoestrogens*, National Center for Toxicological Research, Food and Drug Administration, Little Rock, AR.

Hughes, R. E. (1990). *Dietary Fibre Perspectives—Reviews and Bibliography* (A. R. Leeds, ed.) John Libbey & Co., London: 76.

Hutchins, A. M., Lampe, J. W., Martinit, M. C., Campbell, D. R., and Slavin, J. L. (1995). *J. Am. Diet Assoc., 95*: 769.

Ibrahim, A.-R., and Abul-Hajj, Y. J. (1990). *J. Steroid Biochem. Mol. Biol., 37*: 257.

Imoto, M., Yamashita, T., Sawa, T., Kurasawa, S., Naganawa, H., Takeuchi, T., Bao-quan, Z. and Umezawa, K. (1988). *FEBS Lett., 230*: 43.

Irvine, C., and Fitzpatrick, M. (1995). *NZ Med. J., 108*: 208.

Ito, A., Watanabe, H., and Basaran, N. (1993). *Int. J. Oncol., 2*: 773.

Jing, Y. K., Nakaya, K., and Han, R. (1993). *Anticancer Res., 13*: 1049.

Jing, Y. K., and Waxman, S. (1995). *Anticancer Res., 15*: 1147.

Joannou, G. E., Kelly, G. E., Reeder, A. Y., Waring, M., and Nelson, C. (1995). *J. Steroid Biochem. Mol. Biol., 54*: 167.

Jonas, J. C., Plant, T. D., Gilon, P., Detimary, P., Nenquin, M., and Henquin, J. C. (1995). *Br. J. Pharmacol., 114*: 872.

Jordan, V. C., Koch, R., and Bain, R. R. (1985). *Estrogens in the Environment II. Influences on Development* (J. A. McLachlan, ed.). Elsevier, New York, p. 221.

Kanatani, Y., Kasukabe, T., Hozumi, M., Motoyoshi, K., Nagata, N., and Honma, Y. (1993). *Leuk. Res., 17*: 847.

Kanuck, M. P., and Ellsworth, J. L. (1995) *Life Sci., 57*: 1981.

Kellis, J. T., Jr. and Vickery, L. E. (1984a). *Science, 225*: 1032.

Kelly, G. E., Joannou, G. E., Reeder, A. Y., Nelson, C., and Waring, M. A. (1995). *Proc. Soc. Exp. Biol. Med., 208*: 40.

Kelly, G. E., Nelson, C., Waring, M. A., Joannou, G. E., and Reeder, A. Y. (1993). *Clin. Chim. Acta, 223*: 9.

Kennedy, A. R. (1995). *J. Nutr., 125*: S733.

Keung, W. M., and Vallee, B. L. (1993a). *Proc. Natl. Acad. Sci. USA, 90*: 1247.

Keung, W. M. (1993b). *Alcohol. Clin. Exp. Res., 17*: 1254.

Keung, W. M. (1995). *Biochem. Biophys. Res. Commun., 215*: 1137.

Key, T. J. A., Chen, J., Wang, D. Y., Pike, M. C., and Boreham, J. (1990). *Br. J. Cancer, 62*: 631.

Kiguchi, K., Constantinou, A. I., and Huberman, E. (1990). *Cancer Commun., 2*: 271.

Kirkman, L. M., Lampe, J. W., Campbell, D. R., Martini, M. C., and Slavin, J. L. (1995). *Nutr. Cancer, 24*: 1.

Kitts, W. D., Newsome, F. E., and Runeckles, V. C. (1983). *Can. J. Anim. Sci., 63*: 823.

Kliewer, E. V., and Smith, K. R. (1995). *J. Natl. Cancer Inst., 87*: 1154.

Klus, K., Borgerpapendore, G., and Barz, W. (1993). *Phytochemistry, 34*: 979.

Knight, D. C., and Eden, J. A. (1995). *Maturitas, 22*: 167.

Kondo, K., Tsuneizumi, K., Watanabe, T., and Oishi, M. (1991). *Cancer Res., 51*: 5398.

Korpela, J. T., Adlercreutz, H., and Turunen, M. J. (1988). *Scand. J. Gastroenterol., 23*: 277.

Korpela, J. T., Korpela, R., and Adlercreutz, H. (1992). *Gastroenterology, 103*: 1246.

Krol, W., Czuba, Z. P., Threadgill, M. D., Cunningham, B. D. M., and Pietsz, G. (1995). *Biochem. Pharmacol., 50*: 1031.

Kudou, S., Shimoyamada, M., Imura, T., Uchida, T., and Okubo, K. (1991). *Agric. Biol. Chem., 55*: 859.

Kuo, M. L., Kang, J. J., and Yang, N. C. (1993). *Cancer Lett., 74*: 197.

Kuo, S.-M. and Summers, R. (1995). *Proc. Am. Assoc. Cancer Res. 36*, 594.

Kuriu, A., Ikeda, H., Kanakura, Y., Griffin, J. D., Druker, B., Yagura, H., Kitayama, H., Ishikawa, J., Nishiura, T., Kanayama, Y., Yonezawa, T., and Tarui, S. (1991). *Blood, 78*: 2834.

Kurzer, M. S., Lampe, J. W., Martinit, M. C., and Adlercreutz, H. (1995) *Cancer Epidemiol. Biom. Prev., 4*: 353.

Kyle, E., Bergan, R. C., and Neckers, L. (1995). *Proc. Am. Assoc. Cancer Res., 36*: 388.

Labov, J. B. (1977). *Comp. Biochem. Physiol., 57A*: 3.

Lamartiniere, C. A., Moore, J., Holland, M., and Barnes, S. (1995a). *Proc. Soc. Exp. Biol. Med., 208*: 120.

Lamartiniere, C. A., Moore, J. B., Brown, N. M., Thompson, R., Hardin, M. J., and Barnes, S. (1995b). *Carcinogenesis, 16*: 2833.

Lamartiniere, C. A., Murrill, W., and Brown, N. (1995c). *Proc. Am. Assoc. Cancer Res., 36*: 592.

Lampe, J. W., Martini, M. C., Kurzer, M. S., Adlercreutz, H., and Slavin, J. L. (1994). *Am. J. Clin. Nutr., 60*: 122.

Lampe, J. W., Karr, S. C., Hutchins, A. M., and Slavin, J. L. (1995). *Third International Conference on Phytoestrogens*, National Center for Toxicological Research, Food and Drug Administration, Little Rock, AR.

Landström, M., Zhang, J.-X., Åman, P., Bergh, A., Damber, J.-E., Hallmans, G., Mazur, W., Wähälä, K., and Adlercreutz, H. (1997), submitted.

Le Cam, A. (1991). *Pathol. Biol., 39*: 796.

Le Marchand, L., Kolonel, L. N., Wilkens, L. R., Myers, B. C., and Hirohata, T. (1994). *Epidemiology, 5*: 276.

Leav, I., Merk, F., Kwan, P., and Ho, S. M. (1989). *Prostate, 15*: 23.

Lee, Y. H., Howe, R. S., Sha, S.-J., Teuscher, C., Sheehan, D. M., and Lyttle, C. R. (1989). *Endocrinology, 125*: 3022.

Lee, H. P. Gourley, L., Duffy, S. W., Estève, J., Lee, J., and Day, N. E. (1991). *Lancet, 337*: 1197.

Lehtola, L., Lehväslaiho, H., Koskinen, P., and Alitalo, K. (1992). *Acta Oncol., 31*: 147.

Linassier, C., Pierre, M., Le Pecq, J.-B., and Pierre, J. (1990). *Biochem. Pharmacol., 39*: 187.

Lindner, H. (1976). *Environ. Quality Safety, 5*: 151.

Lindner, H. R. (1967). *Aust. J. Agric. Res., 18*: 305.

Lipworth, L., Adami, H. O., Trichopoulos, D., Carlestrom, K., and Mantzoros, C. (1996). *Epidemiology, 7*: 96.

Lock, M. (1991). *Lancet, 337*: 1270.

Lookhart, G. L., Jones, B. L., and Finney, K. F. (1978). *Cereal Chem., 55*: 967.

Loukovaara, M., Carson, M., and Adlercreutz, H. (1995a). *J. Clin. Endocrinol. Metab., 80*: 160.

Loukovaara, M., Carson, M., and Adlercreutz, H. (1995b). *Biochem. Biophys. Res. Commun. 206*: 895.

Loukovaara, M., Carson, M., and Adlercreutz, H. (1995c). *J. Steroid Biochem. Mol. Biol., 54*: 141.

Loukovaara, M., Carson, M., Palotie, A., and Adlercreutz, H. (1995d). *Steroids, 60*: 656.

Lu, L.-E., W., Grady, J. J., Marshall, M. V., Ramanujam, V. M. S., and Anderson, K. E. (1995a). *Nutr. Cancer, 24*: 311.

Lu, L.-J.W., Broemeling, L. D., Marshall, M. V., and Ramanujam, V. M. S. (1995b). *Cancer Epidemiol. Biomark. Prev., 4*: 497.

Lu, L.-E.W., Anderson, K. E., Grady, J. J., and Nagamani, M. (1996). *Cancer Epidemiol. Biomed. Prev., 5*: 63.

Lubin, F., Wax, Y., and Modan, B. (1986). *J. Natl. Cancer Inst., 77*: 605.

Lundh, T.J.-O., Pettersson, H., and Kiessling, K.-H. (1988). *J. Assoc. Off. Anal. Chem., 71*: 938.

Lyttle, C. R., Medlock, K. L., and Sheehan, D. M. (1984). *J. Biol. Chem., 259*: 2697.

Makishima, M., Honma, Y., Hozumi, M., Sampi, K., Hattori, M., Umezawa, K., and Motoyoshi, K. (1991). *Leukemia Res., 15*: 701.

Makishima, M., Honma, Y., Hozumi, M., Nagata, N., and Motoyoshi, K. (1993). *Biochim. Biophys. Acta, 1176*: 245.

Markaverich, B. M., and Clark, J. H. (1979). *Endocrinology, 105*: 1458.

Markaverich, B. M., Upchurch, S., and Clark, J. H. (1981). *J. Steroid Biochem., 14*: 125.

Markaverich, B. M., Roberts, R. R., Alejandro, M. A., and Clark, J. H. (1984). *Cancer Res., 44*: 1515.

Markaverich, B. M., Roberts, R. R., Alejandro, M. A., and Clark, J. H. (1986). *J. Biol. Chem., 261*: 142.

Markaverich, B. M., Gregory, R. R., Alejandro, M.-A., Clark, J. H., Johnson, G. A., and Middleditch, B. S. (1988a). *J. Biol. Chem., 263*: 7203.

Markaverich, B. M., Roberts, R. R., Alejandro, M., Johnson, G. A., Middleditch, B. S., and Clark, J. H. (1988b). *J. Steroid Biochem., 30*: 71.

Markaverich, B. M., Schauweker, T. H., Gregory, R. R., Varma, M., Kittrell, F. S., Medina, D., and Varma, R. S. (1992). *Cancer Res., 52*: 2482.

Markaverich, B. M., Varma, M. Densmore, C. L., Tiller, A. A., Schauweker, T. H., and Gregory, R. R. (1994). *Int. J. Oncol., 4*: 1291.

Markaverich, B. M., Webb, B., Densmore, C. L., and Gregory, R. R. (1995). *Environ. Health Perspect., 103*: 574.

Markiewicz, L., Garey, J., Adlercreutz, H., and Gurpide, E. (1993). *J. Steroid Biochem. Mol. Biol., 45*: 399.

Markkanen, T., Mäkinen, M. L., Maunuksela, E., and Himanen, P. (1981). *Drugs Exp. Clin. Res., 7*: 711.

Markovits, J., Linassier, C., Fossé, P., Couprie, J., Pierre, J., Jacquemin-Sablon, A., Saucier, J.-M., Le Pecq., J.-B., and Larsen, A. K. (1989). *Cancer Res., 49*: 5111.

Martin, P. M., Horwitz, K. B., Ruyan, D. S., and McGuire, W. L. (1978). *Endocrinology, 103*: 1860.

Mathieson, R. A., and Kitts, W. D. (1980). *J. Endocrinol., 85*: 317.

Matsukawa, Y., Marui, N., Sakai, T., Satomi, Y., Yoshida, M., Matsumoto, K., Nishino, H., and Aoike, A. (1993). *Cancer Res., 53*: 1328.

Matsumoto, P. S., Ohara, A., Duchatelle, P., and Eaton, D. C. (1993). *Am. J. Physiol., 264*: C246.

Mazur, W., Fotsis, T., Wähälä, K., Ojala, S., Salakka, A., and Adlercreutz, H. (1996). *Anal. Biochem., 233*: 169.

McEvily, A. J., Holmquist, B., Auld, D. S., and Vallee, B. L. (1988). *Biochemistry, 27*: 4284.

Medlock, K. L., Branham, W. S., and Sheehan, D. M. (1995a). *Proc. Soc. Exp. Biol. Med., 208*: 67.

Medlock, K. L., Branham, W. S., and Sheehan, D. M. (1995b). *Proc. Soc. Exp. Biol. Med., 208*: 307.

Messina, M., and Barnes, S. (1991). *J. Natl. Cancer Inst., 83*: 541.

Messina, M., Messina, V., and Setchell, K. (1994a). *The Simple Soybean and Your Health*. Avery Publishing Group, Garden City Park, NY.

Messina, M. J., Persky, V., Setchell, K. D. R., and Barnes, S. (1994b), *Nutr. Cancer, 21*: 113.

Messina, M. (1995a). *J. Nutr., 125*: S567.

Messina, M. (1995b). *Am. J. Clin. Nutr., 62*: 645.

Miksicek, R. J. (1993). *Mol. Pharmacol., 44*: 37.

Miksicek, R. J. (1995). *Proc. Soc. Exp. Biol. Med., 208*: 44.

Mills, P. K., Beeson, W. L., Phillips, R. L. and Fraser, G. E. (1989). *Cancer, 64*: 598.

Minami, Y., Staples, M. P., and Giles, G. G. (1993). *Eur. J. Cancer, 29A*: 1735.

Montesano, R., and Orci, L. (1985). *Cell, 42*: 469.

Monti, E., and Sinha, B. K. (1994). *Anticancer Res., 14*: 1221.

Mousavi, Y., and Adlercreutz, H. (1992). *J. Steroid Biochem. Mol. Biol., 41*: 615.

Mousavi, Y., and Adlercreutz, H. (1993). *Steroids, 58*: 301.

Müller, H.-M., Hofmann, J., and Mayr, U. (1989). *Tiernährung, 17*, 47.

Murkies, A. L., Lombard, C., Strauss, B. J. G., Wilcox, G., Burger, H. G., and Morton, M. S. (1995). *Maturitas, 21*: 189.

Murphy, P. A. (1982). *Food Technol.*, 60.

Musey, P. I., Adlercreutz, H., Gould, K. G., Collins, D. C., Fotsis, T., Bannwart, C., Mäkelä, T. Wähälä, K., Brunow, G., and Hase, T. (1995). *Life Sci., 57*: 655.

Mäkelä, S., Pylkkänen, L., Santti, R., and Adlercreutz, H. (1991). *Proceedings of the Interdisciplinary Conference on Effects of Food on the Immune and Hormonal Systems*, Institute of Toxicology, Swiss Federal Institute of Technology and University of Zürich, Schwerzenbach, Switzerland, p. 135.

Mäkelä, S., Davis, V. L., Tally, W. C., Korkman, J., Salo, L., Vihko, R., Santti, R., and Korach, K. S. (1994). *Environ. Health Perspect., 102*: 572.

Mäkelä S. I., Pylkkänen, L. H., Santti, R. S., and Adlercreutz, H. (1995a). *J. Nutr., 125*: 437.

Mäkelä, S., Poutanen, M., Lehtimäki, J., Kostian, M. L. Santti, R., and Vihko, R. (1995b). *Proc. Soc. Exp. Biol. Med., 208*: 519.

Mäkelä, S., Santti, R., Salo, L., and McLachlan, J. A. (1995c). *Environ. Health Perspect., 103*: 123.

Naik, H. R., Lehr, J. E., and Pienta, K. J. (1994). *Anticancer Res., 14*: 2617.

Naim, M., Gestetner, B., Kirson, I., Birk, Y. and Bondi, A. (1973). *Phytochemistry, 22*, 237.

Nakamura, T., Turner, C. H., Yoshikawa, T., Slemenda, C. W., Peacock, M., Burr, D. B., Mizuno, Y., Orimo, H., Ouchi, Y., and Johnston, C. C. (1994). *J. Bone Miner. Res., 9*: 1071.

Nakashima, S., Koike, T., and Nozawa, Y. (1991). *Molc. Pharmacol., 39*: 475.

Newsome, F. E., and Kitts, W. D. (1977). *Can. J. Anim. Sci., 57*: 531.

Newsome, F. E., and Kitts, W. D. (1980). *Animal Reprod. Sci., 3*: 233.

Nilsson, A., Hill, J. L., and Davies, H. L. (1967). *Biochim. Biophys. Act, 148*: 92.

Nilsson, M., Åman, P., Härkönen, H., Hallmans, G., Bach Knudsen, K. E., Mazur, W., and Adlercreutz, H. (1996). *J. Sci. Food Agric. 73*: 143.

Nishio, K., Miura, K., Ohira, T., Heike, Y., and Saijo, N. (1994). *Proc. Soc. Exp. Biol. Med., 207*: 227.

Nomura, A., Henderson, B. E., and Lee J. (1978). *Am. J. Clin. Nutr., 31*: 2020.

Nose, M., Fujimoto, T., Takeda, T., Nishibe, S., and Ogihara, Y. (1992). *Planta Med., 58*: 520.

Nose, M., Fujimoto, T., Nishibe, S., and Ogihara, Y. (1993). *Planta Med., 59*: 131.

Noteboom, W. D., and Gorski, J. (1963). *Endocrinology, 73*: 736.

Nwannenna, A. I., Lundh, T. J. O., Madej, A., Fredriksson, G., and Bjornhag, G. (1995). *Proc. Soc. Exp. Biol. Med., 208*: 92.

Obermeyer, W. R., Warner, C., Casey, R. E., and Musser, S. (1993). *FASEB J., 7* (3 Pt II): Abstract 4985.

Obermeyer, W. R., Musser, S. M., Betz, J. M., Casey, R. E., Pohland, A. E., and Page, S. W. (1995). *Proc. Soc. Exp. Biol. Med., 208*: 6.

Ogawara, H., Akiyama, T., Watanabe, S.-I., Ito, N., Kobori, M., and Seoda, Y. (1989). *J. Antibiotics, XLII*: 340.

Ohno, T., Kato, N., Ishii, C., Shimizu, M., Ito, Y., Tomono, S., and Kawazu, S. (1993). *Endocrine Res., 19*: 273.

Ohta, N., Kuwata, G., Akahori, H., and Watanabe, T. (1979). *Agric. Biol. Chem., 43*: 1415.

Ohta, N., Kuwata, G., Akahori, H., and Watanabe, T. (1980). *Agric. Biol. Chem., 44*: 469.

Okura, A., Arakawa, H., Oka, H., Yoshinari, T., and Monden, Y. (1988). *Biochem. Biophys. Res. Commun., 157*: 183.

Orcheson, L., Rickard, S., Seidl, M., Cheung, F., Luyengi, L., Fong, H., and Thompson, L. U. (1993). *FASEB J., 7* (3 Pt II): Abstract 1686.

Ota, K., and Misu, Y. A. (1958). *GANN, 49* (suppl): 283.

Ozaki, Y., Yatomi, Y., Jinnai, Y., and Kume, S. (1993). *Biochem. Pharmacol., 46*: 395.

Pagliacci, M. C. Smacchia, M., Migliorati, G., Grignani, F., Riccardi, C., and Nicoletti, I. (1994). *Eur. J. Cancer, 30A*: 1675.

Pepper, M. S., and Montesano, R. (1990). *Cell Differ. Dev., 32*: 319.

Perel, E., and Lindner, H. R. (1970). *J. Reprod. Fert., 21*: 171.

Persky, V., and Vanhorn, L. (1995). *J. Nutr., 125*: S709.

Peterson, G., and Barnes, S. (1991). *Biochem. Biophys. Res. Commun., 179*: 661.

Peterson, G., and Barnes, S. (1993). *Prostate, 22*: 335.

Peterson, G., and Barnes, S. (1994) *Mol. Biol. Cell, 5* (suppl): 384a.

Pettersson, H., and Kiessling, K.-H. (1984). *J. Assoc. Off. Anal. Chem., 67*: 503.

Phipps, W. R., Martini, M. C., Lampe, J. W., Slavin, J. L., and Kurzer, M. S. (1993). *J. Clin. Endocrinol. Metabl., 77*: 1215.

Pietinen, P., Rimm, E., Korhonen, P., Hartman, A., Willett, A., Albanes, D., and Virtamo, J. (1996). *Circulation 94*: 2720.

Piontek, M., Hengels, K. J., Porschen, R., and Strohmeyer, G. (1993). *Anticancer Res., 13*: 21193.

Poellinger, L., Whitelaw, M., McGuire, J., Antonsson, C., Pongratz, I., Lindebro, M., Kleman, M., and Gradin, K. (1995). *Third International Conference on Phytoestrogens*. National Center for Toxicological Research, Food and Drug Administration, Little Rock, AR.

Potter, J. D., and McMichael, A. J. (1983). *J. Natl. Cancer Inst., 71*: 703.

Potter, J. D. (1995). *J. Natl. Cancer Inst., 87*: 1039.

Poutanen, M., Isomaa, V., Peltoketo, H., and Vihko, R. (1995) *J. Steroid Biochem. Molc. Biol., 55*: 425.

Pratt, D. E., and Birac, P. M. (1979). *J. Food Sci., 44*: 1720.

Price, K. R., and Fenwick, G. R. (1985). *Food Add. Contamin., 2*: 73.

Pryor, M., Slattery, M. L., Robison, L. M., and Egger, M. (1989). *Cancer Res., 49*: 2161.

Pukkala, E., Gustavsson, N., and Teppo, L. (1987). *Atlas of Cancer Incidence in Finland*. Cancer Society of Finland Publications, Offsett/PKK Oy, Helsinki.

Pylkkänen, L., Santti, R., Newbold, R., and McLachlan, J. A. (1991). *Prostate, 18*: 117.

Rabindran, S. K., Annable, T., Brown, E. B., Collins, K. I., and Greenberger, L. M. (1995). *Proc. Am. Assoc. Cancer Res., 36*: 324.

Raines, E. W., and Ross, R. (1995). *J. Nutr., 125*: S624.

Rannikko, S., and Adlercreutz, H. (1983). *Prostate, 4*: 223.

Rao, C. B. S. (1978). *The chemistry of Lignans*. Andhra University Press and Publications, Waltair, India.

Rastinejad, F., Polverini, P., and Bouck, N. P. (1989). *Cell, 56*: 345.

Rauth, S., Kichina, J., and Green, A. (1995). *Proc. Am. Assoc. Cancer Res., 36*: 351.

Record, I. R., Dreosti, I. E., and McInerney, J. K. (1995). *J. Nutr. Biochem., 6*: 481.

Reddy, B. S., Sharma, C., Mathews, L., Althea, E., Laakso, K., Choi, K., Puska, P., and Korpela, R. (1985). *Mutat. Res., 152*: 97.

Reddy, B. S., and Cohen, L. A. (1986a). *Diet, Nutrition, and Cancer: A Critical Evaluation. Macronutrients and Cancer,* Vol. 1. CRC Press, Boca Raton, FL.

Reddy, B. S., and Cohen, L. A. (1986b). *Diet, Nutrition and Cancer: A Critical Evaluation., Micronutrients, Nonutritive Dieatry Factors, and Cancer,* Vol. 2. CRC Press, Boca Raton, FL.

Reddy, B. S., Sharma, C., Simi, B., Engle, A., Laakso, K., Puska, P., and Korpela, R. (1987). *Cancer Res., 47*: 644.

Reddy, K. B., Mangold, G. L., Tandon, A. K., Yoneda, T., Mundy, G. R., Zilberstein, A., and Osborne, C. K. (1992). *Cancer Res., 52*: 3636.

Rendu, F., Eldor, A., Grelac, F., Bachelot, C., Gazit, A., Gilon, C., Levytoledano, S., and Levitzki,A. (1992). *Biochem. Pharmacol., 44*: 881.

Rocchi, P., Ferreri, A. M., Simone, G., Magrini, E., Cavallazzi, L., and Paolucci, G. (1995). *Anticancer Res., 15*: 1381.

Rokhlin, O. W., and Cohen, M. B. (1995). *Cancer Lett., 98*: 103

Rose, D. P., Boyar, A. P., and Wynder, E. L. (1986). *Cancer, 58*: 2363.

Rose, D. P. (1992). *Nutrition, 8*, 471.

Rose, D. P. (1993). Diet, hormones and cancer. *Ann. Rev. Public Health, 14*: 1.

Rosenblum, E. R., Van Thiel, D. H., Campbell, I. M., Eagon, P. K., and Gavaler, J. S. (1987). *Alcohol Alcohol*. Suppl. 1: 551.

Rosenblum, E. R., Van Thiel, D. H., Campbell, I. M., and Gavaler, J. S. (1991). *Alcohol. Clin. Exp. Res., 15*: 205.

Rosenblum, E. R., Campbell, I. M., Vanthiel, D. H., and Gavaler, J. S. (1992). *Alcohol. Clin. Exp. Res., 16*: 843.

Ross, R. K., Bernstein, L., Lobo, R. A., Shimizu, H., Stanczyk, F. Z., Pike, M. C., and Henderson, B. E. (1992). *Lancet, 339*: 887.

Rutqvist, L. E., Johansson, H., Signomklao, T., Johansson, U., Fornander, T., and Wilking, N. (1995). *J. Natl. Cancer Inst., 87*: 645.

Safe, S. H. (1995). *Environ. Health Perspect., 103*: 346.

Salen, G., Ahrens, E. H., and Grundy, S. M. (1970). *J. Clin. Invest., 49*: 952.

Salo, L., Mäkelä, S., and Santti, R. (1995). *Third International Conference on Phytoestrogens*, National Center for Toxicological Research, Food and Drug Administration, Little Rock, AR.

Sargeant, P., Farndale, R. W., and Sage, S. O. (1993). *FEBS Lett., 315*: 242.

Sathyamoorthy, N., Wang, T. T. Y., and Phang, J. M. (1994). *Cancer Res., 54*: 957.

Scholar, E. M., and Toews, M. L. (1994). *Cancer Lett., 87*: 159.

Schweigerer, L., Christeleit, K., Fleischmann, G., Adlercreutz, H., Wahala, K., Hase, T., Schwab, M., Ludwig, R., and Fotsis, T. (1992). *Eur. J. Clin. Invest., 22*: 260.

Seo, A., and Morr, C. V. (1984). *J. Agric. Food Chem., 32*: 530.

Serraino, M., and Thompson, L. U. (1991). *Cancer Lett., 60*: 135.

Serraino, M., and Thompson, L. U. (1992a). *Nutr. Cancer., 17*: 153.

Serraino, M., and Thompson, L. U. (1992b). *Cancer Lett., 63*: 159.

Setchell, K. D. R., and Adlercreutz, H. (1979). *J. Steroid Biochem., 11*: xv.

Setchell, K. D. R., Lawson, A. M., Mitchell, F. L., Adlercreutz, H., Kirk, D. N., and Axelson, M. (1980a). *Nature, 287*: 740.

Setchell, K. D. R., Bull, R., and Adlercreutz, H. (1980b). *J. Steroid Biochem., 12*: 375.

Setchell, K. D. R., Lawson, A. M., Axelson, M., and Adlercreutz, H. (1980c). *Endocrinological Cancer, Ovarian Function and Disease* (H. Adlercreutz, R. Bulbrook, H. van der Molen, A. Vermeulen and F. Sciarra, eds.). Excerpta Medica. Amsterdam, p. 297.

Setchell, K. D. R., Lawson, A. M., Borriello, S. P., Harkness, R., Gordon, H., Morgan, D. M. L., Kirk, D. N., Adlercreutz, H., Anderson, L. C. and Axelson, M. (1981). *Lancet, 2*: 4.

Setchell, K. D. R., Lawson, A. M., Borriello, S. P., Adlercreutz, H., and Axelson, M. (1982). *Colonic Carcinogenesis: Falk Symposium 31* (R. A. Malt and R. C. N. Williamson, eds.). MTP Press, Lancaster, P. 93.

Setchell, K. D. R., Lawson, A. M., McLaughlin, L. M., Patel, S., Kirk, D. N., and Alexson, M. (1983). *Biomed. Mass Spectrom., 10*: 227.

Setchell, K. D. R., Borriello, S. P., Hulme, P., and Axelson, M. (1984). *Am. J. Clin. Nutr., 40*: 569.

Setchell, K. D. R. (1985). *Estrogens in the Environment II. Influences on Development* (J. A. McLachlan, ed.). Elsevier, New York, p. 69.

Setchell, K. D. R., Gosselin, S. J., Welsh, M. B., Johnston, J. O., Balistreri, W. F., Kramer, L. W., Dresser, B. L. and Tarr, M. J. (1987a). *Gastroenterology, 93*, 225.

Setchell, K. D. R., and Welsh, M. B. (1987b). *J. Chromatogr., 386*, 315.

Setchell, K. D. R., and Adlercreutz, H. (1988). *Role of the Gut Flora in Toxicity and Cancer* (I. Rowland, ed.). Academic Press, London, p. 315.

Setchell, K. D. R., Cassidy, A., and Bingham, S. (1995). *Am. J. Clin. Nutr., 62*: 152.

Severson, R. K., Nomura, A. M. Y., Grove, J. S. and Stemmerman, G. N. (1989). *Cancer Res., 49*: 1857.

Sharma, O. P., Adlercreutz, H., Strandberg, J. D., Zirkin, B. R., Coffey, D. S., and Ewing, L. L. (1992). *J. Steroid Biochem. Mol. Biol., 43*: 557.

Shimizu, H., Ross, R. K., Pike, M. C., and Henderson, B. E. (1990). *Br. J. Cancer, 62*: 451.

Shimizu, H., Ross, R. K., Bernstein, L., Yatani, R., Henderson, B. E., and Mack, T. M. (1991). *Br. J. Cancer, 63*: 963.

Shimokado, K., Yokota, T., Umezawa, K., Sasaguri, T., and Ogata, J. (1994). *Arterioscler. Thromb., 14*: 973.

Shimokado, K., Umezawa, K., and Ogata, J. (1995). *Exp. Cell Res., 220*: 266.

Shirai, T., Imaida, K., Masui, T., Iwasaki, S., Mori, T., Kato, T., and Ito, N. (1994). *Int. J. Cancer, 57*: 224.

Shore, L. S., and Lyttle, C. R. (1986). *Plant Flavonoids in Biology and Medicine: Biochemical, Pharamcological, and Structure-Activity Relationships*. Alan R. Liss, Inc., New York, p. 253.

Shultz, T. D., Bonorden, W. R., and Seaman, W. R. (1991). *Nutr. Res., 11*: 1089.

Shutt, D. A., Weston, R. H., and Hogan, J. P. (1970). *Aust. J. Agric. Res., 21*: 713.

Singh, S., and Langman, M. J. S. (1995). *Gut, 37*: 737.

Shutt, D.A., and Cos, R. I. (1972). *J. Endocrinol., 52*: 299.

Shutt, D. A. (1976). *Endeavour, 35*: 110.

Smith, R. M., Tiesinga, J. J., Shah, N., Smith, J. A., and Jarett, L. (1993). *Arch. Biochem. Biophy., 300*: 238.

Spinozzi, F., Pagliacci, M. C., Migliorati, G., Moraca, R., Grignani, F., Riccardi, C., and Nicoletti, I. (1994). *Leuk. Res., 18*: 431.

Stampfer, M. J., Colditz, G. A., Willett, W. C., Manson, J. E., Rosner, B., Speizer, F. E., and Hennekens, C. H. (1991a). *N. Engl. J. Med., 325*: 756.

Stampfer, M. J., and Colditz, G. A. (1991b). *Prev. Med., 20*: 47.

Stanford, J. L., Herrinton, L. J., Schwartz, S. M., and Weiss, N. S. (1995). *Epidemiology, 6*: 181.

Steele, V. E., Pereira, M. A., Sigman, C. C., and Kelloff, G. J. (1995). *J. Nutr., 125*: S713.

Stephan, E. B., and Dziak, R. (1994). *Calcif. Tissue Int., 54*: 409.

Steusloff, A., Paul, E., Semenchuk, L. A., Disalvo, J., and Pfitzer, G. (1995). *Arch. Biochem. Biophys., 320*: 236.

Stitch, S. R., Smith, P. D., Illingworth, D., and Toumba, K. (1980a). *J. Endocrinol., 85*: 23P.

Stitch, S. R., Toumba, J. K., Groen, M. B., Funke, C. W., Leemhuis, J., Vink, J., and Woods, G. F. (1980b). *Nature, 287*: 738.

Sun, X. Y., Yeh, G. C., and Phang, J. M. (1995). *Proc. Am. Assoc. Cancer Res., 36*: 642.

Supko, J. G., and Malspeis, L. (1995). *Proc. Am. Assoc. Cancer Res., 36*: 382.

Takeda, Y., Nishio, K., Morikage, T., Kubota, N., Kohima, A., Kubo, S., Fujiwara, Y., Niitani, H., and Saijo, N. (1992). *Proc. Am. Assoc. Cancer Res., 33*: 476.

Takeda, Y., Nishio, K., Niitani, H., and Saijo, N. (1994). *Int. J. Cancer, 57*: 229.

Tang, B. Y., and Adams, N. R. (1978). *J. Endocrinol., 78*: 171.

Tang, B. Y., and Adams, N. R. (1980). *J. Endocrinol., 85*: 291.

Tang, G. W. K. (1994). *Maturitas, 19*: 177.

Teppo, L., Pukkala, E., Hakama, M., Hakulinen, A., Herva, A., and Saxén, E. (1980). *Scand. J. Social Med.*, Suppl 19: 14.

Teraoka, H., Ohmura, Y., and Tsukada, K. (1989). *Biochem. Intern., 18*: 1203.

Terpstra, A. H. M., West, C. E., Fennis, J. T. C. M., Schouten, J. A. and van der Veen, E. A. (1984). *Am. J. Clin. Nutr., 39*: 1.

Thompson, L. U., and Serraino, M. (1990). *Proc. Flax Inst. U.S.* Fargo, ND, p. 30.

Thompson, L. U., Robb, P. Serraino, M., and Cheung, F. (1991). *Nutr. Cancer, 16*: 43.

Thompson, L. U. (1995a). *Quality and Safety Aaspects of Food and Nutrition in Europe '95.* World Health Organization, Helsinki, Abstract LQD2.

Thompson, L. U., Rickard, S. E., Orcheson, L. J., and Seidl, M. M. (1995b). *Proc. Am. Assoc. Cancer Res., 36*: 114.

Tikkanen, M. J., Wähälä, K., Ojala, S., Vihma, V., and Adlercreutz, H. (1995). *68th Scientific Sessions of the Am. Heart Assoc.*, Anaheim, CA, p. 82.

Tominaga, S., and Kuroishi, T. (1995). *Cancer Lett. 90*: 75.

Traganos, F., Ardelt, B., Halko, N., Bruno, S., and Darzynkiewicz, Z. (1992). *Cancer Res., 52*: 6200.

Trowell, H. C., and Burkitt, D. P. (1981). *Western Diseases: Their Emergence and Prevention.* Edward Arnold, Ltd, London.

Tsukamoto, C., Shimada, S., Igita, K., Kudou, S., Kokubun, M., Okubo, K., and Kitamura, K. (1995). *J. Agric. Food Chem., 43*: 1184.

Tsunenari, T., Yamada, S., Kawakatsu, M., Negishi, H., and Tsutsumi, M. (1995). *Calcif. Tissue Int., 56*: 5.

Tsutsumi, N. (1995). *Biol. Pharm. Bull., 18*: 1012.

Uckun, F. M., Evans, W. E., Forsyth, C. J., Waddick, K. G., Ahlgren, L. T., Chelstrom, L. M., Burkhardt, A., Bolen, J., and Myers, D. E. (1995). *Science, 267*: 886.

Ullrich, A., and Schlesinger, J. (1990). *Cell, 61*: 203.

Upadhyaya, P., and El-Bayoumy, K. (1995). *Proc. Am. Assoc. Cancer Res., 36*: 597.

Van Thiel, D. H., Galvaoteles, A., Monteiro, E,. Rosenblum, E., and Gavaler, J. S. (1991). *Alcohol. Clin. Exp. Res., 15*: 822.

van't Veer, P., Kolb, C. M., Verhoef, P., Kok, F. J., Schouten, E. G., Hermus, R. J. J., and Sturmans, F. (1990). *Int. J. Cancer, 45*: 825.

Vargas, R., Wroblewska, B., Rego, A., Hatch, J., and Ramwell, P. W. (1993). *Br. J. Pharmacol., 109*: 612.

Veldscholte, J., Berrevoets, C. A., and Mulder, E. (1994). *J. Steroid Biochem. Mol. Biol., 49*: 341.

Verdeal, K., Brown, R. R., Rickardson, T., and Ryan, D. S. (1980). *J. Natl. Cancer Inst., 64*: 285.

Versantvoort, C. H. M., Schuurhuis, G. J., Pinedo, H. M., Eekman, C. A., Kuiper, C. M., Lankelma, J., and Broxterman, H. J. (1993). *Br. J. Cancer, 68*: 939.

Versantvoort, C. H. M., Broxterman, H. J., Lankelma, J., Feller, N., and Pinedo, H. M. (1994). *Biochem. Pharmacol., 48*: 1129.

Verspohl, E. J., Tollkuhn, B., and Kloss, H. (1995). *Cell Signal, 7*: 505.

Vähäpassi, J., and Adlercreutz, H. (1975). *Contraception, 11*: 427.

Walter, E. D. (1941). *J. Am. Chem. Soc., 63*: 32736.

Walz, E. (1931). *Justus Liebigs Annl. Chem., 498*: 118.

Wang, C. F., Mäkela, T., Hase, T., Adlercreutz, H., and Kurzer, M. S. (1994). *J. Steroid Biochem. Mol. Biol., 50*: 205.

Wang, H. J., and Murphy, P. A. (1994a). *J. Agric. Food Chem., 42*: 1674.

Wang, H. J., and Murphy, P. A. (1994b). *J. Agric. Food Chem., 42*: 1666.

Wang, H. L., and Scott, R. E. (1994). *J. Cell Physiol., 158*: 408.

Wang, W. Q., Tanaka, Y., Han, Z. K., and Higuchi, C. M. (1995). *Nutr. Cancer, 23*: 131.

Watanabe, T., Shiraishi, T., Sasaki, H., and Oishi, M. (1989). *Exp. Cell Res., 183*: 335.

Watanabe, S., and Koessel, S. (1993). *J. Epidemiol., 3*: 47.

Waters, A. P. and Knowler, J. T. (1982). *J. Reprod. Fertil., 66*: 379.

Wehling, M. (1994). *Trends Endocrinol. Metab., 5*: 347.

Wei, H. C., Wei, L. H., Frenkel, K., Bowen, R., and Barnes, S. (1993). *Nutr. Cancer, 20*: 1.

Wei, H. C., Bowen, R., Cai, Q. Y., Barnes, S., and Wang, Y. (1995). *Proc. Soc. Exp. Biol. Med., 208*: 124.

Welshons, W. V., Murphy, C. S., Koch, R., Calaf, G., and Jordan, V. C. (1987). *Breast Cancer Res. Treat., 10*: 169.

Whitten, P. L., and Naftolin, F. (1991). *New Biology of Steroid Hormones* (R. B. Hochberg and F. Naftolin, eds.). Raven Press, New York, p. 155.

Whitten, P. L., Russell, E., and Naftolin, F. (1994). *Steroids, 59*: 443.

WHO Study Group. (1994). *Assessment of Fracture Risk and Its Application to Screening for Postmenopausal Osteoporosis*. WHO Technical Report Series 843, Geneva, p. 11.

Willett, W. C., Hunter, D. J., Stampfer, M. J., Colditz, G., Manson, J. E., Spiegelman, D., Rosner, B., Hennekens, C. H., and Speizer, F. E. (1992). *J. Am. Med. Assoc., 21*: 2037.

Willett, W. C., Sacks, F., Trichopoulou, A., Drescher, G., Ferroluzzi, A., Helsing, E., and Trichopoulos, D. (1995). *Am. J. Clin. Nutr., 61*: S1402.

Winter, M. L., Bosland, M. C., Wade, D. R., Falvo, R. E., Nagamani, M., and Liehr, J. G. (1995). *Prostate, 26*: 325.

Xiong, Z. G., Burnette, E., and Cheung, D. W. (1995). *Eur. J Pharmacol.-Molec. Pharm., 290*: 117.

Xu, X., Wang, H. J., Murphy, P. A., Cook, L., and Hendrich, S. (1994a). *J. Nutr., 124*: 825.

Xu, X. M., and Thomas, M. L. (1994b). *Mol. Cell. Endocrinol., 105*: 197.

Xu, X., Harris, K. S., Wang, H. J., Murphy, P. A., and Hendrich, S. (1995a). *J. Nut., 125*: 2307.

Xu, X. M., and Thomas, M. L. (1995b). *Endocrine, 3*: 661.

Yamashita, K., Nohara, Y., Katayama, K., and Namiki, M. (1992). *J. Nutr., 122*: 2440.

Yamashita, Y., Kawada, S., and Nakano, H. (1990). *Biochem. Pharmacol., 39*: 737.

Yanagihara, K., Ito, A., Toge, T., and Numoto, M. (1993). *Cancer Res., 53*: 5815.

Yang, S. G., Saifeddine, M., and Hollenberg, M. D. (1992). *Can. J. Physiol. Pharamcol., 70*: 85.

Yang, S. G., Saifeddine, M., Laniyonu, A., and Hollenberg, M. D. (1993). *J. Pharmacol. Exp. Ther., 264*: 958.

Yatani, R., Chigusa, I., Akazaki, K., Stemmerman, G. N., Welsh, R. A., and Correa, P. (1982). *Int. J. Cancer, 29*: 611.

Yokotsuka, T. (1986). *Adv. Food Res., 30*: 195.

Yuan, J. M., Wang, Q. S., Ross, R. K., Henderson, B. E., and Yu, M. C. (1995). *Br. J. Cancer, 71*: 1353.

15

Ovarian Toxicants: Multiple Mechanisms of Action

Barbara J. Davis and Jerrold J. Heindel
National Institute of Environmental Health Sciences, Research Triangle Park, North Carolina

INTRODUCTION

The ovary plays a central role in the physiology and reproductive status throughout a woman's lifetime. As a reproductive organ, the ovary houses the female germ cells and provides the hormonal milieu needed for successful reproduction. As an endocrine organ, the ovary is the primary source of estradiol and progesterone in nongravid women from birth until menopause. The importance of maintaining normal ovarian function is illustrated by the observations that ovarian dysfunction impairs fertility and increases the chances that a woman will develop gender-associated diseases such as breast, uterine, and ovarian cancer, osteoporosis, and/or autoimmune diseases. Additionally, premature menopause as a result of ovarian dysfunction is also a risk factor for some of these diseases. At least 8–15% of American women (Mosher and Pratt, 1985) and up to 24% of women in the United Kingdom are subfertile (Greenhall and Vessey, 1990), in part due to ovarian dysfunction including polycystic ovarian disease.

Since normal ovarian function is central to women's health, understanding how ovarian function may be impacted by the environment is important in understanding how women's health may be impacted by the environment. Both epidemiological and experimental data demonstrate that exposures to various occupational, environmental, and therapeutic agents can perturb hormonal homeostasis and ovarian function. Hormonal abnormalities and/or menstrual irregularities have been attributed to exposures to carbon disulfide, synthetic estrogens and progesterins, diethylstilbesterol, oil processing, ethylene-based glycol ethers, technical rubber articles, carbamide, carbon monoxide, and methanol in addition to exposures to heavy metals such as lead, arsenic, cadmium, and mercury (Barlow and Sullivan, 1982; Mattison, 1985; Baranski, 1993). In fact, mercury exposures in women have been associated with either hyper- or hypomenorrhea or oligomenorrhea and an increase in anovulatory cycles. Chemotherapeutic agents such as cyclophosphamide or vinblastine cause premature ovarian failure and premature menopause in women (Warne et al., 1973; Chapman et al., 1979). Women exposed to high occupational levels of di-*n*-hexyl phthalate have hypoestrogenic, anovulatory cycles (Aldyreva et al., 1975).

Literally hundreds of chemicals cause reproductive dysfunction in animal models (reviewed in Shardein, 1988; Chapin and Sloane, 1997), and we are just beginning to understand how they cause ovarian dysfunction. In this chapter, we illustrate how toxicants can alter ovarian function, identify vulnerable sites, and discuss what biochemical and molecular pathways may be involved in mediating toxicity. The beginning of the chapter is dedicated to a review of ovarian physiology and morphology, since the first step in understanding how toxicants alter ovarian function is to understand how the ovary functions. The remaining sections discuss the effects of selected ovarian toxicants with particular emphasis on the effects in rodents, since this is the primary animal model used to identify hazards. Use of primate models has been reviewed elsewhere (Sakai and Hodgen, 1987).

ANATOMY

The female reproductive tract forms from mesoderm in the gonadal ridge located on the dorsal coelomic wall, medial to the mesonephric duct. In the adult, the ovaries are paired organs located on the dorsal abdominal wall, lateral and caudal to the kidneys, and embedded in abdominal fat. The round ligament suspends the ovary from the abdominal wall, and the cranial portion of the broad ligament suspends the ovary with the uterus. In the rodent, the ovary is completely enveloped by an ovarian bursa, providing a continuous compartment between the ovary and the oviduct. In the human, the ovary and oviduct are discontinuous.

OVARIAN EMBRYOGENESIS

During embryonic development, the formation of the ovary commences with the arrival of the primordial germ cells from the yolk sac into the gonadal ridge. In rodents, primordial germ cells usually arrive at the gonadal ridge on gestational day 11–11.5 but as early as day 8 (Baker, 1972). In humans, germ cells arrive during the second month of gestation (Baker, 1972; Peters and McNatty, 1980). Concomitant with the migration of the germ cells, the coelomic mesothelium lining the surface of the gondal ridge thickens, proliferates, and then protrudes into the adjacent mesoderm (Peters and McNatty, 1980). This protrusion of surface mesothelium has been thought to give rise to primary and secondary sex cords that surround the primordial germ cells and subsequently develop into follicles. However, other studies suggest that the sex cords arise from the mesonephros tubules and that the coelomic epithelium may not be involved in their formation (Byskov, 1974; Satoh, 1985).

As the cords proliferate to encase the germ cells, the germ cells also proliferate, forming oogonia. Oogonia undergo a series of mitotic divisions, at which point they form oocytes. Subsequently in the rodent, oocytes begin synchronized meiotic divisions and arrest at the diplotene stage around gestational day 17 to postpartum day 5 (Byskov, 1974). In the human the switch from mitosis to meiosis occurs from gestational months 2 to 7 and is not synchronized. Oocytes arrest in the human in the diplotene stage by gestational month 7 (Baker, 1972; Peters and McNatty, 1980). In all species, oocytes remain arrested in the diplotene stage of meiosis and only resume meiosis just prior to ovulation.

In all species, coincident with the onset of meiosis, large numbers of oocytes undergo attrition. The end result of this early proliferation and attrition is a fixed, nonreplenishing population of oocytes in the adult. For example, in the rat, the number of oocytes is reduced from 75,000 to about 27,000 by birth, and by the time of sexual maturity (34–36 days in the rat) rat ovaries contain approximately 5,000 oocytes (Beaumont and Mandl, 1962), although this number varies between strains (Mandl and Zuckerman, 1950). Human ovaries contain about 12 million oocytes at birth and about 400,000 at puberty (Peters and McNatty, 1980). Once the population of oocytes is depleted, the animal (or woman) is no longer fertile.

OVARIAN HISTOLOGY

Follicular Growth

Each oocyte is encircled by support cells within a structure termed a "follicle." The oocyte is maintained and brought to maturity for fertilization through the growth of the follicle and the coordinate interaction of the hormone-producing support cells with the hypothalamic-pituitary-uterine tissues. Because the physiology of the follicle varies with its morphology, follicles are typically classified according to their morphology. The characteristics of the various stages of follicle growth, maturation, and their controls have been described in excellent papers and reviews (e.g., Pedersen and Peters, 1968; Pedersen, 1970; Greenwald, 1974; Hirshfield and Midgley, 1978; Richards, 1980, 1994; Hsueh et al., 1984; Hirshfield and Schmidt, 1987; Hirshfield and Schmidt, 1987; Richards and Hedin, 1988) and will only be highlighted in this chapter. For the mouse, the most commonly used scheme is one proposed by Pedersen and Peters (1968), which classifies follicles based on the size of the oocyte, the size of the follicle and the morphology of the follicle as seen in the mouse. Follicle classifications in the rat are most commonly based on mean cross-section follicular diameter sizes, with descriptive modifications similar to that for the mouse (Hirshfield et al., 1978). While distinctions between the rat and mouse in follicle classifications are necessary in experimental studies, growth and maturation of follicles is a continuous process and shares common morphological and physiological properties in the rat, mouse, and human (Fig. 1).

The primitive oocyte encircled by flattened stromal cells is termed a "primordial" or primary or very small follicle (Hirshfield, 1989). Traditionally, primordial follicles are considered nonproliferating or "resting" follicles that are "recruited" by some factor(s) into growing follicles. Recent evidence, however, suggests that the primordial follicles may not be entirely at "rest," since the supporting cells of these follicles incorporate label after continuous infusions of tritiated thymidine (Hirshfield, 1989) and oocytes and these supporting cells also label for proliferating cell nuclear antigen (PCNA) by immunohistochemistry (B. J. Davis, personal observation). As a percentage of total follicles, these represent the largest pool and represent the population from which all other stages of follicles are derived. It is not clear what triggers the initial growth of the follicles. The oocyte may trigger the growth, or growth may be programmed (Pedersen, 1970). Most evidence suggests that the initiation of growth is independent of pituitary hormones since follicular growth continues in hypophysectomized mice and rats, but it may involve growth factors (for review, see Adashi, 1990). In the mouse, FSH can initiate follicular growth in vitro (Wang et al., 1991). The cumu-

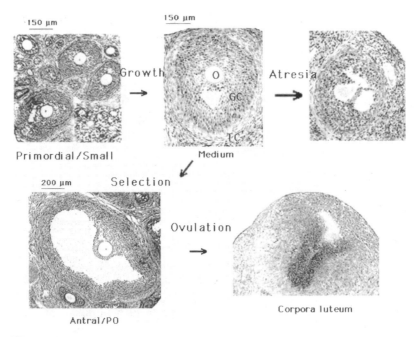

Figure 1 Follicular development in the adult rat. While specific characterization of follicles varies between species, this figure illustrates the similarities of follicular development between all species: follicles are recruited from a pool of primordial oocytes (primordial/small follicles), undergo a period of rapid proliferation of granulosa cell ("GC) and thecal cell ("TC") proliferation (medium follicle: "O" is oocyte) and primarily undergo degeneration ("atresia") while one to several (depending on the species) are "selected" to develop into prevoulatory follicles, respond to an ovulatory surge of LH, release the oocyte, and differentiate into corpora lutea.

lative effect of the various growth-initiating factors is that primordial follicles begin their growth at relatively constant rates (Faddy et al., 1984), and the total number of growing follicles at any one point balances the number of primordial follicles (Krarup et al., 1969).

Once growth is initiated, the oocyte rapidly increases in size and becomes encased by a zona pellucida (Mandl and Zuckerman, 1952a). Concurrently, the supporting stromal cells also begin to proliferate and change their phenotype from flattened cells into cuboidal granulosa cells. This stage is distinguished morphologically from the primordial follicles as "small follicle." In the rat these follicles measure less than 150 μm in diameter; they are termed types 3a and 3b in the mouse. The maximum size of the oocyte is reached concurrent with the development of four granulosa cell layers. Granulosa cells are FSH responsive and are supported by a basement membrane, which separates the granulosa cells from an outer layer of spindle-shaped, LH-responsive thecal cells that are admixed with fibroblasts and vessels. At this stage gonadotropins begin to regulate granulosa cell growth. FSH binding to granulosa cells is associated with the cessation of growth of the oocyte (Richards and Midgley, 1976). FSH increases small follicle granulosa cell proliferation, whereas LH stimulates thecal cell proliferation (Wang et al., 1991). However, proliferation rates of individual cells vary within the same follicles, suggesting that the cell cycles of granulosa cells are not synchronized (Hirshfield, 1985).

The next stage of growth is characterized by a rapid proliferation of granulosa cells and referred to as medium follicles (rat: 151–390 μm; mouse: type 4). Granulosa cells in medium-sized follicles are responsive to FSH (Hirshfield and Midgley, 1978) and contain very low concentrations of estradiol receptors as assessed in gonadotropin-stimulated hypophysectomized rats (Richards, 1975). Both FSH and estradiol stimulate granulosa cell proliferation in these follicles (Goldenberg et al., 1972). Intrinsic factors may also mediate granulosa cell proliferation at this stage since granulosa cell proliferation occurs in hypophysectomized prepubertal rats (Hirshfield and Schmidt, 1987)

After this period of rapid proliferation, rat follicles contain about 1000 granulosa cells in their largest cross-sectional diameter; in both rats and mice, these follicles contain large fluid-filled antrums. These follicles have been referred to by various names including small-antral (Hirshfield, 1987), small-large (Richards et al., 1976), and large follicles (Pedersen and Peters, 1968). They are clearly regulated by FSH. Their granulosa cells have a high content of FSH receptors (Nimrod et al., 1976), and these are the follicles that respond to the proestrus FSH surge and develop into preovulatory, Graafian follicles in the next cycle (Welschen, 1973; Schwartz, 1974; Hirshfield and Midgley, 1978). These follicles also have high levels of estradiol receptors (Richards, 1975) and androgen receptors (Zeleznik et al., 1979). However, they do not have aromatase cytochrome P-450 upon culture, but FSH stimulates aromatase induction after 24 hours of culture (Erickson and Hsueh, 1978; Erickson, 1983). Thus, these follicles lack the ability to synthesize estradiol in vivo.

The final stage of follicular growth is one characterized more by maturation of granulosa and thecal cells into a Graafian or "preovulatory" follicle, which is the follicle that releases its egg in response to an ovulatory surge of LH. In the rodent, these follicles are typically seen only during proestrus. By the time preovulatory follicles are formed in the rat, they contain large antral fluid-filled areas and more than 2000–2500 granulosa cells in their largest cross section. In all species, these follicles are capable of producing estradiol, inhibin, progesterone, and androgens (Hirshfield and Schmidt, 1987). The granulosa cells are a heterogeneous population of cells that are predominantly differentiating and typically are no longer proliferating (Erickson, 1983). Granulosa cells adjacent to the basement membrane ("mural" cells) have aromatase cytochrome P-450 enzymes and a greater abundance of LH receptors, whereas granulosa cells surrounding the oocyte ("cumulus" cells) have a greater abundance of prolactin receptors. Further, granulosa cells also form distinct functional populations in response to FSH-induced steroidogenesis in cell cultures (Rao et al., 1991). In the rat and the mouse, about 8–12 of these follicles develop about every 4 or 5 days in a cycling animal. In the human, only one of these follicles develops about every 28 days.

Follicular Atresia

Of the 5000 primordial follicles in the pubertal rat ovary, only about 1260 complete the growth processes described above leading to ovulation. The rest of the follicles (75%) are lost through a process termed "atresia" (Mandl and Zuckerman, 1950). In comparison, women at the time of sexual maturity have approximately 390,000 oocytes, of which 360 ovulate, and the rest (99%) are lost through atresia. Since follicle numbers are in excess, the process of atresia may play a central role in regulating the number of follicles that ovulate and which follicles ovulate (Hirshfield, 1986). Since FSH and LH are involved in follicular growth, factors that work to antagonize their effects are likely involved in triggering atresia. Thus, a balance is likely between the growth-promoting factors and growth-inhibiting factors that regulates follicle numbers. Some of the endocrine, paparcrine, and autrocrine factors that may play a role in atresia include GnRH, testosterone, IGF-binding proteins, activin, and tumor necrosis factor-alpha (Hsueh et al., 1994; Richards, 1994).

Rates of atresia and the follicle populations that become atretic change with advancing age. For example, in the mouse the rate of atresia decreases after puberty and declines exponentially with advancing age (Jones and Krohn, 1961; Faddy et al., 1984). In mature rats, atresia occurs in growing follicles, whereas in prepubertal rat ovaries, atresia occurs in small follicles. Also, in mature rats rates of atresia in larger follicles vary with the stage of the estrous cycle (Mandl and Zuckerman, 1952; Hirshfield and Midgley, 1978). Moreover, gonadotropins modulate rates of atresia in the rat. Follicular atresia are partially reversible or preventable under certain conditions. For example, exposure to gonadotropins prevents atresia in larger follicles (Hirshfield, 1985, 1986). In addition, the rise in FSH after unilateral ovariectomy induces compensatory follicular development in large healthy follicles and decreases follicular atresia in the remaining ovary. As a consequence, a full complement of follicles ovulate within 24 hours after surgery (Peppler and Greenwald, 1970; Otani and Sasamoto, 1982). Phenobarbital-induced blockade of ovulation also induces atresia in preovulatory follicles in the rat. Thus, the induction of atresia in larger follicles may be due to a lack of hormonal stimulation (Braw and Tsafriri, 1980; Terranova, 1981).

Follicular atresia is the result of granulosa and thecal cell apoptosis or programmed cell death (Hughes and Gorospe, 1991; Tilly et al., 1991; Hurwitz and Adashi, 1992). Apoptosis is characterized morphologically as condensation of nuclear chromatin, cytoplasmic contraction but with intact organelles, separation of cellular fragments into apoptotic bodies, and removal of fragments by phagocytosis. It is an active process initiated by activation of a Ca^{2+}/Mg^{2+}-dependent endogenous endonuclease that cleaves genomic DNA at internucleosomal regions. This cleavage results in distinctive 180- to 200-base -pair DNA oligonucleosomal ladders containing 3'hydroxy tails. Use of in situ labeling of these ends with digoxigenein-ddUTP and a terminal transferase enzyme allows the identification of apoptotic granulosa cells in paraffin-embedded sections of ovary (for review, see Hsueh et al., 1994). Our experience with such in situ methods is that they are not necessarily more sensitive in detecting apoptotic cells than identification of apoptotic bodies on standard hemotoxylin and eosin-stained tissues, and they are certainly more labor intensive and costly. Alternatively, persistent expression of 3β-hydroxysteroid dehydrogenase can be used as an immunohistochemical marker of atresia (Teerds and Dorrington, 1993).

Interstitial Tissue

While thecal cell apoptosis also occurs in follicular atresia (Hughes et al., 1991), most thecal cells undergo hypertrophy, forming the interstitial tissue around the empty lacunae of the degenerative follicle. As the hypertrophy occurs in the interstitial cells, they accumulate large lipid droplets consisting of phospholipids, triglycerides, and cholesterol and its esters (Guraya and Greenwald, 1964). During the preovulatory period, these lipid droplets mobilize from patches of interstitial tissue to developing follicles. Thus, thecal cell hypertrophy during follicular atresia is an important process in the ovary (Welschen et al., 1991). As the ovary ages and follicles continue to degenerate, the interstitial component of the ovary increases in volume. In an aged rat, the ovary may be entirely composed of interstitial tissue (B. J. Davis, personal observation).

Corpora Lutea

If the follicle successfully navigates a course of positive selection, the granulosa and thecal cells prepare to respond to the ovulatory surge of LH, release the oocyte, and differentiate into the cells of the corpus luteum (CL). Terminal differentiation is marked by limited proliferation and extensive cellular hypertrophy. Granulosa cells increase their cytoplasmic volume about 40 times and are characterized by abundant bright eosinophilic cytoplasm filled with discrete vacuoles (lipid/cholesterol/steroid) and centrally located, round, stippled basophilic nuclei with prominent nucleoli. In all species, the newly formed CL are composed of the "large" granulosa-derived cells and the "small" thecal-derived cells which differentially produce progesterone in response to LH. In the cycling rat or mouse, 10–12 follicles ovulate about every 4–6 days, and fresh corpora lutea are readily visualized because of the areas of acute hemorrhage and because the granulosa cells have basophilic cytoplasm and are smaller than in older corpora lutea. Numerous neutrophils, lymphocytes, and macrophages are also distributed throughout the corpora lutea. In all species, it is important to consider that events in follicular growth are important in the formation of the CL since the CL is derived from the follicle. For example, suppression of FSH during the follicular phase results in lower preovulatory estradiol levels and subnormal luteal phase P production and decreased luteal cell mass (Yen, 1986).

While structurally similar in most species, the corpus lutea in the unmated cycling rodent is functionally different from other species since it produces progesterone only transiently and at levels that cannot maintain uterine implantation (Everett, 1961). A secondary surge of LH and diurnal surges of prolactin are necessary to "rescue" the luteal cells of the rat and mouse. Such surges are mediated by mating or cervical stimulation (Smith et al., 1975). In other species, the cells of the corpus luteum maintain progesterone production and can support pregnancy independent of any additional pituitary hormonal stimuli for an initial period. Later in the luteal phase, the CL is dependent on gonadotropin support, most notably chorionic gonadotropin, which acts through the LH receptor to stimulate progesterone. It has also been shown that the chorionic gonadotropin increases estrogen receptor content and stimulates estrogen production. Thus, in the initial phases of CL formation and maintenance, estrogen is luteotropic.

If fertilization or implantation does not occur, the CL regresses with the decreasing production of progesterone and the influence of luteolytic factors such as prostaglandins, oxytocin, tumor necrosis factor-alpha, and estrogen. While CL regression occurs more rapidly in the cycling rat or mouse, many of the same luteolytic factors are involved.

ENDOCRINOLOGY OF THE REPRODUCTIVE CYCLE

Temporal coordination of the dynamic reproductive system in the female manifests as a cycle. Some of the hormones that are involved in the cross-talk between these organs are GnRH, FSH, LH, estradiol, progesterone, inhibin, and follistatin. The pri-

mate (including humans) has a 28-day cycle characterized by a 4- to 7-day menstrua-
tion. Other animals have estrous cycles characterized by limited days of sexual receptivity. The rat and mouse have short estrous cycles (4–5 days) which lack a luteal phase
that could maintain a pregnancy (see Fig. 2). The rodent estrous cycle is defined by
these hormonal variations and resulting morphological changes in reproductive organs
as proestrus, estrus, metestrus, and diestrus 1 and diestrus 2, with ovulation during estrus
(Long and Evans, 1922; Mandl, 1951; LeFevre and McClintock, 1988). Such variations need to be considered when evaluating potential ovarian toxicants.

The endocrine-paracrine interaction between the pituitary and ovarian signaling
systems is described as the "two-cell, two-gonadotropin theory" (Fig. 3) (Hsueh et al.,
1984). In general terms, FSH stimulates the ovarian granulosa cell compartment of
ovarian follicles, and LH stimulates the surrounding thecal cell compartment of ova-

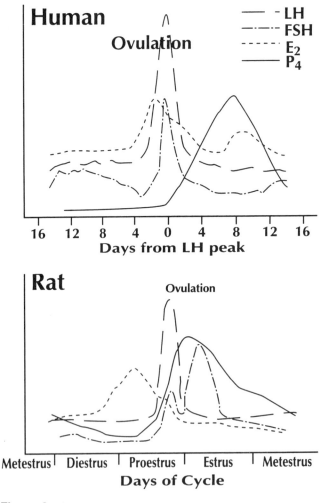

Figure 2 Comparison of the human and rodent female reproductive cycle. While the cycle
characteristics are markedly different between the human and rodent, the endocrinology centering around the LH surge is remarkably comparable as illustrated in this diagram.

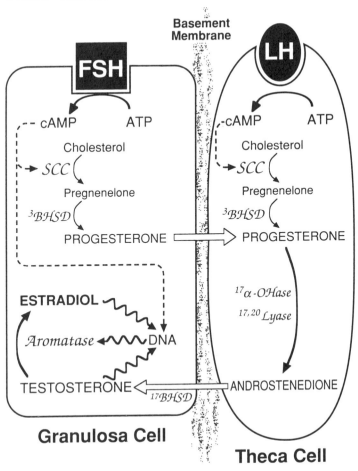

Figure 3 Two-cell, two-gonadotropin theory of estradiol production in the ovary. (Adapted from Hsueh et al., 1984.)

rian follicles. The granulosa and thecal cells are, in turn, dependent upon their own paracrine interaction to produce their hormones. Specifically, slight increases in LH increase the number of LH receptors on the thecal cells. LH stimulates androgen production from the thecal cell compartment and androgens diffuse through the basement membrane to the granulosa cell. Androgens have two actions on the granulosa cell (Hillier and DeZwart, 1981). First, androgens synergize with FSH action to induce aromatase P-450, the enzyme that converts testosterone to estradiol. Second, these androgens are the substrate for aromatase conversion to estradiol. Concomitantly, increasing levels of FSH increases the number of FSH receptors on the granulosa cells. FSH also upregulates aromatase P-450 and cholesterol side chain cleavage P-450, which is the rate-limiting enzyme in granulosa cell progesterone production (Richards and Hedin, 1988). Both androgens (Armstrong and Papkoff, 1976) and increasing levels of estradiol (Richards, 1980) further enhance the progesterone and estradiol-converting pathways. In turn, granulosa cell–produced progesterone diffuses across the basement membrane to the thecal compartment. Progesterone then becomes a substrate for additional thecal cell androgen production. In granulosa and thecal cell co-cultures, progesterone antibody inhibits testosterone by 39% and estrogen by 64% and testosterone antibody inhibits estrogen by 78%, thus illustrating the paracrine dependency of the follicular granulosa and thecal cells (Lui and Hsueh, 1986).

As an autocrine response to this hormonal milieu, granulosa cells acquire their own LH receptors. As granulosa cells become responsive to LH, these cells also produce more estradiol. Thus, aromatase activity, and consequently estradiol production, become linked to the LH surge (Fortune and Hilber, 1986). However, aromatase is rapidly downregulated in the granulosa cell after the LH surge (Richards and Hedin, 1988; Richards, 1994).

OVARIAN-TOXICANT INTERACTIONS

Unquestionably, the ovary is a complex tissue with an intricate set of independent and interdependent processes, each of which is a potential target for perturbation by environmental, occupational, or therapeutic agents. Thus, identifying the cellular and molecular pathways perturbed by toxicants is fundamental to understanding how toxicants can disrupt ovarian function. Although many cellular and molecular pathways are regulated by species-dependent mechanisms, many of the mechanisms are common between animals, including humans. Thus, identifying common pathways that are perturbed by toxicants will greatly enhance our ability to extrapolate effects from one species to another. As such, the examples provided in this chapter are organized by major targets of toxicant action such as follicular growth and differentiation and corpus luteum function with a discussion of what is known about the molecular pathways involved in mediating the toxicty. These toxicants all directly target the ovary, but the reader is reminded that there are reproductive toxicants that act on the hypothalamic/pituitary axis that may manifest as ovarian dysfunction (e.g., Smith, 1983; Vermeulen, 1993).

Mechanism: Depleting Follicle Numbers Through Irreversible Damage to Follicle Populations

Oocytes are a finite population in the female, and they represent a critical target site for chemical destruction (Mattison, 1983). Primordial oocytes in mice appear to be sensitive to a number of agents such as irradiation, polycyclic aromatic hydrocarbons, including benzopyrene and dimethylbenzanthracene (DMBA), cyclophosphamide (Shiromizu et al., 1984; Plowchalk and Mattison, 1992), and 4-vinylcyclohexene (Smith et al., 1990). Since irradiation (Baker, 1971), smoking cigarettes (a source of aromatic hydrocarbons) (Mattison and Thorgeirsson, 1978), and cyclophosphamide treatment (Warne et al., 1973; Koyama et al., 1977) are also associated with premature ovarian failure and sterility in women, it seems plausible that the ovarian toxicity in women exposed to such agents may be due to depletion of primordial germ cells.

Primordial oocyte susceptibility to chemical destruction is species dependent (Mandl, 1964; Baker, 1971; Mattison, 1979; Shiromizu et al., 1984). For example, cyclophosphamide treatment decreases the number of primordial follicles in immature C57BL/6N and DBA/2N mice but not in immature or adult Sprague-Dawley rats (Shiromizu et al., 1984; Jarrell et al., 1987). Similarly, vinylcyclohexene (VCH) decreases primordial follicle numbers in B6C3F1 mice but not in Fisher 322 rats (Smith et al., 1990). It has subsequently been shown that the epoxide of VCH decreases primordial follicle numbers in both rats and mice, suggesting that part of the differences in species susceptibility is due to differences in metabolism of the chemicals. The caveat to this conclusion is that if the metabolism is different between the mouse and rat, then the metabolism may also be different between the mouse, rat, and human. Consequently, without knowing each metabolic pathway in the rodent vs. human, we cannot predict whether human primordial oocytes will be susceptible to destruction by a specific chemical. Additionally, while it is clear that the number of primordial follicles is decreased in mice by such compounds, it is not clear that the cause of this decrease is due to a direct effect on the primordial oocyte or other components of the follicle (Shiromizu et al., 1984) or, if so, whether the human primordial oocyte is similarly susceptible to toxicity. Consequently, in order to establish any correlation of extrapolation of data from the mouse to the human, both the human metabolic pathways and the susceptibility of human primordial oocytes need to be determined for each chemical.

Alternatively, these chemicals may disrupt common pathways involved in follicular growth, and determining these common pathways may be most relevant to understanding how such toxicants cause premature ovarian failure. For example, cyclophosphamide (CP) is an alkylating agent that irreversibly binds DNA alkyl groups and is cytotoxic and mutagenic to proliferating cells (Colvin, 1976). Since both continuous and cyclical cell replication occurs in the ovary of all species, these cells should be particularly vulnerable to CP. Previous studies had already determined that CP treatment decreases granulosa cell numbers in prepubertal rats or immature PMSG-stimulated rats (Ataya et al., 1989) and decreases follicle and corpora lutea numbers in adult rats (Jarrell et al., 1987). CP exposure also blocks granulosa cells at G2 in the cell cycle in mice (Ataya et al., 1989). Thus, we propose that the proliferating cells in growing follicles are targets of CP and that destruction of this population could lead to ovarian failure not only in rats, but also in mice and, most importantly, in women.

To test this hypothesis, CP (200 mg/kg) was given to adult, female virgin Sprague-Dawley rats ($n = 6$–9) all on the same day of the cycle (in this case on proestrus), and then both ovarian morphology and serum hormone levels were temporally evaluated during naturally occurring ovulatory cycles. This experimental design is particularly useful in that it allows the comparison of both morphological and serum hormone changes from a chemical to the changes that naturally occur during the estrous cycle. After a single dose of CP, the initial and specific ovarian lesion in the rat ovary was granulosa cell necrosis and apoptosis in small, medium, and small-antral follicles (Figs. 4–7). Moreover, CP caused greatest damage to medium follicle granulosa cells, since all medium follicles were affected 24 hours after CP exposure and medium follicles contain the greatest proportion of proliferating granulosa cells. In comparison, CP damaged only 25% of small follicles, which are composed of proportionately fewer proliferating cells and have longer cell-cycle times (Hirshfield and Schmidt, 1987). We saw no evidence of CP-induced granulosa cell necrosis or other cellular morphological changes in the primordial and primary follicles. CP also had no effect on differentiated granulosa cells in preovulatory follicles. Preovulatory granulosa cells also responded to the LH surge, since all rats ovulated by 24 hours after treatment, and corpora lutea formed were quantitatively and qualitatively comparable to controls. Thus, the severity of follicular damage directly correlated to the proliferative activity of granulosa cells at the time of CP exposure, while resting or differentiated granulosa cells were spared from CP-induced toxicity. We saw no evidence of oocyte toxicity in rats, but after the

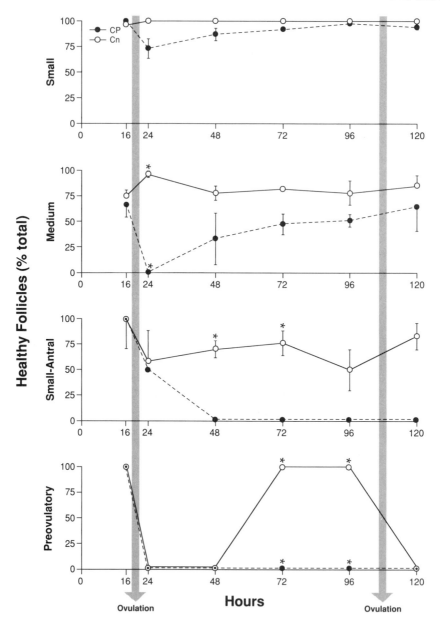

Figure 4 Effect of a single CP exposure on the morphology of specific follicle populations over time. Rats were given a single CP (200 mg/kg) or saline exposure at time 0 (morning of vaginal proestrus), and ovarian morphology was examined at subsequent times after treatment. Follicles were classified by mean cross-section diameter and counted in 20-step sections of one ovary from each of three rats per treatment group from 16 to 120 hours after treatment. Healthy follicles had less than two apoptotic bodies. Total follicles were healthy follicles plus atretic follicles (more than two apoptotic bodies) in controls, or healthy follicles plus follicles with necrotic granulosa cells in CP-treated rats. Healthy follicles were then expressed as a percentage of total follicles counted for each classification. Points represent mean $\pm$ SD (*$p < 0.05$).

(a)

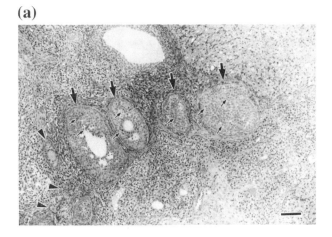

(b)

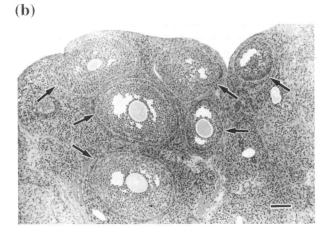

Figure 5 Rat ovary 24 hours after CP (a) or saline (b). After CP, all medium follicles contained granulosa cells with pyknotic nuclei (arrows) and a paucity of mitotic figures in contrast to medium follicles in control rats, which are healthy with numerous mitotic figures and typical of the day of estrus. (Bar = 100 μm.)

same dose given to mice ($n = 3$ per time point) under the same experimental design, medium follicles in mice were similarly affected as in the rats.

The loss of follicles had an immediate and a subsequent effect on serum hormone levels of the ovary and pituitary (Fig. 8). In particular, serum FSH levels increased in a biphasic manner in CP-treated rats compared to control levels. The increase in serum FSH would be predicted because of the interrelationships of the pituitary-ovarian feedback loops (Everett, 1961; Schwartz, 1969). Recall also that the growth of granulosa cells in medium follicles is FSH dependent. The increase in FSH would affect the growth rates of granulosa cells and actually increase their susceptibility to the antiproliferative, cytotoxic effects of CP. More follicles would be damaged, but FSH levels would continue to be increased, recruiting the growth of more follicles, and in that loop allow for the destruction of follicles leading to ovarian failure.

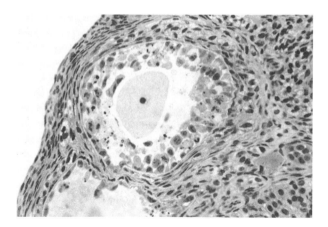

Figure 6 Rat ovary 48 hours after CP, medium follicle. The follicle represents morphology typical of CP-induced granulosa cell toxicity at this time-point. Granulosa cells were characterized by pyknotic, karyorrhectic, and karyolytic nuclei and dissolution of cytoplasm. Other cells were swollen and contained bizarre mitotic figures. (200 × original magnification.)

CP-induced premature ovarian failure has been previously attributed to follicle depletion through FSH-mediated increase in follicle growth (Ataya et al., 1985), although these investigators were unable to document the increase in FSH as we have reported. Given that growing, FSH-dependent follicles are the susceptible population to CP, it has since been hypothesized that antagonizing the action of FSH and slowing the growth of these follicles should protect against CP-induced ovarian injury. In fact, LHRH antagonists that downregulate pituitary responsiveness and hormone secretion effectively protect the ovary from significant follicular loss in both rodents and women (Ataya et al., 1985).

Thus, growing follicles appear to be as important in determining overall follicle numbers and growth as the more numerous primordial follicles. This had been apparent in our studies with CP, since a major consequence of CP-induced granulosa cell

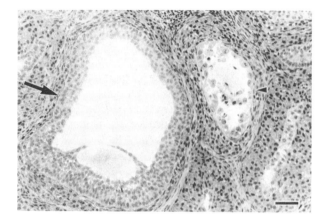

Figure 7 Rat ovary 48 hours after CP. Follicle damaged by CP (arrowhead) compared to atretric follicle (arrow). (Bar = 50 μm.)

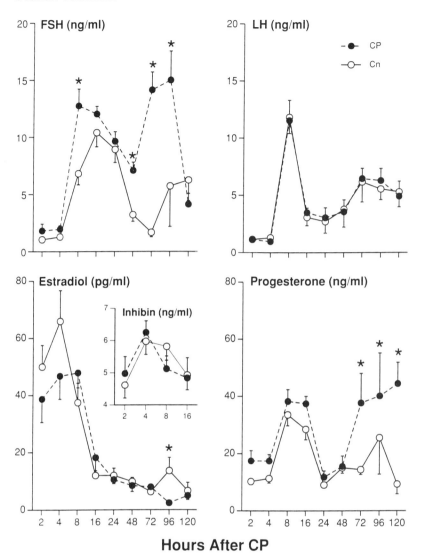

Figure 8 Effect of a single CP exposure on serum hormones over time. Rats were given a single exposure of CP (200 mg/kg) or saline at time 0 (0800 during vaginal proestrus). Serum was analyzed for estradiol, progesterone, FSH, and LH levels 8–120 hours after treatment in 6–9 rats per treatment group. In a separate study, inhibin, estradiol, progesterone, FSH, and LH were analyzed 2–24 hours after the single treatment in 6–9 rats per treatment group. Results were analyzed separately but combined for graphing purposes (*$p < 0.05$).

death had been the disruption of the overall balance of follicle growth (Fig. 9). Specifically, in untreated rats, the ratio of healthy follicles to total follicles had been constant over time, even given cycle-related fluctuations in medium and small antral follicles (Mandl and Zuckerman, 1950, 1952b; Hirshfield, 1978). However, CP killed growing follicles and thereby disrupted this overall healthy-to-total follicle balance. The response to injury had manifested as an increase in the number of growing follicles, as evidenced by significant healthy follicle number increases 24–72 hours after CP

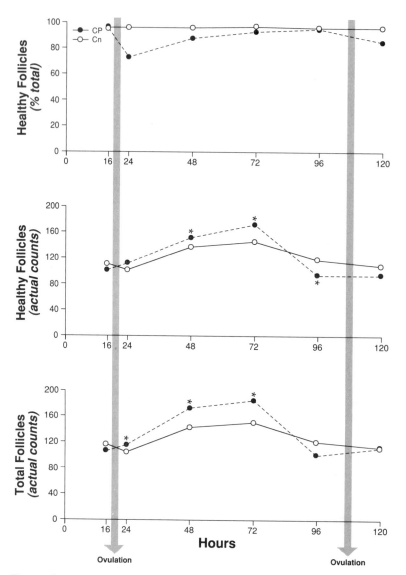

Figure 9 Effect of a single CP exposure on the ratios of healthy to total follicles over time. Rats were given a single CP or saline treatment and follicles were counted 8–120 hours after treatment in 20 sections per ovary in one ovary per rat in three rats per treatment group. Follicle classification and follicle health were determined as described in materials and methods. The percentage of healthy follicles was determined by dividing healthy follicle by total follicles and multiplying by 100. (*$p < 0.05$ refers to significance determined by actual counts.)

exposure. As the healthy follicle balance had been reestablished after insult, follicular growth also slowed, as evidenced by subsequent significant decreases in healthy follicles at 96 hours in treated rats. Thus, the ovary responds to pathological follicle loss by increasing the number of growth healthy follicles at the expense of the total follicle number of reserve. Depletion of total follicles may be manifested as premature ovarian failure.

Other studies also illustrate the importance of the growing follicle populations in determining follicular growth. In the developing prepubertal ovary, there is a "stock-piling" of medium follicles as more follicles enter this stage than leave (Faddy et al., 1984). This stockpiling is more important than initiation of growth from the resting pool in giving flexibility to the follicle system. In the rat unilateral ovariectomy, the mechanism for increasing preovulatory follicles is through an increased recruitment of medium follicles (Hirshfield, 1983).

In summary, chemical agents that deplete follicle numbers can have devastating consequences on ovarian function and the health of women. Studies in mice demonstrate that primordial follicle numbers can be altered by such compounds. Studies across species suggest that growing follicles are a critical target for such compounds and that the pathological effect is mediated both through direct ovarian toxicity and the physiological response of the pituitary.

Mechanism: Altering Steroidogenically Active Follicles

While growing follicles are susceptible to toxicity, steroidogenically active follicles are also susceptible to insult as demonstrated by the effect of di-(2-ethylhexyl) phthalate (DEHP), a ubiquitous environmental contaminant with endocrine-disrupting activity. Occupational exposure to high levels of DEHP has been associated with decreased rates of pregnancy, higher rates of miscarriage, and abnormal gynecological profiles characterized by hypoestrogenism and anovulation when compared to age-matched controls (Aldyreva et al., 1975). We have determined that DEHP specifically affects preovulatory follicles and granulosa cell differentiation, resulting in hypoestrogenic, hypoprogestinic, anovulatory cycles in female rats (Davis et al., 1994a). A primary insult results in a suppression of serum estradiol levels by about 30–40% over the estrous cycle of DEHP-treated Sprague-Dawley rats compared to controls. Since preovulatory (PO) follicle granulosa cells are uniquely composed of differentiated granulosa cells and are the primary source of estradiol in the cycling rat (Richards, 1980), we concluded that DEHP alters the function of these preovulatory granulosa cells. One hallmark of granulosa cell differentiation is a marked increase in cell size. Granulosa cells in PO follicles of DEHP-treated rats are morphologically smaller than granulosa cells of PO follicles in control rats at equivalent days of the cycle, further evidence that DEHP alters the ability of these cells to differentiate. Preceding follicles had been morphologically comparable between DEHP-treated and control rats.

The consequence of the PO follicle granulosa cell toxicity and suppression of serum estradiol is a failed ovulatory LH surge, since increasing estradiol levels are mandatory to induce the ovulatory LH surge (Everett, 1961; Schwartz, 1969). DEHP-treated rats did ovulate after simulation with an LH-like compound (hCG), further evidence that the suppression of estradiol is the primary mediator of the endocrine disruption and that DEHP-treated rats were capable of ovulating if given proper stimulation. As a consequence of the inability of the granulosa cells to differentiate and a failed LH surge, the ovaries of DEHP-treated rats become polycystic. The formation of follicular cysts and the pattern of granulosa cell death after DEHP treatment has also been reported when the LH surge is blocked by phenobarbital (Braw and Tsafriri, 1980).

The hCG experiments demonstrate that DEHP specifically targeted the ovary and not the pituitary. Serum FSH levels are also significantly increased as a result of DEHP-induced suppression of estradiol [recall that estradiol primarily inhibits FSH under physiological conditions (Schwartz, 1969)]. Thus, the disparate changes in FSH (increased) and LH (suppressed) in those experiments can be singularly attributed to DEHP-related suppression of granulosa cell estradiol production at the level of the ovary.

From these in vivo studies, we had determined that the steroidogenically active granulosa cell is the target for DEHP. With the target cell identified, studies were then conducted (and are still in progress) to determine the biochemical and molecular mechanisms by which DEHP alters estradiol production and cell differentiation. Initial studies utilized rat granulosa cell cultures and the active metabolite of DEHP, mono-(2-ethylhexyl) phthalate (MEHP).

Preovulatory granulosa cells produce estradiol from testosterone by aromatase-cytochrome P-450 enzyme, and both FSH from the pituitary and androgens from the neighboring thecal cells are necessary, acting synergistically, for induction and activation of aromatase in these cells (Richards, 1980). FSH acts through cyclic adenosine monophosphate (cAMP) and protein kinase A to regulate and activate aromatase (Richards and Hedin, 1988). Androgens from the thecal cells act as the substrate for aromatase, since granulosa cells lack the 17-α-hydroxylase, 17,20-lyase cytochrome P-450 enzyme complexes necessary to convert progesterone to androstenedione and testosterone, and testosterone binds to its receptor in the granulosa cell to further induce aromatase synthesis (Hillier and DeZwart, 1981). Thus, by supplying FSH and testosterone, cultured granulosa cells can produce estradiol.

Previous studies had determined that MEHP decreases FSH-stimulated cAMP and progesterone in rat granulosa cells (Treinen et al., 1990). To test the hypothesis that MEHP decreased estradiol production as a consequence of its suppression of FSH-stimulated cAMP pathways, estradiol production was compared in granulosa cells exposed to 0- to 100-μM concentrations of MEHP and stimulated by FSH or 8-br-cAMP, a nondegradable cAMP analog (Davis et al., 1994b). Estradiol levels were decreased 20% by 50 μM MEHP and 40% by 100 μM MEHP, regardless of whether estradiol was stimulated by FSH or by 8br-cAMP. Thus, MEHP decreases estradiol independent of its effect on FSH-cAMP, since MEHP decreases estradiol in the presence of an exogenous source of cAMP.

The results suggested that MEHP could have direct effects on aromatase. Consequently, aromatase activity was determined first in culture by measuring granulosa cell estradiol as a product of increasing concentrations of testosterone with or without 100 μM MEHP (Davis et al., 1994b). In cell cultures, MEHP significantly decreased the maximal velocity of aromatase, but not the apparent affinity of aromatase. When aromatase activity was measured directly in isolated granulosa cell microsomes incubated with or without 100 μM MEHP, MEHP had no effect on either the affinity or the velocity of aromatase. Since MEHP decreased the apparent velocity or maximum activity of aromatase in cell cultures but not in isolated microsomes, we concluded that MEHP alters the levels or availability of aromatase in the cell rather than acting as an aromatase inhibitor. Since cellular enzyme levels are regulated by synthesis and degradation, the next step was to examine the effects of MEHP on these processes. When aromatase synthesis was blocked by actinomycin D or cycloheximide, MEHP caused an additional decrease in estradiol. These observations suggested that MEHP affects estradiol by altering aromatase degradation and perhaps cell turnover or metabolism.

Both in vivo studies and in vitro studies of the effects of DEHP are useful to begin to understand how a chemical can impact steroidogenesis in the ovary in an animal model. These studies are also useful to determine the potential effects in humans, particularly since the pathways affected are similar between species. That is, estradiol is the primary endocrine product of the preovulatory follicle, and its release into the peripheral circulation toward the end of follicular development triggers the ovulatory surge of LH in both rats and women (Hillier, 1984). Since DEHP appears to specifically target differentiating granulosa cell estradiol production, the series of events that occurs in rats after DEHP exposure will likely occur in women after similar DEHP exposure. This conclusion is supported by the finding that women occupationally exposed to high concentrations of phthalates have hypoestrogenic anovulatory cycles (Aldyreva et al., 1975) as we described in the rat. Thus, the extrapolation of rodent data on the basis of common physiological and molecular pathways proves to be fundamental in determining how chemicals may impact ovarian function in women.

Mechanism: Altering Corpus Luteal Function

Given the above paradigm of the rat ovarian cycle, it could be anticipated that studying potential luteal cell toxicants in a cycling rat or mouse may be inappropriate with respect to the marked differences between the human and rodent luteal physiology. Consequently, we were surprised by the results of studies we conducted on ethylene glycol monomethyl ether (EGME) and its active metabolite 2-methoxy acetic acid (MAA).

A number of epidemiological studies have incriminated the short-chain ethylene glycol ethers as human reproductive toxicants. Most recently, a series of studies reported increased risks of spontaneous abortion and menstrual and fertility problems among solvent-exposed women in the electronic semiconductor industry (Pastides et al., 1988; Gold et al., 1995) and a dose-response relationship between estimated ethylene glycol exposure and each of these outcomes (Correa et al., 1996). EGME administered in drinking water significantly decreased fertility in female Sprague-Dawley rats (Gulati et al., 1990) and in mice (Gulati et al., 1988) in continuous breeding studies. Additionally, the EGME-treated C3H mice had significantly increased ovarian weights (Gulati et al., 1988). Moreover, both EGME and MAA administration in continuous breeding studies had caused decreased ovarian follicle counts in CD-1 mice (Heindel et al., 1989). Consequently, we wanted to establish the reproductive target site and define mechanisms of EGME toxicity in the nongravid adult, cycling, female rodent, hypothesizing that the follicle would be the target for EGME in the female.

Instead, the results of the studies showed that the ovarian luteal cell is a target of EGME and that the reproductive toxicity in the female rat is manifested morphologically as luteal cell hypertrophy and functionally as progesterone hypersecretion (Davis et al., 1997). The serum hormone levels in the EGME-treated rats suggested that progesterone is increased and corpora lutea hypertrophied independent of any pituitary hormone surges, since FSH, LH, and prolactin all remained at baseline level in EGME-treated rats compared to cycling controls. The morphological and physiological manifestations of EGME treatment mimicked a pregnant or pseudopregnant state in the rat in which high levels of progesterone are produced by hypertrophied corpora lutea and cyclicity is suppressed through progesterone-induced inhibition of FSH and LH surges (Everett, 1961).

To confirm the in vivo findings and determine potential mechanisms of EGME toxicity, we cultured rat luteal cells obtained from hCG-treated immature rats and exposed these cells to MAA. MAA elevated progesterone production at 1 mM compared to controls after 48 hours of culture (Davis et al., 1997). Temporal evaluation suggested that 5 mM MAA caused a continued production of progesterone at 24 and 48 hours of culture, while progesterone production declined in controls.

Luteal cell progesterone production is stimulated by LH- and cAMP-dependent pathways (Richards et al., 1983). Consequently, we hypothesized that MAA maintained progesterone production through LH-stimulated cAMP pathways, but found that MAA increased progesterone whether cells were or were not stimulated with LH. Moreover, levels of cAMP were not significantly altered by any concentration of MAA. These studies suggest that MAA maintains progesterone production independent of LH and cAMP pathways.

Both the in vivo studies and in vitro studies demonstrate that EGME and MAA apparently inhibit luteal cell death and maintain progesterone secretion in the cycling female rat and in the cultured luteal cell. This suggests that EGME may cause menstrual cycle disruption and subfertility in women as well.

However, our rodent data are limited in their direct applicability to predicting EGME reproductive toxicity in women, not only because of the different routes of exposure, but also because of the distinct difference in the function of the corpora lutea in rats and women. Specifically, in the rat, EGME stimulates luteal cell progesterone during a cycle in which the luteal cell does not normally produce progesterone. Women naturally have a luteal phase of their cycle characterized by luteal cell progesterone secretion. Consequently, it is difficult to directly extrapolate how EGME may effect luteal function in women.

In order to better assess the potential hazard of EGME and MAA to women, we then conducted studies to determine whether the same concentrations of MAA increase progesterone in human luteinized granulosa cells as in rat luteal cells (Almekinder et al., unpublished). Human cells were collected from healthy anonymous oocyte donors and treated with 10 IU hCG and 0–5 mM MAA for 6–48 hours. We found that MAA increased progesterone production in cultured human luteal cells at the same concentration as it increased progesterone in rat luteal cells. The in vitro effects of MAA on rat luteal cells are comparable to the in vivo effects of EGME in adult female rats at equivalent concentrations (Davis et al., 1997). Such parallelisms between the human and rat in vitro, and the rat in vitro and in vivo provide compelling evidence that the ovarian luteal cell is a target of EGME in women and may account for a proportion of the reported reproductive dysfunctions in occupationally exposed women.

CONCLUSION

All of these studies serve to illustrate the specific mechanisms by which ovarian toxicants disrupt ovarian function. In other words, ovarian toxicants have multiple modes of action. Further, no single process appears more or less susceptible to the effects of environmental, occupational, or therapeutic agents than any other process. Moreover, chemicals may affect the ovary very differently and still result in a disruption of the endocrine balance, infertility, or premature ovarian failure. Thus, as we continue to identify ovarian toxicants and attempt to determine their potential impact on the health of women, we must continue to recognize common pathways between species, identify cell targets and molecular pathways, and work between in vivo and in vitro assays within rodent systems and human systems. Clearly, more studies are needed to determine molecular mechanims and relate toxicant effects between species in order to advance our knowledge and better predict the risk of exposure to environmental agents.

REFERENCES

Adashi, E. Y. (1990). *Gamete Physiology* (R. H. Asch, J. P. Balmaceda, and I. Johnston, eds.). Norwell, MA.

Aldyreva, M. V., Klimova, R. S., Izyumova, A. S., and Timofievskaya, L.A. (1975). *Gi. Trud. Prof. Zaol., 25*: 25.

Almekinder, J. L., Lenndard, D. E., Walmer, D. K., and Davis, B. J. (1997). *Fund. Appl. Toxicol. 38*: 191–194.

Armstrong, D., and Papkoff, H. (1976). *Endocrinology, 99*: 1144.

Ataya, K., McKanna, J., Weintraub, A., Clark, M., and LeMaire, W. (1985). *Cancer Res., 45*: 3651.

Ataya, K. M., Valeriote, F. A., and Ramahi-Ataya, A. J. (1989). *Cancer Res., 49*: 1660.

Baker, T. (1972). *Reproductive Biology* (H. Balin and S. Glasser, eds.). Excerpta Medica, Amsterdam.

Baker, T. G. (1971). *Am. J. Obstet. Gynecol., 100*: 746.

Baranski, B. (1993). reproductive outcomes. *Environ. Health Perspects., 101*(suppl. 2): 81.

Barlow, S., and Sullivan, F. (1982). *Reproductive Hazards of Industrial Chemicals*. Academic Press, New York.

Beaumont, H. M., and Mandl, A. M. (1962). *Proc. R. Soc. London, 155*: 556.

Braw, R. H., and Tsafriri, A. (1980). *J. Reprod. Fert., 59*: 259.

Byskov, A. (1974). *Nature, 252*: 396.

Chapin, R., and Sloane, R. (1997). breeding and evolving study design and summaries of 88 studies. *Environ. Health Perspect., 105* (suppl.): 199–395.

Chapman, R., Sutcliffe, S., and Malpas, J. (1979). *JAMA, 242*: 1877.

Colvin, M. (1976). *Cancer Res., 36*: 1121.

Correa, A., Gray, R., Cohen, R., Ruthman, N., Shah, F., Seacat, H., and Corn, M. (1996). *Am. J. Epidemiol., 143*: 707.

Davis, B., Maronpot, R., and Heindel, J. (1994a). *Toxicol. Appl. Pharmacol., 128*: 216.

Davis, B., Weaver, R., Gaines, L., and Heindel, J. (1994b). *Toxicol. Appl. Pharmacol., 128*: 224.

Davis, B., Almekinder, J., Flagler, N., Travlos, G., Wilson, R., and Maronpot, R. (1997). *Toxicol. Appl. Pharmacol. 142*: 328–337.

Erickson, G. F. (1983). *Mol. Cell Endocrinol., 29*: 21.

Erickson, G. F., and Hsueh, A. J. W. (1978). *Endocrinology, 102*: 1275.

Everett, J. (1961). *Sex and Internal Secretions* (W. Young, ed.). Williams & Willkins, Baltimore, p. 497.

Faddy, M. J., Jones, E. C., and Edwards, R. G. (1984). *J. Exp. Zool., 197*: 173.

Fortune, J. E., and Hilber, J. L. (1986). *Endocrinology, 143*: 2395.

Gold, E. B., Eskenazi, B., Hammond, S. K., Lasley, B. L., Samuels, S. J., Rasor, M. O., Hines, C. J., Overstreet, J. W., and Schenker, M. B. (1995). *Am. J. Ind. Med., 28*: 799.

Goldenberg, R. L., Vaitukaitis, J. L., and Rose, G. T. (1972). *Endocrinology, 90*: 1492.

Greenhall, E., and Vessey, M. (1990). *Fertil. Steril., 54*: 978.

Greenwald, G. S. (1974). *Handbook of Physiology* (R. O. Greep and E. B. Astwood, eds.). Williams & Wilkins, Baltimore, p. 293.

Gulati, D., Hope, E., Barnes, L., Russell, S., Poonacha, K., Chapin, R., and Morrissey, R. (1988a). *Ethylene Glycol Monomethyl Ether: Reproduction and Fertility Assessment in C3H Mice when Administered in Drinking Water* (Final Study Report). National Toxicology Program, Washington, D.C.

Gulati, D., Hope, E., Barnes, L., Russell, S., Poonacha, K., Chapin, R., and Morrissey, R. (1988b). *Ethylene Glycol Monomethyl Ether: Reproduction and Fertility Assessment in C57BL/6 Mice when Administered in Drinking Water* (Final Study Report). National Toxicology Program, Washington, D.C.

Gulati, D., Hope, E., Barnes, L., Russell, S., Poonacha, K., Chapin, R., and Morrissey, R. (1988c). *Ethylene Glycol Monomethyl Ether: Reproduction and Fertility Assessment in CD-1 Mice when Administered in Drinking Water* (Final Study Report). National Toxicology Program, Washington, D.C.

Gulati, D., Hope, E., Barnes, L., Russell, S., Poonacha, K., and Chapin, R. (1990). *Development of Reproduction and Fertility Assessment Protocol in CD Sprague-Dawley Rats: Litter 5 Design*. National Toxicology Program, Washington, D.C.

Guraya, S. S., and Greenwald, G. S. (1964). *Anat. Rec., 149*: 411.

Heindel, J. J., Thomford, P. J., and Mattison, D. R. (1989). Growth Factors and the Ovary (A. N. Hirshfield, ed.). Plenum Press, New York, 421.

Hillier, S., and DeZwart, F. (1981). *Endocrinology, 109*: 1303.

Hirshfield, A. (1978). *Biol. Reprod., 19*: 606.

Hirshfield, A. (1986). *Biol. Reprod., 35*: 113.

Hirshfield, A. N. (1983). *Biol. Reprod., 28*: 271.

Hirshfield, A. N. (1985). *Biol. Reprod., 32*: 979.

Hirshfield, A. N. (1987). *Reprod. Toxicol., 1*: 71.

Hirshfield, A. N. (1989). *Biol. Reprod., 41*: 309.

Hirshfield, A. N., and Midgley, A. R. (1978). *Biol. Reprod., 19*: 597.

Hirshfield, A. N., and Schmidt, W. A. (1987). *Regulation of Ovarian and Testicular Function* (V. Mahesh, D. Dhindsa, E. Andersone, and S. Kalra, eds.). Plenum Press, New York.

Hsueh, A. J. W., Adashi, P. B. C., and Welsch, T. H. (1984). *Endocrine Rev., 5*: 76.

Hsueh, A., Billig, H., and Tsafriri, A. (1994). *Endocrine Rev., 15*: 707.

Hughes, F., and Gorospe, W. (1991). *Endocrinology, 129*: 2415.

Hurwitz, A., and Adashi, E. (1992). *Mol. Cell Endocrinol., 84*: 19.

Jarrell, J., YoungLai, E., Barr, R., McMahon, A., Belbeck, L., and O'Connell, G. (1987). *Cancer Res., 47*: 2340.

Jones, E. C., and Krohn, P. L. (1961). *Endocrinol., 21*: 469.

Koyama, H., Wada, T., Nishzawa, Y., Iwanaga, T., Yukitoshi, A., Terasawa, T., Kosaki, G., Yamomoto, T., and Wada, A. (1977). *Cancer Res., 39*: 1043.

Krarup, R., Pedersen, T., and Faber, M. (1969). *Nature, 224*: 187.

LeFevre, J., and McClintock, M. K. (1988). *Biol. Reprod., 38*: 780.

Long, J. A., and Evans, H. M. (1922). Mem Univ Calif, *6*: 1.

Lui, Y. S., and Hsueh, A. J. W. (1986). *Biol. Reprod., 35*: 27.

Mandl, A. (1964). *Biol. Rev., 39*: 288.

Mandl, A. M. (1951). *J. Exp. Biol., 28*: 576.

Mandl, A. M., and Zuckerman, S. (1950). *J. Endocrinol., 6*: 426.

Mandl, A. M., and Zuckerman, S. (1952a). *J. Endocrinol., 8*: 126.

Mandl, A. M., and Zuckerman, S. (1952b). *J. Endocrinol., 8*: 341.

Mattison, D. (1983). *Reprod. Toxicol., 117*: 65.

Mattison, D. R. (1979). *Chem-Biol. Interact., 48*: 133.

Mattison, D. R. (1985). *Reproductive Toxicology* (R. L. Dixon, ed.). Raven Press, New York, p. 109.

Mattison, D. R., and Thorgeirsson, S. S. (1978). *Lancet* (Jan 28): 187.

Mosher, W., and Pratt, W. (1985). *Fecundity and Infertility in the United States 1965-1982, National Center for Health Statistics Advance Data*. Vital and Health Statistics. Public Health Service, Washington, D.C.

Nimrod, A., Erickson, G. F., and Ryan, K. J. (1976). *J. Endocrinol., 98*: 56.

Otani, T., and Sasamoto, S. (1982). *J. Reprod. Fert., 65*: 347.

Pastides, H., Calabrese, E., Hosmer, D., and Harris, D. (1988). *J. Occupat. Med., 30*: 543.

Pedersen, T. (1970). *Acta Endocrinol., 64*: 304.

Pedersen, T., and Peters, H. (1968). *J. Reprod. Fert., 17*: 555.

Peppler, R., and Greenwald,, G. S. (1970). *Am. J. Anat., 127*: 1.

Peters, H., and McNatty, K. P. (1980). *The Ovary.* Paul Elek, New York.

Plowchalk, D., and Mattison, D. (1992). *Reprod. Toxicol., 6*: 411.

Rao, I. M., Mill, T. M., Anderson, E., and Mahesh, V. B. (1991). *Anat. Rec., 229*: 177.

Richards, J. (1994). *Endocrine Rev., 15*: 725.

Richards, J. S. (1975). *Endocrinology, 97*: 1174.

Richards, J. S. (1980). *Physiol. Rev., 60*: 51.

Richards, J. S., and Hedin, L. (1988). *Ann. Rev. Physiol., 50*: 441.

Richards, J. S., Ireland, J. J., Rao, M. C., Bernath, G. A., Midgley, A. R., and Reichert, L. E. (1976). *Endocrinology, 99*: 1562.

Richards, J. S., and Midgley, A. R. (1976). *Biol. Reprod., 14*: 82.

Richards, J. S., Sehgal, N., and Tash, J. S. (1983). *J. Biol. Chem., 258*(8): 5227.

Sakai, C., and Hodgen, G. (1987). *Reprod. Toxicol., 1*: 207.

Satoh, M. (1985). *J. Anat., 143*: 17.

Schwartz, N. B. (1969). *Rec. Prog. Horm. Res., 25*: 1.

Schwartz, N. B. (1974). *Biol. Reprod., 10*: 236.

Shardein, J. (1988). *Product Safety Evaluation Handbook* (S. Gad, ed.). Marcel Dekker, New York, p. 291.

Shiromizu, K., Thorgeirsson, S., and Mattison, D. (1984). *Pediatr. Pharmacol., 4*: 213.

Smith, B., Carter, D., and Sipes, I. (1990). *Toxicol. Appl. Pharmacol., 105*: 364.

Smith, C. (1983). *Am. J. Indust. Med., 4*: 107.

Smith, M. S., Freeman, M. E., and Neil, J. D. (1975). *Endocrinology, 96*: 319.

Teerds, K., and Dorrington, J. (1993). *Biol. Reprod., 49*: 989.

Terranova, P. F. (1981). *Endocrinology, 108*: 1885.

Tilly, J., Kowlaski, K., Johnson, A., and Hsueh, J. (1991). *Endocrinology, 129*: 2799.

Treinen, K., Dodson, W., and Heindel, J. (1990). *Toxicol. Appl. Pharmacol., 106*: 334.

Vermeulen, A. (1993). *Environ. Health Perspect., 101*(suppl. 2): 91.

Wang, X., Roy, S. K., and Greenwald, G. S. (1991). *Biol. Reprod., 44*: 857.

Warne, G. L., Fairley, K. F., Hobbs, J. B., and Martin, F. I. R. (1973). *N. Engl. J. Med., 289*: 1159.

Welschen, R. (1973). *Acta Endocrinol., 72*: 137.

Welschen, R., Westof, R., Westof, K. F., Braendle, W. L., and Dizeraga, G. S. (1991). *Biol. Reprod., 44*: 461.

Yen, S. (1986). *Reproductive Endocrinology.* W.B. Saunders Company, Philadelphia.

Zeleznik, A. J., Hillier, S. G., and Ross, G. T. (1979). *Biol. Reprod., 21*: 673.

16

Toxicology of Early Pregnancy, Implantation, and Uterine Function

Audrey M. Cummings
U.S. Environmental Protection Agency, Research Triangle Park, North Carolina

EARLY PREGNANCY AND IMPLANTATION

Background

Early pregnancy is a period wherein a series of complex physiological events are interwoven to yield a successful outcome. From fertilization and embryo transport through early embryonic development and implantation, these events are closely regulated by ovarian steroid hormones. Each phase of early pregnancy is critically related to every ensuing phase as this synchronous course of events is ultimately orchestrated.

Fertilization is accomplished by the union of sperm and ovum in the oviduct (Bedford, 1982). As shown in Figure 1, the sperm initially makes contact with the cumulus oophorus (Fig. 1A). Then (Fig. 1B) the sperm binds to the surface of the zona pellucida. The first polar body was extruded during meiosis I, and the chromosomes are now in meiotic arrest. Next, the sperm fuses with the egg wall after it has penetrated the zona pellucida (Fig. 1C). In Figure 1D, the sperm head has begun to swell and the second polar body is being emitted. In Figure 1E, the second polar body has been emitted and the sperm chromatin has decondensed. Then (Fig. 1F) both pronuclei are formed. In Figure 1G, one sees that the two pronuclei have come into apposition, and in Figure 1H the chromosomes of the male and female are aligned on the mitotic spindle, which begins the first cleavage division. The new embryo now has a full complement of chromosomes (Bedford, 1982).

During the preimplantation period, the fertilized ovum is dividing, developing, and traveling through the oviduct to the uterus (Fig. 2A). Both the rate of embryo transport and the stage of embryonic development must be synchronized with the preparation of the uterus to a stage that is capable of supporting pregnancy. During the transition from the morula stage to the blastocyst stage, the embryo enters the uterus (Johnson and Everitt, 1980a). The zona pellucida is still intact and the embryo receives nourishment from the fluids secreted by the oviduct and uterus. Simultaneously, the endometrium is undergoing preparation to a stage of receptivity for the blastocyst (Fig. 2B). This preparation is strictly regulated by estrogen and progesterone and provides a narrow window of opportunity for implantation (Johnson and Everitt, 1980a). The

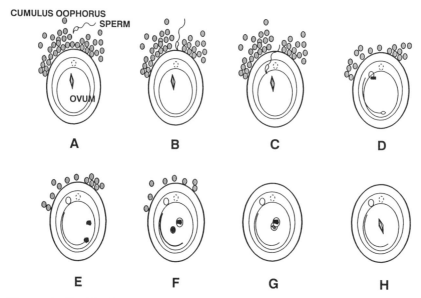

Figure 1 Fertilization of the rat ovum by sperm. See text for details. (Adapted from Bedford, 1982.)

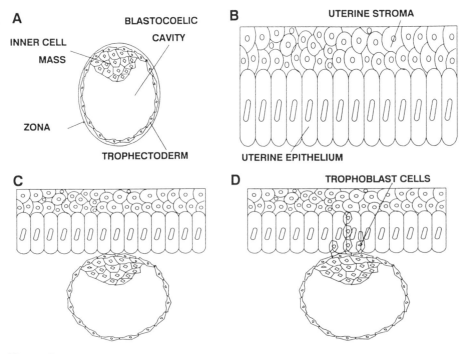

Figure 2 (A) Preimplantation blastocyst. (B) Preimplantation endometrium. (C) Apposition of blastocyst and endometrium. (D) Invasion of endometrium by blastocyst. (Adapted from Johnson and Everitt, 1980.)

synchronization and timing of blastocyst development and uterine preparation for receptivity is critical for successful implantation.

Implantation is a critical stage of early pregnancy when the embryo beings its attachment to the uterine lining. This attachment is the beginning of a process that culminates in the formation of the placenta, a structure that provides essential nutrition for the developing embryo and fetus (Johnson and Everitt, 1980a). At the beginning of the peri-implantation interval, the blastocyst makes contact with the endometrium at the site of implantation (Fig. 2C). The zona pellucida is dissolved, permitting contact between the trophectoderm of the blastocyst and the uterine epithelium. This stage is called apposition. The next step is adhesion, where the trophectoderm and luminal epithelium become adherent. The third stage, attachment of the blastocyst and endometrium, involves very close contact between the trophectoderm and uterine epithelium with interlocking microvilli and changes in cell surface molecules (Johnson and Everitt, 1980a; Weitlauf, 1994).

Events after attachment vary considerably among species. The fourth stage, invasion or penetration of the uterine epithelium, occurs in some species including the human and the rat. The conceptus breaks through the surface epithelium and invades the stroma (Fig. 2D). This invasion can be divided into three phases. The first is a signal that is initiated by the blastocyst, transduced through the epithelium, amplified, and affects the stroma to induce decidualization. There is an initial increase in vascular permeability followed by a differentiation of stromal cells to decidual cells. These decidual cells, which are often polyploid, proliferate rapidly, resulting in a massive growth of decidual tissue. This decidual tissue eventually becomes a maternal component of the placenta and may play a role in the nourishment of the blastocyst as well as the restriction of its invasion. A few hours after the initiation of implantation, the surface epithelium under the blastocyst erodes and trophectodermal processes move between adjacent epithelial cells. Some trophectoderm cells fuse and form a syncytium, while others remain cellular. Uterine glandular tissue in front of the conceptus is destroyed and consumed by the invading trophoblast. The depth of conceptus invasion varies according to species. After the decidua has been invaded, implantation per se is complete. There is a physical hold and a nutritional source for the developing embryo prior to placental formation (Johnson and Everitt, 1980a).

Following, or even simultaneously with, these peri-implantation events, other processes important to pregnancy are occurring, such as the signal to provoke the maternal recognition of pregnancy. This is a signal to convert the cyclic ovary to a noncyclic ovary where the corpus luteum produces sufficient progesterone to maintain pregnancy, at least temporarily. In the human, the signal is human chorionic gonadotropin (hCG), which is secreted by the trophoblast and acts to promote corpus luteum differentiation and progesterone secretion. In the rat, diurnal surges of prolactin from the pituitary during the first week of pregnancy activate the corpus luteum (Johnson and Everitt, 1980a; Weitlauf, 1994). These peaks of prolactin in the rat are generated as a result of the cervical stimulation occurring during coitus.

Due to the difficulty of studying such an intricate process as early pregnancy, the chemical mechanisms involved in implantation and uterine-embryo interactions have been undertaken on a piecemeal basis. Epithelial cell markers, for example, appear to be many and varied. The receptor for colony-stimulating factor-1 (CSF-1) is first detected in cleavage-stage mouse embryos and becomes restricted to trophoblasts (Cross et al., 1994). Estrogen appears to induce the uterine epithelium to secrete cytokines, including members of the epidermal growth factor (EGF) family and leukemia inhibitory factor (LIF) (Cross et al., 1994). EGF, transforming growth factor-alpha (TGFα), heparin-binding EGF (HB-EGF), and amphiregulin are produced in the uterus during the peri-implantation period.

As stated by Edwards (1994), cytokines are major intercellular signals secreted by leukocytes. In the human uterus, IGF-I and IGF-II regulate secretory endometrium; IGF-I receptor transcripts are induced in epithelial cells by estrogens (Edwards, 1994). EGF stimulates epithelial cells and glandular stromal proliferation such that EGF-R mediates mitogenesis, lactoferrin mRNA synthesis, and protein synthesis. The uterine decidua synthesizes ceruloplasmin, α_1-antitrypsin, secretory component, t-piece, complements C3 and C4, LIF, macrophage colony-stimulating factor (MCSF), and IL-6 as the metrial gland is formed (Edwards, 1994). In mice, LIF is expressed in the uterine endometrial glands on day 4 of pregnancy (Stewart et al., 1992). This expression of LIF appears to be under maternal control and precedes implantation of the blastocyst. Female mice lacking a functional LIF gene cannot implant their blastocysts even though they are fertile by other measures such as ovulation (Stewart et al., 1992).

The importance of the various cytokines and growth factors in implantation may be difficult to determine. The gene encoding HB-EGF is expressed only in the luminal epithelium at the site of blastocyst apposition starting 7 hours before attachment (Das et al., 1994). It is induced after estradiol injections in delayed implanting animals. Markoff et al. (1995) have demonstrated, via Western blots, a dramatic increase in IGF-binding protein-4 at the time of embryo implantation (gestation day 4) in the mouse.

Immunocytochemical examination has revealed EGF, TGF-α, and EGF-receptor (EGF-R) in the luminal glandular epithelium of uteri on days 4, 5, and 6 of pregnancy in rats (Johnson and Chatterjee, 1993). EGF-R was also found in the implanting embryo and in decidual cells of the stroma (Johnson and Chatterjee, 1993). the intraluminal injection of EGF into uterine horns induced implantation in the ovariectomized progesterone-treated delayed implanting rat (Tamada et al., 1994), and this induction was inhibited by indomethacin, a prostaglandin synthesis inhibitor. According to Yee et al. (1993), the concentration of prostaglandins of the E, F, and I series is elevated at implantation sites and in the uterus after application of a decidual stimulus. Inhibitors of leukotriene synthesis also inhibit uterine decidualization (Yee et al., 1993).

The cytokine granulocyte-macrophage colony-stimulating factor (GM-CSF) is also found in uterine epithelial cells, is regulated by estrogen and progesterone, and may play a role in the remodeling events in the endometrium during the preimplantation period (Robertson et al., 1996). Experiments reported by Grummer et al. (1994) suggest that the spatial and temporal pattern of the expression of connexin 26 and 43 (gap junctional proteins) in response to implantation and to estrogen and progesterone demonstrate a role for these proteins in early pregnancy.

Vulnerability of Early Pregnancy to Xenobiotics

So how might early pregnancy be affected by xenobiotics? First, the rate of transport of the embryo may be accelerated or retarded, leading to an asynchrony between the conceptus and the uterus. A chemical may affect the preparation of the endometrium such that receptivity is not achieved. Or a chemical may somehow affect the interaction between the conceptus and the uterus and thereby prevent attachment or invasion. The embryo itself may be affected, resulting in embryonic death or malformation. Each of these events during early pregnancy is vulnerable to toxic insult, which can result in pregnancy failure. While such evaluations are not feasible in women, one can assess the vulnerability of each of these sites and stages to specific reproductive toxicants in animal models.

Potential sites of vulnerability thus include, in chronological order, embryo transport, uterine preparation, conceptus-uterine interaction, decidualization, maintenance of pregnancy, and, overall, hormonal support. In this regard, a series of protocols has been assembled and validated in our laboratory for the assessment of toxic effects on early pregnancy in the rat. This panel was developed in order to provide a means to evaluate early pregnancy loss and to probe the point of action at which chemicals may reduce fertility. The protocols include the Early Pregnancy Protocol: Dose Response Evaluation; the Decidual Cell Response Technique; the Pre- versus Postimplantation Protocol; and Embryo Transport Rate Analysis (Cummings, 1990).

Chemicals are often chosen on the basis of the results of subchronic or other studies where reduced fertility has been observed as a consequence of chemical exposure. The chemical is then evaluated in the Early Pregnancy Protocol (Cummings, 1990). This first protocol can assess the effect of toxicants on a range of endpoints during early pregnancy and determine a dose-response curve for reduced fertility. In these experiments (Fig. 3A), female rats are bred with proven fertile, untreated males; a sperm-positive vaginal smear on the day of estrus corresponds to day 0. Animals are dosed during days 1–8 of pregnancy; this dosing regimen begins after fertilization and en-

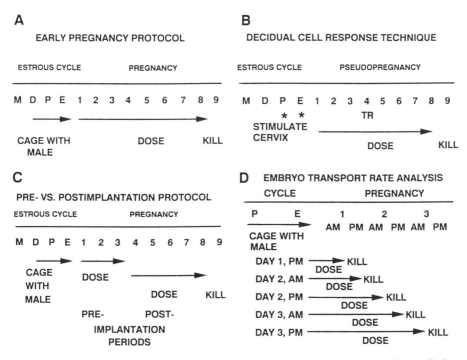

Figure 3 (A) Early Pregnancy Protocol. (B) Decidual Cell Response Technique. (C) Pre- vs. Postimplantation Protocol. (D) Embryo Transport Rate Analysis. M, D, P, and E represent the days of the estrous cycle; 1–9 are the days of early pregnancy. TR = Surgical decidual induction. (Adapted from Cummings, 1990.)

compasses implantation on day 4 and the postimplantation period of decidual growth. Rats are killed on day 9, at which time a range of endpoints is evaluated at necropsy, including the number and weight of the implantation sites, serum steroid hormone levels, ovarian weight, and corpora lutea number. The dose-response evaluation is performed using a range of dosages of the chemical under investigation. If a reduction in the number of implantation sites is seen, the investigation is continued to determine the mechanisms responsible for embryonic loss.

If the number of implantation sites is affected, the Decidual Cell Response (DCR) Technique is applied (Cummings, 1990). The DCR can identify direct uterine effects of toxic chemicals and distinguish between embryotoxic and maternal effects. The DCR is a model for maternal physiological events associated with implantation, it mimics the response of the uterus occurring after embryo implantation, and it can distinguish embryotoxicity from direct effects of chemicals on maternal reproductive physiology. In this protocol (Fig. 3B), manual stimulation of the uterine cervix, performed on proestrus and estrus (or mating of the females with vasectomized males), results in the initiation of pseudopregnancy. This is an induced phase of the rat reproductive cycle that is endocrinologically similar to early pregnancy but lacks fertilized ova. Animals are dosed during days 1–8 of pseudopregnancy. On day 4, the day of receptivity for implantation (if embryos were present), the DCR is induced by surgical uterine trauma. This procedure mimics the effect of implanting blastocysts but affects the entire length of the uterine horn rather than discrete implantation sites. The animals are killed on day 9 of pseudopregnancy, at which time decidual growth is evaluated by weighing the trimmed uterus. Treatment with a chemical during the DCR technique can result in an inhibition of decidual growth along a continuum from barely significant to partially inhibited to 100% blockade of growth. If no effect on the DCR is seen, then toxicity is most likely directed at the embryo and not the uterus. A chemically induced inhibition of the DCR usually signifies a direct uterine effect, potentially mediated via CNS effects. Then further investigation of the molecular mechanisms underlying such an effect may be considered. These investigations will be discussed later in this chapter.

At this stage the Pre- versus Postimplantation Protocol is applied (Cummings, 1990). This protocol can evaluate effects on specific peri-implantation intervals (such as the pre- and postimplantation periods). It can also establish sensitivities of each interval to dosage of administered compounds. In this test (Fig. 3C), rats are bred exactly as in the Early Pregnancy Protocol. The animals are dosed during days 1–3, the preimplantation period, or during days 4–8, the postimplantation interval. All rats are killed on day 9 of pregnancy and multiple endpoints are evaluated, including the number of implantation sites. If a postimplantation effect is seen, the problem is one of continued pregnancy maintenance, which may involve the uterus, the ovary, the embryo, or hormonal support. If a preimplantation loss is detected, an Embryo Transport Rate Analysis is performed.

The Embryo Transport Rate Analysis can measure changes in embryo transport rate and relate these changes to pregnancy failure (Cummings, 1990). In this technique (Fig. 3D), female rats are bred as in the Early Pregnancy Protocol. Animals are dosed at 0900 h during days 1–3 of pregnancy. Groups of rats are killed on the afternoon of day 1 and the mornings and afternoons of days 2 and 3. Embryos are flushed from oviducts and uteri, separately, and the number of embryos found in each segment are counted. An acceleration or retardation of embryo transport rate implicates preimplantation timing as a factor in early pregnancy loss. The finding of no effect on embryo transport rate suggests either uterine or embryonic toxicity, between which the DCR can distinguish.

Each of these tests thus provides information that, when taken together, can be used to delineate some of the toxicological mechanisms by which a chemical may produce early embryonic loss following acute exposure. Some examples of how these evaluations have been used to investigate specific chemicals may clarify the benefit of their use.

Figure 4A shows data from an Embryo Transport Rate Analysis using methoxychlor (MXC) at 0, 100, 200, and 500 mg/kg/day (Cummings and Perreault, 1990). The figure shows the number of embryos found on day 1 in the oviduct and in the uterus as well as the total number found. There was no effect on MXC on these numbers on day 1 when compared with control. On the morning of day 2 (Fig. 4B) there is a different pattern. As the dose of MXC was increased, fewer embryos were found in the oviduct and more embryos were found in the uterus; embryos were ultimately lost from the uterus also. The total number of embryos found on day 2 was reduced at 500 mg/kg/day MXC. By the morning of day 3 (Fig. 4C), all doses of MXC produced a loss of embryos from the oviduct, shifting them to the uterus. At the highest dose, embryos were completely lost from the tract and expelled. Any disruption of the synchrony between the preparation of the uterus for pregnancy and the timely arrival of embryos in the uterus can result in pregnancy failure. Thus, MXC accelerates embryo transport rate through the female tract. The embryos arrive in the uterus too early and the synchronization with uterine preparation is lost. The result is preimplantation embryonic loss.

When bromocryptine was administered during days 1–8 of pregnancy in the Early Pregnancy Protocol (Cummings et al., 1991), there was a reduction in the number of implantation sites found on day 9 at the higher dosages (Fig. 5A). The measurement of progesterone levels on day 9 (Fig. 5B) revealed a reduction in the hormone levels parallel with the loss of embryos at 0.06, 0.08, and 0.10 mg bromocriptine/rat/day. The DCR Technique (Fig. 5C) demonstrated that there was an inhibitory effect of bromocriptine on decidual growth. This effect may be mediated through the change in progesterone levels. When the Embryo Transport Rate Analysis was performed (Fig. 5D), data from days 2 and 3 revealed no effect of bromocriptine on embryo transport rate. The number of embryos found in the oviduct and the uterus were essentially the same for every dose. Bromocriptine is known to block prolactin surges from the pituitary during early pregnancy. In the rat, but not in the human, this blockade of prolactin prevents the development of fully functional corpora lutea and normal progesterone secretion. Without sufficient hormonal support, pregnancy is terminated via the failure of maternal recognition of pregnancy.

The evaluation of 5-azacytidine in the Early Pregnancy Protocol (Cummings, 1994) yielded no effect of the chemical on the number of implantation sites found on day 9 (Fig. 6A). However, one needs to be aware that such results may require further evaluation. When animals were dosed with 5-azacytidine during days 1–8 of pregnancy and killed on day 20, the result was a decline in fetal survival and an increase in resorptions (Fig. 6B). Thus, a day 20 study should follow a study where the results of the Early Pregnancy Protocol are negative on day 9.

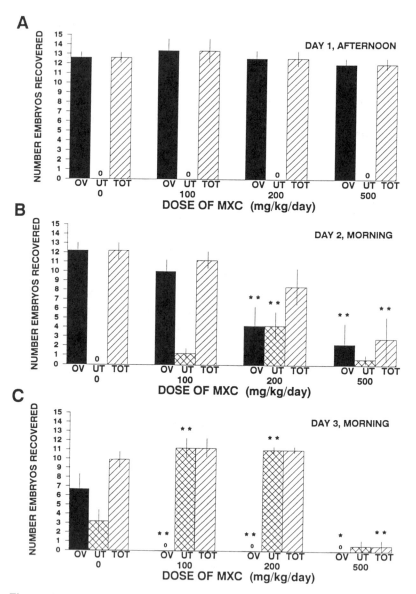

Figure 4 Effect of methoxychlor on Embryo Transport Rate. (A) There is no effect of methoxychlor on day 1. (B) By the morning of day 2, MXC has shifted embryos from the oviduct (OV) to the uterus (UT); the total number found (TOT) was also affected at a dose of 500 mg/kg/day. (C) By the morning of day 3, all doses of MXC affect embryonic location; also, most of the embryos have been expelled from the tract by the highest dose of MXC. (Adapted from Cummings and Perreault, 1990.)

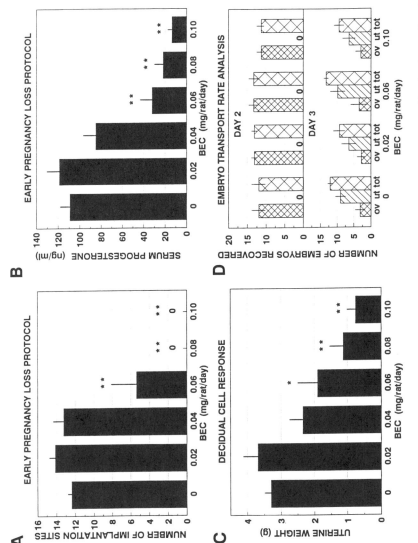

Figure 5 (A) Bromocriptine reduces the number of implantation sites found on day 9 if animals are dosed on days 1–8 of pregnancy. (B) In simultaneous measurements, serum progesterone was found to be reduced. (C) Bromocriptine also blocked the decidual cell response. (D) However, bromocriptine had no effect on embryo transport rate. These data are consistent with the failure of maternal recognition of pregnancy. Abbreviations as for Figure 4. (Adapted from Cummings et al., 1991.)

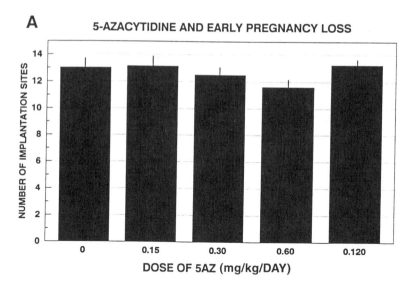

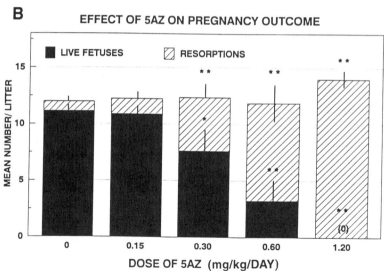

Figure 6 (A) 5-Azacytidine does not reduce the number of implantation sites found on day 9 of early pregnancy. (B) However, dosing during days 1–8 and necropsy on day 20 showed a decrease in the number of fetuses surviving and an increase in the number of resorptions by late pregnancy. (Adapted from Cummings, 1994.)

UTERINE FUNCTION

Background

The uterus is a target organ for the ovarian steroid hormones estrogen and progesterone. Receptors for these hormones are contained within the epithelial, stromal, and myometrial cell compartments, and these receptors bind circulating hormone (Johnson and Everitt, 1980b). The result is a cascade of events wherein the receptor-hormone complex binds the genome at an acceptor site and activates the transcription of specific mRNAs (O'Malley and Tsai, 1992). The mRNAs produced are processed and translated by cytoplasmic ribosomes to form new proteins, which in some way alter the function of the uterine cells (O'Malley and Tsai, 1992).

The cell biology of uterine function is complex, especially when hormone receptors are taken into account. Medlock et al. (1991) showed that 17β-estradiol can downregulate its own receptor in rat uterus. Progesterone was also shown to downregulate the estrogen receptor at high (0.05–5 mg/ml) estradiol levels (Medlock et al., 1994). Progesterone is known to inhibit many estrogen-induced processes in the uterus including protein synthesis, glucose-6-phosphate dehydrogenase activity, peroxidase activity, and complement C3 (Hasty and Lyttle, 1992).

Markers of Xenobiotic Effects

Steroid hormones have been shown to regulate growth factors and cytokines in rat uterus. Estradiol regulated insulinlike growth factor-I (IGF-I) expression and IGF-I receptor expression in the uterus (Murphy and Ghahary, 1990). Data presented by Nelson et al. (1991) suggest that EGF may be a mediator of estrogen action. For example, an antibody specific for EGF inhibited the estrogen-induced uterine growth in mice, and EGF itself also mimicked the effect of estrogen in the induction of uterine lactoferrin mRNA and protein synthesis in mouse uterus (Nelson et al., 1991). Growth factors associated with estrogen action in uterus include TGF-α, TGF-β1, TGF-β2, TGF-β3, and IGF-I (Gray et al., 1995). Data presented by Gray et al. (1995) indicate an estrogen modulation of the expression of platelet-derived growth factors (PDGF) mRNA and protein as well as for the corresponding receptor in mouse uterus.

In general, estrogen promotes the growth of the uterus. Estrogen administered to immature or ovariectomized rats results in rapid uterine tissue growth, imbibition of fluid, and an increase in uterine weight by 6 hours (Astwood, 1938). Estradiol promotes the activity of several uterine enzymes including ornithine decarboxylase (Cohen, et al., 1970), creatine kinase (Notides and Gorski, 1966; Reiss and Kaye, 1981), uterine peroxidase (Lucas et al., 1955; Lyttle and DeSombre, 1977), and glucose-6-phosphate-dehydrogenase (Barker and Warren, 1966). Both uterine growth and the stimulation of these enzyme activities can be considered markers for the potential action of xenobiotics.

MXC is a chlorinated hydrocarbon pesticide that is metabolized to a weakly estrogenic compound HPTE (2,2-bis(p-hydroxyphenyl)-1,1,1-Trichloroethane) (Kupfer and Bulger, 1979). MXC has been shown to stimulate the activity of uterine peroxidase in immature rats, and this stimulation was inhibited by actinomycin D, cycloheximide, and progesterone in a manner similar to that seen with estradiol-stimulated enzyme activity (Cummings and Metcalf, 1994). In a similar manner, MXC has been shown to increase the activity of the estrogen-induced protein ("IP"; creatine kinase) in immature rats; actinomycin D and cycloheximide but not progesterone blocked this effect (Cummings and Metcalf, 1995). These data suggest a parallel action of MXC and estradiol. MXC also stimulated the binding of ^{125}I-EGF to uterine membranes isolated from treated rats, and the data demonstrated that EGF-receptor number was increased by MXC treatment with no changes in binding affinity (K_d) (Metcalf et al., 1995).

In an example of a nonestrogenic effect on uterine function, Sahlin et al. (1994) showed that uterine weight and IGF-I were increased following treatment with either testosterone or 5α-dihydrotestosterone, a steroid not metabolizable to estrogen. Thus, androgenic chemicals may regulate, and adversely affect, uterine function as well.

The Uterus as a Target for Xenobiotics

According to Gellert (1978a), the polychlorinated biphenyl (PCB) Arochlor 1221 induced uterine growth in immature rats at 1000 mg/kg. The chemical also advanced the age of vaginal opening and induced persistent vaginal estrus and anovulation when administered at 10 mg/rat for 2 days to neonatal animals (Gellert, 1978a). Bitman and Cecil (1970) demonstrated the effect of a variety of PCBs on the 18-hour glycogen response of the immature rat uterus. Their results indicate that Arochlors 1221, 1232, 1242, and 1248 were active in this assay (Bitman and Cecil, 1970). Kepone (chlordecone) can produce significant uterine growth in weanling rats at 10 and 100 mg/kg (Gellert, 1978b). Chlordecone has been shown to interact with the estrogen receptor system in rat uterus in vivo and in vitro (Hammond et al., 1979). The chemical induces a stimulation of uterine weight as well as progesterone receptor synthesis and has an affinity for the estrogen receptor approximately 0.01–0.04% that of 17β-estradiol (Hammond et al., 1979). Johnson et al. (1992) demonstrated the estrogenic character of chlordecone in the initiation of implantation and maintenance of pregnancy in progesterone primed delayed implanting hypophysectomized rats. Also, o,p'-DDE (1,1-dichloro-2-p-chlorophenyl)-2-(o-chlorophenyl)ethylene) and o,p'-DDT were shown to affect the uterus in a similar way but to a lesser extent than chlordecone (Johnson et al., 1992).

MXC was shown to increase uterine weight in rodents (Tullner, 1961) using technical grade chemical. Later, the purified chemical was shown to be uterotrophic in immature rats (Welch et al., 1969). Kupfer and Bulger (1979) demonstrated both the proestrogenic nature of MXC (its metabolism to the active form HPTE) and the translocation of the cytosolic estrogen receptor by the active form of the chemical. Bulger et al. (1985) demonstrated the relative binding affinity of MXC and HPTE (among others) with ^{3}H-estradiol in rat uterine cytosol as well as the inhibition of ^{3}H-estradiol binding to rat uterine cytosol. They also demonstrated the capability of these compounds to increase uterine weight and stimulate the activity of ornithine decarboxylase in vivo (Bulger et al., 1985). Johnson et al. (1992) reported that MXC was effective in the initiation of implantation and maintenance of pregnancy in the hypophysectomized, progesterone-treated model of early pregnancy and that this effectiveness is based on the degree to which the various chemicals tested bind the estrogen receptor. When administered during days 1–8 of early pregnancy, MXC doses of 200–500 mg/kg/day reduce the number of implantation sites found on day 9 (Cummings and Gray, 1989). Experiments reported by Cummings and Gray (1987) demonstrated a direct blockade of the uterine decidual cell response by MXC at ≥200 mg/kg/day.

The effect of 2,3,7,8-tetrachlorodibenzo-p-dioxin (TCDD) on the uterus has been evaluated by a number of investigators. According to Safe et al. (1991), TCDD administered to rats inhibited the increase in estradiol-induced uterine wet weight, peroxidase activity, estrogen and progesterone receptor levels, EGF receptor binding, and EGF-R and c-*fos* protooncogene mRNA levels, suggesting that the compound has antiestrogenic activity with respect to uterine function. Investigations reported by DeVito et al. (1992) indicate that a similar downregulation of the estrogen receptor in mice is tissue-specific and appears to involve transcriptional or posttranscriptional mechanisms.

The antiestrogens tamoxifen and clomiphene were shown to inhibit uterine decidual induction in pseudopregnant rats (Barkai et al., 1992). Clomiphene also inhibited the increase in ornithine decarboxylase (an enzyme that normally marks the end of the induction phase and is regulated by estrogen), while no effect of clomiphene on the availability or receptor binding of progesterone was evident (Barkai et al., 1992). The steroidal antiestrogen ICI-182,780 inhibited uterine growth when administered to rats on postnatal days 10–14, but had no effect on uterine luminal epithelial hypertrophy or gland genesis (Branham et al., 1996). ICI-182,780 also reduced estrogen receptor in all cell types while the agonist activity of tamoxifen increased estrogen receptors in uterine cells, increased uterine weight, induced luminal epithelial hypertrophy, and inhibited gland genesis (Branham et al., 1996).

Hydroxyflutamide is an antiandrogen, and experiments reported by Chandrasekhan et al. (1990) were designed to investigate the chemical's potential antiprogestogenic activity in rat uterus. Hydroxyflutamide was shown to delay implantation, fetal development, and parturition in pregnant rats and suppressed decidualization in pseudopregnant rats (Chandrasekhan et al., 1990).

SUMMARY

A variety of chemicals, both environmental and pharmaceutical, can alter uterine function. This alteration can take the form of effects on the receptor systems for the steroid hormones estrogen and progesterone, increases or decreases in the activity and/or expression of specific enzymes, or modifications in the cytokines that regulate and mediate steroid hormone function in uterine tissue.

Alterations in uterine function will obviously ultimately affect early pregnancy and implantation. Among the critical functions are the decidualization of the uterus and its regulation by the ovarian steroid hormones, and this function is vulnerable to toxic effects. In a related manner, the development of sensitivity of the uterus to implantation and the synchronization of that development with the timely arrival of the blastocyst into the uterus can also be compromised by xenobiotics. In the examples described in this chapter, dosages applied are well above those likely for human exposure. The chemicals and their evaluations, however, serve as models for the examination of unknown toxicants that may pose a hazard to the human population.

An understanding of the interrelationships among uterine function, implantation, and early pregnancy is necessary for the prediction, evaluation, and interpretation of the adverse effects of reproductive toxicants. The mechanisms through which such chemicals have their effects have often yet to be elucidated and the investigation of such mechanisms remains a priority in the study of reproductive toxicology. Important future research needs to include the investigation of the molecular mechanisms through which toxicants produce reproductive effects, thus providing a means to extrapolate information gleaned from animals and cells to humans. Such chemical mechanisms may involve steroid receptor systems, cytokines, small molecules such as calcium, calmodulin, and nitric oxide, neurotransmitters such as GABA, or other as yet unrecognized mediators of chemical effects in the reproductive system.

REFERENCES

Astwood, E. B. (1938). *Endocrinology, 23*: 25.

Barkai, U., Kidron, T., and Kraicer, P. F. (1992). *Biol. Reprod., 46*: 733.

Barker, K. L., and Warren, J. C. (1966). *Endocrinology, 78*: 1205.

Bedford, J. M. (1982). *Reproduction in Mammals, Book 1: Germ Cells and Fertilization* (C. R. Austin and R. V. Short, eds.). Cambridge University Press, Cambridge, p. 128.

Bitman, J., and Cecil, H. C. (1970). *J. Agric. Food Chem., 18*: 1108.

Branham, W. S., Fishman, R., Streck, R. D., Medlock, K. L., DeGeorge, J. J., and Sheehan, D. M. (1996). *Biol. Reprod., 54*: 160.

Bulger, W. H., Feil, V. J., and Kupfer, D. (1985). *Mol. Pharmacol., 27*: 115.

Chandrasekhar, Y., Armstrong, D. T., and Kennedy, T. G. (1990). *Biol. Reprod., 42*: 120.

Cohen, S., O'Malley, B. W., and Stastny, M. (1970). *Science, 170*: 336.

Cross, J. C., Werb, Z., and Fisher, S. J. (1994). *Science, 266*: 1508.

Cummings, A. M. (1990). *Fund. Appl. Toxicol., 15*: 571.

Cummings, A. M. (1994). *Fund Appl. Toxicol., 23*: 429.

Cummings, A. M., and Gray, L. E., Jr. (1987). *Toxicol. Appl. Pharmacol., 90*: 330.

Cummings, A. M., and Gray, L. E., Jr. (1989). *Toxicol. Appl. Pharmacol., 97*: 454.

Cummings, A. M., and Metcalf, J. M. (1994). *Reprod. Toxicol., 8*: 477.

Cummings, A. M., and Metcalf, J. M. (1995). *Toxicol. Appl. Pharmacol., 130*: 154.

Cummings, A. M., and Perreault, S. D. (1990). *Toxicol. Appl. Pharmacol., 102*: 110.

Cummings, A. M., Perreault, S. D., and Harris, S. T. (1991). *Fund. Appl. Toxicol., 17*: 563.

Das, S. K., Wang, X.-N., Paria, B. C., Damm, D., Abraham, J. A., Klagsbrun, M., Andrews, G. K., and Dey S. K. (1994). *Development, 120*: 1071.

DeVito, M. J., Thomas, T., Martin E., Umbreit, T. H., and Gallo, M. A. (1992). *Toxicol. Appl. Pharmacol., 113*: 284.

Edwards, R. G. (1994). *Human Reprod., 9* (Suppl. 2): 73.

Gellert, R. J. (1978a). *Environ. Res., 16*: 123.

Gellert, R. J. (1978b). *Environ. Res., 16*: 131.

Gray, K., Eitzman, B., Raszmann, K., Steed, T., Geboff, A., McLachlan, J., and Bidwell, M. (1995). *Endocrinology, 136*: 2325.

Grummer, R., Chwalisz, K., Mulholland, J., Traub, O., and Winterhager, E. (1994). *Biol. Reprod., 51*: 1109.

Hammond, B., Katzenellenbogen, B. S., Krauthammer, N., and McConnell, J. (1979). *Proc. Natl. Acad. Sci. USA, 76*: 6641.

Hasty, L. A., and Lyttle, C. R. (1992). *Biol. Reprod., 47*: 285.

Johnson, D. C., and Chatterjee, S. (1993). *J. Reprod. Fertil., 99*: 557.

Johnson, D. C., Sen, M., and Dey, S. K. (1992). *Proc. Soc. Exp. Biol. Med., 199*: 42.

Johnson, M., and Everitt, B. (1980a). *Essential Reproduction*. Blackwell Scientific Publications, Oxford, p. 247.

Johnson, M., and Everitt, B. (1980b). *Essential Reproduction*. Blackwell Scientific Publications, Oxford, p. 173.

Kupfer, D., and Bulger, W. H. (1979). *Life Sciences, 25*: 975.

Lucas, F. V., Neufeld, H. A., Utterback, J. G., Martin, A. P., and Stotz, E. (1955). *J. Biol. Chem., 214*: 755.

Lyttle, C. R., and DeSombre, E. R. (1977). *Proc. Natl. Acad. Sci. USA, 74*: 3162.

Markoff, E., Henemyre, C., Fellows, J., Pennington, E., Zeitler, P. S., and Cedars, M. I. (1995). *Biol. Reprod., 53*: 1103.

Medlock, K. L., Forrester, T. M., and Sheehan, D. M. (1994). *Proc. Soc. Exp. Biol. Med., 205*: 146.

Medlock, K. L., Lyttle, C. R., Kelekpouris, N., Newman, E. D., and Sheehan, D. M. (1991). *Proc. Soc. Exp. Biol. Med., 196*: 293.

Metcalf, J. L., Laws, S. C., and Cummings, A. M. (1995). *Biol. Reprod., 52* (Suppl. 1): 97.

Murphy, L. J., and Ghahary, A. (1990). *Endocrine Rev., 11*: 443.

Nelson, K. G., Takahashi, T., Bossert, N. L., Walmer, D. K., and McLachlan, J. A. (1991). *Proc. Natl. Acad. Sci. USA, 88*: 21.

Notides, A., and Gorski, J. (1966). *Proc. Natl. Acad Sci. USA, 56*: 230.

O'Malley, B. W., and Tsai, M.-J. (1992). *Biol. Reprod., 46*: 163.

Reiss, N. A., and Kaye, A. M. (1981). *J. Biol. Chem., 256*: 5741.

Robertson, S. A., Mayrhofer, G., and Seamark, R. F. (1996). *Biol. Reprod., 54*: 183.

Safe, S. Astroff, B., Harris, M., Zacharewski, T., Dickerson, R., Romkes, M., and Biegel, L. (1991). *Pharmacol. Toxicol., 69*: 400.

Sahlin, L., Norstedt, G., and Eriksson, H. (1994). *J. Steroid Biochem. Mol. Biol., 51*: 57.

Stewart, C. L., Kaspar, P., Brunet, L. J., Bhatt, H., Gadi, I., Kontgen F., and Abbondanzo, S. J. (1992). *Nature, 359*: 76.

Tamada, H., Kai, Y., and Mori, J. (1994). *Prostaglandins, 47*: 467.

Tullner, W. W. (1961). *Science, 133*: 647.

Weitlauf, H. M. (1994). *The Physiology of Reproduction*, 2nd ed., Vol. I (E. Knobil and J. D. Neill, eds.). Raven Press, Ltd., New York, p. 391.

Welch, R. M., Levin, W., and Conney, A. H. (1969). *Toxicol. Appl. Pharmacol., 14*: 358.

Yee, G. M., Squires, P. M., Cejic, S. S., and Kennedy, T. G. (1993). *J. Lipid Mediators, 6*: 525.

17

TCDD and Uterine Function

Kimberly Silvers and Ellen A. Rorke
Case Western Reserve University, Cleveland, Ohio

INTRODUCTION

The Compound

2,3,7,8-Tetrachloro-dibenzo-p-dioxin (TCDD) is a halogenated aromatic hydrocarbon that has generated significant concern as a potential environmental toxicant. This class of chemicals includes other polychlorinated dioxins, dibenzofurans, and biphenyls. TCDD is produced during the synthesis of 2,4,5-trichlorophenol, a substrate used to manufacture the herbicide and defoliant 2,4,5-trichlorophenoxyacetictrichlorophenol. It is also found as a trace contaminant in many commercial and domestic combustion processes (Bumb et al., 1978, 1980; Crummett, 1982). Environmental contamination with TCDD is widespread, and TCDD has been detected both in the soil and in the food chain. In addition, significant numbers of humans have been exposed to TCDD from industrial accidents and military use (Ayres et al., 1985; Silbergeld and Mattison, 1987). TCDD has been detected in human serum, adipose tissue, and milk (Safe et al., 1978, 1991).

TCDD is one of the most toxic synthetic substances found (Hodgson and Levi, 1987; Peterson et al., 1993); its LD_{50} in guinea pigs, the most sensitive species, is 0.6 μg/kg. The LD_{50} for female rats is 0.045 mg/kg—seven times higher. In these rats it is fetotoxic at 0.1 μg/kg and causes birth defects at 2 ng/kg. TCDD is also a proven hepatic carcinogen in both rats and mice (Hodgson and Levi, 1987; Peterson et al., 1993). As the disparity in the LD_{50} for guinea pigs and rats suggests, significant species variation exists in the toxicity of TCDD. Although transitory effects of TCDD exposure, most notably chloracne, have been documented in humans, no epidemiological study has demonstrated that TCDD causes severe health problems in adult humans (Mofenson et al., 1985; Silbergeld and Mattison, 1987). Initial public concern focused on the carcinogenic activity of TCDD. However, recent attention has shifted to its effects as a reproductive toxicant.

Reproductive Toxicant

TCDD has been reported to cause reproductive toxicity in several mammalian species (Allen et al., 1979; Couture et al., 1990; Peterson et al., 1993). An increased incidence of spontaneous abortions in monkeys, guinea pigs, rabbits, rats, hamsters, and mice from exposure to TCDD has been shown (Kociba and Schwetz, 1982; Peterson et al., 1993). In addition, TCDD exposure results in a disruption of normal estrous cycling in rodents, primates, and humans (Allen et al., 1979; Barsotti et al., 1979; Giavini et al., 1982; Lu and Wong, 1984; Umbreit et al., 1987). An extensive review of the reproductive effects of TCDD reveals a general pattern of fetotoxicity, characterized by thymic hypoplasia, subcutaneous edema, decreased fetal growth, and prenatal mortality, which occur in several different species (Allen et al., 1979; Kociba and Schwetz, 1982; Couture et al., 1990). In addition, species-specific effects of TCDD include cleft palate and hydronephrosis in mice, intestinal hemorrhage in the rat, and additional rib formation in the rabbit (Peterson et al., 1993). Reproductive toxicity in the human has been more difficult to assess, given the fact that industrial and military workers have been exposed to mixtures of compounds rather than pure TCDD (Mofenson et al., 1985; Silbergeld and Mattison, 1987). However, several studies have suggested an increase in the rate of spontaneous abortions and birth defects such as spina bifida and cleft palate in communities exposed to high levels of TCDD (Silbergeld and Mattison, 1987; Peterson et al., 1993).

In addition to such overt signs of fetotoxicity, TCDD may interfere with normal reproductive function by altering the subtle balance between hormones, growth factors, and receptors required to maintain the reproductive systems of both males and females (Astroff et al., 1991; Safe et al., 1991; DeVito et al., 1992; Dohr et al., 1994). That TCDD could affect reproductive tissues first became apparent when it was shown that long-term feeding of TCDD to rats reduced the number of age-related tumors in uterus, pituitary, and breast (Kociba et al., 1978). These three tissues are target sites for the ovarian steroid hormone estradiol. TCDD also promotes cancer in a fourth estrogen target tissue, the liver, and this effect occurs preferentially in female rats (Pitot et al., 1987; Sewall et al., 1993). In addition, several studies have suggested that changes in cell proliferation in reproductive tissues in response to TCDD are mediated by estrogen regulatory pathway (Biegel and Safe, 1990; Safe et al., 1991; Wang et al., 1993; Gierthy et al., 1993). How TCDD might alter uterine function will be the main focus of this review.

THE UTERUS

The primary function of the uterus is to house the developing fetus (Strauss and Gurpide, 1991; Brenner and Slayden, 1994). It can be divided into three parts: the fundus (top), and corpus (body), and the isthmus (canal), which connects to the cervix. These parts are comprised of two tissues: a smooth muscle, the myometrium, which makes up the uterine wall, and the mucosal lining of the uterine cavity, the endometrium, which is the site of implantation for the developing blastocyst. The surface of the endometrium is a columnar epithelium, which is continuous with simple epithelial glands. These glands extend all the way to the myometrium. The endometrial stroma is highly cellular with an amorphous extracellular matrix that has few connective tissue fibers (Strauss and Gurpide, 1991; Brenner and Slayden, 1994). The endometrium displays a complex pattern of growth and differentiation, which occurs on a cyclic basis in the mature adult female and is under the control of the ovaries (Clark and Mani, 1994). Endometrial differentiation and function can be maintained in ovariectomized females upon the replacement of two ovarian steroid hormones, estradiol and progesterone (Strauss and Gurpide, 1991; Brenner and Slayden, 1994; Clark and Mani, 1994).

Cyclic Uterine Changes

Estradiol is the primary hormone required for the development and maintenance of normal uterine function (Reynolds, 1951; Clark and Mani, 1994). During the early proliferative phase of the ovarian cycle, which is dominated by estradiol, the endometrial glands and arterial vessels are small and the surrounding stoma is dense. With time, both the glandular and the stromal cells show increased mitotic activity. The glands and vasculature grow as the stroma becomes edematous and the tissue thickens. In the later part of the proliferative phase, the glands and blood vessels become coiled as they continue to grow. During the luteal phase, which follows ovulation and is under control of progesterone, glycogen accumulates first at the base of the glands and with time moves toward the apex and into the lumina. The stroma continues to be highly edematous. If pregnancy ensues, the endometrium is maintained and ready for implantation; if not it is shed and the cycle is repeated (Strauss and Gurpide, 1991; Brenner and Slayden, 1994; Clark and Mani, 1994).

Estrogen Action

Estradiol modulates uterine growth and differentiation via its interaction with its intracellular receptor (Jordan et al., 1985; O'Malley et al., 1991; Orti et al., 1992). The estrogen receptor belongs to a large superfamily of ligand-dependent transcription factors, which are characterized by zinc finger domains (Giguere et al., 1988; Murdoch et al., 1990; Predki and Sarkar, 1992; Ikeda et al., 1993; Inoue et al., 1993; Mader et al., 1993; Luisi et al., 1994; Zilliacus et al., 1994). The zinc fingers promote the interaction of receptors with specific enhancers (O'Malley et al., 1991; Kato et al., 1992; Nigro et al., 1992; Predki and Sarkar, 1992; Vamvakopoulos and Chrousos, 1993; Luisi et al., 1994). Other domains of the receptor include the ligand-binding domain, which is found in the C-terminal, and two transcriptional activation sites—one in the ligand-binding domain, the other in the N-terminal region (Giguere et al., 1988; O'Malley et al., 1991; Wilson et al., 1992; Gorski et al., 1993; Luisi et al., 1994). Unliganded receptors are found as part of large complexes formed by noncovalent association of receptor proteins with the 90 kDa heat-shock protein (HSP90) (Catelli et al., 1985; Chambraud et al., 1990; Schlatter et al., 1992; Scherrer et al., 1993). Two other proteins, HSP70 and P59, have also been found in many of these complexes (Ratajczak et al., 1990, 1993; Takahashi et al., 1994; Elledge et al., 1994). These complexes, which are found in the nucleus, cannot bind DNA (Catelli et al., 1985; Ratajczak et al., 1990; 1991; Scherrer et al., 1993). Once ligand is bound, the receptors undergo a conformation change, dissociate from the large complexes, and are transformed to activated species that bind to higher-affinity hormone-response elements on specific genes for DNA (O'Malley et al., 1991; Nigro et al., 1992; Luisi et al., 1994). Receptors can dimerize prior to interacting with their response elements, and receptor dimers bind with greater affinity and stability than receptor monomers (Hirst et al., 1992; Wilson et al., 1993; Furlow et al., 1993). Higher-order interactions between receptor dimers at different response elements lead to tetrameric complexes with 10- to 100-fold greater affinity for their response elements (Gaub et al., 1990; O'Malley et al., 1991; Luisi et al., 1994). In addition, protein-protein interactions among receptor complexes and other transcription factors facilitate the initiation of transcription at regulated genes (Gaub et al., 1990; Schuh and Mueller, 1993; Burbach et al., 1994; Jacq et al., 1994; Smidt et al., 1994). It appears that steroid receptors enhance the formation of rapid-start complexes by RNA polymerases and enhance the assembly of a network of transcription factors at the TATA site (Bagchi et al., 1990; Schild et al., 1993; Ying and Gorski, 1994).

A number of genes expressed in the uterus are regulated by estradiol. These include epidermal growth factor (EGF) and its receptor (Stancel et al., 1987; DiAugustine et al., 1988; Lingham et al., 1988; Gardner et al., 1989), insulinlike growth factors (IGFs) (Murphy et al., 1987a; Sahlin et al., 1990), and transforming growth factors-α and -β (TGF-α, TGF-β) (Dickson et al., 1986; Lippman et al., 1987; Anzai et al., 1992; Beck and Garner, 1992; Gong et al., 1994). Estradiol also enhances the expression of various proto-oncogenes, including c-*myc*, c-*fox*, and c-*ras* (Murphy et al., 1987b; Travers and Knowler, 1987; Loose-Mitchell et al., 1988). Uterine responsiveness to progesterone increases following estradiol treatment via the upregulation of the progesterone receptor (Leibl et al., 1981; Nardulli et al., 1988; Aronica and Katzenellenbogen, 1991). Progesterone, on the other hand, counteracts the effects of estrogens by reducing estrogen receptor levels (Coulson and Pavlik, 1977; Evans and Leavitt, 1980; Okulicz et al., 1981; Sumida et al., 1981) and by increasing the rate of estradiol metabolism to less active metabolites (Clark, 1980; MacDonald et al., 1982; Quarmby and Martin, 1982). Consequently, uterine estrogen receptor levels are high during the proliferative phase and decline after ovulation, whereas progesterone receptors peak at the time of ovulation (Clark et al., 1980; Senekjian et al., 1989; Berman et al., 1992; Hild Petitio et al., 1992).

TCDD AS AN ANTIESTROGEN

Estrogens are known to promote cancers of the breast, uterus, and pituitary (Li and Li, 1990; Zhu and Liehr, 1993; Liehr, 1990; Newbold et al., 1990). Therefore, the observation by Kociba et al. (1978) that rats fed TCDD had a lower incidence of cancers in these target tissues led investigators to study its effects on these estrogen-responsive tissues. TCDD has been shown to cause a number of antiestrogenic effects. Initially it was found to block the estradiol-induced uterine weight gain in rats and mice (Gallo et al., 1986; Romkes et al., 1987; Umbreit et al., 1988). Prepubertal females treated with TCDD have smaller uteri than untreated females, suggesting that TCDD is blocking the uterotrophic effects of endogenous estrogens (Astroff and Safe, 1988; Umbreit et al., 1988). Interestingly, when the uterine responsiveness of rats to TCDD is evaluated in ovariectomized and intact rats, it appears that uteri from ovariectomized females are less responsive to TCDD then uteri from intact females (Romkes et al., 1987; DeVito et al., 1992). Furthermore, uteri from 21-day-old female rats are far less responsive to TCDD than uteri from 28-day-old rats (Safe et al., 1991; White et al., 1995). This difference may be due in part to the increasing ovarian production and circulating levels of estradiol found in the older females as they approach puberty (Lax et al., 1983; Docke et al., 1984). These data suggest that circulating levels of estradiol help delineate the uterine responsiveness to TCDD.

TCDD can also modulate uterine responsiveness to high levels of exogenously administered estradiol. When 1–5 μg of estradiol is administered to immature females, uterine weight increases two- to threefold (Hisaw, 1959; Pollard and Martin, 1968; Hayes et al., 1981). When co-administered with estradiol, TCDD blocks the uterine weight gain (Gallo et al., 1986; Umbreit et al., 1987; Astroff and Safe, 1988). Since estradiol stimulates water imbibition as well as cell proliferation in the uterus, it is unclear if there is a preferential inhibition of these two responses by TCDD. Furthermore, the uterus has a number of important cell types—epithelial, stromal, and smooth muscle—all of which are modulated by estradiol (McCormack and Glasser, 1980). There are no data currently available that indicate whether these cell types are all equally responsive to TCDD.

A number of biochemical endpoints known to be regulated by estradiol (Coulson and Pavlik, 1977; Hayes et al., 1981; Leibl et al., 1981; Murphy et al., 1987b; Travers and Knowler, 1987; DiAugustine et al., 1988; Loose-Mitchell et al., 1988; Gardcner et al., 1989; Aronica and Katzenellenbogen, 1991) have now been measured, which confirm TCDD's antiestrogenic activity in the uterus. These include uterine peroxidase activity (Astroff and Safe, 1990, 1991), epidermal growth factor receptor levels (Astroff et al., 1990; Astroff and Safe, 1991), progesterone receptor levels (Astroff and Safe, 1988; Romkes and Safe, 1988), and oncogene expression (Astroff et al., 1991). When TCDD and estradiol are co-administered, estrogen-dependent responses are blocked. These antiestrogenic effects of TCDD show the same kinetic parameters, suggesting a common pathway of inhibition. In addition to the observation that TCDD is anti-estrogenic in the uterus, TCDD has been shown to have antiestrogenic activity in other estrogen-responsive systems, including human breast cancer cells lines (Gierthy et al., 1987; Gierthy and Lincoln, 1988; Vickers et al., 1989; Harris et al., 1990b; Fernandez and Safe, 1992) and human liver cell lines (Randerath et al., 1990; Lin et al., 1991; Zacharewski et al., 1991).

Ah Receptor

Like steroid hormones, TCDD binds to an intracellular receptor, which translocates to the nucleus and alters transcription in the target cell (Poland and Knutson, 1982; Jones et al., 1985; Landers and Bunce, 1991; Nebert et al., 1993). The receptor for TCDD, named AhR for aryl hydrocarbon receptor, is a heterodimer found in the cytosol complexed with heat-shock protein and other proteins (Perdew et al., 1993; de Marais et al., 1994). The ligand-receptor complex isolated from the nucleus is also found complexed with a third protein, ARNT, or AhR nuclear translocator protein (Kohn et al., 1993; Matsushita et al., 1993; Chan et al., 1994; Hankinson, 1994). Both AhR and ARNT contain a basic helix-loop-helix motif (Struhl, 1989; Matsushita et al., 1993), which allows binding to DNA at sites known as xenobiotic-responsive elements (XRE) [sometimes referred to as dioxin-responsive elements (DRE)] (Jaiswal et al., 1987; Denison et al., 1988; Denison et al., 1989; Whitlock et al., 1989; Landers and Bunce, 1991; Whitlock 1993). XRE differs from steroid hormone–binding sites in that it lacks a requirement for zinc, and direct or inverted repeats are not present in the sequence, as in the steroid hormone–response elements (Saatcioglu et al., 1990; Nebert et al., 1993). Several genes are regulated by AhR, including the cytochrome P450 genes Cyp1A1 and Cyp1A2 (Jaiswal et al., 1985; Cresteil et al., 1987; Nebert and Jones,

1989; Whitlock et al., 1989; Thomsen et al., 1991) and genes involved in Phase II metabolism such as NAD(P)H:menadione oxidoreductase (Henry and Gasiewicz, 1986; Gasiewicz et al., 1986), aldehyde dehydrogenase (Simpson et al., 1985; Tank et al., 1986; Dunn et al., 1988), UDP-glucuronosyltransferase (Umbreit et al., 1989), and glutathione transferase (Telakowski Hopkins et al., 1988; Nebert, 1991; Nebert et al., 1993). The induction of cytochrome P-450 1A1, measured as aryl hydrocarbon hydroxylase (AHH) activity, is considered to be the prototypical response to TCDD treatment (Whitlock, 1990; Roberts et al., 1991).

One of the key questions concerning the role of TCDD in reproductive toxicity is how the interaction between TCDD and estrogen-regulated effects might occur. Several structure-activity studies, in which the affinity of TCDD-like congeners for the AhR is correlated with their ability to elicit antiestrogenic effects, suggest that the AhR is involved in the effects of TCDD in both uterus and breast cancer cell lines (Harris et al., 1989; Harris et al., 1990a,b; Astroff and Safe, 1991; Zacharewski et al., 1994). In addition, age and strain differences in the antiestrogenic activity of TCDD have been correlated with differences in Ah receptor levels (Umbreit et al., 1988; Umbreit et al., 1989; Lin et al., 1991; Safe et al., 1991; White et al., 1995). Further evidence for the role of AhR in mediating the antiestrogenic effects of TCDD comes from in vitro studies using cells with wild-type and mutant AhR activity. In Ah-responsive Hepa 1c1c7 cells, pretreatment with TCDD followed by estradiol caused a decrease in nuclear ER levels and ER protein and PS2 expression (Zacharewski et al., 1991; Zacharewski et al., 1994). In Ah-nonresponsive mutant cells, however, pretreatment with TCDD does not affect these parameters (Zacharewski et al., 1991; Safe et al., 1991; Zacharewski et al., 1994). Similar results have been shown using human ovarian carcinoma cell lines (Rowlands et al., 1993). In addition, in another estrogen-dependent cell type, the MCF-7 human breast cancer cell line, pretreatment with TCDD followed by estradiol caused a decrease in nuclear ER levels (Harris et al., 1990b), and this decrease was blocked by an AhR antagonist, α-napthoflavone (Wang et al., 1993). These studies strongly suggest that AhR is a necessary participant in the antiestrogenic effects of TCDD.

TCDD does not compete with estradiol for the estrogen receptor (Poland and Knutson, 1982; Romkes et al., 1987; Astroff and Safe, 1988; Romkes and Safe, 1988). Alternative explanations have been raised to explain the antiestrogenic effects of TCDD, all of which involve the binding of TCDD to its receptor and the formation of a complex with the nuclear translocation protein (Matsushita et al., 1993; Hankinson, 1994). In the first paradigm, TCDD upregulates the target cell metabolism of estradiol to less active products, thereby creating a local hormone deficiency in which the estrogen receptors are not activated (Namkung et al., 1985; Spink et al., 1990, 1992, 1994). In the second paradigm, it is proposed that TCDD acts via a downregulation of estrogen receptor (Umbreit et al., 1988; Astroff and Safe, 1988; Romkes and Safe, 1988; Safe et al., 1991). The diminished estrogen receptor levels would leave the cell less responsive to circulating levels of estradiol. Expression of growth factors and their receptors is also believed to be an essential intermediate in the uterine response to estradiol (Stancel et al., 1987; Lingham et al., 1988; Gong et al., 1994). TCDD could block the effects of estradiol by interfering with the expression of a local growth factor (Madhukar et al., 1984; Astroff et al., 1990; Dohr et al., 1994). Finally, since uterine growth is the culmination of a complex set of responses to ovarian hormones, which are themselves regulated by pituitary gonadotropin hormones and hypothalamic growth factors, changes in uterine growth in vivo may reflect not only direct effects of TCDD on the uterus but also indirect effects of TCDD on pituitary and/or ovarian function (Moore et al., 1989; Bookstaff et al., 1990; Li et al., 1995). These mechanisms are nonexclusive and may all play a significant role in TCDD action.

Postreceptor Activity

Estrogen Metabolism and TCDD

Metabolism of estradiol in the human liver has been shown to occur through hydroxylation at the 2, 4, 6a, 15a, and 16a positions (Breuer et al., 1966; Fishman, 1983). These reactions are mediated by the cytochrome P-450 enzymes, present mainly in liver and some extrahepatic tissue. The predominant cytochromes P-450 involved in hydroxylation of estradiol are cytochromes P-450 1A1, 1A2, 3A3, and 3A4 (Guengerich, 1990; Aoyama et al., 1990; Spink et al., 1992). Since TCDD leads to the induction of cytochrome P-450 1A1 and 1A2 (Nebert and Jones, 1989), one of the first theories concerning the reproductive toxicity of TCDD was that it led to increased metabolism of estrogen (Gierthy et al., 1988; Spink et al., 1990, 1992; Thomsen et al., 1991; Savas et al., 1993; Kupfer et al., 1994). This effect might lead to decreased estrogen levels circulating in the plasma, or at a specific target site, which would then interfere with normal reproductive function. Increased metabolism of estradiol at the cellular level, and thus less estradiol available to bind to the estrogen receptor, could also offer one explanation for the observed decrease in estrogen receptor binding and decreased occupied nuclear receptor without necessitating a downregulation of total estrogen receptors. TCDD has been shown to increase 2-hydroxylation of estradiol fourfold in rat liver microsomes and has been shown to increase estradiol metabolism in cultured MCF-7 cells (Gierthy et al., 1988; Spink et al., 1990, 1992). Decreased circulating levels of estrogen in TCDD-treated primates have been reported (Barsotti et al., 1979). However, it has never been determined whether this increased metabolism of estradiol is physiologically relevant at estrogen-responsive sites. In addition, although treatment of pregnant rats with TCDD has been reported to increase the cytochrome P-450 levels in maternal hepatic tissue, the serum levels of estradiol remained unchanged (Shiverick and Muther, 1983). A similar study also showed that an increase in cytochrome P-450 activity in maternal hepatic, fetal hepatic, and placental tissue did not greatly increase the metabolism of estradiol to its 2-hydroxy or 4-hydroxy catechol metabolites (Namkung et al., 1985). Several studies have also shown that doses of TCDD lower than those required to induce cytochrome P-450 1A1 activity still elicit antiestrogenic effects (Romkes et al., 1987; Astroff and Safe, 1988; Safe et al., 1991; Wang et al., 1993). Furthermore, studies have shown that the main cytochrome P-450 present in mouse endometrial stromal cells is a different cytochrome P-450 enzyme, P-450 EF (Savas et al., 1993, 1994). Both immunoblot studies and immunoinhibition of endometrial cell microsomal metabolism suggest that cytochrome P-450 1A is not involved in endometrial cell metabolism (Savas et al., 1993). Although cytochrome P-450 EF shares homology with cytochrome P-450 1A1 and 1A2 and is also inducible by TCDD, this enzyme possesses a different substrate specificity and does not seem to be involved in steroid metabolism (Savas et al., 1994). At present it is unknown which P-450 is present in human endometrial cells. Thus, although TCDD does induce cytochrome P-450 in mouse endometrial stromal cells, the specific enzyme induced is different from that in MCF-7 cells and may not affect uterine estrogen metabolism to the same degree as in the MCF-7 breast cancer cell. These results suggest that an induction in estradiol metabolism, although important, does not completely account for the effects of TCDD on the uterus.

Role of Estrogen Receptor Function in TCDD Toxicity

Studies in a variety of estrogen-responsive cell lines suggest a role for the ER in cell responsiveness to TCDD treatment (Harris et al., 1989, 1990a; Vickers et al., 1989; Biegel and Safe, 1990). While AhR was detectable in each of the cell lines studied, the level of TCDD binding to these receptors did not correlate with levels of cytochrome P450 1A1 mRNA or AHH activity. Instead, AHH activity correlated with the estrogen receptor content of these cells (Vickers et al., 1989). In another study, TCDD inhibited estradiol-induced cell proliferation and estrogen-mediated secretion of several proteins only in breast cancer cell lines that were both ER and AhR positive (Biegel and Safe, 1990). When breast cancer cells positive for the Ah receptor but negative for the estrogen receptor are treated with TCDD, the toxicant forms the nuclear AhR complex and binds to a DRE; however, there is no induction of AHH activity (Thomsen et al., 1994). Similarly, when these estrogen receptor–negative cells are transfected with an Ah-responsive plasmid containing the 5' flanking region from the CYP 1A1 gene upstream of a CAT reporter, there is no induction of activity by TCDD. However, when the cells are co-transfected with a human estrogen receptor and the Ah-responsive plasmid, CAT activity is increased 10-fold (Thomsen et al., 1994). Co-transfection with a progesterone receptor expression plasmid does not increase CAT activity, indicating that the response is specific for the estrogen receptor (Thomsen et al., 1994). These studies indicate that although estrogen receptors are not necessary for the formation of a complex TCDD-AhR with its response element, the estrogen receptor is necessary for biological activity.

TCDD and Transcription of the Estrogen Receptor

Just how the AhR and ER interact to cause the antiestrogenic effects of TCDD is not known. Cell responsiveness to hormonal stimulation is determined in part by the hormone receptors present, and factors that change the expression of the estrogen receptors change the cells' ability to recognize and respond to estradiol (Coulson and Pavlik, 1977; Evans and Leavitt, 1980; Okulicz et al., 1981; Sumida et al., 1981; Read et al., 1989). Initially it was thought that the effects of TCDD on uterine tissue may be due to its modulation of the uterine estrogen receptors (Romkes et al., 1987; Astroff and Safe, 1988; Romkes and Safe, 1988; Safe et al., 1991). This view was strengthened when it was found that treatment of animals with a single dose of TCDD decreases both hepatic and uterine levels of ER (Romkes et al., 1987; Hruska and Olson, 1989; Goldstein et al., 1990; Lin et al., 1991; DeVito et al., 1992). Normally estradiol causes an increase in uterine estrogen receptor levels (Coulson and Pavlik, 1977; Clark et al., 1980). TCDD was found to block this effect of estradiol, much like its ability to block uterine weight increases (Romkes et al., 1987; Astroff and Safe, 1988; Safe et al., 1991). These results suggest that the antiestrogenic effects observed for TCDD might be due to a downregulation in the estrogen receptor. Furthermore, it has been shown that the 5' region and coding region of exon one of the human ER gene contains six partial and two full-length DREs (White and Gasiewicz, 1993). Although the activity of this promoter has not been tested, activated Ah receptor complexes from mouse and from MCF-7 cells have been shown to bind to the full-length DRE, according to gel shift mobility assays (Moore et al., 1993). These results suggest that TCDD bound to AhR might interfere with transcription of the estrogen receptor. However, when total ER levels were measured in cells exposed to TCDD, it was found that there was no change in their levels for 16–24 hours following treatment (Gierthy et al., 1987; DeVito et al., 1992). Total estrogen receptor levels are inhibited by TCDD after 24 hours (DeVito et al., 1992), the time in which one can measure the inhibition of uterine weight, peroxidase activity, and the other estrogen-dependent biological endpoints (Safe et al., 1991). Northern analysis as well as mRNA stability experiments reveal a similar time course (Wang et al., 1993; Lu et al., 1994). Thus, the modulation of estrogen receptor appears to be a long-term response to TCDD, consistent with a transcriptional mechanism, but not primarily responsible for the antiestrogenic effects on the uterus. Both estrogen and TCDD induce a large number of genes. TCDD might directly inhibit estrogen-induced transcription or induce other proteins that interfere with estrogen-induced transcription (Zacharewski et al., 1994).

TCDD and Estrogen Receptor Protein Levels

When cells are treated with estradiol, there is a time-dependent accumulation of occupied nuclear ER complex, which peaks at 1–3 hours following the addition of hormone. Following this, nuclear ER complex levels diminish, and this loss of estrogen receptors is referred to as processing (Clark et al., 1980; Eckert and Katzenellenbogen, 1982; Clark and Mani, 1994). Pretreatment with TCDD 6 or 12 hours prior to the addition of estradiol significantly decreases the accumulation of the occupied nuclear ER complexes (Harris et al., 1990b; Zacharewski et al., 1991; Wang et al., 1993). This decrease is accompanied by a decrease in immunodetectable ER protein (Harris et al., 1990b; Zacharewski et al., 1991). Paralleling the effects on nuclear receptor levels, TCDD diminishes the binding of proteins in the nuclear extract of MCF-7 cells to radiolabeled estrogen response element (ERE) in gel mobility shift assays (Wang et al., 1993; Lu et al., 1994). Since no supershift assays were performed in these studies, it is unknown if the changes in the gel mobility shift assays are due to differences in the estrogen receptor level or if they reflect the modulation of a different transcription factor by TCDD (Wang et al., 1993). Based on these results, it has been suggested that TCDD may downregulate the level of the estrogen receptor by posttranslational mechanism.

TCDD AND GROWTH FACTORS

Epidermal Growth Factors

Another way TCDD modulates uterine growth may be via its effects on growth factors and their receptors (Dohr et al., 1994; Vogel and Abel, 1995). Uterine cells are growth stimulated by EGF (Irwin et al., 1991; Beck and Garner, 1992). These cells express not only the receptor for EGF but also express and respond to several ligands known to interact with it (Gardner et al., 1987; DiAugustine et al., 1988; Gardner et al., 1989; Beck and Garner, 1992). These include EGF (DiAugustine et al., 1988; Bhattacharyya et al., 1994), TGF-α (Anzai et al., 1992; Gong et al., 1994), and heparin-binding EGF (HB-EGF) (Brigstock et al., 1990; Das et al., 1994). Based on numerous observations showing that estrogen upregulates uterine EGFR and its ligands, it has been suggested that the EGF pathway is an important intermediate for estrogenic stimulation of uterine cell growth (Gardner et al., 1987, 1989; DiAugustine et al., 1988; Beck and Garner, 1992; Nelson et al., 1992; Das et al., 1994). TCDD causes a downregulation of epidermal growth factor receptors (EGFR) in several tissues and inhibits its induction by estradiol (Madhukar et al., 1984; Astroff et al., 1990; Goldstein et al., 1990; Astroff and Safe, 1991; Safe et al., 1991; Sewall et al., 1993). The decrease in EGFR is accompanied by a decrease in EGFR mRNA, and no change in affinity of EGF for its receptor is seen (Astroff et al., 1990; Goldstein et al., 1990). The modulation has been attributed both to direct effects on the expression of EGFR (Astroff et al., 1990; Goldstein et al., 1990) and to the enhanced secretion of TGF-α, a ligand for EGFR (Choi et al., 1991; Dohr et al., 1994; Vogel and Abel, 1995). In those model systems in which TCDD induces TGF-α, the EGFR is activated and internalized and TCDD produces EGF-like effects (Madhukar et al., 1984; Choi et al., 1991; Dohr et al., 1994; Vogel and Abel, 1995) There is no indication that the uterine EGFR is activated following TCDD administration; furthermore, EGF is estrogenic and TCDD is antiestrogenic.

In addition to its effects on growth, the EGFR has other effects in the uterus. The interaction of uterine stromal cell EGFR with the HB-EGF on the surface of the blastocyst is believed to be one of the first interactions during implantation (Brigstock et al., 1990; Das et al., 1994). In the myometrium, EGF has been shown to induce contractions (Gardner et al., 1987). It is unclear to what degree, if any, these EGFR-dependent parameters are affected by TCDD.

Other Growth Factors

A number of growth factors are involved in uterine function (Bhattacharyya et al., 1994; Werner et al., 1994). Studies in several nonuterine target cells have shown that TCDD stimulates TGF-ß mRNA and protein accumulation (Abbott and Birnbaum, 1990; Gaido et al., 1992; Vogel and Abel, 1995). Uterine epithelial cells are both growth inhibited (Anzai et al., 1992) and stimulated to undergo apoptosis, a coordinated program of cell death characterized by DNA fragmentation by TGFβ (Rotello et al., 1991). Although some emphasis has been placed on TCDD downregulation of EGFR levels to explain its antiuterotrophic activity, It is not known if the inhibition of estrogen-induced uterine growth is also associated with TCDD modulation of TGF-ß and changes in apoptosis.

Another growth factor involved in uterine function is IGF, which regulates the growth and differentiation of a wide variety of cells via its interaction with its cell surface receptor (Rosenfeld et al., 1990; Werner et al., 1994). The interaction of IGF with its receptor is controlled by IGF-binding proteins (IGFBPs), which, depending upon the cell type and hormonal milieu, can potentiate or inhibit the binding of IGF to its receptor (Rosenfeld et al., 1990). IGF I and II, the type I receptor for IGF, and several IGFBPS are expressed in the uterus (Murphy et al., 1987a; Tang et al., 1994; Werner et al., 1994). IGF potentiates the effects of estrogen on uterine growth and is believed to be an essential intermediate for the uterine growth-promoting effects of estrogen (Irwin et al., 1991; Werner et al., 1994). In addition, estrogen stimulates endometrial IGF-I production during the peri-implantation process (Murphy et al., 1987a; Sahlin et al., 1990), and IGF-I and IGF-I receptor levels fluctuate during the ovarian cycle in response to estrogens (Werner et al., 1994). TCDDs antiestrogenic effects have been correlated with an inhibition of IGF-I receptor levels in MCF-7 breast cancer cells (Liu et al., 1992). However, it has yet to be determined whether TCDD directly affects uterine IGF pathway.

EXTRAUTERINE EFFECTS OF TCDD

Since the control of estrogen secretion requires both positive and negative feedback at the level of the hypothalamus and pituitary, alterations in CNS function by TCDD could interfere with normal uterine cyclic activity. Studies in rats have shown that TCDD can alter the brain's control over reproductive function. Normally, castrated male animals will not increase LH secretion if given estrogen and progesterone as female animals will. However, if male rats are exposed to TCDD perinatally and the same experiment is performed, LH secretion in males will be enhanced (Mably et al., 1992a). This increased responsiveness of the pituitary in males effectively "feminizes" their response to estrogen, suggesting that TCDD may interfere with normal differentiation of the CNS. TCDD has also been shown to alter the ability of the pituitary to respond to feedback from androgens. TCDD treatment of male animals results in a net androgen-deficient state. Normally, this would cause an increase in GnRH receptors in the pituitary in an effort to restore the steady state of androgens. However, in TCDD-treated animals, no increase in GnRH receptors is seen, and thus no compensatory increase in serum LH occurs. This prevents the animals from compensating for their decreased androgen levels (Mably et al., 1992b). While other studies have shown that in utero and lactational exposure to TCDD feminizes male sexual behavior, these changes could not be attributed to decreased perinatal androgen or estradiol levels, since the volumes of hormone-responsive brain nuclei are unaffected. In addition, in utero or lactational exposure to TCDD had no effect on the concentration of ER in brain nuclei in adulthood (Bjerke et al., 1994). Thus, the mechanism by which TCDD alters reproductive behavior is not known. These experiments do, however, suggest the possibility that TCDD can interfere with reproduction at the level of the CNS, offering another explanation for the altered reproductive function seen in animals treated with TCDD.

SUMMARY

TCDD blocks the trophic activity of estrogens in the uterus. This antiestrogenic activity is associated with enhanced hepatic and uterine metabolism of estradiol, diminished uterine nuclear estrogen receptor levels, and downregulation of growth factor receptors, which act as intermediates in the estrogen regulatory pathway. These three changes create a local hypoestrogenic state by diminishing the levels of active hormone as well as the uterine ability to respond to estradiol. Furthermore, TCDD modulates the ovaries as well as the hypothalamic pituitary axis. Since the uterus is under the control of hormones secreted by these glands, changes in their activity produce indirect effects on uterine function. To what degree these individual changes are responsible for antiuterotrophic effects of TCDD is unclear. The literature on TCDD and the uterus has focused primarily on the immature female, and it is unknown if TCDD will modulate uterine function in the adult cycling female. Clearly research is needed to determine if the consequences of TCDD's antiestrogenic activity include increased reproductive toxicity.

ACKNOWLEDGMENTS

We wish to thank Drs. Eleana McCoy and Richard L. Eckert for their suggestions during the preparation of the manuscript. This work was supported by NIH grant ES05227.

REFERENCES

Abbott, B. D., and Birnbaum, L. S. (1990). *Toxicol. Appl. Pharmacol., 106*: 418.

Allen, J. R., Barsotti, D. A., Lambrecht, L. K., and VanMiller, J. P. (1979). *Ann. NY Acad. Sci., 320*: 419.

Anzai, Y., Gong, Y., Holinka, C. F., Murphy, L. J., Murphy, L. C., Kuramoto, H., and Gurpide, E. (1992). *J. Steroid Biochem. Mol. Biol., 42*: 449.

Aoyama, T., Korzekwa, K., Nagata, K., Gillette, J., Gelboin, H. V., and Gonzalez, F. J. (1990). *Endocrinology, 126*: 3101.

Aronica, S. M., and Katzenellenbogen, B. S. (1991). *Endocrinology, 128*: 2045.

Astroff, B., Eldridge, B., and Safe, S. (1991). *Toxicol. Lett., 56*: 305–315.

Astroff, B., Rowlands, C., Dickerson, R., and Safe, S. (1990). *Mol. Cell Endocrinol., 72*: 247–252.

Astroff, B. and Safe, S. (1988). *Toxicol. Appl. Pharmacol., 95*: 435–443.

Astroff, B. and Safe S. (1991). *Toxicology, 69*: 187–197.

Ayres, S. M., Webb, K. B., Evans, R. G., and Miles, J. (1985). *Environ. Health Perspect., 62*: 329–335.

Bagchi, M. K., Tsai, S. Y., Weigel, N. L., Tsai, M. J., and O'Malley, B. W. (1990). *J. Biol. Chem., 265*: 5129–5134.

Barsotti, D. A., Abrahamson, L. J., and Allen, J. R. (1979). *Bull. Envir. Contam. Toxicol., 21*: 463–469.

Beck, C. A. and Garner, C. W. (1992). *Mol. Cell Endocrinol. 84*: 109–118.

Bergman, M. D., Schachter, B. S., Karelus, K., Combatsiaris, E. P., Garcia, T., and Nelson, J. F. (1992). *Endocrinology, 130*: 1923–1930.

Bhattacharyya, N., Ramsammy, R., Eatman, E., Hollis, V. W., and Anderson, W. A. (1994). *J. Submicrosc. Cytol. Pathol., 26*: 147–162.

Biegel, L. and Safe, S. (1990). *J. Steroid Biochem. Mol. Biol., 37*: 725–732.

Bjerke, D. L., Brown, T. J., MacLusky, N. J., Hochberg, R. B., and Peterson, R. E. (1994). *Toxicol. Appl. Pharmacol., 127*: 258–267.

Bookstaff, R. C., Kamel, F., Moore, R. W., Bjerke, D. L., and Peterson, R. E. (1990). *Toxicol. Appl. Pharmacol., 105*: 87–92.

Bookstaff, R. C., Moore, R. W., and Peterson, R. E. (1990). *Toxicol. Appl. Pharmacol., 104*: 212–224.

Brenner, R. M. and Slayden, O. D. (1994). *The Physiology of Reproduction* (E. Knobil and J. D. Neil, eds.). Raven Press, New York, pp. 541–569.

Breuer, H, Knupper, R., and Haupt, M. (1966). *Nature, 212*: 76–78.

Brigstock, D. R., Heap, R. B., Baaarker, P. J., and Brown, K. D. (1990). *Biochem. J., 266*: 273–282.

Bumb, R. R., Crummett, W. B., Coutie, S. S., Gledhill, J. R., Hummel, R. H., Kagel, R. O., Lamparski, L. L. Luoma, E. V., Miller, D. L., Nestrick, T. J., Shadoff, L. A., Stegl, R. H., and Woods, J. S. (1980). *Science, 210*: 385–390.

Catelli, M. G., Binart, N., Jung Testas, I., Renoir, J. M., Baulieu, E. E., Feramisco, J. R., and Welch, W. J. (1985). *EMBO J., 4*: 3131.

Chambraud, B., Berry, M., Redeuilh, G., Chambon, P., and Baulieu, E. E. (1990). *J. Biol., Chem., 265*: 20686.

Chan, W. K., Chu, R., Jain, S., Reddy, J. K., and Bradfield, C. A. (1994). *J. Biol. Chem., 269*: 26464.

Chen, H. T. (1987). *Endocrinology, 120*: 247.

Choi, E. J., Toxcano, D. G., Ryan, J. A., Riedel, N., and Toscano, W. A. (1991). *J. Biol. Chem., 266*: 9591.

Clark, B. F. (1980). *J. Endocrinol., 85*: 155.

Clark, J. H., and Mani, S. K. (1994). *The Physiology of Reproduction* (E. Knobil and J. D. Neil, eds.). Raven Press, New York, p. 1011.

Clark, J. H., Markaverich, B., Upchurch, S., Eriksson, H., Hardin, J. W., and Peck, E. J. J. (1980). *Recent Prog. Horm. Res., 36*: 89.

Coulson, P. B., and Pavlik, E. J. (1977). *J. Steroid Biochem., 8*: 205.

Couture, L. A., Abbott, B. D., and Birnbaum, L. S. (1990). *Teratology, 42*: 619.

Cresteil, T., Jaiswal, A. K., and Eisen, H. J. (1987). *Arch. Biochem. Biophys., 253*: 233.

Crummett, W. B. (1982). *Chlorinated Dioxins and Related Compounds: Impact on the Environment* (O. Hutzinger, R. W. Frei, E. Merian, and F. Pocchiaari, eds.). Pergamon Press, Oxford.

Dalman, F. C., Sturzenbecker, L. J., Levin, A. A., Lucas, D. A., Perdew, G. H., Petkovitch, M., Chambon, P., Grippo, J. F., and Pratt, W. B. (1991). *Biochemistry, 30*: 5605.

Das, S. K., Tsukamura, H., Paria, B. C., Andrews, G. K., and Dey, S. K. (1994a). *Endocrinology, 134*: 971.

Das, S. K., Wang, X. N., Paria, B. C., Damm, D., Abraham, J., A., Klgasbrun, M., Andrews, G. K., and Dey, S. K. (1994b). *Development, 120*: 1113.

de Morais, S. M., Giannone, J. V., and Okey, A. B. (1994). *J. Biol. Chem., 269*: 12129.

Denison, M. S., Fisher, J. M., and Whitlock, J. P. J. (1988). *Proc. Natl Acad Sci USA, 85*: 2528.

Denison, M. S., Fisher, J. M., and Whitlock, J. P. J. (1989). *J. Biol. Chem., 264*: 16478.

DeVito, M. J., Thomas, T., Martin, E., Umbreit, T. H., and Gallo, M. A. (1992). *Toxicol. Appl. Pharmacol., 113*: 284.

DiAugustine, R. P., Petrusz, P., Bell, G. I., Brown, C. F., Korach, K. S., McLachlan, J. A., and Teng, C. T. (1988). *Endocrinology, 122*: 2355.

Dickson, R. B., Bates, S. E., McManaway, M. E., and Lippman, M. E. (1986). *Cancer Res., 46*: 1707.

Docke, F., Rohde, W., Stahl, F., Smollich, A., and Dorner, G. (1984). *Exp. Clin. Endocrinol., 83*: 6.

Dohr, O., Vogel, C., and Abel, J. (1994). *Exp. Clin. Immunogenet., 11*: 142.

Dunn, T. J., Lindahl, R., and Pitot, H. C. (1988). *J. Biol. Chem., 263*: 10878.

Eckert, R. L., and Katzenellenbogen, B. S. (1982). *Cancer Res., 42*: 139.

Elledge, R. M., Clark, G. M., Fuqua, S. A., Yu, Y. Y., and Allred, D. C. (1994). *Cancer Res., 54*: 3752.

Evans, R. W., and Leavitt, W. W. (1980). *Proc. Natl. Acad. Sci. USA, 77*: 5856.

Fernandez, P., and Safe, S. (1992). *Toxicol. Lett., 61*: 185.

Fishman, J. (1983). *Annu. Rev. Physiol., 45*: 61.

Furlow, J. D., Murdoch, F. E., and Gorski, J. (1993). *J. Biol. Chem., 268*: 12519.

Gaido, K. W., Maness, S. C., Leonard, L. S., and Greenlee, W. F. (1992). *J. Biol. Chem., 267*: 24591.

Gallo, M. A., Hesse, E. J., Macdonald, G. J., and Umbreit, T. H. (1986). *Toxicol. Lett., 32*: 123.

Gardner, R. M., Lingham, R. B., and Stancel, G. M. (1987). *FASEB J., 1*: 224.

Gardner, R. M., Verner, G., Kirkland, J. L., and Stancel, G. M. (1989). *J. Steroid Biochem., 32*: 339.

Gasiewicz, T. A., Rucci, G., Henry, E. C., and Baggs, R. B. (1986). *Biochem. Pharmacol., 35*: 2737.

Gaub, M. P., Bellard, M., Scheuer, I., Chambon, P., and Sassone Corsi, P. (1990). *Cell, 63*: 1267.

Giavini, E., Prati, M., and Vismara, C. (1982). *Environ. Res., 29*: 185.

Gierthy, J. F. and Lincoln, D. W. (1988). *Breast Cancer Res. Treat., 12*: 227.

Gierthy, J. F., Lincoln, D. W., Gillespie, M. B., Seeger, J. I., Martinez, H. L., Dickerman, H. W., and Kumar, S. A. (1987). *Cancer Res., 47*: 6198.

Gierthy, J. F., Lincoln, D. W., Kampcik, S. J., Dickerman, H. W., Bradlow, H. L., Niwa, T., and Swaneck, G. E. (1988). *Biochem. Biophys. Res. Commun., 157*: 515.

Geirthy, J. F., Bennett, J. A., Bradley, L. M., and Cutler, D. S. (1993). *Cancer Res., 53*: 3149.

Giguere, V., Yang, N., Segui, P., and Evans, R. M. (1988). *Nature, 331*: 91.

Goldstein, J. A., Lin, F. H., Stohs, S. J., Graham, M., Clarke, G., Birnbaum, L., and Lucier, G. (1990). *Prog. Clin. Biol. Res., 331*: 187.

Gong, Y., Murphy, L. C., and Murphy, L. J. (1994). *J. Steroid Biochem. Mol. Biol., 50*: 13.

Gorski, J., Furlow, J. D., Murdoch, F. E., Fritsch, M., Kaneko, K., Ying, C., and Malayer, J. R. (1993). *Biol. Reprod., 48*: 8.

Guengerich, F. P. (1990). *Life Sci., 47*: 1981.

Hankinson, O. (1994). *Adv. Enzyme Regul., 34*: 159.

Harris, M., Zacharewski, T., Piskorska Pliszczynska, J., Rosengren, R., and Safe, S. (1990a). *Toxicol. Appl. Pharmacol., 105*: 243.

Harris, M., Zacharewski, T., and Safe, S. (1990b). *Cancer Res., 50*: 3579.

Harris, M., Piskorska-Pliszcznska, J., Zacharewski, T., Romkes, M., and Safe, S. (1989). *Cancer Res., 49*: 4531.

Hayes, J. R., Rorke, E. A., Roberston, D. W., Katzenellenbogen, B. S., and Katezenellenbogen, J. A. (1981). *Endocrinology, 108*: 164.

Henry, E. G., and Gasiewicz, T. A. (1986). *Chem. Biol. Interact., 59*: 29.

Hild Petito, S., Verhage, H. G., and Fazleabas, A. T. (1992). *Endocrinology, 103*: 2343.

Hirst, M. A., Hinck, L., Danielsen, M., and Ringold, G. M. (1992). *Proc. Natl. Acad. Sci. USA, 89*: 5527.

Hisaw, F. L. (1959). *Endocrinology, 64*: 276.

Hodgson, E., and Levi, P. E. (1987). *A Textbook of Modern Toxicology.* Elsevier, New York.

Hruska, R. E. and Olson, J. R. (1989). *Toxicol. Lett., 48*: 289.

IARC. (1978) *Monographs on the Evaluation of Carcinogenic Risk to Humans.* IARC 18,

Ikeda, M., Ogata, F., Curtis, S. W., Lubahn, D. B., French, F. S., Wilson, E. M., and Korach, K. S. (1993). *J. Biol. Chem., 268*: 10296.

Inoue, S., Orimo, A., Hosoi, T., Kondo, S., Toyoshima, H., Kondo, T., Ikegami, A., Ouchi, Y., Orimo, H., and Muramatsu, M. (1993). *Proc. Natl. Acad. Sci. USA, 90*: 11117.

Irwin, J. C., Utian, W. H., and Eckert, R. L. (1991). *Endocrinology, 129*: 2385.

Jacq, X., Brou, C., Lutz, Y., Davidson, I., Chambon, P., and Tora, L. (1994). *Cell, 79*: 107.

Jaiswal, A. K., Gonzalez, F. J., and Nebert, D. W. (1985). *Science, 228*: 80.

Jaiswal, A. K., Gonzalez, F. J., and Nebert, D. W. (1987). *Mol. Endocrinol., 1*: 312.

Jones, P. B., Galeazzi, D. R., Fisher, J. M., and Whitlock, J. P. J. (1985). *Science, 227*: 1499.

Jordan, V. C., Mittal, S., Gosden, B., Koch, R., and Lieberman, M. E. (1985). *Environ. Health Perspect., 61*: 97.

Kato, S., Tora, L., Yamauchi, J., Masushige, S., Bellard, M., and Chambon, P. (1992). *Cell, 68*: 731.

Kociba, R. J., and Schwetz, B. A. (1982). *Drug Met. Rev., 13*: 387.

Kociba, R. J., Keyes, D. G., Beger, J. E., Carreon, R. M., Wade, C. E., Dittenber, D. A., Kalnins, R. P., Frauson, L. E., Park, C. L., Barnard, S. D., Hummel, R. A., and Humiston, C. G. (1978). *Toxicol. Appl. Pharmacol., 46*: 279.

Kohn, M. C., Lucier, G. W., Clark, G. C., Sewall, C., Tritscher, A. M., and Portier, C. J. (1993). *Toxicol. Appl. Pharmacol., 120*: 138.

Kupfer, D., Mani, C., Lee, C. A., and Rifkind, A. B. (1994). *Cancer Res., 54*: 3140.

Landers, J. P., and Bunce, N. J. (1991). *Biochem. J., 276*: 273.

Lax, E. R., Tamulevicius, P., Muller, A., and Schriefers, H. (1983). *J. Steroid Biochem., 19*: 1083.

Leibl, H., Bieglmayer, C., and Spona, J. (1981). *Endocrinol. Exp., 15*: 35.

Li, J. J., and Li, S. A. (1990). *Endocr. Rev., 11*: 524.

Li, X., Johnson, D. C., and Rozman, K. K. (1995). *Toxicol. Appl. Pharmacol., 133*: 321.

Liehr, J. G. (1990). *Mutat. Res., 238*: 269.

Lin, F. H., Stohs, S. J., Birnbaum, L. S., Clark, G., Lucier, G. W., and Goldstein, J. A. (1991). *Toxicol. Appl. Pharmacol., 108*: 129.

Lingham, R. B., Stancel, G. M., and Loose Mitchell, D. S. (1988). *Mol. Endocrinol., 2*: 230.

Lippman, M. E., Dickson, R. B., Gelmann, E. P., Rosen, N., Knabbe, C., Bates, S., Bronzert, D., Huff, K., and Kasid, A. (1987). *J. Cell Biochem., 35*: 1.

Liu, H., Biegel, L., Narasimhan, T. R., Rowlands, C., and Safe, S. (1992). *Mol. Cell Endocrinol., 87*: 19.

Loose-Mitchell, D. S., Chaippettta, C., and Stancel, G. M. (1988). *Mol. Endocrinol., 2*: 946.

Lu, Y., Wang, X., and Safe, S. (1994). *Toxicol. Appl. Pharmacol., 127*: 1.

Lu, Y. C., and Wong, P. N. (1984). *Am. J. Ind. Med., 5*: 81.

Luisi, B. F., Schwabe, J. W., and Freedman, L. P. (1994). *Vitam. Horm., 49*: 1.

Mably, T. A., Moore, R. W., Goy, R. W., and Peterson, R. E. (1992a). *Toxicol. Appl. Pharmacol., 114*: 108.

Mably, T. A., Moore, R. W., and Peterson, R. E. (1992b). *Toxicol. Appl. Pharmacol., 114*: 97.

MacDonald, R. G., Gianferrari, E. A., and Leavitt, W. W. (1982). *Steroids, 40*: 465.

Mader, S., Chambon, P., and White, J. H. (1993). *Nucleic Acids Res., 21*: 1125-1132.

Madhukar, B. V., Brewster, D. W., and Matsumura, F. (1984). *Proc. Natl. Acad. Sci. USA, 81*: 7407.

Matsushita, N., Sogawa, K., Ema, M., Yoshida, A., and Fujii Kuriyama, Y. (1993). *J. Biol. Chem., 268*, 21002-21006.

McCormack, S. A., and Glasser, S. R. (1980). *Endocrinology, 106*: 1634.

Mofenson, H., Becker, C., Kimbrough, R., Lawrence, R., Lovejoy, F., Winters, W., Carracio, T. R., Hardel, L. K., Rumack, B. H., and Spyker, D. (1985). *Vet. Hum. Toxicol., 27*: 434.

Moore, R. W., Parsons, J. A., Bookstaff, R. C., and Peterson, R. E. (1989). *J. Biochem. Toxicol., 4*: 165.

Moore, M., Narasimhan, T. R., Wang, X., Krishnan, V., Safe, S., Williams, H. J., and Scott, A. I. (1993). *J. Steroid Biochem. Mol. Biol., 44*: 251.

Murdoch, F. E., Meier, D. A., Furlow, J. D., Grunwald, K. A., and Gorski, J. (1990). *Biochemistry, 29*: 8377.

Murphy, L. J., Murphy, L. C., and Friesen, H. G. (1987a). *Mol. Endocrinol., 1*: 445.

Murphy, L. J., Murphy, L. C., and Friesen, H. G. (1987b). *Endocrinology, 120*: 1882.

Namkung, M. J., Porubek, D. J., Nelson, S. D., and Juchau, M. R. (1985). *J. Steroid Biochem., 22*: 563.

Nardulli, A. M., Greene, G. L., O'Malley, B. W., and Katzenellenbogen, B. S. (1988). *Endocrinology, 122*: 935.

Nebert, D. W. (1991). *Mutat. Res., 247*: 267.

Nebert, D. W. (1993). *J. Natl. Cancer Inst., 85*: 1888.

Nebert, D. W., and Jones, J. E. (1989). *Int. J. Biochem., 21*: 243.

Nebert, D. W., Puga, A., and Vasiliou, V. (1993). *Ann. NY Acad. Sci., 685*: 624.

Nelson, K. G., Takahashi, T., Lee, D. C., Luetteke, N. C., Bosseeert, N. L., Ross, K., Eitzman, B. E., and McLachlan, J. A. (1992). *Endocrinology, 131*: 1657.

Newbold, R. R., Bullock, B. C., and McLachlan, J. A. (1990). *Cancer Res., 50*: 7677.

Nigro, V., Molinari, A. M., Armetta, I., de Falco, A., Abbondanza, C., Medici, N., and Puca, G. A. (1992). *Biochem. Biophys. Res. Commun., 186*: 803.

Okulicz, W. C., Evans, R. W., and Leavitt, W. W. (1981). *Steroids, 37*: 463.

O'Malley, B. W., Tsai, S. Y., Bagchi, M., Weigel, N. L., Schrader, W. T., and Tsai, M. J. (1991). *Recent. Prog. Horm. Res., 47*: 1.

Orti, E., Bodwell, J. E., and Munck, A. (1992). *Endocr. Rev., 13*: 105.

Perdew, G. H., Hord, N., Hollenback, C. E., and Welsh, M. J. (1993). *Exp. Cell Res., 209*: 350.

Peterson, R. E., Theobald, H. M., and Kimmel, G. L. (1993). *Crit. Rev. Toxicol., 23*: 283.

Pitot, H. C., Goldsworthy, T. L., Moran, S., Kennan, W., Glauert, H. P., Maronpot, R. R., and Campbell, H. A. (1987). *Carcinogenesis, 8*: 149.

Poland, A., and Knutson, J. C. (1982). *Annu. Rev. Pharmacol., Toxicol., 22*: 517.

Pollard, I., and Martin, L. (1968). *Steroids, 11*: 897.

Predki, P. F., and Sarkar, B. (1992). *J. Biol. Chem., 267*: 5842.

Quarmby, V. E., and Martin, L. (1982). *Mol. Cell Endocrinol., 27*: 317.

Randerath, K., Putman, K. L., Randerath, E., Zacharewski, T., Harris, M., and Safe, S. (1990). *Toxicol. Appl. Pharmacol., 103*: 271.

Ratajczak, T., Hlaing, J., Brockway, M. J., and Hahnel, R. (1990). *J. Steroid Biochem., 35*: 543.

Ratajczak, T., Carrello, A., Mark, P. J., Warner, B. J., Simpson, R. J., Moritz, R. L., and House, A. K. (1993). *J. Biol. Chem., 268*: 13187.

Read, L. D., Greene, G. L., and Katzenellenbogen, B. S. (1989). *Mol. Endocrinol., 3*: 295.

Reynolds, S. R. M. (1951). *Physiol. Rev., 1*: 244.

Roberts, E. A., Johnson, K. C., and Dippold, W. G. (1991). *Biochem. Pharmacol., 42*: 521.

Romkes, M., and Safe, S. (1988). *Toxicol. Appl. Pharmacol., 92*: 368.

Romkes, M., Piskorska Pliszczynska, J., and Safe, S. (1987). *Toxicol. Appl. Pharmacol., 87*: 306.

Rosenfeld, R. G., Lamson, G., Pham, H., Oh, Y., Conover, C., De Leon, D. D., Donovan, S. M., Ocrant, I., and Giudice, L. (1990). *Recent Prog. Horm. Res., 46*: 99.

Rotello, R. J., Lieberman, R. C., Purchio, A. F., and Gerschenson, L. E. (1991). *Proc. Natl. Acad. Sci. USA, 88*: 3412.

Rowlands, C., Krishnan, V., Wang, X., Santostefano, M., Safe, S., Miller, W. R., and Langdon, S. (1993). *Cancer Res., 53*: 1802.

Roy, D., Weisz, J., and Leihr, J. G. (1990). *Carcinogenesis, 11*: 459.

Saatcioglu, F., Perry, D. J., Pasco, D. S., and Fagan, J. B. (1990). *J. Biol. Chem., 265*: 9251.

Safe, S., Astroff, B., Harris, M., Zacharewski, T., Dickerson, R., Romkes, M., and Biegel, L. (1991). *Pharmacol. Toxicol., 69*: 400.

Sahlin, L., Rodriguez Martinez, H., Stanchev, P., Dalin, A. M., Norstedt, G., and Eriksson, H. (1990). *Zentralbl. Veterinarmed. A, 37*: 795.

Savas, U., Christou, M., and Jefcoate, C. R. (1993). *Carcinogenesis, 14*: 2013.

Savas, U., Bhattacharyya, K. K., Christou, M., Alexander, D. L., and Jefcoate, C. R. (1994). *J. Biol. Chem., 269*: 14905.

Scherrer, L. C., Picard, D., Massa, E., Harmon, J. M., Simons, S. S. J., Yamamoto, K. R., and Pratt, W. B. (1993). *Biochemistry, 32*: 5381.

Schild, C., Claret, F. X., Wahli, W., and Wolffe, A. P. (1993). *EMBO J., 12*: 423.

Schlatter, L. K., Howard, K. J., Parker, M. G., and Distelhorst, C. W. (1992). *Mol. Endocrinol., 6*: 132.

Schuh, T. J., and Mueller, G. C. (1993). *Receptor, 3*: 125.

Scientific Review Committee of the American Academy of Clinical Toxicology (1985). *J. Toxicol. Clin. Toxicol, 23*: 191.

Senekjian, E. K., Press, M. F., Blough, R. R., Herbst, A. L., and DeSombre, E. R. (1989). *Am. J Obstet. Gynecol., 160*: 592.

Sewall, C. H., Lucier, G. W., Tritscher, A. M., and Clark, G. C. (1993). *Carcinogenesis, 14*: 1885.

Shiverick, K. T., and Muther, T. F. (1983). *Biochem. Pharmacol., 32*: 991.

Silbergeld, E. K., and Mattison, D. R. (1987). *Am. J. Ind. Med., 11*: 131.

Simpson, V. J., Baker, R., and Deitrich, R. A. (1985). *Toxicol. Appl. Pharmacol., 79*: 193.

Smidt, M. P., Wijnholds, J., Snippe, L., van Keulen, G., and Ab, G. (1994). *Biochim. Biophys. Acta, 1219*: 115.

Spink, D. C., Lincoln, D. W., Dickerman, H. W., and Gierthy, J. F. (1990). *Proc. Natl. Acad. Sci. USA, 87*: 6917.

Spink, D. C., Eugster, H. P., Lincoln, D. W. I. I., Schuetz, J. D., Schuetz, E. G., Johnson, J. A., Kaminsky, L. S., and Gierthy, J. F. (1992). *Arch. Biochem. Biophys. 293*: 342.

Spink, D. C., Hayes, C. L., Young, N. R., Christou, M., Sutter, T. R., Jefcoate, C. R., and Gierthy, J. F. (1994). *J. Steroid Biochem. Mol. Biol., 51*: 251.

Stancel, G. M., Gardner, R. M., Kirkland, J. L., Lin, T. H., lingham, R. B., Loose Mitchell, D. S., Mukku, V. R., Orengo, C. A., and Verner, G. (1987). *Adv. Exp. Med Biol., 230*: 99.

Strauss III, J. F., and Gurpide, E. (1991). *Reproductive Endocrinology* (R. B. Jaffe and S. S. C. Yen, eds.). W. B. Saunders and Co., Philadelphia, p. 309.

Struhl, K. (1989). *Trends Biochem. Sci., 14*: 137.

Sumida, C., Gelly, C., and Pasqualini, J. R. (1981). *J. Recept. Res., 2*: 221.

Takahashi, S., Mikami, T., Watanabe, Y., Okazaki, M., Okazaki, Y., Okazaki, A., Sato, T., Asaishi, K., Hirata, K., Narimatsu, E., et al. (1994). *Am. J. Clin. Pathol., 101*: 519.

Tang, X. M., Rossi, M. J., Massterson, B. J., and Chegini, N. (1994). *Biol. Reprod., 50*: 1113.

Tank, A. W., Deitrich, R. A., and Weiner, H. (1986). *Biochem Pharmacol., 35*: 4563.

Telakowski Hopkins, C. A., King, R. G., and Pickett, C. B. (1988). gene: identification of regulatory elements required for basal level and inducible expression. *Proc. Natl. Acad. Sci. USA, 85*: 1000.

Thomsen, J. S., Nissen, L., Stacey, S. N., Hines, R. N., and Autrup, H. (1991). *Eur. J. Biochem., 197*: 577.

Thomsen, J. S., Wang, X,. Hines, R. N., and Safe, S. (1994). *Carcinogenesis, 15*: 933.

Travers, M. T., and Knowler, J. T. (1987). *FEBS Lett., 211*: 27.

Umbreit, T. H., Hesse, E. J., and Gallo, M. A. (1987). *Arch. Environ. Contam. Toxicol., 16*: 461.

Umbreit, T. H., Hesse, E. J., Macdonald, G. J., and Gallo, M. A. (1988). *Toxicol. Lett., 40*: 1.

Umbreit, T. H., Engles, D., Grossman, A., and Gallo, M. A. (1989). *Toxicol. Lett., 48*: 29.

Vamvakopoulos, N. C., and Chrousos, G. P. (1993). *J. Clin. Invest., 92*: 1896.

Vickers, A. E., and Lucier, G. W. (1991). *Carcinogenesis, 12*: 391.

Vickers, P. J., Dufresne, M. J., and Cowan, K. H. (1989). *Mol. Endocrinol., 3*: 157.

Vogel, C., and Abel, J. (1995). *Arch. Toxicol., 69*: 259.

Wang, X., Rosengren, R., Morrison, V., Santostefano, M., and Safe, S. (1992). *Biochem. Pharmacol., 43*: 1635.

Wang, X., Porter, W., Kirshnan, V., Narasimhan, T. R., and Safe, S. (1993). *Mol. Cell Endocrinol., 96*: 159.

Werner, H., Adamo, M., Roberts, C. T. J., and LeRoith, D. (1994). *Vitam. Horm., 48*: 1.

White, T. E., and Gasiewicz, T. A. (1993). *Biochem. Biophys. Res. Commun., 193*: 956–962.

White, T. E. K., Rucci, G., Liu, Z., and Gasiewicz, T. A. (1995). *Toxicol. Appl. Pharmacol., 133*: 313–320.

Whitlock, J. P. J. (1990). *Annu. Rev. Pharmacol. Toxicol., 30*: 251–277.

Whitlock, J. P. J. (1993). *Chem. Res. Toxicol., 6*: 754–763.

Whitlock, J. P. J., Denison, M. S., Fisher, J. M., and Shen, E. S. (1989). *Mol. Biol. Med., 6*: 169–178.

Wilson, T. E., Fahrner, T. J., and Milbrandt, J. (1993). *Mol. Cell Biol., 13*: 5797–5804.

Wilson, T. E., Paulsen, R. E., Padgett, K. A., and Milbrandt, J. (1992). *Science, 256*: 107–110.

Ying, C. and Gorski, J. (1994). *Mol. Cell Endocrinol. 99*: 183–192.

Zacharewski, T., Harris, M., and Safe, S. (1991). *Biochem. Pharmacol., 41*: 1931.

Zacharewski, T. R., Bondy, K. L., McDonell, P., and Wu, Z. F. (1994). *Cancer Res., 54*: 2707.

Zhu, B. T., and Liehr, J. G. (1993). *Arch. Biochem. Biophys., 304*: 248.

Zilliacus, J., Carlstedt Duke, J., Gustafsson, J. A., and Wright, A. P. (1994). DNA-binding specificities within the nuclear receptor family of transcription factors. *Proc. Natl. Acad. Sci. USA, 91*: 4175.

18

Antiestrogenic Activity of TCDD and Related Compounds

Timothy Zacharewski
Michigan State University, East Lansing, Michigan

Stephen H. Safe
Texas A&M University, College Station, Texas

INTRODUCTION

Origins and Environmental Impact of PCDDs and PCDFs

Polychlorinated dibenzo-*p*-dioxins (PCDDs) and dibenzofurans (PCDFs) are formed as by-products in a number of industrial processes and during combustion (Hutzinger and Fiedler, 1993; Rappe, 1993b; Fiedler, 1996). For example, PCDDs/PCDFs (Fig. 1) have been identified as contaminants in the production of chlorinated phenols and their derived products as well as other chlorinated industrial compounds. These compounds have also been detected as by-products in sewage sludge, pulp and paper mill effluents, in diverse high-temperature industrial processes, and as by-products of incineration of diverse organic materials including municipal and industrial waste, wood, and coal. The mass balance of industrial and anthropogenic inputs of PCDDs/PCDFs into the environment is unknown and depends on specific local and regional sources. Regulatory decisions in most industrialized countries have resulted in significantly decreased emissions of these compounds from many of the identified industrial sources.

PCDDs and PCDFs are stable to both chemical and thermal degradation, and this stability contributes to their persistence in the environment. These compounds are also lipophilic and environmental residues preferentially bioaccumulate in the food chain and PCDDs/PCDFs are routinely detected in fish, wildlife, human adipose tissue, serum, and milk (Rappe, 1993a,b; U.S. Environmental Protection Agency, 1995). The identification of parts per trillion residues of toxic PCDDs/PCDFs in diverse environmental samples and in food has raised considerable scientific and public concern regarding the potential adverse impacts of these chemicals on human health (U.S. Environmental Protection Agency, 1995). In addition to low-level environmental exposure to PCDDs/PCDFs, several more highly exposed groups have also been identified, and these include industrial workers involved in the manufacture of products that contain PCDDs/PCDFs as contaminants (e.g., herbicides), individuals such as herbicide sprayers who handled PCDD/PCDF-containing pesticides, and several groups who were accidentally

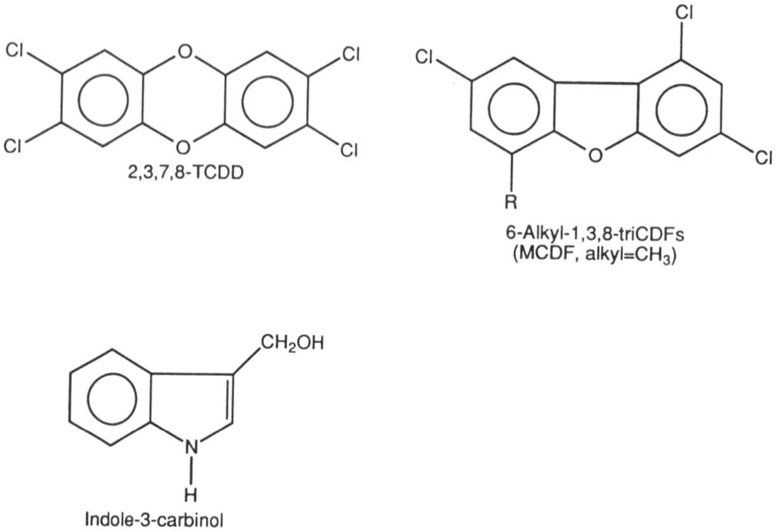

Figure 1 Structures of Ah receptor agonists that exhibit antiestrogenic activity.

exposed to these compounds (U.S. Environmental Protection Agency, 1995). For example, an industrial accident in Seveso, Italy, released trichlorophenate contaminated primarily with 2,3,7,8-tetrachlorodibenzo-*p*-dioxin (TCDD) (Fig. 1) resulting in the subsequent exposure of the population living in the vicinity of the factory (Mocarelli et al., 1991; Bertazzi et al., 1993). The widespread human exposure and environmental contamination of PCDDs and PCDFs has spurred research on the biochemical and toxic responses and mechanisms of action elicited in animal models, mammalian cells in culture, and exposed human populations.

Biochemical and Toxic Responses

There are 75 and 135 possible PCDD and PCDF congeners, respectively, and the effects of only a relatively small number of these compounds have been investigated. Most studies show that PCDDs and PCDFs elicit common biochemical and toxic responses; however, there are significant differences in their potencies (Poland and Knutson, 1982; Goldstein and Safe, 1989; Safe, 1990, 1994). TCDD, the most toxic member of this class of chemicals, has been extensively used as a prototype to investigate the various activities of these compounds and their mechanism of action (Landers and Bunce, 1991; Lucier et al., 1993; Swanson and Bradfield, 1993; Whitlock, Jr., 1993; Okey et al., 1994; Safe, 1995). The diverse spectrum of biochemical and toxic responses elicited by TCDD and related compounds include a wasting syndrome, immunosuppressive effects, hepatotoxicity and porphyria, carcinogenic and anticarcinogenic activity, hypo- and hyperplastic effects, reproductive and developmental activity, tumor-promotion activity, neurotoxicity, chloracne and other dermal effects, and the disruption of several endocrine-response pathways. TCDD also modulates expression of several genes, including the activities of some phase I and phase II drug-metabolizing enzymes, cytokines, growth factors and hormones, and their receptors (Table 1) (Safe, 1995). The effects of TCDD on a specific target organ or cell are highly variable and depen-

Table 1 Inhibition of Mammary Tumor Growth by TCDD in Rodent Models

Animal species	Rodent model	Effects
Sprague-Dawley rat	Spontaneous tumor formation	TCDD (0.1, 0.01, and 0.001 µg/kg/day) inhibited spontaneous tumor formation (53, 27, and 20% inhibition, respectively)
Sprague-Dawley rat	DEN initiation, TCDD promotion	No tumor observed in animals treated with TCDD (2 × 0.125 µg/kg/day × 30) vs. 36% (4/11) in controls
Sprague-Dawley rat	DMBA initiation, TCDD inhibitor	TCDD (3 × 10 µg/kg/week) total inhibition of tumor growth (decreased tumor volumes to < control levels)
B6C2F1 mice	MCF-7 cell solid tumor xenografts plus E2	Inhibition of tumor growth by TCDD (2–8 µg/kg)

dent on the age, sex, species, and strain of animal. For example, chloracne is one of the hallmarks of relatively high exposure to TCDD and related compounds and is observed in humans, rabbits (ears), and some inbred strains of mice, but not in most other species. Several studies show that female Sprague-Dawley rats develop liver tumors after long-term dietary exposure to TCDD (Kociba et al., 1978) or after exposure in a shorter-term initiation/promotion protocol (Tritscher et al., 1995). However, In the same animals that develop liver tumors, there is a significant decrease in formation of mammary tumors, which form spontaneously in female Sprague-Dawley rats. In contrast, male Sprague-Dawley rats are resistant to TCDD-induced hepatocarcinogenesis (Kociba et al., 1978).

The Aryl Hydrocarbon Receptor as an Intracellular Target for TCDD

Based on the results of genetic studies in inbred strains of mice and the structure-dependent effects of TCDD and related compounds, it was hypothesized that the initial target for TCDD was an intracellular protein designated as the Ah receptor (Poland and Glover, 1975; Poland et al., 1975). Poland and coworkers first identified the Ah receptor in hepatic cytosol derived from C57BL/6 mice using radiolabeled TCDD as a ligand (Poland et al., 1976). Their studies showed that the Ah receptor was a low-capacity, high-affinity binding protein which bound saturably with [3H]TCDD. Subsequent research in several laboratories has identified the Ah receptor in both target and non-target tissues/cells using [3H]TCDD, other radiolabeled PCDDs/PCDFs, and polynuclear aromatic hydrocarbons (PAHs) as radioligands (Okey et al., 1994; Safe, 1995). The properties of crude Ah receptor preparations were highly variable and dependent on techniques used in individual laboratories; however, photoaffinity labeling studies showed that the apparent molecular weight of the cytosolic Ah receptor varied between species and within strains of mice (Poland et al., 1976; Poland and Glover, 1987, 1990). For example, the Mr values for the cytosolic Ah receptor in mouse, chicken, guinea pig, rabbit, rat, human, monkey, and hamster liver were 95, 101, 103, 104, 106, 106,

113, and 125 kDa, respectively (Poland and Glover, 1987). Subsequent studies have reported the primary structure of the Ah receptor from several species based on the sequence of their corresponding genes, which have been cloned (Burbach et al., 1992; Ema et al., 1992, 1994; Dolwick et al., 1993; Schmidt et al., 1993; Swanson and Bradfield, 1993). The Ah receptor is a member of the basic helix-loop-helix (bHLH) family of proteins and contains several domains within the protein which are responsible for DNA binding, ligand binding, and interaction with other proteins required for ligand-induced transactivation (Swanson and Bradfield, 1993). Inter- and intraspecies differences in the molecular weight of the Ah receptor are due primarily to variability in the carboxyl terminus of these proteins. For example, Dolwick and coworkers (1993) reported that comparison of the amino acid sequences of the human and murine receptors indicated that the basic regions, HLH and PAS domains in the amino-terminal region of these proteins were highly conserved (100, 98, 87%, respectively), whereas there was <60% homology in the carboxyl termini. Ema and coworkers (1994) compared the structure and properties of the Ah receptor from the Ah-responsive C57BL/6 and less responsive DBA/2 mouse and showed that two critical alternations were primarily responsible for the differential Ah responsiveness. An alanine[375]-to-valine change and an elongated carboxy-terminal sequence accounted for the reduced ligand-binding affinity of the Ah receptor from the DBA/2 mice. Similar amino acid changes in the ligand-binding domain were also observed in the human Ah receptor, which also exhibits lower binding affinity for TCDD.

Molecular Mechanism of Action-Induction of CYP1A1 Gene Expression

The Ah receptor is primarily cytosolic, and after addition of TCDD, a 290- to 300-kDa complex is formed at low temperature. This complex contains heat-shock protein (hsp) 90 and exhibits low DNA-binding affinity (Perdew, 1988; Pongratz et al., 1992). Increased temperature or high-salt treatment results in formation of a 180- to 200-kDa Ah receptor complex, which is identical with the nuclear Ah complex isolated from cells or organs after treatment with TCDD (Prokipcak and Okey, 1988; Wang et al., 1991, 1992; Perdew, 1992). This lower molecular weight complex exhibits higher binding affinity for DNA and consists of the Ah receptor and a second protein, designated the Ah receptor nuclear translocator (Arnt) protein. The *arnt* gene has been cloned and identified as another bHLH protein (Hoffman et al., 1991). Subsequent studies on the 5′-flanking region of the CYP1A1 gene have identified *cis*-acting genomic dioxin or xenobiotic responsive elements (XREs or DREs), which interact with the nuclear Ah receptor resulting in transactivation of the CYP1A1 gene (Gonzalez and Nebert, 1985; Jones et al., 1985, 1986a,b; Fujisawa-Sehara et al., 1986, 1988; Neuhold et al., 1989). Functional DREs/XREs have also been identified in the 5′-flanking sequences of the CYP1A2, CYP1B1, glutathione S-transferase (Ya), NADPH:quinone oxidoreductase, and aldehyde-3-dehydrogenase genes and play a role in the regulation of gene expression (Rushmore et al., 1990; Asman et al., 1993; Jaiswal, 1994; Quattrochi et al., 1994). The nuclear Ah receptor is a unique ligand-induced bHLH transcription factor, which is not a member of the steroid/thyroid hormone receptor superfamily (Evans, 1988). However, like other members of the superfamily, the activity of the nuclear Ah receptor as a transcription factor may involve interactions with other nuclear proteins and binding to other DNA motifs (Pimental et al., 1993; Vasiliou et al., 1995).

Ah RECEPTOR AGONISTS AS ANTIESTROGENS—IN VIVO LABORATORY STUDIES

TCDD and related compounds modulate several endocrine responses that may be important for the toxic effects elicited by these compounds. For example, TCDD induces plasminogen activator inhibitor-2, transforming growth factor alpha (TGF-α), TGF-β, and interleukin-1β mRNA levels (Choi et al., 1991; Sutter et al., 1991; Gaido et al., 1992) and decreases circulating thyroid hormone levels and binding capacities of the estrogen, epidermal growth factor, and glucocorticoid receptors (Bastomsky, 1977; Madhukar et al., 1984; McKinney et al., 1985; Rozman et al., 1985; Henry and Gasiewicz, 1987; Jones et al., 1987; Ryan et al., 1989; Sunahara et al., 1989; Lin et al., 1991). TCDD also causes cell-specific induction or inhibition of steroid hormone metabolism, and this may contribute to the developmental and reproductive toxicity of TCDD (Peterson et al., 1993). The linkages between TCDD-induced endocrine disruption and specific toxic responses have not been established for most responses (Safe, 1995). However, research in our laboratories and others have characterized the antiestrogenic activity of Ah receptor agonists and some of the mechanisms associated with these responses.

Antitumorigenic Activity in Rodents

Kociba and coworkers (1978) utilized Sprague-Dawley rats for investigating the dose-dependent carcinogenic activity of TCDD in the diet (for 2 years). Although TCDD was hepatocarcinogenic, there was a significant decrease in spontaneous mammary (Table 1) and uterine tumor formation. The effects of TCDD as an inhibitor of spontaneous mammary carcinogenesis in Sprague-Dawley rats was confirmed in a recent study that utilized diethylnitrosamine as an initiator and TCDD as a promoter of hepatocarcinogenesis (Tritscher et al., 1995). Female Sprague-Dawley rats were also utilized in the 7,12-dimethylbenzanthracene (DMBA) mammary tumor model (Holcomb and Safe, 1994). Rats were treated with a carcinogenic dose of DMBA (20 mg/rat) and when mammary tumors were first detected, the animals were treated with corn oil (vehicle control) or 10 µg/kg/wk TCDD in corn oil. After 3 weeks, there was 3.9-fold increase in tumor volume in rats that received corn oil alone, whereas in the TCDD-treated animals there was a 72% decrease in tumor volume compared to tumor volumes at the beginning of the experiment. Moreover, in some of the animals, the original small mammary tumors were no longer detectable 3 weeks after treatment with TCDD. Several relatively nontoxic analogs of TCDD have been developed for potential clinical use in treatment of breast cancer, and one of these compounds, 6-methyl-1,3,8-trichlorodibenzofuran (MCDF), has been tested in the DMBA-mammary tumor model. At a dose as low as 5 mg/kg/wk ($\times$3), MCDF significantly inhibited ($>75\%$) tumor growth, and other alkyl-substituted PCDFs are currently being investigated in this assay system (McDougal et al., 1996). Gierthy and coworkers (1993) have reported that in immunosuppressed B6D2F1 mice treated with 17β-estradiol (E2) and bearing MCF-7 cell xenografts, TCDD also inhibited tumor growth. The results of the rodent bioassays indicate that TCDD and at least one other analog (i.e., MCDF) inhibit mammary tumor formation and growth.

At least one other class of Ah receptor agonists has been characterized as inhibitors of mammary tumor growth in rodents. Indole-3-carbinol (I3C) (Fig. 1) is a component of brassica vegetables, and this compound binds to the Ah receptor and is a weak Ah receptor agonist (Bjeldanes et al., 1991; Jellinck et al., 1991). Moreover, in the acidic environment of the gut, I3C rearranges to a number of hetero-PAHs, including indolo[3,2-b]carbazole, which exhibit higher binding affinity for the Ah receptor (Bjeldanes et al., 1991; De Kruif et al., 1991). Several studies have demonstrated that I3C or vegetables inhibit mammary tumor formation and growth (Wattenberg, 1978; Stoewsand et al., 1988; Bradlow et al., 1991; Grubbs et al., 1995), and these antiestrogenic effects may be Ah receptor-mediated. However, since I3C and related compounds also induce phase I and II drug-metabolizing enzymes, it is possible that metabolic detoxification of mammary tumor initiators is also important for the in vivo antiestrogenic activity of I3C.

Antiestrogenic Activity in the Rodent Uterus

The antiestrogenic activity of TCDD and related compounds has been extensively investigated in the mouse and rat uterus. TCDD inhibited E2-induced uterine wet weight increase and decreased cytosolic and nuclear estrogen receptor (ER) levels in several strains of mice (Gallo et al., 1986; Umbriet et al., 1988, 1989; Umbriet and Gallo, 1988; DeVito et al., 1992). Although TCDD exhibited antiestrogenic activity in the mouse uterus, cotreatment with TCDD plus E2 or tamoxifen potentiated the toxicity of TCDD, and this was due in part to increased hepatic retention of TCDD (MacKenzie et al., 1992).

The effects of TCDD and related PCDD and PCDF congeners were extensively investigated in the immature female Sprague-Dawley rat uterus. Initial studies compared the effects of TCDD and progesterone on uterine wet weight and progesterone and estrogen receptor (PR and ER) levels (Romkes et al., 1987; Romkes and Safe, 1988). In 25-day old female rats, TCDD alone decreased uterine wet weight and cytosolic ER and PR levels, and in rats treated with E2 plus TCDD, there was a significant decrease in E2-induced uterine wet weight and ER and PR binding. In parallel experiments utilizing the same response, both progesterone and TCDD exhibited comparable antiestrogenic activity; however, TCDD was >30 times more potent and the induced effects were more persistent than those observed for progesterone (Romkes and Safe, 1988). Both the Ah receptor and PR are expressed in rat uteri; however, TCDD did not competitively bind to the PR and progesterone did not competitively bind to the Ah receptor. The antiestrogenic activity of TCDD in the 25-day-old female rat uterus was also observed in ovariectomized female rats (Romkes et al., 1987). Subsequent studies showed that TCDD also inhibited basal and E2-inducible uterine peroxidase activity, c-*fos*, and EGF receptor mRNA levels (Astroff and Safe, 1990; Astroff et al., 1990, 1991). Moreover, results of structure-activity studies showed that there was a correlation between the antiestrogenic potency of specific PCDD/PCDF congeners and their binding affinity for the Ah receptor. For inhibition of E2-induced uterine peroxidase activity and wet weight, TCDD ≥ 2,3,4,7,8-pentaCDF > 1,2,4,7,8-pentaCDD, and this rank order of potency is similar to the Ah receptor–binding affinities of these congeners and their activities for other Ah receptor–mediated responses (Astroff and Safe, 1990). Other studies have confirmed the antiestrogenic activity of the coplanar PCB congeners 3,3′,4,4′-tetraCB in the rat (Jansen et al., 1993) and 3,3′,4,4′,5,5′-hexaCB in the mink

(Patnode and Curtis, 1994) uterus, respectively. A recent study reported that after administration of TCDD to 19-day-old weanling female Sprague-Dawley rats, no antiestrogenic activities were observed (White et al., 1995), and this contrasted to results reported in immature and ovariectomized rats and mice. Similar age-dependent results were observed in this laboratory (Dickerson and Safe, 1992). Subsequent studies showed that uterine Ah receptor levels undergo significant age-dependent changes during early development; between 7 and 28 days of age, levels increased form 4.4 to 11.3 fmol/mg, which was not significantly different than observed in ovariectomized 70-day-old rats (12.9 fmol/mg). Therefore, the insensitivity of 19- to 21-day-old mice to the antiestrogenic activity of TCDD may be related, in part, to lower Ah receptor levels.

The development of Ah receptor–mediated antiestrogens for use in the clinical treatment of breast cancer requires agonists with low toxicity but high antiestrogenic activity. MCDF was initially developed in this laboratory as an Ah receptor antagonist, and the results of several studies showed the MCDF inhibited TCDD-induced CYP1A1/CYP1A2 induction in vivo and in vitro (Astroff et al., 1988; Astroff and Safe, 1988, 1989; Harris et al., 1989b). MCDF also inhibited TCDD-induced immunotoxicity, fetal cleft palate, and hepatic porphyria in mice (Bannister et al., 1989; Lorenzen and Okey, 1990). However, MCDF did not inhibit TCDD-induced antiestrogenicity in the female rat uterus and exhibited Ah receptor agonists activity for this response. Several alkyl PCDFs have been investigated in the rat uterine assay, and their potential utility or efficacy as antiestrogens was determined from the ratio of their CYP1A1/antiestrogenicity potencies (ED_{50} values). Induction of hepatic CYP1A1-dependent activity in rats by halogenated aromatics correlates with their toxic potencies (Safe, 1990), and therefore compounds with the highest ratios will be the most effective antiestrogens (i.e., antiestrogens with relatively low toxicity). The results illustrated in Figure 2 compare the values of these ratios for MCDF and TCDD for inhibiting E2-induced uterine cytosolic (c) and nuclear (n) ER and PR levels. The ratios for MCDF vary from 3.2 to 5.9, whereas the corresponding values for TCDD were 0.007 to 0.013, and the low values for TCDD reflect the high toxicity of this compound relative to its antiestrogenic

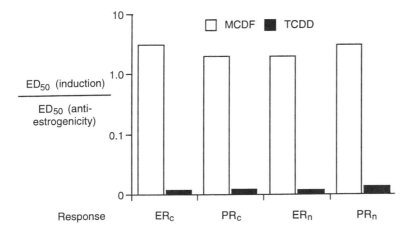

Figure 2 Comparative ED_{50} (induction) / ED_{50} (antiestrogenicity) ratios for MCDF (□) and TCDD (■). The dose-response studies and ED_{50} values were determined in 25-day-old female Sprague-Dawley rats as described. (From Astroff and Safe, 1988.)

potency. In contrast, although MCDF is less potent than TCDD as an antiestrogen in the female rat uterus, the high ED_{50} (CYP1A1 induction)/ED_{50} (antiestrogenicity) ratio is primarily due to the low activity of MCDF as an inducer of CYP1A1-dependent activity (Astroff and Safe, 1988). The low activity of MCDF as an inducer of CYP1A1 is paralleled by the low toxicity of the compound (Bannister et al., 1989; Astroff and Safe, 1988). A recent study investigated the structure-dependent activity of several alkyl PCDDs and PCDF as antiestrogens in the female rat uterus (Dickerson et al., 1995). The antiestrogenic activities of a series of 6-alkyl-1,3,8-substituted alkyl PCDFs that are substituted on only two of the four lateral positions (2, 3, 7, and 8) were compared to alkyl PCDDs/PCDFs that contained three or four lateral substituents. The antiestrogenic potencies for most of the compounds were similar; however, the alkyl PCDFs substituted on three or four lateral substituents were more potent inducers of CYP1A1 than the 1,3,6,8-substituted alkyl PCDFs. The ED_{50} (induction)/ED_{50} (antiestrogenicity) ratios for 6-isopropyl-1,3,8-triCDF, 6-methyl-2,3,4,8-tetraCDF, and 8-methyl-2,3,4,7-tetraCDF were 17,100, 1.69, and 0.69 for uterine wet weight and 13,990, 3.34, and 0.97 for PR binding, respectively. The high ratio observed for 6-isopropyl-1,3,8-triCDF was primarily due to the low activity of this compound as an inducer of CYP1A1. 1,3,6,8-Substituted alkyl PCDFs are Ah receptor agonists which exhibit low toxicity and antiestrogenic activity in the female rat uterus and DMBA-induced mammary tumor model, and their potential utility as chemotherapeutic drugs for treatment of breast cancer is being further investigated.

ANTIESTROGENIC EFFECTS OF Ah RECEPTOR AGONISTS IN HUMAN BREAST CANCER CELL LINES

Several human breast cancer cell lines have been used to investigate the antiestrogenic activities of TCDD and related compounds. These cell lines include the estrogen-responsive MCF-7, ZR-75-1, and T47-D cells and the estrogen-nonresponsive MDA-MB-231 cells. MCF-7 and T47-D cells express the Ah receptor and exhibit induced AHH and EROD or XRE-regulated reporter gene activities following treatment with TCDD (Harris et al., 1989a; Vickers et al., 1989; Thomsen et al., 1991). Although MDA-MB-231 cells also express the Ah receptor and Ah receptor complexes have been measured in nuclear extracts, TCDD does not induced AHH and EROD activities. However, transient transfection of a human estrogen receptor expression plasmid into MDA-MB-231 cells was found to restore TCDD-mediated induction of an XRE-regulated reporter gene, suggesting that Ah receptor and ER responsiveness were linked (Thomsen et al., 1994). In contrast, a unique BaP-resistant MCF-7 cell line that lacks Ah receptor-responsiveness did not exhibit compromised ER-responsiveness, even though these cells express Ah receptor and ER (Moore et al., 1994). Therefore, there do not appear to be simple correlations between Ah receptor expression/TCDD responsiveness and ER expression/ER responsiveness.

TCDD and related compounds exhibit a broad range of antiproliferative, antiestrogenic, and antimitogenic activities in a number of human breast cancer cell lines. Treatment of MCF-7 cells with TCDD significantly inhibited E2-induced cell proliferation and postconfluent focus development (Gierthy and Lincoln, 1988; Biegel and Safe, 1990; Gierthy et al., 1993; Merchant et al., 1993; Spink et al., 1994b). It has also been demonstrated that TCDD and related compounds inhibit the E2 induction of tissue plasminogen activator, cathepsin D, pS2, glucose-to-lactate conversion, and PR levels (Gierthy et al., 1987; Biegel and Safe, 1990; Narasimhan et al., 1991; Krishnan and Safe, 1993; Moore et al., 1993; Harper et al., 1994; Zacharewski et al., 1994; Krishnan et al., 1995). Studies have shown that these antiestrogenic effects occur at both the protein and gene expression levels and have been attributed to induced E2-hydroxylase activities, downregulation of ER levels, and inhibition of E2-induced gene expression (Safe, 1995). In addition, TCDD also inhibited MCF-7 and T47-D cell proliferation induced by growth factors such as IGF-I, TGF-α, and EGF (Fernandez and Safe, 1992; Liu et al., 1992; Fernandez et al., 1994). Structure-activity relationship studies and experiments conducted in mouse Hepa 1c1c7 wild-type and mutant cell lines have conclusively demonstrated that these effects, including downregulation of ER levels, are mediated by the Ah receptor.

In addition to TCDD, MCDF also exhibited antiestrogenic activities in MCF-7 cells. MCDF is a partial Ah receptor agonist which competitively binds to the Ah receptor and elicits antiestrogenic effects comparable to TCDD (Zacharewski et al., 1992). For example, MCDF induces ER downregulation and inhibits E2-induced cell proliferation and cathepsin D secretion. However, it exhibits relatively weak activity for other Ah receptor–mediated biochemical and toxic responses including lethality, porphyria, teratogenicity, immunotoxicity, and P4501A1 induction. Moreover, in cotreatment studies, MCDF partially antagonized several TCDD-induced responses (Astroff et al., 1988; Gierthy and Lincoln, 1988; Yao and Safe, 1989). Therefore, MCDF, which does not exhibit any affinity for the ER, may be the prototype of a novel class of Ah receptor–mediated antiestrogens that could be used in the treatment of human breast cancer. Other Ah receptor–mediated antiestrogens currently being investigated include the natural products I3C and ICZ as well as other alkyl dibenzofuran and benzocoumarin derivatives (Bjeldanes et al., 1991; Liu et al., 1993, 1994; Tiwari et al., 1994).

MECHANISM OF ACTION OF Ah RECEPTOR AGONISTS AS ANTIESTROGENS

Three different mechanisms have been proposed to explain the antiestrogenic activities of TCDD and related compounds, namely (1) increased oxidative metabolism of E2 by Ah receptor–mediated induction of mixed function oxidases, (2) downregulation of ER-binding activity and protein levels, and (3) Ah receptor–mediated suppression of E2-induced gene expression (Safe, 1995).

The Induced Oxidative Metabolism Hypothesis

The "increased metabolism" hypothesis is based on observations that induction of aryl hydrocarbon hydroxylase (AHH) and ethoxyresorufin O-deethylase (EROD) activities by TCDD correlates with increased E2 hydroxylase activities. Studies have shown that free E2 levels decrease from 10 to 0.029 nM within 12 hours after incubation of MCF-7 cells pretreated with 10 nM TCDD for 72 hours. Intracellular E2 and estrone levels were also depleted in TCDD pretreated cells. In vitro studies with microsomes from TCDD-treated MCF-7 cells exhibited increased levels of E2 hydroxylation at positions C-2, C-4, C-15a, and C-16a (Spink et al., 1990, 1992, 1994a,b). The C-2, C-15a, and C-16a E2-hydroxylase activities have been attributed to P4501A1, while P4501A2 and

P4501B1 catalyzed hydroxylation of E2 at the C-2 and C-4 positions, respectively. Therefore, the antiestrogenic activities of TCDD and related compounds have been attributed to the induced oxidative metabolism of E2 to less potent ER agonists.

However, several experimental paradigms indicate that induced E2 metabolism cannot fully account for all of the observed antiestrogenic activities of TCDD and related compounds. For example:

1. Circulating levels of E2 were unaffected in animals exhibiting antiestrogenic effects following treatment with TCDD (Shiverick and Muther, 1982; DeVito et al., 1992). However, it has been argued that in the microenvironment of a cell, E2 is degraded to less active metabolites as it transverses the cytoplasm (Gierthy et al., 1993).

2. TCDD and related compounds, including the partial agonists MCDF, I3C, and ICZ, inhibit E2-induced responses at time points and at concentrations for which there is no measurable AHH or EROD activity (Biegel and Safe, 1990; Harris et al., 1990; Harasimhan et al., 1991; Chaloupka et al., 1992; Krishnan et al., 1992, 1995; Krishnan and Safe, 1993; Moore et al., 1993; Liu et al., 1994; Tiwari et al., 1994). TCDD also elicited antiestrogenic activities in Hepa 1c1c7 C1 mutant cells, which do not exhibit induction of AHH and EROD activities due to a mutation within the CYP1A1 gene (Zacharewski et al., 1994).

3. E2-induced ERE-regulated reporter gene activity in transiently transfected MCF-7 cells was not inhibited by TCDD, thus suggesting that induced oxidative metabolism of E2 is not associated with the antiestrogenic effects of TCDD and related compounds (Zacharewski et al., 1994).

4. Transient transfection experiments using promoter-regulated reporter genes indicate that, in addition to the ERE, DNA motifs within the regulatory region are required for TCDD-mediated suppression of E2-induced activity (Zacharewski et al., 1994).

5. Transient transfection experiments using promoter-regulated reporter genes and ligand-independent chimeric receptor activators have shown that TCDD-mediated suppression of induced gene expression was independent of E2 metabolism (Zacharewski et al., 1994).

Ah Receptor–Mediated Downregulation of ER Levels

Downregulation of ER-binding activity and protein levels may also contribute to the antiestrogenic activities of TCDD. Decreases in ER levels resulting from enhanced receptor turnover and inhibition of receptor replenishment have been proposed as mechanisms explaining the antagonistic activities of other treatments including progestins and some antiestrogens (Leavitt, 1985; Dauvois et al., 1992). In vitro studies demonstrate that treatment of MCF-7 and wild-type Hepa 1c1c7 cells results in a 60% reduction of nuclear ER-binding activity (Harris et al., 1990; Zacharewski et al., 1991). This effect is ligand- and structure-dependent and requires functional Ah receptor complexes. In Hepa 1c1c7 cells, downregulation of ER-binding activity was accompanied by a concomitant decrease in immunoreactive ER protein levels. However, the effect could be blocked by the protein synthesis inhibitors, cycloheximide and actinomycin D, thus implying that in Hepa 1c1c7 cells, ER downregulation is protein synthesis dependent and that other factors are required in addition to the Ah receptor (Zacharewski et al., 1991).

Downregulation of ER levels by TCDD and related compounds may involve transcriptional and posttranscriptional mechanisms. Preliminary studies in mice have found that TCDD decreases ER mRNA suggesting either decreased ER gene expression or altered mRNA turnover or stability as possible mechanisms (DeVito et al., 1990; Colella and Gallo, 1993). Moreover, the identification of dioxin response elements within the ER gene sequence has prompted speculation that Ah receptor complexes may also act as repressors by inhibiting the binding of transcriptional activators required for ER gene expression or by interfering with the transcriptional machinery (White and Gasiewicz, 1993).

Although downregulation of ER levels may be a contributing factor, there have not been any conclusive studies demonstrating its role in the antiestrogenic activities of TCDD and related compounds. For example, transient transfection studies have shown that TCDD-mediated downregulation of ER levels did not affect the E2 induction of an ERE-regulated reporter gene (Zacharewski et al., 1994). Furthermore, TCDD treatment of MCF-7 cells did not significantly affect steady-state ER mRNA levels as determined by Northern analysis or the rate of ER gene transcription in nuclear run-on assays (Wang et al., 1993). Therefore, the remaining ER levels ($\sim 40\%$) following TCDD treatment were sufficient to maximally induce the ERE-regulated reporter gene, thus suggesting that ER downregulation is not cardinal in order to elicit the antiestrogenic effects of TCDD and related compounds.

Ah Receptor–Mediated Suppression of E2-Induced Gene Expression

The ability of TCDD to inhibit the induction of estrogen-responsive genes that may promote breast tumor cell promotion has been proposed as a possible mechanism of action to explain some of the antiestrogenic effects of TCDD and related compounds. In order to further investigate this hypothesis, the effect of TCDD and related compounds on the expression of two estrogen-inducible genes, pS2 and cathepsin D, has been examined in MCF-7 human breast cancer cells (Zacharewski et al., 1994; Krishnan et al., 1995; Gillesby et al., 1997). The presence of pS2 and cathepsin D in breast tumor biopsies has been clinically used as prognostic markers to assess the status of a tumor and to identify patients that are more likely to respond to antihormonal therapy and/or who are at high risk for recurrence and death (Rio and Chambon, 1990; Rochefort, 1990). Estrogen-inducible expression of both genes is regulated by distinct response elements located within their respective 5' regulatory regions (Berry et al., 1989; Krishnan et al., 1994). Consequently, pS2 and cathepsin D are excellent models to investigate possible Ah receptor–ER interactions that contribute to the antiestrogenic activities of TCDD and related compounds.

Studies have demonstrated that TCDD significantly decreases E2-induced secretion of pS2 and cathepsin D protein levels in MCF-7 human breast cancer cells (Biegel and Safe, 1990; Zacharewski et al., 1994; Krishnan et al., 1995). Furthermore, it has also been reported that TCDD inhibits E2-induced (1) cathepsin D mRNA levels, (2) rate of cathepsin D gene transcription, and (3) pS2- and cathepsin D–regulated reporter gene activity in transiently transfected cells. Collectively, these studies indicate that TCDD-mediated suppression of E2-induced responses occurs at the gene-expression level and may contribute to the antiestrogenic effects of TCDD and related compounds.

In order to investigate the role of Ah receptor complexes in TCDD-mediated suppression, Hepa 1c1c7 wild-type and mutant cells were transiently transfected with pS2- and cathepsin D–regulated reporter genes. These cells have been previously used to investigate the role of Ah receptor complexes in the induction of aryl hydrocarbon hydroxylase (AHH) activity and the downregulation of ER-binding activity and protein levels (Hankinson et al., 1991; Zacharewski et al., 1991; White and Gasiewicz, 1993). The mutant cell lines contain mutations in the genes that encode for the Ah receptor (i.e., Group B-, C12) or ARNT (Group C-, C4, or BPrc1, class II), and therefore, TCDD does not induce the formation of nuclear complexes or AHH activity in these cells (Miller et al., 1983; Hankinson et al., 1991). Significant induction of reporter gene activity was observed in Hepa 1c1c7 wild-type and mutant cells using pS2 and cathepsin D constructs following treatment with E2 (Zacharewski et al., 1994; Krishnan et al., 1995). However, TCDD-mediated suppression of E2-induced reporter gene activity was only observed in wild-type cells, while no effect was observed in transiently transfected Ah receptor– and ARNT-deficient mutant cells, thus demonstrating the requirement of functional Ah receptor complexes. The necessity of functional Ah receptor complexes was further illustrated when TCDD-mediated suppression of E2-induced pS2-regulated reporter gene activity was reestablished following the co-transfection of cDNAs for either Ah receptor or ARNT, which complemented the deficiency of the mutant cell (Zacharewski et al., 1994).

Several possible mechanisms could be responsible for TCDD-mediated suppression of estrogen-induced gene expression at the transcriptional level, including (1) inhibition of transcriptional factor DNA binding, (2) blocking of transcriptional initiation, and (3) silencing (Renkawitz, 1990). Inhibition of DNA binding can result from either the obstruction of a particular positive transcription factor binding to its enhancer or by interfering with the transcriptional machinery binding to the start site (i.e., TATA box). Most examples of inhibition involve competition between transacting factors binding to the same specific motif or to unique, adjacent sequences, which create steric hindrance. Transcriptional repression by inhibition of DNA binding can also result from formation of inactive dimers or from inactivation of a transactivating complex via interaction with an inhibitor. The initiation of transcription can also be suppressed by protein-protein interactions that block or mask activation domains within the positive transcription factor that is responsible for inducing gene expression. In addition, transmission of the signal from the transactivator can be repressed by squelching or titrating intermediate factors that are required to transduce the signal to the transcription machinery. Moreover, inhibition of gene expression can involve specific DNA elements, referred to as silencers, which bind factors that repress gene expression. Specific arrangements of activators, repressors, and the initiation complex are not critical, since repressor factors bound to silencers lock the promoter into configurations that are inaccessible to the transcriptional machinery.

An examination of the effects of TCDD on the induction of estrogen-responsive genes by E2 suggests that inhibition of induced gene expression may involve a number of novel mechanisms. For example, deletion analysis of the 5' regulatory region of cathepsin D has identified a specific region within –296 to + 59 that is required for E2 induction and the TCDD-mediated inhibitory response. Previous studies have established that an ERE-Sp1 complex, GGGCGGn23ACGGG (–199 to –165), mediates induction of cathepsin D expression by E2 (Krishnan et al., 1994). Further analysis of this region has identified an XRE-like motif (GCGCGTGCCC) located at positions

−181 to −172 within the ERE-Sp1 response element, which differs from the consensus DRE sequence (T/GnGCGTGA/CG/CA) at position 9 (differences indicated in bold-face, lower case type) (Lusska et al., 1993) (Fig. 3). Electrophoretic mobility shift assays using oligonucleotides containing wild-type and mutated ER-Sp1 response elements con-clusively demonstrated that TCDD blocked the formation of a ERE-Sp1/DNA complex (Krishnan et al., 1995). However, direct binding of Ah receptor complexes to the wild-type ERE-Sp1 oligonucleotide was not observed in gel electrophoretic mobility shift as-says, possibly due to increased protein-DNA dissociation resulting in the loss of the retarded band in the assay system. Subsequent UV cross-linking studies demonstrated that Ah receptor complexes isolated from cells treated with E2 and TCDD could di-rectly bind to the XRE-like motif located within the ERE-Sp1 response element. These results suggest that the E2-induced formation of ERE-Sp1 complexes is required to pro-mote Ah receptor complex binding with the XRE-like site. The formation of ER-Sp1/DNA complexes may assist Ah receptor complex binding by increasing accessibility to the XRE and through protein-protein interactions, which may then destabilize or fa-cilitate ERE-Sp1 complex dissociation from the response element. Therefore, the mecha-nism of TCDD-mediated suppression of E2-induced cathepsin D gene expression in-volves inhibition of ERE-Sp1 complex binding to its response element.

Studies suggest that TCDD-mediated inhibition of E2-induced pS2 expression involves a comparable mechanism of action (Gillesby et al., 1996). A similarly XRE-like motif, which also differs from the consensus DRE sequence (caGCGTGAGc) at positions 1 and 9, has been identified with the pS2 5′ regulatory region and is essen-tial for the TCDD inhibitory response. However, the pS2 XRE-like motif, located at positions −523 to −514, is approximately 100 bp upstream of the ERE. The role of the pS2 XRE-like motif was confirmed by introducing a T-to-C point mutation at position −518, which abolished TCDD-mediated suppression of E2-induced reporter gene ac-tivity. These results were consistent with previous studies, which demonstrated that comparable mutations to the XRE core were incapable of forming AhR/DNA complexes and were ineffective in competitive electrophoretic mobility shift assays (Bank et al., 1992; Shen and Whitlock, 1992; Yao and Denison, 1992). As observed with the cathe-psin D XRE-like motif, direct binding of Ah receptor complexes to the pS2 XRE-like motif was not observed in band shift assays, although unlabeled oligonucleotides were found to effectively compete with wild-type XRE probes in competitive assays. Fur-thermore, UV cross-linking studies demonstrate that Ah receptor complexes can bind to the pS2 XRE-like motif (Gillesby et al., 1997). Although the inability to show di-rect binding between Ah receptor complexes and the pS2 XRE-like motif is problem-

Sequence									Description
T/G	n	G	C	G	T	G	A/C	G/C	A	consensus XRE
G	c	G	G	G	T	G	C	C	**C**	cathepsin D XRE
C	a	G	C	G	T	G	A	G	**C**	pS2 XRE

Figure 3 Comparison of the consensus XRE to the cathepsin D and pS2 XRE-like motifs. These motifs are required for TCDD-mediated inhibition of E2-induced gene expression. Nucleotides in bold, lowercase indicate deviations from the consensus XRE.

atic, competitive electrophoretic mobility shift assays and UV cross-linking studies indicate that an interaction does occur between TCDD-transformed cytosol and the pS2 XRE-like motif.

Studies examining the effect of changes to XRE positions 1 or 9 reveal abolition of enhancer function in transient transfection assays. These results indicate that DNA binding and enhancer function can be dramatically affected by bases adjacent to the XRE core and that in vitro examination of receptor:enhancer binding affinity using gel electrophoretic mobility shift assays is not predictive of function (Lusska et al., 1993). Consistent with these results, TCDD does not induce a cathepsin D or pS2 XRE-like motif-regulated reporter gene. Therefore, nucleotide changes at positions 1 and 9 may contribute to the inhibitory nature of these regulatory elements. Together, the results from cathepsin D and pS2 suggest that XRE-like motifs play a prominent role in TCDD-mediated inhibition of E2-induced gene expression.

ACKNOWLEDGMENTS

The financial assistance of the National Institutes of Health (CA64081), the Texas Agricultural Experiment Station, the Canadian Network of Toxicology Centres, and the Natural Sciences and Engineering Research Council of Canada (Strategic Grant) is gratefully acknowledged. S. Safe is a Sid Kyle Professor of Toxicology and T. Zacharewski is supported by a PMAC/HRF/MRC Research Career Award in Medicine.

REFERENCES

Asman, D. C., Takimoto, K., Pitot, H. C., Dunn, T. J., and Lindahl, R. (1993). *J. Biol. Chem., 268*: 12530.

Astroff, B., and Safe, S. (1988). *Toxicol. Appl. Pharmacol., 95*: 435.

Astroff, B., and Safe, S. (1989). *Toxicology, 59*: 285.

Astroff, B., and Safe, S. (1990). *Biochem. Pharmacol., 39*: 485.

Astroff, B., Zacharewski, T., Safe, S., Arlotto, M. P., Parkinson, A., Thomas, P., and Levin, W. (1988). *Mol. Pharmacol., 33*: 231.

Astroff, B., Rowlands, C., Dickerson, R., and Safe, S. (1990). *Mol. Cell. Endocrinol., 72*: 247.

Astroff, B., Eldridge, B., and Safe, S. (1991). *Toxicol. Lett., 56*: 305.

Bank, P. A., Yao, E. F., Phelps, C. L., Harper, P. A., and Denison, M. S. (1992). *Eur. J. Pharmacol., 228*: 85.

Bannister, R., Biegel, L., Davis, D., Astroff, B., and Safe, S. (1989). *Toxicology, 54*: 139.

Bastomsky, C. H. (1977). *Endocrinology, 101*: 292.

Berry, M., Nunez, A., and Chambon, P. (1989). *Proc. Natl. Acad. Sci. USA, 86*: 1218.

Bertazzi, P. A., Pesatori, A. C., Consonni, d., Tironi, A., Landi, M. T., and Zocchetti, C. (1993). *Epidemiology, 4*: 398.

Biegel, L., and Safe, S. (1990). *J. Steroid Biochem. Mol. Biol., 37*: 725.

Bjeldanes, L. F., Kim, J. Y., Grose, K. R., Bartholomew, J. C., and Bradfield, C. A. (1991). *Proc. Natl. Acad. Sci. USA, 88*: 9543.

Bradlow, H. L., Michnovicz, J. J., Telang, N. T., and Osborne, M. P. (1991). *Carcinogenesis, 12*: 1571.

Burbach, K. M., Poland, A. B., and Bradfield, C. A. (1992). *Proc. Natl. Acad. Sci. USA, 89*: 8185.

Chaloupka, K., Krishnan, V., and Safe, S. (1992). *Carcinogenesis, 13*: 2223.

Choi, E. J., Toscano, D. G., Ryan, J. A., Riedel, N., and Toscano, W. A. (1991). *J. Biol. Chem.,* 266: 9591.

Colella, T. A., and Gallo, M. A. (1993). *Toxicologist, 13*: 102.

Dauvois, S., Danielian, P. S., White, R., and Parker, M. G. (1992). *Proc. Natl. Acad. Sci. USA,* 89: 4037.

De Kruif, C. A., Marsman, J. W., Venekamp, J. C., Falke, H. E., Noordhoek, J., Blaauboer, B. J., and Wortelboer, H. M. (1991). *Chem. -Biol. Interact., 80*: 303.

DeVito, M. J., Thomas, T., Martin, E., Umbreit, T., and Gallo, M. (1990). *Toxicologist, 10*: 246.

DeVito, M. J., Thomas, T., Martin, E., Umbreit, T. H., and Gallo, M. A. (1992). *Toxicol. Appl. Pharmacol., 113*: 284.

Dickerson, R., and Safe, S. (1992). *Toxicol. Appl. Pharmacol., 113*: 55.

Dickerson, R., Howie-Keller, L., and Safe, S. (1995). *Toxicol. Appl. Pharmacol., 135*: 287.

Dolwick, K. M., Schmidt, J. V., Carver, L. A., Swanson, H. I., and Bradfield, C. A. (1993). *Mol. Pharmacol., 44*: 911.

Ema, M., Sogawa, K., Watanabe, N., Chujoh, Y., Matsushita, N., Gotoh, O., Funae, Y., and Fujii-Kuriyama, Y. (1992). *Biochem. Biophys. Res. Comm., 184*: 246.

Ema, M., Ohe, N., Suzuki, M., Mimura, J., Sogawa, K., Ikawa, S., and Fujii-Kuriyama, Y. (1994). *J. Biol. Chem., 269*: 27337.

Evans, R. M. (1988). Science, *240*: 889.

Fernandez, P., and Safe, S. (1992). *Toxicol. Lett., 61*: 185.

Fernandez, P., Burghardt, R., Smith, R., Nodland, K., and Safe, S. (1994). *Eur. J. Pharmacol., 270*: 53.

Fiedler, H. (1996). *Chemosphere, 32*: 55.

Fujisawa-Sehara, A., Sogawa, K., Nishi, C., and Fujii-Kuriyama, Y. (1986). *Nucleic Acids Res., 14*: 1465.

Fujisawa-Sehara, A., Yamane, M., and Fujii-Kuriyama, Y. (1988). *Proc. Natl. Acad. Sci. USA,* 85: 5859.

Gaido, K. W., Maness, S. C., Leonard, L. S., and Greenlee, W. F. (1992). *J. Biol. Chem., 267*: 24591.

Gallo, M. A., Hesse, E. J., MacDonald, G. J., and Umbreit, T. H. (1986). *Toxicol. Lett., 32*: 123.

Gierthy, J. F., and Lincoln, D. W. (1988). *Breast Cancer Res., 12*: 227.

Gierthy, J. F., Lincoln, D. W., Gillespie, M. B., Seeger, J. I., Martinez, H. L., Dickerman, H. W., and Kumar, S. A. (1987). *Cancer Res., 47*: 6198.

Gierthy, J. F., Bennett, J. A., Bradley, L. M., and Cutler, D. S. (1993). *Cancer Res., 53*: 3149.

Gillesby, B., Santostefano, M., Porter, W., Safe, S., Wu, Z. F., and Zacharewski, T. (1997). *Biochemistry, 36*: 6080.

Goldstein, J. A., and Safe, S. (1989). *Halogenated Biphenyls, Naphthalenes, Dibenzodioxins and Related Compounds* (R. D. Kimbrough and A. A. Jensen, eds.). Elsevier-North Holland, Amsterdam, p. 239.

Gonzalez, F. J., and Nebert, D. W. (1985). *Nucleic Acids Res., 13*: 7269.

Grubbs, C. J., Steele, V. E., Casebolt, T., Juliana, M. M., Eto, I., Whitaker, L. M., Dragnev, K. H., Kelloff, G. J., and Lubet, R. L. (1995). *Anticancer Res., 15*: 709.

Hankinson, O., Brooks, B. A., Weir-Brown, K. I., Huffman, E. C., Johnson, B. S., Nanthur, J., Reyes, H., and Watson, A. J. (1991). *Biochimie, 73*: 61.

Harper, N., Wang, X., Liu, H., and Safe, S. (1994). *Mol. Cell Endocrinol., 104*: 47.

Harris, M., Piskorska-Pliszczynska, J., Zacharewski, T., Romkes, M., and Safe, S. (1989a). *Cancer Res., 49*: 4531.

Harris, M., Zacharewski, T., Astroff, B., and Safe, S. (1989b). *Mol. Pharmacol., 35*: 729.

Harris, M., Zacharewski, T., and Safe, S. (1990). *Cancer Res., 50*: 3579.

Henry, E. C., and Gasiewicz, T. A. (1987). *Toxicol. Appl. Pharmacol., 89*: 165.

Hoffman, E. C., Reyes, H., Chu, F., Sander, F., Conley, L. H., Brooks, B. A., and Hankinson, O. (1991). *Science, 252*: 954.

Holcomb, M., and Safe, S. (1994). *Cancer Lett., 82*: 43.

Hutzinger, O., and Fiedler, H. (1993). *Chemosphere, 27*: 121.

Jaiswal, A. K. (1994). *J. Biol. Chem., 269*: 14502.

Jansen, H. T., Cooke, P. S., Porcelli, J., Liu, T.-C., and Hansen, L. G. (1993). *Reprod. Toxicol., 7*: 237.

Jellinck, P. H., Michnovicz, J. J., and Bradlow, H. L. (1991). *Steroids, 56*: 446.

Jones, M. K., Weisenburger, W. P., Sipes, G., and Russell, D. H. (1987). *Toxicol. Appl. Pharmacol., 87*: 337.

Jones, P. B., Galeazzi, D. R., Fisher, J. M., and Whitlock, J. P., Jr. (1985). *Science, 227*: 1499.

Jones, P. B., Durrin, L. K., Fisher, J. M., and Whitlock, J. P., Jr. (1986a). *J. Biol. Chem., 261*: 6647.

Jones, P. B., Durrin, L. K., Galeazzi, D. R., and Whitlock, J. P., Jr. (1986b). *Proc. Natl. Acad. Sci. USA, 83*: 2802.

Kociba, R. J., Keyes, D. G., Beger, J. E., Carreon, R. M., Wade, C. E., Dittenber, D. A., Kalnins, R. P., Frauson, L. E., Park, C. L., Barnard, S. D., Hummel, R. A., and Humiston, C. G. (1978). *Toxicol. Appl. Pharmacol., 46*: 279.

Krishnan, V., and Safe, S. (1993). *Toxicol. Appl. Pharmacol., 120*: 55.

Krishnan, V., Narasimhan, T.R., and Safe, S. (1992). *Anal. Biochem., 204*: 137.

Krishnan, V., Wang, X., and Safe, S. (1994). *J. Biol. Chem., 269*: 15912.

Krishnan, V., Porter, W., Santostefano, M., Wang, X., and Safe, S. (1995). *Mol. Cell. Biol., 15*: 6710.

Landers, J. P., and Bunce, N. J. (1991). *Biochem. J., 276*: 273.

Leavitt, W. W. (1985). *Molecular Mechanism of Steroid Hormone Action* (V. K. Moudgil, ed.). De Gruyter, Berlin, p. 437.

Lin, P. H., Selinfreund, R., and Wharton, W. (1991). *Peptide Growth Factors, Part. C., 198*: 251.

Liu, H., Biegel, L., Narasimhan, T. R., Rowlands, C., and Safe, S. (1992). *Mol. Cell. Endocrinol., 87*: 19.

Liu, H., Santostefano, M., Lu, Y., and Safe, S. (1993). *Arch. Biochem. Biophys., 306*: 223.

Liu, H., Wormke, M., Safe, S., and Bjeldanes, L. F. (1994). *J. Natl. Cancer Inst., 86*: 1758.

Lorenzen, A., and Okey, A. B. (1990). *Toxicol. Appl. Pharmacol., 106*: 53.

Lucier, G. W., Portier, C. J., and Gallo, M. A. (1993). *Environ. Health Perspect., 101*: 36.

Lusska, A., Shen, E., and Whitlock, J. P., Jr. (1993). *J. Biol. Chem., 268*: 6575.

MacKenzie, S. A., Thomas, T., Umbreit, T. H., and Gallo, M. A. (1992). *Toxicol. Appl. Pharmacol., 116*: 101.

Madhukar, B. V., Brewster, D. W., and Matsumura, F. (1984). *Proc. Natl. Acad. Sci. USA, 81*: 7407.

McDougal, A., Howell, J., and Safe, S. (1996). *Toxicologist, 16*: 1040.

McKinney, J. D., Fawkes, J., Jordan, S., Chae, K., Oatley, S., Coleman, R. E., and Briner, W. (1985). *Environ. Health Perspect., 61*: 41.

Merchant, M., Krishnan, V., and Safe, S. (1993). *Toxicol. Appl. Pharmacol., 120*: 179.

Miller, A. G., Israel, D., and Whitlock, J. P., Jr. (1983). *J. Biol. Chem., 258*: 3523.

Mocarelli, P., Needham, L. L., Marocchi, A., Patterson, D. G., Jr., Brambilla, P., Gerthoux, P. M., Meazza, L., and Carreri, V. (1991). *J. Toxicol. Environ. Health, 32*: 357.

Moore, M., Narasimhan, T. R., Wang, X., Krishnan, V., Safe, S., Williams, H. J., and Scott, A. I. (1993). *J. Steroid Biochem. Mol. Biol., 44*: 251.

Moore, M., Wang, X., Lu, Y.-F., Wormke, M., Craig, A., Gerlach, J., Burghardt, R., and Safe, S. (1994). *J. Biol. Chem., 269*: 11751.

Narasimhan, T. R., Safe, S., Williams, H. J., and Scott, A. I. (1991). *Mol. Pharmacol., 40*: 1029.

Neuhold, L. A., Shirayoshi, Y., Ozato, K., Jones, J. E., and Nebert, D. W. (1989). *Mol. Cell. Biol., 9*: 2378.

Okey, A. B., Riddick, D. S., and Harper, P. A. (1994). *Toxicol. Lett., 70*: 1.

Patnode, K. A., and Curtis, L. R. (1994). *Toxicol. Appl. Pharmacol., 127*: 9.

Perdew, G. H. (1988). *J. Biol. Chem., 263*: 13802.

Perdew, G. H. (1992). *Biochem. Biophys. Res. Commun., 182*: 55.

Peterson, R. E., Theobald, H. M., and Kimmel, G. L. (1993). *C.R.C. Crit. Rev. Toxicol., 23*: 283.

Pimental, R. A., Liang, B., Yee, G. K., Wilhelmsson, A., Poellinger, L., and Paulson, K. E. (1993). *Mol. Cell. Biol., 13*: 4365.

Poland, A., Glover, E., and Kende, A. S. (1976). *J. Biol. Chem., 251*: 4936.

Poland, A., and Glover, E. (1975). *Mol. Pharmacol., 11*: 389.

Poland, A., and Glover, E. (1987). *Biochem. Biophys. Res. Commun., 146*: 1439.

Poland, A., and Glover, E. (1990). *Mol. Pharmacol., 38*: 306.

Poland, A., and Knutson, J. C. (1982). *Annu. Rev. Pharmacol. Toxicol., 22*: 517.

Poland, A., Glover, E., Robinson, J. R., and Nebert, D. W. (1975). *J. Biol. Chem., 249*: 5599.

Pongratz, I., Mason, G. G. F., and Poellinger, L. (1992). *J. Biol. Chem., 267*: 13728.

Prokipcak, R. D., and Okey, A. B. (1988). *Arch. Biochem. Biophys., 267*: 811.

Quattrochi, L. C., Vu, T., and Tukey, R. H. (1994). *J. Biol. Chem., 269*: 6949.

Rappe, C. (1993a). *Organohalogen Compounds, Dioxin '93, 12*: 163.

Rappe, C. (1993b). *Chemosphere, 27*: 211.

Renkawitz, R. (1990). *Trends Genet., 6*: 192.

Rio, M. C., and Chambon, P. (1990). *Cancer Cells, 2*: 269.

Rochefort, H. (1990). *Semin. Cancer Biol., 1*: 153.

Romkes, M., and Safe, S. (1988). *Toxicol. Appl. Pharmacol., 92*: 368.

Romkes, M., Piskorska-Pliszczynska, J., and Safe, S. (1987). *Toxicol. Appl. Pharmacol., 87*: 306.

Rozman, K., Hazelton, G. A., Klaassen, C. D., Arlotto, M. P., and Parkinson, A. (1985). *Toxicology, 37*: 51.

Rushmore, T. H., King, R. G., Paulson, K. E., and Pickett, C. B. (1990). *Proc. Natl. Acad. Sci. USA, 87*: 3826.

Ryan, R. P., Sunahara, G. I., Lucier, G. W., Birnbaum, L. S., and Nelson, K. G. (1989). *Toxicol. Appl. Pharmacol., 98*: 454.

Safe, S. (1990). *C.R.C. Crit. Rev. Toxicol., 21*: 51.

Safe, S. (1994). *C.R.C. Crit. Rev. Toxicol., 24*: 87.

Safe, S. (1995). *Pharmacol. Ther., 67*: 247.

Schmidt, J. V., Carver, L. A., and Bradfield, C. A. (1993). *J. Biol. Chem., 268*: 22203.

Shen, E. S., and Whitlock, J. P. (1992). *J. Biol. Chem., 267*: 6815.

Shiverick, K. T., and Muther, T. F. (1982). *Toxicol. Appl. Pharmacol., 65*: 170.

Spink, D. C., Lincoln, D. W., Dickerman, H. W., and Gierthy, J. F. (1990). *Proc. Natl. Acad. Sci. USA, 87*: 6917.

Spink, D. C., Eugster, H., Lincoln, D. W., II, Schuetz, J. D., Schuetz, E. G., Johnson, J. A., Kaminsky, L. S., and Gierthy, J. F. (1992). *Arch. Biochem. Biophys., 293*: 342.

Spink, D. C., Hayes, C. L., Young, N. R., Christou, M., Sutter, T. R., Jefcoate, C. R., and Gierthy, J. F. (1994a). *J. Steroid Biochem. Mol. Biol., 51*: 251.

Spink, D. C., Johnson, J. A., Connor, S. P., Aldous, K. M., and Gierthy, J. F. (1994b). *J. Toxicol. Environ. Health, 41*: 451.

Stoewsand, G. S., Anderson, J. L., and Munson, L. (1988). *Cancer Lett., 39*: 199.

Sunahara, G. I., Lucier, G. W., McCoy, Z., Bresnick, E. H., Sanchez, E. R., and Nelson, K. G. (1989). *Mol. Pharmacol., 36*: 239.

Sutter, T. R., Guzman, K., Dold, K. M., and Greenlee, W. F. (1991). *Science, 254*: 415.

Swanson, H. I., and Bradfield, C. A. (1993). *Pharmacogenetics, 3*: 213.

Thomsen, J. S., Nissen, L., Stacey, S. N., Hines, R. N., and Autrup, H. (1991). *Eur. J. Biochem., 197*: 577.

Thomsen, J. S., Wang, X., Hines, R. N., and Safe, S. (1994). *Carcinogenesis, 15*: 933.

Tiwari, R. K., Guo, L., Bradlow, H. L., Telang, N. T., and Osbone, M. P. (1994). *J. Natl. Cancer Inst., 86*: 126.

Tritscher, A. M., Clark, G. C., Sewall, C., Sills, R. C., Maronpot, R., and Lucier, G. W. (1995). *Carcinogenesis, 16*: 2807.

Umbriet, T. H., and Gallo, M. A. (1988). *Toxicol. Lett., 42*: 5.

Umbriet, T. H., Hesse, E. J., MacDonald, G. J., and Gallo, M. A. (1988). *Toxicol. Lett., 40*: 1

Umbriet, T. H., Scala, P. L., MacKenzie, S. A., and Gallo, M. A. (1989). *Toxicology, 59*: 163.

U.S. Environmental Protection Agency (1995). *Health Assessment Document for 2,3,7,8-Tetrachlorodibenzo-p-dioxin (TCDD) and Related Compounds.* Office of Research and Development, Washington, DC.

Vasiliou, V., Puga, A., Chang, C.-Y., Tabor, M. W., and Nebert, D. W. (1995). *Biochem. Pharmacol., 50*: 2057.

Vickers, P. J., Dufresne, M. J., and Cowan, K. H. (1989). *Mol. Endocrinol., 3*: 157.

Wang, X., Narisimhan, T. R., Morrison, V., and Safe, S. (1991). *Arch. Biochem. Biophys., 287*: 186.

Wang, X., Santostefano, M., Lu, Y., and Safe, S. (1992). *Chem. -Biol. Interact., 85*: 79.

Wang, X., Porter, W., Krishnan, V., Narasimhan, T. R., and Safe, S. (1993). *Mol. Cell. Endocrinol., 96*: 159.

Wattenberg, L. W. (1978). *J. Natl. Cancer Inst., 60*: 11.

White, T. E. K., and Gasiewicz, T. A. (1993). *Biochem. Biophys. Res. Commun., 193*: 956.

White, T. E. K., Rucci, G., Liu, Z., and Gasiewicz, T. A. (1995). *Toxicol. Appl. Pharmacol., 133*: 313.

Whitlock, J. P., Jr. (1993). *Chem. Res. Toxicol., 6*: 754.

Yao, C., and Safe, S. (1989). *Toxicol. Appl. Pharmacol., 100*: 208.

Yao, E. F., and Denison, M. S. (1992). *Biochemistry, 31*: 5060.

Zacharewski, T., Harris, M., and Safe, S. (1991). *Biochem. Pharmacol., 41*: 1931.

Zacharewski, T., Harris, M., Biegel, L., Morrison, V., Merchant, M., and Safe, S. (1992). *Toxicol. Appl. Pharmacol., 13*: 311.

Zacharewski, T., Bondy, K., McDonell, P., and Wu, Z. F. (1994). *Cancer Res., 54*: 2707.

19

Reproductive and Developmental Toxicity of Metals: Issues for Consideration in Human Health Risk Assessments

Rafael A. Ponce and Elaine M. Faustman
University of Washington, Seattle, Washington

INTRODUCTION

Since the 1980s, risk assessments have become an important tool used by industries and state and federal agencies to systematically identify and characterize potential public health hazards resulting from exposure to environmental compounds. The use of risk assessment has grown, at least in part, because it provides a common framework for negotiation and decision making and attempts to create objectivity in judgment (Hornstein, 1992). A framework for regulatory decision making that utilizes the characterization of risk (Fig. 1) was proposed in the early 1980s (Calkins et al., 1980). This framework was used by the National Research Council to establish its "redbook" risk assessment principles (NRC, 1983) wherein the goal of risk characterization is achieved through a combined qualitative and quantitative analysis of risk. The most recent version of the NRC framework is shown in Figure 2 (NRC, 1994). Qualitative information used in risk analysis includes the strength of evidence for an adverse outcome, the nature of the outcomes resulting from an exposure or situation, exposure routes, the identification of sensitive populations, information on areas of uncertainty (and their influence on the risk estimate), and descriptive information on other factors that influence the risk estimate. Quantitative information includes exposure assessments, estimates of the potential magnitude of the risks, and uncertainty information regarding the risk estimates.

Historically, cancer as an adverse endpoint has been the primary focus of quantitative risk assessment. However, more recently noncancer endpoints such as potential adverse reproductive or developmental outcomes have gained prominence. This change is a result, in part, of the increasing recognition of a broad range of potential public health impacts and of the new methodologies for calculating noncancer risks.

A significant area in environmental risk assessment involves the assessment of risks to human health from exposure to heavy metals. Inorganic compounds, primarily metals, had been found at approximately 65% of the hazardous waste sites in the United States as of 1992 (ATSDR, 1992), and of the top 20 substances on the 1993 ATSDR

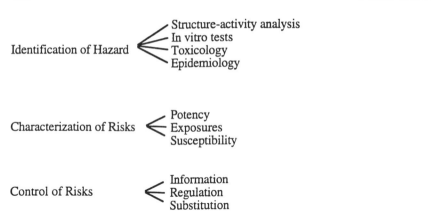

Figure 1 Framework for regulatory decision making: risk assessment and risk management.

Priority List of Hazardous Substances, lead, arsenic, and metallic mercury were the first three, cadmium was number six, and hexavalent chromium was number nineteen. Of particular concern are the noncancer hazards that are posed by heavy metals. In this chapter, we will focus on an assessment of the potential adverse effects of metals to reproduction and development. The large number of children born in the United States each year with malformations or impairments (4–8% of live births) of unknown etiology [66% of all malformations (Slikker, 1994)] and the recognized effects of metals

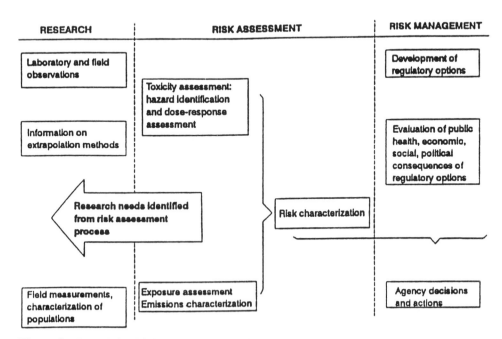

Figure 2 The relationship between research, risk assessment, and risk management is presented. While information is typically presented as flowing from research to risk assessment and, ultimately, to risk management, the figure highlights the significant influence that the identification of data gaps in the risk assessment process has on driving new basic research. Through such iterations between research and risk assessment, uncertainties in exposure assessment and hazard potency can be reduced, and risk management decisions can be refined.

(e.g., methylmercury, lead) on both development and reproduction speak to the need for further research on this topic.

This chapter will highlight the issues and concerns specific to the assessment of potential risks to human reproduction and development following exposures to heavy metals. Various properties or effects of metals will be used to illustrate risk assessment principles for reproductive and developmental effects. The risk assessment model as defined by the National Research Council (NRC, 1983) consists of *hazard identification*, *dose-response assessment*, *exposure analysis*, and *risk characterization*. However, in the U.S. EPA's guidelines for risk assessment of developmental toxicants (U.S. EPA, 1991) and the National Research Council (NRC, 1994), hazard identification and dose-response assessment are combined under the term *toxicity assessment*. This is because the toxic effects are difficult to assess without simultaneous consideration of dose, therefore, this chapter will follow the risk assessment model for developmental and reproductive toxicants as *toxicity assessment*, *exposure assessment*, and *risk characterization*, as shown in Figure 2. This model demonstrates the iterative nature of risk assessment with emphasis on toxicity assessment and the identification of additional research needs to refine risk characterization. We will use risk to refer to the probability of an adverse reproductive or developmental outcome.

TOXICITY ASSESSMENT

The assessment of metals for their potential adverse reproductive and developmental effects involves the joint consideration of the nature of the potential hazards from exposure and an evaluation of dose-response information. Toxicity information may be obtained from an analysis of chemical and physical characteristics, short-term tests, in vitro bioassays, and an evaluation of epidemiological information. Commonly referenced metals with regard to potential reproductive and developmental toxicity are listed in Table 1.

Use of Metal Chemistry and Structure-Activity Information in Toxicity Assessment

Metals have limited chemical or physical information that is specific to reproductive and developmental toxicity assessments. Compound structure, solubility, stability, pH sensitivity, electrophilicity, and chemical reactivity can provide valuable information for understanding and predicting potential metal toxicity. For example, chemical reactivity information such as the preferred metal ion–binding sites for metals on cellular macromolecules can be generally predicted by examining the strength of association

Table 1 Metals Reported to Affect Reproduction and Development

Male Reproductive Systems: aluminum, boranes, boron, cadmium, cobalt, lead, inorganic mercury, methylmercury, molybdenum, nickel, silver, uranium

Female Reproductive Systems: arsenic, lead, lithium, inorganic mercury, methylmercury, molybdenum, nickel, selenium, thallium

Developmental Systems: arsenic, boron, cadmium, lead, lithium, inorganic mercury, methylmercury, thallium

Information presented in this table is not meant to be inclusive. Effects for all metals are functions of dose response and bioavailability, and thus care must be taken in extrapolating these reported effects for prediction of effects in exposed humans.
Source: Schardein, 1993; Rogers and Kavlock, 1996; Thomas, 1996.

between the metal ion and a variety of ligands. Generally speaking, metal ions that bind to O-donor sites include the alkali, alkaline earth, and lanthanide metal ions and also Al^{3+}, Cr^{3+}, and Fe^{3+}. Dipositive metal ions in the first transition row between Mn^{2+} and Cu^{2+} favor N donors. Cd^{2+}, CH_3Hg^+, Hg^{2+}, and Pb^{2+} favor binding to S groups. Zn^{2+} is transitional in binding to N and S donors (Martin, 1988).

The affinity of the metal for protein ligands and the chemical form of the metal (i.e., valence state, alkylation state, etc.) will affect both the distribution (pharmacokinetics) and the toxic potency (pharmacodynamics) of the metal. For example, the affinity of mercuric mercury (Hg^{2+}) and methylmercury (CH_3Hg^+) for the sulfhydryl group of cysteine underlies their potency as general (nonspecific) enzyme poisons; cysteines involved in disulfide bond formation commonly stabilize protein structure. Despite the affinity of these two mercury compounds for sulfhydryl groups, the nonalkylated Hg^{2+} is primarily a renal toxicant, while methylmercury is a significant developmental and adult neurotoxicant. The difference in their site of action appears to be due to the inability of Hg^{2+} to cross the blood-brain barrier and the ability of methylmercury, when bound to cysteine, to mimic methionine and utilize the neutral amino acid transport system to gain access to the central nervous system (Aschner and Clarkson, 1988, 1989; Dock et al., 1994; Vahter et al., 1994).

Chemical form plays an important role in determining metal bioavailability. This presents a challenge to the risk assessor because environmental levels of metals are usually regulated without considering bioavailability and metals are frequently regulated as a class of compounds rather than specific salts or oxides. To illustrate, investigations into lead bioavailability using rodent bioassays demonstrated that the bioavailability of the lead ore from Skagway, AK, was low relative to that of other ores (Dieter et al., 1993). This information helped explain the relatively low blood lead levels (<15 µg/dl) observed in Skagway residents despite high residential soil lead levels (>500 ppm) and demonstrated a lower adverse health risk from environmental lead than might have been predicted based on environmental monitoring and the use of default assumptions of bioavailability (Middaugh et al., 1989).

Cadmium acetate and cadmium chloride have been tested and, as soluble salts, shown to be bioavailable and produce adverse reproductive effects (NAS, in press). However, because cadmium oxide is minimally soluble it does not appear to have systematic effects outside of its direct effects on lung tissue following inhalation and deposition.

Bioavailability is not only a function of chemical form, but of physical form as well. For example, the risks to human reproduction and development from processing ore will vary according to the stage in ore processing. While the inspiration of hard metal dusts may lead to localized respiratory effects, including asthma, pulmonary irritation, and edema (Chiappino, 1994), metal roasting or welding leads to the production of metal fumes (e.g., lead, arsenic, cobalt, copper, and zinc fumes), which can have systemic effects such as metal fume fever (Burgess, 1978; Rolfe et al., 1992; Gordon and Fine, 1993).

Relatively little is known about the nature of the interactions between metals and the potential adverse risks from metal mixtures. Because nonessential trace metals may mimic or compete with essential trace metals, the standard default assumption of additivity of risk may not be justifiable for certain metal mixtures, especially during development. Some simple metal mixtures, the nature of their interactivity, and the potential risk to health are discussed later in this chapter. A number of other factors that influence the toxic potency of metals, including nutritional deficiency and the presence of protective systems, will also be discussed later in this chapter.

Use of In Vitro Assays in Toxicity Assessment

The literature is replete with experimental investigations into the developmental effects of trace metals utilized in in vitro test systems, and standard in vitro test systems for investigating developmental toxicants have been discussed in relation to their utility for risk evaluation (Faustman, 1988; Whittaker and Faustman, 1994; Moore et al., 1995). In general, in vitro studies are used to investigate one or more of the mechanistic pathways proposed for the adverse effects of metals on reproduction and development. Examples of these mechanisms include energy depletion, interrupted cell proliferation, altered cell membrane function and intercellular communication, intracellular oxidation, altered gene expression, and inhibited enzyme function.

Because of the general, nonspecific nature of the interactions between metals and cellular macromolecules, there are no standard in vitro tests used to evaluate metal effects. Table 2 shows some examples of the types of in vitro/short-term assays that have been used to evaluate metals for reproductive or developmental toxicity.

Several approaches have been taken for the development of in vitro systems. One approach has been to model in vitro systems for developmental toxicants after "like" events that occur in development (e.g., regeneration, reaggregation, and cell-cell communication). Examples of these types of assays include the hydra reaggregation assay and the cell attachment assay using mouse ovarian tumor cells (Faustman, 1988). Another approach of in vitro systems is to examine a narrow window of the differentiation process in development or gametogenesis. This approach is seen in the use of whole embryo culture of pre- or postimplantation embryos as a model of development or in germ-Sertoli cell or oocyte cultures as models of reproductive processes. In these models, only a narrow window of observation is possible due to difficulties in maintaining in vitro–like conditions in vitro for an extended period of time. Other examples of in vitro studies that may be used for the evaluation of reproductive toxicants include the acrosome reaction assessment, which measures the ability of sperm to undergo an acrosome reaction), and the sperm penetration assay, which measures the ability of sperm to penetrate and initiate fertilization in a hamster egg.

Some issues specific to the application of in vitro tests for metals involve metal solubility, light reactivity, equilibrium chemistry, and volatility. For example, elemental mercury has a relatively high vapor pressure, methylmercury is sensitive to light, and the speciation of metals in aqueous media will change as a function of pH. These factors may influence the effective in vitro exposure concentration over time, making it difficult to meaningfully evaluate exposure-response information, and thus careful consideration should be given to the design of experimental protocols for investigating the toxicology of metals in vitro. For example, significant shifts in the exposure-response curve may be observed when evaluating methylmercury cytotoxicity in vitro depending on media composition. Because serum proteins in culture media are rich in cysteine and thus in protein sulfhydryl groups, the presence of serum will effectively bind methylmercury and decrease the "free" methylmercury to the cell. Also, different serum lots may contain different quantities of sulfhydryl-containing proteins, making it difficult to standardize exposure protocols; the experimenter should use one serum lot per experimental series to minimize this effect or utilize serum-free alternatives as culture

Table 2 Examples of In Vitro and Short-Term Assays for Development and Reproduction Applied to the Assessment of Metal Effects

System	Assay	Endpoint
Development		
Intact embryo/fetus	Chernoff-Kavlock Test	Viability, malformation, and growth retardation
	Rodent preimplantation	Early embryo development
	Rodent postimplantation	Early embryo organogenesis
	Chick embryo	Avian development
	Xenopus	Frog development and FETAX assay
Invertebrate	Fish	Pician development
	Drosophila	Fruit fly development
	Hydra	*Hydra* reaggregation and regeneration
Organ culture	Limb culture	Limb bud development
	Palatal shelves	Palatal shelf closure
Cell culture		
Primary cells	Micromass	Evaluation of limb and neuronal cell differentiation
Established lines	Mouse ovarian	Cell attachment
	Palatal cells	Growth inhibition
Reproduction		
Male	Sperm	Viability, motility, and biochemical assessment
	Germ/Sertoli cell culture	Morphological, biochemical, and functional assessments
	Germ cell assays	Dominant lethal mutations, heritable translocations, and specific loci variants
Female	Hamster cell penetration assay	Ability of sperm to penetrate and initiate fertilization
	Oocyte cultures	Viability, functional biochemical assays
	Fertilization assay	Fertilization of superovulated eggs

Source: Faustman, 1988; Olshan and Faustman, 1993; Whittaker and Faustman, 1994; Rogers and Kavlock, 1996; Thomas, 1996.

media. Similar problems may arise when examining the mechanisms of lead toxicity as laboratory reagents are often contaminated with lead and because lead may bind to ligands present in media or elsewhere, reducing the effective free lead concentration. The effect of such opportunistic binding is proportionally greater as the concentration of lead in solution decreases (Simons, 1993).

In vitro tests have been most frequently used for mechanistic evaluations, however, in some selected examples they have been applied as prescreens before full-scale, in vivo assessment, as guidance for the development of in vivo experiments (including the choice of appropriate measurement endpoints), and for the prioritization of compounds for pharmaceutical development or regulatory action. Although in vitro systems provide useful information that may assist risk evaluations, they are not sufficient for risk assessment (Moore et al., 1995). This is partially due to validation issues that still require resolution, but also because most short-term tests only evaluate a select set of all possible developmental or reproductive processes. One area where progress is being made is the use of in vitro information in mechanistically based risk models. For example, a biologically based dose-response model has been developed to assess risks to fetal growth and development from exposures to methylmercury (Leroux et al., 1996). Values for model parameters, which include cell cycle progression rates, cell differentiation rates and cell death rates, were obtained from short-term, in vitro experiments carried out on embryonic rat midbrain cells exposed to methylmercury (Ponce et al., 1994). This information was combined with in vivo information on normal neuronal cell kinetics, and dose-response curves were predicted for methylmercury-induced brain abnormalities. Model predictions were similar to dose-response relationships predicted from in vivo malformation results.

Examples of in vivo short-term tests are also found in Table 2. These tests are designed to evaluate mortality, growth alterations, structural abnormalities, and functional deficits. The Chernoff-Kavlock assay is an example of a short-term in vivo system used for prescreening of developmental toxicants. In this assay, pregnant rodent females are exposed during organogenesis to a limited range of dose levels close to those producing maternal toxicity. Offspring are examined for viability, growth, and malformations, and the results can be used to prioritize chemicals for further developmental testing.

In Vivo Tests for Developmental and Reproductive Toxicity Assessment

The use of animal models to evaluate potential risks to humans from developmental and reproductive toxicants continues to be a central tool in toxicology. The extrapolation of risks from animals to humans is based on the "the phylogenetic continuity of species" (Winneke and Lilienthal, 1992). This continuity is observed at the microscopic level of cell structure and function and includes similarities in energy metabolism, signal transduction, transmission of genetic material, and other aspects of cell structure and function that underlie the most basic processes of cellular activity. While developmental processes have striking similarities in the patterns of organogenesis across species, reproductive similarities are less clear, and with regard to similarities in toxic response there are disparities across test species. For example, of the human developmental toxicants tested in hamsters, 54% tested as false negative for causing growth retardation. These data contrast with a finding that only 10% of the human developmental toxicants tested in mice tested as false negatives for growth retardation (Rogers and Kavlock, 1996). While care should be taken when making animal-to-human extrapolations of toxicity or risk based on some of the in vivo tests of developmental or reproductive toxicity provided below, these tests are extremely valuable toward gaining an understanding of mechanisms of toxicity, evaluating ameliorative approaches, and estimating potential risks.

Standardized in vivo toxicological testing protocols for reproductive and developmental toxicity analysis have been established by various groups, including the U.S. Food and Drug Administration (FDA), U.S. EPA, and the National Toxicology Program (NTP) (U.S. EPA, 1991, 1996; Rogers and Kavlock, 1996; Thomas, 1996). The protocols established by FDA include Segment I tests of fertility and reproductive function, Segment II studies of developmental toxicity and malformation, Segment III studies of postnatal and perinatal toxicity, and multigenerational studies of fertility, growth, and development. EPA has also developed two-generation studies of reproductive function, and the NTP conducts the Fertility Assessment by Continuous Breeding. While multigenerational studies are technically difficult and expensive to conduct, they provide important behavioral mechanistic information, which would not otherwise be obtained. A listing of endpoints used to assess maternal and developmental toxicity is provided in Table 3. These endpoints include the measurement of physical characteristics, clinical examinations, tissue-specific evaluations, and behavioral testing. The analyses of chemically induced developmental toxicity include the evaluation of reproductive capabilities of both sexes over one complete spermatogenic cycle and several estrous cycles. Examples of reproductive toxicity test endpoints that are specific for either male or female toxicity are provided in Table 4.

Use of Epidemiological Data in Toxicity Assessment

Well-performed epidemiological studies that positively associate an exposure to a disease are generally the most convincing evidence of potential risk to humans from an exposure. However, such associations are difficult to obtain because of limited information on exposure parameters (e.g., timing, duration, and concentration of exposure), variability in susceptibility to disease, and the presence of other factors that may "confound" study interpretation.

Common epidemiology study designs fall into two major classes: experimental and observational. In experimental studies, individuals are randomly assigned to either a treatment group or a control group and are prospectively studied. Experimental studies are generally the domain of medical research, including pharmaceutical clinical trials. Observational studies include case reports and diagnostic tests, ecological studies, cross-sectional studies, case-control studies, and cohort studies. Case reports usually involve a small number of people and present findings on specific, often unusual diseases associated with a particular exposure, while diagnostic test reports usually present results from a new medical procedure. Ecological studies utilize routine data to evaluate the association of a disease with a cause among a group of groups (e.g., countries or races). Because such studies do not evaluate individuals, they are usually used as exploratory or preliminary studies, which will require further, refined evaluation. Cross-sectional studies evaluate groups at one time to establish risk factors for a disease. While cross-sectional studies can make an association, they are not useful for establishing a cause-effect relationship. Case-control studies examine differences in exposure between groups with a disease (the cases) and without disease (the controls). Such studies are often used

Table 3 Examples of In Vivo Measurement Endpoints for the Assessment of Maternal and Developmental Toxicity

Endpoints used to assess maternal toxicity:
- Mortality
- Mating index   (no. with seminal plugs or sperm)/(no. mated) $\times$ 100
- Fertility index   (no. with implants)/(no. of matings) $\times$ 100
- Gestation length   Useful when animals are allowed to deliver pups
- Body weight   Day 0, during gestation, day of necropsy
- Body weight change   Throughout gestation, during treatment, posttreatment to sacrifice, corrected maternal
- Organ weights   Absolute, relative to body weight, relative to brain weight
- Clinical evaluations   Types, incidence, degree, and duration of clinical signs; enzyme markers; clinical chemistries
- Gross necropsy and histopathology
- Food and water consumption

Endpoints used to determine developmental toxicity:

Endpoints typically measured at terminal phase of pregnancy
- Implantation sites
- Preimplantation loss
- Affected (nonlive and malformed)
- Live offspring with malformations and variations
- Corpora lutea
- Resorptions and fetal deaths
- Fetal weight

Endpoints that can be measured postnatally
- Stillbirths
- Offspring growth (birth, postnatally)
- Neurobehavioral development and function[a]
     Reflex development
     Sensory function
     Cognitive function
- Reproductive system development and function[a]
     Ovarian cyclicity
     Fertility
- Other organ system function (e.g., renal, cardiovascular)[a]
- Offspring viability (birth, first week, at weaning, etc.)
- Physical landmarks of development
     Motor activity
     Social/Reproductive behavior
     Neuropathology and brain weights
     Sperm measures (e.g., morphology, motility, number)
     Pregnancy outcome

[a]Actual endpoints measured depend on the function or organ system being studied.
Source: Adapted from U.S. EPA, 1991; Moore et al., 1995.

to examine the occurrence of a rare disease. Cohort studies usually involve following groups of individuals with differences in their exposure status over time to determine whether there are differences in the occurrence of disease. There are a number of potential "flaws and fallacies" in the interpretation of epidemiological studies, and these are usually dependent on study design (Michael et al., 1984).

    Because of the environmental and industrial prevalence of heavy metals, there have been many epidemiological investigations of the effects of metal exposure on reproduction and development. These epidemiological investigations have studied environmentally exposed populations, occupational cohorts, and accidental exposures to evaluate the effects of metals on reproduction and development. Generally, epidemiological in-

Table 4 In Vivo Endpoints for the Evaluation of Reproductive Toxicity

Male endpoints	Female endpoints
Testis: weight; size in situ; spermatid reserves;	Body weight
Gross and histological evaluation; nonfunctional tubules (%); tubules with lumen sperm (%); tubule diameter; counts of leptotene spermatocytes	Ovary: organ weight; histology; number of oocytes; rate of follicular atresia; follicular steroidogenesis; follicular maturation; oocyte maturation; ovulation; luteal function
Epididymis: weight and histology; number of sperm in distal half; motility of sperm, distal end (%); gross sperm morphology, distal end (%); detailed sperm morphology, distal end (%)	Hypothalamus: histology; altered synthesis and release of neurotransmitters, neuromodulators, and neurohormones
Biochemical assays	Pituitary: histology; altered synthesis and release of tropic hormones
Accessory sex glands: histology; gravimetric	Endocrine: gonadotropin; chorionic gonadotropin levels; estrogen and progesterone
Semen: total volume; gel-free volume; sperm concentration; total sperm/ejaculate; total sperm/day of abstinence; sperm motility (various tests); gross sperm morphology; detail sperm morphology	Oviduct: histology; gamete transport; fertilization; transport of early embryo
Endocrine: luteinizing hormone; follicle-stimulating hormone; testosterone; gonadotropin-releasing hormone	Uterus: cytology and histology; luminal fluid analysis (xenobiotics, proteins); decidual response; dysfunctional bleeding
Fertility	Cervix/Vulva/Vagina: cytology; histology; mucus production; mucus quality (sperm penetration test)
Ratio exposed:pregnancy females	Fertility: ratio exposed:pregnant females; number of embryos or young per pregnant female; ratio viable embryos:corpora lutea; number 2- to 8-cell eggs
Number embryos per pregnant female	
Sperm per ovum	
Other tests considered: tonometric measurement of testicular consistency; qualitative testicular histology; state of cycle at which spermiation occurs; quantitative testicular histology	

Source: Adapted from Thomas, 1996.

vestigations of environmental exposures have been hampered by difficulties in assessing exposure profiles, especially those exposures involving mixtures.

In many epidemiology studies the composition of the metal exposures will not be known, and in occupational settings, the metals are usually present as a mixture of multiple forms of the same metals with multiple metal species. Also, the chemical form of the ores used in a smelter can vary over time as production economics may shift the source of the crude ores, and very different metal profiles will be present across metal industries. Table 5 shows a comparison of epidemiology studies conducted for lead with an evaluation of chromosomal aberrations as an endpoint. It is unknown whether differences in metal exposures with regards to form and/or species composition could account for any of the differences among study results. Minimal details were provided on study design or power of detection, thus, true comparisons of these studies are not possible. As another example of disparate findings among studies, Table 6 presents a comparison of studies that evaluated paternal exposure to lead and the related reproductive outcomes (both positive and negative studies are presented for most endpoints). Results from this comparison demonstrate the difficulties in observing consistent findings among human populations, even among those who are generally at most risk because of their occupation and within one set of endpoints. In general, the uncertainty associated with evaluating exposures to mixtures of different metal species and forms will be greater in nonoccupational epidemiology studies because more information is usually available in occupational settings with regards to exposure composition, refining methods, worker exposure concentrations, human biomonitoring information, etc.

While high concentration exposures to metals have been strongly associated with developmental or reproductive toxicity, there is considerable scientific debate regard-

Table 5 Occupational Exposure to Lead and Its Relationship to Chromosomal Aberrations

Exposed subjects	Type of aberration
Positive findings	
Lead oxide workers	Chromatid and chromosome breaks
Chemical factory workers	Chromatid gaps, breaks
Zinc plant workers	Gaps, fragments, rings, exchanges, dicentrics
Blast-furnace workers, metal grinders, scrap metal workers	Gaps, breaks, hyperploidy, structural abnormalities
Battery plant workers and lead foundry workers	Gaps, breaks, fragments
Lead oxide factory workers	Chromatid and chromosome aberrations
Battery melters, tin workers	Dicentrics, rings, fragments
Ceramic, lead, and battery workers	Breaks, fragments
Smelter workers	Gaps, chromatid and chromosome aberrations
Battery plant workers	Chromatid and chromosome aberrations
Negative findings	
Policemen	
Lead workers	
Shipyard workers	
Smelter workers	
Volunteers (ingested lead)	
Children (near a smelter)	

Source: Thomas, 1996.

Table 6 Association Between Paternal Exposures to Metals and Male Reproductive Dysfunction

Metal	Sperm abnormalities +	Sperm abnormalities −	Hormonal imbalance +	Hormonal imbalance −	Reduced fertility +	Reduced fertility −	Spontaneous abortion +	Spontaneous abortion −	Birth defect +	Birth defect −	Childhood malignancy +	Childhood malignancy −
Cadmium		1,2		3		1,2 5 5		4				
Manganese					15							
Lead	6–11		12,13	14	5,6 8,11	16	4,17	18,19	20,21 28	19,22	23	24–27
Organic mercury	29		30									
Inorganic mercury					15,31 32		31,32		33,31			
Boron	34				35							

[1]Kazantzis et al., 1963; [2]Saaranen et al., 1989; [3]Mason, 1990; [4]Lindbohm et al., 1991; [5]Gennart et al., 1992; [6]Lancranjan et al., 1975; [7]Cullen et al., 1984; [8]Assenato et al., 1987; [9]Lerda, 1992; [10]Telisman et al., 1990; [11]Chowdhury, 1993; [12]McGregor and Mason, 1990; [13]Ng et al., 1991; [14]Gennart et al., 1992; [15]Lauwerys et al., 1985; [16]Coste et al., 1991; [17]Al-Hakkak et al., 1986; [18]Lindbohm et al., 1991; [19]McDonald et al., 1989; [20]Boue et al., 1975; [21]Olshan et al., 1991; [22]Kristensen et al., 1993; [23]Kantor et al., 1979; [24]Wilkins III and Sinks, 1984; [25]Olshan et al., 1990; [26]Wilkins III and Koutras, 1988; [27]Olsen et al., 1991; [28]Sallmen et al., 1992; [29]Popescu, 1978; [30]Barregard et al., 1994; [31]Alcser et al., 1989; [32]Cordier et al., 1991; [33]Matte et al., 1993; [34]Tarasensko et al., 1972; [35]Whorton et al., 1994.
Source: Adapted from Tas et al., 1996.

ing the levels of safe exposure (where no adverse developmental effects are expected to occur). In particular, debate exists over the interpretation of epidemiological studies of neurodevelopmental effects from low-level, developmental lead and methylmercury exposures. Epidemiological studies of low-level trace metal exposures test the limits of current scientific capabilities. These limits arise from the extremely subtle nature of the behavioral alterations under study, the need for rigorous evaluation of potential confounders, and the large population sample required to draw meaningful conclusions. Moreover, there is debate over the utility and predictiveness of the behavioral test batteries currently used to evaluate subtle differences in neurodevelopment. For example, the World Health Organization currently relies on an epidemiological study of humans exposed in utero to methylmercury as a basis for their food consumption recommendations (Cox et al., 1989; WHO, 1990). However, important issues remain with regard to the interpretation of dose-response information (Cox et al., 1995), the potential for recall bias (uncertainty in age of birth and age of walking), and the applicability of data from short-term, high-concentration exposures from a fungicide applied to grain to derive recommendations for long-term, low-concentration exposures from consumption of fish and marine mammals, which naturally contain methylmercury. Two large-scale epidemiological investigations in the Seychelles (Myers et al., 1995) and the Faroe Islands (Weihe et al., 1996) have been recently undertaken in order to address some of the questions raised in the study of the Iraqi children. In particular, the risks to child neurodevelopment among high consumers of fish and marine mammals (i.e., at the upper range of normal environmental exposures) will be clarified. These data will also allow for reevaluation of the assumptions used to estimate levels of safe exposure by the World Health Organization, FDA, and other regulatory agencies.

Confidence in the causative association between exposure to a metal and an adverse developmental outcome comes from satisfying standard criteria for causation: strength of association, dose-response relationship, consistency, temporality, biological plausibility, and specificity (Hill, 1965). The establishment of a causal relationship is often very clear following high-concentration, accidental exposures to human populations, however, the determination of risks from low-concentration (near background), long-term exposures is exceedingly difficult. This is because the strength of association is often weak (relative risk < 2) and "more likely to be affected by classification biases, confounding, case or control selection, and selective subgroup analysis than would be the case for large order associations" when exposure concentrations are near background (Wynder, 1996). Despite the problems inherent in their use, epidemiological investigations that demonstrate a significant association between exposures to a metal and an adverse developmental or reproductive outcome "may assume a dominant role in human risk assessment" (Moore et al., 1995). This dominance exists because the assumptions required to extrapolate toxicity data between human populations have a lower inherent uncertainty than the assumptions required to extrapolate toxicity data across species (i.e., from animals to humans).

In addition to considerations of statistical power, bias, confounding, validity of exposure/outcome measures, and the ability to generalize to other populations, epidemiological investigations of the effects of metals on development or reproduction have unique considerations. Because of the temporal nature underlying both development and reproduction, considerations of timing of exposure are important in epidemiological studies, and different adverse outcomes would be expected to occur with exposure depending on the stage of development or reproduction. To approach this issue, for example, segmental analysis of hair mercury content was used to evaluate the time-course of fetal exposure to mercury during the 1971–72 Iraqi poisoning incident (March et al., 1980).

Evaluation of Dose-Response Information for Toxicity Assessment

The evaluation of dose-response information involves a "determination of the relationship between the magnitude of exposure and the probability of occurrence of the health effects in question" (NRC, 1983). Such dose-response information is rarely available for humans and is more commonly derived from animal investigations rather than epidemiological studies. In animal investigations, toxicity endpoints are characterized and parameters that affect response identified. Extrapolations from animal experiments to predict human response are made for route of exposure, high-dose to low-dose exposures, short-term to long-term exposures, and for other parameters that affect response. Of particular concern with regard to evaluating the dose-response relationship for metals in humans is the nutritional status of the population, the potential existence of sensitive subpopulations (e.g., pregnant women and young children), tissue protection and repair, and potential interactions between metals.

Factors Affecting Individual Dose Response

Metal chemistry has been exploited in evolution in the development of a diverse array of enzymes that rely on binding metal ions (in which the metals may act as ligands) to achieve normal activity; trace metals required for normal biochemistry are termed essential trace metals and include iron, zinc, cobalt, manganese, and chromium (Martin, 1988); nonessential metals, such as mercury, lead, and cadmium, serve no currently recognized biological purpose. For the essential trace metals, there is a U-shaped dose-effect relationship (Eaton and Klassen, 1996). At the low end of exposure, there may be nutritional deficiency, and at the upper end there is toxicity; between these two extremes lies an optimal range of exposures; this range may be narrow or wide depending on the nutrient. While low-dose exposures of humans to nonessential metals may be innocuous, increasing exposures only lead to toxicity. The toxic manifestations of exposure to higher concentrations of metals "is usually unrelated to its essentiality" (Martin, 1988). Normal reproduction and development is extremely sensitive to the maintenance of an adequate nutritional balance of essential trace elements. Disruption of this balance by metal deficiency or by metal-induced perturbation can significantly increase susceptibility to adverse reproductive or developmental effects.

Nutritional deficiency of essential trace metals may exacerbate the adverse effects of metal exposure and may also affect metal bioavailability. For example, selenium deficiency has been correlated with decreased birth weight following methylmercury exposure in mice, an effect not observed under conditions of normal dietary selenium (however, selenium status did not appear to affect the incidence of cleft palate) (Nishikido et al., 1988), and antagonism between selenium and methylmercury has been observed to varying degrees under experimental conditions (Ganther et al., 1972; Ridlington and Whanger, 1981; Chang, 1983; Svensson et al., 1992); selenium also apparently protects against cadmium toxicity (Ganther, 1980; Rana and Boora, 1992). Additionally, iron deficiency appears to increase the intestinal absorption of manganese (ATSDR, 1992), and lead toxicity is influenced by dietary calcium, iron, vitamins A and D, protein, and alcohol (ATSDR, 1993). Exposure to trace metals may also alter normal metabolism of essential trace metals during development. For example, a decreased incorporation of iron into heme in embryonic mouse liver and impaired fetal growth was observed following lead exposure (Gerber and Maes, 1978). Trace metals may also mimic the biochemistry of other elements. For example, lead is incorporated into bone, is transported across the membrane, and is bound to calmodulin, protein kinase C and other proteins because of its ability to mimic calcium (Clarkson, 1993; Simons, 1993); during development and pregnancy, times of high bone turnover and resorption, bone lead may be mobilized leading to increases in blood lead levels (Silbergeld et al., 1993; O'Flaherty, 1994, 1995). Finally, caloric restriction has also been shown to affect the response of animals to chemical exposures, including metals (Hart et al., 1995).

Because of the toxicity of trace metals, both homeostatic mechanisms and inducible protective enzymes have evolved that regulate intracellular trace metal concentrations. While homeostatic adaptation allows the cell to handle a variety of chemical exposures over a range of concentrations, this adaptation can be overwhelmed. Thus, for many endpoints there are thresholds of exposures above which cellular adaptation can no longer protect tissues from injury. An example of cellular adaptation is the induction of the γ-glutamyl cysteine synthetase gene in response to methylmercury exposure (Woods et al., 1992). Other genes that can be induced by metals include metallothionein, DNA repair, antioxidant defenses, and the glutathione-*S*-transferases. These inducible systems defend against both overt toxicity and chronic cell injury following metal exposures. In addition to these changes at the molecular level, changes can also be induced in response to chemical exposure at the cell and tissue level. These include chemically induced hyperplasia and tissue repair (induced cell replication). However, compensatory repair during the course of development may be limited because of the immature nature of these systems, the inability of these systems to conduct repair fast enough during the dynamic processes of growth, and the unidirectional nature of differentiation and development. Two important, inducible protective systems with respect to metals are metallothionein and glutathione.

Metallothionein, with a large content of sulfhydryl groups, is an inducible protein with a high affinity for group II metals (Haerslev et al., 1995). During pregnancy, metallothionein appears to be involved in the placental transport of essential trace metals (i.e., copper and zinc) from the dam to the fetus in rats and may protect the fetus from exposure to potentially harmful trace metals (Chan et al., 1993). For example, injection of rats with cadmium (1.0 mg/kg/day) for 8 days immediately prior to mating led to an increase in metallothionein and cadmium concentration in placenta and an increase in maternal renal cadmium concentrations; metallothionein, cadmium, zinc, and copper concentrations in newborn rats of exposed dams were unaltered when compared to controls (Chan and Cherian, 1993). These data support a role for metallothionein in protecting the newborn against elevated cadmium and in sequestering cadmium in the kidney. These data also demonstrate a potential risk of cadmium-induced renal dysfunction in the dam due to pregnancy-related mobilization of cadmium from the liver—a relatively specific effect observed during female reproduction (Chan and Cherian, 1993).

Glutathione (GSH), a cysteine-containing tripeptide, is another example of an intracellular defense against metal toxicity. GSH is involved in both the maintenance of intracellular thiol status (often perturbed in the presence of metals with affinity for protein sulfhydryl groups) and the excretion of metals out of the cell and the body (Stein et al., 1988; Tanaka et al., 1992; Dutczak and Ballatori, 1994). For example, Miura et al. (1994a,b) have demonstrated that cell lines shown to be resistant to methylmercury have elevated levels of intracellular GSH; cells treated with BSO (an inhibitor of GSH production) had diminished methylmercury resistance. The authors demonstrated that resistant cells had lower intracellular methylmercury levels as a result of slow uptake and rapid efflux of methylmercury, an effect ascribed to GSH-mediated binding and excretion of methylmercury (Miura et al., 1994a,b). The intracellular production of GSH can be induced (Woods et al., 1992; Woods and Ellis, 1995), and the distribution of GSH and glutathione-*S*-transferases in the nervous system is developmentally regulated (Beiswanger et al., 1995).

The induction of protective systems and induced cell repair in response to metal exposures will significantly impact the shape of the dose-response curve. In particular, a lack of consideration of tissue repair and hormesis may lead to the overprediction of risks in the low-dose range of exposures if a linear extrapolation of risks is made from responses identified under high concentration exposures (Hart and Frame, 1996).

Physiologically Based Pharmacokinetic Models

In order to provide improved dose-response information when extrapolating between species, investigators and risk assessors are turning to the use of physiologically based pharmacokinetic (PBPK) models. PBPK models use physiological parameters (e.g., cardiac output, organ perfusion rats, ventilation rate, etc.) to analyze chemical kinetics and metabolism. Because the relevant physiological parameters can be scaled and adjusted between species, PBPK models are used to predict chemical kinetics and metabolism between species. These types of studies also allow investigators to address the assumption that the target organ dose is proportional to the amount inhaled or ingested, an assumption that may lead to inaccurate estimates of low-dose risk if a chemical exhibits nonlinear pharmacokinetics (Graham, 1995).

Pregnancy dramatically alters maternal physiology (including changes to the cardiovascular, gastrointestinal, respiratory, and endocrine systems), and these alterations can significantly impact chemical toxicokinetics. For example, renal blood flow, glomerular filtration rate, plasma volume, and body fat are all increased during pregnancy. These physiological changes result from "resetting maternal homeostatic mechanisms to deliver essential nutrients to the fetus, and remove heat, carbon dioxide, and waste products from the fetus," however, the physiological mechanisms vary among species (Juchau et al., 1985; Mattison et al., 1991). Because PBPK models utilize species-specific physiological parameters (which can be obtained during pregnancy for both the maternal and fetal compartments), PBPK models allow for an evaluation of distribution kinetics and an estimation of target organ concentrations during various stages of pregnancy. Improved PBPK models attempt to evaluate the impacts of individual variability and parameter uncertainty through the use of Monte Carlo simulation (Hetrick et al., 1991; Thomas et al., 1996); however, these techniques have not yet been applied to metals.

A limited number of PBPK models have been developed for both lead and mercury during pregnancy and early development (Farris et al., 1993; Gray, 1995; O'Flaherty, 1995). For example, PBPK models have been developed to investigate the kinetics of methylmercury distribution between the mother and the fetus during various stages of pregnancy and development (Farris et al., 1993; Gray, 1995) and the kinetics of lead distribution during human growth (O'Flaherty, 1995); in general, these models rely extensively on data derived in vivo (Mattison et al., 1991; O'Flaherty, 1994, 1995). However, there do not appear to be PBPK models specific to pregnancy or early development for other metals. Because of their utility, PBPK models will certainly continue to grow as a risk assessment tool.

EXPOSURE ASSESSMENT

Exposure assessment is the process of determining the nature of environmental exposure by exposure type, intensity, frequency, duration, and source. With regard to reproductive and developmental toxicity, timing of exposure has special significance. Because gametogenesis and development occur sequentially, with one event preceding another, a disruption in this sequence can have significant impact downstream, and exposures that may be safe during one developmental or reproductive stage may be detrimental at others (Kamrin et al., 1994). While recovery or adaptation may be possible under some circumstances, loss of function may be unavoidable, leading to reproductive or developmental deficits.

A significant component of exposure assessment involves the identification of exposure pathways and estimating the overall contribution of exposures from these pathways to overall exposure. Generally, a lack of information on the historic presence of the material in the environment and on individual characteristics that may affect exposure make uncertainty in exposure assessment a significant contributor to overall uncertainty in risk assessment. One tool for the improvement of exposure assessment is the use of biomarkers of exposure.

Biomarkers of exposure involve the use of biological material to estimate the extent of an environmental exposure. Such biomarkers may include specific analyses of blood, urine, hair, or other material for the presence of the compound (or its metabolite). Knowledge of pharmacokinetics combined with information on potential time of exposure can be used to estimate the magnitude of the original exposure. Biomarkers of exposure are extremely useful because they can significantly reduce the number of assumptions used in estimating individual exposure. For example, biomarkers of exposure provide an integrated estimation of exposure (i.e., all exposure routes are combined), and there is no need to account for potential differences in matrix effects on bioavailability. While some biomarker assays for human metal exposures have become standard (e.g., blood lead testing, hair mercury analysis), the evaluation of exposures for other metals remains problematic (e.g., cadmium). Because of the need for information on time of exposure to evaluate potential risks to development and reproduction, traditional exposure measures for metals have limited application for reproductive and developmental risk assessment. The limitations arise because traditional measures of exposure often only reflect cumulative or recent exposures, making it difficult or impossible to determine temporal trends in exposure. Because risks to development or reproduction vary according to reproductive or developmental stage, a lack of temporal exposure information will add uncertainty to the risk evaluation.

An example of an ideal biomarker of exposure to trace metals is the segmental analysis of hair for methylmercury. Segmental hair methylmercury analysis is ideal because it has high specificity, that is, it reflects dietary exposures as there is no exogenous contamination of hair with methylmercury (not true for total mercury hair analysis). Segmental analysis also allows for the determination of the time-course of exposure (Al-Shahristani and Al-Haddad, 1973; Lind et al., 1988; Lasorsa and Citterman, 1991; Suzuki et al., 1993). This time-course information provides an exposure profile (including information on peak exposure), which can then be compared against gestational age. In general, there is a great need for the development of new biomarkers of exposure for metals for the evaluation of reproductive and developmental risks.

RISK CHARACTERIZATION

Risk characterization is "the process of estimating the incidence of a health effect under the various conditions of human exposure described in exposure assessment" (NRC, 1983). Risk characterization allows for a quantitative estimate of the risks to the population of concern through consideration of dose-response and exposure information.

For chemicals that cause systemic toxicity (i.e., noncarcinogens), the U.S. EPA has traditionally used an approach wherein the Lowest Observed Adverse Effect Level (LOAEL) or the No Observed Adverse Effect Level (NOAEL) obtained from a study is used to derive the allowable exposure level in humans. This level is termed the reference dose (RfD) or the reference concentration (RfC). As defined by Barnes and Dourson (1988), RfD/C is "an estimate (with uncertainty spanning perhaps an order of magnitude) of a daily exposure to the human population (including sensitive subgroups) that is likely to be without an appreciable risk of deleterious effects during a lifetime." RfD/C is calculated by using the NOAEL (or LOAEL) derived from a critical study (identified on the basis of such factors as choice of animal model, route of exposure, tested endpoints, overall study quality, sensitivity of response, etc.) and dividing the effect level by uncertainty factors and modifying factors. Uncertainty factors with default values of 10 are often included for intraspecies and interspecies (animal to human) variation, for extrapolation of short-term tests to long-term exposures, or for other experimental limitations. In total, these uncertainty factors may result in a 100- to 1000-fold decrease in the LOAEL to obtain the RfD (Graham, 1995). Guidelines for the choice of critical studies and use of uncertainty factors are provided in the U.S. EPA guidelines for developmental toxicity (U.S. EPA, 1991).

Criticisms regarding the applicability of the RfD/C approach in risk assessment are that the shape of the dose-response curve is ignored, that guidelines have not been developed to take into account study reliability (Barnes and Dourson, 1988), that there is no mechanism to estimate risk probabilities above the RfD/C (Kimmel, 1990), that the NOAEL is not set at a consistent effect level and can vary from 1 to 20% effect (Leisenring and Ryan, 1992; Allen et al., 1994a), and that the NOAEL must be one of the experimental doses (Crump, 1984). Some of the criticisms of the RfD/C are addressed with the use of the benchmark dose (BMD). The BMD (Fig. 3) is defined as the upper statistical confidence limit "for the dose corresponding to a specific increase in level of health effect over the background level" (Crump, 1984). Using a confidence interval for BMDs improves the description of the uncertainty in the shape of the dose-response curve. Estimates of the BMD reflect the shape of the dose-response curve and are not as constrained by experimental design as they do not have to be one of the experimental doses tested. Using the BMD approach it is possible to estimate risks in the full range of the dose-response curve. The validity of the BMD as an improved alternative to the NOAEL-based RfD has been reviewed for developmental toxicants (Faustman, 1988; Allen et al., 1994a,b; Kavlock et al., 1995), and the U.S. EPA has recently begun to utilize BMD methodology in risk assessments published in IRIS. To

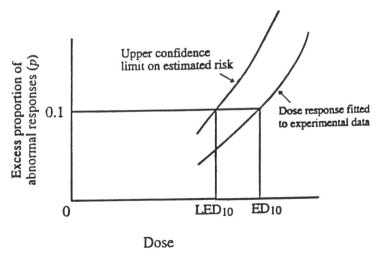

Figure 3 The typical model for obtaining a benchmark dose is presented. In this model, a maximum likelihood dose-response curve is fit to the experimental data. The risk assessor can use either the maximum likelihood estimate (MLE) or an upper confidence limit (typically a 90- or 95-percentile upper confidence limit) to derive an estimate of exposure expected to result in an excess proportion of adverse response above the background rate. As presented, the benchmark dose expected to result in a 10% increase in adverse response above background (BMD_{10}) is shown for both the MLE dose (ED_{10}) and the dose based on the upper confidence limit (LED_{10}).

date, BMDs have been used to calculate three RfCs and one RfD (U.S. EPA, 1996). The RfD for methylmercury has been set using a BMD based on human delayed postnatal development, and an RfD is under development for boron based on reduced body weight observed in rodent studies (U.S. EPA, 1996). The Benchmark approach has also been applied for assessment of reproductive effects (Pease et al., 1991; Auton, 1994).

Another application of the NOAEL is in the derivation of the margin of exposure (MOE). Because NOAELs delineate a level at which no adverse effects are observed to occur in an animal bioassay, they may be used as screening guides by public health and regulatory officials. The MOE is calculated as the ratio of the NOAEL to the observed exposure concentration expressed, where both factors are expressed as mg/kg/day. For example, if the NOAEL for a chemical is 1 mg/kg/day and the observed exposure concentration is estimated at 0.01 mg/kg/day, the MOE is 100. MOEs below 100 are common screening limits below which further evaluation is recommended (Faustman and Omenn, 1996).

UNCERTAINTY IN RISK ASSESSMENTS OF METALS

Uncertainty in risk assessments is usually classified as either variability or state-of-knowledge uncertainty (IAEA, 1989; Hoffman and Hammonds, 1994). In the previous sections of this chapter we have identified key areas of uncertainty and identified numerous factors that influence the potential of metals to cause reproductive or developmental effects. These factors have included nutritional status, timing of exposure relative to gestational organogenesis events, and metal pharmacokinetics and bioavailability. Such uncertainty dramatically impacts our ability to accurately model dose-response relationships for metals. Risk assessments evaluating different chemicals have identified the uncertainty in evaluating dose-response relationships as one of the major contributors to overall uncertainty in risk assessment (McKone, 1994; Cullen, 1995). Obviously, these observations suggest a critical area for research investigation. In practice, risk assessments for reproductive and developmental effects make use of safety factors to adjust for uncertainty, although these adjustments are in one direction and do not provide information on the likelihood of effect or extent of uncertainty (Weil, 1972; Barnes and Dourson, 1988; Kimmel, 1990; U.S. EPA, 1993).

Recent EPA guidelines for both noncancer and cancer risk assessments put the burden on the risk assessor to defend the use of default assumptions in RfD calculations, and the risk assessor is challenged to directly address the uncertainty associated with the "state-of-knowledge." For toxicologists, one of the most fruitful paths to overcome default assumptions is to define more clearly the mechanisms of action for the toxicant under assessment. Table 7 shows example guidelines that can be used to assess proposed mechanisms of developmental toxicity for metals, and these guidelines may be extended to include the evaluation of metals as potential reproductive toxicants.

One approach to analyzing the influence of model parameters (including assumed or default values) on the calculated risk estimate and the uncertainty associated with the estimated risk is Monte Carlo simulation. Monte Carlo simulations utilize random sampling from among probability density functions, which define input parameters in a risk model (IAEA, 1989; Morgan and Henrion, 1990; Law and Kelton, 1991). Because Monte Carlo analysis provides the risk manager with quantitative information on the level of uncertainty associated with the risk estimate rather than a single point estimate of risk, risk managers may make more informed decisions. Monte Carlo analysis is most commonly used for analyzing the effect of various exposure scenarios on risk, however, the development of probability distributions for effect has also been proposed (Gray et al., 1993; Evans et al., 1994a,b), and uncertainty analysis is now being ap-

Table 7 Guidelines for Assessment of Proposed Mechanistic Pathways in Chemically Induced Teratogenesis

Temporal association	Does dysmorphogenesis precede, occur simultaneously with, or follow the critical event?[a]
Dosage relationship	Does the potential mechanistic event occur at or below those doses that result in the dysmorphogenesis?
	Is there a dose-response relationship between exposure and severity of the developmental outcome?
Structure-activity relationships	Do chemicals with similar structures cause similar developmental outcomes?
Strength and consistency of association	Is the proposed mechanistic process strongly or weakly linked with the appearance of the developmental outcome?
	Is the proposed mechanistic process required for the appearance of the developmental outcome?
	How reproducible (in time and space) is the teratogenic outcome if the proposed mechanistic event occurs?
	Does modification of the mechanistic event, or one step in the process, alter or eliminate the adverse developmental outcome?
Coherence	Is there a molecular/cellular basis for the proposed mechanism of action by the chemical or physical agent that elicits the initial event?

[a]Critical event in an action of the chemical or physical agent that is producing developmental toxicity.
Source: Faustman et al., 1996.

plied to PBPK modeling (Hetrick et al., 1991; Thomas et al., 1996). Methods and suggestions for the selection of input distributions for simulation modeling have been described (Finley et al., 1994; Hattis and Burmaster, 1994; Lipton et al., 1995). Alternative analytical and simulation methods exist for propogation of uncertainty (Ferson and Kuhn, 1992; Ferson and Long, 1994), but these methods have not seen wide use.

SUMMARY

This chapter has used the NRC framework for risk assessment (Fig. 2) to organize and evaluate the data needed to conduct a risk assessment of the reproductive and developmental effects of metals. The intent has been to illustrate what quantitative and qualitative information is used and woven together in an overall risk assessment. One of the most useful references for this chapter has been an article by Moore et al. (1995), who built on the U.S. EPA guidelines for reproductive and developmental risk assessment. This combined effort from a diverse group of academic, private sector, and governmental experts described in detail an "evaluative process" for assessing human reproductive and developmental toxicity of agents. One highlight from this process was the description of the evaluative process as combining scientific data with professional judgment. It uses a weight-of-evidence approach to summarize, in a narrative statement, the strength and relevancy of the scientific data for assessing human risks and the degree of confidence that scientists have in the available information. Consistent with the need to consider mechanistic rationale for assessing human risks, it pushes the scientific community to defend default assumptions (e.g., a need to critically evaluate the relevancy of rodent studies that are the bases for deriving potential human health risks from an exposure). Such an effort is consistent with the philosophy of the NRC's report on the need for risk assessment to utilize both science and judgment (NRC, 1994). In defining the "degree of confidence" that scientists have with a risk assessment, it also delineates a process whereby "critical" data needs are described—not as "wouldn't it be nice to know," but as those gaps in information that significantly impact the evaluation (i.e., have the greatest associated uncertainty). A description of critical data gaps was recognized as an important component of risk assessment to be responsive to the needs of the public health community, the regulatory community, and the private sector.

The objectives of risk assessment are shown in Table 8 and can vary based on the responsibilities of risk-management groups. Critical information needs in risk as-

Table 8 Objectives of Risk Assessment

1. Balance risks and benefits
   Drugs
   Pesticides
2. Set target levels of risk
   Food contaminants
   Water pollutants
3. Set priorities for program activities
   Regulatory agencies
   Manufacturers
   Environmental and consumer organizations
4. Estimate residual risks and extent of risk reduction after steps are taken to reduce risk

sessments include both qualitative and quantitative risk information, information on level of certainty, and the identification of data needs. The risk-assessment process as shown in Figure 2 provides an organized framework whereby continuous iterations lead to refined risk information, and identified data needs drive research. In this process, when research provides improved information, the findings replace default assumptions, decrease uncertainty, and improve risk assessment. The challenge to those of us interested in metal toxicity is to respond to these requests for additional data and to improve the scientific basis of our risk-assessment decisions.

REFERENCES

Al-Hakkak, Z. S., Hamamy, H. A., Murad, A. M. B., and Hussain, A. F. (1986). *Mutat. Res., 171*: 53.

Al-Shahristani, H., and Al-Haddad, I. K. (1973). *J. Radioanal. Chem., 15*: 59.

Alcser, K. H., Brix, K. A., Fine, L. J., Kallenbach, R., and Wolfe, R. A. (1989). *Am. J. Ind. Med., 15*: 517.

Allen, B. C., Kavlock, R. J., Kimmel, C. A., and Faustman, E. M. (1994a). *Fundam. Appl. Toxicol., 23*: 487.

Allen, B. C., Kavlock, R. J., Kimmel, C. A., and Faustman, E. M. (1994b). *Fundam. Appl. Toxicol., 23*: 496.

Aschner, M., and Clarkson, T. W. (1988). *Brain Res., 462*: 31.

Aschner, M., and Clarkson, T. W. (1989). *Pharmacol. Toxicol., 64*: 293.

Assenato, G., Paci, C., Baser, M. E., Molinini, R., Candela, R. G., Altamura, B. M., and Giorgino, R. (1987). *Arch. Environ. Health, 42*: 124.

ATSDR (1992). *Toxicological Profile for Manganese*. U.S. Department of Health and Social Services, Atlanta.

ATSDR (1993). *Toxicological Profile for Lead*. U.S. Department of Health and Social Services, Atlanta.

Auton, T. R. (1994). *Reg. Toxicol. Pharmacol., 19*: 152.

Barnes, D. G. and Dourson, M. (1988). *Reg. Toxicol. Pharmacol., 8*: 471.

Barregard, L., Lindsedt, G., Schutz, A., and Sallsten, G. (1994). *Occup. Environ. Med., 51*: 536.

Beiswanger, C. M., Diegmann, M. H., Novak, R. F., Philbert, M. A., Graessle, T. L., Reuhl, K. R., and Lowndes, H. E. (1995). *Neurotoxicology, 16*: 425.

Boue, J., Boue, A., and Lazar, P. (1975). *Teratology, 12*: 11.

Burgess, W. A. (1978). *Patty's Industrial Hygeine and Toxicology* (G. D. Clayton and F. E. Clayton, eds.). John Wiley and Sons, New York, p. 1149.

Calkins, D. R., Dixon, R. L., Gerber, C. R., Zarin, D., and Omenn, G. S. (1980). *J. Natl. Can. Inst., 64*: 169.

Chan, H. M. and Cherian, M. G. (1993). *Toxicol. Appl. Pharmacol., 120*: 308.

Chan, H. M., Tamura, Y., Cherian, M. G., and Goyer, R. A. (1993). *Proc. Soc. Exp. Biol. Med., 202*: 420.

Chang, L. W. (1983). *Exp. Pathol., 23*: 143.

Chiappino, G. (1994). *Sci. Total Environ., 150*: 65.

Chowdhury, R. (1993). *NICE*, Sept. 26–Oct. 1.

Clarkson, T. W. 91993). *Annu. Rev. Pharmacol. Toxicol., 32*: 545.

Cordier, S., Deplan, F., Mandereau, L., and Hemon, D. (1991). *Br. J. Ind. Med., 48*: 375.

Coste, J., Mandereau, L., Pessione, F., Bregu, M., Faye, C., Hemon, D., and Spira, A. (1991). *Eur. J. Epidemiol., 7*: 154.

Cox, C., Clarkson, T. W., Marsh, D. O., Amin-Zaki, L., Tikriti, S., and Myers, G. G. (1989). *Environ. Res., 49*: 318.

Cox, C., Marsh, D., Myers, G., and Clarkson, T. (1995). *Neurotoxicology, 16*: 727.

Crump, K. S. (1984). *Fundam. Appl. Toxicol., 4*: 854.

Cullen, A. C. (1995). *J. Air Waste Manage. Assoc., 45*: 538.

Cullen, M. R., Kayne, R. D., and Robins, J. M. (1984). *Arch. Environ. Health., 39*: 431.

Dieter, M. P., Matthews, H. B., Jeffcoat, R. A., and Moseman, R. F. (1993). *J. Toxicol. Environ. Health, 39*: 79–93.

Dock, L., Mottet, K., and Vahter, M. (1994). *Pharmacol. Toxicol., 74*: 158.

Dutczak, W. J. and Ballatori, N. (1994). *J. Biol. Chem., 269*: 9746.

Eaton, D. L. and Klassen, C. D. (1996). *Casarett and Doull's Toxicology: The Basic Science of Poisons* (C. D. Klassen, M. O. Amdur, and J. Doull, eds.). McGraw-Hill, New York, p. 13.

Evans, J. S., Graham, J. D., Gray, G. S., and Sielken, R. L. (1994a). *Risk Anal., 14*: 25.

Evans, J. S., Gray, G. M., Sielken, J., R.L., Smith, A. E., Valdez-Flores, C., and Graham, J. D. (1994b). *Regul. Toxicol. Pharmacol., 20*: 15.

Farris, F. F., Dedrick, R. L., Allen, P. V., and Smith, J. C. (1993). *Toxicol. Appl. Pharmacol., 119*: 74.

Faustman, E. M. (1988). *Mutat. Res., 205*: 355.

Faustman, E. M. and Omenn, G. S. (1996). *Casarett and Doull's Toxicology: The Basic Science of Poisons* (C. D. Klassen, M. O. Amdur, and J. Doull, eds.). McGraw-Hill, New York, p. 75.

Faustman, E. M., Ponce, R. A., Seeley, M. S., and Whittaker, S. G. (1996). *Handbook of Developmental Toxicity* (R. Hood, ed.). CRC Press, Boca Raton, FL, p. 13.

Ferson, S., and Kuhn, R. (1992). *Computer Techniques in Environmental Studies* (P. Zannetti, ed.). Elsevier, London, p.

Ferson, S. and Long, T. (1994). *Environmental Toxicology and Risk Assessment* (J. Hughes, G. Biddinger, and E. Mones, eds.). American Society for Testing and Materials, Philadelphia.

Finley, B., Proctor, D., Scott, P., Harrington, N., Paustenbach, D., and Price, P. (1994). *Risk Anal., 14*: 533.

Ganther, H. E. (1980). *Ann. NY Acad. Sci., 355*: 212.

Ganther, H. E., Goudie, C., Sunde, M. L., Kopecky, M. J., and Wagner, P. (1972). *Science, 175*: 1122.

Gennart, J. P., Bernard, A., and Lauwerys, R. (1992a). *Int. Arch. Occup. Environ. Health, 64*: 49.

Gennart, J. P., Buchet, J. P., Roels, H., Ghyselen, P., Ceulemans, E., and Lauwerys, R. (1992b). *Am. J. Epidemiol., 135*: 1208.

Gerber, G. B. and Maes, J. (1978). *Toxicology, 9*: 173.

Gordon, T. and Fine, J. M. (1993). *Occup. Med., 8*: 504.

Graham, J. D. (1995). *Toxicology, 102*: 29.

Gray, D. G. (1995). *Toxicol. Appl. Pharmacol., 132*: 91.

Gray, G. M., Cohen, J. T., and Graham, J. D. (1993). *Environ. Health Perspect., 6*: 203.

Haerslev, T., Jacobsen, G. K., Horn, N., and Damsgaard, E. (1995). *Apmis, 103*: 568.

Hart, R. W. and Frame, L. T. (1996). *Belle Newslett.*, 5.

Hart, R. W., Keenan, K., Abdo, K. M., Leakey, J., and Lyn-Cook, B. (1995). *Fundam. Appl. Toxicol., 25*: 195.

Hattis, D. and Burmaster, D. E. (1994). *Risk Anal., 14*: 713.

Hetrick, D., Jarabek, A., and Travis, C. (1991). *J. Pharmacokinet. Biopharm., 19*: 1.

Hill, A. B. (1965). *Proc. R. Soc. Lond. Med., 58*: 295.

Hoffman, F. O. and Hammonds, J. S. (1994). *Risk Anal., 14*: 707.

Hornstein, D. (1992). *Columbia Law Rev., 92*: 501.

IAEA (1989). *Evaluating the Reliability of Predictions Made Using Environmental Transfer Models.* International Atomic Energy Agency, Vienna.

Juchau, M. R., Giachelli, C. M., Fantel, A. G., Greenaway, J. C., Shepard, T. H., and Faustman Watts, E. M. (1985). *Toxicol. Appl. Pharmacol., 80*: 137.

Kamrin, M. A., Carney, E. W., Chou, K., Cummings, A., Dostal., L. A., Harris, C., Henck, J. W., Loch Caruso, R., and Miller, R. K. (1994). *Toxicol. Lett., 74*: 99.

Kantor, A. F., Curnen, M. G., Meigs, J. W., and Flannery, J. T. (1979). *J. Epidemiol. Commun. Health, 33*: 253.

Kavlock, R. J., Allen, B. C., Faustman, E. M., and Kimmel, C. A. (1995). *Fundam. Appl. Toxicol., 26*: 211.

Kazantzis, G., Flynn, F. V., Spwage, J. S., and Trott, D. G. (1963). *O. J. Med., 32*: 165.

Kimmel, C. A. (1990). *Neurotoxicology, 11*: 189.

Kristensen, P., Irgens, L. M., Daltveit, A. K., Andersen, A. (1993). *Am. J. Epidemiol., 137*: 134.

Lancranjan, I., Popescu, H. I., Gavanescu, O., Klepsch, I., and Serbanescu, M. (1975). *Arch. Environ. Health, 30*: 396.

Lasorsa, B. K. and Citterman, R. J. (1991). Segmental analysis of mercury in hair in 80 women of Nome, Alaska. *OCS Study MMS 91-0065*. Minerals Management Service, Anchorage.

Lauwerys, R., Roels, H., Genet, P., Toussaint, G., Bouckaert, A., and De Cooman, S. (1985). *Am. J. Ind. Med., 7*: 171.

Law, A. M. and Kelton, W. D. (1991). *Simulation Modeling and Analysis*. McGraw-Hill, Inc., New York.

Leisenring, W. and Ryan, L. (1992). *Regul. Toxicol. Pharmacol., 15*: 161.

Lerda, D. (1992). *Am. J. Ind. Med., 22*: 567.

Leroux, B. G., Leisenring, W. M., Moolgavkar, S. H., and Faustman, E. M. (1996). *Accepted with revisions*.

Lind, B., Bigras, L., Cernichiari, E., Clarkson, T. W., Friberg, L., Hellman, M., Kennedy, P., Kirbride, J., Kjellstrom, T., and Ohlin, B. (1988). *Fresenius Z. Anal. Chem., 332*: 620.

Lindbohm, M. L., Hemminki, K., Bonhomme, M. G., Anttila, A., Rantala, K., Heikkila, P., and Rosenberg, M. J. (1991a). *Am. J. Ind. Med., 81*: 1029.

Lindbohm, M. L., Sallmen, M., Anttila, A., Taskinen, H., and Hemmniki, K. (1991b). *Scand. J. Work Environ. Health, 17*: 95.

Lipton, J., Shaw, W. D., Holmes, J., and Patterson, A. (1995). *Regul. Toxicol. Pharmacol., 21*: 192.

Marsh, D. O., Myers, G. J., Clarkson, T. W., Amin-Zaki, L., Tikriti, S., and Majeed, M. A. (1980). *Ann. Neurol., 7*: 348.

Martin, R. B. (1988). *Handbook on Toxicity of Inorganic Compounds* (H. G. Seiler and H. Sigel, eds.). Marcel Dekker, Inc., New York, p. 9.

Mason, H. J. (1990). *Hum. Exp. Toxicol., 9*: 91.

Matte, T. D., Mulinare, J., and Erickson, J. D. (1993). *Am. J. Ind. Med., 24*: 11.

Mattison, D. R., Blann, E., and Malek, A. (1991). *Fundam. Appl. Toxicol., 16*: 215.

McDonald, A. D., McDonald, J. C., Armstrong, B., Cherry, N. M., Nolin, A. D., and Robert, D. (1989). *Br. J. Ind. Med., 46*: 329.

McGregor, A. J. and Mason, M. J. (1990). *Hum. Exp. Toxicol., 9*: 371.

McKone, T. E. (1994). *Risk Anal., 14*: 449.

Michael, W., Boyce, W. T., and Wilcox, A. J. (1984). *Biomedical Bestiary: An Epidemiologic Guide to Flaws and Fallacies in the Medical Literature*. Little, Brown and Company, Boston.

Middaugh, J. P., Li, C., and Jenkerson, S. A. (1989). Health hazard and risk assessment from exposure to heavy metal ore in Skagway, AK (Final Report). Alaska Department of Health and Social Services, Anchorage.

Miura, K., Clarkson, T. W., Ikeda, K., Naganuma, A., and Imura, N. (1994a). *Environ. Health Perspect., 3*: 313.

Miura, K., Ikeda, K., Naganuma, A., and Imura, N. (1994b). *In Vitro Toxicol., 7*: 59.

Moore, J. A., Daston, G. P., Faustman, E., Golub, M. S., Hart, W. L., Hughes, Jr., C., Kimmel, A., Lamb, J. C. t., Schwetz, B. A., and Scialli, A. R. (1995). *Reprod. Toxicol., 9*: 61.

Morgan, M. G., and Henrion, M. (1990). *Uncertainty: A Guide to Dealing with Uncertainty in Quantitative Risk and Policy Analysis.* Cambridge University Press, Cambridge.

Myers, G. J., Marsh, D. O., Davidson, P. W., Cox, C., Shamlaye, C. F., Tanner, M., Choi, A., Cernichiari, E., Choisy, O., and Clarkson, T. W. (1995). *Neurotoxicology, 16*: 653.

NAS (in press). *Review of the Army's Zinc-Cadmium Risk Assessment.* Subcommittee on Zinc-Cadmium Sulfide, National Academy of Sciences, Washington, DC.

Ng, T. P., Goh, H. H., Ng, Y. L., Ong, H. Y., Ong, C. N., Chia, K. S., Chia, S. E., and Jeryaratnam, J. (1991). *Br. J. Ind. Med., 48*: 485.

Nishikido, N., Satoh, Y., Naganuma, A., and Imura, N. (1988). *Toxicol. Lett., 40*: 153.

NRC (1983). *Risk Assessment in the Federal Government: Managing the Process.* National Academy Press, Washington, DC.

NRC (1994). *Science and Judgment in Risk Assessment.* National Academy Press, Washington, DC.

O'Flaherty, E. J. (1994). *Environ. Health Perspect. Suppl., 102*: 103.

O'Flaherty, E. J. (1995a). *Toxicol. Lett., 82*: 367.

O'Flaherty, E. J. (1995b). *Toxicol. Appl. Pharmacol., 131*: 297.

Olsen, J. H., Brown, P. N., Schlugen, G., and Jensen, O. M. (1991). *Eur. J. Cancer, 27*: 958.

Olshan, A. F. and Faustman, E. M. (1993). *Annu. Rev. Public Health, 14*: 159.

Olshan, A. F., Breslow, N., Daling, J. R., Falletta, J. M., Grufferman, S., Robinson, L. L., Waskerwitz, M., and Hammond, G. D. (1990). *Cancer Res., 50*: 3212.

Olshan, A. F., Teschke, K., and Baird, P. A. (1991). *Am. J. Ind. Med., 20*: 447.

Pease, W., Vandenberg, J., and Hopper, K. (1991). *Environ. Health Perspect., 91*: 141.

Ponce, R. A., Kavanagh, T. J., Mottet, N. K., Whittaker, S. G., and Faustman, E. M. (1994). *Toxicol. Appl. Pharmacol., 127*: 83.

Popescu, H. I. (1978). *Br. Med. J., 1*: 1347.

Rana, S. V. and Boora, P. R. (1992). *Bull. Environ. Contam. Toxicol., 48*: 120.

Ridlington, J. W. and Whanger, P. D. (1981). *Fundam. Appl. Toxicol., 1*: 368.

Rogers, J. M. and Kavlock, R. J. (1996). *Casarett and Doull's Toxicology: The Basic Science of Poisons* (C. D. Klassen, M. O. Amdur, and J. Doull, eds.). McGraw-Hill, New York, p. 301.

Rolfe, M. W., Paine, R., Davenport, R. B., and Strieter, R. M. (1992). *Am. Rev. Respir. Dis., 146*: 1600.

Saaranen, M., Kantola, M., Saarikoski, S., and Vanha-Perttula, T. (1989). *Andrologia, 21*: 140.

Sallmen, M., Lindbohm, M. L., Anttila, A., Taskinen, H., and Hemminki, K. (1992). *J. Epidemiol. Commun. Health, 42*: 519.

Schardein, J. L. (1993). *Chemically Induced Birth Defects.* Marcel Dekker, New York.

Silbergeld, E. K., Sauk, J., Sommerman, M., Todd, A., McNeill, F., Fowler, B., Fontaine, A., and van Buren, J. (1993). *Neurotoxicology, 14*: 225.

Simons, T. J. B. (1993). *Neurotoxicology, 14*: 77.

Slikker, J., W. (1994). *Neurotoxicology, 15*: 6.

Stein, A. F., Gregus, Z., and Klassen, C. D. (1988). *Toxicol. Appl. Pharmacol., 93*: 351.

Suzuki, T., Hongo, T., Yoshinaga, J., Imai, H., Nakazawa, M., Matsuo, N., and Akagi, H. (1993). *Arch. Environ. Health, 48*: 221.

Svensson, B. G., Schutz, A., Nilsson, A., Akesson, I., Akesson, B., and Skerfving, S. (1992). *Sci. Total Environ., 126*: 61.

Tanaka, T., Naganuma, A., and Imura, N. (1992). *Eur. J. Pharmacol., 228*: 9.

Tarasensko, N. Y., Kasparov, A. A., and Strongina, O. M. (1972). *Gig. Tr. Prof. Zabol, 11*: 13.

Tas, S., Lauwerys, R., and Lison, Đ. (1996). *Crit. Rev. Toxicol., 26*: 261.

Telisman, S., Cvitkovic, P., Gavella, M., and Pongracic, J. (1990). *Peking. People's Republic of China*, 29.

Thomas, J. A. (1996). *Casarett and Doull's Toxicology: The Basic Science of Poisons* (C. D. Klassen, M. O. Amdur and J. Doull, eds.). McGraw-Hill, New York, p. 547.

Thomas, R., Bigelow, P., Keffe, T., and Yang, R. (1996). *Am. Ind. Hyg. Assoc. J., 57*: 23.

U.S. EPA (1991). *Guidelines for Developmental Toxicity Risk Assessment.* U.S. Environmental Protection Agency.

U.S. EPA (1993). EPA/625/4-89/024.

U.S. EPA (1996). *Draft Benchmark Dose Technical Guidance Document.* U.S. EPA, Washington, DC.

Vahter, M., Mottet, N. K., Friberg, L., Lind, B., Shen, D. D., and Burbacher, T. (1994). *Toxicol. Appl. Pharmacol., 124*: 221.

Weihe, P., Grandjean, P., and White R. (1996). Tenth International Congress on Circumpolar Health, Anchorage, AK.

Weil, C. S. (1972). *Toxicol. Appl. Pharmacol., 21*: 454.

Whittaker, S. G., and Faustman, E. M. (1994). *In Vitro Toxicology* (S. Cox, ed.). Raven Press, Ltd., New York, p. 97.

WHO (1990). *Environmental Health Criteria 101: Methylmercury.* World Health Organization, Geneva.

Whorton, M. D., Haas, J. L., Trent, L., and Wong, O. (1994). *Occup. Environ. Med., 51*: 761.

Wilkins III, J. R. and Koutras, R. H. (1988). *Am. J. Ind. Med., 14*: 299.

Wilkins III, J. R. and Sinks, T. H. J. (1984). *J. Occup. Med., 26*: 427.

Winneke, G. and Lilienthal, H. (1992). *Toxicol. Lett., 64/65*: 239.

Woods, J. S. and Ellis, M. E. (1995). *Biochem. Pharmacol., 50*: 1719.

Woods, J. S., Davis, H. A., and Baer, R. P. (1992). *Arch. Biochem. Biophys., 296*: 350.

Wynder, E. L. (1996). *Am. J. Epidemiol., 143*: 747.

20

The Biology of the Male Reproductive Tract

Michael J. McPhaul
University of Texas Southwestern Medical Center, Dallas, Texas

INTRODUCTION

A great deal has been learned in recent years that permits many aspects of the male reproductive tract to be defined in more than merely descriptive terms. Owing to the breadth of information available, nothing more than a broad outline is possible in a single chapter. Within this limited forum, the goal of this chapter is to outline the process of normal male development as well as the factors and cell types involved in the normal function of the male reproductive tract and to present an overview of the range of disease states that results when these processes are disturbed. Although this discussion draws heavily on information gleaned from the study of genetic mutations that have disrupted the development or function of the male reproductive tract, it is likely that environmental influences could alter these systems at any level to cause disease as well.

DIFFERENTIATION OF THE TESTES AND SEX DETERMINATION

The recognition that the differentiation of the testes is the central event in sexual development dates to the 1940s when gonadectomy and gonadal transplantation experiments in developing animals established that the presence of a functional testes during a specific developmental window was necessary and sufficient for normal male sexual development (reviewed in Jost, 1953). The subsequent recognition of the importance of the inheritance of the Y chromosome to mammalian testicular development led to the concept that the expression of specific genes—determined by the chromosomal composition—determines whether the indifferent gonad will differentiate as a testes or an ovary. In placental mammals, the subsequent steps of sexual development are mediated by the hormones produced as a result of the type of gonadal differentiation that occurs (Jost, 1953; Wilson et al., 1995).

A number of genes were entertained as potential regulators of gonadal differentiation, most notably the HY male transplantation antigen (Wachtel, 1983; McLaren et al., 1984). It was findings from investigations of the pathogenesis of specific forms of sex reversal, however, that led to the first major insights into the molecular mechanisms underlying gonadal differentiation. The advent of sensitive techniques capable of identifying Y-encoded DNA sequences even when Y-chromosomal segments were undetectable by karotypic analyses led to the recognition that the X chromosomes of subjects with a rare syndrome—46,XX males—contained segments of DNA normally located on the Y chromosome. The detailed study of patients with this disorder—and its conceptual antithesis, 46,XY females—using positional cloning techniques led to the identification of a region of the Y chromosome that was consistently present on one of the X chromosomes of 46,XX males and absent or defective in 46,XY females. Such investigations led to the discovery of a gene, termed Sry, that is postulated to be central initiator of testicular differentiation. In transgenic experiments, it has been shown that the expression of a functional Sry gene is both necessary and sufficient to effect testicular differentiation in XX mice (Gubbay et al., 1990; Koopman et al., 1990, 1991; Page et al., 1990; Sinclair et al., 1990).

How the expression of the Sry protein causes the indifferent gonad to differentiate into a testis has not yet been defined. Studies of Sry mRNA expression in rodents indicate that Sry expression precedes morphological differentiation of the gonad into a structure that is recognizable as a testis and that the initial expression of Sry is within Sertoli cells (Koopman et al., 1990). The sequence of the Sry gene predicts a molecule with motifs ("zinc fingers") that suggests a direct role in the modulation of genes important to the process of sexual development. In current models, the presence of the Y chromosome leads to the expression of the Sry protein, which then initiates a cascade of genetic events culminating in testicular differentiation (Lovell-Badge and Hacker, 1995). Inhibition of P-450 aromatase expression and activation of the gene encoding Müllerian inhibiting substance have been proposed as target genes modulated by the action of Sry (Haqq et al., 1993).

In recent years, a number of other genes (in addition to Sry) have been implicated in the development of the testis. In some instances, the effects are specific for the effects on male development, but in others alterations of male phenotypic development are only a component of more global effects.

Steroidogenic factor-1 (SF-1, also known as Ad4BP), an orphan member of the steroid hormone receptor family, was identified as a molecule important for the expression of adrenal steroidogenic P-450s (Morohashi et al., 1992; Honda et al., 1993; Ikeda et al., 1994). Targeted disruption of the SF-1 gene revealed that this molecule plays crucial roles in the differentiation of hormone-producing cells, as mice homozygous for the disrupted SF-1 allele not only showed defects of steroidogenesis, but also demonstrated a virtual absence of a variety of steroidogenic tissues, including the testis (Luo et al., 1994).

While the importance of SF-1 followed from molecular studies performed in cell culture, the importance of the DAX-I gene (DSS-AHC, critical region of the X chromosome, gene 1) was deduced from investigations of a rare form of X-linked congenital adrenal hypoplasia that is often associated with a form of hypogonadotropic hypogonadism. This clinical entity was traced by positional cloning studies to defects of an unusual protein that contains regions with a greater than 30% identity to several members of the steroid hormone receptor family of genes. Despite these homologies, this protein is unusual in that, although it is capable of binding to DNA, it lacks the type of DNA-binding domain that is characteristic of other members of this family. Defects in this gene are clearly responsible for the observed defects of both gonadal and adrenal function (Muscatelli et al., 1994; Zanaria et al., 1994). It is likely that the DAX-1 protein, like SF-1, serves as an important regulator of gonadal development and function.

A number of other molecules have been suggested to play important roles in testicular differentiation, including SOX-9, WT-1, c-kit, and steel factor. The exact roles played by each in the genetic cascade of sexual development are less clear and are the subject of active research. Table 1 summarizes some of what is known about these molecules.

DIFFERENTIATION OF THE MALE GENITAL TRACT

Prior to the eighth week of gestation, the urogenital tracts of male and female human fetuses cannot be discriminated morphologically (Fig. 1). With the functional differentiation of the testes, however, the indifferent urogenital tract is subjected to the actions of two classes of hormones: androgens and Müllerian inhibiting substance (MIS, also known as anti-Müllerian hormone, or AMH).

The actions of androgens on the urogenital tract are complex and present several intriguing biological problems. Both testosterone and 5α-dihydrotestosterone have been shown to bind to the same receptor protein, the androgen receptor (Fig. 2), a member of the steroid hormone receptor family of genes that is encoded on the X chromosome (Chang et al., 1988; Trapman et al., 1988; Tilley et al., 1989). In common with other members of this family, the androgen receptor is comprised of distinct domains (Evans, 1988). The carboxyl-terminal segment of the receptor is necessary and sufficient to bind ligands with high affinity. The central segment of the receptor mediates the sequence-specific binding of the receptor to DNA and is required for the activation of responsive genes. The precise role of the amino terminus of the AR in the activation of genes has not been defined, but it is clear that its presence is required for the ability to maximally stimulate many androgen-responsive genes. Whether this is due to steric effects (e.g., influencing the ability of the receptor to bind to DNA) or to additional points of contact for the interaction of the AR with the transcription apparatus or with distinct coactivators is not yet clear.

The major androgens produced by the Leydig cells of the testes are testosterone and its 5α reduced metabolite, 5α-dihydrotestosterone (DHT) (see below). Although testosterone is the more abundant androgen in blood, in many target tissues 5α-dihydrotestosterone is the principal androgen that is identified bound to the receptor protein in vivo (Anderson and Liao, 1968; Bruchovsky and Wilson, 1968). Moreover, the formation of 5α-dihydrotestosterone is required for normal development of the external genitalia and the prostate.

The interactions of testosterone and DHT with the androgen receptor are somewhat different. The AR binds testosterone with a somewhat lower affinity than it does DHT. In addition, once formed, AR-testosterone complexes dissociate more rapidly than AR-DHT complexes. Nonetheless, these subtle differences seem inadequate to explain the observed biological differences in the actions of these two androgens. At least three different mechanisms are possible explanations for the dichotomy of androgen action. The first is that although DHT and testosterone (T) can act to modulate the activities of many genes to a similar fashion, when complexed to the androgen receptor, they cause the androgen receptor to assume conformations that are distinct enough that some

Table 1 Molecules Participating in the Process of Testicular Differentiation

Molecule	Class	Result of gene defect	Ref.
AR	TF, SHR	Defective virilization of all androgen target tissues; Degree of defect depends on the residual level of AR function	McPhaul, 1995; Quigley et al., 1995
5α-RII		Variable defects in the virilization of selected tissues (e.g., of external genitalia); preserved virilization of Wolffian duct derivatives (e.g., epididymis)	Wilson et al., 1993
Sry	HMG TF	Failure of testicular development in 46XY individuals	Berta et al., 1990; Pontiggia et al., 1994
SF-1	TF	Failure of testicular differentiation; differentiation of hormone-producing cells globally affected	
DAX	TF	Defects result in a form of congenital adrenal hypoplasia; a form of hypogonadotropic hypogonadism often coexists	
MIS	GF	Persistence of Müllerian duct derivatives (uterus, fallopian tubes) in affected 46XY males	Imbeaud et al., 1994
MIS receptor	TmRec	Phenotype similar to subjects with defects of MIS production	Imbeaud et al., 1995
WT-1	TF	Defects in the Wilm's tumor (WT) locus cause two congenital malformation syndromes WAGR (WT, aniridia, genitourinary abnormalities, and mental retardation) and Denys-Drash syndrome (WT, intersex disorders, nephropathy) associated with a failure of urogenital development	Kriedberg et al., 1993; Rauscher, 1993
Steel	GF	Defects in production of steel factor lead to abnormalities in melanocytes, hematopoietic precursors, and germ cells; germ cell defect is believed caused by a reduction of germ cell survival	Besmer et al., 1993
c-Kit	TmRec	Defects of this TM receptor mirror those observed in animals carrying defective steel alleles	Besmer et al., 1993
SOX 9	HMG TF	Defects cause syndrome Campomelic dysplasia, a rare autosomal syndrome of skeletal defects. Abnormal male development often accompanies this disorder and characterized by defects of testicular differentiation	Kwok et al., 1995

TF = Transcription factor; SHR = factor is member of the steroid hormone receptor family of genes; GF = growth factor; TmRec = transmembrane receptor; HMG = high mobility group.

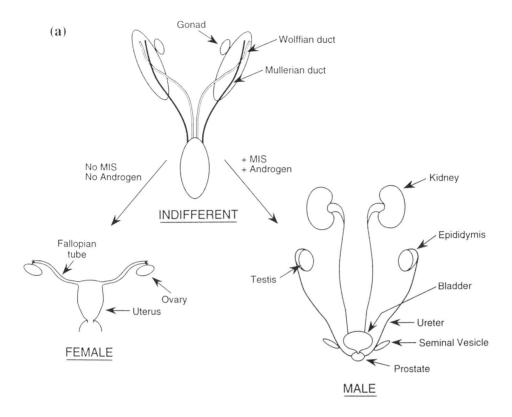

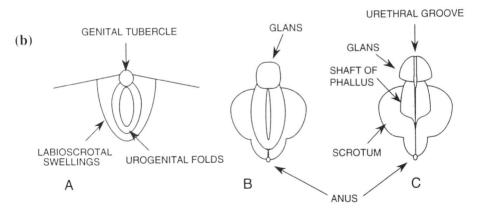

Figure 1 Development of the urogenital tract: (Top) Prior to the differentiation of the testes, the urogenital tracts of male and female embryos are indistinguishable (indifferent state). As the result of the actions of MIS secreted by the testes, the Müllerian ducts involute. In response to the effects of testicular androgens, the Wolffian ducts develop into the epididymis, vas deferens, and seminal vesicles. 5α-DHT formation (catalyzed by 5α-reductase II) is required for the normal development of the prostate. For the sake of clarity, the bladder, uterus, and kidneys have been excluded from the diagram of the developed female urogenital tract. (Bottom) As in the case of the internal structures, the external genitalia of male and female embryos begin as identical rudiments. In response to androgens, the urogenital tubercle enlarges to form the glans, the urogenital folds form the shaft of the phallus, and the labioscrotal swellings fuse to form the scrotum. The normal development of these external structures is dependent on the formation of 5α-DHT by the 5α-reductase II isozyme. (Adapted from Wilson et al., 1995.)

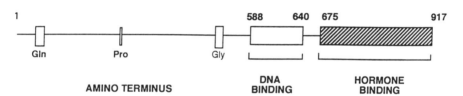

Figure 2 The predicted structure of the human androgen receptor: The androgen receptor, like other steroid receptors, is comprised of distinct domains that mediate the specific binding of DNA and ligand. Although the amino terminus of the receptor is required for full activity in assays of receptor function, how this effect is exerted has not been defined. The relative positions of the three repeated amino acid motifs are indicated. Variations in the lengths of the glycine (Gly) and glutamine (Gln) repeats account for the slightly different sizes of the open reading frames predicted from the nucleotide sequences of the AR cDNAs isolated by different researchers. Expansion of the glutamine repeat has been linked to the pathogenesis of Kennedy's disease (LaSpada et al., 1991), and contraction of this same element has been associated with an increased risk of developing prostate cancer.

genes are regulated selectively by T or DHT. No gene that fits this description has yet been identified. A second theory holds that in some tissues, the tissue concentrations of androgen are limiting. In these tissues, the formation of DHT—which is bound more avidly by the androgen receptor and forms more stable complexes with it (Griño et al., 1990)—leads to a higher concentration of active hormone-receptor complexes. This idea predicts that when studied in cell types in which tissue metabolism is minimal, T and DHT will have similar maximal potencies. Most genes that have been examined show responsiveness to both T and DHT, and the only example studied carefully in cells devoid of metabolism demonstrated a behavior consistent with this type of regulation (Deslypere et al., 1992). Finally, testosterone and DHT differ in the metabolic conversions that they can undergo (e.g., testosterone can be aromatized, DHT cannot). It is conceivable that differences in the conversion of one of these steroids is at the root of these biological differences.

An additional intriguing feature of the action of androgen during the development of the male urogenital tract is the interplay between the stromal and epithelial components of several tissues of the urogenital tract, most notably the prostate and seminal vesicle. Elegant studies by Cunha and coworkers have demonstrated that many of the effects of androgen in stimulating the development of such tissues is indirect. That is, in addition to the signals mediated by the androgen receptor within the cells themselves, important cellular signals pass between the stroma and epithelium (Cunha et al., 1987).

The result of the combined effects of T and DHT on the male urogenital tract is to virilize the Wolffian duct structures and the external genitalia. With respect to the Wolffian ducts, this translates into the development of the Wolffian ducts into the epididymis, vas deferens, and seminal vesicles. In the case of the external genitalia, the actions of androgen (principally 5α-dihydrotestosterone) are required to mediate the enlargement of the genital tubercle to form the phallus and the genital swelling to form the scrotum into which the testes descend. Normal prostatic development, like that of the external genitalia, is dependent on the formation of 5α-dihydrotestosterone by steroid 5α-reductase II.

The second hormone that has been shown to be crucial to normal male phenotypic development is Müllerian inhibiting substance. Jost first postulated the existence of a hormone derived from the testes that mediated the regression of structures derived from the Müllerian ducts during normal male sexual development (Jost, 1953). In the absence of this hormone, the Müllerian ducts would not involute, but instead grow and develop into the uterus and fallopian tubes. Through the use of bioassays that employed organ cultures of fetal urogenital ridges to detect MIS bioactivity, MIS was purified and cDNAs encoding it cloned (Cate et al., 1986; also reviewed in Josso et al., 1993; Lee and Donohoe, 1993). The mature human MIS is a 536-amino-acid protein derived from a larger precursor following cleavage of a signal peptide. This protein undergoes dimerization, glycosylation, and further proteolytic processing. Although the exact roles of the different molecular forms of MIS are not clear, assays of the different cleavage products in bioassays suggest that the bulk of the biological activity resides in the carboxyl-terminal domain of the protein. As deduced from studies of its role in mediating the involution of the Müllerian ducts, mRNA encoding MIS is first detected in the Sertoli cells at the time of testicular differentiation. In the developing male gonad, MIS mRNA levels decrease postnatally but are detectable in the granulosa cells during follicular development postnatally, suggesting an additional role in the development of germ cells (Hirobe et al., 1992, 1994).

AMH exerts its actions via a plasma membrane receptor that is most closely related to the type II receptor for TGF-β (diClemente et al., 1994; Imbeaud, 1995). The deduced amino acid sequence of the receptor predicts a molecule with a single transmembrane domain, which utilizes a cytoplasmic serine/threonine kinase as its intracellular signaling mechanism.

ANDROGENS AND ANDROGEN-BINDING PROTEINS

In addition to the classic intracellular androgen receptor protein, two other proteins were found to be capable of binding androgens with high affinity: one from human serum and the other from the rat epididymis. The first of these was discovered in the late 1960s when a number of groups demonstrated that proteins in human plasma were capable of binding androgens and estrogens with high affinity (reviewed in Westphal, 1971). This binding activity, referred to as sex hormone–binding globulin (SHBG), was found to have affinities and physical properties that distinguished it from the classic intracellular AR. The second of these proteins, androgen-binding protein (ABP), was identified in a completely different source—the rat testis and epididymis—but was found to possess many physical properties that were similar to the human SHBG. While several characteristics suggested a close relationship between these two binding activities, other evidence, particularly immunological, suggested that they were not identical.

The determination of the protein sequence of human SHBG (Walsh et al., 1986) and the determination of the deduced amino acid sequences of rat ABP, human ABP, and rat SHBG (Gershagen et al., 1987; Joseph et al., 1987; Sullivan et al., 1991) clarified the relationships between these proteins. Within a species, it appears that the binding activities (SHBG and ABP) are in fact derived from the same gene and due to proteins with identical primary amino acid sequences (true for both rat and human ABP/SHBG). From the predicted amino acid sequences, rat and human proteins are predicted to be about 68% identical. The difference in physical properties evident between the proteins derived from the epididymis (ABP) and the SHBG from serum (synthesized in the liver) are caused in large part by different oligosaccharide modifications. It is believed that in both molecules a dimer of the protein subunits is the species that binds the steroid, although the exact stoichiometry of this binding is still somewhat uncertain.

The exact physiological role played by these binding proteins remains the subject of considerable interest and uncertainty (discussed in detail in Joseph, 1994). Two major roles have been proposed. The first is that the formation of these steroid-protein complexes influences the availability of steroid hormone for delivery to target tissues. In this model, SHBG/ABP plays a role that is essentially one of sequestration. Considerable evidence exists to support this role, although most of it is of an indirect nature. The second role that has been postulated is that such complexes play a distinct role in the delivery of a signal to specific cell types or tissues. According to this view, SHBG and ABP modulate the activities of the steroids that they bind, either by delivering the steroid hormone to specific cell types or by binding to a specific cell surface receptor that is itself coupled to intracellular second messenger systems. Recent studies using cultured cell model systems suggest that such a receptor for SHBG exists and may signal via the action of cAMP (Nakhla et al., 1994; Fortunati et al., 1996). It is possible that combinations of both of these mechanisms may be operative.

CELLULAR COMPONENTS OF THE ADULT TESTES

The normal testis is a complex organ composed of multiple distinct cell types. On histological grounds, three major cell types have been recognized: Leydig cells, Sertoli cells, and germ cells. The roles of other cell types in the function of the normal testes have been less well studied but have been inferred from more recent studies (Saez et al., 1991). The major histological cell types and their configurations are depicted schematically in Figures 3 and 4.

Leydig Cells

Although Leydig cells were identified first in histological sections in the late 1800s, it was not until the study of partially purified cellular components of the testes became possible that the steroidogenic capacities of the Leydig cells were unambiguously demonstrated. The development of gradient techniques to yield fractions highly enriched in Leydig cells has permitted detailed studies of Leydig cell function to be pursued.

The principal function of Leydig cells is the synthesis of the male hormones, androgens. In mammalian species, the principal steroid produced is testosterone, and in normal men approximately 5 mg of testosterone is secreted by the testes daily. Smaller quantities of 5α-dihydrotestosterone and estradiol are secreted directly by the testes. The remaining fraction of these latter two hormones is found in the blood is derived from extragonadal conversion by the 5α-reductase and P-450 aromatase enzymes, respectively (Ito and Horton, 1971; MacDonald et al., 1971; Walsh et al., 1974).

As depicted in Figure 5, the mature Leydig cell expresses all of the enzymatic machinery required to synthesize testosterone (Payne and Youngblood, 1995). The initial delivery of cholesterol into mitochondria, is a regulated, rate-limiting step in steroidogenesis, which is facilitated by the StAR protein (*St*eroidogenic *A*cute *R*egulatory pro-

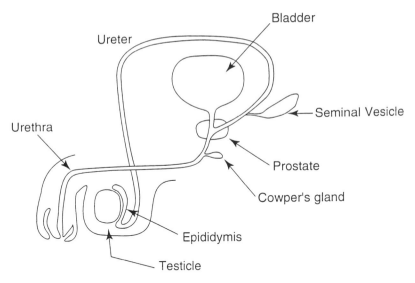

Figure 3 Schematic of the male reproductive tract: The diagram illustrates the locations and the relative positions of the components of the male reproductive tract. The drawing is not drawn to scale.

tein) (Clark et al., 1994). This transfer precedes the initial enzymatic cleavage of cholesterol in the mitochondria to form pregnenolone, the first step in the testosterone synthetic pathway. This and several subsequent reactions are catalyzed by members of the cytochrome P-450 family of enzymes (Waterman et al., 1986). Two of the steps in the synthesis of testosterone are catalyzed by members of different gene families, however. 3β-Hydroxysteroid dehydrogenase is critical for the formation of the 3-one-4-ene structure of the A-ring of testosterone (Keeney and Mason, 1992; Simard et al., 1995). The final step, conversion of androstenedione to testosterone, is catalyzed by a specific form of 17β-hydroxysteroid dehydrogenase, 17β-HSD III (Geissler et al., 1994). The critical role of this specific form of 17-HSD in the synthesis of androgens has been demonstrated by the analysis of patients with male pseudohermaphroditism caused by defects in the enzyme (see below).

Sertoli Cells

The Sertoli cell (Fig. 4) represents a highly specialized cell type that serves a critical role in spermatogenesis. The mature Sertoli cell has a complicated morphology that can be fully appreciated only by analyzing serial sections using electron microscopy. The intricate branching structure of individual cells extends from the basal lamina to the luminal surface of the seminiferous tubule and is crucial to the delicate relationship that exists between the Sertoli cells and the developing germ cells that they envelop and support. The shape of the Sertoli cell is not static, and the structure of individual cells can be seen to vary with the maturation state of the spermatogonia and spermatocytes with which they are associated. An important feature of the Sertoli cell is the existence of tight junctional complexes (occluding junctional complexes) that are localized toward the basal segments of the cells. These junctions ring the seminiferous tubule, are im-

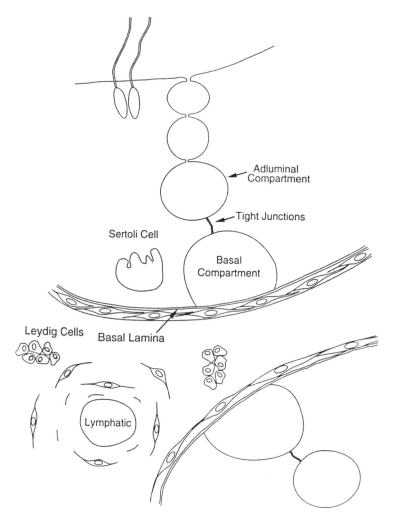

Figure 4 Schematic of the cell types of the testis: The diagram depicts several of the major cell types of the testis. The interstitial spaces are filled with loose connective tissue, Leydig cells, lymphatics, and blood vessels, as indicated. To the upper and lower right, portions of seminiferous tubules are shown, illustrating the adluminal and luminal compartments created by the tight junctional complexes.

permeable to a variety of macromolecules, and serve to divide the tubule into basal and adluminal compartments. The separation serves to sequester the developing spermatocytes and spermatozoa in both a nutritional and immunological sense (Dym and Fawcett, 1970).

Sertoli cells clearly provide more than a structural scaffold and interactions between the Sertoli cells, and the germ cells play a central role in regulating the process of spermatogenesis (Fritz and Murphy, 1993; Griswold, 1995). Owing to this central role, the proteins produced by Sertoli cells have been the object of great interest, and a number of such molecules have been characterized. The functions of many of these proteins are known or suspected.

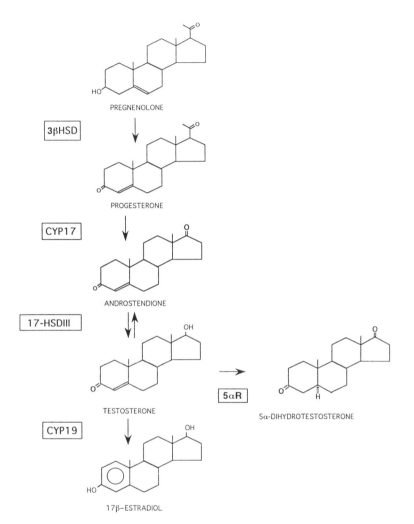

PREGNENOLONE

3βHSD

PROGESTERONE

CYP17

ANDROSTENDIONE

17-HSDIII

TESTOSTERONE

5αR

5α-DIHYDROTESTOSTERONE

CYP19

17β−ESTRADIOL

Figure 5 Steroid synthetic pathways: The steps involved in the biosynthesis of testosterone, 5α-dihydrotestosterone, and estradiol are shown. The reactions are catalyzed by the enzymes indicated: 3β-HSD is 3β-hydroxysteroid dehydrogenase, "17α" is 17α-hydroxylase/17,21-lyase, "17HSDIII" is 17β-hydroxysteroid dehydrogenase type III, "5αR" is steroid 5α-reductase, "Aromatase" is the cytochrome P450 aromatase. Single arrows indicate reactions that are irreversible, and dual arrows represent reversible reactions.

One of the first proteins that was demonstrated to be produced by cultured Sertoli cells was transferrin. Cloning studies established that the transferring produced by the Sertoli cell is identical at the amino acid level to that produced by the liver. Owing to the sequestration of the developing germ cell from the systemic circulation by the occluding junctions, the germ cells are dependent on the Sertoli cells for their supply of iron. Iron is taken up by the Sertoli cells from the blood by receptor-mediated processes, delivered to lysosomes, and repackaged into transferrin synthesized within the Sertoli cell. Iron is then delivered to the developing spermatocytes via the Sertoli-derived transferrin (Sylvester and Griswold, 1994). A similar cycle is believed to control the delivery of copper to the developing germ cells.

Other proteins have been identified that are known or are postulated to play a role in the transport of nutrients and hormones. Androgen-binding protein (ABP) represents an example of this class of proteins that has been carefully studied. As noted above, ABP is a dimeric molecule that is capable of binding both androgens (testosterone or 5α-dihydrotestosterone) or estrogen with high affinity. In contrast to transferrin, ABP is secreted by the Sertoli cell and becomes localized within the luminal compartment of the seminiferous tubule. Within this compartment, ABP has been suggested to play important roles. In model systems, ABP has been shown to enhance the delivery of androgen to selected sites in the male reproductive tract, particularly the caput of the epididymis (Turner and Roddy, 1990; Torres et al., 1995). Immunohistochemical studies have demonstrated the selective internalization of ABP by the epididymal epithelial cells (Pelliniemi et al., 1981). Although such findings suggest that ABP may play important roles in the physiology of sperm maturation of transport, these functions have not been precisely defined.

The synthesis of a number of hormones by the Sertoli cells themselves has also been demonstrated. Among these, MIS and inhibin have been the most intensively studied. The synthesis of MIS is restricted to the Sertoli cell and has been defined on the basis of its role in promoting the regression of the Müllerian duct–derived structures. In addition to this well-established function, MIS may play distinct roles in the reproductive tract (see above). By contrast, inhibin production is not specific for Sertoli cells and is produced by a wide range of gonadal and nongonadal cell types. In the context of the Sertoli cell, however, the production of inhibin—particularly inhibin-B—is believed to be an important component of the feedback loop by which FSH secretion and spermatogenesis are controlled (Illingsworth et al., 1996).

Finally, a number of molecules have been identified that serve functions that are still somewhat uncertain. SGP-1 and SGP-2 serve as examples of such proteins. Sertoli cell–sulfated glycoprotein-1 (SGP-1) is a heavily glycosylated and sulfated protein that was first detected as a major polypeptide secreted by cultured Sertoli cells. Analyses of cDNAs encoding SGP-1 demonstrated a sequence similarity of 67% to human prosaponin (O'Brien and Kishimoto, 1991), a ganglioside binding and transport protein (Hiraiwa et al., 1992). Studies have demonstrated that SGP-1 exists in several forms. One form (approximately 70 kDa) is secreted into the seminiferous tubule lumen, where it is found complexed to spermatozoa, and a second form is localized to lysosomes. Analyses of the patterns of tissue expression revealed that a wide range of tissues express SGP-1. These patterns of expression suggest that both the intracellular and extracellular forms may play a role in the transport of lipids, particularly glycolipids.

If the role of SGP-1 has not been clearly defined, then SGP-2 (also known as "clusterin") is enigmatic because so many different roles have been ascribed to this single protein. Like SGP-1, SGP-2 expression is not unique to Sertoli cells and the mature, heterodimeric protein has been identified in a number of different tissues. This protein is synthesized as a single polypeptide chain precursor that is subsequently cleaved to yield its mature form. This protein is the major protein secreted by cultured rat Sertoli cells and is present in high concentrations in the seminal fluid of a number of species, including human. In the diverse tissues in which it has been identified, it has been suggested to play roles as diverse as participating in the process of apoptosis and facilitating the transport of lipids (Fritz and Murphy, 1993). Although its role in the reproductive tract remains uncertain, roles in the facilitation of fertilization and the maturation of spermatozoa have been suggested.

Germ Cells

In the mammalian embryo, testicular differentiation begins as the result of the interactions of the germs cells with mesenchymal cells of the mesonephros and cells derived from the coelomic epithelium. Importantly, these primordial germ cells are not initially localized to the gonadal ridges, but migrate to these regions from the yolk sac during the fifth week of gestation. During and subsequent to this migration, the germ cells divide and begin to interact with other cell types within the gonadal ridge. It is certain that multiple steps are necessary for this migration to occur normally. While many such influences remain to be defined, it is now clear that the transmembrane receptor, c-kit, and its ligand (steel factor) are necessary for the normal migration and survival of the germ cell population (Besmer et al., 1993).

One of the earliest events in testicular development is the interaction that occurs between the primordial germ cells and the Sertoli cell precursors to form aggregates that fuse to form cord-like structures that later become the seminiferous tubules (Magre and Jost, 1991). As noted above, the differentiation of the Sertoli cell appears to be an early event and may in fact be central to the tubular organization, to the subsequent differentiation of the Leydig cell, and to the onset of steroidogenesis (Burgoyne et al., 1988). After their incorporation into the seminiferous tubules, the primordial germ cells are termed gonocytes. Following a period of mitosis during the differentiation of the testis, the gonocytes divide only very slowly until the onset of puberty (Müller and Skakkebaek, 1983).

Spermatogenesis

Spermatogenesis is the process by which diploid germ cell precursors develop into the highly specialized haploid cells, spermatozoa, that are capable of fertilizing ova. This process involves a balance of meiotic and mitotic cell divisions that ensure the maintenance of adequate numbers of germ cell precursors, while at the same time permitting the production of upward of 2×10^8 haploid spermatozoa daily.

The different germ cells derivatives have been classified on the basis of their morphology as well as their nuclear organization (Burgoyne et al., 1988). As shown in Figure 6, in humans this process can be viewed as a progression of germ cells through a series of six stages. Progression through this cycle 5½ times is required for development and release as a mature spermatozoa, a process taking some 70 days in humans. During this process, cells which begin as the least differentiated precursors (A_d type spermatogonia) are transformed into progressively more mature spermatogonia (A_p and B). After entering the meiotic phase, each preleptotene spermatocyte undergoes a series of meiotic divisions, resulting in the production of four haploid round spermatids. In the final phase of spermiogenesis, the DNA is compacted and structures specialized for motility (flagellum) and for oocyte penetration (acrosome) develop.

In many animals such as rat, the seminiferous epithelium exhibits a "regular" morphology in which the germ cells and Sertoli cells within a tubule segment are all

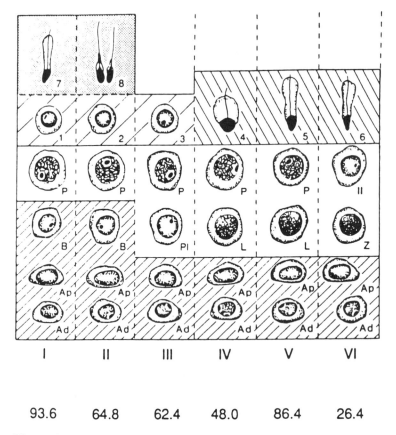

93.6   64.8   62.4   48.0   86.4   26.4

Figure 6 Stages of spermatogenesis: A diagrammatic representation of the different stages of the human spermatogenic cycle. Each vertical column represents the complement of developing germ cells associated with a Sertoli cell of the stage indicated (I to VI). The duration of each of the stages is shown below (in hours). Each germ cell traverses the cycle 5½ times prior to release (at stage II) as a mature spermatozoon. Ad, Ap, and B represent the different types of human spermatogonia (Clermont, 1972). Preleptotene spermatocytes represent the stage at which the developing spermatogonia enter the meiotic phase. L, Z, and P are leptotene, zygotene, and pachytene primary spermatocytes. Following the second meiotic division (II, secondary spermatocyte), the haploid spermatids progress through the three phases of spermatid development shown. Reproduced with permission from R. M. Sharpe in *Physiology of Reproduction* (Knobil and Neill, eds.). Raven Press, New York, 1994.

in approximately the same developmental stage. In humans, a more complex arrangement has been described in which the seminiferous epithelium is organized in such a fashion that waves of cells in the different phases of development are arranged in a coiled or helical fashion (Schulze and Rehder, 1984).

EPIDIDYMIS

The epididymis is a single duct that connects the testes and the vas deferens. The efferent ductules exit the superior pole of the testis and connect to the initial segment of the epididymis. Although referred to as a single entity, the epididymis varies in structure continuously along its length, and at least three regions are often distinguished (caput, corpus, and cauda). The morphology and histology of the epithelium and the thickness of the muscular coat vary dramatically from region to region (Fig. 7) (Yeung et al., 1991).

The epididymis performs at least three functions. Its most basic role is in the transport of the sperm to the urogenital outflow tracts. This transport (estimated to require 2–6 days in the human epididymis) is accomplished by the bulk flow of fluid and by the contraction of the muscular layer that invests the epididymis. Although considerable variability exists, it appears that an inverse relationship exists between the rate of sperm production and the rate of transit through the epididymis (Johnson and Varner, 1988).

Equally fundamental is its role in the storage of sperm. The production rate of spermatozoa by the testes is relatively constant compared to the frequency with which they are needed for reproduction. As such, the epididymis represents a major site of sperm storage, storing as many as 2×10^7 spermatozoa in each epididymis (Haqq et al., 1993). In animal models, sperm has been shown to remain viable within the epididymis for periods measured in weeks.

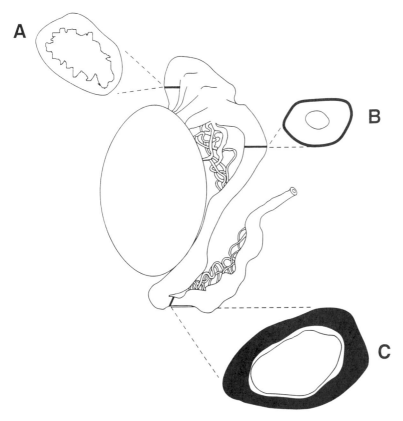

Figure 7   The structure of the epididymis: The structure of the epididymis shows marked regional variations. Examples of these differences are depicted by presentation of the cross sections from the caput (A), corpus (B), and cauda (C). As shown in these cross sections, the thickness of the outer muscular coat varies considerably (depicted in solid black). (Adapted from Baumgarten et al., 1971.)

By far the most intriguing aspect of epididymal function is its role in sperm maturation. Investigations in a number of animal species have clearly established that spermatozoa undergo numerous functional changes as they move through the epididymis, acquiring progressively increased motility and an increased capacity to fertilize ova. Although the basis for this maturation process has not been defined, the secretory products of the epididymal epithelium have been the object of much study, under the assumption that the environment created by the epididymal epithelium actively contributes to this process (Hinton and Palladino, 1995). Investigations have focused on specific protein products (Boue et al., 1992) and molecules that act to alter the composition of the epididymal microenvironment (Cornwall and Hann, 1995). Some of the strongest evidence to support the critical role of the epididymis in the maturation of sperm has come from unexpected quarters. The studies of Sonnenberg-Riethmacher and coworkers (Sonnenberg-Riethmacher et al., 1996) demonstrate that selected alterations in the expression of genes expressed in the epithelium lead to infertility, results that parallel the outcomes of experiments performed using ligation techniques.

Although such investigations leave little doubt that the properties of spermatozoa isolated from different parts of the epididymis possess different functional properties, some investigators have questioned the degree that results from animal studies can be extrapolated directly to humans (Bedford, 1994). This issue has been underscored by the description of men with obstructive azoospermia who have been treated with anastomosis of the vas deferens to the vasa efferentia or even to a seminiferous tubule with resulting fertility. The fertility (albeit reduced) that is observed in such cases may indicate that in humans other parts of the male reproductive tract (e.g., vas deferens or vasa efferentia) may be capable of performing functions that remove an absolute dependence of sperm maturation on epididymal function.

MALE ACCESSORY SEX GLANDS

General

The secondary sex glands of the male reproductive tract (Fig. 3) are an intriguing aspect of the male reproductive tract. Although every species has a series of glands that empty into the principal duct structures of the reproductive tract, the number, structure, and anatomy of these is highly variable from species to species (Price, 1962).

The inconsistent nature of the structure of these glands is paralleled by the variable nature of many of the protein products and composition of their secretions. This variability suggests that these organs subserve roles that are principally mechanical and or nutritive in nature. These inferences are supported by observations from the clinical and research arenas that suggest that these structures (or their effects) are not absolutely required for fertility (see above).

Seminal Vesicles

The seminal vesicles are the most cephalad of the secondary sex glands (Fig. 3). These structures lie posterior to the urinary bladder and superior to the prostate. These paired glands are derived from evaginations of the mesonephric ducts prior to their entry into the urethra.

The histology of these glands reveals a complex branching architecture of the mucosa, which is lined by columnar epithelium. The thin layer of lamina propria upon which the epithelium rests also invests the mucosa and separates it from the muscularis, which is the outermost layer (Aumuller and Adler, 1979).

It is clear from a consideration of the histological features of the seminal vesicles that they are designed for a purpose that is principally secretory in nature. In keeping with this expectation, over 70% of the human ejaculate volume is comprised of the secretory products of the seminal vesicles. A great deal is now known regarding the composition of the seminal vesicle secretions. Despite this knowledge, little is known regarding the role that the components play, other than as components of an environment that is conducive to sperm storage, motility, and viability (Aumuller and Seitz, 1990).

An interesting exception in this regard is semenogelin, a high molecular weight protein that is one of the most abundant in seminal plasma. This protein is believed to play an important role in the coagulation and liquefaction of semen that occurs immediately following ejaculation. The semenogelin protein that is a major component of this clot is a substrate for the action of prostatic specific antigen (PSA), produced within the prostatic epithelial cells and secreted into the seminal fluid by the prostate. This protein is related to kallekreins and enzymatically cleaves the semenogelin, which leads to its subsequent liquefaction (Lilja et al., 1987, 1989). While defects in fertility have not been traced to abnormalities of these steps, these events may serve distinct roles in facilitating sperm movement or fertilization.

Prostate

The prostate is the largest of the human accessory sex glands (Fig. 3). Unlike the seminal vesicle, the prostate is not truly a single gland, but instead is a collection of glands that enter the urethra at different points.

The morphogenesis of the prostate is complex and, like the morphogenesis of the seminal vesicles, involves an interplay between the stromal and epithelial components of the gland. In contrast to the seminal vesicles, however, the normal development of the prostate is dependent on the formation of 5α-DHT, as shown by the absent prostatic development in patients with genetic deficiencies of steroid 5α-reductase (Imperato-McGinley et al., 1974; Walsh et al., 1974; Wilson et al., 1993) or in animals following the administration of 5α-reductase inhibitors (Imperato-McGinley et al., 1985; George and Peterson, 1988).

As in the case of the seminal vesicles, a great deal is known about the prostate and its secretions. Of the proteins produced by the prostate, three have been most carefully studied, however: PSA, prostate membrane specific antigen (PMSA), and prostate specific acid phosphatase (PAP). Although specific roles have been attributed to PSA and PMSA [semen coagulation (Lilja et al., 1987, 1989) and folate hydrolase activity (Heston et al., 1997, respectively), the largest role played by these proteins is as a marker of prostatic cancer.

CONTROL OF TESTICULAR FUNCTION

Normal testicular function involves the coordinated regulation of the major cell types that comprise the testes. This regulation, which is coordinated in a hierarchical fashion, has the hypothalamus as its seat. Cell bodies of the GnRH-secreting neurons are localized to specific regions of the central nervous system, particularly the medial basal hypothalamus and preoptic areas (MBH/POA). Processes of these neurons contact and release GnRH into the long portal vessels, which deliver it to the cells of the anterior pituitary gland. These pulses of GnRH stimulate the gonadotropes of the anterior pituitary to secrete FSH and LH in a pulsatile fashion (Marshall et al., 1992).

As noted below, the activity of the male hypothalamic-pituitary-testicular axis changes dramatically at different stages of development. Following its reactivation during puberty, the hypothalamic pulse generator effects the pulsatile release of the mature GnRH decapeptide into the hypophyseal portal vessels. These pulses of GnRH are delivered to the gonadotropes of the anterior pituitary and activate the high-affinity GnRH receptor. The predicted structure of this receptor [as deduced from cDNAs encoding the human GnRH receptor (Kakar et al., 1992) and other species, such as the rat (Eidne et al., 1992; Kaiser et al., 1992)] indicates that it is composed of seven transmembrane-spanning regions, similar to other G-protein–coupled receptors. The binding of GnRH activates the production of cAMP by adenylate cyclase and stimulates the pulsatile release of FSH and LH into the systemic circulation (Stojilkovic and Catt, 1995). FSH and LH are carried through the blood to target cells localized principally within the testes. It is important to note that the development of the hypothalamus and pituitary are absolute requirements for normal gonadal function. Diseases or conditions that impair the integrity of the hypothalamus or pituitary function can result in the complete absence of gonadal function.

The gene encoding the LH receptor has been characterized and shown to encode a G-protein–coupled receptor predicted to possess seven transmembrane-spanning segments (reviewed in Morales et al., 1995). LH receptors are localized to the surfaces of the Leydig cells, where they can be detected early in embryogenesis (Zhang et al., 1994). The binding of LH to its receptor leads to a stimulation of adenylate cyclase and an increase of intracellular cyclic AMP. Although it is believed that cAMP is a principal intracellular messenger of LH action, the LH receptor is coupled to phospholipase C as well, and the binding of ligand results in the mobilization of intracellular calcium and the accumulation of inositol phosphates. This dual signal is apparently due to the coupling of the LH receptor to both Gs and Gi (Herrlich et al., 1996). The principal effect of the binding of LH to the LH receptor on Leydig cells is the stimulation of testicular steroidogenesis (Luo et al., 1994). The pulsatile release of LH from the pituitary results in the pulsatile secretion of testosterone.

FSH receptors are localized to the Sertoli cells of the testes (Heckert and Griswold, 1991; Bockers et al., 1994). The predicted amino acid sequence of the FSH receptor indicates that it is closely related to the LH receptor. As with the LH receptor, the FSH receptor is believed to transduce signals via both calcium and cAMP-dependent pathways (Gorczynska et al., 1994). FSH stimulation has effects both on Sertoli cell proliferation and on the production of differentiated products (Griswold, 1988).

Feedback inhibition of the hypothalamic-pituitary-testicular axis is effected at the level of both the pituitary and the hypothalamus. Steroid hormones (testosterone either directly or through conversion to estradiol) act to inhibit LH production. Inhibin produced by Sertoli cells—principally inhibin-B—are thought to participate in the regulation of FSH levels (Illingsworth et al., 1996).

PHASES OF TESTICULAR FUNCTION

Testicular function in humans and other mammals varies at different ages and can be divided into several distinct phases, as depicted in Figure 8 (Winter et al., 1976). The

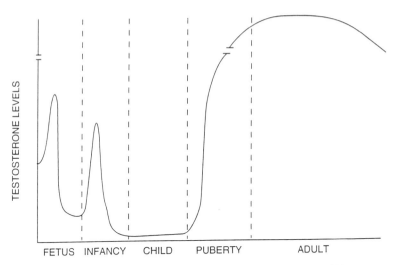

Figure 8   The phases of testicular function: Testis function at different ages as represented by the levels of testosterone measured at each phase of development. (Adapted from Winter et al., 1976.)

first phase occurs during the first weeks of gestation and follows the differentiation of the testes and the maturation the HPG axis. The activity of the testes at this time is responsible for the surge of androgens that effects the differentiation of the male urogenital tract.

During the latter months of gestation it is believed that the rising levels of placental-derived steroids causes the feedback inhibition of gonadotropin secretion, resulting in a decrease in gonadotropin to very low levels by the time of birth. Following birth, the levels of gonadotropins rise again, leading to a neonatal surge of androgen production. This is believed to be due to the removal (at birth) of the inhibitory influences of the maternal steroids upon the fetal GnRH pulse generator. The consequences of this surge in human development are not clear, although in other vertebrates this surge is important for imprinting of sexual behavior.

The neonatal surge ends as inhibitory influences are exerted within the CNS to inhibit the activity of the HPG axis. The nature of this inhibition has remained the subject of some debate, but has been postulated to result from an increased sensitivity of the hypothalamus to gonadal steroids, although some authors have postulated that additional inhibitory mechanisms may be involved. Whatever the mechanisms, the end result is that during this period gonadotropin levels remain very low in the face of the low sex steroid levels that characterize the prepubertal state.

At approximately 8 years of age, the activity of the HPG axis begins to increase. This is first manifested as nocturnal gonadotropin surges, but these gradually increase in amplitude and frequency until adult levels are attained throughout the day. In response to these trophic influences, testicular volume increases, serum testosterone rises into the normal adult range, secondary sexual characteristics develop, and active spermatogenesis can be detected.

Although testicular function does not normally cease abruptly as does ovarian function in women, it is clear that in many normal men a gradual decline in testicular function is observed with advancing age. Histologically, it can be demonstrated that the number of Leydig cells gradually declines. In the studies of Neaves et al. (1984) Leydig cell numbers decreased nearly 50% in older men. These changes in cell number can be correlated with discernible effects, as the levels of serum testosterone decline in older men compared to healthy age-matched control subjects (Bremner et al., 1983; Deslypere and Vermeulen, 1984).

ALTERATIONS OF MALE REPRODUCTIVE TRACT FUNCTION

Investigations of gene defects that alter the development or function of components of the male reproductive tract have been instrumental in defining many of the genes that are critical to this process. Such defects can be selective or global in their effects. The following discussion is not intended to be comprehensive, but is intended instead to illustrate the range of diseases and syndromes that can be caused by defects in specific genes that control or participate in the function or development of the male reproductive tract. As suggested in the final section, it is likely that each level demonstrated to be affected by genetic lesions would also be vulnerable to perturbation by environmental exposures as well.

Disease States

Androgen Resistance (Androgen Insensitivity)

Androgen resistance due to defects in the androgen receptor encompasses a wide spectrum of phenotypes, ranging from individuals in which the androgen receptor is completely inactive to patients in which receptor function is only minimally deranged. Clinical characteristics of each of these different categories are listed in Table 2.

A large number of genetic mutations have been defined in patients with different forms of androgen resistance (McPhaul, 1995; Quigley et al., 1995). As indicated, several different types of genetic defect can give rise to an androgen-resistant pheno-

Table 2 Disorders Caused by AR Defects

Syndrome	Defects of virilization	Types of genetic lesion
Complete testicular feminization (complete androgen insensitivity)	Defects of virilization of Wolffian ducts, prostate, and external genitalia	D, T, AA
Reifenstein syndrome	Defects of external genitalia	AA
Infertility, undervirilization	Normal male development	AA

An overview of the types of syndromes associated with defects of the androgen receptor. Incomplete testicular feminization has been used to describe a phenotype that is intermediate to those that of complete testicular feminization and the Reifenstein syndrome. The term partial androgen insensitivity has been used to describe phenotypes less severe than complete androgen insensitivity. Recently, Quigley and coworkers (1995) proposed a scheme to more precisely codify the degree of defect. D, T, AA denote defects caused by deletions, termination codons, and amino acid substitutions, respectively. Note that to date, only amino acid substitutions have been described to cause the entire range of androgen-resistant phenotypes.

In addition to these classic syndromes caused by defects of the AR, Kennedy syndrome (SBMA) has been traced to an expansion of the glutamine repeat within the amino terminus of the AR (LaSpada et al., 1991). The pathogenesis of this neurodegenerative disorder in which patients also exhibit signs of mild, late onset androgen resistance is not known.

Recent studies of androgen receptor structure in prostate tumors have demonstrated that amino acid substitutions within the hormone binding domain of the AR can also cause that receptor to respond to adrenal hormones or antiandrogens as if to agonists (reviewed in McPhaul, 1995).

type, including mutations that cause premature termination of the receptor protein or single amino acid substitutions within an otherwise intact receptor. A few general rules can now be discerned. First, mutations that cause interruption of the primary amino acid sequence invariably result in the most severely affected phenotype, that of complete testicular feminization. By contrast, amino acid substitutions within the androgen receptor open reading frame can impair androgen receptor function to a variable extent and can result in the entire spectrum of androgen-resistant phenotypes.

Amino acid substitutions are the most frequent type of mutation encountered in the AR gene that cause androgen resistance, and these alterations invariably localize to either the hormone-binding domain or the DNA-binding domain of the receptor protein. The mechanisms by which such amino acid substitutions cause androgen resistance appear to fall into two major categories. Substitutions within the DNA-binding domain of the androgen receptor impair the binding of the receptor to DNA and do not allow the stimulation of responsive genes. The precise mechanism(s) by which amino acid substitutions in the hormone-binding domain impair receptor function are less clear, but the amino acid substitutions that cause androgen resistance appear to do so by blocking the ability of the hormone-binding domain to bind hormone or by destabilizing the hormone-receptor complex. This active hormone-receptor complex is required for the activation of responsive genes, likely by contacting components of the transcription apparatus (Marcelli et al., 1994).

In the androgen resistance syndromes caused by defects in the androgen receptor, testicular differentiation occurs normally. As such, Müllerian duct regression occurs normally and such patients do not have a uterus or fallopian tubes. Histological examination of the testes of patients with more complete forms of testicular feminization reveals seminiferous tubules containing Sertoli cells and immature germ cells but no evidence of spermatogenesis. In less complete forms (such as Reifenstein syndrome) tubules with spermatogenesis are occasionally visualized. In patients with complete forms, the Wolffian ducts do not develop, but in the less severely affected subjects the Wolffian duct derivatives show variable degrees of virilization (O'Leary, 1965; Justrabo et al., 1978). These findings indicate that weak androgen signals are sufficient to permit the development of the Wolffian duct derivatives. In keeping with the central importance of the action of testosterone within the testes in regulating normal spermatogenesis, azoospermia and infertility are major features of all but the least severe forms of androgen resistance.

Defects in the Action of Müllerian Inhibiting Substance

Patients with persistent Müllerian duct syndrome are usually identified during surgery to repair cryptorchidism or an inguinal hernia. This syndrome is characterized by the persistence of structures derived from the Müllerian ducts (uterus, fallopian tubes) in otherwise normal phenotypic men.

The pedigrees of affected families is consistent with an autosomal recessive mode of inheritance. Analysis of patients has suggested a heterogeneous etiology. Assays of serum and measurement of testicular MIS (AMH) mRNA and bioactive MIS levels have demonstrated that defects of MIS production account for the majority of affected patients. Analysis of the MIS gene in these individuals have identified a number of defects in the MIS gene that result in the synthesis of a protein product that is inactive in bioassays (Imbeaud et al., 1994).

These same studies, however, indicate the existence of a number of individuals with an apparently identical phenotype that produce biologically active MIS. Analyses of such patients has permitted the identification of defects within the gene that encodes the receptor for MIS. In fact, the identification of such an affected individual and the correlation of the mutant with the phenotype was instrumental in confirming the identity of the MIS receptor (Imbeaud et al., 1995).

Defects of Leuteinizing Hormone Receptor Function—Leydig Cell Hypoplasia

This disorder is a rare form of male pseudohermaphroditism. Affected patients demonstrate a female phenotype, producing low basal levels of testosterone that do not rise following stimulation with hCG. Although it has been recognized for many years that this entity is caused by a disorder that affects the normal differentiation of the Leydig cell (Berthezene et al., 1976; Brown et al., 1978; Schwartz et al., 1981; Martinez-Mora et al., 1991; Koopman et al., 1991; Kremer et al., 1995), only recently have researchers identified the etiology of this disorder. These studies have established that mutations within specific regions of the LH receptor are the cause of the defects in steroidogenesis. These molecular defects result in the synthesis of defective LH receptors that block the differentiation of the Leydig cell and result in a virtual absence of testosterone synthesis. It is of interest that the antithesis of this syndrome (familial male-limited precocious puberty) is also caused by mutations of the LH receptor (Kremer et al., 1993; Kosugi et al., 1995; Laue et al., 1995), but these mutations lead to an activation of the receptor in the absence of ligand.

Defects of the Follicle-Stimulating Hormone Receptor

While inactivating mutations of the LH receptor lead to dramatic alterations of gonadal function, only a single pedigree has been described in which patients carry inactivating mutations of the FSH receptor. In this family, the biological effects of the mutation on male reproductive function were quite subtle. Aittomaki and coworkers (1995) encountered this family during studies of a form of ovarian dysgenesis. These workers were able to show that the presumptive cause for the ovarian dysfunction was that the affected member carried a mutation that resulted in a single amino acid substitution in the FSH receptor. In vitro studies of the mutant receptor established that it resulted in a demonstrable defect of FSH receptor function. With regard to male reproductive function, these investigators were able to show that several, but not all, of the male patients that harbored this mutant allele were infertile. These results suggest that a functional FSH receptor is not required for fertility, results that are consistent with data from studies that indicate that spermatogenesis is not absolutely dependent on FSH in the adult (Tapanainen et al., 1997).

Disorders in the Synthesis or Metabolism of Hormones

The normal differentiation and function of the testes requires the participation of the products of a number of different genes, only a few of which have been defined. These genes include those involved in the determination of testicular development such as SF-1 and DAX (see above), as well as genes that encode products necessary for normal testicular function.

A number of syndromes have been traced to defects of steroidogenesis. Such defects may result in alteration of steroid metabolism that are tissue selective or which are manifested in all steroidogenesis tissues. Depending on the degree of deficiency and the enzyme affected, there may also be symptoms associated with the steroid precursors that accumulate as a result of the metabolic block. In each of the diseases, the phenotype observed is in keeping with the nature of the hormonal abnormalities that result. Lesions that interfere with the synthesis of androgen in genetically male embryos will lead to forms of male pseudohermaphroditism ranging in severity from minor defects of virilization to a complete female phenotype. Blocks in the synthesis of estrogen (and that leave the synthesis of androgen intact) will lead to the development of a virilized fetus. Alterations in steroidogenesis that lead to the accumulation of androgens or androgen precursors can cause isosexual precocious male development or result in the virilization of females (reviewed in Donohoue et al., 1995).

In addition to blocks in steroid hormone synthesis caused by defects in steroidogenic P-450s, defects in two enzymes, 5α-reductase II and 17β-hydroxysteroid dehydrogenase type 3 (17β-HSD III), have provided important insight into the mechanisms of male development and gonadal function. The discussion below focuses on lesions in these two enzymes.

5-α Reductase

The studies of Anderson and Liao (1968) and Bruchowsky and Wilson (1968) led to the recognition that 5α-DHT was the principal androgen bound to the AR in androgen target tissues, such as the prostate. This observation raised the possibility that specific androgens might play specific roles in normal androgen physiology. These experiments were placed into a physiological context with the description of a form of male pseudohermaphroditism associated with a paucity of urinary 5α-reduced steroid metabolites and reduced 5α-reductase activity in cultured fibroblasts (Imperato-McGinley et al., 1974; Walsh et al., 1974). The genetics of this disorder was consistent with a recessive inheritance, suggesting that the pathogenesis was caused by a deficiency of the 5α-reductase enzyme. The demonstration that the administration of 5α-reductase inhibitors caused a phenocopy of this disorder in rats (Imperato-McGinley et al., 1985; George and Peterson, 1988) led to the concept that the conversion of testosterone to 5α-DHT by steroid 5α-reductase was essential to the normal development of the external genitalia and the prostate in humans and in animals.

The successful cloning of a cDNA encoding a 5α-reductase by Andersson et al. (1989) initiated a new era of androgen physiology. It quickly became apparent that two different enzymes were expressed in the tissues of mammalian species and that these two enzymes differed substantially in their physical properties. The demonstration that these enzymes differed in their distribution of expression led to the expectation that they might serve different physiological functions (Russell and Wilson, 1994).

Subsequent investigations have examined the molecular basis of 5α-reductase deficiency in a large number of patients (Wilson et al., 1993). These studies have established that the clinical phenotype classically described as 5α-reductase deficiency is due to defects in the gene encoding steroid 5α-reductase II. This finding suggests that the steroid 5α-reductase I enzyme plays a separate role in the physiology of androgens, one unrelated to sex determination. Recent studies of mice in which gene encoding steroid 5α-reductase I had been disrupted suggested that this isozyme plays a role in the process initiation of parturition, at least in rodents (Mahendroo et al., 1996). Studies to extend these results to humans have not yet been reported.

17β-Hydroxysteroid Dehydrogenase Type 3

The clinical syndrome that was later traced to a deficiency of 17β-HSD III was recognized in the early 1970s as a rare form of male pseudohermaphroditism caused by a block in the ability to convert androstendione to testosterone (Saez et al., 1971, 1972).

As a result of this enzymatic defect, severe defects of male development are common and affected individuals are usually identified as phenotypic females at birth. Clinical evaluations, however, reveal a 46,XY karotype, testes that are histologically normal in appearance, absent Müllerian duct structures, and virilized Wolffian duct–derived structures.

Cloning studies again provided the basic tools needed to analyze the etiology of this disease. Using an expression-cloning strategy, Geissler and colleagues isolated a third form of 17β-HSD (Geissler et al., 1994). Although clearly related to other isozymes that had been isolated previously (17β-HSD I and II), this enzyme possessed distinctive cofactor requirements, showed a marked tissue specificity of expression (testis), and preferentially catalyzed the conversion of androstendione to testosterone. Importantly, a variety of inactivating mutations could be demonstrated in patients afflicted with this disorder.

Two aspects of this disorder are particularly intriguing. The first is the apparently normal virilization of Wolffian duct–derived structures (epididymis, seminal vesicles, and vas deferens) that is observed in affected individuals, even those in whom the determined genotype predicts a complete or near complete absence of functional 17β-HSD III enzyme. Under such circumstances, the levels of testosterone derived from the pathways accounting for testosterone synthesis by the normal testis should be virtually nonexistent. Such findings suggest either that androstendione itself is capable of virilizing the Wolffian duct structures directly or that androstendione is converted by other 17β-HSD isozymes—locally or in other tissues—to testosterone. That these same structures do not virilize in patients affected with complete defects of the androgen receptor would seem to favor the latter possibility.

The second puzzle presented by this disorder is the age-related alterations that occur in the phenotype of affected individuals. As noted above, virtually all patients are initially identified as female. Despite these assignments, many subjects show appreciable virilization and substantial levels of serum T following the time of puberty. The results of in vitro studies of the expressed mutant 17β-HSD enzyme predicted for such patients and the evaluation of the available clinical data suggest that the increase in serum testosterone that occurs is caused by an increased conversion of androstendione to testosterone in extraglandular sites. Whether this reflects changes in the expression of one of the other 17β-HSD isoenzymes or an increase in substrate availability to the other isozymes remains to be determined (Andersson et al., 1996).

Syndromes of Male Hypogonadism Associated with Low or Inappropriately Normal Levels of Gonadotropins

Although mechanistically distinct, defects in male development could also occur as the result of defects in the hormones or mechanisms controlling gonadal function. Such defects would include mutations that affect the synthesis of GnRH or of the β subunits of FSH or LH.

Strictly speaking, Kallman syndrome is not caused by defects of hormone synthesis or action, but to a disorder that alters the development of the GnRH-secreting neurons. The GnRH neurons migrate from the olefactory placode into several regions of the hypothalamus, particularly the MBH/POA. Kallman syndrome, as classically described in 1944, is the association of hypogonadotropic hypogonadism and anosmia (Kallman et al., 1944). Genetic studies have demonstrated that this disorder is heterogeneous. Forms that show X-linked, autosomal dominant, and autosomal recessive forms have been described (Chaussain et al., 1988). Clues to the types of defect that might be encountered in this disease came from the recognition that the hypothalamic neurons that secrete GnRH migrate from their sites of origin in the olfactory epithelium and that the GnRH-secreting neurons migrate along the olfactory nerves (or closely associated pathways). This concept allowed for explanations that could potentially explain the often dual defects observed in Kallman patients as due to lesions that affected the development of the olfactory system and the GnRH-secreting neurons. Limited pathological studies supported models that postulated lesions causing alterations in the normal migration of the GnRH-secreting neurons (Schwanzel-Fukuda et al., 1989).

Because of its higher frequency, the X-linked form has been the most carefully studied to date. Positional cloning studies localized the defective gene to the long arm of the X-chromosome, and subsequent experiments identified a gene, termed KAL1 (Legouis et al., 1994), that was found to be mutated in patients with the X-linked form of Kallman syndrome (Hardelin et al., 1993).

The predicted protein product encoded by this gene is intriguing in that it predicts a large, presumably extracellular protein (680 amino acids) that contains motifs (four contiguous fibronectin type III repeats) suggesting a role in both cell adhesion and morphogenesis. Further, the predicted amino acid sequence also encodes a motif (WHEY acidic protein 4-disulfide core or WAP) that is contained in a number of proteins with serine protease activity. A similar motif has recently been identified in a factor involved in the differentiation of the urogenital sinus (Rowley et al., 1995). Consistent with a presumed role in its participation in the development and migration of the GnRH neurons, the KAL1 gene is expressed in the CNS of animal models (Legouis et al., 1993) and of humans (Duke et al., 1995) at the appropriate developmental stages. Current models suggest that at these developmental times, the KAL1 gene product plays a critical role in the cells within the CNS that are the targets of the olfactory and terminalis neurons and that the absence of the nerve fibers disrupts the migration of the GnRH neurons (Hardelin and Petit, 1995). It seems likely that the genes mutated in the other genetically distinct forms of Kallman syndrome (i.e., autosomal forms) may affect—directly or indirectly—distinct components involved in the migration and differentiation of the GnRH neurons.

Selective Defects of Spermatogenesis

Selective defects of spermatogenesis can be caused by abnormalities that affect the production, maturation, or transport or of spermatozoa. Examples of the latter two types of disturbance are discussed below.

The Cystic Fibrosis Transmembrane Conductance Regulator

Defects in the transport of spermatozoa are most commonly encountered when the outflow tracts from the testes are blocked. While this is most commonly encountered following ligation of the vas deferens for voluntary sterilization, it is clear that a small number (1–2%) of infertile men have congenital abnormalities of the reproductive tract [congenital bilateral absence of the vas deferens (CBAVD)] that can result in obstructive azoospermia. The observation that this condition was frequently present in men with cystic fibrosis (CF) led to an examination of the gene that causes classic forms of CF (the cystic fibrosis transmembrane conductance regulator (Riordan et al., 1989) to ascertain whether abnormalities of this gene could be identified in CBAVD, even in the absence of a classic CF clinical picture.

Initial studies of patients with CBAVD identified mutations in the CF gene, but in many instances only a single mutant allele was identified. More recent analyses have analyzed a large number of patients with CBAVD and have examined such individuals for allelic variants (particularly the "5T" allele) that have been correlated with the expression of lowered levels of the CFTR mRNA (Chillon et al., 1995).These studies established that a large proportion of affected individuals carry combinations of defective CFTR alleles. As such, it appears that this disorder is a tissue-selective manifestation of defects in the cystic fibrosis transmembrane regulator (CFTR) and represents the mildest clinical expression of defects in the CFTR gene.

DAZ

In 1976, Tiepolo and Zuffardi reported the presence of deletions of the long arm of the Y chromosome in a group of six men with a normal male habitus but with azoospermia (Tiepolo and Zuffardi, 1976). On the basis of these findings, these authors suggested that a specific gene or genes within this region were required for normal spermatogenesis. The establishment of a molecular map of the Y chromosome permitted the identification of genes within these intervals absent in patients with small deletions associated with azoospermia, and several candidate genes were quickly identified. RBM (Ma et al., 1993; Laval et al., 1995) and DAZ (Reijo et al., 1995) are two such genes that have been the object of additional studies, as both have attributes that suggested that they might be integral to specific steps in spermatogenesis. In humans, both genes are encoded on the Y chromosome, expressed in the testis, and are believed—on the basis of motifs within the predicted amino acid structure—to function in the sequestration or processing of RNA.

In many instances, the identification of a gene that is believed to perform a specific function is quickly followed by attempts to demonstrate such a function by targeted expression or deletion in experimental animals. Such experiments have proven difficult to perform with respect to both RBM and DAZ for two reasons. First, both are members of a gene family of related proteins that appear to function as RNA-binding proteins. Second, the Y-chromosomal location that was instrumental in identifying DAZ and RBM in the pathogenesis of azoospermia does not appear to be conserved in other species and may in fact be unique to humans. Instead, a closely related autosomal gene, DAZLA (*DAZ-Like, Autosomal*) has been identified in the mouse (Cooke et al., 1996).

Despite these obstacles, evidence is accumulating that molecules such as DAZ may play important roles in the process of spermatogenesis. In this regard, the strongest evidence has come from the studies of Eberhart et al., who have demonstrated that a gene that appears to be the *Drosophila* homolog of DAZ, *boule*, plays a critical role in spermatogenesis in flies (Eberhart et al., 1996). Although autosomally encoded in the fly, mutations in this protein lead to an inability of germ cells to traverse the checkpoints that control entry into the meiotic metaphase. These investigators were also able to show that this inability to undergo meiotic cell divisions in flies carrying mutant alleles of the *boule* locus could be overcome by the germline transformation using a *boule* transgene expressed under the control of a β_2-tubulin promoter, which is active in postmitotic germ cells. Since such direct experiments have not yet been reported for RBM or related genes, their functions in the process of spermatogenesis remain to be formally demonstrated (Burgoyne, 1996).

Disorders of the Male Reproductive Tract Resulting from Environmental Exposures

Cells of any lineage can be susceptible to selective forms of toxicity, and such damage can occur in response to at least two types of agents. In one instance, the apparent selective toxic effects are the reflection of the action of an agent that has toxicity for many cell types, such as DNA-damaging agents. In these instances, the damage may be more evident in specific cell types owing to the high rate of cell turnover or because the pharmacokinetics of drug delivery or excretion cause selected tissues to be exposed to higher compound concentrations. In other circumstances, a molecule appears to have a selective effect in tissues that reflect specific aspects of their biology. In addition to these more "classic" forms of toxicity, the dependence of testicular function on the endocrine system adds an additional form of damage in which the function of selected tissues can be altered—either permanently or reversibly—by the action of compounds to perturb reproductive function. Examples of each type have been described for the male reproductive tract.

Toxic Effects

One of the best characterized environmental exposures affecting the function of the male reproductive tract was reported by Whorton et al. in 1977. These authors investigated the association of oligospermia and azoospermia in pesticide workers exposed to a variety of chemicals in a manufacturing facility. Review of the exposures and clinical status of the workers suggested that individuals with severely depressed sperm counts differed from those with normal sperm counts in that the former had longer exposures to the pesticide dibromochloropropane (DBCP). Clinical studies suggested that the azoospermic men had higher levels of LH and FSH, but that their levels of testosterone were indistinguishable. The results of testicular biopsies indicated the selective loss of spermatogonia with preservation of Sertoli and Leydig cells (Biara et al., 1978; Potashnik et al, 1979). In humans, these changes are at best only partially reversible (Potashnik and Porath, 1995). Although similar changes have been observed in rats exposed to this compound (Torkelson et al., 1961), the mechanism for this selective toxicity has never been identified.

A number of additional agents, including heavy metals, have been described to have discernible effects on the male reproductive tract. Such toxicities are discussed in detail in subsequent chapters.

Direct Hormonal Effects—Antiandrogens

Recent studies have explored an additional type of environmental exposure with selective effects on the male reproductive tract that has an endocrine basis. In 1950, Burlington and Lindeman (1950) observed that DDT injected into male chickens inhibited the development of the comb and wattle and caused a marked reduction of testicular size. Owing to the structural similarity of DDT to compounds with estrogenic activity (particularly to diethylstilbesterol), these studies were interpreted as reflecting the estrogenic properties of DDT or its metabolites. Subsequent experiments by Wakeling and Visek (1973) suggested that another mechanism might be operative, in that their studies demonstrated that the binding of labeled androgen by the androgen receptor of the rat prostate could be effectively competed by o,p-DDT, a DDT metabolite.

These issues have now been clarified. Kelce and coworkers (1995) have shown that naturally occurring metabolites of DDT possess substantial activities as anti-androgens. This effect appears to be mediated directly via competition at the level of the AR itself and to occur at concentrations that are biologically achievable. As such, environmental exposures of this type may be contributors to several phenomena, including effects that have been observed in avian populations, a role in mediating the decline in sperm counts that have been reported in recent decades, and the increased frequency of developmental defects of male differentiation (Kelce et al., 1995; Wong et al., 1995).

Direct Hormonal Effects—Androgenic

As noted above, both male and female embryos possess androgen receptors. As such, female embryos can be virilized by exposure to compounds that possess androgenic activities. Although virilization could result from exposure to androgens, in recent times such circumstances are unusual owing to the duration and level of exposure required. For this reason, such developmental defects are usually caused by drug exposures (such as danazol) Brunskill, 1992, androgen-producing tumors, or defects of steroidogenesis that result in the accumulation of steroid metabolites with androgenic properties or that can be converted to androgens (Donohoue et al., 1995).

While only rarely encountered as an inadvertent exposure in adults, anabolic steroids are commonly abused in order to enhance body-building efforts or to improve athletic performance. A variety of agents—both parenteral and oral—are available through either legal means or via clandestine sources (Wilson, 1988). Most drugs used for this purpose are androgens that are substituted at various positions on the steroid nucleus (primarily the 1 or 17 positions) to disrupt the metabolism that would occur if no modifications were present. Although the extent to which such drugs enhance body building remains controversial, the compounds are believed to exert most—if not all—of their effects via the same androgen receptor protein that mediates the effects of the physiological concentrations of androgen present in normal males. In addition to this more classic mechanism of action for this class of androgenic compounds, some authors have noted that 17-substituted androgens have the capacity to compete for glucocorticoid binding at the level of the glucocorticoid receptor and have suggested that some component of their actions might be caused by their capacity to inhibit the catabolic effects of gluocorticoids (Raaka et al., 1989).

The untoward effects noted as a result of these compounds are generally related to the effects of supraphysiological levels of androgen. Excessive dosing with drugs of this class that are capable of being aromatized may cause gynecomastia and breast tenderness as the result of excessive estrogen formation. A far more difficult problem is encountered when patients stop using these drugs. Owing to the inhibition of the hypothalamic-pituitary axis effected by the phamacological doses of these agents, testicular atrophy occurs. Furthermore, the secondary hypogonadism induced by these potent drugs can be quite profound and may result in prolonged periods of symptomatic hypogonadism.

SUMMARY

It is clear that considerable progress has been made in defining the mechanisms that contribute to the normal development of the male phenotype and the influences that govern the normal function of the male reproductive tract. It is certain that the coming years will see a continuation of this discovery process and a wealth of information that can be applied to the diagnosis and treatment of diseases that affect male development or male reproductive function.

When viewed from the perspective of the toxicologist, however, this proliferation of information can be quite daunting. The definition of the myriad components that comprise or influence the function of the male reproductive tract make it clear that exposures may exert their effects in a number of different ways and at a number of different levels. The development of methods to identify such potentially harmful effects prior to observing the end results in human populations will present an increasingly challenging task.

REFERENCES

Aittomaki, K., Lucena, J. L., Pakarinen, P., Sistonen, P., et al. (1995). *Cell, 82*: 959.

Anderson, K. M., and Liao, S. (1968). *Nature, 219*: 277.

Andersson, S., Bishop, R. W., and Russell, D. W. (1989). *J. Biol. Chem., 264*: 167249.

Andersson, S., Russell, D. W., and Wilson, J. D. (1996). *Trends Endocrinol. Metab., 7*: 121.

Aumuller, G., and Adler, G. (1979). *Cell Tissue Res., 198*: 145.

Aumuller, G., and Seitz, J. (1990). *Int. Rev. Cytol., 121*: 127.

Baumgarten, H. G., Holstein, A. F., and Rosengren, E. (1971). *Z. Zellforsch., 120*: 37.

Bedford, J. M. (1988). *Ann. NY Acad. Sci., 541*: 284.

Bedford, J. M. (1994). *Hum. Reprod., 9*: 2187.

Berta, P., Hawkins. J. R., Sinclair, A. H., Taylor, A., Griffiths, B. L., Goodfellow, P. N., and Fellous, M. (1990). *Nature, 348*: 448.

Berthezene, F., Forest, M. G., Grimaud, J. A., Claustrat, B., and Mornex, R. (1976). *N. Engl. J. Med., 295*: 969.

Besmer, P., Manova, K., Duttlinger, R., Huang, E. J., Packer, A., Gyssler, C., and Bachvarova, R. F. (1993). *Development* (suppl.): 125.

Biava, C. G., Smuckler, E. A., and Whorton, D. (1978). *Exp. Mol. Pathol., 29*: 448.

Blaquier, J. A., Cameo, M. S., Cuasnicu, P. S., Gonzalez Echeverria, M. F., Pineiro, L., and Tezon, J. G. (1988). *Ann. NY Acad. Sci., 541*: 292.

Bockers, T. M., Nieschlag, E., Kreutz, M. R., and Bergmann, M. (1994). *Cell Tiss. Res., 278*: 595.

Boue, F., Lassalle, B., Duquenne, C., Villaroya, S., Testart, J., Lefevre, A., and Finaz, C. (1992). *Mol. Reprod. Dev., 33*: 470.

Bowen, P., Lee, C. S. N., Migeon, C. J., Kaplan, N. M., Whalley, P. J., McKusick, B. A., and Reifenstein Jr., E. C. (1965). *Squibb Inst. Med. Res. 62*(2): 252.

Bremner, W. J., Vitiello, M. V., and Prinz, P. N. (1983). *J. Clin. Endocrinol. Metab., 56*: 1278.

Breton, S., Smith, P. J., Lui, B., and Brown, D. (1996). *Nature Med., 2*: 470.

Brown, D. M., Markland, C., and Dehner, L. P. (1978). *J. Clin. Endocrinol. Metab., 46*: 1.

Bruchovsky, N. and Wilson, J. D. (1968). *J. Biol. Chem., 243*: 2012.

Brunskill, P. J. (1992). *Br. J. Obstet. Gynecol., 99*: 212.

Burgoyne, P. S. (1996). *Nature, 381*: 740.

Burgoyne, P. S., Buehr, M., Koopman, P., Rossant, J., and McLaren, A. (1988). *Development, 102*: 443.

Burlington, H., and Lindeman, V. F. (1950). *Proc. Soc. Exp. Biol. Med., 74*: 48.

Casals, T., Bassas, L., Romero, J. R., Chillon, M., Gimenez, J., Ramos, M. D., Tapia, G., Narvaez, H., et al. (1995). *Hum. Genet., 95*: 205.

Cate, R. L., Mattaliano, R. J., Hession, C., Tizard, R., et al. (1986). *Cell, 45*: 685.

Chang, C., Kokontis, J., and Liao, S. (1988). *Proc. Natl. Acad. Sci. USA, 85*: 7211.

Chaussain, J. L., Toublanc, J. E., Feingold, et al. (1988). *Hormone Res., 29*: 202.

Chillon, M., Casals, T., Mercier, B., Bassas, L., Lissens, W., Silber, S., Romey, M., Ruiz-Romero, J., et al. (1995). *N. Engl. J. Med., 332*: 1475.

Clark, B. J., Wells, J., King, S. R., and Stocco, D. M. (1994). *J. Biol. Chem., 269*: 28314.

Clermont, Y. (1972). *Physiol. Rev., 52*: 198.

Cooke, H. J., Lee, M., Kerr, S., and Ruggiu, M. (1996). *Hum. Mol. Genet., 5*: 513.

Cornwall, G. A. and Hann, S. R. (1995). *Mol. Reprod. Dev., 41*: 37.

Cunha, G. R., Donjacour, A. A., Cooke, P. S., et al. (1987). *Endo. Rev., 8*: 338.

Deslypere, J. P., and Vermeulen, A. (1984). *J. Clin. Endocrinol. Metab., 59*: 955.

Deslypere, J. P., Young, M., Wilson, J. D., and McPhaul, M. J. (1992). *Mol. Cell Endocrinol., 88*: 15.

diClemente, N., Wilson, C., Faure, E., et al. (1994). *Mol. Endocrinol., 8*: 1006.

Donohoue, P. A., Parker, K., and Migeon, C. J. (1995). *The Metabolic Basis of Inherited Disease* (C. R. Scriver, et al., eds.). McGraw-Hill, New York, p. 2929.

Duke, V. M., Winyard, P. J., Thorogood, P., Soothill, P., Bouloux, P. M., and Woolf, A. S. (1995). *Mol. Cell Endocrinol., 110*: 73.

Dym, M., and Fawcett, D. W. (1970). *Biol. Reprod., 3*: 308.

Eberhart, C. G., Maines, J. Z., and Wasserman, S. A. (1996). *Nature, 381*: 783.

Eidne, K. A., Sellar, R. E., Couper, G., Anderson, L., and Taylor, P. L. (1992). *Mol. Cell Endocrinol., 90*: R5.

Evans, R. M. (1988). *Science, 240*: 889.

Feldman, M., Lea, O. A., Petrusz, P., Tres, L. L., Kierszenbaum, A. L., and French, F. S. (1981). *J. Biol. Chem., 256*: 5170.

Fortunati, N., Fissore, F., Fazzari, A., Becchis, M., et al. (1996). *Endocrinology, 137*: 686.

Fritz, I. B. (1994). *Ciba Found. Symp., 182*: 271.

Fritz, I. B., and Murphy, B. (1993). *Trends Endocrinol. Metab., 4*: 41.

Geissler, W. M., Davis, D. L., Wu, L. et al. (1994). *Nature* Genetics 7: 34.

George, F. W., and Peterson, K. (1988). *Endocrinology, 122*: 1159.

Gershagen, S., Fernlund, P., and Lundwall, A. (1987). *FEBS Lett., 220*: 129.

Gorczynska, E., Spaliviero, J., and Handelsman, D. J. (1994). *Endocrinology, 134*: 293.

Gottlieb, B. (1997). http: www.mcgill.ca/androgenb

Griño, P. B., Griffin, J. E., and Wilson, J. D. (1990). *Endocrinology, 126*: 1165.

Griswold, M. D. (1988). *Int. Rev. Cytol., 110*: 133.

Griswold, M. D. (1995). *Biol. Reprod., 52*: 211.

Gubbay, J., Collignon, J., Koopman, P., Capel, B., Economou, A., Munsterberg, A., Vivian, N., Goodfellow, P., Lovell-Badge, R. (1990). *Nature, 346*: 245.

Haqq, C. M., King, C. Y., Donahoe, P. K., and Weiss, M. A. (1993). *Proc. Natl. Acad. Sci. USA, 90*: 1097.

Hardelin, J. P., Levilliers, J., Blanchard, S., Carel, J. C., Leutenegger, M., Pinard-Bertelletto, J. P., Bouloux, P., and Petit, C. (1993). *Hum. Mol. Genet., 2*: 373.

Hardelin, J. P., and Petit, C. *Bailliere's Clin. Endocrinol. Metab., 9*: 489.

Heckert, L. L., and Griswold, M. D. (1991). *Mol. Endocrinol., 5*: 670.

Herrlich, A., Kuhn, B., Grosse, R., Schmid, A., Schultz, G., and Gudermann, T. (1996). *J. Biol. Chem., 271*: 16764.

Heston, W. D. (1997). *Urology, 49* (suppl. 3A): 104.

Hinrichsen, M. J., and Blaquier, J. A. (1980). *J. Reprod. Fertil., 60*: 291.

Hinton, B. T., and Palladino, M. A. (1995). *Microsc. Res. Technique, 30*: 67.

Hiraiwa, M., Soeda, S., Kishimoto, Y., and O'Brien, J. (1992). *Proc. Natl. Acad. Sci.* USA, 89: 11254.

Hirobe, S., He, W-W., Lee, M. M., and Donahoe, P. K. (1992). *Endocrinology, 131*: 854.

Hirobe, S., He, W. W., Gustafson, M. L., MacLaughlin, D. T., and Donahoe, P. K. (1994). *Biol. Reprod., 50*: 1238.

Honda, S., Morohashi, K., Nomura, M., Takeya, H., Kitajima, M., and Omura, T. (1993). *J. Biol. Chem., 268*: 7494.

Igdoura, S. A., Hermo, L., Rosenthal, A., and Morales, C. R. (1993). *Anat. Rec., 235*: 411.

Igdoura, S. A., and Morales, C. R. (1995) *Mol. Reprod. Dev., 40*: 91.

Ikeda, Y., Shen, W., Ingraham, H. A., and Parker, K. L. (1994). *Mol. Endocrinol., 8*: 654.

Illingsworth, P. J., Groome, N. P., Byrd, W., Rainey, W. E., McNeilly, A. S., Mather, J. P., and Bremner, W. J. (1996). *J. Clin. Endocrinol., Metab., 81*: 1321.

Imbeaud, S., Carre-Eusebe, D., Rey, R., Belville, C., Josso, N., and Picard, J.-Y. (1994). *Hum. Mol. Gen., 3*: 125.

Imbeaud, S., Faure, E., Lamarre, I. et al. (1995). *Nature Genet., 11*: 382.

Imperato-McGinley, J., Guerrero, L., Gautier, T., and Peterson, R. E. (1974). *Science, 186*: 1213.

Imperato-McGinley, J., Binienda, Z., Arthur, A., et al. (1985). *Endocrinology, 116*: 807.

Ito, T., and Horton, R. (1971). *J. Clin. Invest., 50*: 1621.

Jenkins, E. A. P., Andersson, S., Imperato-McGinley, J., Wilson, J. D., and Russell, D. W. (1992). *J. Clin. Invest., 89*: 293.

Johnson, L., and Varner, D. D. (1988). *Biol. Reprod., 39*: 812.

Joseph, D. R. (1994). *Vitam. Horm., 49*: 197.

Joseph, D. R., Hall, S. H., and French, F. S. (1987). *Proc. Natl. Acad. Sci. USA, 84*: 339.

Josso, N., Cate, R. L., Picard, J. Y., Vigier, B., di Clemente, N., Wilson, C., Imbeaud, S., Pepinsky, R. B., Guerrier, D., Boussin, L., et al. (1993). *Recent Prog. Horm. Res., 48*: 1.

Jost, A. (1953). *Recent Progress in Hormone Research.* Academic Press, New York, p. 379.

Justrabo, E., Cabanne, F., Michiels, R., Bastien, H., Dusserre, P., Pansiot, F., and Cayot, F. (1978). *Pathology, 126*: 165.

Kaiser, U. B., Zhao, D., Cardona, G. R., and Chin, W. W. (1992). *BBRC, 189*: 1645.

Kakar, S. S., Musgrove, L. C., Devor, D. C., Sellers, J. C., and Neill, J. D. (1992). *BBRC, 189*: 289.

Kallman, F. J., Schoenfeld, W. A., and Barrera, S. E. (1944). *Am. J. Mental Def., XLVIII*: 203.

Karsch-Mizrachi, I., and Haynes, S. R. (1993). *Nucl. Acids Res., 21*: 2229.

Keeney, D. S., and Mason, J. I. (1992). *Endocrinology, 130*: 2007.

Kelce, W. R., Stone, C. R., Laws, S. C., Gray, L. E., Kemppalnen, J. A., and Wilson, E. M. (1995). *Nature, 375*: 581.

Kipen, H. M., and Zuber, C. (1994). *Ann. NY Acad. Sci., 736*: 58.

Koopman, P., Munsterberg, A., Capel, B., Vivian, N., and Lovell-Badge, R. (1990). *Nature, 348*: 450.

Koopman, P., Gubbay, J., Vivian, N., Goodfellow, P. N., Lovell-Badge, R. (1991). *Nature, 351*: 117.

Kosugi, S., Van Dop, C., Geffner, M. E., Rabi, W., Carel, J.-C., Chaussain, J.-L., Mori, T., Merendino, Jr., J. J., and Shenker, A. (1995). *Hum. Molec. Gen., 4*: 183.

Krasovskii, G. N., Varshavskaya, S. P., and Borisov, A. I. (1976). *Environ. Health Perspect, 13*: 69.

Kremer, H., Mariman, E., Otten, B. J., Moll, G. W., Stoelinga, G. B. A., et al. (1993). *Hum. Molec. Gen., 2*: 1779.

Kremer, H., Kraaij, R., Toledo, S., Post, M., Fridman, J. B., Hayashida, C. Y., van Reen, M., Milgrom, E., Ropers, H.-H., et al. (1995). *Nature Genet., 9*: 160.

Kriedberg, J. A., Sariola, H., Loring, J. M., Maeda, M., Pelletier, J., Housman, D., and Jaensisch, R. (1993). *Cell, 74*: 679.

Kwok, C., Weller, P. A., Guioli, S., Fosterr, J. W., Mansour, S., et al. (1995). *Am. J. Hum. Genet., 57*: 1028.

Lancranjan, I., Popescuu, H. I., Gavanescu, O., Klepsch, I., and Serbanescu, M. (1975). *Arch. Environ. Health, 30*: 396.

LaSpada, A. R., Wilson, E. M., Lubahn, D. B., et al. (1991). *Nature, 352*: 77.

Latronico, A. C., Anasti, J., Arnhold, I. J. P., Rapaport, R., Mendonca, B. B., Bloise, W., Castro, M., Tsigos, C., and Chrousos, G. P. (1996). *N. Engl. J. Med., 334*: 507.

Laue, L., Chan, W.-Y., Hsueh, A. J. W., Kudo, M., et al. (1995). *Proc. Natl. Acad. Sci. USA,* *92*: 1906.

Laval, S. H., Glenister, P. H., Raspberry, C., Thornton, C. E., Mahadevaiah, S. K., Cooke, H. J., et al. (1995). *Proc. Natl. Acad. Sci. USA, 92*: 10403.

Lee, M. M., and Donahoe, P. K. (1993). *Endocrinol. Rev., 14*: 152.

Legouis, R., Lievre, C. A., Leibovici, M., Lapointe, F., and Petit, C. (1993) *Proc. Natl. Acad. Sci. USA, 90*: 2461.

Legouis, R., Cohen-Salmon, M., Del Castillo, I., and Petit, C. (1994). *Biomed. Pharmacother., 48*: 241.

Lilja, H., Oldbring, J., Rannevik, G., and Laurell, C.-B. (1987). *J. Clin. Invest., 80*: 281.

Lilja, H., Abrahamsson, P.-A., and Lundwall, A. (1989). *J. Biol. Chem., 26*: 1894.

Lovell-Badge, R., and Hacker, A. (1995). *Phil. Trans. R. Soc. London, Series B: Biol Sci., 350*: 205.

Lubahn, D. B., Joseph, D. R., Sar, M., et al. (1988). *Mol. Endocrinol., 2*: 1265.

Luo, X., Ikeda, Y., Parker, K. L. (1994). *Cell, 77*: 481.

Ma, K., Inglis, J. D., Sharkey, A., Bickmore, W. A., Hill, R. E., et al. (1993). *Cell, 75*: 1287.

MacDonald, P., Grodin, J., and Siiteri, P. (1971). *Control of Gonadal Steroid Secretion* (D. Baird and Strong, eds.). Williams and Wilkins, Baltimore, p. 158.

Magre, S., and Jost, A. (1991). *J. Electron. Micro. Technol., 19*: 172.

Mahendroo, M. S., Cala, K. M., and Russell, D. W. (1996). *Mol. Endocrinol., 10*: 380.

Marcelli, M., Zoppi, S,. Wilson, C. M., Griffin, J. E., and McPhaul, M. J. (1994). *J. Clin. Invest., 94*: 1642.

Marshall, J. C., Dalkins, A. C., Haisenleder, D. J., and Griffin, M. L. (1992). *Am. Clin. Climatol. Assoc., 104*: 31.

Martinez-Mora, J., Saez, M. J., Toran, N., Isnard, R., Perez-Iribarne, M. M., Egozcue, J., and Audi, L. (1991). *Clin. Endocrinol., 34*: 485.

McLaren, A., Simpson, E., Tomonari, K., Chandler, P., and Hohh, H. (1984). *Nature, 312*: 552.

McPhaul, M. J. (1995). *Molecular Endocrinology* (Weintraub, B. D., ed.). Raven Press, New York, p. 411.

Morales, C. R., el-Alfy, M., Zhao, Q., and Igdoura, S. A. (1995). *Histol. Histopathol., 10*: 1023.

Morales, C. R., el-Alfy, M., Zhao, Q., and Igdoura, S. A. (1996). *J. Histochem. Cytochem., 44*: 327.

Morohashi, K., Honda, S., Inomata, Y., Handa, H., and Omura, T. (1992). *J. Biol. Chem., 267*: 17913.

Morse, H. C., Horike, N., Rowley, M. J., Heller, C. G. (1973). *J. Clin. Endocrinol. Metab., 37*: 882.

Müller, J., and Skakkebaek, N. E. (1983). *Int. J. Androl., 6*: 143.

Münsterberg, A., and Lovell-Badge, R. (1991). *Development, 113*: 613.

Muscatelli, F., Strom, T. M., Walker, A. P., et al. (1994). *Nature, 372*: 672.

Nakhla, A. M., Khan, M. S., Romas, N. P., and Rosner, W. (1994). *Proc. Natl. Acad. Sci. USA, 91*: 5402.

Neaves, W. B., Johnson, L., Porter, J. C., Parker, C. R., Jr., and Petty, C. S. (1984). *J. Clin. Endocrinol. Metab., 59*: 756.

O'Brien, J. S., and Kishimoto, Y. (1991). *FASEB J., 5*: 301.

O'Leary, J. A. (1965). *Fertil. Steril., 16*: 813.

Page, D. C., Fisher, E. M. C., McGillivray, B., and Brown, L. G. (1990). *Nature, 346*: 279.

Payne, A. H., and Youngblood, G. L. (1995). *Biol. Reprod., 52*: 217.

Pelliniemi, L. J., Dym, M., Gunsalus, G. L., Musto, N. A., Bardin, C. W., and Fawcett, D. W. (1981). *Endocrinology, 108*: 925.

Pontiggia, A., Rimini, R., Harley, V. R., Goodfellow, P. N., Lovell-Badge, R., and Bianchi, M. E. (1994). *EMBO J., 13*: 6115.

Potashnik, G., and Porath, A. (1995). *J. Occ. Environ. Med., 37*: 1287.

Potashnik, G., Yanai-Ibar, I., Sacks, M. I., and Israeli, R. (1979). *Isr. J. Med., 15*: 438.

Price, D. (1962). *Natl. Cancer Inst. Monogr., 12*: 1.

Quigley, C. A., De Bellis, A., Marschke, K. B., el-Awady, M. K., Wilson, E. M., and French, F. S. (1995). *Endocr. Rev., 16*: 271.

Raaka, B. M., Finnerty, M., and Samuels, H. H. (1989). *Mol. Endocrinol., 3*: 332.

Rauscher, F. J. (1993). *FASEB J., 7*: 896.

Reijo, R., Lee, T. Y., Salo, P., Alagappan, R., Brown, L. G., Rosenberg, M., Rozen, S., Jaffe, T., Straus, D., Hovatta, O., et al. (1995). *Nature Genet., 10*: 383.

Reijo, R., Alagappan, R. K., Patrizio, P., and Page, D. C. (1996). *Lancet 347*: 1290.

Reijo, R., Seligman, J., Dinulos, M. B., Jaffe, T., Brown, L. G., Distecge, C. M., and Page, D. C. (1996). *Genomics 35*: 346.

Riordan, J. R., Rommens, J. M., Kerem, B., et al. (1989). *Science, 245*: 1066.

Rowley, D. R., Dang, T. D., Larsen, M., Gerdes, M. J., McBride, L., and Lu, B. (1995). *J. Biol. Chem., 270*: 22058.

Russell, D. W., and Wilson, J. D. (1994). *Ann. Rev. Biochem., 63*: 25.

Russell, L. D. (1980). *Gamete Res., 3*: 179.

Saez, J. M., dePeretti, E., Morera, A. M., David, M., Bertrand, J. (1971). *J. Clin. Endocrinol. Metab., 32*: 604.

Saez, J. M., Morera, A. M., dePeretti, E., and Bertrand, J. (1972). *J. Clin. Endocrinol. Metabl., 34*: 598.

Saez, J. M., Avallet, O., Lejeune, H., and Chatelain, P. G. (1991). *Horm. Res., 36*: 104.

Savitz, D. A., and Chen, J. (1990). *Environ. Health Perspect., 88*: 325.

Schulze, W., and Rehder, U. (1984). *Cell Tissue Res., 237*: 395.

Schwanzel-Fukuda, M., Bick, D., and Pfaff, D. W. (1989). *Mol. Brain Res., 6*: 311.

Schwartz, M., Imperato-McGinley, J., Peterson, R. E., Cooper, G., Morris, P. L., MacGillivray, M., and Hensle, T. (1981). *J. Clin. Endocrinol. Metab., 53*: 123.

Segaloff, D. L., and Ascoli, M. (1993). *Endocrinol. Rev., 14*: 324.

Shemi, D., Sod-Moriah, U. A., Abraham, M., Friedlander, M., Potashnik, G., and Kaplanski, J. (1989). *Andrologia, 21*: 229.

Shen, W., Moore, C., Ikeda, Y., Parker, K. L., and Ingraham, H. A. (1994). *Cell, 77*: 651.

Simard, J., Sanchez, R., Durocher, F., Rheaume, E., Turgeon, C., Labrie, Y., Luu-The, V., Mebarki, F., Morel, Y., de Launoit, Y., et al. (1995). *J. Steroid Biochem. Mol. Biol., 55*: 489.

Sinclair, A. H., Berta, P., Palmer, M. S., Hawkins, J. R., Griffiths, B. L., Smith, M. J., Foster, J. W., Frischuf, A., Lovell-Badge, R., and Goodfellow, P. N. (1990). *Nature, 346*: 240.

Sonnenberg-Riethmacher, E., Walter, B., Riethmacher, D., Godecke, S., and Birchmeier, C. (1996). *Genes Dev., 10*: 1184.

Stojilkovic, S. S., and Catt, K. J. (1995). *Rec. Prog. Horm. Res., 50*: 161.

Sullivan, P. M., Petrusz, P., Szpirer, C., and Joseph, D. R. (1991). *J. Biol. Chem., 266*: 143.

Sylvester, S. R., and Griswold, M. D. (1994). *J. Androl., 15*: 381.

Tapanainen, J. S., Aittomaki, K., Min, J., Vaskivuo, T., Huhtaniemi, I. T. (1997). *Nature Genet., 15*: 205.

Tiepolo, L., and Zuffardi, O. (1976). *Hum. Genet., 34*: 119.

Tilley, W. D., Marcelli, M., Wilson, J. D., and McPhaul, M. J. (1989). *Proc. Natl. Acad. Sci. USA, 86*: 327.

Torkelson, T. R., Sadek, S. E., and Rowe, V. K. (1961). *Toxicol. Appl. Pharm., 3*: 545.

Torres, M., Gomez-Pardo, E., Dressler, G. R., and Gruss, P. (1995). *Development, 121*: 4057.

Trapman, J., Klaassen, P., Kuiper, G. G., et al. (1988). *Biochem. Biophys. Res. Comm., 153*: 241.

Turner, T. T., and Roddy, M. S. (1990). *Biol. Reprod., 43*: 414.

Turner, T. T., and Yamamoto, M. (1991). *Biol. Reprod., 45*: 358.

Vogt, P., Keil, R., Kohler, M., Lengauer, C., Lewe, D., and Lewe, G. (1991). *Hum. Genet., 86*: 341.

Wachtel, S. (1983). *H-Y Antigen and the Biology of Sex Determination*. Grune and Stratton, New York.

Wakeling, A. E., Visek, W. J. (1973). *Science, 181*: 659.

Walsh, K. A., Titani, K., Takio, K., Kumar, S., Hayes, R., Petra, P. H. (1986). *Biochemistry, 25*: 7584.

Walsh, P. C., Madden, J. D., Harrod, M. J., Goldstein, J. L., MacDonald, P. C., and Wilson, J. D. (1974) *N. Engl. J. Med., 291*: 944.

Waterman, M. R., John, M. E., and Simpson, E. R. (1986). *Cytochrome P-450. Structure, Mechanism, and Biochemistry*. Plenum Press, New York, p. 450.

Westphal, U. (1971). *Steroid Protein Interactions*. Springer-Verlag, New York, p. 356.

Whorton, D., Krauss, R. M., Marshall, S., and Milby, T. H. (1977). *Lancet* (Dec 17): 1259.

Wong, C., Kelce, W. R., Sar, M., and Wilson, E. M. (1995). J. Biol. Chem., *270*: 19998.

Wilson, J. D. (1988). *Endocrinol. Rev., 9*: 181.

Wilson, J. D., Harrod, M. J., Goldstein, J. L., Hemsell, D. L., and MacDonald, P. C. (1974). *N. Engl. J. Med., 290*: 1097.

Wilson, J. D., Griffin, J. E., and Russell, D. W. (1993). *Endocrinol. Rev., 14*: 577.

Wilson, J. D., George, F. W., and Renfree, M. B. (1995). *Recent Prog. Horm. Res., 50*: 349.

Winter, J. S. D., Hughes, I. A., Reyes, F. I., and Faiman, C. (1976). *J. Clin. Endocrinol. Metab., 42*: 679.

Yeung, C. H., Cooper, T. G., Bergmann, M., and Schulze, H. (1991). *Am. J. Anat., 191*: 261.

Zanaria, E., Muscatelli, F., Bardoni, B., Strom, T. M., et al. (1994). *Nature, 372*: 635.

Zhang, F. P., Hamalainen, T., Kaipia, A., Parkarinen, P., and Huhtaniemi, I. (1994). *Endocrinology, 134*: 2206.

21

Reciprocal Mesenchymal-Epithelial Interactions in the Development of the Male Urogenital Tract

Gerald R. Cunha, Annemarie A. Donjacour, and Simon W. Hayward
University of California, San Francisco, San Francisco, California

DEVELOPMENT OF THE MALE REPRODUCTIVE SYSTEM

The development of the male reproductive system is a complex process, which occurs during fetal, neonatal, prepubertal, and pubertal periods. In the initial ambisexual stage, male and female embryos are indistinguishable anatomically, having a pair of undifferentiated gonads and two sets of mesodermally derived gonoducts, the Wolffian and Müllerian ducts, within the gonadal ridge. Sexual development in rats and mice proceeds in a cranial-to-caudal direction along the ducts over a period of approximately 5 days in the latter half of gestation (Price and Ortiz, 1965). The gonadal primordia can develop into either testes or ovaries depending on the presence or absence of the Y chromosome. In the male the Wolffian ducts (WD) and in the female the Müllerian ducts are maintained, while the other set of ducts is normally destroyed. Müllerian-inhibiting substance (MIS) is secreted by Sertoli cells of the fetal testes and causes the Müllerian ducts to regress, first at the cranial end and progressing caudally (Josso et al., 1977; Donahoe et al., 1982). Transgenic female mice that inappropriately express MIS lack structures derived from the Müllerian duct, i.e., the oviduct, uterus, and upper portion of the vagina (Behringer et al., 1990), while male mice that lack a functional MIS gene develop uteri and oviductal structures in conjunction with male urogenital organs (Behringer et al., 1994). The observation that MIS inhibited proliferation of isolated Müllerian duct mesenchyme but not epithelium in vitro (Tsuji et al., 1992) suggests the possibility that Müllerian duct mesenchyme may mediate the destructive effects of MIS on the Müllerian duct. This idea is supported by the discovery of a candidate MIS receptor, which is localized in the mesenchyme surrounding the Müllerian duct (Baarends, et al., 1994; di Clemente, et al., 1994).

Androgens from the fetal testis prevent regression of the mesodermal WD and stimulate its development into epididymis and ductus deferens. In some species, seminal vesicles (SV) and ampullary glands (derived from the lower end of the WD) also develop (Jost, 1953; Jost, 1965; Wilson et al., 1981; Cunha et al., 1987). The mesonephric tubules at the cranial end of the WD are also maintained by androgens and form the efferent ductules that connect the testis with the epididymis. In the absence of testosterone, the mesonephric tubules and the WDs degenerate, as in females or in males subjected to fetal castration (Raynaud and Frilley, 1947; Jost, 1953). This is also the case in the absence of functional androgen receptors (AR) in mice and humans with the testicular feminization mutation (Tfm) (Lyons and Hawkes, 1970; He et al., 1990; Quigley et al., 1995). In double mutant mice that lack both AR and a functional MIS gene, genetic male animals that have testes and produce testosterone retain their Müllerian ducts, which form oviducts, uteri, and a vagina. In such transgenic mice WDs degenerate and male secondary sex organs do not develop (Behringer et al., 1994).

Organogenesis of WD structures proceeds after androgens prevent the degeneration of the WDs. In the case of the ductus deferens, the initially round ductal profile develops three to four invaginations that project into the ductal lumen, giving the duct a slightly folded appearance in transverse section. A thick coat of smooth muscle develops from the mesenchyme surrounding the ductus deferens. In the case of the epididymis, which develops from the cranial end of the Wolffian duct, extensive proliferation lengthens the duct manyfold. The elongating duct then coils and folds on itself to form a compact structure having an enormous length of duct packed into a relatively small organ. The SV forms as a diverticulum of the lower end of the WD. As it elongates, the SV curves to form a canelike structure, whose epithelium then undergoes a unique process of evagination and branching morphogenesis to form one of the most elaborately folded mucosas in the body (Cunha and Lung, 1979; Lung and Cunha, 1981). Budding of the ampullary gland from the lower end of the WD occurs during early postnatal life in the few species that have this small gland (Price and Williams-Ashman, 1961).

The urogenital sinus develops into adult structures in both sexes. In females the urogenital sinus forms the lower portion of the vagina, while in males the urogenital sinus forms the prostate, periurethral glands, and bulbourethral glands (Gittinger and Lasnitzki, 1972; Lasnitzki, 1974; Cunha, 1986). Androgens from the fetal testis stimulate budding of the endodermally derived urogenital sinus epithelium just caudal to the bladder neck (Sugimura et al., 1986b; Timms et al., 1994). In rats and mice the location of prostatic buds around the circumference of the urogenital sinus delineates the lobes of the prostate that are clearly distinguishable in adulthood. These are designated ventral, anterior, and dorsolateral prostate (Jesik et al., 1982; Sugimura et al., 1986b; Hayashi et al., 1991). Numerous prostatic cords elongate into the urogenital sinus mesenchyme (UGM), and some, especially in the ventral prostate, begin to undergo branching morphogenesis prior to birth. In other species the lobar subdivision of the prostate is less complex or distinct. In baboons and macaques the prostate is divided into cranial and caudal lobes, while in the human true lobes are impossible to distinguish (Lewis et al., 1981; McNeal, 1983a; Lewis, 1984). The cranial lobe of the primate prostate seems to be involved in coagulation of semen. The homologous rat anterior prostate produces secretions capable of replacing those of the cranial prostatic lobe of the rhesus monkey in inducing coagulation of monkey SV fluid (Van Wagen, 1936). It has been proposed that the cranial lobe of the monkey has the same embryonic history as the central zone of the human prostate (McNeal, 1983b).

While the dependence of male reproductive development upon androgens is undisputed (Jost, 1953; Wilson et al., 1981; Cunha et al., 1987), the precise roles for testosterone and its more active metabolite, 5α-dihydrotestosterone (DHT), are still under investigation. Testosterone is the major androgen secreted by the testis, but is metabolized to DHT by the enzyme Δ^4-3-ketosteroid-5α-reductase (5α-reductase) in various androgen target tissues such as the prostate (Wilson and Lasnitzki, 1971; Lasnitzki and Franklin, 1972; Wilson et al., 1981, 1993; Shima et al., 1990; Russell and Wilson, 1994). DHT has a 10-fold greater affinity for AR than testosterone (Carlson and Katzenellenbogen, 1990; Deslypere et al., 1992). This is accomplished by DHT binding to AR for much longer than testosterone (Zhou et al., 1995). This occurs because binding rates are about the same but dissociation rates are much lower for DHT versus testosterone. Also, DHT is the major hormone identified in nuclei of the adult prostate (Anderson and Liao, 1968; Bruchovsky and Wilson, 1968; Gloyna and Wilson, 1969; Wilson and Gloyna, 1970). For many species, 5α-reductase activity is generally lower in WD structures than in the urogenital sinus (Wilson and Lasnitzki, 1971; Siiteri and Wilson, 1974). However, when cranial, middle, and caudal segments of the fetal WD are assayed separately, 5α-reductase activity increases from low levels cranially to high levels caudally in the developing SV in which 5α-reductase levels are equivalent to that of the urogenital sinus (Tsuji et al., 1994b). In situations in which 5α-reductase activity is greatly impaired or lost, for example, with naturally occurring mutations in the gene encoding 5α-reductase in the human population or following administration of 5α-reductase inhibitors to rats, the prostate is more severely inhibited in its growth than the epididymis or SV (Imperato-McGinley, et al., 1974; Imperato-McGinley, 1984; Imperato-McGinley et al., 1984). This has been interpreted to indicate that DHT is the active androgen in masculinization of the embryonic urogenital sinus, while testosterone alone is sufficient to support the growth of WD derivatives. This is an oversimplification, as the developing SV has high 5α-reductase activity (Tsuji et al., 1994b), and postnatal growth and development of the SV appear to be dependent upon DHT (Shima et al., 1990). Thus, in a strict sense testosterone is not the sole active androgen for the WD, as DHT is important for the development of the SV from the caudal end of the WD.

Organs of WD and urogenital sinus origin appear to differ in their overall sensitivity to androgens. In studies that compared the ability of a 5α-reductase inhibitor, finastride, and an antiandrogen, flutamide, to suppress growth of male reproductive organs of the rat, the prostate was the most sensitive to both types of agents. The epididymis was not affected by finastride and required a very large dose of flutamide for its growth to be affected. The SV had an intermediate response to both compounds (Imperato-McGinley et al., 1992). There is an important caveat in interpreting these results; size of urogenital organs was not evaluated immediately following inhibitor treatment, i.e., neonatally, but rather in adulthood. Therefore, final organ weights reflected pubertal growth in response to androgens as well as the interference of inhibitors during early androgen-dependent development. In this regard, neonatal castration of rats leads not only to smaller prostates and SVs in adulthood, but also to glands that are unable to respond fully when stimulated by exogenous androgens in adulthood (Chung and MacFadden, 1980). This situation contrasts with that found in animals castrated in adulthood in which exogenous androgens can induce full regrowth of the regressed prostate and SV. Observations that neonatal "priming" of the rat sex accessory glands with androgens enhances the responsiveness to androgens in adulthood suggest that there are long-term developmental effects resulting from fetal and neonatal hormonal environments (Chung and Ferland-Raymond, 1975). These effects may be the result of direct steroidal actions on the epithelium or they may be mediated by the mesenchyme of the developing organ.

Androgen action in target organs is mediated by AR (Evans, 1988; Quigley et al., 1995). The fetal male reproductive organs show an interesting pattern of AR expression. AR are detectable in the mesenchyme of all male genital rudiments during the ambisexual stage and do not appear in the epithelia of these organs for several days (Shannon and Cunha, 1983; Takeda et al., 1985; Murakami, 1987; Cooke et al., 1991a; Husmann et al., 1991; Takeda and Chang, 1991). As smooth muscle differentiates from mesenchyme, AR remain high in the muscular layer of the male genital tract throughout development and into adulthood (Hayward et al., 1996b; Prins and Birch, 1993). Epithelial AR appear first in the efferent ductules at day 16 of gestation in the mouse and are subsequently expressed in a cranial to caudal sequence throughout the urogenital tract. In the embryonic mouse, AR appear in the epithelium of the epididymis and ductus deferens at 19 days of gestation, in the SV at 1–2 days postnatal, and in the prostate at approximately 4–6 days postnatal (Cooke et al., 1991a; Cunha et a., 1991a). This pattern is similar in the rat but with the expression of epithelial AR occur somewhat earlier (Prins and Birch, 1995; Hayward et al., 1996a). Therefore, most of the androgen-dependent developmental processes that occur in fetal life affect organs that have high AR levels in the mesenchyme and undetectable to low levels of AR in the epithelium. This suggests a paracrine mechanism of androgen action in the developing male reproductive tract as will be discussed below.

The neonatal and prepubertal periods are characterized by low levels of circulating androgens (Pointis et al., 1980; Corpechot et al., 1981; Winter et al., 1981), even though significant morphogenesis and cytodifferentiation occur in the male reproductive organs during these periods (Donjacour and Cunha, 1988a). The epithelial cords of the neonatal prostate undergo extensive branching morphogenesis. In the mouse, in which this has been quantitated, 85% of the adult number of ductal tips and branchpoints are generated by 15 days of age (Sugimura et al., 1986b). Morphogenesis in the neonatal SV involves extensive epithelial evagination and branching. The simple cane-shaped epithelial tube at birth is converted into a complex highly infolded epithelium by one week after birth during periods when serum androgens are particularly low (Lung and Cunha, 1981).

Both the epithelial and stromal components of the prostate and SV undergo cytodifferentiation in the neonatal period in rats and mice. In the SV a narrow lumen is present from the earliest stages. SV epithelial cells are low columnar with nuclei elongated perpendicular to the basement membrane. Basal cells of the SV epithelium are identifiable at approximately 11–13 days postnatal in the mouse (Deane and Wurzelmann, 1965). At birth the prostatic epithelium is organized as solid cords in rats and mice. Epithelial proliferation occurs preferentially at the ductal tips throughout postnatal development (Sugimura et al., 1986c). Lumen formation begins proximally near the urethra at about 1–2 days postnatal (Hayward et al., 1996a). This process appears to involve apoptosis of centrally located epithelial cells and polarization and differentiation of epithelial cells along the basal lamina. Epithelial cells of the solid cords co-express cytokeratins that are normally found in fully differentiated luminal cells (keratins 8 and 18) as well as cytokeratins normally found in mature basal cells (keratins 5, 14). As canalization occurs, peripheral epithelial cells of the solid cords segregate into morphologically identifiable luminal and basal epithelial cells, each expressing their characteristic sets of cytokeratins (Hayward et al., 1996a). Prostatic luminal cells are columnar and have basally located, oval-shaped nuclei (Evans and Chandler, 1987). Prostatic basal cells are traditionally described as being triangular in shape. However, recent descriptions of these cells in the rat prostate have noted that basal cells have a very complex shape with long cellular processes extending both around the epithelial ducts and between luminal cells towards the lumen (Timms et al., 1976; Soeffing and Timms, 1995). When wholemounts of prostatic ducts are stained with antibodies to cytokeratins 5 or 14, basal cells can be seen to possess long cell processes and present a rather large "footprint" on the basal lamina (Hayward et al., 1996c). The close co-incidence of basal cell differentiation with ductal canalization strongly suggests that the prostatic basal cells may play an architectural role, perhaps supporting the ductal structure of the organ.

Androgen receptor expression in rodent prostatic epithelial cells is initiated at around birth (Prins and Birch, 1995; Hayward et al., 1996a). The intensity of immunocytochemical staining for epithelial AR increases after birth on a per cell basis, especially in luminal epithelial cells. In addition, the proportion of epithelial cells that are positive for AR also increases over time until by 15 days postnatal all luminal epithelial cells have high AR expression (Takeda and Chang, 1991; Prins and Birch, 1995). Prostatic basal cells, especially in proximal regions, are reported to be mostly AR-negative late in postnatal development (Prins and Birch, 1995), although some AR protein and message are detectable by electron microscopy in basal cells of the adult rat prostate (Soeffing and Timms, 1995).

The mesenchymal cells of both the prostate and SV differentiate during the first 3 weeks of postnatal life. In the prostate the mesenchymal cells close to the epithelium become elongated and closely apposed to the ducts, and gradually and sequentially express the proteins that characterize smooth muscle cells (Table 1) (Flickinger, 1970, 1972; Hayward et al., 1996b). Once formed, these smooth muscle cells maintain a high level of AR expression (Prins and Birch, 1995; Hayward et al., 1996b). The interductal stromal cells differentiate into fibroblasts, express vimentin, and decrease their expression of AR until by 3 weeks of age most interductal fibroblasts appear AR-negative (Prins et al., 1991; Prins and Birch, 1995; Hayward et al., 1996b). Smooth muscle differentiation in the male urogenital tract is dependent upon androgens. This was demonstrated in experiments in which human fetal prostatic rudiments were divided into right and left halves and grafted underneath the renal capsule of intact and castrated male athymic nude rats (Hayward et al., 1996d). In intact male hosts development of prostatic smooth muscle progressed throughout the graft. In contrast, in castrated hosts prostatic smooth muscle failed to develop. The small amount of smooth muscle observed in these grafts was mostly restricted to blood vessels. Given the fact that prostatic epithelium was present in human fetal prostatic grafts to both intact and castrated hosts prior to grafting and was visible in tissue sections, the failure of smooth muscle differentiation in grafts to castrated hosts could be attributed not to a lack of prostatic epithelium, but instead to a lack of androgenic stimulation.

The arrangement of smooth muscle and fibroblastic connective tissue in the SV is different from that of the prostate. The SV develops a lamina propria adjacent to the epithelium; this consists of extracellular matrix and fibroblastic cells with low AR expression. The thick smooth muscle layer of the SV is organized outside of the lamina propria. SV smooth muscle cells express the same pattern of marker proteins (smooth muscle α-actin, desmin, myosin, and vinculin), as do prostatic smooth muscle cells (Hayward et al., 1996b). Maintenance of smooth muscle differentiation in the adult male genital tract is also dependent upon androgens. Following castration, smooth muscle in the prostate and SV dedifferentiate in an orderly fashion by losing smooth muscle

Table 1 Differentiation of Smooth Muscle in the Developing Rat Ventral Prostate

Marker	19dE UGS	0d postnatal	5d postnatal	10d postnatal	15d postnatal	Adult
α-Actin	++	++	++	+++	+++	+++
Myosin	−	−	+	+	+	+++
Laminin	−	−	+/−	+/−	+	+++
Vinculin	+/−	+	++	++	++	+++
Vimentin	++	++	++	++	++	+

Vimentin is initially widely expressed and during development, becomes localized to the interductal connective tissue and is excluded from the differentiating muscle. The markers of smooth muscle differentiation are expressed sequentially in a proximal to distal manner along the growth ducts.
Source: Hayward et al., 1996b.

markers in the reverse order in which they were expressed during development (Hayward et al., 1996b).

Secretory protein production by rat prostatic epithelium begins during the second to third week after birth. In the rat ventral prostate, supranuclear clear areas in luminal cells have been observed as early as 12 days postnatal and are thought to represent the early stages of secretory activity (Price, 1936). Secretory proteins of the rat dorsolateral prostate and coagulating gland are first detectable by Western blotting of prostatic homogenates between 14 and 16 days of age (Lopes et al., 1996). SV secretion is first detectable by biochemical methods at 25 days postnatal (Fawell and Higgins, 1986), however, secretory granules are visible at approximately 18 days postnatal (Deane and Wurzelmann, 1965). Epithelial AR are required for full secretory cytodifferentiation in both the prostate and SV (Cunha and Young, 1991; Donjacour and Cunha, 1993), however, it is likely that stromal factors also play a role in this process, possibly indirectly by maintaining the proper epithelial cell morphology.

Puberty, as defined by an elevation in serum androgens at approximately 28 days of age in mice (Selmanoff et al., 1977; Corpechot et al., 1981), does not have a dramatic qualitative effect on the development of the SV and prostate, since (a) ductal morphogenesis of these glands is essentially complete, (b) epithelial and stromal cytodifferentiation are well underway, and (c) secretory proteins are present prior to puberty. The major changes in these organs during puberty are substantial increases in wet weight and DNA content. In the mouse this occurs rapidly between 30 and 60 days of age, when wet weight and DNA content plateau (Donjacour and Cunha, 1988a). Serum androgens remain high throughout adulthood, but secondary sex organs do not continue to grow. Proliferation does not stop entirely, however. The mitotic index of adult rat prostatic epithelium is approximately 0.3% (Evans and Chandler, 1987). It has been reported that prostatic epithelial cells undergo apoptosis in the proximal ducts in adulthood, while a low level of epithelial proliferation continues at the ductal tips. Thus, a balance between cell division and cell death is maintained under normal circumstances in the adult prostate (Lee et al., 1990; Sensibar et al., 1990).

In summary, the adult organ represents the endpoint of the developmental process. During the development of the prostate, both epithelium and stroma develop in concert spatially and temporally. The epithelial cords grow, branch, and canalize with segregation of the epithelial cells to the luminal and basal phenotypes. Concurrent with, and immediately adjacent to, the prostatic epithelium, the fetal mesenchyme differentiates from an undifferentiated loose connective tissue into dense sheathes of smooth muscle surrounding the epithelial ducts. The fully differentiated prostatic smooth muscle has been postulated to act under the influence of androgens to inhibit epithelial proliferation and maintain epithelial differentiation in the adult gland (Tenniswood, 1986; Cunha et al., 1986b; Hayward et al., 1996d; Nemeth and Lee, 1996). The idea that differentiated smooth muscle acts as an inhibitor of epithelial proliferation raises the possible role of smooth muscle in prostatic disease. The idea of a renewal of fetal inductive ability in adult prostatic stroma associated with areas of human benign prostatic hyperplasia has been suggested (McNeal, 1983a, 1984). We have recently proposed that prostatic smooth muscle may inhibit growth of microscopic foci of prostatic cancer and may be responsible for the low level of clinical cancer as compared to the high level of latent disease found at autopsy (Cunha et al., 1996b; Hayward et al., 1996d). These ideas suggest that a knowledge of prostatic smooth muscle–derived signaling molecules may be of therapeutic importance.

MESENCHYMAL-EPITHELIAL INTERACTIONS

As indicated above, the male genital tract develops from two anlage: the mesoderm-derived WDs and the endoderm-derived urogenital sinus. Both are regionally specialized in a cranial-caudal sequence to form the epididymis, ductus deferens, and SV (all WD derivatives) and the bladder, prostate, and bulbourethral gland (all derived from the urogenital sinus). Within the WD and urogenital sinus, the regional differentiation of these organs is specified by mesenchymal-epithelial interactions. For example, prostatic and SV development are induced by specific mesenchymal populations, which specify epithelial differentiation. Organogenesis is absolutely dependent upon these mesenchymal-epithelial interactions, as differentiation of both the epithelium and mesenchyme is abortive if the epithelium and mesenchyme are grown by themselves. Appropriate epithelial and mesenchymal differentiation only occurs if the two tissues are allowed to interact. This concept applies not only to the male urogenital tract, but also to all other organ rudiments composed of an epithelial parenchyma and mesenchyme, for example, the female urogenital tract, intestine, lung, bladder, and mammary gland (Cunha, 1976a; Kratochwil, 1987; Haffen et al., 1989).

In males, UGM induces prostatic ductal morphogenesis, regulates epithelial proliferation, and specifies expression of prostatic secretory proteins provided the mesenchyme is associated with an epithelium of endodermal origin (Cunha et al., 1987, 1992a). Likewise, seminal vesicle mesenchyme (SVM) induces SV epithelial development and differentiation provided the mesenchyme is associated with an epithelium of WD origin (Higgins et al., 1989b; Cunha et al., 1991b). [In the female genital tract, uterine mesenchyme induces uterine development from Müllerian epithelium, and vaginal mesenchyme induces vaginal epithelial differentiation (Cunha, 1976b; Boutin et al., 1992).] When both male and female organogenetic inductions are examined as a whole, it is evident that virtually all aspects of epithelial differentiation are determined and specified by the mesenchyme. Table 2 lists the types of epithelial differentiation markers specified by the mesenchyme.

As noted above, the outcome of epithelial-mesenchymal interactions is dependent upon not only the source of the inducing mesenchyme, but also the germ layer origin and responsiveness of the epithelium (see Table 3 for examples). Thus, while the mesenchyme induces and specifies the differentiation of the epithelium, developmental endpoints induced by the mesenchyme are strictly limited by the developmental reper-

Table 2 Mesenchymal Effects on Epithelial Development in the Urogenital Tract

Mesenchymal effect	Ref.
Induces epithelial morphogenesis	Cunha et al., 1983a; Kratochwil, 1987
Specifies epithelial morphology	Cunha et al., 1983a; Kratochwil, 1987
Specifies expression of epithelial cytokeratins	Bigsby and Cunha, unpublished
Regulates epithelial proliferation	Sugimura et al., 1986a
Specifies epithelial secretory proteins	Cunha et al., 1991b; Donjacour and Cunha, 1993
Induces expression of epithelial androgen receptors	Cunha et al., 1980b
Determines expression of epithelial heparan sulfate proteoglycans	Boutin et al., 1991b
Determines expression of epithelial hox genes	Pavlova et al., 1994

toire of the germ layer or the anlage from which the epithelium is derived. For example, prostatic differentiation has only been elicited from endoderm-derived epithelia, specifically endoderm of the urogenital sinus. Attempts to induce prostatic differentiation from foregut endoderm have bene unsuccessful (G. R. Cunha, unpublished). Likewise, SV differentiation has only been induced from mesodermal epithelium derived from the WD (Table 3). Thus, the origin of the responding epithelium plays a critical role in determining developmental outcome of experimental tissue recombinants.

As indicated above, the rodent prostate is composed of well-defined lobar subdivisions (anterior, dorsal, lateral, and ventral prostates), each having unique patterns of ductal branching and also expressing lobe-specific secretory proteins (Sugimura et al., 1986b; Hayashi et al., 1991; Kinbara and Cunha, 1995). These lobe-specific differences in prostatic differentiation appear to be induced by regionally distinct subpopulations of mesenchyme within the urogenital sinus (Sugimura et al., 1985; Takeda et al., 1990; Timms et al., 1995). Recently, it has been appreciated that the dorsal and ventral portions of UGM have different inductive properties. This is based upon the observation that only ventral UGM is an effective inducer of ventral prostatic differentiation (Takeda et al., 1990; Timms et al., 1995). Presumably, dorsal UGM induces dorsal-lateral prostate, although this has not been tested. Curiously, in situ SVM is an inducer of both the SV and also the anterior lobe of the prostate (coagulating gland) as both of these glands develop within a common mass of mesenchyme surrounding the lower end of the WD. When SVM is used as a glandular inducer in experimental tissue recombinants, the developmental outcome is restricted by the germ layer origin of the epithelium. For example, prostatic differentiation occurs when SVM is associated with endodermal epithelia (Table 3), while SV epithelial differentiation occurs when SVM is associated with epithelium of WD origin (Boutin et al., 1991a; Cunha et al., 1991b; Tsuji et al., 1994a; Donjacour and Cunha, 1995). Since SVM is physically in close proximity to the dorsal-lateral prostate, the lobar identity of the prostate induced by SVM is dorsal-lateral prostate (Donjacour and Cunha, 1995; Kinbara et al., 1995). In reality, because SVM and UGM are so intimately related, they can be considered to represent a common glandular inductive field with slight regional differences cranially-caudally and dorsal-ventrally. Thus, UGM is a prostatic inducer when associated with endodermal epithelia and an inducer of SV when associated with WD epithelia. Whereas UGM possesses regional subpopulations of prostatic inducers that induce the entire prostatic complex (ventral, lateral, and dorsal lobes), SVM is purely a dorsal-lateral prostatic inducer provided endodermal epithelia are used.

While mesenchyme induces epithelial differentiation, epithelial-mesenchymal interactions are in fact reciprocal in that epithelium induces smooth muscle differentiation in the mesenchyme. Evidence in support of a role of the epithelium as an inducer of smooth muscle differentiation can be found in studies on the intestine, urinary bladder, uterus, and male genital tract. The role of epithelium as an inducer of smooth muscle differentiation is particularly evident in the mouse uterus. When undifferentiated uterine mesenchyme obtained from newborn mice was grafted beneath the renal capsule of syngeneic female hosts and grown for 1 month, only small amounts of smooth muscle differentiated, most of which was associated with blood vessels. In contrast, when uterine mesenchyme was grafted in combination with uterine epithelium, large amounts of myometrium developed (Cunha et al., 1989, 1992b). By the same token, differentiation and morphological patterning of smooth muscle in the male urogenital tract is regulated via cell-cell interactions with epithelium. For example, UGM grafted

Table 3 Instructive and Permissive Inductions in the Urogenital Tract

Mesenchyme	Epithelium	Epithelial germ layer	Result	Type of induction	Ref.
SVM	Prostate	Endoderm	Prostate	Permissive	Kimbara et al., 1992
VM	UGS	Endoderm	Prostate	Permissive	Cunha, 1975
UGM	UGS	Endoderm	Prostate	Permissive	Cunha, 1972
UGM	Prostate	Endoderm	Prostate	Permissive	Neubauer, et al., 1986; Norman, et al., 1986 Hayashi, et al., 1993; Kinbara, et al., 1995
UGM	Bladder	Endoderm	Prostate	Instructive	Cunha et al., 1980a, 1983b
UGM	Urethra	Endoderm	Prostate	Instructive	Boutin, et al., 1991a
UGM	Sinus vagina	Endoderm	Prostate	Instructive	Boutin, et al., 1991a
SVM	Bladder	Endoderm	Prostate	Instructive	Donjacour and Cunha, 1988b
SVM	UGS	Endoderm	Prostate	Permissive	Cunha, 1972
BUG-M	Prostate	Endoderm	Prostate	Permissive	Kimbara, et al., 1992
SVM	SV	Mesoderm	SV	Permissive	Higgins, et al., 1989a
SVM	WD	Mesoderm	SV	Instructive	Higgins, et al., 1989b
SVM	Epididymis	Mesoderm	SV	Instructive	Turner, et al., 1989
SVM	Ureter	Mesoderm	SV	Instructive	Cunha, et al., 1991b
SVM	Vas deferens	Mesoderm	SV	Instructive	Cunha, et al., 1991b
UGM	SV	Mesoderm	SV	Permissive	Cunha, et al., 1991b
UGM	Epididymis	Mesoderm	SV	Instructive	Cunha and Young, in preparation
UGM	Ureter	Mesoderm	SV	Instructive	Cunha and Young, in preparation
UGM	Vas deferens	Mesoderm	SV	Instructive	Cunha and Young, in preparation
BUG-M	SV	Mesoderm	Ductal	Not known	Tsuji, et al., 1994a

SVM, Seminal vesicle mesenchyme; UGM, urogenital sinus mesenchyme; VM, vaginal mesenchyme; BUG-M, bulbourethral gland mesenchyme; WD, Wolffian duct; UGS, urogenital sinus; SV, seminal vesicle.

and grown for 1 month in male hosts formed little if any smooth muscle. In contrast, when UGM was grafted with epithelium of either adult prostate, bladder, or embryonic urogenital sinus, prostatic ducts developed which were surrounded by actin-positive smooth muscle cells organized into thin sheaths as is appropriate for the rat prostate (Cunha et al., 1992b). It is also noteworthy that rat UGM formed thick sheets of smooth muscle surrounding the epithelial ducts in tissue recombinants composed of rat UGM plus human prostatic epithelium (Hayward et al., 1996d). This pattern of smooth muscle is characteristic of human prostate and demonstrates that the human prostatic epithelium induced the rat mesenchyme to take on human stromal patterning. These observations suggest that smooth muscle differentiation is induced and spatially patterned by epithelium in both male and female genital tracts. Similar findings have been made for the urinary bladder and intestine (Haffen et al., 1987; Baskin et al., 1996). Model systems utilizing subcutaneous grafts of collagen gels containing human prostatic epithelium have shown analogous determination and patterning effects of the stroma by the transplanted epithelium. Transplants of adult human prostatic epithelium were able to reprogram adult mouse subcutaneous fibroblasts to form smooth muscle sheaths surrounding the epithelium (Hayward et al., 1992). In a similar model subcutaneous grafts of collagen gels containing fetal rat gut epithelium were able to induce smooth muscle differentiation from subcutaneous fibroblasts of the host. In this case it was clear that the smooth muscle was appropriately oriented, demonstrating epithelial determination of both type and patterning of stromal differentiation (Del Buono et al., 1992).

While mesenchymal-epithelial interactions play a fundamental role in development of the male urogenital tract, the overall developmental process is elicited by androgens, which regulate development, growth, and function within androgen target organs via AR. During fetal and/or neonatal periods, androgens prevent programmed cell death of the WDs, elicit elongation and coiling of the epididymis, induce ductal branching in the prostate, and elicit the unique epithelial morphogenesis characteristic of the SV (Cunha et al., 1992a). Since prostatic and SV development is androgen-dependent and is induced by mesenchyme, it is not surprising that androgenic effects on epithelial development are mediated via mesenchyme. Analysis of tissue recombinants prepared with wild-type and androgen-insensitive Tfm tissues has clarified the relationship between mesenchymal-epithelial interactions, AR, and androgen action. Analysis of chimeric prostates and SVs composed of wild-type mesenchyme plus Tfm epithelium has demonstrated that AR-deficient Tfm epithelium can undergo androgen-dependent ductal morphogenesis, epithelial proliferation, and columnar cytodifferentiation (Cunha and Young, 1991; Cunha et al., 1992a). However, it should be noted that tissue recombinants containing Tfm epithelium are unable to express secretory activity, demonstrating that epithelial AR are required to express androgen-dependent secretory proteins (Cunha and Young, 1991; Donjacour and Cunha, 1993). Findings of this nature demonstrate that certain "androgenic effects" on epithelium may be independent of androgens in a strict sense, instead being elicited by paracrine factors produced by AR-positive mesenchyme. Clearly, the results of the Tfm/wild-type tissue recombination experiments in conjunction with the mesenchymal expression of AR during development suggest that many so-called androgenic effects expressed in epithelial cells are elicited via paracrine influences from mesenchyme and ъo not require the presence of epithelial AR. Tfm/wild-type recombination experiments further demonstrate that in the adult prostate epithelial morphology is maintained by androgenic action on the prostatic stroma. Since prostatic smooth muscle is the primary stromal tissue expressing AR, this strongly implicates smooth muscle as the mediator of androgen action in the adult.

While the prostate is primarily considered to be an androgen target organ, it is also sensitive to estrogens from fetal to adult periods (Huggins and Webster, 1948; Leav et al., 1971; Mawhinney and Neubauer, 1979; Merk et al., 1982; Naslund and Coffey, 1986; Zhao et al., 1992; Prins et al., 1993). Effects of estrogen on the prostate are thought to be mediated via estrogen receptors (ER), which are known to be expressed in the prostate specifically in the stroma (Jung-Testas et al., 1981; Purvis et al., 1985; Schulze and Barrack, 1987; West et al., 1988; Mobbs et al., 1990; Schulze and Claus, 1990). The paracrine action of androgen on the prostate via stromal AR raises the possibility that estrogenic effects on the prostate may be similarly elicited via ER in the stroma. While direct evidence for a role of stromal ER is not yet available for the prostate, recent studies in the uterus are consistent with this concept. As for androgens, 17β-estradiol (E2) regulates postnatal uterine and vaginal epithelial growth and differentiation. Estradiol elicits its effects via ER, which are expressed in both epithelial and stromal cells of the adult uterus. Effects of estradiol on target epithelial cells have been assumed to be mediated directly through epithelial ER. However, based upon analysis of ER expression and E2 responsiveness in neonatal mouse uteri (Bigsby and Cunha, 1986, 1987; Yamashita et al., 1989; Yamashita et al., 1990), mitogenic effects of E2 on neonatal uterine epithelium (UtE) appear to be mediated by estrogen-receptor–positive (ER+) stromal cells, a conclusion further supported by failure of estrogens to elicit proliferation in isolated UtE in vitro (Casimiri et al., 1980; Iguchi et al., 1985; Julian et al., 1992). To unequivocally determine the roles of epithelial versus stromal ER in epithelial proliferation, a transgenic ER knockout (ERKO) mouse (Lubahn et al., 1993) has been used to produce tissue recombinants in which epithelium, stroma, or both are devoid of functional ER. For this purpose the following tissue recombinants were prepared with UtE and uterine stroma (UtS) from wild-type (wt) and ERKO mice: wt-Uts + wt-UtE, wt-UtS + ERKO-UtE, ERKO-UtS + ERKO-UtE, and ERKO-UtS + wt-UtE. All tissue recombinants were grown in intact female nude mice for 4 weeks, at which time the hosts were ovariectomized. Ten days after ovariectomy, the hosts were injected with oil or E2 followed 16 hours later by injection of BrdU and sacrifice 2 hours later. Epithelial labeling index was determined in tissue sections. In tissue recombinants prepared with wt-UtS (wt-UtS + wt-UtE, wt-UtS + ERKO-UtE), epithelial labeling index was induced about several-fold by E2 over oil-treated controls. By contrast, in tissue recombinants prepared with ERKO-UtS (ERKO-UtS + ERKO-UtE and ERKO-UtS + wt-UtE), epithelial labeling index was low and similar in E2- versus oil-treated specimens (Cooke et al., 1997a). These data clearly demonstrate that E2 induction of epithelial proliferation is a paracrine event requiring ER-positive stroma. Moreover, epithelial ER are not required for E2-induced epithelial proliferation. These studies demonstrating the paracrine action of estradiol on uterine epithelial proliferation are certainly relevant to male genital tract structures, which are known to express ER and to be responsive to natural and xenobiotic estrogen.

The effects of estrogens on prostatic development have been examined in a number of studies. Exposure to exogenous estrogens during fetal and neonatal development leads to a range of phenotypic abnormalities, which can be seen in adulthood. Exposure of the neonatal rat to high levels of estrogens and progestagens can permanently perturb both development and adult androgenic response of the prostate (Rajfer and Coffey, 1978; Higgins et al., 1981; Naslund and Coffey, 1986). Examination of the prostate demonstrates ER exclusively in the stroma of proximal regions of the prostatic ducts (Schulze and Barrack, 1987; Cooke et al., 1991b). The effects of fetal exposure to estrogens are particularly important because of the widespread use of diethylstilbestrol (DES) by pregnant women between 1948 and 1971. Between 4 and 6 million women were exposed to DES to prevent spontaneous abortion (Herbst and Bern, 1981). While an extensive literature now exists on the daughters of women exposed to DES during pregnancy, there has been very little research to examine the effects of this compound on human males. It is known that human fetal prostates grown in athymic rodent hosts underwent severe squamous metaplasia in response to DES, which persisted in some of cases even after withdrawal of the estrogen (Sugimura et al., 1988; Yonemura et al., 1995). These data suggest that estrogens can have profound effects on human prostatic development, and further that these changes, which are most profound in the epithelium, are probably mediated by estrogenic action on the prostatic mesenchyme.

GROWTH FACTORS AS MEDIATORS OF MESENCHYMAL-EPITHELIAL INTERACTIONS

Most studies on the role of growth factors in the male genital tract have focused on the adult prostate (Story, 1991; Steiner, 1993; Cunha et al., 1995). Such studies have yielded the following information:

1. Growth factors are expressed by prostatic stromal and/or epithelial cells.
2. Isolated cells (usually epithelium) cultured in vitro respond to growth factors.
3. Response of target cells to a growth factor is mediated through appropriate receptors.

Two major problems are generally encountered in most studies. Growth factors generally have been investigated in cells derived from the growth-quiescent adult prostate, and highly artificial culture systems involving the growth of cells on plastic substrata have been employed. While past studies have identified many families of growth factors in the prostate, it is impossible to establish the relevance of such studies to actual prostatic growth when growth-quiescent adult prostates are used as the source of experimental tissue. It is our firm belief that the role of growth factors in the prostate *must* be investigated during periods of actual prostatic growth so that the relevance of a given growth factor to actual prostatic growth and development can be established. For these reasons we have focused on the role of growth factors in the developing prostate. A basic outline of the ways in which mesenchyme and epithelium may interact in the developing and adult prostate to control gene expression was described by Tenniswood (1986). Advances in our understanding of the molecules involved in intercellular signaling, combined with the rapidly developing field of transgenic mice (gene knockout or overexpression), now allows analysis of some of the individual signaling molecules involved in the processes. One growth factor ideally suited as a potential paracrine mediator of mesenchymal-epithelial interactions is keratinocyte growth factor (KGF).

KGF is produced by mesenchymal and fibroblastic cells and has a high degree of specificity for epithelial cells (Aaronson et al., 1991). Based upon homology at the RNA level with other fibroblast growth factors (FGFs), KGF has been designated FGF-

7 (Finch et al., 1989; Yan et al., 1991). The KGF receptor (KGFR) is a splice variant of the fibroblast growth factor receptor 2 (FGFR2) gene and is expressed principally by epithelial cells (Miki et al., 1992). In skin, KGF transcripts are found exclusively in dermis, while KGFR message is exclusively expressed in epidermis (Finch et al., 1989; Bottaro et al., 1990; Orr-Urtreger et al., 1993). KGF promotes epidermal proliferation (Marchese et al,. 1990) and is induced to particularly high levels following skin wounding (Werner et al., 1992). Given the expression of KGF in mesenchyme and concomitant expression of KGFR in epithelium (Orr-Urtreger et al., 1993; Mason et al., 1994; Finch et al., 1995), KGF is well suited as a paracrine regulator of mesenchymal-epithelial interactions in development. In this regard KGF is also expressed during ductal branching morphogenesis in mesenchyme of the developing salivary gland, mammary gland, and lung (Mason et al., 1994; Finch et al., 1995). KGF has also been identified in the adult rat and human prostate (Rubin et al., 1992; Yan et al., 1992) and plays a role in androgen-dependent growth and development of the SV in neonatal mice (Alarid et al., 1994).

To determine the possible role of KGF in prostatic development, a serum-free organ culture system has been developed using ventral prostates (VPs) from newborn rats (Sugimura et al., 1996). As shown in Fig. 1a, the newborn rat VP is composed of three to four main ducts, each have three to six branches surrounded by undifferentiated mesenchyme. In the presence of testosterone (T) neonatal rat, VPs underwent

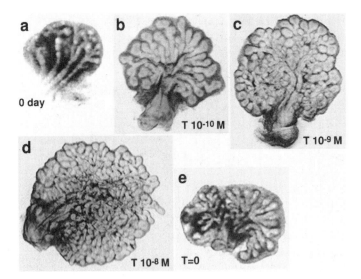

Figure 1 (a) Rat VP prior to culture. Note that each main duct has three to six distal tips and that epithelial ducts are surrounded by mesenchyme. Neonatal rat VPs cultured for 6 days in vitro under serum-free conditions in the presence of testosterone at (b)10^{-10} M, (c) 10^{-9} M, and (d) 10^{-8} M. Note increased ductal growth and branching morphogenesis at higher testosterone concentrations. (e) Neonatal rat VP cultured for 6 days in vitro under serum-free conditions in the absence of testosterone. While development has proceeded somewhat relative to glands before culture (compare with Fig. 1a), ductal growth and branching morphogenesis is far less than that elicited by an optimal dose of testosterone.

extensive ductal branching in a dose-dependent manner during 6 days of culture (Fig. 1b–d). Maximal ductal branching morphogenesis was achieved with T at 10^{-8} M. Growth (DNA content) of VPs cultured with T increased in a dose-dependent manner with maximal response observed at 10^{-8} M (Sugimura et al., 1996). Ductal branching morphogenesis was greatly reduced in VPs cultured without T (Fig. 1e), even though VPs cultured in the absence of exogenous T grew somewhat and developed relative to glands at the start of culture. This minimal growth response is attributed to the carryover of endogenous T.

In the absence of T, KGF (50 or 100 ng/ml) by itself was able to elicit extensive ductal growth and branching morphogenesis. The pattern of ductal branching induced by KGF was comparable and similar in magnitude to that elicited by T (Fig. 2). DNA content of VPs cultured with 100 ng/ml KGF alone versus T at 10^{-8} M was equivalent, 1.8 ± 0.46 mg versus 1.88 ± 0.09 mg, respectively, which represents an increase in DNA content of about 85% that of VPs before culture and an measure of about 64% that of VPs cultured without T.

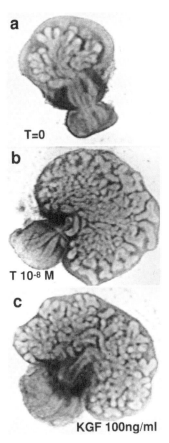

Figure 2 Representative images of rat VPs cultured for 6 days in serum-free medium (a) without testosterone, (b) with testosterone (10^{-8} M), and (c) without testosterone + KGF 100 ng/ml. Note that ductal growth and branching morphogenesis elicited by KGF is comparable to that elicited by testosterone.

During the neonatal period the rat VP undergoes cytodifferentiation to form ducts lined by tall simple columnar epithelial cells. At birth the rat VP consists of solid epithelial cords surrounded by an undifferentiated mesenchyme. After 6 days of culture in the presence of T (10^{-8} M), many of the solid epithelial cords canalized, and in such ducts the epithelium differentiated into tall columnar cells. These differentiation events were greatly retarded in newborn rat VPs cultured for 6 days without T. However, in VPs cultured for 6 days in the absence of T but with KGF at either 50 or 100 ng/ml, ductal canalization occurred and a tall columnar epithelium differentiated.

To further evaluate the role of KGF in prostatic development, newborn rat VPs were cultured as explants for 6 days in the presence of T (10^{-9} M) plus either the 1G4 anti-KGF monoclonal antibody (Mab) (Bottaro et al., 1993) or peptide 412, a soluble portion of the third IgG-like loop of the KGF-R which binds KGF and thus competes with active receptors for endogenous KGF (Bottaro et al., 1993). Both Mab 1G4 and peptide 412 greatly reduced T-induced ductal branching morphogenesis of the neonatal rat VP (Fig. 3) during 6 days of culture. The inhibitory effects of these agents were

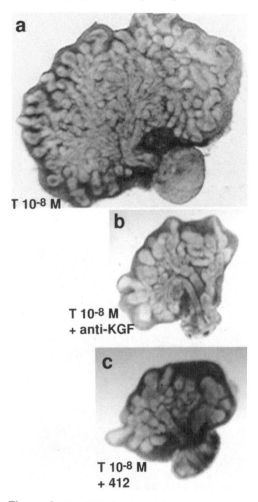

Figure 3 Rat VPs incubated for 6 days in the presence of testosterone (10^{-9} M) alone (a) or in combination with anti-KGF Mab (b) or peptide 412 (c). Note dramatic inhibition of growth and ductal branching morphogenesis elicited by anti-KGF and peptide 412.

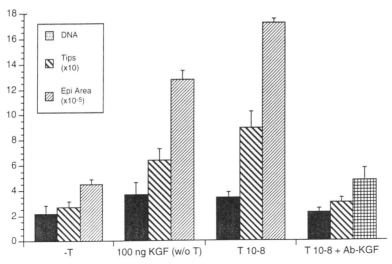

Figure 4 Effect of testosterone, KGF, anti-KGF, and peptide 412 on DNA content per rat VP, the number of ductal tips per rat VP, and epithelial area per rat VP folowing 6 days of incubation in serum-free medium.

also reflected in a reduction in (a) the number of ductal tips per explant (b) epithelial area per explant, and (c) DNA content per explant (Fig. 4) (Sugimura et al., 1996).

Since organ culture experiments implicated KGF in T-induced ductal branching morphogenesis, RT-PCR analysis was performed to assess transcripts for KGF and KGF-R in neonatal rat VPs. KGF and KGF-R transcripts were detected in rat VPs cultured with and without T. The KGF transcripts detected in the whole VP were shown to be derived from the mesenchyme. By in situ hybridization KGF-R transcripts were observed in the epithelial ducts and not in the surrounding mesenchyme (Sugimura et al., 1996). Thus, the effects of androgens on prostatic epithelial development are mediated at least in part by KGF, which is produced by mesenchyme and in turn regulates epithelial development via cell surface receptors.

While KGF is ideally suited as a paracrine factor mediating mesenchymal-epithelial interactions, it is only one of many growth factors present in the prostate (Story, 1991; Stiner, 1993; Cunha et al., 1994). Indeed development, growth, and maintenance of adult male urogenital tract structures are surely regulated by multiple interacting autocrine and paracrine factors, which are regulated either directly or indirectly by androgens or other hormones. Moreover, there is the distinct possibility that cell surface growth factor receptor pathways intersect at some level with the AR pathway, as has been recently reported (Culig et al., 1994). By using transgenic mice and tissue recombination technology in combination, it should be possible to unravel the complex web of interacting pathways that regulate growth and development of the male urogenital tract.

ACKNOWLEDGMENTS

This work was supported by NIH grants DK 32157, CA 59831, DK 47517, DK 45861, AG 13784, DK 52708, and CA 64872.

REFERENCES

Aaronson, S. A., Bottaro, D. P., Miki, T., Ron, D., Finch, P. W., Fleming, T. P., Ahn, J., Taylor, W., G., and Rubin, J. S. (1991). *Ann. N.Y. Acad. Sci., 638*: 62.

Alarid, E. T., Rubin, J. S., Young, P., Chedid, M., Ron, D., Aaronson, S. A., and Cunha, G. R. (1994). *Proc. Natl. Acad. Sci. USA, 91*: 1074.

Anderson, K. M., and Liao, S. (1968). *Nature, 219*: 277.

Baarends, W. M., van Helmond, M. J. L., Post, M., van der Schoot, J. C. M., Hooerbrugghe, J. W., de Winter, J. P., Uilenbroek, J. T. J., Karels, B., Wilming, L. G., Meijers, J. H. C., Themmen, A. P. N., and Grootegoed, J. A. (1994). *Development, 120*: 189.

Baskin, L. S., Young, P., Hayward, S., and Cunha, G. (1996). *J. Urol., 156*: 1820

Behringer, R. R., Cate, R. L., Froelick, G. J., Palmiter, R. D., and Brinster, R. L. (1990). *Nature, 345*: 167.

Behringer, R. R., Finegold, M. J., and Cate, R. L. (1994). *Cell, 79*: 415.

Bigsby, R. M., and Cunha, G. R. (1986). *Endocrinology, 119*: 390.

Bigsby, R. M., and Cunha, G. R. (1987). *Endocrinology, 120* (Suppl.): 64.

Bottaro, D. P., Rubin, J. S., Ron, D., Finch, P. W., Florio, C., and Aaronson, S. A. (1990). *J. Biol. Chem., 265*: 12767.

Bottaro, D. P., Fortney, E., Rubin, J. S., and Aaronson, S. A. (1993). *J. Biol. Chem., 268*: 9180.

Boutin, E. L., Battle, E., and Cunha, G. R. (1991a). *Differentiation, 48*: 99.

Boutin, E. L., Sanderson, R. D., Bernfield, M., and Cunha, G. R. (1991b). *Dev. Biol., 148*: 63.

Boutin, E. L., Battle, E., and Cunha, G. R. (1992). *Differentiation, 49*: 101.

Bruchovsky, N., and Wilson, J. D. (1968). *J. Biol. Chem., 243*: 2012.

Carlson, K. E., and Katzenellenbogen, J. A. (1990). *J. Steroid Biochem., 36*: 549.

Casimiri, V., Rath, N. C., Parvez, H., and Psychoyos, A. (1980). *J. Steroid Biochem., 12*: 293.

Cung, L. W., and Ferland-Raymond, G. (1975). *Endocrinology, 97*: 145.

Chung, L. W., and MacFadden, D. K. (1980). *Invest. Urol., 17*: 337.

Cooke, P. S., Young, P., and Cunha, G. R. (1991a). *Endocrinology, 128*: 2867.

Cooke, P. S., Young, P., Hess, R. A., and Cunha, G. R. (1991b). *Endocrinology, 128*: 2874.

Cooke, O. A., Buchanan, D. L., Young, P., Setiawan, T., Brody, J., Korach, K. S., Taylor, J., Lubahn, D. B., and Cunha, G. R. (1997). *Proc. Natl. Acad. Sci., 94*: 6535.

Corpechot, C., Baulieu, E. E., and Robel, P. (1981). *Acta Endocrinol. (Copenh.), 96*: 127.

Culig, Z., Hobisch, A., Cronauer, M. V., Radmayr, C., Trapman, J., Hittmair, A., Bartsch, G., and Klocker, H. (1994). *Cancer Res., 54*: 5474.

Cunha, G. R. (1972). *Anat. Rec., 172*: 179.

Cunha, G. R. (1975). *Endocrinology, 95*: 665.

Cunha, G. R. (1976a). *Int. Rev. Cytol., 47*: 137.

Cunha, G. R. (1976b). *J. Exp. Zool., 196*: 361.

Cunha, G. R. (1986). *Urologic Endocrinology* (Rajfer, J., ed.). WB Saunders Co, Philadelphia, p. 6.

Cunha, G. R., and Lung, B. (1979). *Accessory Glands of the Male Reproductive Tract* (Spring-Mills, E., and Hafez, E. S. E., eds.). Ann Arbor Science., Ann Arbor, MI, p. 1.

Cunha, G. R., and Young, P. (1991). *Endocrinology, 128*: 3293.

Cunha, G. R., Alardi, E. T., Turner, T., Donjacour, A. A., Boutin, E. L., and Foster, B. A. (1992a). *J. Androl., 13*: 465.

Cunha, G. R., Battle, E., Young, P., Brody, J., Donjacour, A., Hayashi, N., and Kinbara, H. (1992b). *Epithelial Cell Biol., 1*: 76.

Cunha, G. R., Chung, L. W. K., Shannon, J. M., Taguchi, O., and Fujii, H. (1983a). *Recent Prog. Horm. Res., 39*: 559.

Cunha, G. R., Cooke, P. S., Bigsby, R., and Brody, J. R. (1991a). *The Structure and Function of Nuclear Hormone Receptors* (Parker, M. G., ed.). Academic Press, New York, p. 235.

Cunha, G. R., Donjacour, A. A., Cooke, P. S., Mee, S., Bigsby, R. M., Higgins, S. J., and Sugimura, Y. (1987). *Endocrine Rev., 8*: 338.

Cunha, G. R., Foster, B., Thomson, A., Sugimura, Y., Tanji, N., Tusji, M., Terada, N., Finch, P. W., and Donjacour, A.A. (1995). *World J. Urol., 13*: 264.

Cunha, G. R., Hayward, S. W., Dahiya, R., and Foster, B. A. (1996b). *Acta Anat., 155*: 63.

Cunha, G. R., Lung, B., and Reese, B. (1980a). *Invest. Urol., 17*: 302.

Cunha, G. R., Reese, B. A., and Sekkingstad, M. (1980b). *Endocrinology, 107*: 1767.

Cunha, G. R., Sekkingstad, M., and Meloy, B. A. (1983b). *Differentiation, 24*: 174.

Cunha, G. R., Sugimura, Y., Foster, B. A., Rubin, J., and Aaronson, S. (1994). *Biomed. Pharacother., 48*: S9.

Cunha, G. R., Young, P., and Brody, J. R. (1989). *Biol. Reprod., 40*: 861.

Cunha, G. R., Young, P., Higgins, S. J., and Cooke, P. S. (1991b). *Development, 111*: 145.

Danielpour, D., Kadomatsu, K., Anzano, M. A., Smith, J. M., and Sporn, M. B. (1994). *Cancer Res., 54*: 3413.

Deane, H. W., and Wurzelmann, S. (1965). *Am. J. Anat., 117*: 91.

Del Buono, R., Fleming, K. A., Morey, A. L., Hall, P. A., and Wright, N. A. (1992). *Development, 114*: 67.

Deslypere, J. P., Young, M., Wilson, J. D., and McPhaul, M. J. (1992). *Mol. Cell Endocrinol., 88*: 15.

di Clemente, N., Wilson, C., Faure, E., Boussin, L., Carmillo, P., Tizard, R., Picard, J. Y., Vigier, B., Josso, N., and Cate, R. (1994). *Mol. Endocrinol., 8*: 1006.

Donahoe, P. K., Budzik, G., Trelstad, R., Mudgett-Hunter, M., Fuller, A. F. J., Hutson, J., Ikawa, H., Hayashi, A., and MacLaughlin, D. (1982). *Rec. Prog. Horm. Res., 38*: 279.

Donjacour, A. A., and Cunha, G. R. (1988a). *Dev. Biol., 128*: 1.

Donjacour, A. A., and Cunha, G. R. (1988b). *J. Cell Biol., 107*: 609a.

Donjacour, A. A., and Cunha, G. R. (1993). *Endocrinology, 131*: 2342.

Donjacour, A. A., and Cunha, G. R. (1995). *Development, 121*: 2199.

Evans, G. S., and Chandler, J. A. (1987). *Prostate, 10*: 163.

Evans, R. M. (1988). *Science, 240*: 889.

Fawell, S. E., and Higgins, S. J. (1986). *Mol. Cell. Endocrinol., 48*: 39.

Finch, P. W., Rubin, J. S., Miki, T,. Ron, D., and Aaronson, S. A. (1989). *Science, 245*: 752.

Finch, P. W., Cunha, G. R., Rubin, J. S., Wong, J., and Ron, D. (1995). *Dev. Dynam., 203*: 223.

Flickinger, C. J. (1970). *Z. Zellforsch., 109*: 1.

Flickinger, C. J. (1972). *Am. J. Anat., 134*: 107.

Gittinger, J. W., and Lasnitzki, I. (1972). *J. Endocrinol., 52*: 459.

Gloyna, R., and Wilson, J. (1969). *J. Clin. Endocrinol., 29*: 970.

Haffen, K., Kedinger, M., and Simon-Assmann, P. (1987). *J. Pediatr. Gastroenterol. Nutr., 6*: 14.

Haffen, K., Kedinger, M., and Simon-Assmann, P. (1989). *Human Gastrointestinal Development* (Lebenthal, E., ed.). Raven Press, New York, p. 19.

Hayashi, N., Sugimura, Y., Kawamura, J., Donjacour, A. A., and Cunha, G. R. (1991). *Biol. Reprod., 45*: 308.

Hayashi, N., Cunha, G. R., and Parker, M. (1993). *Epithel. Cell Biol., 2*: 66.

Hayward, S. W., Del Buono, R., Hall, P. A., and Deshpande, N. (1992). *J. Cell Sci., 102*: 361.

Hayward, S. W., Dahiya, R., Cunha, G. R., Bartek, J., Despande, N., and Narayan, P. (1995). *In Vitro, 31A*: 14.

Hayward, S. W., Baskin, L. S., Haughney, P. C., Cunha, A. R., Foster, B. A., Dahiya, R., Prins, G. S., and Cunha, G. R. (1996a). *Acta Anatomica, 155*: 81.

Hayward, S. W., Baskin, L. S., Haughney, P. C., Foster, B. A., Cunha, A. R., Dahiya, R., Prins, G. S., and Cunha, G. R. (1996b). *Acta Anatom., 155*: 94.

Hayward, S. W., Brody, J. R., and Cunha, G. R. (1996c). *Differentiation, 60*: 219.

Hayward, S. W., Cunha, G. R., and Dahiya, R. (1996d). *Ann. N.Y. Acad. Sci., 84*: 50.

He, W. W., Young, C.Y.-F., and Tindall, D. J. (1990). *Endocrinology, 126*(Suppl.): 240.

Herbst, A., and Bern, H. (1981). *Developmental Effects of DES in pregnancy*. Thieme Stratton, New York.

Higgins, S. J., Brooks, D. E., Fuller, F. M., Jackson, P. J., and Smith, S. E. (1981). *Biochem. J., 194*: 895.

Higgins, S. J., Young, P., Brody, J. R., and Cunha, G. R. (1989a). *Development, 106*: 219.

Higgins, S. J., Young, P., and Cunha, G. R. (1989b). *Development, 106*: 235.

Huggins, C., and Webster, W. O. (1948). *J. Urol., 59*: 258.

Husmann, D. A., McPhaul, M., and Wilson, J. D. (1991). *Endocrinology, 128*: 1902.

Iguchi, T., Uchima, F.-D.A., Ostrander, P. L., Hamamoto, S. T., and Bern, H. A. (1985). *Proc. Jpn. Acad., 61*: 292.

Imperato-McGinley, J. (1984). *Prog. Cancer Res. Ther., 31*: 491.

Imperato-McGinley, J., Guerrero, L., Gautier, T., and Peterson, R.E. (1974). *Science, 186*: 1213.

Imperato-McGinley, J., Peterson, R. E., and Gautier, T. (1984). *Sexual Differentiation: Basic and Clinical Aspects* (Serio, M., et al., eds.). Raven Press, New York, p. 233.

Imperato-McGinley, J., Sanchez, R. S., Spencer, J. R., Yess, B,. and Vaughan, E. D. (1992). *Endocrinology, 131*: 1149.

Jesik, C. J., Holland, J. M., and Lee, C. (1982). *Prostate, 3*: 81.

Josso, N., Picard, J. Y., and Tran, D. (1977). *Rec. Prog. Hormone Res., 33*: 117.

Jost, A. (1953). *Rec. Prog. Horm. Res., 8*: 379.

Jost, A. (1965). *Organogenesis* (Urpsrung, R. L., and DeHaan, H., eds.). Holt, Rinehart and Winston, New York, p. 611.

Julian, J., Carson, D. D., and Glasser, S. R. (1992). *Endocrinology, 130*: 68.

Jung-Testas, I., Groyer, M. T., Bruner-Lorand, J., Hechter, O., Baulieu, E.-E., and Robel, P. (1981). *Endocrinology, 109*: 1287.

Kinbara, H., and Cunha, G. R. (1995). *Prostate, 28*: 58.

Kinbara, H., Cunha, G.R., Boutin, E., Hayashi, N. and Kawamura, J. (1995). *Prostate,*

Kratochwil, K. (1987). *Developmental Biology: A Comprehensive Synthesis* (Gwatkin, R. B. L., ed.). Plenum Press., New York, p. 315.

Lasnitzki, I. (1974). *Male Accessory Sex Organs: Structure and Function* (Brandes, D., ed.). Academic Press, New York, p. 348.

Lasnitzki, I., and Franklin, H. R. (1972). *J. Endocrinol., 54*: 333.

Leav, I., Mofrin, R. F., Ofner, P., Cavazos, L. F., and Leeds, E. B. (1971). *Endocrinology, 89*: 465.

Lee, C., Sensibar, J. A., Dudek, S. M., Hiipakka, R. A., and Liao, S. (1990). *Biol. Reprod., 43*: 1079.

Lewis, R. W. (1984). *Prog. Clin. Biol. Res., 145*: 235.

Lewis, R. W., Kim, J. C. S., Irani, D., and Roberts, J. A. (1981). *Prostate, 2*: 51.

Lopes, E. S., Foster, B. A., Donjacour, A. A., and Cunha, G. R. (1996). *Endocrinology, 137*: 4225.

Lubahn, D. B., Moyer, J. S., Golding, T. S., Couse, J. F., Korach, K. S., and Smithies, O. (1993). *Proc. Natl. Acad. Sci. USA, 90*: 11162.

Lung, B., and Cunha, G. R. (1981). *Anat. Rec., 199*: 73.

Lyons, M. F., and Hawkes, S. G. (1970). *Nature, 227*: 1217.

Marchese, C., Rubin, J., Ron, D., Faggioni, A., Torrisi, M. R., Messina, A., Frati, L., and Aaronson, S. A. (1990). *J. Cell Physiol., 144*: 326.

Mason, I. J., Pace, F. F., Smith, R., and Dickson, C. (1994). *Mech. Dev., 15*: 15.

Mawhinney, M. G., and Neubauer, B. L. (1979). *Invest. Urol., 16*: 409.

McNeal, J. E. (1983a). *Monogr. Urol., 4*: 3.

McNeal, J. E. (1983b). *Benign Prostatic Hypertrophy* (Hinman, F. J., ed.). Springer-Verlag, New York, p. 152.

McNeal, J. E. (1984). *Prog. Clin. Biol. Res., 145*: 27.

Merk, F. B., Ofner, P., Kwan, P. W. L., Leav, I., and Vena, R. L. (1982). *Lab. Invest., 47*: 437.

Miki, T., Bottaro, D. P., Fleming, T. P., Smith, C. L., Burgess, W. H., Chan, A.M.-L., and Aaronson, S. A. (1992). *Proc. Natl. Acad. Sci. USA, 89*: 246.

Mobbs, B. G., Johnson, I. E., and Liu, I. (1990). *Prostate, 16*: 235.

Murakami, R. (1987). *J. Anat., 151*: 209.

Narayan, P., and Dahiya, R. (1992). *J. Urol., 148*: 1600.

Naslund, M. J., and Coffey, D. S. (1986). *J. Urol., 136*: 1136.

Nemeth, J. A., and Lee, C. (1996). *Prostate, 28*: 124.

Neubauer, B. L., Best, K. L., Hoover, D. M., Slisz, M. L., Van, F. R. M., and Goode, R. L. (1986). *Fed. Proc., 45*: 2618.

Norman, J. T., Cunha, G. R., and Sugimura, Y. (1986). *Prostate, 8*: 209.

Orr-Urtreger, A., Bedford, M. T., Burakova, T., Arman, E., Zimmer, Y., Yayon, A., Givol, D., and Lonai, P. (1993). *Dev. Biol., 158*: 475.

Pavlova, A., Boutin, E., Cunha, G. R., and Sassoon, D. (1994). *Development, 120*: 335.

Pointis, G., Latreille, M. T., and Cedard, L. (1980). *J. Endocrinol., 86*: 483.

Price, D. (1936). *Am. J. Anat., 60*: 79.

Price, D., and Ortiz, E. (1965). *Organogenesis* (Ursprung, R. L., and DeHaan, H., eds.). Holt, Rinehart and Winston, New York, p. 629.

Price, D., and Williams-Ashman, H. G. (1961). *Sex and Internal Secretions*, 3rd ed. (Young, W. C., ed.). Williams and Wilkins, Baltimore, p. 366.

Prins, G. S., and Birch, L. (1993). *Endocrinology, 132*: 169.

Prins, G. S., and Birch, L. (1995). *Endocrinology, 136*: 1303.

Prins, G., Birch, L., and Greene, G. (1991). *Endocrinology, 129*: 3187.

Prins, G. S., Woodham, C., Lepinske, M., and Birch, L. (1993). *Endocrinology, 132*: 2387.

Purvis, K., Morkas, L., Rui, H., and Attramadal, A. (1985). *Arch. Androl., 15*: 143.

Quigley, C. A., De Bellis, A., Marschke, K. B., El-Awady, M. K., Wilson, E. M., and French, F. S. (1995). *Endocr. Rev., 16*: 271.

Rajfer, J., and Coffey, D. S. (1978). *Invest. Urol, 16*: 186.

Raynaud, A,. and Frilley, M. (1947). *C. R. Soc. Biol., 141*: 658.

Rubin, J. S., Peehl, D. M., Chedid, M., Alardi, E. T., Cunha, G. R., Ron, D., and Aaronson, S. A. (1992). 2nd International Symposium on the Biology of Prostatic Growth, *NIH publication 14*:

Russell, D. W., and Wilson, J. D. (1994). *Annu. Rev. Biochem., 63*: 25.

Schulze, H., and Barrack, E. R. (1987). *Endocrinology, 121*: 1773.

Schulze, H., and Claus, S. (1990). *Prostate, 16*: 331.

Selmanoff, M. K., Goldman, B. D., and Ginsburg, B. E. (1977). *Endocrinology, 100*: 122.

Sensibar, J. A., Liu, X., Patai, B., Alger, B., and Lee, C. (1990). *Prostate, 16*: 263.

Shannon, J. M., and Cunha, G. R. (1983). *Prostate, 4*: 367.

Shima, H., Tsuji, M., Young, P. F., and Cunha, G. R. (1990). *Endocrinology, 127*: 3222.

Siiteri, P. K., and Wilson, J. D. (1974). *J. Clin. Endocrinol. Metab., 38*: 113.

Soeffing, W. J., and Timms, B. G. (1995). *J. Androl., 16*: 197.

Steiner, M. S. (1993). *Urology, 42*: 99.

Story, M. T. (1991). *Prostate Cancer: Cell and Molecular Mechanisms in Diagnosis and Treatment* (Isaacs, J. T., ed.). Cold Spring Harbor Laboratory Press, Cold Spring Harbor, NY, p. 123.

Sugimura, Y., Norman, J. T., Cunha, G. R., and Shannon, J. M. (1985). *Prostate, 7*: 253.

Sugimura, Y., Cunha, G. R., and Bigsby, R. M. (1986a). *Prostate, 9*: 217.

Sugimura, Y., Cunha, G. R., and Donjacour, A. A. (1986b). *Biol. Reprod., 34*: 961.

Sugimura, Y., Cunha, G. R., Donjacour, A. A., Bigsby, R. M., and Brody, J. R. (1986c). *Biol. Reprod., 34*: 985.

Sugimura, Y., Cunha, G. R., Yonemura, C. U., and Kawamura, J. (1988). *Human Path., 19*: 133.

Sugimura, Y., Foster, B. A., Hom, Y. H., Rubin, J. S., Finch, P. W., Aaronson, S. A., Hayashi, N., Kawamura, J., and Cunha, G. R. (1996). *Int. J. Develop. Biol., 40*: 941.

Takeda, H., and Chang, C. (1991). *J. Endocrinol., 129*: 83.

Takeda, H., Mizuno, T., and Lasnitzki, I. (1985). *J. Endocrinol., 104*: 87.

Takeda, H., Suematsu, N., and Mizuno, T. (1990). *Development, 110*: 273.

Tenniswood, M. (1986). *Prostate, 9*: 375.

Timms, B., Lee, C., Aumuller, G., and Seitz, J. (1995). *Microsc. Res. Technique, 30*: 319.

Timms, B. G., Chandler, J. A., and Sinowatz, F. (1976). *Cell Tiss Res, 173*: 542.

Timms, B. G., Mohs, T. J., and DiDio, J. A. (1994). *J. Urol., 151*: 1427.

Tsuji, M., Shima, H., Yonemura, C. Y., Brody, J., Donahoe, P. K., and Cunha, G. R. (1992). *Endocrinology, 134*: 1481.

Tsuji, M., Shima, H., Boutin, G., Young, P., and Cunha, G. R. (1994a). *J. Andrology, 15*: 565.

Tsuji, M., Shima, H., Terada, N., and Cunha, G. R. (1994b). *Endocrinology, 134*: 2198.

Turner, T., Young, P., and Cunha, G. R. (1989). *J. Cell Biol., 109*: 69a.

Van Wagen, G. (1936). *Anat. Rec., 66*: 411.

Werner, S., Peters, K. G., Longaker, M. T., Fuller, P. F., Banda, M. J., and Williams, L. T. (1992). *Proc. Natl. Acad. Sci. USA, 89*: 6896.

West, N. B., Roselli, C. E., Resko, J. A., Greene, G. L., and Brenner, R. M. (1988). *Endocrinology, 123*: 2312.

Wilson, J., George, F., and Griffin, J. (1981). *Science, 211*: 1278.

Wilson, J., Griffin, J., and Russell, D. W. (1993). *Endocr. Rev., 14*: 577.

Wilson, J. D., and Gloyna, R. E. (1970). *Rec. Prog. Horm. Res., 26*: 309.

Wilson, J. D., and Lasnitzki, I. (1971). *Endocrinology, 89*: 659.

Winter, J. S. D., Faiman, C., and Reyes, F. (1981). *Mechanisms of Sex Differentiation in Animals and Man* (Austin, C. R., and Edwards, R. G., ed.). Academic Press, New York, p. 205.

Yamashita, S., Newbold, R. R., McLachlan, J. A., and Korach, K. S. (1989). *Endocrinology, 125*: 2888.

Yamashita, S., Newbold, R. R., McLachlan, J. A., and Korach, K. S. (1990). *Endocrinology, 127*: 2456.

Yan, G. C., Nikolaropoulos, S., Wang, F., and McKeehan, W. L. (1991). *In Vitro Cell Dev. Biol., 27A*: 437.

Yan, G., Fukabori, Y., Nikolaropoulos, S., Wang, F., and McKeehan, W. L. (1992). *Mol. Endocrinol., 6*: 2123.

Yan, G., Fukabori, Y., McBride, G., Nikolaropolous, S., and McKeehan, W. (1993). *Mol. Cell Biol., 13*: 4513.

Yonemura, C. Y., Cunha, G. R., Sugimura, Y., and Mee, S. L. (1995). *Acta Anatom., 153*: 1.

Zhao, G. Q., Holterhus, P. M., Dammshäuser, I., Hoffbauer, G., and Aumüller, G. (1992). *Prostate, 21*: 183.

Zhou, Z. X., Lane, M. V., Kemppainen, J. A., French, F. S., and Wilson, E. M. (1995). *Mol. Endocrinol., 9*: 208.

22

Influence of Estrogenic Agents on Mammalian Male Reproductive Tract Development

Retha R. Newbold

National Institute of Environmental Health Sciences, Research Triangle Park, North Carolina

INTRODUCTION

Estrogens are involved in normal growth and differentiation of the mammalian female genital tract. In male reproductive tract tissues, the role of estrogens has not been emphasized, mainly because of the overwhelming importance of testosterone, the androgen responsible for male reproductive tract structure and function. In early sex differentiation, the role of estrogens, of either maternal or fetal origin, in male reproductive tract development remains unclear. Although both human and rodent fetuses are bathed in endogenous estrogens, compounds with estrogenic activity are thought to be relatively inactive in the fetal genital system because of their binding to extracellular proteins, conjugation, and metabolism to inactive forms. Therefore, the focus has remained on androgens, not estrogen, as the hormone responsible for growth and development of male reproductive tissues.

The role of estrogens in normal functioning of the male reproductive tract should not be overlooked since it has been recognized for years that the testis synthesizes estrogen as well as testosterone and that testosterone can be aromatized to estrogen (de Jong et al., 1974; Dorrington et al., 1978). Acknowledging the presence of estrogen synthesis in the male reproductive system raises the possibility that there is a specific function for estrogens and further suggests the potential of male-specific estrogen target tissues, a concept that remains under investigation (Hess et al., 1995).

Although the role of estrogens in normal male reproductive tract development and function is uncertain, it has been shown that exposure to exogenous estrogens during critical stages of sex differentiation profoundly alters male sexual development. For decades experimental studies have described the adverse effects of estrogens on male reproductive tract growth and differentiation, but only recently has the potential for environmental estrogens and other endocrine-disrupting compounds to alter male sexual differentiation raised much concern (Colborn and Clement, 1992; Colborn et al., 1995). Since the early 1980s, studies have reported "feminization" of wildlife exposed to es-

trogenic chemicals in the environment. Investigators from the University of Florida
describe altered sex differentiation of male alligators that live in polluted waters in South
Florida (Guillette et al., 1995). DDT has been implicated as the culprit; DDT or its
metabolites may act as estrogens or they may block the effects of androgens. Still other
examples exist where sex differentiation in various wildlife species is dramatically affected by exposure to exogenous estrogens: feminization of male western gulls in southern California (Fry and Toone, 1981), induction of vitellogenin synthesis in male fish
growing in the estrogenic effluent from sewage plants (Purdom et al., 1994; Sumpter
and Jobling, 1995), and the feminization of male turtles or change in sex ratio of turtles
hatching in PCB-contaminated areas (Crews et al., 1995) demonstrate the adverse effects of estrogenic compounds on male reproductive development. Whether low-level
environmental pollutants are causing adverse effects in humans remains a topic of much
controversy.

Accidental exposures of humans to high levels of estrogenic compounds, however, provide evidence of the profound effects of estrogens on the developing male
reproductive tract. The well-documented, long-lasting effects of prenatal exposure to
diethylstilbestrol (DES) on both male and female offspring is a reminder. DES, a potent synthetic estrogen, was prescribed to a large population of pregnant women to
prevent miscarriage and other pregnancy complications. Subsequently, it was shown
that their offspring exposed to DES in utero exhibited a wide range of reproductive tract
abnormalities, including a low, but significant increased incidence of vaginal adenocarcinoma (for review, see Herbst and Bern, 1981).

Taken together, these examples provide ample basis for concern that chemicals
with estrogenic activity are adversely affecting reproductive tract development and function if exposure occurs during critical stages of differentiation.

ADVERSE EFFECTS OF ESTROGENIC SUBSTANCES:
DES AS AN EXAMPLE

Although it is a relatively recent concern that chemicals in the environment like kepone,
methoxychlor, DDT, and its metabolites with estrogenic potential are causing adverse
effects, the scientific community has a long history with synthetic estrogenic compounds.
It has been recognized for centuries that the ovary controlled the estrous cycle, but not
until the twentieth century was the biologically active substance produced by the ovary
described. The term "estrogens" was coined for these substances because of their ability
to induce estrus in animals. The physiological, chemical, and pharmacological properties of estrogens were studied by many laboratories; the search for a synthetic substi-
tute followed immediately. DES, a nonsteroidal compound with properties similar to
the natural female sex hormone, estradiol, was first synthesized in 1938 in London by
Sir Edward Charles Dodds and his colleagues. DES was specifically designed for its
estrogenic activity. Like many of today's environmental estrogens, DES was structurally different from the natural estrogens. In fact, Dodd's research with DES provided
early demonstration that compounds with diverse structures could exhibit similar biological functions associated with estrogens. DES also demonstrated another significant
lesson, the potential toxic effects of estrogens.

Soon after its discovery, DES was marketed for numerous uses including as a
growth promoter for poultry and livestock. Its clinical uses were not overlooked; by

1946, DES was prescribed for numerous gynecological conditions including prevention of threatened miscarriages in high-risk pregnancies. With its increasing popularity, DES was soon widely prescribed for use even during normal pregnancies, similar to vitamin pills. In 1971, however, a report associated DES with a rare form of cancer termed "vaginal adenocarcinoma" detected in a small percentage of adolescent daughters of women who had taken the drug while pregnant. DES was also linked to more frequent benign problems in the oviduct, uterus, cervix, and vagina of DES-exposed offspring including structural and functional changes in the reproductive tract. Similarly, the DES-exposed male offspring reported structural, functional, and cellular abnormalities following prenatal exposure; hypospadias, microphallus, retained testes, inflammation, and decreased fertility were reported (Bibbo et al., 1975, 1977; Gill et al., 1976, 1978, 1979, 1981; Hoefnagel, 1976; Cosgrove et al., 1977; Driscol and Taylor, 1980; Loughlin et al., 1980; Whitehead and Leiter, 1981; Henderson et al., 1982; Depue et al., 1983; Conley et al., 1984; Pottern et al., 1985).

Thus, DES became one of the first examples of an estrogenic toxicant in humans; DES was shown to cross the placenta and induced a direct effect on the developing fetus. (For a detailed review of the history of DES, its uses, and the resulting human experience, see Newbold and McLachlan, 1996.)

Questions on the mechanisms involved in the DES-induced toxic effects have prompted the development of an animal model to study the adverse effects of estrogens on the sexual differentiation of both male and female mammals. A description follows of a murine animal model that has been used to duplicate and predict adverse effects in humans with similar DES exposure. Some of the most significant effects in males are discussed with the expectation that these findings might provide information pertinent to the evaluation of the possible adverse effects of other environmental estrogenic compounds.

DES ANIMAL MODEL TO STUDY HUMAN DISEASE

The DES experience has focused much attention on the role of estrogenic compounds in genital tract development and especially their role in inducing neoplasia when exposure occurs during sensitive developmental periods. Assessment of the abnormalities resulting from developmental exposure to DES and other estrogenic substances requires detailed knowledge of the reproductive tract. Thus, an animal model was developed to study structural, functional, and long-term effects of DES exposure during development. Outbred CD-1 mice were treated subcutaneously with DES on days 9–16 of gestation. This time period encompasses the major period of organogenesis of the genital tract in the mouse. DES doses ranged from 0.01 to 100 µ/kg maternal body weight; the highest dose was equal to or less than that given therapeutically to pregnant women, and the lower DES doses are comparable to weaker estrogenic compounds found in the environment. As seen in Table 1, the effects of prenatal exposure to DES on male mice and humans are comparable. Abnormalities such as undescended and hypoplastic testes, infertility, epididymal cysts, sperm abnormalities, hypospadias, microphallus, Müllerian duct remnants, prostatic inflammation, and squamous metaplasia of the prostatic ventricle have been reported in both species. The similarity of effects recommends the mouse model as a good comparative study. Thus, treatment with DES during critical stages of differentiation permanently alters the developing male reproductive tract,

Table 1 Comparative Developmental Effects of Prenatal Exposure to
Diethylstilbestrol in Male Offspring of Mice and Humans

Developmental effect	Mouse	Human
Subfertility	+	+
Sperm abnormalities	+	+
Decreased sperm count	+	+
Epididymal cysts	+	+
Hypoplastic and cryptorchid testes	+	+
Testicular tumors:		+
Interstitial cell tumors	+	?
Seminoma	+	+
Rete tumors	+	?
Anatomical feminization	+	+
Microphallus	+	+
Hypospadias	+	+
Retention of Müllerian duct remnants	+	+
Tumors in retained Müllerian duct remnants	+	?
Seminal vesicle tumors	+	?
Prostatic inflammation	+	+
Prostatic tumors	+	?
Immune dysfunction	+	+

Source: Adapted from Newbold, 1995.

as well as that of females, in both species. In this chapter only the developmental effects in males will be discussed.

EMBRYOLOGICAL TARGET TISSUES

Early in the normal fetal development of the reproductive tract, an undifferentiated stage
exists in which the sex of the embryo cannot be determined (Fig. 1, top panel). At this
stage, the gonads have not developed into either testis or ovary, and all embryos have
two sets of genital ducts, Müllerian (paramesonephric) and Wolffian (mesonephric)
ducts. In the female, as sex differentiation occurs, the gonad differentiates into an ovary
and the Müllerian duct develops into oviduct, uterus, cervix, and upper vagina, while
the Wolffian (mesonephric) duct regresses. In the male, under the influence of testicular secretions from the developing gonad, the mesonephric ducts are maintained and
develop into epididymis, vas deferens, seminal vesicles, and ejaculatory duct, while the
Müllerian duct regresses (Fig. 1, lower panel). Exposure to DES during critical periods of sex differentiation results in altered differentiation of the gonad and alterations
in both Müllerian (female) and Wolffian (male) duct systems. In fact, the regression
of the Müllerian ducts in males exposed to DES during gestation was incomplete, with
remnants located mainly in the area of the testis (appendix testis) and the seminal
colliculus (prostatic ventricle). These two anatomical positions are consistent with the
epididymal cysts and nodular masses of the coagulating gland and metaplasia of the
seminal colliculus observed in aged prenatal DES-exposed male mice (McLachlan et
al., 1975). To examine the early effects of DES on the developing duct systems (Fig.
2), epithelial cell height and duct diameter was determined in the Müllerian and Wolf-

Undifferentiated Stage of Reproductive Tract Development

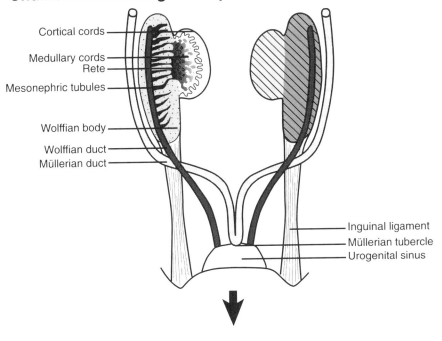

Cortical cords
Medullary cords
Rete
Mesonephric tubules

Wolffian body
Wolffian duct
Müllerian duct

Inguinal ligament
Müllerian tubercle
Urogenital sinus

Differentiated Male

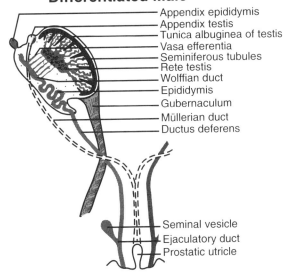

Appendix epididymis
Appendix testis
Tunica albuginea of testis
Vasa efferentia
Seminiferous tubules
Rete testis
Wolffian duct
Epididymis
Gubernaculum
Müllerian duct
Ductus deferens

Seminal vesicle
Ejaculatory duct
Prostatic utricle

Figure 1 Normal differentiation of the fetal male reproductive tract: (top) undifferentiated stage of reproductive tract development; (bottom) differentiated stage of development. (Adapted from Tuchmann-Duplessis and Haegel, 1982.)

fian ducts and the mesonephric tubule (rete) regions. Data on the diameter of structures in control and DES-treated (days 9–16 of gestation) male mice are summarized in Table 2; similar findings of increased epithelial cell height in DES-treated male mice as compared to controls were seen in each structure (data not presented). It is clearly

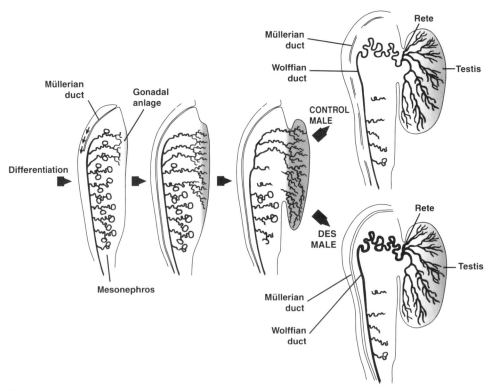

Figure 2 Detailed differentiation of cranial portion of the developing fetal reproductive tract. Note development and differentiation of Müllerian, Wolffian, and rete systems and the effects of DES during development. The Müllerian duct, which normally regresses during reproductive tract development in the male, is maintained in the prenatal DES treated mouse. In addition, mesonephric (Wolffian) duct and mesonephric tubules (rete) are also affected by exposure to DES at critical stages of differentiation. The diameter of ductal systems and cell height of the epithelium lining the ducts are increased in response to estrogen. This early histological response is manifested by long-term changes in these tissues seen later in life and discussed in the text.

illustrated that estrogens affect these developing tissues even after DES treatment is terminated on day 16 of gestation. The relationship of these early histological changes with subsequent changes, like functional defects observed later in life, is discussed.

REPRODUCTIVE TRACT DYSFUNCTION

Subfertility has been reported in both humans and mice following developmental exposure to DES (Table 1). For mice, this parameter was assessed in the animal model by breeding prenatal DES-exposed male mice to control females of the same strain. In the DES-exposed mouse, a slight reduction in fertility was seen at the DES-10 dose, but the two higher doses (50–100 μg/kg) were associated with noticeable decreases in reproduction; only 50% of DES-male offspring treated with 50 μg/kg and 40% treated with 100 μg/kg were fertile (Fig. 3). Rodent models have been reported to be insensi-

Table 2 Early Effects of DES on the Developing Reproductive Tract Tissues[a]

A. Diameter of mesonephric tubules (rete)[b]

	Control	DES
Fetal day 13	24.3 ± 1.0[c]	21.3 ± 0.0
Fetal day 14	26.3 ± 2.8	29.5 ± 1.0
Fetal day 16	24.8 ± 0.5	31.0 ± 1.0
Neonatal day 1	30.8 ± 2.5	36.5 ± 3.0
Neonatal day 5	32.5 ± 2.5	36.5 ± 3.0

B. Diameter of Müllerian duct[b]

	Control	DES
Fetal day 13	37.3 ± 1.5[c]	33.3 ± 3.0
Fetal day 14	20.3 ± 1.0	29.3 ± 2.0
Fetal day 16	0	49.5 ± 0.0
Neonatal day 1	0	189.5 ± 2.0
Neonatal day 5	0	223.3 ± 10.0

C. Diameter of Wolffian (mesonephric) duct[b]

	Control	DES
Fetal day 13	32.8 ± 2.8[c]	38.3 ± 2.8
Fetal day 14	35.5 ± 5.3	51.0 ± 2.3
Fetal day 16	47.3 ± 1.5	65.5 ± 4.5
Neonatal day 1	44.0 ± 5.0	54.8 ± 5.3
Neonatal day 5	37.3 ± 1.5	46.5 ± 2.8

[a]Male mice were exposed prenatally to DES (100 μg/kg maternal body weight) on days 9–16 of gestation unless sacrifice day occurred earlier. Males were sacrificed at ages indicated and reproductive tract tissues fixed and sectioned en bloc. Measurements were obtained from 3 separate readings per section, 3 sections per fetus, and a minimum of 3 fetuses, each from a different litter.
[b]Diameter is measured in μm from the areas indicated in schematic.
[c]Numbers in mean ± SEM. For location of each measurement, see Figure 2.

tive to toxic insult with chemicals, requiring almost 10-fold sperm count reduction to occur before fertility is affected (Meistich, 1989). Therefore, the decline in fertility that is seen following prenatal DES exposure is probably biologically significant. The extrapolation to humans, however, is difficult. Even in unexposed humans, sperm counts differ widely; moreover, some individuals have counts that make them subfertile, and some have counts in a range at which any reduction would shift them into a subfertile category. Thus, these difficulties may help explain some of the inconsistencies in fertility reported in the prenatal DES-exposed human data (Gill et al., 1976; Wilcox et al., 1995) as well as humans exposed to environmental estrogenic contaminants (Carlson et al., 1992; Bromwich et al., 1994; Irvine et al., 1994; Auger et al., 1995; Olsen et al., 1995).

In summary, in the rodent model, multiple factors appear to be related to the observed decreased fertility, including (1) retained testes and Müllerian remnants, (2)

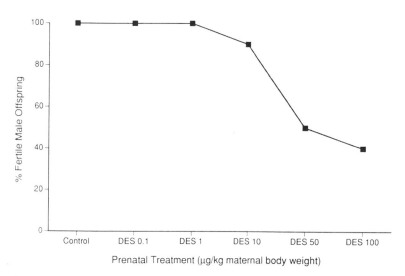

Figure 3 Fertility of adult male mice exposed prenatally to varying doses of DES. Outbred CD-1 male mice were exposed prenatally to DES on days 9–16 of gestation to varying doses of DES as described. Prenatally DES-exposed males were bred at 3 months of age to control females of the same strain. The DES-10 dose caused a slight reduction in fertility, while the two highest doses (DES-50 and 100) were associated with marked decreases in reproduction.

abnormal sperm morphology and motility, (3) lesions in reproductive tract tissues, (4) abnormal reproductive tract secretions, and (5) inflammation. Alterations in the testis, retained Müllerian ducts, and differentiation of Wolffian duct structures will be further discussed.

GONADAL EFFECTS

It has already been discussed that prenatal exposure to DES at the highest dose tested (100 µg/kg of maternal body weight) renders 60% of the male offspring sterile; the number of sterile males decreases with decreasing prenatal DES doses. The majority of the sterile males had cryptorchid testes ranging in position from firmly attached to the posterior pole of the kidney to high in the scrotal sac. Studies from other laboratories have also described undescended testes and testicular hypoplasia in 70% of male offspring of mice treated with DES (10 mg/kg) on gestational days 17 or 19 (Normura and Kanzaki, 1977). In the current study, doses lower than prenatal DES-10 were not associated with retained testes and fertility did not appear to be compromised. Table 3 summarizes some of the effects of DES on the testis; testis location and weights were determined and testis weight to body weight ratios were calculated. To determine if fertility problems observed at the 50 and 100 µg/kg doses of DES were solely related to cryptorchid testis, mice with descended or retained testes were examined for sperm abnormalities. As seen in Table 4, prenatal DES exposure caused an increase in sperm abnormalities and a decrease in the number of motile sperm in descended as well as retained testes. However, sperm samples collected from cryptorchid testes showed a higher proportion of morphologically abnormal sperm and nonmotile sperm, and in 80%

Table 3 Effects of Prenatal DES Exposure on Testis[a]

Treatment	Body weight (g)	Testis location[b]	Testis weight (g)		Testis Weight:Body Weight Ratio	
			Left	Right	Left	Right
Control	39.05 ± 2.83	DT	0.12 ± 0.003	0.13 ± 0.003	0.35 ± 0.02	0.37 ± 0.03
DES-0.01	38.49 ± 1.75	DT	0.11 ± 0.006	0.12 ± 0.006	0.29 ± 0.02	0.32 ± 0.02
DES-0.1	34.19 ± 0.69	DT	0.13 ± 0.007	0.14 ± 0.008	0.39 ± 0.03	0.41 ± 0.03
DES-1	33.33 ± 1.31	DT	0.12 ± 0.007	0.13 ± 0.008	0.36 ± 0.03	0.38 ± 0.04
DES-2.5	34.09 ± 1.64	DT	0.11 ± 0.007	0.11 ± 0.009	0.32 ± 0.03	0.33 ± 0.03
DES-5	33.12 ± 0.63	DT	0.12 ± 0.003	0.13 ± 0.003	0.37 ± 0.02	0.41 ± 0.02
DES-10	31.07 ± 1.07	DT	0.11 ± 0.013	0.13 ± 0.009	0.37 ± 0.04	0.41 ± 0.03
DES-50	33.16 ± 2.15	RT	0.07 ± 0.008	0.07 ± 0.004	0.21 ± 0.02	0.19 ± 0.02
		DT	0.12 ± 0.003	0.13 ± 0.001	0.35 ± 0.01	0.38 ± 0.01
DES-100	32.84 ± 2.52	RT	0.08 ± 0.008	0.08 ± 0.003	0.24 ± 0.03	0.24 ± 0.02
		DT	0.11 ± 0.007	0.13 ± 0.002	0.33 ± 0.02	0.35 ± 0.01

[a]Male mice were exposed prenatally to DES on days 9–16 of gestation to varying doses of DES. Males were sacrificed at 3 months of age and body weight, testis location, and weights determined.

[b]Prenatal DES-treated males exhibited cryptorchidism (retained testis at high doses; location of testis ranged from firmly attached to the posterior pole of the kidney, high in the abdominal cavity to high in the scrotal sac).

DT = Descended testis; RT = retained testis.

Numbers are the mean ± SEM from 5 animals per treatment group except DES-50 and DES-100, which were represented by 14 animals in each group.

Table 4 Sperm Abnormalities in Mice Exposed Prenatally to DES

Prenatal treatment	No. of mice	Sperm	
		% Abnormal	% Motile
Control	10	5.9 (1.5–13.9)	86.4 (43–98)
DES-50			
Descended	8	17.1 (3.5–48.0)	22.6 (0–96)
Retained	10	56.2 (35.1–100)	15.3 (0–45.1)
DES-100			
Descended	7	18.7 (2.6–40.0)	21.5 (0–95)
Retained	10[a]	39; 100	1; 0

Sperm were collected from the vas deferens of sexually mature CD-1 male mice and analyzed. Males were the offspring of mice treated subcutaneously with DES (50 or 100 µg/kg/day) on days 9–16 of gestation; values in parentheses are ranges.

[a]Eight of these animals had no detectable sperm that could be analyzed. Mice exposed to lower doses with DES prenatally were not analyzed for sperm abnormalities since their reproductive capacity was not significantly different from control animals.

of these cases no sperm could be collected from the vas deferens (McLachlan, 1981). This suggests an adverse affect of estrogens on the testis, in addition to cryptorchidism.

To examine the long-term effects on the testis and to follow up on reports suggesting an association of prenatal DES exposure and the development of testicular seminoma (Gill et al., 1978, 1979, 1981; Conley et al., 1984), male mice exposed to DES in utero were followed to 10–18 months of age. In addition to nonmalignant abnormalities (retained testis) already described in this mouse model and described in men exposed prenatally to DES, there were degenerative changes in the testes of 82% of the mice prenatally exposed to DES, ranging from mild to severe; both retained and descended testis were involved (Table 5). Degenerative changes were described by four

Table 5 Testicular Lesions Following Prenatal Exposure to Diethylstilbestrol[a]

Lesion	Number
Retained[b]	252/277 (91)
Inflammation	23/277 (8)
Degenerative changes[c]	226/277 (82)
Interstitial cell tumor	2/277 (1)
Interstitial cell carcinoma	5/277 (2)

[a]Lesions in the reproductive tract of male mice exposed prenatally to DES. Males were the 10- to 18-month offspring of CD-1 mice treated with DES (100 µg/kg, subcutaneously) on days 9–16 of gestation. Number in parentheses is %. Similar abnormalities were not observed in control animals in this study.

[b]This number represents animals with at least one retained testis.

[c]These data include degenerative changes categories 2, 3, and 4 described in the text.

categories of increasing severity: (1) relatively normal-looking but smaller than control testis, (2) marked reduced spermatogenesis with giant cells within the lumen of the seminiferous tubules, (3) atrophied seminiferous tubules with hyalinized basement membrane and thickened arterioles, and (4) mostly necrotic or scar tissue comprising the testis. Categories 2, 3, and 4 are included in Table 5 as degenerative changes of the testis. Mineralization was seen in the testes of 9% of the DES-exposed mice. This was primarily in degenerative testes and often within reminiscent seminiferous tubules. These types of degenerating changes and the occurrence of mineralization of the testis were not observed in any of the corresponding control animals in this study.

Marked inflammation of the testis was seen in 8% of prenatally exposed DES animals (Table 5). Inflammatory changes appeared to be secondary to either escape of sperm from seminiferous tubules into interstitial tissues and/or necrosis; primary inflammatory disease in the testes was not apparent. Sperm granulomas were also frequently seen in degenerating testes. These were typical granulomatous reactions with recognizable sperm, mixed with or surrounded by neutrophils in a collar of foamy macrophages.

Interstitial cell tumors were seen in 7 out of 227 DES-treated males. These lesions mainly occurred in the older animals. Interstitial cell hyperplasia and carcinoma have been produced experimentally in certain strains of mice after prolonged treatment with various estrogenic compounds; in fact, Huseby and colleagues (1976) have studied many aspects of interstitial cell tumor formation. The significance of the finding in the prenatal study is the fact that there are so many interstitial cell carcinomas in proportion to interstitial cell tumors; out of seven tumors, five are malignant. In addition, interstitial cell tumors are not common in this strain of mouse, and these mice were treated only during prenatal development, not long term as reported by other investigators (Huseby, 1976).

These data suggest that mice exposed to DES have an increased risk of developing testicular abnormalities including hypotropic testes and interstitial cell tumors. Although the incidence of testicular tumors is low in the corpus testis, the additional finding of rete testis adenocarcinoma (discussed later) raises the combined incidence of DES-exposed animals with tumors to 8%. Prior to these findings, most of the testicular abnormalities have been subtle, but these reports suggest that the changes in the testis, including neoplasia, are significant.

Induction of interstitial cell tumors in mouse testis has been reported after prolonged administration of various estrogenic compounds; however, the strain of mouse, the specific agent, and the duration of treatment affect the results. Spontaneously occurring interstitial cell tumors have been reported in hybrid strains of mice as well as in experimentally induced cryptorchid testis of Balb/c mice. Other than the findings in the DES mouse model described in this laboratory, the only demonstrated association between malignant growths in the testis and DES exposure has been in an adult mouse injected repeatedly with larger doses of the drug (Hooker, 1940).

Although seminomas have been experimentally induced in dogs, such germ cell tumors of the testis are rare in rodents. Cryptorchidism is considered to be a predisposing factor to seminoma in men and dogs, but in spite of the high incidence of retained testis in the mouse model, we were able to demonstrate this particular testicular lesion in only one case in all our historical DES-treated animals.

The association of prenatal DES exposure and the development of testicular tumors in men has become a subject of much controversy over the last few years as attention has shifted to identify the DES-exposed males. Some reports, specifically addressing factors for cancer of the testis, list prenatal DES exposure as a risk factor, while other studies show no relationship to hormonal treatment during pregnancy. The data in the experimental mouse model support the contention that DES-treated males are at a greater risk for testicular tumors than unexposed males.

MÜLLERIAN DUCT EFFECTS

Early in genital tract development, the Müllerian (paramesonephric) ducts in the male fetus regress in response to Müllerian-inhibiting substance (MIS) produced by the fetal testes (Fig. 1). The mechanism by which this substance acts is uncertain; however, MIS or an interaction of MIS and testicular secretions seems to be essential for Müllerian duct regression during the critical phase of sexual development. Since 1939, it has been known that exogenous estrogens interfere with the regression of Müllerian ducts in the male (Wolf, 1939). Studies from our laboratory have also shown that prenatal treatment of mice with DES results in the persistence of Müllerian duct derivatives in adult males, which are homologous to the female differentiated duct structures such as oviduct, uterus, and upper vagina (McLachlan et al., 1975; McLachlan, 1981; Newbold and McLachlan, 1985, 1988: Newbold et al., 1987). Since a similar report by Driscol and Taylor (1980) described persistent Müllerian duct remnants in humans exposed in utero to DES, retention of Müllerian tissue in adults may be a general DES-induced biological phenomenon.

Several studies in our laboratory have suggested that DES exerts its effect mainly by altering the response of the Müllerian duct tissue itself rather than by suppression of MIS production by the fetal testes or a structural effect in the MIS protein (Newbold et al., 1984). By using an organ culture assay system for MIS activity where undifferentiated reproductive tract tissues are co-cultured with fetal testes, mechanisms of DES inhibition of MIS-induced regression of the Müllerian ducts were studied (Fig. 4). In this culture system, DES-treated or control undifferentiated ducts (fetal reproductive tracts) were recombined along with treated or control fetal testes. Prenatal DES exposure was by subcutaneous injection to the mother (100 μg/kg body weight) on days 9–12 of gestation. All embryonic tissues were removed on day 13 of gestation and cultured for 72 hours. In organ culture, Müllerian duct regression, comparable to that seen in vivo, occurred when control reproductive tracts were associated with control testes (Fig. 4A). However, maintenance of the Müllerian ducts was observed in 100% of the tissues when DES-treated testes and DES-treated reproductive tracts were cultured together (Fig. 4B). When recombinations of control reproductive tracts and DES-treated testes were formed, regression of the Müllerian ducts was seen in 87% (Fig 4C). But in the combinations of control testes and DES-treated reproductive tracts, 41% of the cultured tissue showed partial regression of the Müllerian duct and 59% showed no regression (Fig. 4D). These data support in vivo studies, suggesting that prenatal exposure to DES inhibits Müllerian duct regression, but also further suggest that this inhibitory effect is mainly due to a decrease in responsiveness of the treated fetal Müllerian duct (Newbold et al., 1984).

The long-term effects on retained Müllerian duct remnants in male offspring were followed. Persistent Müllerian remnants in the cranial portion of the male reproductive tract were studied; caudal persistent Müllerian remnants (prostatic utricle) have been

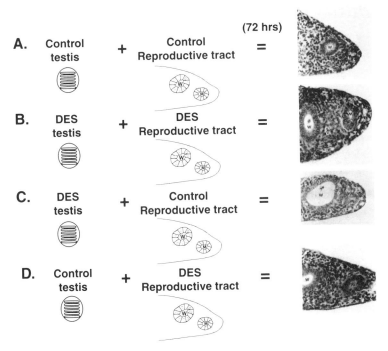

Figure 4 Organ culture of fetal testis and undifferentiated reproductive tracts. Pregnant female mice were exposed to corn oil (control) or DES dissolved in corn oil (DES 100 h/kg maternal body weight) on days 9–12 of gestation. Embryonic tissues were removed on day 13 of gestation and cultured for 72 hours. Undifferentiated ducts (Müllerian and Wolffian) were removed from fetuses on day 13 of gestation and recombined with fetal testis of the same age. (A) The Müllerian duct (M) regresses in organ culture while the Wolffian duct is maintained: (control[testis] + control[reproductive tract]). (B) The Müllerian duct (M) persists in culture and appears as it did at the original time of explant. The Wolffian duct (W) is also stimulated: (DES[testis] + DES[reproductive tract]). (C) The Müllerian duct (M) lumen is significantly reduced and the mesenchyme has started to condense around the duct. This diagram demonstrates partial regression of the Müllerian duct: (DES[testis] + Control[reproductive tract]). (D) There is little regression of the Müllerian duct (M). It persists as it did at the original time of culture. The Wolffian duct (W) also appears to be stimulated: (Control[testis] + DES[reproductive tract]). These data support previous in vivo results that prenatal exposure to DES has an inhibitory effect on Müllerian duct regression and further suggest that this inhibitory effect is mainly due to a decrease in responsiveness of the treated embryonic Müllerian duct. (From Newbold et al., 1984.)

previously described (McLachlan, 1981), as well as other genital abnormalities induced in other male rodent models treated prenatally or neonatally with DES or other estrogenic compounds (Greene et al., 1940; Thorburg, 1948; Dunn and Greene, 1963; Mori, 1967; Arai, 1970, 1984; Arai et al., 1977, 1979; Nomura and Kanzaki, 1977; Rustia, 1979; Vorherr et al., 1979). At 10–18 months of age, prenatally DES-exposed males were examined for reproductive tract abnormalities in the area of the testis, epididymis, and vas deferens. Prominent Müllerian remnants were observed in 268 out of 277 (97%) of the DES-exposed males but not in any of the control animals (Table 6). These remnants demonstrated the ability to differentiate into "femalelike structures" homologous to oviduct and uterus. The Müllerian remnants were often enlarged and cystic

Table 6 Abnormalities in Male Mice Exposed Prenatally
to Diethylstilbestrol[a]

	Number of Animals	
	Out of 277	Percent
Müllerian remnants[b]		
Prominent structure	268	97[c]
Cystic structure	121	44[c]
Cystic hyperplasia	18	6[c]
Hyperplasia	85	31[c]
Inflammation	27	10[c]
Tumors[d]	23	8[c]
Wolffian derivatives[e]		
Cystic epididymis	32	12[c]
Inflammation	87	29[c]
Sperm granuloma of epididymis	16	6[c]
Hyperplasia of epididymal duct	1	0
Tumors[f]	1	0

[a]Males were 10- to 18-month-old offspring of CD-1 mice treated subcutaneously
with DES (100 µg/kg) on days 9–16 of gestation.
[b]Squamous metaplasia and gland formation in the male Müllerian remnants are
not included in the table because of low incidence.
[c]Significantly different from corresponding controls: Fisher Exact Test $p < 0.05$.
[d]Tumors of the Müllerian remnants were distributed and cytologically characterized as follows: 2 adenomas, 15 cystadenomas, 1 carcinoma, 2 stromal tumors,
1 stromal sarcoma, 1 complex stromal sarcoma, and 1 adenocarcinoma.
[e]At 18 months of age, the control animals had minor inflammatory changes in
Wolffian derivatives. One 18-month-old control animal had severe inflammation
of the epididymis and another animal had a sperm granuloma. No other lesions
were observed in the control animals at any age in this study.
[f]Tumor in the Wolffian derivatives was adenoma of the epididymis.
Source: Newbold et al., 1987.

(44%) and shared supporting connective tissue with adjacent male structures. Previously
reported lesions (McLachlan et al., 1975), termed "epididymal cysts," were histologically determined to be cystic oviductlike structures and were, therefore, considered a
Müllerian duct abnormality. Pathological changes in these male oviductal homologs
included diverticuli or "gland formation" similar to those reported in DES-exposed
female mice. Squamous metaplasia in the uterine homolog and cystadenomas were seen
in the Müllerian remnants of DES-treated animals ranging from 3 to 18 months of age.
In addition, there were some malignant lesions such as carcinoma, stromal tumors, stromal sarcoma, complex stromal sarcoma, and adenocarcinoma.

These data demonstrate that transplacental exposure to the estrogenic substance
DES affects the differentiation and normal development of the male genital tract involving retained Müllerian duct–derived tissues. The long-term changes in this tissue include both benign and malignant tumors.

MESONEPHRIC (WOLFFIAN) DUCT EFFECTS

Studies from our laboratory have pointed to the fact that the Wolffian (mesonephric)
duct, as well as the Müllerian duct, is a target for DES (McLachlan et al., 1980;

Newbold et al., 1983; Haney et al., 1984). In fact, in adult female mice and humans exposed to DES, hyperplastic mesonephric remnants are a common finding (Haney et al., 1984). The relationship of these dysmorphic mesonephric structures to the pathogenesis of hyperplastic or neoplastic disease in the female is still being studied. The findings, however, raise the possibility that structures derived from the mesonephric ducts or tubules in male offspring may be dysplastic. In fact, as seen in Table 6, epididymal structures were cystic (12%) and exhibited inflammation (29%) and sperm granulomas (6%) in animals 10–18 months of age. Epididymal cysts specifically of mesonephric duct and an adenoma of the epididymal duct in separate animals were observed. No comparable abnormalities were noted in 122 control males of corresponding ages.

In addition to effects on epididymal structures, changes were also noted in the rete testis. Byskov (1978, 1981) has attributed the differentiated rete system to mesonephric tubular origin (Fig. 2). Therefore, the rete testis, a clearly definable adult structure apparently derived from mesonephric tubules, was specifically evaluated in males exposed prenatally to DES. In a group of animals ranging from 10 to 18 months of age, 56% of the male mice exposed to DES during gestation had various degrees of papillary proliferation and hyperplasia of the epithelium of the rete testis. The simplest form of this lesion consisted of knoblike overgrowth of the cuboidal epithelium and diffuse hyperplasia of the epithelium. These overgrowths had coalesced to form papillomas consisting of small cuboidal cells with increased hyperchromatism and vacuolated cells in other DES-treated animals. Control males, 10–18 months of age, had mild focal epithelial hyperplasia of the rete testis in 24% of mice, but papillary proliferation of the epithelium was never observed.

A more severe lesion resembling adenocarcinoma of the rete testis was observed in mice after in utero exposure to DES. This lesion was not found in any controls but was found in 5% of the prenatally DES-exposed mice. Although distant metastases were rarely found, the tumors often infiltrated into the seminiferous tubules; the histological pattern of these tumors was suggestive of either papillary adenocarcinoma or tubulopapillary adenocarcinoma. Detailed reports of these animals are given in Newbold et al. (1985, 1986, 1996) and Newbold and McLachlan (1988).

Adenocarcinoma of the rete testis in humans and experimental animals is extremely rare. Criteria were first formulated by Felk and Hunter (1945) and have been accepted by most investigators as the basis for identifying this neoplasm in humans. The diagnostic criteria for tumors of the rete testis are (1) involvement centering on the mediastinum testis rather than in the testis proper, (2) lack of direct extension through the parietal tunica, (3) transition from normal epithelial structure to neoplastic structures in the rete testis, (4) no evidence of teratoma, and (5) lack of any other primary tumor.

In the mouse, as in humans, tumors were confined to the mediastinum testis. Maximal involvement of the rete testis was seen with minimal involvement of the corpus. Tumor cells appeared to be transformed from cytologically normal rete epithelium to neoplastic epithelium. Sections were examined from the mouse testis containing the lesion and from the opposite testis, but no teratomatous elements were found. In one mouse, both an interstitial cell tumor in the corpus testis and an adenocarcinoma of the rete testis were observed. Thus, the lesions in mice conform closely to the criteria established for rete adenocarcinoma in humans.

The rare occurrence of adenocarcinoma of the rete testis in humans shows no age preference (age range 20–80 years), and tumors were reported almost equally divided between right and left testis. Likewise, in the animal model there was no statistical difference between the prevalence of rete testis lesions in mice 10–18 months of age, although older animals have not been examined. In humans, the tumors usually were present as a testicular mass often associated with a hydrocele. It is of special interest that, in the group of human cases reported, there were at least three tumors in maldescended testes. Cryptorchidism have been implicated as a predisposing factor for testicular cancer, like seminomas. The previously reported high incidence of retained testes in mice following prenatal DES exposure and the occurrence of this specific rare lesion of the testis resembling rete adenocarcinoma, described in the present report, raise the possibility of an association between cryptorchidism, prenatal DES exposure, and testicular cancer. In fact, researchers studying risk factors for cancer of the testis have evidence supporting this association (Henderson et al., 1979). Although cryptorchidism results in decreased or lack of spermatogenesis in male mice, this cannot solely account for the higher prevalence of rete lesions since 36% of the animals diagnosed within the rete lesion had spermatogenesis occurring in the testis.

The demonstration of an extremely rare lesion such as rete testis adenocarcinoma is unique in its prevalence alone since it appears in 5% of the prenatal DES animals. Yoshitomi and Morii (1984) reported the spontaneous occurrence of a rete adenocarcinoma in a 23-month-old mouse from a colony of 500 aged mice (0.2% incidence). Therefore, our data suggest prenatal exposure to DES results in at least a 20-fold increase in the prevalence of adenocarcinoma. In fact, if these males were allowed to age beyond 18 months, they might have a higher incidence of abnormalities of the rete.

In an attempt to increase the incidence of rete tumors in the prenatally DES-exposed male mice, 5-mg pellets of estradiol were subcutaneously implanted into the animals for 2 months prior to their sacrifice. This type of treatment, secondary exposure during adult life to an estrogen, did not increase the tumor incidence at this site. The rete lesions did not appear to be responsive to estrogen stimulation (data unpublished).

Although the induction of interstitial cell tumors by DES in adult mice has been well studied, no studies of estrogen-treated adult mice have reported abnormalities of the rete. The increased incidence of lesions in the rete testis seen in this mouse model suggests that this lesion is associated with developmental exposure to DES. An increased incidence of a rare tumor, vaginal adenocarcinoma, focused attention on the adverse effects on female offspring of women given DES while pregnant. The occurrence of rete testis tumors in the male offspring of mice given DES during pregnancy suggests that this may be an analogous situation since naturally occurring rete testis lesions are extremely rare.

To date, no reports of rete hyperplasia or adenocarcinoma in humans have been attributed to prenatal exposure to DES, although three cases of seminomas have been described in prenatally DES-exposed men, suggesting an association of prenatal DES treatment with subsequent development of testicular tumors (Gill et al., 1978, 1979; Conley et al., 1983; Yoshitomi and Morii, 1984). It is interesting to note that a recent report states that rete adenocarcinoma can be misdiagnosed as seminoma and seminoma must be ruled out before a diagnosis of rete adenocarcinoma can be made; thus, caution should be taken in diagnosing any testicular lesions associated with prenatal DES exposure.

Together these data demonstrate that exposure to DES during critical stages of sex differentiation affects the differentiation and development of the Müllerian (para-mesonephric) and Wolffian (mesonephric) ducts and tubules. The long-term changes in these tissues include lesions, some of which are neoplastic, although the natural history of the lesions is not known.

MECHANISMS

Numerous studies demonstrate that prenatal exposure to DES inhibits Müllerian duct regression; studies with the prenatal DES exposed mouse model suggest that this in-hibitory effect is due to a decrease in responsiveness of the treated embryonic duct rather than a defect in the testicular produced MIS. It remains uncertain whether the male Müllerian remnants developed tumors spontaneously rather than as a direct consequence of DES treatment. Further, the possibility of factors from the testis stimulating tumor formation needs to be determined. The fact that DES-treated males had higher levels of circulating estrogens may also increase the incidence of tumors in these remnants, although implantation of estrogen pellets in 10-month-old prenatally DES-exposed male mice did not increase the incidence of Müllerian duct tumors. Although tumor incidence could not be accelerated with additional estrogen treatment, these Müllerian duct rem-nants were demonstrated to exhibit a strong immunostaining reaction for the estrogen receptor and demonstrated the ability to produce lactoferrin, an estrogen-responsive uterine protein in females, in response to estrogen treatment.

The role of the estrogen receptor (ER) in the induction of abnormalities and tu-mors following developmental exposure to DES is being further studied using transgenic mice that overexpress ER (MT-mER). Transgenic ER mice were developmentally treated with DES and followed as they aged. Preliminary data suggest that the mice that are overexpressing the ER are at a higher risk of developing abnormalities including rete adenocarcinoma as compared to DES-treated wild-type mice and, further, that these abnormalities occur at an earlier age. These transgenic mouse data suggest that the ER may play a role in DES-induced toxic effects in the mouse.

Another mechanism being further investigated involves altered protein synthesis in reproductive tract tissues exposed prenatally to DES. Lactoferrin, an estrogen-induc-ible uterine glycoprotein in the female, has been identified in the seminal vesicle of the prenatally DES-exposed male mouse. Seminal vesicle tissues of the male mouse do not normally synthesize lactoferrin, however, following prenatal DES treatment, the seminal vesicle constitutively synthesizes lactoferrin; it is regulated by estrogen as it is in the uterus (Pentecost et al., 1988; Newbold et al., 1989; Beckman et al., 1994). Thus, the prenatal DES seminal vesicle tissue (Wolffian-duct derived) has been permanently "femi-nized" to produce lactoferrin. Other protein patterns that are typically female (Newbold et al., 1984) continue to be studied.

Certainly, the retained Müllerian duct remnants in the male have hyperplastic and dysplastic potential and appear to be more susceptible to tumor formation than adja-cent Wolffian-derived male structures, although a higher incidence of epididymal cysts and rete adenocarcinoma were observed in DES-treated males as compared to control males. Accumulating experimental data support the contention that DES-treated male mice are at a higher risk for tumor or cyst formation than normal males.

A clear direct relationship between prenatal DES exposure and neoplasia of the Müllerian remnants and Wolffian-derived structures in men has not been established. Increased numbers of genital and urinary tract abnormalities were found among sons of women treated with DES during pregnancy, but to date no strong evidence suggests a link to cancer except for the reports suggesting an association with testicular seminoma (Gill et al., 1978, 1979: Conley et al., 1983). The animal model raises the possibility that remnants of Müllerian duct structures may pose a risk of developing benign or malignant changes in humans as they age. Furthermore, Wolffian duct–derived structures may also be affected; screening for rete testicular adenocarcinoma should be considered as well as continued careful follow-ups for testicular abnormalities.

SUMMARY AND CONCLUSION

Accumulating evidence suggests that exposure of the developing male fetus to exogenous estrogen adversely affects the differentiation of the genital tract. Some of the alterations are summarized in Table 7. Data thus far suggest that reproductive tract structure and function are altered, and long-term changes include neoplasia. Although fertility is decreased in mice after developmental estrogen exposure, this parameter does not appear to be the most sensitive one, since low-dose estrogen–exposed males do not experience infertility or subfertility. The same low-dose males do, however, have an increased incidence of abnormalities of the reproductive tract, including inflammation and tumors, later in life. This suggests that estrogenic substances occurring in the environment at low-dose amounts may not necessarily adversely affect fertility, but they may have additional long-term consequences that warrant concern.

Although animal studies may be considered carefully if extrapolation to humans is to follow, the prenatal DES mouse model has provided some interesting comparisons to similarly exposed humans. In fact, early experimental studies of cryptorchid

Table 7 Summary of Alterations of the Male Reproductive
Tract After Developmental Estrogen Exposure

Fertility (subfertile or sterile)
Testis
Location (cryptorchid)
Function (sperm # and % motile ↓; % abnormal sperm ↑)
Histology (↑ inflammation; rete testis hyperplasia and neoplasia; interstitial cell tumors; seminomas)
Müllerian-derived structures
Retained
Histology—tumors
Prostatic utricle—hyperplastic and tumors
Mesonephric-derived structures (generalized inflammation)
Epididymis—inflammation, sperm granulomas, tumors
Prostate—tumors
Seminal vesicle—feminization; tumors
Coagulating gland—tumors
Ampullary glands—excessive
Rete testis—tumors

testes and epididymal cysts were predictive of findings in DES-exposed humans. Therefore, considering the data presented in this report, close surveillance of DES-exposed men is essential, especially for potential long-term adverse effects. In addition, although DES is a potent estrogen, it provides markers of the adverse effects of exposure to estrogenic substances during development, whether these exposures come from naturally occurring chemicals, synthetic or environmental contaminants, or pharmaceutical agents.

ACKNOWLEDGMENTS

The author is greatly indebted to Ms. Wendy Jefferson for skillful technical expertise and her critical editorial comments and typing of this manuscript. The author also thanks Dr. Michael Shelby and Ms. Kristine Witt for their helpful comments and review of the manuscript. Finally, the long-standing association and contribution to the field of environmental estrogens of Dr. John McLachlan is acknowledged.

REFERENCES

Arai, Y. (1970). *Endocrinology, 86*: 918.

Arai, Y. (1984). *Experimentia, 24*: 180.

Arai, Y., Suzuki, Y., and Nishizuka, Y. (1977). *Virchows Arch.* [A] *376*: 21.

Arai, Y., Chen, C. Y., and Nishizuka, Y. (1978). *Gann, 69*: 861.

Auger, J., Kunstman, J., Czyglik, and Jouannet, P. (1995). *N. Engl. J. Med., 332*(5): 281.

Beckman, W. C., Newbold, R. R., Teng, C. T., and McLachlan, J. A. (1994). *J. Urol., 151*: 1370.

Bibbo, M., Al-Naqeeb, M., Baccarini, I., Gill, W., Newton, M., Sleeper, K., Sonek, M., and Wied, G. L. (1975). *J. Reprod. Med., 15*: 29.

Bibbo, M., Gill, W. B., Azizi, F., Blough, R., Fang, V. S., and Rosenfield, R. L. (1977). *Am. J. Obstet. Gynecol., 49*(1): 103.

Bromwich, P., Cohen, J., Stewart, I., and Walker, A. (1994). *Br. Med. J., 309*: 19.

Byskov, A. G. (1978). *Biol. Reprod., 19*: 720.

Byskov, A. G. (1981). *Mechanisms of Sex Differentiation in Animals and Man* (C. R. Austin and R. G. Edwards, eds.). Academic Press, New York, p. 145.

Carlsen, E., Giwereman, A., Keiding, N., and Skakkeback, N. (1992). *Br. Med. J., 305*: 609.

Colborn, T., and Clement, C. eds. (1992). *Chemically-Induced Alterations in Sexual and Functional Development: The Wildlife/Human Connection*. Princeton Scientific Publishing, Princeton, NJ.

Colborn, T., Dumanoski, D., and Meyers, J. P. (1995). *Our Stolen Future*. Penguin Books, New York.

Conley, G. R., Sant, G. R., Ucci, A. A., and Mitcheson, H. D. (1984). *JAMA, 249*: 1325.

Cosgrove, M. D., Benton, B., and Henderson, B. E. (1977). *J. Urol., 117*: 220.

Crews, D., Bergeron, J. M., and McLachlan, J. A. (1995). *Environ. Health Perspect., 103*(7): 73.

de Jong, F. H., Hey, A. H., and Van der Molen, H. J. (1974). *J. Endocrinol., 60*: 409.

Depue, R. H., Pike, M. C., and Henderson, B. E. (1983). *J. Natl. Cancer Inst., 71*: 1151.

Dorrington, J. M., Fritz, B., and Armstrong, D. T. (1978). *Biol. Reprod., 18*: 55.

Driscol, S. G., and Taylor, S. H. (1980). *Obstet. Gynecol., 56*: 537.

Dunn, T. B,. and Greene, A. W. (1963). *J. Natl. Cancer Inst., 31*: 425.

Feek, J. D., and Hunter, W. C. (1945). *Arch. Pathol., 40*: 399.

Fry, M., and Toone, C. K. (1981). *Science, 213*: 922.

Gill, W. B., Schumacher, G. F. B., and Bibbo, M. (1976). *J. Reprod. Med., 16*: 147.

Gill, W. B., Schumacher, G. F. B., and Bibbo, M. (1978). *Intrauterine Exposure to Diethylstil-bestrol in the Human* (A. L. Herbst, ed.). American College of Obstetrics and Gynecology, Chicago.

Gill, W. B., Shumacher, G. F. B., Bibbo, M., Straus, F. H., II, and Shoenberg, H. W. (1979). *J. Urol., 122*: 36.

Gill, W. B., Schumacher, G. F. B., and Hubby, M. H. (1981). *Developmental Effects of Diethylstil-bestrol (DES) in Pregnancy* (A. L. Herbst and H. A. Beru, eds.). Thieme Stratton, New York, p. 103.

Greene, R. R., Burrill, M. W., and Ivy, A. C. (1940). *Am. J. Anat., 67*: 305.

Guillette, L. J., Crain, D. A., Rooney, A. A., and Pickford, D. B. (1995). *Environ. Health Perspect., 103*(Suppl. 7): 151.

Haney, A. F., Newbold R. R., Fetter, B. F., and McLachlan, J. A. (1984). *Biol. Reprod., 30*: 471.

Henderson, B. E., Benton, B., Jing, J., Yu, M. C., and Pike, M. C. (1979). *Int. J. Cancer, 23*: 598.

Henderson, B. E., Ross, R. K., Pike, M. C., and Casagrande, J. T. (1982). *Cancer Res., 42*: 3232.

Herbst, A. L., and Bern, H. A., eds. (1981). *Developmental Effects of Diethylstilbestrol (DES) in Pregnancy*. Thieme-Stratton, New York.

Hess, R. A., Bunick, D., and Bahr, J. M. (1995). *Environ. Health Perspect., 103*(7): 59.

Hoefnagel, D. (1976). *Lancet, 1*: 152.

Hooker, C. W., Gardner, W. U., and Pfeiffer, C. A. (1940). *JAMA, 115*: 443.

Huseby, R. A. (1976). *J. Tox. Environ. Health, 1*(1): 177.

Irvine, D. (1994). *Br. Med. J., 309*: 131.

Loughlin, J. E., Robboy, S. J., and Morrison, A. S. (1980). *N. Engl. J. Med.*, July 10: 122.

McLachlan, J. A. (1981). *Developmental Effects of Diethylstilbestrol (DES) in Pregnancy* (A. L. Herbst and H. A. Bern, eds.). Thieme-Stratton, New York, p. 148.

McLachlan, J. A., and Korach, K. S. (1995). *Environ. Health Perspect., 103*(7): 3.

McLachlan, J. A., Newbold, R. R., and Bullock, B. C. (1975). *Science, 190*: 991.

McLachlan, J. A., Newbold, R. R., and Bullock, B. C. (1980). *Cancer Res., 40*: 3988.

Meistrich, M. L. (1989). *Toxicology of the Male and Female Reproductive Systems* (P. K. Warking, ed.). Hemisphere Publishing Corporation, New York, p. 303.

Mori, T. (1967). *J. Fac. Sci. Univ. Tokyo Sect. 4, 11*: 243.

Newbold, R. R. (1995). *Environ. Health Perspect., 103*(Suppl 7): 83.

Newbold, R. R., Bullock, B. C., and McLachlan, J. A. (1983). *Biol. Reprod., 28*: 735.

Newbold, R. R., Bullock, B. C., and McLachlan, J. A. (1985). *Cancer Res., 45*: 5145.

Newbold, R. R., Bullock, B. C., and McLachlan, J. A. (1986). *Am. J. Pathol., 125*: 625.

Newbold, R. R., Bullock, B. C., and McLachlan, J. A. (1987). *Tert. Ca. Mut., 7*: 377.

Newbold, R. R., Carter, D. B., Harris, S. E., and McLachlan, J. A. (1984). *Biol. Reprod., 30*: 459.

Newbold, R. R., and McLachlan, J. A. (1978). *J. Toxicol. Environ. Health, 4*: 491.

Newbold, R. R., and McLachlan, J. A. (1985). *Estrogens in the Environment* (J. A. McLachlan, ed.). Elsevier, New York, p. 288.

Newbold, R. R., and McLachlan J. A. (1988). *Toxicology of Hormones in Prenatal Life* (T. Mori and Nagasawa, eds.). CRC Press, Inc., Boca Raton, FL, p. 89.

Newbold, R. R., and McLachlan, J. A. (1996). *Cellular and Molecular Mechanisms of Hormonal Carcinogenesis: Environmental Influences* (J. Huff, J. Boyd, and J. C. Barrett, eds.). Wiley-Liss, New York, p. 131.

Newbold, R. R., Pentecost, B. T., Yamashita, S., Lum, K., Miller, J. V., Nelson, P., Blair, J., Kong, H., Teng, C. T., and McLachlan, J. A. (1989). *Endocrinology, 124*: 2568.

Newbold, R. R., Suzuki, Y., and McLachlan, J. A. (1984). *Endocrinology, 115*: 1863.

Nomura, T., and Kanzaki, T. (1977). *Cancer Res., 37*: 1099.

Olsen, G., Bodner, K., Ramlow, J., Ross, C., and Lipshultz, L. (1995). *Fertil. Steril., 63*(4): 887.

Pentecost, B. T., Newbold, R. R., Teng, C. T., and McLachlan, J. A. (1988). *Mol. Endocrinol., 2*: 1243.

Pottern, L. M., Brown, L. M., Hoover, R. N., Javadpour, N., O'Connell, K. J., Stutzman, R. E., and Blattner, W. A. (1985). *J. Natl. Cancer Inst., 74*: 377.

Purdom, C. E., Hardiman, P. A., Bye, V. J., Eno, N. C., Tyler, C. R., and Sumpter, J. P. (1994). *Chem. Ecol., 8*: 275.

Rustia, M. (1979). *J. Natl. Cancer Inst. Monogr., 51*: 77.

Sharpe, R., and Skakkeback, N. (1993). *Lancet, 341*: 1392.

Sumpter, J. P., and Jobling, S. (1995). *Environ. Health Perspect., 103*(7): 173.

Throberg, J. V. (1948). *Acta Endocrinol. Copenh. Suppl., 2*: 1.

Tuchmann-Duplessis, H., and Haegel, P. (1982). *Illustrated Human Embryology*. Springer Verlag, New York.

Vorherr, H., Messer, R. H., Vorherr, U. F., Jordon, S. W., and Kornfeld, M. (1979). *Biochem. Pharmacol., 28*: 1865.

Whitehead, E. D., and Leiter, E. (1981). *J. Urol., 125*: 47.

Wilcox, A. J., Baird, D. D., Weiunberg, C. R., Hornsby, P. P., and Herbst, A. L. (1995). *N. Engl. J. Med., 332*(21): 1411.

Wolf, E. (1939). *C.R. Acad. Sci.* [D] (Paris), *208*: 1532.

Yoshitomi, K., and Morii, S. (1984). *Vet. Pathol., 21*: 300.

23

Toxicology of the Male Excurrent Ducts and Accessory Sex Glands

Gary R. Klinefelter
U.S. Environmental Protection Agency, Research Triangle Park, North Carolina

Rex A. Hess
College of Veterinary Medicine, University of Illinois, Urbana, Illinois

INTRODUCTION

Toxicology of the excurrent ducts in the male is a relatively new topic, as past efforts have emphasized an understanding of the mechanisms of direct and indirect effects of chemical exposure on the testis. It is difficult to separate direct versus indirect effects on the excurrent ducts because functional impairment of either interstitial cells, Sertoli cells, or germ cells may indirectly affect the function of epithelium downstream. For example, since the epididymis is strongly dependent upon androgen stimulation, any compound that compromises Leydig cell function will decrease androgen concentrations in the blood or rete testis fluid, which in turn will have dramatic effects on epididymal function and thus sperm maturation and fertility. Of course, the overwhelming long-term effect of androgen depletion will be testicular injury, which results in a decline in spermatogenesis and eventually in testicular atrophy. However, short-term effects will be on epididymal function and sperm maturation. Despite the inherent difficulty, it is important that we begin to separate direct from indirect effects of toxicants on the male reproductive tract and to understand how these effects interfere with testicular function as well as impact posttesticular sperm maturation, i.e., fertilizing capability. This chapter will address many of these questions and provide a general overview of structure/function and toxic response of rete testis, efferent ductules, epididymis, seminal vesicles, and prostate. Although the reproductive tract shows considerable similarity between species, it must be emphasized that there can be major differences in structure and function. There are also major regulatory differences between the various regions of the tract. In general, hormonal regulation of the epididymis and male accessory sex organs appears to be under the control of androgens. However, the epithelium

---

Although the research described herein has been funded in part by the U.S. Environmental Protection Agency, it does not necessarily reflect the views of the agency and no official endorsement should be inferred.

of the rete testis, ductuli efferentes, and initial segment of the epididymis does not appear to be regulated entirely by the androgens, but appears to be under the control of other testicular factors present in the luminal fluid. Thus, it is important in male reproductive toxicology to understand the mechanism of toxicant action on the affected organs, and any species differences in normal structure and function, before attempting to extrapolate data obtained in animal studies to human exposure conditions.

STRUCTURE AND FUNCTION OF THE EXCURRENT DUCTS

Rete Testis

The rete testis in many large domestic animals is located within a central zone of the testis forming the mediastinum (Amann et al., 1977; Orsi et al., 1983; Hees et al., 1987). However, in both rats and humans the rete testis is found at the margin of the testis and usually forms an extratesticular portion that empties into the ductuli efferentes (Reid and Cleland, 1957; Dym, 1976). The rete testis consists of flattened channels or interconnecting lacunae that are lined by low cuboidal to columnar epithelium. The luminal surface is covered by short microvilli, and each cell contains a single cilium (Dym, 1976). Tight junctions between cells exclude the passage of intercellular tracer molecules, and basolateral junctions contain elaborate interdigitations and desmosomes (Dym, 1976; Morales et al., 1984). Both vimentin and cytokeratin intermediate filaments are expressed by the epithelial cells (Tung et al., 1987). Although physiological data indicate that the rete testis modifies seminiferous tubular fluid and exhibits morphological characteristics of an absorptive epithelium (Morales et al., 1984; Morales and Hermo, 1986; Tung et al., 1987), the significance of rete participation in the removal of luminal fluids in male reproduction is unclear (Dym, 1976; Viotto et al., 1993).

Efferent Ductules (Ductuli Efferentes)

The efferent ductules arise separately from the rete testis near the tunica albuginea and form a series of small tubules between the rete testis and epididymis (Fig. 1). They may number between 2 and 33 depending on the species (see Ilio and Hess, 1994, for an extensive review). Near the rete, the ductuli have a wider lumen and are embedded in the superior epididymal ligament and a thick layer of fat. In the rat, coni vasculosa forms highly tortuous tubules near the epididymis. Within the conus, the ductules anastomose and become invested with a connective tissue capsule. As the terminal ductule enters the head of the epididymis, it is smaller in diameter and highly coiled. In mammals, there are two basic designs of efferent ductules (Ilio and Hess, 1994). The first pattern is that of a funnel in which the ductules anastomose into a single tubule that changes abruptly into the initial segment of the epididymis, which is typical for the rat, mouse, and some guinea pigs (Jones and Jurd, 1987; Ilio and Hess, 1994). The second type of organization involves parallel coils of ductules that form multiple entries into the head of the epididymis, as seen in most guinea pigs and large mammals, including man (Hemeida et al., 1978). Unlike other mammals, the caput epididymidis in man is occupied mostly by the efferent ducts that leave the testis as parallel straight tubules, which then become coiled tortuously into lobules that fold over one another before emptying into the epididymis.

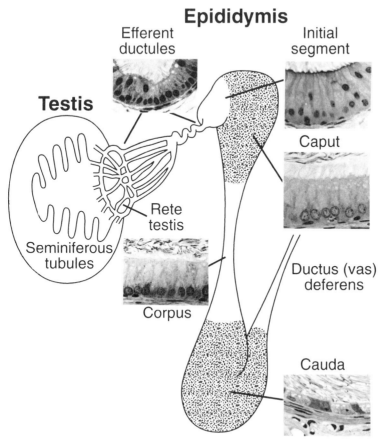

Figure 1 The male reproductive tract showing rete testis, efferent ductules, initial segment of the epididymis, caput, corpus, and cauda epididymidis, and the vas deferens. Insets of the corresponding epithelia are indicated (x335).

Histology

In general, there are relatively few differences in the light microscopic appearance of ductuli efferentes between the species studies. However, the relative amounts of vacuoles and granules found within the nonciliated cells varies both between species as well as along the length of the ductules within a species (Morita, 1966; Goyal and Hrudka, 1981; Gray et al., 1983; Goyal et al., 1992; Ilio and Hess, 1994). In addition to the nonciliated cell, all efferent ductal epithelia contain a population of ciliated cells. Detailed descriptions of the histological feature of the efferent ductules for a number of diverse species, such as rodents, dog, buffalo, elephant, and man, are available (Hamilton, 1975; Jones, 1977; Robaire and Hermo, 1988; Ilio and Hess, 1994).

The ductules are lined by a single layer of columnar epithelium supported by a thin layer of smooth muscle and connective tissue. In cross section, the lumen of the ductules contains few spermatozoa, except in the terminus, where spermatozoa become more concentrated (Talo, 1981). In the rat, the initial zone of the ductules is wide and the epithelium contains numerous lysosomal granules (Ilio and Hess, 1994). Moreover, the volume of vacuoles in the nonciliated cells in the initial zone is six to seven times that found in the terminal zone (Jones and Jurd, 1987). Ductules in the conus and terminal zones are narrower, and their epithelia contain fewer lysosomal granules.

Histochemical, cytochemical, and immunocytochemical studies on the epithelium of the ductuli efferentes have shown that the epithelium is rich in numerous enzymes including acid phosphatase and carbonic anhydrase (Cohen et al., 1976; Goyal et al., 1980). Sodium/Potassium ATPase has also been found with high activity in the efferent ductules (Byers and Graham, 1990; Ilio and Hess, 1992). Additionally, lipids (Ilio and Hess, 1994), glycogen, and many novel proteins such as immobilin (Hermo et al., 1992a), glutathione S-transferase (Veri et al., 1993), and sulfated glycoprotein-1 (SGP-1) (Hermo et al., 1992b; Igdoura et al., 1993) are present in the ductal epithelium. Lysosomal localization of acid phosphatase and thiamine pyrophosphatase enzymes is consistent with the endocytotic activity of cells in which endocytosed material such as testicular proteins is degraded (Hermo and Morales, 1984). Indeed, the lysosomes within the nonciliated cells disappear following ligation at the level of the rete testis (Hermo and Morales, 1984; Robaire and Hermo, 1988).

In all species examined, the nonciliated cells have a well-developed endocytotic system specialized for the uptake of particular material and fluid from the ductal lumen, a microvillus (i.e., brush) border, a profusion of apical canaliculi, vesicles, and a variety of large vacuoles and membrane-bound bodies of different shapes, sizes, and staining intensities (Hermo et al., 1988; Robaire and Hermo, 1988; Ilio and Hess, 1994). Coated pits, apical tubules, endosomes, multivesicular bodies, and lysosomes are components of an elaborate endocytotic apparatus capable of fluid phase, adsorptive (Morales and Hermo, 1983; Hermo and Morales, 1984; Hermo et al., 1985), and receptor-mediated endocytosis (Byers et al., 1985; Veeramachaneni et al., 1990; Veeramachaneni and Amann, 1991). Testicular fluid is taken up sequentially by the endocytotic apparatus from coated pits to multivesicular bodies then to lysosomes and broken down by means of hydrolytic enzymes (Yokoyama and Chang, 1971; Wrobel, 1972; Hermo and Morales, 1984). Other organelles such as mitochondria, rough endoplasmic reticulum, and Golgi apparatus are also found in the supranuclear region of nonciliated cells. In contrast to large mammals (Goyal and Hrudka, 1980, 1981; Gray et al., 1983), the Golgi apparatus shows no evidence of secretory granule formation (Robaire and Hermo, 1988), so secretory function is doubtful (Hoffer, 1972; Robaire and Hermo, 1988).

The ciliated cells possess reduced amounts of vesicular structures associated with endocytosis (Hermo et al., 1985) and are connected to one another via tight junctions. In some regions the nucleus is displaced more apically than in the nonciliated cells. The cilia contain typical $9+2$ doublet microtubules (Ilio and Hess, 1994). Ciliated and nonciliated cells are attached at their boundaries to adjacent cells by tight junctional complexes that are segmented or incomplete, suggesting that the permeability barrier of this epithelium is weak and may facilitate bulk fluid movement (Pudney and Fawcett, 1984). However, studies using tracer molecules have demonstrated egress of tracer from blood to lumen (Suzuki and Nagano, 1978) rather than from lumen to blood (Morales and Hermo, 1983; Hermo and Morales, 1984). This suggests that the efferent ductules may be the primary site of antibody invasion along the excurrent duct, which can lead to sperm agglutination and ductal occlusion (Dym and Romrell, 1975; Tung and Alexander, 1980). The lateral aspects of the plasma membranes of efferent ductule cells form a well-localized "tubular network" (Ramos and Dym, 1977; Jones and Jurd, 1987; Robaire and Hermo, 1988; Ilio and Hess, 1992), which become dilated when absorption is active (Pudney and Fawcett, 1984). While this is consistent with active fluid transport (Suzuki and Nagano, 1978), the dilation is considerably less than that normally found in resorptive epithelium such as the proximal convoluted tubules of the kidney (Ilio and Hess, 1992).

The epithelium of the ductuli efferentes rests on a basement membrane consisting of an amorphous ground substance and connective tissue fibrils (Ilio and Hess, 1994). Beneath this lies collagen and elastic fibers and one or more layers of smooth muscle cells (Lopez and Breuker, 1986). Blood capillaries near the epithelium have thin endothelium and are equipped with fenestrations and vesicles in conformation with the absorptive capacity of the ductules (Montorzi and Burgos, 1967; Suzuki, 1982). A rich sympathetic innervation of short adrenergic neurons, particularly in the smooth muscle layer of the coni vasculosa, has been demonstrated in marsupials (Maruch et al., 1989). The nerve supply in rats was described as double adrenergic/cholinergic innervation forming perivascular, subepithelial, and muscle plexuses (Garnacho et al., 1989). Adrenergic innervation, however, appears not to be a constant feature in higher mammals (El-Badawi and Schenk, 1967).

Hormonal Regulation

Regulation of ductuli efferentes is not well understood. A few studies have focused on the effects of ductal ligation or androgen withdrawal on the structure and function of the efferent ducts. Fawcett and Hoffner (1979) found that in the rat ligation of the ductules at the level of the rete had little effect on the distal segment of the efferent ductules, but did induce epithelial regression in the initial segment of the epididymis. This regression could not be prevented by exogenous androgens. Recently, Hess et al. (unpublished results) found that ligation at the rete does have a profound effect on specific histological features of the efferent ductal epithelium; with the loss of rete testis fluid the ductules showed a decrease in the number of dense granules, similar to that of blind-ending tubules (Guttroff et al., 1991).

The activity of Na^+,K^+-ATPase in the different regions of the rat efferent ductules and epididymis appears to be stimulated by circulating androgens (Ilio and Hess, 1994). However, factors within the ductal lumen are important in modulating the enzymatic activity. Castration or ligation of the duct near the testis decrease enzyme activity in both efferent ductules and the initial segment (Ilio and Hess, 1994). Neither exogenous nor endogenous androgens (i.e., testosterone or dihydrotestosterone) prevent this decrease in activity completely; castration plus androgen replacement showed slightly increased activity over castration in efferent ductules and caput epididymidis. Interestingly, the administration of 17β-estradiol to intact animals also lowers enzyme activity in the proximal segments of the efferent ductules as well as in the caput epididymidis, but the conus region of the ductules and initial segment were not affected (Ilio and Hess, 1994). These results suggest that there may be a selective modulation of enzyme activity by estrogen in different regions of the excurrent ducts.

There is considerable evidence to suggest that the efferent ductules are a major target for hormones. Receptors for several hormones have been identified in the efferent ductules, suggesting that the ductuli may be a possible site of hormone action. Androgen receptors have been localized mainly in the principal cells of the epithelium of the ductules and epididymis (Schleicher et al., 1984; Tekpetey and Amann, 1988; Roselli et al., 1991; Cooke et al., 1991a). However, estrogen receptors (ER) have also been localized in the ductuli efferentes (Schleicher et al., 1984; Sapino et al., 1987; West and Brenner, 1990; Cooke et al., 1991b; Iguchi et al., 1991; Greco et al., 1992, 1993; Gist et al., 1996). The functional significance of ER in the efferent ductal epithelium remains to be determined. However, it was recently discovered that developing spermatids are capable of converting androgens to estrogens because they contain active aromatase (Nitta et al., 1993; Kwon et al., 1995). These new data raise the possibility that luminal estrogen is targeted toward the ductal epithelium via apical transport systems for the purpose of modifying androgen responsiveness of these cells proportional to the number of sperm being transported.

Vitamin D_3–binding sites have also been shown to have their greatest intensity in the epithelium of the efferent ductules (Stumpf et al., 1987; Schleicher et al., 1989). Its labeling pattern was found to overlap with [^{3}H]-dihydrotestosterone– and [^{3}H]-estradiol–binding sites. Since vitamin D is involved in the homeostasis of calcium and phosphorus in the kidney, the presence of vitamin D receptors in the efferent ductules raises the possibility of hormonal regulation of fluid absorption via flux in local calcium and phosphorus concentrations in the efferent ductules. Other hormones or receptors present in the efferent ductules include opioid receptors (Wolfe et al., 1989), proenkephalin (Garrett et al., 1990; Douglass et al., 1991), oxytocin (Veeramachaneni and Amann, 1990), and inhibin (Veeramachaneni et al., 1989). Their functions in this region are unknown.

Function of the Efferent Ductules

Known functions of the efferent ductules include sperm transport, water resorption, ion transport, protein resorption, steroid metabolism, and spermiophagy (reviewed in Ilio and Hess, 1994). The time interval required for spermatozoa to travel the length of the ductuli efferentes is approximately 45 minutes in the rat (English and Dym, 1981), but little is known for other species. There is controversy over the mechanism of sperm transport. Talo (1981) has observed that although ciliary beat moves spermatozoa and round cells, the beat is not exclusively in the epididymal direction; cilia situated in opposite sides of the lumen beat in opposite directions. The cilia have been postulated to reduce the flow of fluids by creating a reflux (Winet, 1980). Hess (unpublished observation), using video microscopy of microdissected ductules, has observed ciliary beat that appeared to serve a function of simply stirring the fluid, possibly for homogeneous resorption of fluids.

Many physiological and micropuncture studies on the proximal segments of the excurrent ducts in different species have confirmed the original findings of Crabo (1965) that more than 90% of the fluids secreted by the seminiferous epithelium is reabsorbed in the ductuli efferentes. The reported values nonetheless vary between 50 and 96% (Djakiew and Jones, 1983; Howards et al., 1975a,b; Jones, 1980, 1987; Jones and Jurd, 1987; Levine and Marsh, 1971; Turner, 1984). Although the efferent ductules are now recognized as the major site for rete testis fluid absorption, the underlying mechanisms for absorption remain unresolved. However, the work of several laboratories suggests that the primary mechanism of fluid (i.e., water) transport in the ductuli efferentes involves the coupling of water and active ion transport (Levine and March, 1971; Hamilton, 1975; Jesse and Howards, 1976; Wong and Yeung, 1977a,b,c,d; Au et al., 1978; Wong et al., 1978a,b, 1979; Turner, 1979, 1984; Hohlbrugger, 1980; Jones and Jurd, 1987; Hinton and Turner, 1988; Wong, 1990; Ilio and Hess, 1992).

One model used to explain water and ion movement in ductuli efferentes is the "standing osmotic gradient" model (Diamond and Bossert, 1967; Hamilton, 1975; Byers and Graham, 1990). The basic tenet of this model predicts the creation of a local hypertonic environment in the apical regions of the epithelium. This model is based upon the active transport of Na^+ (and Cl^-) from the luminal fluid into the apical cytoplasm by apical pumps located adjacent to cellular tight junctions. Contrary to this, Ilio and Hess (1994) showed that Na^+,K^+-ATPase is located primarily at the base and not in the apical regions of the epithelium. This finding is consistent with a generalized model of Na^+ transport (Koefoed-Johnsen and Ussing, 1958; Mills et al., 1977), which requires an active Na^+ transport only in the basal epithelium. Therefore, a unified model for water transport in the efferent ductules has been developed to accommodate these data (Ilio and Hess, 1994). In this model, the transport of Na^+ in the ductuli efferentes would occur in two phases: (1) a carrier or channel-mediated entry across the luminal surface (either through Na^+ channels or by an electrochemically neutral Na^+/H^+ exchange mechanism) coupled to (2) its ATP-dependent extrusion across basal surfaces in exchange of K^+ through the activity of Na^+,K^+-ATPase found in the deep basolateral regions of the epithelium (Ilio and Hess, 1992).

The movement of water from the lumen to the interstitium could proceed by two paths: the paracellular pathway (across the junctional complexes) and/or the transcellular pathway (through the cells). These two pathways are not independent, as the basolateral intercellular spaces become their final common path. The paracellular pathway is made possible by the apical junctional complexes that have been characterized as "leaky" (Suzuki and Nagano, 1978; Nagano and Suzuki, 1980; Pudney and Fawcett, 1984). Because of low electrical resistance across these junctions, a small driving force generated by Na^+,K^+-ATPase in the basal aspects of the cells can selectively draw solutes (e.g., Cl^-) and large quantities of water through the junctional complexes similar to that described in the cells of the proximal convoluted tubules (Ullrich, 1990). The transcellular pathway can be mediated in two ways:

1. By the passive movement of water that follows active Na^+ transport. This is most likely attributed to high-permeability water channels found in high concentration in the apical brush border of nonciliated cells in efferent ductules (Brown et al., 1993).
2. By the incorporation of water in the carbonic anhydrase–mediated reaction, $H_2O + CO_2 \, 'H_2CO_3 \, 'H^+ + HCO_3^-$, which is consistent with the abundance of carbonic anhydrase found in the nonciliated cells of the efferent ductules (Cohen et al., 1976; Goyal et al., 1980).

The absorption of protein in the efferent ductules has been demonstrated by the disappearance of certain bands of proteins from the rete testis fluid between the ductuli efferentes and the initial segment of the epididymis owing to their absorption in the ductuli and/or the initial segment (Koskimies and Kormano, 1975; Olson and Hinton, 1985; Jones, 1987). It has been calculated that about half of the total protein leaving the testis was absorbed in the ductuli efferentes (Jones and Jurd, 1987). The capacity of the efferent ductal epithelium to resorb molecules both through fluid-phase, adsorptive endocytosis and receptor-mediated endocytosis has been confirmed by several studies (Pelliniemi et al., 1981; Morales and Hermo, 1983; Hermo and Morales, 1984; Hermo et al., 1985; Veeramachaneni and Amann, 1991).

After fixation by vascular perfusion, secretory blebs have bene observed in certain regions of the ductuli efferentes in goat (Gray et al., 1983), bull (Goyal, 1985),
and dog (Ilio and Hess, 1994). The secretory blebs were also present in the ductal lumen distal to the site of secretion. Thus, in some species apocrine secretions do occur
in the epithelium of the efferent ductules. Several protein bands that are absent in rete
testis fluid have been identified in fluid from the initial segment of the epididymis. It
is thought that some of these proteins were secreted in either the efferent ductules or
the initial segment of the epididymis (Olson and Hinton, 1985).

Epididymis

It is now well accepted that sperm acquire the ability to fertilize eggs during their transit
through the epididymal duct and that this process of maturation is orchestrated via complex interactions between the epithelium lining the duct, the luminal fluid, and the sperm
(Bedford, 1975). Given the ultimate goal, the functional maturation of testicular sperm,
it is not surprising that the epididymal duct possesses great diversity with respect to
both structure and function along its length.

As described above, the network of efferent ducts converge to form this single
highly, convoluted duct at the level of the initial segment—the first morphologically
distinguishable region of the epididymis (Fig. 1). A columnar, pseudostratified epithelium persists throughout the length of the epididymis; but in the initial segment the
columnar principal cells are significantly taller than in other regions of the epididymis.
In addition to the principal, basal, and halo cells found in all regions of the epididymis, the initial segment of the rat contains a unique, deeply stained cell type known as
the narrow cell. This cell type is strikingly similar to the mitochondria-rich cell found
throughout the length of the epididymis in other species, including humans (Robaire
and Hermo, 1988). The initial segment also plays a major role in the absorption of
specific proteins and fluid. Both androgen-binding protein (ABP) and transferrin are
removed by endocytosis from the luminal fluid in the initial segment to a greater extent than any other region of the epididymis (Veeramachaneni et al., 1991), albeit not
nearly to the extent these proteins are endocytosed in the efferent ducts. While it is
speculated that fluid phase endocytosis may follow the receptor-mediated endocytosis
of specific macromolecules such as ABP, transferrin, and α_2-macroglobulin, most of
the fluid absorption occurs via passive diffusion, following the uptake of chloride ion
(Wong et al., 1978) and carnitine (Hinton and Hernandez, 1985). This significant removal from the luminal fluid explains how sperm are increasingly concentrated as they
pass through the proximal regions of the duct. The luminal milieu of the epididymis is
altered not only by absorptive functions, but also be secretory functions. While the
mechanism of protein secretion in the epididymis remains controversial, there is no
question that proteins are secreted by the epithelium within the initial segment, and many
of these proteins appear identical to those secreted in more distal regions of the epididymis (Brooks, 1983).

The epithelium in the caput epididymidis is shorter than that in the initial segment, but it is otherwise morphologically similar, containing principal cells, basal cells,
and occasional halo cells. The principal cells contain 5α-reductase activity responsible
for the conversion of testosterone to dihydrotestosterone (Klinefelter and Amann, 1980),
the androgen that acts via the androgen receptor to stimulate protein synthesis and secretion and promote the fertilizing ability of sperm (Orgebin-Crist and Jahad, 1978).

In the rat, the levels of dihydrotestosterone are approximately 10-fold higher than those of testosterone in the caput and corpus epididymidis, but the levels of these androgens are similar in the cauda epididymidis (Vreeburg, 1975; Turner et al., 1984). The increased levels of dihydrotestosterone in the caput/corpus epididymidis results directly from the increased activity of 5α-reductase in these proximal regions of the duct (Robaire and Hermo, 1988). While a significant amount of testosterone enters the lumen of the epididymis via the rete testis and efferent ductules, most of the testosterone within the epithelium appears to be taken from the blood by the epithelium against a concentration gradient (Turner, 1988). It has been speculated that ABP in the lumen of the epididymis may be responsible for this antigrade uptake of testosterone (Turner, 1991). Interestingly, recent data suggest that the testosterone entering the epididymis via the lumen is associated with the transport of ABP from the seminiferous tubule to the epididymis (Danzo, 1995). The notion that the testosterone within the lumen of the epididymis may be secondary to the transport of ABP is consistent with findings that epididymal sperm maturation can only be maintained for several days in castrated rats if exogenous testosterone is administered (Dyson and Orgebin-Crist, 1973; Klinefelter et al., 1994a).

It has been demonstrated, both in vivo (Holland et al., 1992) and in vitro (Klinefelter and Hamilton, 1985; Klinefelter et al., 1990), that specific proteins synthesized by the epididymal epithelium are secreted into the lumen of the caput and become associated with sperm. While a few proteins synthesized within the epididymis have been shown to be androgen-dependent (Brooks, 1983), we are only beginning to examine the extent to which androgens regulate the secretion of specified proteins and the association of specific proteins with the sperm membrane. In general, however, androgen-dependent protein synthesis and secretion in the epididymis appears to be less dynamic than androgen-dependent protein synthesis and secretion in the seminal vesicles and prostate, i.e., the half-life for androgen-dependent protein mRNA is relatively long-lived in the epididymis (Brooks, 1987), and androgen-dependent proteins comprise a relatively minor percentage of the total protein (Brooks, 1983). Indeed, the predominant secretory protein in the proximal epididymis is sulfated glycoprotein-2 (SGP-2) or clusterin, a protein that does not appear to be under the direct control of androgens (Brooks, 1983; Mattmüeller and Hinton, 1992; Cyr and Robaire, 1992). Other epididymal proteins such as retinoic acid–binding protein (Zwain et al., 1992) and 5α-reductase (Robaire et al., 1977; Brown et al., 1983) appear to be regulated by factors entering the epididymal lumen from the testis.

The corpus epididymidis is characterized by an epithelium comprised of principal cells that contain abundant infranuclear lipid (Hamilton, 1975), basal cells, and halo cells. In addition, clear cells are frequently observed in distal corpus of the rat. As indicated above, the corpus epididymidis is involved in androgen-dependent protein synthesis; but in addition to androgen involvement in protein synthesis, androgens also play a role in lipid biosynthesis and turnover. Glycerol phosphocholine is synthesized within the proximal regions of the epididymis from lecithin derived from bloodborne lipoprotein and secreted into the lumen (Hammerstedt and Rowan, 1979), and this process is androgen-dependent. Since epithelial cells in the corpus have abundant deposits of lipid, these cells might play a unique role in the turnover of lipid in the plasma membrane of sperm as they transit the epididymis (Hammerstedt and Parks, 1987; Hall et al., 1991).

In the rat, the epithelium in the cauda epididymidis is characterized by the presence of numerous clear cells. These cells are responsible for the phagocytosis of cytoplasmic droplets shed from sperm during epididymal transit (Hermo et al., 1988). In this region of the epididymis one also finds the highest concentration of inositol as this small organic molecule becomes concentrated along the length of the epididymis. The inositol within the lumen of the cauda is derived from active uptake by the epithelium (Cooper, 1982), as well as synthesis within the epithelial cells (Robinson and Fritz, 1979); these processes appear to be under the control of androgens (Pholpramool et al., 1982).

In both rats and men, sperm are fully capable of successfully fertilizing an egg by the time they reach the proximal cauda epididymidis (Robaire and Hermo, 1988). Collectively, the biochemical changes (i.e., protein acquisition and modification, lipid turnover, thiol reduction) that sperm undergo during their transit to the proximal cauda epididymidis confer increases in the velocities of progressively motile sperm (Yeung et al., 1992) and fertilizing ability (Dyson and Orgebin-Crist, 1973), but the mechanisms underlying these functional changes are not yet understood.

The journey from the testis to the cauda takes approximately 4 days in rats (Robb et al., 1978) and 2 days in men (Johnson and Varner, 1988). Transit of sperm through the epididymis presumably is mediated by the rich adrenergic innervation (Eliasson and Risly, 1968), although cholinergic innervation also exists (El-Badawi and Schenk, 1967). The degree of adrenergic innervation increases along the length of the epididymis, with the greatest innervation in the cauda (Kaleczyc et al. 1993)—a feature that undoubtedly contributes to ejaculatory response. The timely transport of sperm through the epididymis plays a critical role in the acquisition of successful fertilizing ability. Indeed, sperm that are allowed to age within the cauda epididymidis lose their fertilizing ability (Cooper and Orgebin-Crist, 1977), as do sperm whose transit is accelerated through the epididymis (discussed below).

While the dependence of the epididymis on androgen is clear, other hormones appear to play roles in the functional maintenance of the epididymis as well. Estrogen has been shown to accelerate the rate of sperm transit through the epididymis (Meistrich et al., 1975). Moreover, it has been suggested that estrogen may regulate clear cell function (Schleicher et al., 1984). This is consistent with data that demonstrate that there are more high-affinity estrogen-binding sites in the cauda than in more proximal regions of the epididymis (Danzo and Eller, 1979). In addition to the presumptive role for estrogen, there also appears to be a role for prolactin, vitamin D, and vitamin A in normal epididymal function (Robaire and Hermo, 1988).

The Seminal Vesicles, Prostate, and Bulbourethral Glands

The seminal vesicles are not present in all species, and structurally these glands vary considerably among those species in which they are present. In the rat, the seminal vesicles consist of paired vesicular glands that converge, along with the prostate and bladder, into the pelvic urethra. The mucosa of these glands forms deep, intricate folds of pseudostratified, columnar epithelium (Setchell et al., 1994). As discussed below, the function of these androgen-dependent glands, particularly the secretion of fluid, is influenced dramatically by alterations in androgen status. Thus, change in seminal vesicle weight has been a valuable diagnostic parameter in male reproductive toxicology. Unlike the epididymis, the epithelium of the seminal vesicles is responsible for the secretion of only a few proteins (Higgins et al., 1976; Higgins and Bruchell, 1978), and secretion of each appears to be highly androgen dependent (Higgins et al., 1978).

The seminal vesicles contribute significantly to the sperm-free fluid volume of the ejaculate. The total volume of secreted seminal vesicle fluid appears to be essential for transport of a sufficient number of sperm into the uterus (Pietz and Olds Clarke, 1986). In rodents, surgically removing incremental amounts of the gland results in smaller vaginal plugs and fewer uterine sperm, presumably the result of vaginal leakage (Carballada and Esponda, 1992). This most likely results from insufficient availability of coagulating proteins such as semenogelin (Aumuller and Riva, 1992). In addition to protein, the seminal vesicles produce abundant amounts of fructose, which presumably serves as a source of energy for sperm survival in the female reproductive tract (Mann and Mann, 1981).

The prostate is present in all mammals, but like the seminal vesicles, there is considerable structural heterogeneity between species. The prostate is a tubuloalveolar gland, which also is lined by a pseudostratified, columnar epithelium. The prostate lies anterior to both the seminal vesicles and bladder; and in the rat it consists of ventral and dorsolateral segments. The prostate secretes large amounts of citric acid and polyamines, but roles for these products have not yet been determined. Like the seminal vesicles, the rat prostate is highly androgen dependent (Parker and Mainwaring, 1977), and almost half of the total protein secreted by the prostate represents a single protein (Lea et al., 1979; Parker et al., 1982). This protein, referred to as prostatein or prostatic binding protein, is comprised of two heterodimers: one contains the C1 and C3 polypeptides and the other contains the C2 and C3 polypeptides (Parker et al., 1980; Parker et al., 1983). The expression of the C3 mRNA has been a valuable diagnostic measure for androgen-regulated gene transcription in the prostate (Parker et al., 1980). Both the size and appearance of the prostate epithelium are highly responsive to changes in androgen status in the male rat. Three secreted proteins have been identified in the human prostate, i.e., prostatic acid phosphatase, prostate-specific antigen, and prostate-specific protein; the serum level of prostate-specific antigen is a useful indicator of prostate cancer (Luke and Coffey, 1994).

Within the prostate, testosterone is converted to dihydrotestosterone via the action of 5α-reductase, and it is the latter androgen that is primarily responsible for the androgen-dependent growth and differentiation of the prostate (Luke and Coffey, 1994). Indeed, concentrations of DHT are fivefold greater than T within the prostate—5 vs. 1 ng/g. While the prostate's use of DHT is clear, the maintenance of normal androgen-dependent function of the prostate appears to involve a complex interaction between the stromal and epithelial compartments, between growth factors and hormones, and between the extracellular matrix and the prostate cell (Luke and Coffey, 1994).

The bulbourethral glands, which are not found in dogs or bears, also exhibit considerable structural variability between species. Typically they consist of a single pair of glands, but as many as three pair can be observed in certain marsupials (Setchell et al., 1994). In the rat and man, these small tubuloalveolar glands flank the urethra near the root of the penis, and their secretions drain into the urethra via tiny connecting ducts. Because of their obscure location in the rat, the bulbourethral glands are seldom examined in male reproductive toxicology. However, as discussed below, these glands are highly responsive to estrogens and estrogenic chemicals.

EMBRYONIC DEVELOPMENT AND ANOMALIES OF THE MALE EXCURRENT DUCTS

Normal Development

The contribution of the mesonephros to the development of Sertoli cells and rete testis epithelium remains controversial (Satoh, 1985; George and Wilson, 1988; Rabinovici and Jaffe, 1990; Dinges et al., 1991; Takeuchi, 1992; Merchant-Larios et al., 1993). However, the dual expression of cytokeratin and vimentin in the rete testis epithelium has led some to conclude that rete cells are of mesonephric tubule origin (Tung et al., 1987; Dinges et al., 1991). There is also considerable evidence to support the theory the efferent ductules, as well as the initial segment of the epididymis, arise from mesonephric tubules within the mesonephros rather than from the adjacent tissue comprising the Wolffian duct (Linder, 1971; Marshall et al., 1979; Croisille, 1981; George and Wilson, 1988; Hinton and Turner, 1988; Takeuchi, 1992).

The remainder of the epididymis, the vas deferens, and the seminal vesicles are derived from the Wolffian duct, an evagination of the mesonephros which lies next to the Müllerian duct (George and Wilson, 1988). The epididymis is formed from the upper aspect of the Wolffian duct, which lies adjacent to the testis. The vas deferens arises from the middle of the Wolffian duct, while the seminal vesicles originate from the lower aspect of the duct before entering the urogenital sinus. The prostate is unique in that it is derived from the pelvic portion of the urogenital sinus.

The timely appearance and action of two hormones that arise from the fetal testis, Müllerian inhibiting hormone and androgen (i.e., testosterone, dihydrotestosterone), are requisite to the normal phenotypic development of the male excurrent duct system. Müllerian inhibiting hormone induces the regression of the Müllerian duct, which in turn is permissive for the development of the testis and Wolffian duct derivatives. However, in addition to playing a permissive role in the development of the excurrent duct system, there is evidence suggesting a more active role, specifically promoting the structural and functional development/differentiation of the tissue (George and Wilson, 1994). For example, female transgenic mice that overexpress Müllerian inhibiting hormone develop gonads containing spermatogenic cords (Behringer et al., 1990). Thus, the mechanism of action of Müllerian inhibiting hormone is unclear. Moreover, testosterone accelerates the regression of Müllerian ducts by Müllerian inhibiting hormone in vitro (Ikawa et al., 1982), suggesting synergism between these two hormones.

Testosterone that is produced shortly after the formation of the spermatogenic cords in the male embryo also has profound effects on the phenotype of the male excurrent ducts. Testosterone presumably exerts its influence via a classical receptor-mediated mechanism. However, involvement of other receptor-independent mechanisms have been speculated since the androgen receptor has a 10-fold greater affinity for the 5α-reduced metabolite, dihydrotestosterone, compared to its affinity for testosterone. Moreover, the dihydrotestosterone-receptor complex activates reporter genes at much lower concentrations than testosterone (Deslypere et al., 1992). Finally, 5α-reductase activity is not detectable in the Wolffian duct derivatives until after differentiation is complete (Wilson and Lasnitzki, 1971).

Early studies using mice expressing the testicular feminization mutation (Tfm), and thus failing to produce the androgen receptor, clearly demonstrated that the androgen receptor is required for normal differentiation of the Wolffian duct derivatives (Lyon and Hawkes, 1970). Thus, regardless of the exact mechanism(s), testosterone promotes differentiation of the epididymis, vas deferens, and seminal vesicles. By contrast, 5α-reductase activity is present in the urogenital sinus before differentiation of the prostate (Wilson and Lasnitzki, 1971). Dihydrotestosterone, produced primarily by the epithelial cells of the urogenital sinus via the action of 5α-reductase activity, is believed to bind to androgen receptor within adjacent stromal cells, induce the transcription of specific growth factor(s), and promote the differentiation of the prostate (Luke and Coffey, 1994).

Estrogens and Abnormal Development

The effects of diethylstilbestrol (DES) exposure on the development and function of the male reproductive tract include cryptorchidism, testicular atrophy, epididymal cysts, sperm abnormalities, sperm granulomas, adenocarcinoma, prostatic inflammation, and other changes that could affect fertility and male reproductive function (Table 1). One interesting discovery has been the persistence of Müllerian remnants of portions of the female reproductive tract adjacent to the male tract, which apparently interferes with the development of the male tract by decreasing its response to Müllerian inhibiting hormone (Bullock et al., 1988).

Additionally, DES has been shown to induce adenocarcinoma of the rete testis (Newbold et al., 1985). The presence of rete testis neoplasms in mice exposed to DES and the induction of epididymal granulomas in both hamsters (Wilson et al., 1986) and mice (Arai et al., 1983) strongly suggest abnormal growth of the mesonephric tubules during development of the reproductive system. However, a precise biochemical mechanism to account for this abnormal development is lacking. Recently, Sato et al. (1994) demonstrated that DES exposure in utero does induce an early appearance of estrogen receptor in the male tract and that this appearance is coincident with an increase in cell division. Thus, it appears that DES may be acting directly on the developing ductal tissues in the male. One hypothesis would be that blind-ending tubules are formed, which become filled with stagnant sperm leading to epididymal sperm granulomas. DES stimulation of cell division in the mesonephric tubules may also lead to abnormal growth of the excurrent ducts.

Similar to DES, the administration of estradiol perinatally also results in abnormalities of the male reproductive system (Table 1). A recent study also showed that when the estrogen receptor is genetically "knocked out," male mice develop normally but are essentially infertile (Korach, 1994). The study of animals lacking functional estrogen receptor should provide a model for separating nonestrogenic from estrogenic-like effects of chemicals such as DES and also for determining the precise role of estrogen in the development and function of the male reproductive tract.

TOXICOLOGY OF THE EXCURRENT DUCTS

Rete Testis and Efferent Ductules

The rete testis and efferent ductules are probably the least studied tissues in male reproductive toxicology. Due to their small size and location within the fat pad in rodents, the efferent ductules are typically discarded when the testis and epididymis are removed and weighed. Because there is little mention of the rete testis or efferent ductules in the toxicology literature, it is difficult to know whether these organs are

Table 1 Effects of Exposure to Estrogenic or Estrogenlike Chemicals During Fetal and Perinatal Periods on Development and Function of the Male Reproductive Tract

Effect	Ref.
Cryptorchid testis	Nomura and Kanzaki, 1977; McLachlan et al., 1981; Bullock et al., 1988; Walker et al., 1990
Atrophic seminiferous tubules or hypogonadism	McLachlan, 1977; Bibbo et al., 1977; Gill et al., 1976; 1977; 1979; Bullock et al., 1988; Guisti et al., 1995; Mori, 1967; Jones, 1980; Arai et al., 1983; Bibbo et al., 1977; Elias et al., 1990; Bellido et al., 1985; Deschamps et al., 1987
Persistent Müllerian duct remnants	McLachlan, 1977; Newbold et al., 1987; Bullock et al., 1988;
Increases in estrogen receptors	Sato et al., 1994
Epididymal cysts	Conley et al., 1983; Gill et al., 1976; 1977; 1979; Jones, 1980; Arai et al., 1983; McLachlan, 1977; McLachlan et al., 1975; Bullock et al., 1988; Gray et al., 1985; Dunn and Greene, 1963; Warner et al., 1979
Nodules or hyperplasia of rete testis, efferent ductules, or epididymis	Sapino et al., 1987; Bullock et al., 1988; Arai et al., 1983; Newbold et al., 1987; Mori, 1967
Sperm granulomas in epididymis	Wilson et al., 1986; Bullock et al., 1988; Deschamps et al., 1987; Rustia and Shubik, 1976
Adenocarcinoma/adenomas or rete testis or epididymis	Newbold et al., 1985; 1986; 1987; Bullock et al., 1988; Connell and Donjacour, 1985; Yasuda et al., 1988; Deschamps et al., 1987
Decrease in daily sperm production or sperm concentration	Gill et al., 1976; Bibbo et al.,1977; Schumacher et al., 1981; Wilson et al., 1986; Giusti, 1995
Sperm abnormalities or decreased motility	Bibbo et al., 1977; Bush et al., 1986; Giusti, 1995

relatively insensitive to toxicants or if they have simply been overlooked. Therefore, this section of the chapter is organized by toxicants known to induce effects within these tissues.

Direct effects of toxicants on the rete testis have not been reported. However, developmental effects (see DES) and indirect effects are recognized. One physical response of the rete is it ability to enlarge due to fluid back-pressure whenever the efferent ductules become occluded. This response has been observed following exposures to α-chlorohydrin (Cooper et al., 1974) and benzimidazole carbamates (Hess et al., 1991; Nakai et al., 1992). In both cases, the occlusions are primarily located in the efferent ductules and initial segment of the epididymis (Ericsson, 1970; Cooper et al., 1974; Nakai et al., 1992). Although blockage of the efferent ducts induces long-term

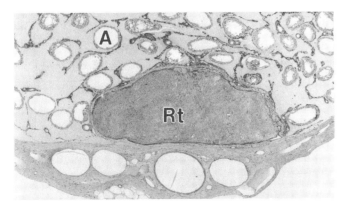

Figure 2 Day 70 following a single dose exposure to carbendazim (400 mg/kg) in the rat. The rete testis (Rt) is swollen with compacted sperm and the seminiferous tubules are atrophic (A) with edematous interstitial space (x25).

seminiferous tubular atrophy (Fig. 2), there appears to be an attempt by the epithelium of the rete testis to regenerate new passages for fluid exit (Fig. 3). It is doubtful if such regeneration aids in reestablishment of functional seminiferous epithelium following complete tubular atrophy. However, when occlusions are more distal, i.e., in the caput epididymidis, the efferent ductules appear to function in the removal of ductal fluids and sperm and thus prevent or reduce the back-pressure atrophy (Kuwahara, 1976).

α-Chlorohydrin

α-Chlorohydrin (3-chloro-1,2-propanediol; U-5897) is a well-known chemical that targets the male reproductive tract, particularly the efferent ductules and proximal regions of the epididymis. At high dosages, it induces distinct histopathological lesions in the efferent ductules and initial segment of the epididymis, but at lower dosages α-chlorohydrin appears to have direct effects on sperm (Crabo et al., 1975; Mohri et al., 1975) and moderate effects on the ductal epithelium (Samojlik and Chang, 1970) as discussed

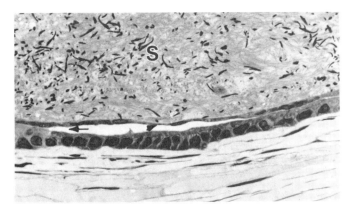

Figure 3 Higher magnification of Figure 2. The rete testis epithelium (arrow) has regrown along the luminal edge of the compacted sperm (S) in an attempt to form a new lumen (x300).

below. High dosages of α-chlorohydrin induce the formation of intratubular spermatoceles in the ductuli efferentes, with larger formations near the rete testis than in the more distal regions (Cooper and Jackson, 1972; Jones, 1978). The epithelium of these spermatoceles is attenuated and fragmented; sometimes spermatozoa are seen escaping into the connective tissue resulting in the formation of granulomata. Eventually, hyalinization and ductal fibrosis supervenes. A higher dosage in rats produces obstructive lesions in the initial segment of the epididymis and stimulates the nonciliated cells of the efferent ductules to form pseudopodia and to phagocytize sperm (Hoffer and Hamilton, 1974; Hoffer et al., 1975). However, in contrast to earlier findings of Cooper and Jackson (1972), the epithelium of the ductules is otherwise normal in appearance.

There is considerable variation between species in response to α-chlorohydrin. The rat efferent ductules and initial segment regions respond rapidly with irreversible infertility due to ductal occlusions, but no effect has been observed in the mouse (Ericsson, 1970) or rabbit (Back et al., 1975). Moderate responses have been observed in the guinea pig (Ericcson and Baker, 1970) and the dog (Dixit et al., 1975). In the dog, α-chlorohydrin caused an increase in luminal diameter in the proximal epididymidis, the loss of luminal sperm, and regression of the seminiferous tubules. However, the testes recovered within 100 days posttreatment, indicating that either the ductal occlusions were repaired or new canals were formed.

The mechanism of action by α-chlorohydrin within the ductal epithelium appears to be mediated through the blood vasculature, although there does not appear to be endothelial damage, which can cause ductal occlusions (Mason and Shaver, 1952). The effect on the blood vascular system remains controversial. One study reported no effect on epididymal blood flow (Brown-Woodman and White, 1976), while another study found that sodium nitrite (a vasodilator) given 30 minutes prior to α-chlorohydrin prevented the epididymal lesions (Kalla and Singh, 1979).

The similarity between proximal convoluted tubules of the kidney and efferent ductules is well established and includes a major function in the resorption of luminal fluids, including water and proteins (Hinton and Turner, 1988; Ilio and Hess, 1994). Thus, it is not surprising that oxalic acid, a metabolite of α-chlorohydrin, causes diuresis and acute glomerular nephritis as well as spermatocoeles in the male (Jones et al., 1978). It is possible that the metabolite of α-chlorohydrin produces calcium oxalate crystals within the ductal epithelial cells, leading to cell death or denuding of the epithelium. Desquamation of epithelial cells has been reported to occur within hours of exposure (Hoffer et al., 1973). It is thought that the denuded epithelial cells form a luminal plug that induces inflammation and subsequent sperm granulomas (Cooper et al., 1974; Kalla and Chohan, 1976). It is interesting to note that long-term ethylene glycol exposure also causes testicular edema and eventual seminiferous tubular atrophy (Jones et al., 1978). It is likely that the mechanism is the same as α-chlorohydrin, although the reported study did not examine the epididymis.

Gunn et al. (1969) suggested that α-chlorohydrin acts on the mechanisms of fluid resorption and that the response is dependent upon the presence of luminal fluids. However, a later study showed that ligation of the efferent ductules prior to exposure did not prevent the epithelial sloughing (Cooper et al., 1974). Evidence that α-chlorohydrin may interfere with fluid resorption comes from the work of Hinton et al. (1983) showing that α-chlorohydrin or a metabolite reduces sugar transport in the rat caput epithelium; and interference with glucose metabolism would indirectly inhibit energy-dependent processes such as ion transport (Wong et al., 1979). Because the efferent ductules are responsible for resorption of nearly 90% of the fluids entering from the rete testis (Jones and Jurd, 1987), inhibition of the Na^+/K^+ ATPases located along the basolateral membranes of the ductal epithelium (Ilio and Hess, 1992) also would greatly reduce transepithelial movement of water.

Regardless of the biochemical mechanisms of α-chlorohydrin action, the morphological response of rete testis and efferent ductules is well established. The first response is the ductal occlusion, followed by stagnation of sperm within the lumen of the efferent ductules and subsequent luminal dilation. The dilation appears to induce an inflammatory response and sperm granuloma formation. This sequence of events causes the rete testis to become swollen with seminiferous tubular fluids and if the distal occlusions are not resolved, the swelling continues into the seminiferous tubules (Ericsson, 1975), leading to back-pressure atrophy of the testis (Ilio and Hess, 1994).

Ethane Dimethanesulfonate

A single dose of ethane dimethanesulfonate (EDS) causes hyperplasia of ductuli efferentes epithelium and the formation of small intratubular sperm retention cysts (Cooper and Jackson, 1972). These later coalesce between 5 and 7 days posttreatment, forming large spermatoceles, which rupture inducing the formation of permanent sperm granulomas (Cooper and Jackson, 1973). In contrast to the permanent lesions induced by α-chlorohydrin, EDS-induced occlusions appear to be resolved either by recanalization or some other mechanism (Cooper and Jackson, 1973) and fertility recovers in some males. However, as discussed below, the effects on the efferent ductules and proximal regions of the epididymis are most likely separate from those affecting the fertilizing ability of sperm.

Benzimidazole Carbamates

Occlusion of the efferent ductules (Fig. 4) is common following exposure to the fungicide benomyl and its metabolite carbendazim (Hess et al., 1991; Nakai et al., 1992).

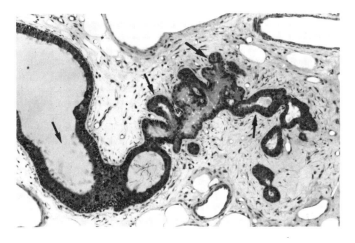

Figure 4 Ductuli efferentes (DE) 70 days posttreatment with a single dose of the benomyl metabolite carbendazim (400 mg/kg). Note the abnormal growth of the ductal epithelium (arrows at the former site of occlusion (x120). (From Ilio and Hess, 1994.)

These fungicides have been important environmental chemicals for controlling mold, fungus, and other organisms that infest lawns and garden and orchard plants. A single dose of these compounds induces rapid testicular effects, detectable within hours as an increase in testis weight, but having long-term effects that lead to testicular atrophy and infertility (Carter et al., 1987; Hess et al., 1991; Nakai et al., 1992). As with α-chlorohydrin, occlusion of the efferent ductules produces a fluid back-pressure that causes seminiferous tubular swelling, loss of germ cells, and eventual atrophy of the testis. Thus, the earlier reports of long-term atrophy of the testis following subchronic and acute multiple exposures to carbendazim (Carter et al., 1987; Rehnberg et al., 1989; Gray et al., 1990) are explained entirely by the pathophysiological mechanism of ductal occlusions. The more subtle effects detected in epididymal sperm, at the lower dosages (Gray et al., 1990), are explained by the direct effects of carbendazim on the seminiferous epithelium (Nakai et al., 1992).

The sequential pathological events that occur in the efferent ductules following benzimidazole exposures are not understood entirely. Testicular swelling was observed as early as 8 hours and occlusions were evident in 2 days postexposure, which suggests that the initial lesion responsible for the long-term infertility (Carter et al., 1987) occurs shortly after dosing. A comparison of the 70-day data (Nakai et al., 1992) with that from the 245-day study (Carter et al., 1987) is noteworthy because the percentage of testes with "total seminiferous tubular atrophy" was 21% on day 70 and 50% on day 245. This suggests that either seminiferous tubular regression continues long after treatment has stopped, or that the additional doses used were effective in spreading the damage to a greater number of efferent ductules. This latter explanation is consistent with the finding that subchronic exposures (Rehnberg et al., 1989; Gray et al., 1990) induce smaller testes than a single exposure (Hess et al., 1991). Thus, it appears that higher dosages and/or repeated doses may occlude more ductules or induce a greater number of permanent lesions.

Long-term atrophic effects on the testis were more severe after carbendazim treatment than with benomyl. This difference appears to be related to the ability of carbendazim to induce more severe occlusions in the efferent ductules (Hess et al., 1991). It remains to be determined whether the occlusions are independent of the massive sloughing of germ cells that become highly compacted within the distended ductules. The inflammatory response associated with occluded ductules may be the first reaction, causing subsequent damage to the ductal epithelium. Differences in the degree of inflammation could contribute to the large variation seen in testicular responses at 70 and 245 days postexposure (Carter et al., 1987). It is likely that severe inflammations are followed by destruction of the efferent ductal epithelium and then by fibrosis. With a medium inflammatory response, surviving epithelial cells with intact basement membrane can regenerate the ductule with hyperplastic growth around the occlusions (Nakai et al., 1993), as seen in other tubular systems (Siegel and Bulger, 1975).

The response of efferent ductal epithelium to injury induced by ductal occlusions appears to be dependent upon the degree of inflammation caused by the trauma. An acute inflammatory reaction appears to be induced by the compacted luminal contents that cause swelling of the ductal lumen, which in turn facilitates the release of a chemotactic substance, possibly a cytokine of the interleukin-8 superfamily (Colditz et al., 1989), and subsequent recruitment of neutrophils. Moreover, leakage of sperm antigens may then draw neutrophils toward the ductal lumen and stimulate phagocytic activity (Vogelpoel and Verhoef, 1984). This is similar to other organ systems in which the indirect damage caused by neutrophil emigration into the interstitium and through the epithelium promotes granuloma formation and stimulates fibroblasts (Bowden and Adamson, 1985).

Epithelia undergoing a mild inflammatory response often exhibit irregular epithelial growth along the edge of luminal contents or the formation of multiple abnormal ductules (Fig. 4). These abnormal ductules form from epithelial cells, which migrate from the periphery of the occluded lumen, indicating attempted recanalization. At later time periods, the occluded lumen fills with fibrotic connective tissue and is surrounded by several abnormal ductules. The abnormal ductules are characterized by fewer cytoplasmic organelles such as lysosomal granules associated with absorptive function and reduced epithelial cell height. Epithelial cells of the abnormal ductules differentiate but the morphological features of these ductules are similar to those of blind ending tubules (Guttroff et al., 1991). Moreover, there is no evidence that occluded ductules form patent connections via the abnormal ductules.

Other chemicals that cause sloughing of the seminiferous epithelium apparently do not induce ductal occlusions; therefore, the mechanism for blockage will likely be a factor acting directly on the efferent ductules. Since the normal epithelium of the efferent ductules is "leaky" (Nagano and Suzuki, 1980) and capable of reabsorbing much of the water in rete testis fluids (Jones and Jurd, 1987), damage to the epithelium could exacerbate the rate of water loss from the lumen and the resulting dehydration would further compact an already crowded lumen of sloughed germ cells. However, the precise pathobiochemical mechanism that induces ductal occlusions remains to be determined.

6-Chloro-6-Deoxyglucose

This compound also produces histopathological lesions in the efferent ductules and proximal portions of the epididymis (Ford and Waites, 1981). Similar to the effects seen with the benzimidazole carbamates, 6-chloro-6-deoxyglucose (CDG) exposure at high dosages induces testicular swelling by day 3 postexposure followed by testicular atrophy by day 77. These changes are associated with occluded efferent ductules and epididymal spermatocoeles with severe inflammation. The mechanism of action appears to be similar to that of α-chlorohydrin, in that CDG also reduces the transport of glucose into the ductal epithelium (Hinton et al., 1983).

Quinazolinone

2,3-Dihydro-2-(1-naphthyl)-4-(1H), Quinazolinone, was used as an antiarthritic drug and found to be a potent antitumor drug, but it also turns out to be an antifertility drug and induces sterility in numerous animals (Ericsson, 1975). The chemical induces occlusions of the efferent ductules and initial segment of the epididymis, swollen ductal lumens, spermatocoeles, sperm granulomas, and transient swelling of the testis prior to total atrophy (Ericsson, 1971). Its mechanism of action on the male tract has not been reported.

Metals

Cadmium exposure causes massive necrosis of the initial segment and other regions of the epididymis by damaging the vascular endothelium (Mason and Young, 1967; Nagy,

1985). This disruption of blood flow to the testis and epididymis is thought to be responsible for these effects. Lesions in the ductuli efferentes have also been reported (Mason and Young, 1967). The pathological changes following cadmium exposure included sperm stasis and impaction, ciliary loss, epithelial hyperplasia, edema, and phagocytosis of sperm. Future studies of this metal should examine the efferent ductules for biochemical changes. Cadmium alters the activity of several enzymes (Singhal et al., 1985), including that of carbonic anhydrase, which has its highest concentration in the epithelial cells lining the efferent ductules (Cohen et al., 1976) and, as indicated above, is probably involved in water and H^+ movement in this region (Ilio and Hess, 1992).

There are conflicting reports on the effects of lead on the male reproductive system (Thomas and Brogan, 1983). Although lead disrupts the oxidoreductase enzymes of the midpiece in epididymal sperm (Marchlewicz et al., 1993), there is no evidence that this metal has a direct effect on epididymal function. However, Marchlewicz et al. (1993) clearly showed a reduction in sperm concentration in the epididymis without evidence of testicular effects. These data suggest that lead may have an effect on the ability of the ductal epithelium to resorb fluids.

Mercury is a potent neurotoxin and therefore not normally considered a reproductive toxicant. However, there is evidence that Young's syndrome, which causes failure of normal efferent ductule function and infertility, is associated with mercury toxicity (Hendry et al., 1990). In this syndrome, the luminal fluid shows increased viscosity. Moreover, the epithelium of efferent ductules exhibits an abnormal accumulation of lipid, which also may indicative of decreased flow of fluid through the ductules (Mitchinson et al., 1975).

Dinitrobenzene

Dinitrobenzene, a chemical widely used in the manufacture of dyes, plastics, and explosives, has also been reported to cause a limited number of obstructions in the ductuli efferentes, some eventually proceeding to granulomatous formation, mineralization, and fibrosis (Linder et al., 1988). However, there has been no detailed description of the changes in the ductal epithelium following exposure.

Other Compounds

Smoking has been implicated in epithelial disruption by causing detachment of ciliary tufts in the efferent ductules or initial segment (Bornman et al., 1989). High dosages of fluoride also produce the loss of cilia as well as loss of epithelial lining in the ductuli efferentes and initial segment (Susheela and Kumar, 1991). Moreover, a deficiency in vitamin E affects efferent ductule epithelium (Mason and Shaver, 1952), and uranyl nitrate, which causes specific injury to the proximal convoluted tubules of the kidney, also induces testicular swelling similar to that observed following efferent ductular occlusions (Mason and Shaver, 1952).

In conclusion, there are basically two responses by the efferent ductules to toxic insult. Either the ductules fail to absorb the luminal fluids, which results in a diluted semen and infertility, or the ductules become occluded by some mechanism yet to be discovered. Regardless of how the ductules become occluded, once they are blocked, the long-term result is the same in most cases. The blockage induces a build-up of fluid in the rete testis and seminiferous tubules, which leads to testicular swelling, ischemia consequent to pressure build-up, and ultimately atrophy. Future studies should address the mechanisms of ductular occlusions and other factors that interfere with the normal function of this region of the excurrent duct system.

Epididymis

Mechanisms of Toxicity

Ascertaining and understanding toxicity within the epididymis is not a straightforward matter. As indicated above, the maturation of sperm within the epididymis requires interaction between the epididymal epithelium, the luminal fluid, and sperm. In addition, it is difficult to determine whether toxicity observed at the level of epididymal sperm is a consequence of direct toxicant action on the epididymis or represents a lesion that originated within the testis. Moreover, toxicity in the epididymis can be manifest as toxicity to sperm alone, toxicity that involves the epididymal epithelium or lamina propria, or toxicity to sperm that is mediated via toxic insult to the epididymal epithelium or lamina propria.

While virtually all male reproductive toxicity tests include an evaluation of epididymal sperm quantity and quality, the effects observed on epididymal sperm are frequently linked to a concomitant insult in the testis. For example, exposure to chemicals such as dibromo-chloropropane (Whorton et al., 1977), dinitrobenzene (Hess et al., 1988; Linder et al., 1988), dibromoacetic acid (Linder et al., 1995), cadmium (Saksena and Lau, 1979; Gray et al., 1992), ethylene glycol monoethyl ether (Clegg and Zenick, 1988), and ethane dimethanesulfonate (Jackson, 1964; Jackson and Morris, 1977) has been shown to disrupt spermatogenesis, decrease sperm reserves in the cauda epididymidis significantly, and thereby reduce fertility.

Aside from those testicular toxicants that decrease fertility by decreasing the number of epididymal sperm available upon ejaculation, low-dose exposure to certain mutagenic agents such as ethane methanesulfonate are likely to compromise the genetic integrity of testicular germ cells and thereby compromise fertility without effecting a decrease in sperm numbers. In a study of cyclophosphamide, an immunosuppressive agent known to produce dominant lethal chromosome mutations, when males were dosed for only 4 days and then mated, significant postimplantation losses occurred (Qui et al., 1992). Moreover, when the efferent ducts were ligated prior to cyclophosphamide administration, postimplantation losses persisted, indicating that the effect was exerted on sperm already present in epididymis. Whether this effect represents direct or indirect action has not been determined. Interestingly, cyclophosphamide also was shown to alter the incidence of specific epididymal epithelial cell types (Trasler et al., 1988).

An example of indirect-acting dominant lethal toxicity to epididymal sperm was provided by dominant lethal studies of methyl chloride (Working et al., 1985). In this study postimplantation losses were observed one week after acute inhalation exposure to methyl chloride. Subsequent studies demonstrated that the damage to sperm DNA was secondary to an inflammatory response in the cauda epididymidis (Chapin et al., 1984) as treatment with anti-inflammatory agents prevented both the inflammatory response and the postimplantation loss (Chellman et al., 1986).

Numerous other chemicals have been shown to alter measures of epididymal sperm quality such as sperm motility, sperm morphology, and sperm protein composition. These endpoints appear to contribute significantly to the overall fertility (i.e., fertilizing ability and normal embryo development) of epididymal sperm. While α-chlorohy-drin was one of the first chemicals considered to be an epididymal toxicant, the mechanism of its toxicity remains unclear. Both α-chlorohydrin and its structural analog, epichlorohydrin, reduce the fertilizing ability of sperm within a few days of exposure (Ericsson and Baker, 1970; Cooper et al., 1974; Tsunoda and Chang, 1976). Since both chemicals produce sperm granulomas in the epididymis, it was speculated that the reduced fertilizing ability was secondary to granuloma formation, much like the methyl chloride scenario.

Studies using lower doses of α-chlorohydrin (Tsunoda and Chang, 1976) and epichlorohydrin (Toth et al., 1989) failed to observe granuloma formation, but fertility was still reduced. It also has been suggested that this reduction in fertility is the result of chemical-specific effects on sperm motility (Toth et al., 1989; Slott et al., 1990). Studies showed that α-chlorohydrin inhibits glyceraldehyde-3-phosphate dehydrogenase activity (Jones and O'Brien, 1980) and alters the membrane protein composition (Tsang et al., 1981) of cauda epididymal sperm. Recent work in our laboratory (Klinefelter et al., 1997a) demonstrates that epichlorohydrin compromises the fertility of cauda epididymal sperm at doses as low as 3 mg/kg, a dose that does not alter sperm motion parameters. Moreover, the decrease in fertility was strongly correlated with a diminution in a specific sperm protein. Interestingly, 6-chloro-6-deoxyglucose, a chemical proposed for use as a male contraceptive, was also shown to alter the protein profile of epididymal sperm (Tsang et al., 1981). These authors speculated that this might be attributed to inability of proteins secreted by the epididymis to associate with the sperm membrane. The lack of an appropriate in vitro model to test this hypothesis precluded further study.

Ornidizole, a drug with trichomonacidal activity, is another chemical that exerts qualitative effects on epididymal sperm. Both the motility and fertilizing ability of epididymal sperm are reduced within one week of exposure (McClain and Downing, 1988; Oberlander et al., 1994). The short onset of these effects strongly suggests a direct effect on the epididymis; however, there are not yet definitive data to support this notion. Moreover, while data demonstrating that in vivo, but not in vitro, exposure of sperm to ornidizole decreases the velocity of sperm suggest that the effect may be mediated through the epithelium, this too remains to be determined.

In addition to epididymal toxicity manifested by insult to the epididymal epithelium and/or sperm, chemicals that modify the innervation of the epididymal duct also are likely to alter both the quantity and quality of sperm within the epididymis. Indeed, partial sympathetic denervation of the epididymis recently has been shown to increase the number of sperm stored in the cauda epididymidis (Billups et al., 1991a), decrease the swimming velocities of these sperm (Billups et al., 1991b), and alter the protein composition of cauda epididymal luminal fluid (Ricker et al., 1996). Treatment with the antihypertensive drug guanethidine has been shown to selectively destroy the adrenergic innervation of the epididymis, vas deferens, seminal vesicles, and prostate (Evans et al., 1972). This chemical also produces significant increases in both the weight of the cauda epididymidis, the number of cauda epididymal sperm that are stored (Lamano Carvalho et al., 1993), and sperm transit time through the cauda epididymidis (Kempinas et al., 1997).

There are recent data to suggest that chemicals that produce epididymis-specific reductions in cauda epididymal sperm, i.e., no change in testicular sperm numbers, also compromise fertility (Klinefelter et al., 1997). Indeed, a toxicant-induced, epididymis-specific reduction in the number of sperm stored in the cauda epididymidis is likely to perturb the process of epididymal sperm maturation if the decrease in cauda sperm reserves is caused by accelerated epididymal sperm transit. Interestingly, the chemicals that produce this epididymis-specific decrease in cauda epididymal sperm are structurally quite diverse and include the weakly estrogenic pesticide methoxychlor (Gray et al., 1989), the antiandrogenic fungicide vinclozolin (Gray et al., 1994), the polychlorinated chemicals 2,3,7,8-tetrachlorodibenzo-*p*-dioxin (TCDD; Mably et al., 1992; Gray et al., 1995a) and TCDD-like cogener PCB169 (Gray et al,. 1995b), and two alkylating derivatives of ethylene glycol, ethane dimethanesulfonate (EDS), and chloroethylmethanesulfonate (CEMS) (Klinefelter et al., 1994b).

Recent data suggest that at least one of these chemicals, CEMS, does indeed accelerate the transit of sperm through the epididymis (Klinefelter and Suarez, 1997), but the mechanism of its action remains elusive. It will be important to determine the cellular and biochemical mechanisms underlying toxicant-induced decreases in epididymal sperm transit, as a significant decrease in transit time in humans is likely to have an adverse consequence on the fertility of sperm in the human ejaculate.

Models to Assess Direct-Acting Toxicants

As suggested above, it is inherently difficult to establish that a toxicant exerts its direct action on the epididymis due to the fact that the testis can contribute, via either the luminal fluid or the systemic circulation, to toxicity manifested in the epididymis. Thus, while many chemicals have been considered epididymal toxicants, relatively few have been studied in sufficient fashion to indicate direct action. Over the past several years, studies from our laboratory have focused on putative epididymal toxicants and methodology to identify direct action.

To control experimentally for any testicular factor contributing to the observed toxicity in the epididymis, it is necessary to apply our understanding of epididymal sperm maturation. In the rat, sperm maturation occurs within the 4 days required for sperm to travel from the androgen-dependent caput region to the proximal cauda epididymidis (Robb et al., 1978), where fertilizing ability is first observed (Dyson and Orgebin-Crist, 1973). Thus, an evaluation of sperm in the proximal cauda epididymidis 4 days following the onset of toxicant exposure avoids any possible contribution of testicular sperm. Moreover, sperm within the proximal cauda are more homogeneously aged compared to those in the distal cauda, thereby affording greater sensitivity. Finally, to circumvent toxicity in the epididymis that might result from alterations in testicular fluid or decreased testosterone production, efferent duct ligation and exogenous testosterone supplementation may be utilized, respectively (Klinefelter et al., 1990).

We used this methodology to determine whether EDS, well known for its Leydig cell toxicity, and CEMS, a structurally related alkane sulfonate, were capable of exerting effects in the rat epididymis that were independent of circulating testosterone levels or testicular fluid (Klinefelter et al., 1990; Klinefelter et al., 1994b). Both chemicals were shown to elicit similar, testosterone-independent effects in the epididymis including a disappearance of clear cells from the cauda epididymidis (Fig. 5), an epididymis-specific decrease in cauda epididymal sperm, and an alteration of the protein profile of sperm from the proximal cauda epididymidis.

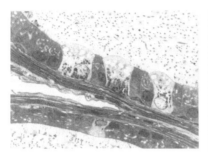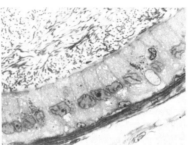

Figure 5   (Left) The epithelium of the proximal cauda epididymidis in a normal adult rat. Note the presence of numerous clear cells (arrow) (x1100). (Right) The epithelium of the proximal cauda epididymidis in a rat treated with 4 daily doses of chloroethylmethanesulfonate (12.5 mg/kg). Note that distinguishable clear cells have disappeared; cells containing lysosomes (arrowhead) presumably represent degenerating clear cells (x1100).

While the results of these studies suggested that EDS and CEMS act directly on the epididymis, the possibility remained that the epididymis was being exposed to nonandrogenic, testicular manifestations via the systemic circulation, which in turn perturbed structure and function of the epididymis. Thus, both in vivo and in vitro methods were developed to establish that sperm maturation was compromised by direct action on the epididymis. The effect of EDS exposure in vivo on the ultimate phase of sperm maturation, i.e., fertilizing ability, was evaluated using in utero insemination. Although in utero insemination of epididymal sperm in rats is not novel (Dyson and Orgebin-Crist, 1973; Tsunoda and Chang, 1976) it has not been applied to toxicology despite the suggestion that it would increase the sensitivity of detecting toxicant-induced changes in fertility (Amann, 1986). This notion is supported by the observation that a normal rat ejaculates an excess of qualitatively normal sperm; ejaculated sperm counts need to be decreased 1000-fold before a decrease in fertility is observed (Gray et al., 1992).

We first determined the number of proximal cauda epididymal sperm which, when inseminated in utero, would achieve 75% fertility, a submaximal value in the linear range of the dose-response curve (Klinefelter et al., 1994a). An equivalent number of sperm from the proximal cauda of control and EDS-treated rats were inseminated in utero 4 days after exposure, and fertilizing ability was evaluated. To make the test for qualitative effects more rigorous, a lower dose of EDS, one that failed to alter sperm motion parameters, was chosen. Once significant EDS-induced decreases in fertility were observed, treatment groups were added to determine whether the reduced fertilizing ability resulted from direct action on the epididymis. For this, a castrate, testosterone (T)-implanted rat model was used. These rats have completely normal fertility for several days (Dyson and Orgebin-Crist, 1973). However, when EDS was administered to these animals, fertilizing ability was reduced significantly to a level that was not statistically different from that of intact rats treated with EDS. This in vivo study demonstrated conclusively that EDS is capable of compromising sperm maturation, i.e., fertilizing ability, in the absence of the testis, presumably via direct influence on the epididymis.

To verify direct action of EDS on the epididymis, an in vitro coculture system was established (Klinefelter, 1992, 1993; Klinefelter et al., 1992b). To retain morphological integrity and differentiated function, plaques of epididymal epithelial cells, rather than single cells, were plated on a customized extracellular matrix in a bicameral chamber (Fig. 6). While the epithelial cells do not form a monolayer, they do spread out to form sheets of contiguous cells. Importantly, cultured epithelial cells from the caput epididymidis synthesize and secrete proteins identical or similar to those found in luminal fluid obtained by micropuncture, particularly the heavy and light chains of SGP-2 or clusterin (Mattmueller and Hinton, 1992). Once the epithelial cells are established in culture, a second, coculture insert can be added to evaluate the effects of toxicants on cocultured sperm. In the absence of EDS, cocultured caput sperm acquire some of the proteins, including clusterin, secreted by the underlying epithelial cells. Moreover, during the 24-hour coculture period the progressive motility and straight-line velocity of these sperm increase significantly; with values approximating those reported for sperm from the corpus epididymidis (Yeung et al., 1992). While several facets of sperm maturation occur in untreated cocultures, treatment with EDS results in dose-related decreases in the association of secreted protein by sperm as well as significant decreases in progressive motility and straight-line velocity (Klinefelter et al., 1992a). Thus, EDS does compromise facets of epididymal sperm maturation directly.

More recently we have used cultured epithelial cells from the caput epididymidis of the castrate, testosterone-implanted animal to identify alterations in protein synthesis and secretion induced by antiandrogens (Klinefelter and Kelce, 1994c). Previous work reported difficulty obtaining antiandrogen-induced alterations in the epididymis when intact rats were used (Dhar et al., 1982; de las Heras, 1988). In the castrated, testosterone-implanted animal, the testosterone concentration in the epididymis is approximately 50% of control, i.e., decreased from 30 to 15 nM (Klinefelter et al., 1994a), thereby lowering the threshold for competitive binding of antiandrogen to the androgen receptor. Similarly, androgen supplementation in culture medium also was reduced to 15 nM. While the profile of proteins synthesized and secreted by epithelial cells in untreated cultures appeared identical to that seen for cells from intact animals, both the

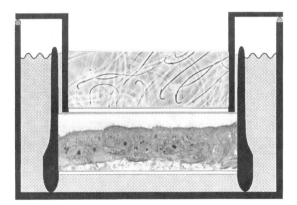

Figure 6   (Left) Sheets of contiguous caput epididymal epithelial cells and attached extracellular matrix can be recovered from culture 48 hours after plating epithelial cell plaques. (Right) Diagram illustrating the configuration of the epididymal epithelial cell-sperm coculture system. The epithelial cell plaques are plated on a customized extracellular matrix spread across a 0.4-mm porous Millicell-CM insert. After 2 days immature sperm from the caput epididymidis are placed in a 3.0-mm porous Transwell insert suspended just above the underlying epithelial cells.

known antiandrogen, hydroxyflutamide, and metabolites of the antiandrogenic pesticide, vinclozolin (Gray et al. 1994), inhibited androgen-dependent protein synthesis and secretion in a dose-related manner. Interestingly, proteins previously not identified as androgen dependent also were diminished by antiandrogens in this study. Since previous work to identify androgen-dependent proteins utilized long-term castration and androgen replacement paradigms, the acutely regulated, androgen-dependent proteins may deserve more attention in future studies concerning androgen regulation of epididymal function.

Potential Biomarkers of Fertility

Given the increasing number of chemicals that await toxicological testing and the increasing number of infertile couples reported by human fertility clinics, there is a clear need to develop biomarkers of fertility. In the study described above to determine whether EDS can act directly to alter the fertilizing ability of a cauda epididymal sperm, we observed a significant correlation between a specific sperm protein and fertilizing ability. It seemed reasonable, however, to suspect that multiple factors act in concert to confer normal fertility (Amann et al., 1993) and that exposure to different toxicants might compromise fertility by differentially targeting these requisite factors. In a subsequent multivariant study, four epididymal toxicants (i.e., EDS, CEMS, epichlorohydrin, and hydroxyflutamide) were evaluated, each at two dose levels (Klinefelter et al., 1997). In this study each recipient female was stimulated with a vasectomized teaser male prior to insemination and fertility was evaluated 9 days after insemination. Each toxicant decreased the fertility of proximal cauda sperm significantly compared to sperm from controls. While endpoints such as serum testosterone, epididymal tissue testosterone, progressive motility, and cauda reserves were reduced significantly by treatment, the only endpoint that was highly correlated with fertility was an acidic 22 kDa protein recovered in detergent extracts of the inseminated sperm. Using discriminant analysis, quantitation of this protein following two-dimensional gel electrophoresis enabled correct identification of 90% of the animals with greater than 50% fertility and 94% of the animals with less than 50% fertility. The results suggest that this sperm protein (SP22) may serve as a useful biomarker in predicting fertility following a toxic insult. Recent data indicate that SP22 is identical to human DJ1 protein, a protein for which function is undetermined (Welch et al., 1997). The fact that this protein appears in two-dimensional gels of detergent-extracted rete testis sperm (unpublished observation) indicates testicular origin and suggests destabilization during epididymal transit following toxicity to the epididymis (Fig. 7). If this indeed proves to be the case, additional important epididymal biomarkers of fertility await discovery.

Alterations Following Developmental Exposures

While the aforementioned studies demonstrate toxicity within the epididymis following adult (i.e., postpubertal) exposure to certain chemicals, an increasing number of studies are providing data to support the notion that the developing epididymis is extremely sensitive to toxic insult. Among the chemicals that perturb the developing epididymis is the antiandrogenic fungicide vinclozolin, 2,3,7,8-tetrachlorodibenzo-p-dioxin (TCDD or dioxin), and the dioxinlike congener PCB 169. Moreover, when male rat

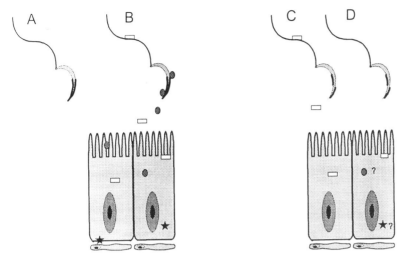

Figure 7 A hypothetical representation of how toxicants might alter sperm fertilizing ability and epididymal transit time. (A) Sperm entering the lumen of the epididymis from the testis acquire both androgen-independent proteins such as clusterin (open rectangle) as well as androgen-dependent protein (solid circles). (B) The association of these androgen-dependent proteins with the sperm membrane acts to modify and/or stabilize SP22 and confers fertilizing ability. Androgen-dependent protein secreted basally (star) may serve to regulate smooth muscle contractility and regulate, i.e., decelerate, transit. (C) Antiandrogenic chemicals diminish the synthesis of androgen-dependent proteins and thereby destabilize SP22 and thereby decrease fertility, while androgen-independent proteins continue to be secreted and associated with the sperm membrane. Antiandrogens would also accelerate sperm transit. (D) Other epididymal toxicants such as ethane dimethanesulfonate and chloroethylmethanesulfonate decrease the overall secretion and sperm-association of both androgen-independent and androgen-dependent protein; protein synthesis may be compromised as well. Again, fertilizing ability would be compromised and transit time would be accelerated.

offspring are dosed with the estrogenic pesticide methoxychlor beginning from weaning through adulthood, puberty is delayed, the weight of the epididymis is decreased, and the number of sperm in the ejaculate is reduced while the number of sperm in the testis remains unchanged (Gray et al., 1989; Anderson et al., 1995). When prepubertal bulls were exposed to either estradiol or zeranol, and estrogenic mycotoxin derivative, normal histological differentiation of the epididymis at sexual maturation was compromised significantly (Veeramachaneni et al., 1988). In light of the recent concern over estrogens in the environment, more definitive research on the bioactivity of environmental estrogens in the male is warranted.

The effects of environmental antiandrogens in male rats have been more thoroughly characterized. Indeed, it is well known that when the fetus is exposed to antiandrogenic chemicals such as flutamide during critical periods of reproductive development, Wolffian duct derivatives develop abnormally (Schardein, 1993). When the fungicide vinclozolin was administered during sexual differentiation, i.e., gestation day 14 to postnatal day 3, male offspring had epididymal granulomas, significantly reduced ejaculated sperm counts, and were infertile (Gray et al., 1994). The reduction in ejaculated sperm counts and fertility was attributed to the hypospadias that was observed in these animals; however, when adult animals were exposed to similar doses of vinclozolin, there was a reduction in cauda sperm reserves and epididymal weight (Fail et al., 1995). While the epididymis was adversely affected, it is noteworthy that the decreases in seminal vesicle weight and prostate weight were more conspicuous and sensitive, confirming the inherent refractoriness of the epididymis to antiandrogens relative to the other androgen-dependent organs.

Studies designed to determine what, if any, qualitative alterations gestational and perinatal antiandrogen exposure causes in the epididymis are warranted. As indicated above, when castrate, T-implanted adults are exposed to hydroxyflutamide, both fertility and sperm protein composition are affected; exposure of the developing epididymis to the relatively weaker, environmental antiandrogens such as vinclozolin and p,p'-DDE (Kelce et al., 1994, 1995) may produce similar dysfunction in the adult. Regardless, it is clear that if sufficiently high exposure to these chemicals occurs during the period of reproductive development, qualitative alterations in epididymal function will be inconsequential since anomalies such as hypospadias an ectopic testes will preclude normal mating (Gray and Kelce, 1996).

After binding to the Ah receptor, TCDD alters estrogen-dependent, androgen-dependent, and thyroid-dependent processes as well as growth factor–mediated action. An early multigenerational study demonstrated that dietary exposure to only 0.01 μg/kg/day resulted in decreased fertility of males and females mated in the F_2 and F_3 generations, but did not affect the males in the parental (F_0) generation, males that were not exposed during development (Murray et al., 1979). This suggested that developing reproductive system is uniquely sensitive to this chemical. More recent studies have shown that a single gestational exposure of 1 μg/kg significantly decreases epididymal weight and epididymal sperm numbers (Mably et al., 1992; Gray et al. 1995a). The decreases in epididymal (Mably et al., 1992) and ejaculated (Gray et al., 1995a) sperm were much greater than the observed decrease in testicular sperm, suggesting an epididymis-specific effect. Recently, this decrease has been attributed to accelerated transit of sperm through the cauda epididymidis (Wilker et al., 1996). These effects appear to occur in the absence of significant changes in androgen status, i.e., circulating T and epididymal androgen receptor content are normal (Gray et al., 1995a), and therefore may reflect modification of an androgen-independent event or postnuclear modification of an androgen-dependent event. The significant decrease in ejaculated sperm numbers undoubtedly accounted, at least in part, for the infertility. However, a recent evaluation of sperm protein profiles in adults following gestational or perinatal exposure to TCDD reveals multiple changes (unpublished observations).

Interestingly, a low-dose exposure of the dioxinlike congener PCB169 during gestation also reduces cauda sperm reserves as well as ejaculated sperm counts in offspring (Gray et al., 1995b); the epididymis-specific decrease in sperm numbers is even more dramatic than that observed following TCDD exposure. Finally, lactational exposure of males to Arochlor 1254 (Sager et al., 1991) also decreases fertility in adulthood. Thus, like TCDD, perinatal exposure to certain PCBs appears to imprint the developing epididymis such that sperm quantity, and perhaps sperm quality, is compromised in the adult epididymis. Unfortunately, the mechanism by which this might occur remains elusive.

The Seminal Vesicles, Prostate, and Bulbourethral Glands

As indicated above, the seminal vesicles and prostate have proved to be invaluable diagnostic indicators of antiandrogenic action in the male. Antiandrogen-induced decreases in prostate weight seem to be irreversible and highly sensitive when exposure occurs in utero, but not when exposure occurs in later life. For example, exposure to vinclozolin during puberty reduces the weights of the seminal vesicle and prostate slightly, while exposure to similar doses between gestation day 14 and postnatal day 3 results in agenesis of the prostate (Gray et al., 1994). Thus, it is apparent that the earlier an exposure to antiandrogen occurs, the more dramatic the manifestation of regressive changes in this androgen-dependent gland. This refractoriness in adult life is presumably due in part to the fact that the weaker environmental antiandrogens bind with less affinity to the androgen receptors than more potent synthetic antiandrogens (Kelce et al., 1995) and the natural substrate for the androgen receptor, i.e., testosterone and dihydrotestosterone, increase dramatically during reproductive development, thereby reducing the efficacy of competitive binding. However, Kelce and coworkers (1995) demonstrated that p,p'-DDE is capable of exerting significant effects on the prostate of adult castrate, T-implanted rats. Aside from decreased prostate weight, these prostates express significantly less C3 mRNA, the androgen-regulated subunit of prostatein, compared to control rats.

The seminal vesicles and prostate are influenced less by gestational and perinatal exposure to TCDD than by antiandrogen exposure. A relatively high dose of TCDD produces only modest decreases in seminal vesicle weight (Gray et al., 1995a) and prostate weight (Roman et al., 1995). When observed, the decrease in organ weight does not appear to be associated with any alteration in androgen status (Gray et al., 1995a; Roman et al., 1995). Gestational exposure to PCB169, on the other hand, decreases seminal vesicle and prostate weights by 50% compared to control (Gray et al., 1995a). As with the TCDD- and PCB169-induced alterations in the epididymis, the mechanism underlying these decreases in accessory sex gland weight remains elusive.

The accessory glands are influenced not only by the action of antiandrogens that act via the androgen receptor, but also by inhibitors of 5α-reductase activity. In utero exposure to the potent 5α-reductase inhibitor, finasteride produces irreversible alterations of the seminal vesicles and prostate (Imperato-McGinley et al., 1992). Moreover, when adult male rats were exposed to finasteride, fertility was reduced significantly (Cukierski et al., 1991; Wise et al., 1991). This decrease in fertility had an earlier onset when rats were exposed during the prepubertal period than when rats were exposed only as adults (Wise et al., 1991). An in utero insemination study confirmed that the reduction in fertility was not attributed to a change in the quality of cauda epididymal sperm (Cukierski et al., 1991). Moreover, mating behavior was unaffected. The observed reduction in fertility was, however, associated with decreases in the weights of the seminal vesicles and prostate and failure to form copulatory plugs. Thus, by inhibiting the action of 5α-reductase, there is less DHT available within the seminal vesicle and prostate to maintain normal differentiated function. As a consequence, the formation of copulatory plugs was compromised.

As mentioned above, the bulbourethral glands are highly sensitive to exposure to estrogenic chemicals. Metaplastic and cystic lesions of the bulbourethral glands have been observed in sheep maintained on clover pastures, a rich source of phytoestrogens (Bennetts, 1946; Bennetts et al., 1946). Moreover, pastures containing *Trifolium*

subterraneium, *T. protense*, and *T. repens* plants provide an enriched source of estrogenic activity that is likely to result in alteration of the bulbourethral glands. Metaplastic
lesions of these glands were also observed in bulls implanted neonatally with zeranol
(Deschamps et al., 1987). Zeranol, a derivative of the estrogenic mycotoxin zearalenone
found in moldy cereals, is frequently used as an anabolic agent in cattle and sheep. A
Canadian study estimated that the daily exposure of young children to zearalenone via
food consumption was in the $\mu M/kg$ body wt/day range (Kuiper-Goodman et al., 1987).
Considering that humans are also exposed, albeit inadvertently, to phytoestrogens and
estrogenic mycotoxicns, it may be quite important to carefully evaluate estrogen-responsive tissues such as the bulbourethral glands in toxicology studies, even when the rat
is used as the animal model.

RELEVANCE TO HUMAN HEALTH

It seems reasonable in some cases to evaluate human male reproductive lesions based
upon mechanisms of known functions in laboratory animals, such as the kidneylike
function of the efferent ductules and initial segment of the epididymis. For example,
in a study of 218 autopsies, it was observed that whenever there was a chronic renal
failure with hemodialysis, there was also a cystic dilation of the rete tetis and efferent
ductules (Nistal et al., 1989). The rete and ductules also contained calcium oxalate
crystals and proteinaceous material, and the testes exhibited hypospermatogenesis. These
pathological changes are quite similar to those following α-chlorohydrin exposure in
the rat, and thus could be interpreted as a generalized response of proximal regions
epididymis to substances or conditions that affect ion and fluid transport.

The role that the initial segment plays in seminiferous tubular atrophy is relevant
to human health. Experimental data clearly show that occlusions of the rete testis and
efferent ductules and possibly the initial segment of the epididymis will cause a build-
up of fluid within the seminiferous tubules and subsequent atrophy within a few days
(Hess et al., 1991; Ilio and Hess, 1994). What is not known is the number of ductules
that must be occluded to induce atrophy or if atrophy is proportional to the number of
occlusions. Finally, the presence of focal atrophy and oligospermia may be better understood by improving our knowledge of tubule function in the head of the epididymis.

Today there is increased public concern that sperm counts are declining in the
human population as a consequence of exposure to environmental pollutants (Giwercman
and Skakkebaek, 1992; Sharpe, 1993). While all attention to date has been focused on
the possibility of compromise within the testis, it is important to recognize that the
epididymis regulates the transit of sperm from the testis to the distal cauda epididymidis
and vas deferens, where they are stored until ejaculation or emission via the urine. If
a toxicant accelerates sperm transit through the epididymis, it is quite likely that significantly fewer sperm will be stored within the cauda epididymidis and a reduction in
ejaculated sperm counts will be observed. Moreover, under conditions of accelerated
epididymal transit, sperm have less time to mature. Since a significant percentage of
sperm in a fertile male's ejaculate is qualitatively abnormal (Wyrobek et al., 1982) and
the time required for sperm to transit through the caput/corpus epididymidis in humans
is roughly half that of rats (Robb et al., 1978; Johnson and Varner, 1988), a toxicant-
induced decrease in sperm quality could render an individual infertile. Thus, acceler-

ated epididymal transit is likely to compromise both the quantity and quality of ejaculated sperm.

Just as toxicant-induced changes in epididymal transit are likely to have a bearing on fertility in humans, chemicals that perturb the hormone-dependent processes involved in epididymal sperm maturation also are likely to affect fertility in humans, provided exposure is sufficient to compromise the normal mechanism of hormone action. Given this consideration, it is disconcerting that select populations of infants and prepubertal boys may be exposed to levels of antiandrogenic pesticides that can alter androgen receptor–mediated gene transcription (Kelce et al., 1995). While the mechanism of action of TCDD-like chemicals on the male reproductive system is not clear, these chemicals have long half-lives and bioaccumulate. Thus, it is feasible that long-term prepubertal exposure may indeed be sufficient to mediate changes in fertility at the level of the epididymis and/or the acessory glands (Gray and Kelce, 1996).

CONCLUDING REMARKS

The male reproductive tract consists of several unique tubular organs linked together by a common ductal lumen. Functions differ significantly for the tissues comprising the excurrent ducts, and therefore it is inappropriate to assume a common response to toxicants. Toxic responses of the efferent ductules have been overlooked traditionally because of their obscure location outside the epididymal connective tissue capsule. Morphological and biochemical studies, however, support the opinion that this part of the ductal system is physiologically important, particularly in the reabsorption of rete testis fluid and for the maintenance of proper sperm concentrations in the lumen of the epididymis. It remains to be seen whether the statement by Wakeley in 1943 that most epididymal cysts in man arise from dysfunction of the efferent ductules holds true. However, in rodent studies, if occlusions occur during the early period of a subchronic or subacute test, then all subsequent exposures to the toxicant will have little to do with the resulting testicular atrophy because regression will be a direct result of the rapid fluid back-pressure. In light of this mechanism of toxicity, many studies in which testicular atrophy is found may need to be reexamined for terminal lesions in the efferent ductules and initial segment of the epididymis.

Toxicology of the epididymis also is a relatively new area in reproductive toxicology. Recent work demonstrates that multiple chemicals appear to exert similar effects on the epididymis and that many of these effects, i.e., disappearance of clear cells, loss of specific sperm protein, and decreased cauda sperm reserves, are independent of the testis. In vivo and in vitro methodologies now exist to allow us to determine whether epididymal toxicants compromise epididymal function directly. Moreover, receptor-binding assays, transfection assays, and gene knockout models will become increasingly important to distinguish the relative potency and mode of action for the increasing number of putative endocrine-disruptive chemicals. The availability of such methods should foster new research to better understand the mechanism of action of environmental chemicals on epididymal function, as well as to develop useful cellular and biochemical markers of toxic insult which are likely to compromise fertility in exposed humans. Finally, a significant new effort must be directed toward understanding toxicant-induced changes in epididymal transit time as well as understanding the unique sensitivity of developing Wolffian duct derivatives to toxic insult.

REFERENCES

Amann, R. P. (1986). *Environ. Health Perspect., 70*: 149.

Amann, R. P., Johnson, L., and Pickett, B. W. (1977). *Am. J. Vet. Res., 38*: 1571.

Amann, R. P., Hammerstedt, R. H., and Veeramachaneni, D. N. R. (1993). *Reprod. Fertil. Dev.,* 5: 361.

Anderson, S. A., Pearce, S. W., Fail, P. A., and McTaggart, B. T. (1995). *Toxicologist, 15*: 164A.

Arai, Y., Mori, T., Suzuki, Y., and Bern, H. (1983). *Int. Rev. Cytol., 84*: 235.

Au, C. L., Ngai, H. K., Yeung, C. H., and Wong, P. Y. D. (1978). *J. Endocrinol., 77*: 265.

Aumuller, G., and Riva, A. (1992). *Andrologia, 24*: 183.

Back, D. J., Glover, T. D., Shenton, J. C., and Boyd, G. P. (1975). *J. Reprod. Fertil., 45*: 117.

Bedford, J. M. (1975). *Handbook of Physiology*, Section 7 *Endocrinology*, Vol. V. *The Male Reproductive System* (D. W. Hamilton and R. O. Greep, eds.). American Physiology Society, Washington, DC, p. 303.

Behringer, R. R., Cate, R. L., Froelick, G. J., Palmiter, R. D., and Brinster, R. L. (1990). *Nature, 345*: 167.

Bennetts, H. W. (1946). *Aust. Vet. J., 22*: 70.

Bennetts, H. W., Underwood, E. J., and Shier, F. L. (1946). *Aust. Vet. J., 22*: 2.

Billups, K. L., Tillman, S. L., and Chang, T. S. K. (1991a). *J. Urol., 143*: 625.

Billups, K. L., Tillman, S. L., and Chang, T. S. K. (1991b). *Fertil. Steril., 56*: 1076.

Bornman, M. S., Kok, E. L., du Plessis, D. J., and Otto B. S. (1989). *Andrologia, 21*: 18.

Bowden, D. H., and Adamson, I. Y. R. (1985). *J. Pathol., 147*: 257.

Brooks, D. E. (1983). *Mol. Cell. Endocrinol., 29*: 255.

Brooks, D. E. (1987). *Mol. Cell. Endocrinol., 53*: 59.

Brown, D. V., Amann, R. P., and Wagley, L. M. (1983). *Biol. Reprod., 28*: 1257.

Brown, D., Verbavatz, J.-M., Valenti, G., Lui, B., and Sabolic, I. (1993). *Eur. J. Cell Biol.,* 61: 264.

Brown-Woodman, P. D. C., and White, J. G. (1976). *Aust. J. Biol. Sci., 29*: 545.

Bullock, B. C., Newbold, R. R., and McLachlan, J. A. (1988). *Environ. Health Perspect., 77*: 29.

Byers, S., and Graham, R. (1990). *Am. J. Anat., 188*: 31.

Byers, S. W., Musto, N. A., and Dym, M. (1985). *J. Androl., 6*: 271.

Carballada, R., and Esponda, P. (1992). *J. Reprod. Fert., 95*: 639.

Carter, S. D., Hess, R. A., and Laskey, J. W. (1987). *Biol. Reprod., 37*: 709.

Chapin, R. E., White, R. D., Morgan, K. T., and Bus, J. S. (1984). *Toxicol. Appl. Pharmacol.,* 76: 328.

Chellman, G. J., Morgan, K. T., Bus, J. S., and Working, P. K. (1986). *Toxicol. Appl. Pharmacol., 85*: 365.

Clegg, E. D., and Zenick, H. (1988). *Toxicologist, 8*: 119.

Cohen, J. P., Hoffer, A. P., and Rosen, S. (1976). *Biol. Reprod., 14*: 339.

Colditz, I., Zwahlen, R., Dewald, B., and Baggioli, M. (1989). *Am. J. Pathol., 134*: 755.

Cooke, P. S., Young, P., and Cunha, G. R. (1991a). *Endocrinology, 128*: 2867.

Cooke, P. S., Young, P., Hess, R. A., and Cunha, G. R. (1991b). *Endocrinology, 128*: 2867.

Cooper, T. G. (1982). *J. Reprod. Fertil., 64*: 373.

Cooper, T. G., and Orgebin-Crist, M.-C. (1977). *Biol. Reprod., 16*: 258.

Cooper, E. R. A., and Jackson, H. (1972). *J. Reprod. Fertil., 28*: 317.

Cooper, E. R. A., and Jackson, H. (1973). *J. Reprod. Fertil., 34*: 445.

Cooper, E. R. A., Jones, A. R., and Jackson, H. (1974). *J. Reprod. Fertil., 38*: 379.

Crabo, B. (1965). *Acta Vet. Scand., 6*: 8.

Crabo, B. G., Zimmerman, K. J., Gustafsson, B., Holtman, M., Koh, T. J. P., and Graham, E. F. (1975). *Int. J. Fertil., 20*: 87.

Croisille, Y. (1981). *Prog. Reprod. Biol., 8*: 1.

Cukierski, M. A., Sina, J. L., Prahalada, S., Wise, L. D., Antonello, J. M., MacDonald, J. S., and Robertson, R. T. (1991). *Reprod. Toxicol., 5*: 353.

Cyr, D. G., and Robaire, B. (1992). *Endocrinology, 130*: 2160.

Danzo, B. J. (1995). *Endocrinology, 136*: 4004.

Danzo, B. J., and Eller, B. C. (1979). *Endocrinology, 105*: 1128.

de las Heras, M. A., Suescun, M. O., and Calandra, R. S. (1988). *J. Reprod. Fertil., 83*: 177.

Deschamps, J. C., Ott, R. S., McEntee, K., Health, E. H., Heinrichs, R. R., Shanks, R. D., and Hixon, J. E. (1987). *Am. J. Vet. Res., 48*: 137.

Deslypere, J.-P., Young, M., Wilson, J. D., and McPhaul, M. J. (1992). *Mol. Cell Endocrinol., 88*: 15.

Dhar, J. D., Srivastava, S. R., and Setty, B. S. (1982). *Andrologia, 14*: 55.

Diamond, J. M., and Bossert, W. (1967). *J. Gen. Physiol., 50*: 2061.

Dinges, H. P., Zatloukal, K., Schmid, C., Mair, S., and Wirnsberger, G. (1991). *Virchows Archiv. A Pathol. Anat., 418*: 119.

Dixit, V. P,. Lohiya, N. K., Arya, M., and Agrawal, M. (1975). *Acta Biol. Med. Germ., 34*: 1851.

Djakiew, D., and Jones, R. C. (1983). *J. Reprod. Fertil., 68*: 445.

Douglass, J., Garrett, S. H., and Garrett, J. E. (1991). *Ann. N.Y. Acad. Sci., 637*: 384.

Dym, M. (1976). *Anat. Rec., 186*: 493.

Dym, M., and Romrell, L. J. (1975). *J. Reprod. Fertil., 42*: 1.

Dyson, A. L., and Orgebin-Crist, M. C. (1973). *Endocrinology, 93*: 391.

El-Badawi, A., and Schenk, E. A. (1967). *Am. J. Anat., 121*: 1.

Eliasson, R., and Risley, P. L. (1968). *Acta. Physiol. Scand., 73*: 311.

English, H. F. and Dym, M. (1981). *Ann. N.Y. Acad. Sci., 383*: 445.

Ericsson, R. J. (1970). *J. Reprod. Fertil., 22*: 213.

Ericsson, R. J. (1971). *Proc. Sec. Exp. Biol. Med., 137*: 532.

Ericsson, R. J. (1975). *Chemosterilants* (J. J. Sciarra, C. Markland, and J. J. Speidel, eds.). Harper & Row, New York, p. 262.

Ericsson, R. J., and Baker, V. F. (1970). *J. Reprod. Fertil., 21*: 267.

Evans, B., Gannon, B. J., Heath, J. W., and Burnstock, G. (1972). *Fertil. Steril., 23*: 657.

Fail, P. A., Pearce, S. W., Anderson, S. A., Tyl, R. W., and Gray, L. E. (1995). *The Toxicologist, 15*: 293A.

Fawcett, D. W., and Hoffer, A. P. (1979). *Biol. Reprod., 20*: 162.

Ford, W. C. L., and Waites, G. M. H. (1981). *Contraception, 24*: 577.

Garnacho, S. S., Vega, J. A., Arenal, A. A. and Casa, A. P. (1989). *Arch. Esp. de Urol., 42*: 727.

Garrett, J. E., Garrett, S. H., and Douglass, J. (1990). *Mol. Endocrinol., 4*: 108.

George, F. W., and Wilson, J. D. (1988). *The Physiology of Reproduction*, Vol. 1 (E. Knobil and J. Neill, eds.). Raven Press, New York, p. 3.

George, F. W., and Wilson, J. D. (1994). *The Physiology of Reproduction*, Vol. 1 (E. Knobil and J. Neill, eds.). Raven Press, New York, p. 3.

Gist, D. H., Hess, R. A., Bahr, J., and Bunick, D. (1996). *J. Androl., 16* (Suppl. 1).

Giwercman, A., and Skakkebaek, N. E. (1992). *Int. J. Androl., 15*: 373.

Goyal, H. O. (1985). *Am. J. Anat., 172*: 155.

Goyal, H. O., and Hrudka, F. (1980). *Andrologia, 12*: 404.

Goyal, H. O., and Hrudka, F. (1981). *Andrologia, 13*: 292.

Goyal, H. O., and Hutto V., and Robinson, D. D. (1992). *Anat. Rec., 233*: 53.

Gray, B. W., Brown, B. G., Ganjam, V. K., and Whitesides, J. F. (1983). *Biol. Reprod., 29*: 525.

Gray, L. E., Otsby, J., Ferrell, F., Rehnberg, G., Linder, R., Cooper, R., Goldman, J., Slott, V., and Laskey, J. (1989). *Fundam. Appl. Toxicol., 12*: 92.

Gray, L. E., Ostby, J., Linder, R., Goldman, J., Rehnberg, G., Cooper, R. (1990). *Fundam. Appl. Toxicol., 15*: 281.

Gray, L. E., Marshall, R., Ostby, J., and Setzer, R. W. (1992). *Toxicologist, 12*(1): 433A.

Gray, L. E., Ostby, J. S., and Kelce, W. R. (1994). *Toxicol. Appl. Pharmacol., 129*: 46.

Gray, L. E., Kelce, W. R., Monosson, E., Ostby, J. S., and Birnbaum, L. S. (1995a). *Toxicol. Appl. Pharmacol., 131*: 108.

Gray, L. E., Ostby, J., Wolf, C., Miller, D. B., Kelce, W. R., Gordon, C. J., and Birnbaum, L. (1995b). *Toxicology-Ecotoxicology-Mechanism of Action-Metabolism* (L. Birnbaum et al., eds.). Edmonton/Alberta, Canada, p. 33.

Gray, L. E., and Kelce, W. R. (1996). *Tox. Ind. Health*, (in press).

Greco, T. L., Furlow, J. D., Duello, T. M., and Gorski, J. (1992). *Endocrinology, 130*: 421.

Greco, T. L., Duello, T. M., and Gorski, J. (1993). *Endocrinol. Rev., 14*: 59.

Gunn, S. A., Gould, T. C., and Anderson, W. A. D. (1969). *Proc. Soc. Exp. Biol. Med., 132*: 656.

Guttroff, R. F., Cooke, P. S., and Hess, R. A. (1991). *Anat. Rec., 232*: 423.

Hall, J. C., Hadley, J., and Doman, T. (1991). *J. Androl., 12*: 76.

Hamilton, D. W. (1975). *Handbook of Physiology*, Section 7 *Endocrinology*, Vol. V. *The Male Reproductive System* (D. W. Hamilton and R. O. Greep, eds.). American Physiology Society, Washington, DC, p. 259.

Hammerstedt, R. H., and Parks, J. E. (1987). *J. Reprod. Fertil., 34*: 133.

Hammerstedt, R. H., and Rowan, W. A. (1979). *Biochem. Biophys. Acta, 710*: 370.

Hemeida, N. A., Sack, W. O., and McEntee, K. (1978). *Am. J. Vet. Res., 39*: 1892.

Hendry, W. F., Parslow, J. M., Levison, D. A., Royle, M. G., and Parkinson, M. C. (1990). *Ann. Roy. Coll. Surg. Engl., 72*: 396.

Hermo, L., and Morales, C. (1984). *Am. J. Anat., 171*: 59.

Hermo, L., Clermont, Y., and Morales, C. (1985). *Anat. Rec., 211*: 285.

Hermo, L., Spier, N., and Nadler, N. J. (1988). *Am. J. Anat., 182*: 107.

Hermo, L., Morales, C., and Oko, R. (1992a). *Anat. Rec., 232*: 401.

Hermo, L., Oko, R., and Robaire, R. (1992b). *Anat. Rec., 232*: 202.

Hees, H., Wrobel, K.-H., Kohler, T., Leiser, R., and Rothbacher, I. (1987). *Cell Tissue Res., 248*: 143.

Hess, R. A., Linder, R. E., Strader, L. F., and Perreault, S. D. (1988). *J. Androl., 9*: 327.

Hess, R. A., Moore, B. J., Forrer, J., Linder, R. E., and Abuel-Atta, A. A. (1991). *Fundam. Appl. Toxicol., 17*: 733.

Higgins, S. J., and Burchell, J. M. (1978). *Biochem. J., 174*: 543.

Higgins, S. J., Burchell, J. M., and Mainwaring, W. I. P. (1976). *Biochem. J., 158*: 271.

Higgins, S. J., Burchell, J. M., Parker, M. G., and Herries, D. G. (1978). *Eur. J. Biochem., 91*: 327.

Hinton, B. T., and Hernandez, H. (1985). *J. Androl., 6*: 300.

Hinton, B. T., and Turner, T. T. (1988). *News Physiol. Sci., 3*: 28.

Hinton, B. T., Hernandez, H., and Howards, S. S. (1983). *J. Androl., 4*: 216.

Hoffer, A. P. (1972). *Anat. Rec., 172*: 331.

Hoffer, A. P., and Hamilton, D. W. (1974). *Anat. Rec., 178*: 376.

Hoffer, A. P., Hamilton, D. W., and Fawcett D. W. (1973). *Anat. Rec., 175*: 203.

Hoffer, A. P., Hamilton, D. W., and Fawcett, D. W. (1975). *J. Reprod. Fertil., 44*: 1.

Hohlbrugger, G. (1980). *Fertil. Steril., 34*: 50.

Holland, M. K., Vreeburg, J. T. M., and Orgebin-Crist, M.-C. (1992). *J. Androl., 13*: 266.

Howards, S. S., Johnson, A, and Jessee, S. (1975a). *Fertil. Steril., 26*: 20.

Howards, S. S., Johnson, A., and Jessee, S. (1975b). *Fertil. Steril., 26*: 13.

Igdoura, S. A., Hermo, L., Rosenthal, A., and Morales, C. R. (1993). *Anat. Rec., 235*: 411.

Iguchi, T., Uesugi, Y., Sato, T., Ohta, Y., and Tagasugi, N. (1991). *Mol. Androl., 3*: 109.

Ikawa, H., Hutson, J. M., Budzik, G. P., MacLaughlin, D. T., and Donahoe, P. K. (1982). *J. Pediatr. Surg., 17*: 453.

Ilio, K. Y., and Hess, R. A. (1992). *Anat. Rec., 234*: 190.

Ilio, K. Y., and Hess, R. A. (1994). *Microsc. Res. Tech., 29*: 432.

Imperato-McGinley, J., Sanchez, R. S., Spencer, J. R., Yee, B., and Vaughan, E. D. (1992). *Endocrinology, 131*: 1149.

Jackson, C. M., and Morris, I. M. (1977). *Andrologia, 9*: 29.

Jackson, H. (1964). *Br. Med. Bull., 20*: 107.

Jessee, S. J., and Howards, S. S. (1976). *Biol. Reprod., 15*: 631.

Johnson, L., and Varner, D. D. (1988). *Biol. Reprod., 39*: 812.

Jones, A. R. (1978). *Life Sci., 23*: 1625.

Jones, A. R., and O'Brien, R. W. (1980). *Xenobiotica, 10*: 365.

Jones, A. R., Milton, D. H., and Murcott, C. (1978). *Xenobiotica, 8*: 573.

Jones, L. A. (1980). *Proc. Soc. Exp. Biol. Med., 165*: 17.

Jones, R. C. (1977). *Immunological Influence on Human Fertility* (B. Boeucher, ed.). Academic Press, Inc., Sydney, p. 67.

Jones, R. C. (1987). *J. Reprod. Fertil., 89*: 193.

Jones, R. C., and Jurd, K. M. (1987). *Aust. J. Biol. Sci., 40*: 79.

Jones, R. C., and Holt, W. V. (1981). *J. Anat., 133*: 79.

Kaleczc, J., Majewski, M., Calka, J., and Lakomy, M. (1993). *Folia Histochem. Cytobiol., 31*: 117.

Kalla, N. R., and Chohan, K. S. (1976). *Exp. Path., Bd., 12*: 19.

Kalla, N. R., and Singh, B. (1979). *J. Reprod. Fertil., 56*: 149.

Kelce, W. R., Monosson, E., Gamcsik, M. P., Laws, S. C., and Gray, L. E. (1994). *Toxicol. Appl. Pharmacol., 126*: 276.

Kelce, W. R., Stone, C. R., Laws, S. C., Gray, L. E., Kemppainen, J. A., and Wilson, E. M. (1995). *Nature, 375*: 581.

Kempinas, W. G., Suarez, J., Roberts, N., Ferrell, J., Strader, L., McElroy, W. K., Goldman, J., and Klinefelter, G. (1977). *Biol. Reprod., 56*: Suppl. 473A.

Klinefelter, G. R. (1993). *Methods in Toxicology* (R. E. Chapin and J. J. Heindel, eds.). Academic Press, New York, p. 274.

Klinefelter, G. R., and Amann, R. P. (1980). *Biol. Reprod., 22*: 1149.

Klinefelter, G. R., and Hamilton, D. W. (1985). *Biol. Reprod., 33*: 1017.

Klinefelter, G. R., and Kelce, W. R. (1994). *Biol. Reprod., 51*: 207A.

Klinefelter, G. R., Laskey, J. W., Roberts, N. L., Slott, V., and Suarez, J. D. (1990). *Toxicol. Appl. Pharmacol., 105*: 271.

Klinefelter, G. R., Roberts, N. L., and Suarez, J. D. (1992a). *J. Androl., 13*: 409.

Klinefelter, G. R. (1992b). *J. Tiss. Cult. Meth., 14*: 195.

Klinefelter, G. R., Laskey, J. W., Perreault, S. D., and Ferrell, J. (1994a). *J. Androl., 15*: 318.

Klinefelter, G. R., Laskey, J. W., Kelce, W. R., Ferrell, J., Roberts, N. L., Suarez, J. D., and Slott, V. (1994b). *Biol. Reprod., 51*: 82.

Klinefelter, G. R., Laskey, J. L., Ferrell, J., Suarez, J. D., and Roberts, N. L. (1997a). *J. Androl. 18*: 139.

Klinefelter, G. R. and Suarez, J. D. (1977b). *Reprod. Toxicol., 11*: 511.

Koefoed-Johnsen, V., and Ussing, H. H. (1958). *Acta Physiol. Scand., 42*: 298.

Korach, K. S. (1994). *Science, 266*: 1524.

Koskimies, A. I., and Kormano, M. (1975). *J. Reprod. Fertil., 43*: 345.

Kuiper-Goodman, T., Scott, P. M., and Watanabe, H. (1987). *Reg. Toxicol. Pharmacol., 7*: 253.

Kuwahara, M. (1976). *Tohoku J. Exp. Med., 120*: 251.

Kwon, S., Hess, R. A., Bunick, D., Nitta, H., Janulis, L., Osawa, Y., and Bahr, J. M. (1995). *Reprod., 53*: 1259.

Lamano-Carvalho, T. L., Favaretto, A. L. V., Petenusci, S. O., and Kempinas, W. G. (1993). *Brazil. J. Med. Biol. Res., 26*: 639.

Lea, O. A., Petrusz, P., and French, F. S. (1979). *J. Biol. Chem., 254*: 6196.

Levine, N., and Marsh, D. J. (1971). *J. Physiol., 213*: 557.

Linder, E. (1971). *Ann. NY Acad. Sci., 177*: 204.

Linder, R. E., Hess, R. A., Perreault, S. D., Strader, L. F., and Barbee, R. R. (1988). *J. Androl., 9*: 317.

Linder, R. E., Klinefelter, G. R., Strader, L. F., Suarez, J. D., Roberts, N. L., and Perreault, S. D. (1995). *Fundam. Appl. Toxicol., 28*: 9.

Lopez, M. L., and Breuker, H. (1986). *Andrologia, 18*: 133.

Luke, M. C., and Coffey, D. S. (1994). *The Physiology of Reproduction* (E. Knobil and J. Neill, eds.). Raven Press, New York, p. 1435.

Lyon, M. F., and Hawkes, S. G. (1970). *Nature, 227*: 1217.

Mably, T. A., Bjerke, D. L., Moore, R. W., Fitzpatrick, A. G., and Peterson, R. E. (1992). *Toxicol. Appl. Pharmacol., 114*: 118.

Mann, T., and Lutwak-Mann, C. (1981). *Male Reproductive Function and Semen.* Springer-Verlag, Berlin, p. 495.

Marchlewicz, M., Protasowicki, M., Rózewicka, L., Piasecka, M., and Laszczynska, M. (1993). *Folia Histochem. Cytobiol., 31*: 55.

Marshall, F. F., Reiner, W. G., and Goldberg, B. S. (1979). *Invest. Urol., 17*: 78.

Maruch, S. M., Alves, H. J., and Machado, C. R. (1989). *Acta Anat., 134*: 257.

Mason, K. E., and Shaver, S. L. (1952). *Ann. NY Acad. Sci., 55*: 585.

Mason, K. E., and Young, J. O. (1967). *Anat. Rec., 159*: 311.

Mattmueller, D. R., and Hinton, B. T. (1992). *Mol. Reprod. Dev., 32*: 73.

McClain, R. M., and Downing, J. C. (1988). *Toxicol. Appl. Pharmacol., 92*: 488.

Meistrich, M. L., Hughes, T. J., and Bruce, W. R. (1975). *Nature, 258*: 145.

Merchant-Larios, H., Moreno-Mendoza, N., and Buehr, M. (1993). *Int. J Dev. Biol., 37*: 407.

Mills, J. W., Ernst, S. A., and DiBona, D. R. (1977). *J. Cell Biol., 73*: 103.

Mitchinson, M. J., Sherman, K. P., and Stainer-Smith, A. M. (1975). *J. Pathol., 115*: 57.

Mohri, H. D., Sutter, D. A. I., Brown-Woodman, P. D. C., White, I. G., and Ridley, D. D. (1975). *Nature, 255*: 75.

Montorzi, N. M., and Burgos, M. H. (1967). *Z. Zellforsch., 83*: 58.

Morales, C., and Hermo, L. (1983). *Cell Tissue Res., 230*: 503.

Morales, C., and Hermo, L. (1986). *Cell Tissue Res., 245*: 323.

Morales, C., Hermo, L., and Clermont, Y. (1984). *Anat. Rec., 209*: 185.

Morita, J. (1966). *Arch. Histol. Jpn., 26*: 341.

Murray, F. J., Smith, F. A., Nitschke, K. D., Humiston, C. G., Kociba, R. J., and Schwetz, B. A. (1979). *Toxicol. Appl. Pharmacol., 50*: 241.

Nagano, T., and Suzuki, F. (1980). *Arch. Histol. Jpn., 43*: 185.

Nagy, F. (1985). *Arch. Androl., 15*: 91.

Nakai, M., Hess, R. A., Moore, B. J., Guttroff, R., Strader, L. F., and Linder, R. E. (1992). *J. Androl., 13*: 507.

Nakai, M., Moore, B. J., and Hess, R. A. (1993). *Anat. Rec., 235*: 51.

Newbold, R. R., Bullock, B. C., and McLachlan, J. A. (1985). *Cancer Res., 45*: 5145.

Niemi, M., and Kormano, M. (1965). *Anat. Rec., 151*: 159.

Nistal, M., Santamaria, L., and Paniagua, R. (1989). *Human Pathol., 20*: 1065.

Nitta, H., Bunick, D., Hess, R. A., Janulis, L., Newton, S. C., Osawa, Y., Shizuta, Y., Toda, K., and Bahr, M. J. (1993). *Endocrinology, 132*: 1396.

Nomura, T., and Kanzaki, T. (1977). *Cancer Res., 37*: 1099.

Oberlander, G., Yeung, C. H., and Cooper, T. G. (1994). *J. Reprod. Fertil., 100*: 551.

Olson, G. E., and Hinton, B. T. (1985). *J. Androl., 6*: 20.

Orgebin-Crist, M.-C., and Jahad, N. (1978). *Endocrinology, 103*: 46.

Orsi, P. A. M., Dias, S. M., Seullner, G., Guazzelli Filho, J. and Vicentini, C. A. (1983). Anat. Anz., 153:249–254.

Parker, M. G. (1982). *Nature, 298*: 92.

Parker, M. G., and Mainwaring, W. I. P. (1977). *Cell, 12*: 401.

Parker, M. G., White, R., and Williams, J. G. (1980). *J. Biol. Chem., 255*: 6996.

Parker, M. G., White, R., Hurst, H., Needham, M., and Tilly, R. (1983). *J. Biol Chem., 258*: 12.

Pelliniemi, L. J., Dym, M., Gunsalus, G. L., Musto, N. A., Bardin, C. W. and Fawcett, D. W. (1981). *Endocrinology, 108*: 925.

Pholpramool, C., White, R. W., and Setchell, B. P. (1982). *J. Reprod. Fertil., 66*: 547.

Pietz, B., and Olds-Clark, E. P. (1988). *Biol. Reprod., 35*: 608.

Pudney, J., and Fawcett, D. W. (1984). *Anat. Rec., 208*: 383.

Qui, J., Hales, B. F., and Robaire, B. (1992). *Biol. Reprod., 46*: 926.

Rabinovici, J., and Jaffe, R. B. (1990). *Endocrinol. Rev., 11*: 532.

Ramos Jr., A. S., and Dym, M. (1977). *Biol. Reprod., 17*: 339.

Rehnberg, G. L., Cooper, R. L., Goldman, J. M., Gray, L. E., Hein, J. F., and McElroy, W. K. (1989). *Toxicol. Appl. Pharmacol., 101*: 55.

Reid, B. L., and Cleland, K. W. (1957). *Aust. J. Zool., 5*: 223.

Ricker, D. D., Chamness, S. L., Hinton, B. T., and Chang, T. S. K. (1996). *J. Androl., 17*: 117.

Robaire, B., and Hermo, L. (1988). *The Physiology of Reproduction* (E. Knobil and J. Neill, eds.). Raven Press, New York, p. 999.

Robaire, B., Ewing, L. L., Zirkin, B. R., and Irby, D. C. (1977). *Endocrinology, 101*: 1379.

Robb, G. W., Amann, R. P., and Killian, G. J. (1978). *J. Reprod. Fertil., 54*: 103.

Robinson, R., and Fritz, I. B. (1979). *Can. J. Biochem., 57*: 962.

Roman, B. L., Sommer, R. J., Shinomiya, K., and Peterson, R. E. (1995). *Toxicol. Appl. Pharmacol., 134*: 241.

Roselli, C. E., West, N. B., and Brenner, R. M. (1991). *Biol. Reprod., 44*: 739.

Sager, D., Girard, D., and Nelson, D. (1991). *Environ. Toxicol. Chem., 10*: 737.

Sakena, S. K., and Lau, I. F. (1979). *Endokrinologie, 74*: S.6.

Samojlik, E., and Chang, M. C. (1970). *Biol. Reprod., 2*: 299.

Sapino, A., Pagani, A., Godano, A. and Bussolati, G. (1987). *Virchows Arch. A, 411*: 409.

Sato T., Chiba, A., Hayashi, S., Okamura, H., Ohta, Y., Takasugi, N., and Iguchi, T. (1994). *Reprod. Toxicol., 8*: 145.

Satoh, M. (1985). *J. Anat., 143*: 17.

Schardein, J. L. (1993). *Chemically Induced Birth Defects*. Marcel Dekker, New York, p. 271.

Schleicher, G., Drews, U., Stumpf, W. E., and Sar, M. (1984). *Histochemistry, 81*: 139.

Schleicher, G., Privette, T. H., and Stumpf, W. E. (1989). *J. Histochem. Cytochem., 37*: 1083.

Setchell, B. P., Maddocks, S., and Brooks, D. E. (1994). *The Physiology of Reproduction* (E. Knobil and J. Neill, eds.). Raven Press, New York, p. 1063.

Sharpe, R. M. (1993). *J. Endocrinol., 136*: 357.

Sharpe, R. M., and Skakkebaek, N. E. (1993). *Lancet, 341*: 1392.

Siegei, F. L., and Bulger, R. E. (1975). *Virchows Arch. B Cell Path., 18*: 243.

Singhal, R. L., Vijayvargiya, R., and Skukla, G. S. (1985). *Endocrine Toxicology* (J. A. Thomas, K. S. Korach, and J. A. McLachlan, eds.). Raven Press, New York, p. 149.

Slott, V. L., Suarez, J. D., Simmons, J. E., and Perreault, S. D. (1990). *Fundam. Appl. Toxicol., 15*: 597.

Stumpf, W. E., Sar, M., Chen, K., Morin, J., and DeLuca, H. F. (1987). *Cell Tissue Res., 247*: 453.

Susheela, A. K., and Kumar, A. (1991). *J. Reprod. Fertil., 92*: 353.

Suzuki, F. (1982). *Am. J. Anat., 163*: 309.

Suzuki, F., and Nagano, T. (1978). *Anat. Rec., 191*: 503.

Takeuchi, S.-I. (1992). *Nippon Hinysk. Gak. Zus., 7*: 1043.

Talo, A. (1981). *J. Reprod. Fertil., 63*: 17.

Tekpetey, F. R., and Amann, R. P. (1988). *Biol. Reprod., 38*: 1051.

Thomas, J. A., and Brogan, W. C. (1983). *Am. J. Ind. Med., 4*: 127.

Toth, G. P., Zenick, H., and Smith, M. K. (1989). *Fundam. Appl. Toxicol., 13*: 16.

Trasler, J. M., Hermo, L., and Robaire, B. B. (1988). *Biol. Reprod., 38*: 463.

Tsang, A. Y. F., Lee, W. M., and Wong, P. Y. D. (1981). *Int. J. Androl., 4*: 703.

Tsunoda, Y., and Chang, M. C. (1976). *J. Reprod. Fertil., 46*: 401.

Tung, K. S. K., and Alexander, N. J. (1980). *Am. J. Pathol., 101*: 17.

Tung, P. S., Rosenior, J., and Fritz, I. B. (1987). *Biol. Reprod., 36*: 1297.

Turner, T. (1979). *Invest. Urol., 16*: 311.

Turner, T. (1984). *J. Reprod. Fertil., 72*: 509.

Turner, T. T. (1988). *Biol. Reprod., 39*: 399.

Turner, T. T. (1991). *The Male Germ Cell: Spermatogonium to Fertilization* (B. Robaire, ed.). New York Academy of Sciences, New York, p. 364.

Turner, T. T., Jones, C. E., Howards, S. S., Ewing, L. L., Zegeye, B., and Gunsalus, G. L. (1984). *Endocrinology, 115*: 1925.

Ullrich, K. J. (1990). *Methods in Enzymology*, Vol. 191 (S. Fleischer and B. Fleischer, eds.). Academic Press, New York, p. 1.

Veeramachaneni, D. N. R., and Amann, R. P. (1990). *Endocrinology, 126*: 1156.

Veeramachaneni, D. N. R., and Amann, R. P. (1991). *J. Androl., 12*: 288.

Veeramachaneni, D. N. R., Sheerman, G. B., Floyd, J. G., Ott, R. S., and Hixon, J. E. (1988). *Fundam. Appl. Toxicol., 10*: 73.

Veeramachaneni, D. N. R., Schanbacher, B. D., and Amann, R. P. (1989). *Biol. Reprod., 41*: 499.

Veeramachaneni, D. N., Amann, R. P., Palmer, J. S., and Hinton, B. T. (1990). *J. Androl., 11*: 140.

Veri, J.-P., Hermo, L., and Robaire, B. (1993). *J. Androl., 14*: 23.

Viotto, M. J. S., Orsi, A. M., Vicentini, C. A., Mello Dias, S., and Gregório, E. A. (1993). *Anat. Histol. Embryol., 22*: 114.

Vogelpoel, F. R., and Verhoef, J. (1984). *Arch. Androl., 14*: 123.

Vreeburg, J. T. M. (1975). *J. Endocrinol., 67*: 203.

Welch, J. E., Klinefelter, G. R., Suarez, J. D., Roberts, N. L., and Barbee, R. R. (1997). *Mol. Biol. Cell, 8*: Suppl. (In press).

West, N. B., and Brenner, R. M. (1990). *Biol. Reprod., 42*: 533.

Whorton, D., Krauss, R. M., Marshall, S., and Milby, T. H. (1977). *Lancet, 2*: 1259.

Wilker, C. E., Safe, S. H., and Johnson, L. (1996). *Toxicologist, 30*: 142A.

Wilson, J. D., and Lasnitzki, I. (1971). *Endocrinology, 89*: 659.

Wilson, T. M., Therrien, A., and Harkness, J. E. (1986). *Lab. Anim. Sci., 36*: 41.

Winet, H. (1980). *J. Androl., 1*: 303.

Wise, L. D., Minsker, D. H., Cukierski, M. A., Clark, R. L., Prahalada, S., Antonello, J. M., MacDonald, J. S., and Robertson, R. T. (1991). *Reprod. Toxicol., 5*: 337.

Wolfe Jr., S. A., Culp, S. G., and De Souza, E. B. (1989). *Endocrinology, 124*: 1160.

Wong, P. Y. D. (1990). *Reprod. Fertil. Dev., 2*: 115.

Wong, P. Y. D., and Yeung, C. H. (1977a). *J. Reprod. Fertil., 49*: 77.

Wong, P. Y. D., and Yeung, C. H. (1977b). *Endocrinology, 101*: 1391.

Wong, P. Y. D., and Yeung, C. H. (1977c). *J. Endocrinol., 72*: 12P.

Wong, P. Y. D., and Yeung, C. H. (1977d). *J. Reprod. Fertil., 51*: 469.

Wong, P. Y. D., and Yeung, C. H. (1978). *J. Physiol., 275*: 13.

Wong, P. Y. D., Au, C. L., and Ngai, H. K. (1978a). *Int. J. Androl., Suppl., 2*: 608.

Wong, Y. C., Wong, P. Y. D., and Yeung, C. H. (1978b). *Experientia, 34*: 485.

Wong, P. Y. D., Au, C. L., and Ngai, H. K. (1979). *The Spermatozoon* (D. W. Fawcett and J. M. Bedford, eds.). Urban & Schwarzenberg, Inc., Baltimore-Munich, p. 57.

Working, P. K., Bus, J. S., and Hamm, T. E. (1985). *Toxicol. Appl. Pharmacol., 77*: 133.

Wrobel, K.-H. (1972). *Z. Zellforsch., 135*: 129.

Wyrobek, A. J., Gordon, I. A., Watchmaker, G., and Moore, D. H. H. (1982). *Banbury Report: Indicators of Genotoxic Exposure* (B. A. Bridges, B. E. Butterworth, and I. B. Weinstein, eds.). Cold Spring Harbor, p. 527.

Yeung, C. H., Oberlander, G., and Cooper, T. G. (1992). *J. Reprod. Fertil., 96*: 427.

Yokoyama, M., and Chang, J. P. (1971). *J. Histochem. Cytochem., 19*: 766.

Zwain, I. H., Grima, J., and Cheng, C. Y. (1992). *Endocrinology, 131*: 1511.

24

Developmental Male Reproductive Toxicology of 2,3,7,8-Tetrachlorodibenzo-*p*-dioxin (TCDD) and PCBs

Beth L. Roman and Richard E. Peterson
University of Wisconsin, Madison, Wisconsin

INTRODUCTION

Environmental Exposure to Polychlorinated Dibenzo-*p*-dioxins, Dibenzofurans, and Biphenyls

Polychlorinated dibenzo-p-dioxins (PCDDs), dibenzofurans (PCDFs), and biphenyls (PCBs) are persistent environmental contaminants. The latter class of chemicals, the PCBs, were formerly used in the manufacture of dielectrics, hydraulic fluids, plastics, and paints. Although PCB production has ceased in the United States, the continued use and ultimate disposal of PCB-containing products serve as current sources of PCB input into the environment. In contrast, PCDDs and PCDFs were never produced for commercial purposes, but instead have been and continue to be introduced into the environment as by-products of a number of processes, including incineration of municipal and hospital wastes, smelting and steel production, burning of coal, wood, and petroleum products for fuel, manufacturing of chlorinated compounds, and bleaching of pulp and paper (Zook and Rappe, 1994). PCDDs, PCDFs, and PCBs are primarily found sorbed to particulate matter; they are highly resistant to degradation and therefore persist in the environment. Because these chemicals are highly lipophilic and resistant to metabolism by phase I and phase II enzymes, they bioaccumulate in fish, wildlife, domestic animals and ultimately humans. The primary source of human exposure to PCDDs, PCDFs and PCBs is diet, with meat, dairy products, and fish containing the majority of the dietary intake of these contaminants (Startin, 1994). Of the PCDD, PCDF and PCB congeners, the most potent and thus prototype congener is 2,3,7,8-tetrachlorodibenzo-p-dioxin (TCDD).

Mechanism of Action of TCDD

In laboratory mammals, TCDD exposure causes a wide variety of adverse health effects including, but not limited to, delayed-onset mortality, immune dysfunction, tumor promotion, epithelial hyperplasia, developmental toxicity, and reproductive toxicity (Poland and Knutson, 1982; Peterson et al., 1993). Two compelling lines of evidence suggest that most, if not all, toxic effects of TCDD and TCDD-like halogenated aromatic hydrocarbons are mediated by binding of these compounds to a cytosolic protein known as the aryl hydrocarbon receptor (AhR). First, PCDD, PCDF, and PCB congeners exhibit a structure-activity relationship with respect to AhR binding that correlates very well with the rank order of toxic potency of these congeners (Poland and Knutson, 1982). In general, congeners that are planar or can assume a nearly planar configuration bind to the receptor with the highest affinity and are the most potent toxicants. Second, the ability of TCDD to cause toxicity segregates with the *Ahr* locus. For example, C57BL/6 mice, which express a high-affinity AhR (*Ahr*[b] allele), are more sensitive to TCDD-induced lethality, thymic atrophy, and cleft palate than DBA2 mice, which express a low-affinity AhR (*Ahr*[d] allele; Poland and Glover, 1980). Similarly, *Ahr*[-/-] C57BL/6 mice generated by gene targeting are refractory to TCDD exposure in terms of thymic atrophy and alterations in liver histology, both of which are observed in their wild-type (*Ahr*[+/+]) and heterozygous (*Ahr*[+/-] counterparts (Fernandez-Salguero et al., 1996). However, although it is generally agreed that AhR binding is necessary for manifestation of TCDD toxicity, the events that follow receptor binding that actually lead to toxicity are not well established.

Three possible scenarios are presented below and diagrammed in Figure 1. The mechanisms which TCDD upregulates transcription of a number of genes is very well-understood at the molecular level (Fig. 1; for review, see Schmidt and Bradfield, 1996). TCDD is very lipophilic and easily enters cells by passive diffusion. Once in the cytoplasm, TCDD binds to the AhR, a member of the basic-helix-loop-helix (bHLH) family of transcription factors (Burbach et al., 1992). In its unliganded state, the AhR is bound in the cytoplasm to two 90 kDa heat shock proteins (hsp90). TCDD binding facilitates release of hsp90 and allows translocation of the liganded AhR to the nucleus, where it dimerizes with another bHLH protein, the AhR nuclear translocator (ARNT). This liganded AhR/ARNT complex is a heterodimeric transcription factor that enhances transcription of a number of genes encoding xenobiotic metabolizing enzymes (CYP1A1, CYP1A2, glutathione-S-transferase-Ya, quinone reductase, and class 3 aldehyde dehydrogenase) by binding to enhancer elements (dioxin response elements, or DREs) in the 5' regulatory regions of these genes, achieving signaling via chromatin disruption as well as interaction with the basal cellular transcriptional machinery (Schmidt and Bradfield, 1996, and references therein; Rowlands et al., 1996). However, although it is possible that activation of these xenobiotic metabolizing enzymes could lead to toxicity, for example by metabolizing an endogenous compound important in normal cell function, induction of these enzymes is generally considered to be more of a biomarker of TCDD exposure than a mechanism of toxicity. Other members of the *Ahr* gene battery that exhibit increases in mRNA abundance in response to TCDD exposure include CYP1B1, interleukin-1β, plasminogen activator inhibitor-2 (Sutter et al., 1991) and transforming growth factor-α (Choi et al., 1991), although the mechanism by which these increases occur is unknown, as functional DREs have not yet been identified in these genes. But again, none of these or any other TCDD-responsive genes identified to date have been directly linked to toxicity. However, it remains possible that tissue-specific enhancement of transcription of as yet unidentified genes plays a key role in mediating the toxic effects of TCDD.

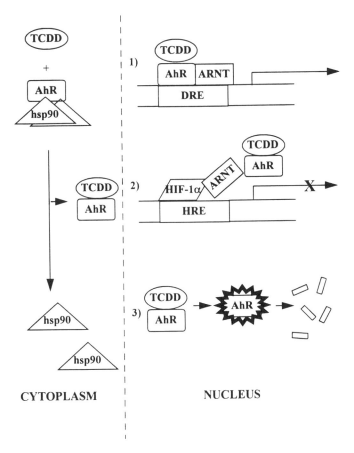

Figure 1 Three possible mechanisms by which TCDD could cause toxicity. TCDD diffuses into the cell and binds to the aryl hydrocarbon receptor, facilitating release of hsp90 and allowing translocation of the liganded AhR to the nucleus. In the nucleus: (1) The liganded AhR could dimerize with ARNT, and this complex could then bind to dioxin response elements (DREs) in the 5′ regulatory region of genes and enhance transcription via chromatin disruption and intraction with the basal cellular transcriptional machinery. (2) The liganded AhR could compete with other PAS proteins for dimerization with ARNT, resulting in downregulation of genes dependent on ARNT and an alternative partner for transcription. For example, the liganded AhR could compete with HIF-1α for ARNT, resulting in the downregulation of a battery of hypoxia-inducible genes. (HRE = hypoxia response element.) (3) The ligand-activated AhR could be rapidly depleted from cells, perhaps decreasing the pool of AhR available for binding to an endogenous ligand and/or activating transcription of genes important in normal cellular function.

Both the AhR and ARNT are members of a growing family of regulatory proteins defined by the presence of a PAS (*Per*, *A*RNT, *S*im) domain, a stretch of 200–300 amino acids containing two 51- amino acid repeats that is involved in dimerization (Nambu et al., 1991), and, in the case of AhR, in ligand binding and hsp90 association (Dolwick et al., 1993; Whitelaw et al., 1993). Most of these proteins also contain an N-terminal bHLH motif that functions in DNA binding. To date, family members include *Drosophila* proteins Per (Jackson et al., 1986), Sim (Crews et al., 1988) and trachealess (Wilk et al., 1996), as well as the mammalian proteins AhR (Burbach et al., 1992), ARNT and ARNT 2 (Hoffman et al., 1991; Hirose et al., 1996), Sim1 and Sim2 (Ema et al., 1996), hypoxia inducible factor-1α (HIF-1α, Wang et al., 1995), steroid receptor coactivator-1 (SRC-1, Yao et al., 1996), endothelial PAS protein (EPAS, Tian et al., 1997; also known as HIF-1α-like factor, Ema et al., 1997; or MOP-2, Hogenesch et al., 1997), neuronal PAS domain proteins 1 and 2 (NPAS1,2, Zhou et al., 1997; NPAS2 also known as MOP4, Hogenesch et al., 1997), Clock (King et al., 1997), AIB1 (Anzick et al., 1997), and MOPs 3 and 5 (Hogenesch et al., 1997). Interestingly, the AhR is the only PAS protein known to bind ligand, although the physiological role of the AhR in cellular function is not understood, and no endogenous AhR ligand has yet been identified. The AhR has been shown to dimerize in a ligand-dependent manner only with ARNT (Reyes et al., 1992) and ARNT2 (Hirose et al., 1996). Conversely, ARNT is able to dimerize with a number of other PAS proteins, including Sim1 and Sim2 (Probst et al., 1997), HIF-1α (Wang et al., 1995) and EPAS (Ema et al., 1997; Tian et al., 1997; Hogenesch et al. 1997), with the latter two forming active transcriptional complexes recognizing hypoxia response elements.

The observation that ARNT has multiple dimerization partners suggests a second mechanism by which TCDD toxicity might occur (Fig. 1). That is, TCDD toxicity might be the result of downregulation of transcription of genes dependent on ARNT and an alternative partner (i.e., not AhR), as opposed to upregulation of transcription of genes dependent on AhR/ARNT. For example, if the liganded AhR could deplete the pool of ARNT available for binding to HIF-1α, then downregulation of genes responsive to hypoxia, such as erythropoietin (Semenza and Wang, 1992) and vascular endothelial growth factor (Forsythe et al., 1996), would result, perhaps leading to toxicity. It should be noted that one in vitro study demonstrated an interference by activated or overexpressed HIF-1α with the AhR signaling pathway, although the converse was not true, i.e., the activated AhR did not interfere with the HIF-1 signaling pathway (Gradin et al., 1996). Still, liganded AhR-mediated interference with HIF-1 signaling in vivo or with other PAS protein signaling pathways remains a possibility.

A third mechanism by which TCDD might cause toxicity is by downregulating AhR protein expression (Fig. 1). TCDD exposure has been shown to rapidly deplete AhR protein levels in a variety of cultured cells to 10–20% of control (Giannone et al., 1995; Pollenz, 1996). Similar effects of TCDD exposure on AhR protein levels have been noted in adult male rat reproductive organs after exposure to 25 µg TCDD/kg for 24 hours (Roman et al., 1998a) as well as in female rat liver, thymus, spleen, and lung after exposure to 15 µg TCDD/kg (Pollenz et al., 1998). In the latter study, significant decreases (10–40% of control) were observed as early as 8 hours after dosing and persisted to a lesser degree as late as 2 weeks after dosing (Pollenz et al., 1998). Examination of $Ahr^{-/-}$ mice suggests that the AhR is important in hepatic and immune system development (Fernandez-Salguero et al., 1995; Schmidt et al., 1996), and further examination will most likely reveal other organs in which the AhR is important for normal morphogenesis and/or function. Therefore, it follows that severe TCDD-induced depletions in the amount of AhR available for binding of a putative endogenous ligand and/or activating transcription of genes important in normal cellular function could very well lead to toxicity, particularly during early development, when even transient interference with signal transduction pathways could result in permanent changes in organ structure or function.

EFFECTS OF IN UTERO AND LACTATIONAL TCDD EXPOSURE ON THE MALE RAT REPRODUCTIVE SYSTEM

Rationale for Studies, Overview of Results, and Experimental Design

In laboratory mammals, TCDD exposure causes a wide variety of adverse health effects, including male reproductive toxicity (Peterson et al., 1993). In male rats, TCDD exposure during adulthood results in decreased testis, seminal vesicle, and ventral prostate weights, altered testicular and epididymal morphology, increased incidence of epididymal sperm granulomas, and decreased fertility (Khera and Ruddick, 1973; Moore et al., 1985; Rune et al., 1991; Johnson et al., 1992). These adverse reproductive effects are contributed to by TCDD-induced decreases in plasma androgen concentrations, a phenomenon due, in part, to reduced Leydig cell number, volume, and steroidogenic enzyme activity, as well as impaired Leydig cell luteinizing hormone (LH) responsiveness (Moore et al,. 1985, 1991; Mebus et al., 1987; Kleeman et al., 1990; Rune et al., 1991; Johnson et al., 1992, 1994). The ED_{50} for androgenic deficiency associated with adult TCDD exposure is approximately 15,000 ng/kg (Moore et al., 1985), a dose that is in the overtly toxic in terms of decreased body weight gain. In addition, this dose is extremely high compared to background levels of human exposure to TCDD-like compounds. When the concentration of PCDD, PCDF and PCB AhR agonists is normalized with respect to the toxic potency of TCDD, the background body burden of AhR agonists can be expressed in terms of TCDD equivalents, or TEQs. In humans in Western industrialized countries, the average background body burden of TEQs is estimated to be 8–13 ng/kg (DeVito et al., 1995), which is three orders of magnitude below the ED_{50} for androgenic deficiency in response to adult TCDD exposure (Fig. 2).

The sensitivity of the fetal and neonatal male reproductive system to many endocrine-disrupting chemicals is greater than that of the adult male reproductive system (Peterson et al., 1997). Because adult TCDD exposure reduced circulating androgen concentrations in male rats, it was hypothesized that in utero and lactational TCDD exposure might produce similar effects on androgenic status during early development and that these effects might occur in response to a very low maternal dose of TCDD. And because androgens are required during early development for proper morphogenesis of male accessory sex organs, for imprinting responses of male accessory sex organs to later androgenic stimulation, for initiating spermatogenesis, and for directing a male pattern of sexual differentiation of the central nervous system (CNS) (Chowdhury and Steinberger, 1975; Chung and Ferland-Raymond, 1975; Rajfer and Coffey, 1979; MacLusky and Naftolin, 1981), it seemed likely that profound effects on reproductive system development and sexual behavior of male offspring might occur. Indeed, in the Holtzman rat in utero and lactational exposure to a single low dose of TCDD (64–1000 ng/kg maternal body weight) affected external indicators of androgenic status, decreased androgen-dependent accessory sex organ weights throughout development, decreased the responsiveness of the adult ventral prostate to androgenic stimulation, decreased daily sperm production (DSP) and cauda epididymal sperm number (CESN), partially demasculinized and partially feminized sexual behavior, and partially feminized the regulation of LH secretion (Mably et al., 1992a,b,c; Bjerke and Peterson, 1994; Bjerke et al., 1994a,b; Roman et al., 1995). The ED_{50} for most of these effects was 160 ng TCDD/kg maternal body weight, about 1/100 of the ED_{50} for androgenic deficiency following adult exposure, and the LOAEL for the most sensitive effects was 64 ng TCDD/kg, not much higher than the estimated average background body burden of

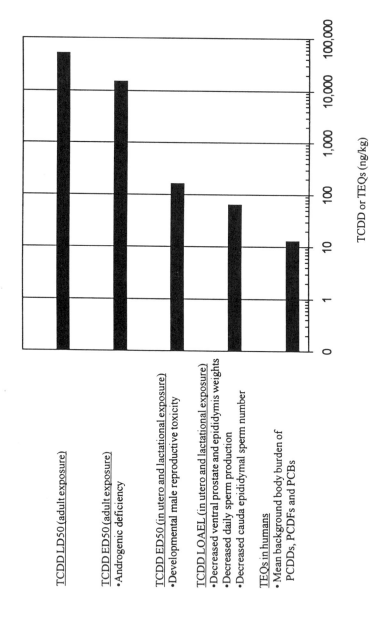

Figure 2 Comparison of TCDD doses required to produce male reproductive system toxicity in the Holtzman rat when exposure occurs during adulthood or in utero and via lactation. Note that the LOAEL for the most sensitive developmental male reproductive effects is two orders of magnitude lower than the ED_{50} for androgenic deficiency in response to adult TCDD exposure, and less than one order of magnitude greater than the mean background body burden of TEQs in humans in Western industrialized countries.

TEQs in humans (Fig. 2). However, several recent studies have provided both direct as well as compelling indirect evidence that the spectrum of effects of in utero and lactational TCDD exposure on the male rat reproductive system cannot be explained by decreased testicular androgen production or plasma androgen concentrations during early development. Nor can they be entirely explained by decreased testicular androgen production or plasma androgen concentrations later in development, when most of the adverse male reproductive effects of perinatal TCDD exposure have been observed. Thus, although the guiding hypothesis that in utero and lactational TCDD exposure would interfere with testicular steroidogenesis was not proven true, the developing male reproductive system has indeed been shown to be extremely sensitive to in utero and lactational TCDD exposure. The mechanisms underlying these effects are the subject of current investigations but to date remain largely unknown.

In the studies described below, unless stated otherwise, pregnant Holtzman rats were orally dosed with TCDD (64–1000 ng/kg) or vehicle on gestation day (GD) 15, a time when most organogenesis is complete and the testis is just beginning to synthesize androgens (Warren et al., 1975, 1984; Feldman and Bloch, 1978). On the day of birth, postnatal day (PND) 0, litters were normalized to 10 pups and lactational exposure continued until sacrifice or weaning (PND 21). In other studies, pregnant Long Evans rats were orally dosed with TCDD (1000 ng/kg) or vehicle on GD 8 or 15. On PND 0, litters were normalized to eight pups and lactational exposure continued until sacrifice or weaning (PND 27). Because GD 8 dosing in Long Evans rats was either ineffective (accessory sex organ weights, DSP, CESN, sexual behavior) or less effective (time to preputial separation, ejaculated sperm number) in producing male reproductive toxicity than GD 15 dosing (Gray et al., 1995), results of GD 8 dosing are not presented in this review. An overview of results from these studies, and from similar studies in Syrian hamsters and ICR mice, is presented in Table 1.

It is important to note that the lowest maternal dose of TCDD administered in these studies was less than 1/100 of the ED_{50} for producing an androgenic deficiency with adult exposure and that the amount of TCDD to which each male offspring was exposed, although this has not been directly determined, was only a small fraction of the maternal dose. A recent study has reported that, in Sprague-Dawley rats, 0.01% of a maternal TCDD dose of 5600 ng/kg administered on GD 18 was present in the liver of each fetus after 1 or 2 days of transplacental exposure, whereas 2–3% was present in the liver of each pup after transplacental and 4 days of subsequent lactational exposure (Li et al., 1995).

Perinatal Plasma Androgen Concentrations and Testicular Androgen Production

Although decreases in prenatal (GD 17–21) and early postnatal (2 h after birth) plasma testosterone concentrations have been reported in male Holtzman rats exposed to 1000 ng TCDD/kg maternal body weight on GD 15 (Mably et al., 1992c), a more recent study could not reproduce these decreases on GD 18 or 20 or at 2 hours after birth (Chen et al., 1993). In addition, TCDD exposure did not decrease intratesticular testosterone content or impair the ability of the LH analog, human chorionic gonadotropin (hCG), to stimulate the production of testosterone from bisected testis preparations at these times (Chen et al., 1993). Combined with the fact that in utero and lactational TCDD exposure did not alter plasma testosterone concentrations on PND 1, 3, or 5

Table 1 Summary of Developmental Male Reproductive Toxicity of TCDD in Rats, Hamsters, and Mice[a]

	Holtzman rats	Long Evans rats	Syrian hamsters	ICR mice
Plasma androgen concentrations				
Perinatal	No effect[b]	No effect	—[c]	—
Adult	No effect[b]	No effect	No effect	No effect
Testicular androgen production				
Perinatal	No effect	No effect	—	—
Adult	No effect[b]	No effect	—	—
External indicators of androgenic status				
Anogenital distance (relative)	No effect[b]	No effect	No effect	No effect
Testis descent	Delayed	—	—	No effect
Preputial separation	Delayed	Delayed	Delayed	No effect
Flank gland development	NA[d]	NA	No effect	NA
Accessory sex organ weights (relative)				
Ventral prostate	Decreased[e]	No effect	—	Decreased
Dorsolateral prostate	Decreased	—	—	No effect
Seminal vesicle	Decreased	Decreased	Decreased	No effect
Coagulating gland	Decreased	—	—	Decreased
Accessory sex organ androgen responsiveness				
Ventral prostate	Decreased	No effect	—	—
Seminal vesicle	No effect	Decreased	—	—
Testis weight				
Absolute	Decreased	Decreased	Decreased	No effect
Relative	No effect[b]	—	—	—
Epididymis weight				
Absolute	Decreased	Decreased	Decreased	No effect
Relative	No effect	—	—	—
Sperm numbers and fertility				
Daily sperm production	Decreased	No effect	No effect	No effect
Cauda epididymal sperm number	Decreased	Decreased	Decreased	Decreased
Ejaculated sperm number	Decreased	Decreased	Decreased	—
Fertility	No effect	Decreased	—	—
Sexual behavior				
Masculine	Demasculinized	Demasculinized	No effect	—
Feminine	Feminized	No effect	—	—
Regulation of LH secretion	Feminized	—	—	—

[a]Male Holtzman rats were born to dams dosed on GD 15 with 64–1000 ng TCDD/kg. Male Long Evans rats were born to dams dosed on GD 15 with 1000 ng TCDD/kg. Male Syrian hamsters were born to dams dosed on GD 11 with 2000 ng TCDD/kg. Male ICR mice were born to dams dosed with 15,000–60,000 ng TCDD/kg on GD 14.
[b]No statistically significant effect in the majority of studies.
[c]Not measured.
[d]Not applicable.
[e]Decreased = Significantly decreased at one or more time points.

(Mably et al., 1992c), it seems likely that the mechanism by which TCDD affects the developing male rat reproductive system does not involve impaired perinatal testicular androgen production and reduced perinatal plasma androgen concentrations. This conclusion is upheld by studies in Long Evans rats, in which there was no effect of GD 15 TCDD exposure on plasma testosterone concentrations or in vitro LH-stimulated testicular testosterone production on PND 0 (Gray et al., 1995).

Postweaning Plasma Androgen Concentrations and Testicular Androgen Production

Although reductions in plasma testosterone concentrations in the male rat during perinatal development cannot explain effects of in utero and lactational TCDD exposure on the reproductive system and sexual behavior, reductions in plasma androgen concentrations later in life might contribute to some of these effects. With a TCDD dose of 1000 ng/kg, significant decreases in plasma testosterone concentrations were noted on PND 63 with in utero exposure alone and lactational exposure alone but not with combined in utero and lactational exposure (Bjerke and Peterson, 1994). In utero and lactational TCDD exposure (64–1000 ng/kg) tended to decrease plasma testosterone and 5α-dihydrotestosterone (DHT) concentrations in male Holtzman rats in a dose-dependent manner between PNDS 32 and 120 (Mably et al., 1992c), although the statistical significance of these decreases could not be established due to large interanimal variability. A more recent study (Roman et al., 1995) also showed a trend for decreased plasma testosterone concentrations from PND 21 through PND 63; plasma concentrations of another prevalent circulating androgen, 5α-androstane-3α, 17β-diol (3α-diol), tended to be decreased only from PND 21 through PND 49. In this study, decreases in plasma androgen concentrations were not statistically significant at any time and were not as severe (average 83% of control) as those previously reported (average 52% of control) (Mably et al., 1992c). In addition, intratesticular testosterone content was not significantly decreased by in utero and lactational TCDD exposure, although intratesticular 3α-diol content was depressed at early times (PNDs 21 and 32). In vitro testosterone production by decapsulated testes in response to a range of hCG concentrations was not affected by in utero and lactational TCDD exposure, although significant decreases in hCG-stimulated-3α-diol production (PNDs 32 and 49) as well as a significant increase in hCG-stimulated 3α-diol production (PND 63) were observed.

It is clear from results presented here that, in male Holtzman rats, testosterone and 3α-diol were not equally affected by in utero and lactational TCDD exposure. Because intratesticular 3α-diol content (PNDs 21 and 32) and in vitro hCG-stimulated 3α-diol production (PNDs 32 and 49) were more severely depressed than intratesticular testosterone content and in vitro hCG-stimulated testosterone production, it is possible that activities of the testicular enzymes (5α-reductase, 3α-hydroxysteroid oxidoreductase) involved in converting testosterone to 3α-diol could be reduced by TCDD exposure. However, the physiological significance of decreases in intratesticular 3α-diol content and in vitro hCG-stimulated testicular 3α-diol production is not clear, as decreases in these parameters did not result in statistically significant decreases in plasma 3α-diol concentration or correlate temporally with observed effects on androgen-dependent accessory sex organ weights (Roman et al., 1995).

Similar to findings in Holtzman rats, in utero and lactational TCDD exposure did not decrease plasma testosterone concentration, intratesticular testosterone concentration, or LH-stimulated testicular testosterone production in Long Evans rats on PND 49 or PND 270 (Gray et al., 1995), strengthening the hypothesis that the adverse effects of in utero and lactational TCDD exposure on the male rat reproductive system cannot be fully explained by an androgenic deficiency.

External Indicators of Androgenic Status

The androgenic status of male rats can be indirectly measured in a number of ways. Anogenital distance (AGD), or the distance from the base of the genital tubercle to the anus, is longer in males than in females and can be used as a gauge of circulating androgen concentration and/or perineum androgen responsiveness (Neumann et al., 1970). Male Holtzman rats exposed to TCDD in utero and via lactation exhibited decreased AGD on PNDs 1, 3, 4, and 5 (Mably et al., 1992c; Bjerke and Peterson, 1994; Bjerke et al., 1994b; Roman et al., 1995). However, in most cases, when AGD was normalized with respect to crown-rump length, TCDD-induced decreases in AGD were no longer apparent (Bjerke and Peterson, 1994; Bjerke et al., 1994b; Roman et al., 1995). Similar results have been seen in the Long Evans rat, in which GD 15 exposure decreased absolute but not relative AGD on PNDs 3, 8, 15, and 22 (Gray et al., 1995). These data provide strong indirect evidence that perinatal androgenic status is not altered by in utero and lactational TCDD exposure.

Two additional easily measured external indicators of androgenic status are time to testis descent and time to preputial separation (Rajfer and Walsh, 1977; Korenbrot et al., 1977), which occur in control rats between PND 20–23 and PND 42–45, respectively. In the Holtzman strain, in utero and lactational TCDD exposure delayed testis descent by as many as 1.7 days (Mably et al., 1992c; Bjerke et al., 1994b) and preputial separation by as many as 3.2 days (Bjerke et al,. 1994b; Bjerke and Peterson, 1994; Roman et al., 1995). Preputial separation, an indication of pubertal development, was delayed by 3.6 days in the Long Evans strain (Gray et al., 1995). It is possible that decreases in circulating androgen concentrations observed at these times could have contributed to the significant delay in the onset of these developmental milestones, but because body weights were significantly decreased in these animals during lactation and juvenile development (Mably et al., 1992c; Bjerke and Peterson, 1994; Bjerke et al., 1994b; Roman et al., 1995b; Gray et al., 1995), a generalized developmental delay cannot be ruled out.

Accessory Sex Organ Weights and Androgen Responsiveness

In the male Holtzman rat, one of the most sensitive effects of in utero and lactational TCDD exposure is a reduction in ventral prostate weight, which occurs with a maternal TCDD dose as low as 64 ng/kg (Mably et al., 1992c). At a maternal TCDD dose of 1000 ng/kg, decreased ventral prostate weights have been observed as early as PND 14 (Roman and Peterson, 1998) and as late as PND 120 (Mably et al., 1992c). These results demonstrate that TCDD exposure impairs ventral prostate growth and morphogenesis early in development, resulting in a lesion that cannot be compensated for later in life. Recent results have shown that the dorsolateral prostate is also particularly sensitive to in utero and lactational TCDD exposure (Roman et al., 1995). Between PNDs 21 and 63, TCDD-induced decreases in relative weights of the ventral prostate and dorsolateral prostate were of similar magnitude (16–43%), with the magnitude of relative weight reduction decreasing with age. Decreases in ventral prostate and dorsolateral prostate weights were reflected in both DNA and protein content (Bjerke and Peterson, 1994; B. Roman, unpublished results).

The mechanism by which TCDD impairs prostate growth and development is unknown at this time. Although in utero and lactational TCDD exposure does not significantly decrease circulating androgen concentrations, it is possible that it might interfere with androgen action within the ventral prostate and dorsolateral prostate. Circulating testosterone and 3α-diol are converted within the fetal and postnatal prostate to DHT, the active intracellular androgen, by 5α-reductase and 3α-hydroxysteroid oxidoreductase, respectively (Wilson and Lasnitzki, 1971; Orlowski et al., 1988). In utero and lactational TCDD exposure significantly decreased ventral prostate DHT concentration at weaning (PND 21) to 62% of control, but had no effect on DHT concentration at later times, suggesting that impaired DHT synthesis (or increased DHT catabolism) might contribute to early decreases in ventral prostate weight. In support of this hypothesis, acute, high-dose TCDD exposure has been shown to decrease hepatic 5α-reductase activity in rats (Gustafsson and Ingelman-Sundberg, 1979). However, dorsolateral prostate DHT concentration was not decreased by in utero and lactational TCDD exposure at or subsequent to weaning (Roman et al., 1995), implying that, at least in the dorsolateral prostate and probably in the ventral prostate as well, TCDD interferes with growth either (1) by inhibiting androgen action at some point distal to DHT synthesis or (2) by interfering with an androgen-independent growth controlling mechanism.

Because the effects of in utero and lactational TCDD exposure on ventral prostate development could not be fully explained by decreased circulating androgen concentrations or by impaired ability of the ventral prostate to convert circulating androgens to DHT, responsiveness of the ventral prostate to testosterone stimulation was examined (Bjerke et al., 1994b). Male offspring of TCDD-exposed and vehicle-exposed dams were castrated in adulthood and implanted with graded lengths of silastic tubing containing crystalline testosterone, resulting in plasma testosterone concentrations ranging from subphysiological to supraphysiological. Over this range of implant lengths, plasma testosterone concentrations and ventral prostate testosterone and DHT concentrations were not different in TCDD-exposed and control animals, implying that peripheral testosterone metabolism was not altered by in utero and lactational TCDD exposure. However, both the weight and protein content of the ventral prostate in TCDD-exposed rats were significantly reduced over all implant lengths, suggesting that the responsiveness of the ventral prostate to androgens was impaired by TCDD.

The observation that decreased androgen production could not account for effects of in utero and lactational TCDD exposure on prostate development implied that the prostate might be a primary target of TCDD exposure. The developing ventral and dorsolateral prostates express both the AhR and ARNT and are responsive to in utero and lactational TCDD exposure in terms of CYP1A1 induction (Roman and Peterson, 1998). In addition, a number of androgen-regulated mRNAs that code for markers of functional differentiation (secretory activity) of luminal epithelial cells exhibited transient decreases in response to in utero and lactational TCDD exposure (Roman and Peterson, 1998). However, there was no effect of acute TCDD exposure (25 μg TCDD/kg, 24 hour) of adult males on androgen-regulated prostatic mRNA abundance, suggesting that the transient decreases observed in the developing prostate were not the result of direct action of the liganded AhR on these genes but instead were reflective of a TCDD-induced delay in luminal epithelial cell differentiation.

Morphometric analysis revealed that very early growth of the prostate was impaired by TCDD exposure, as evidenced by decreased number of prostatic epithelial buds emerging from the fetal urogenital sinus on GD 20 and decreased ventral prostate cell proliferation index on PND 1 (Roman et al., 1998b). However, there was no evidence of decreased proliferation at later times or increased apoptosis at any time. Immunohistochemical analysis (Roman et al., 1998b) further revealed that differentiation of both smooth muscle cells and luminal epithelial cells was delayed in ventral prostates from TCDD-exposed animals, with the latter observation correlating well with observed delays in the onset of secretory activity. In addition, striking alterations in ventral prostate histology were noted in approximately 40% of TCDD-exposed animals as late as the juvenile stage of development (PND 32). The severity of histological abnormalities correlated precisely with the severity of the decrease in tissue wet weight, and alterations in differentiation and histological arrangement of cell types could not be explained by a developmental delay. Typical abnormalities included epithelial hyperplasia, decreased abundance of fully differentiated luminal epithelial cells, increased density (or continuous layer) of basal epithelial cells, altered spatial distribution of androgen receptor expression and increased thickness of the periductal smooth muscle sheath. Taken together, results suggest that effects of in utero and lactational TCDD exposure on prostate weight were contributed to by impaired growth at very early times, and delayed and/or impaired differentiation that could be of permanent consequence.

Seminal vesicle weight is also decreased by in utero and lactational TCDD exposure, although in Holtzman rats, the sensitivity of this organ to TCDD is not as great as that of the ventral prostate. A dose as low as 160 ng TCDD/kg maternal body weight significantly decreased seminal vesicle weight in Holtzman rats during the juvenile (PND 32) and postpubertal (PND 63) stages of sexual development, although doses up to 1000 ng TCDD/kg did not affect seminal vesicle weight at sexual maturity (PND 120) (Mably et al., 1992c). One subsequent study has confirmed decreases in seminal vesicle weight on PND 63 (Bjerke and Peterson, 1994), although data from another study (Roman et al., 1995), while exhibiting a trend for TCDD-induced decreases in seminal vesicle weight between PNDs 32 and 63, could not confirm these decreases with any statistical significance. In this latter study, seminal vesicle and coagulating gland weights of TCDD-exposed animals ranged from 81 to 104% of control, with the magnitude of relative weight reduction increasing with age. This trend was directly opposite to that for the magnitude of reduction in ventral prostate and dorsolateral prostate weights, which decreased with age. Also unlike the ventral prostate, responsiveness of the seminal vesicle to testosterone stimulation in adulthood was not impaired by in utero and lactational TCDD exposure (Bjerke et al., 1994b), implying that slight decreases in plasma androgen concentrations might explain the relatively minor decreases in seminal vesicle weight. The observation that the time course of effects on growth of the ventral and dorsolateral prostates is very different from the time course of effects on growth of the seminal vesicle and coagulating gland lends further support to the idea that TCDD-induced reductions in ventral prostate and dorsolateral prostate weights cannot be attributed to an androgenic deficiency or a generalized developmental delay.

Accessory sex organ sensitivity to in utero and lactational TCDD exposure seems to be rat strain–specific. In the Long Evans strain, the weight of the combined seminal vesicle/coagulating gland was decreased in TCDD-exposed animals at 11 months of age, although the weight of the ventral prostate was not different from control (Gray et al., 1995). The decrease in seminal vesicle weight probably reflected a decrease in androgen-stimulated secretory protein production, because the weight of the expressed seminal vesicle (devoid of secretions) from TCDD-exposed animals was not different from control. In addition, testosterone responsiveness (in terms of increased weight) of the seminal vesicle was decreased by in utero and lactational TCDD exposure, a phenomenon that could not be explained by decreased androgen receptor number. In the Long Evans strain, testosterone responsiveness of the ventral prostate was unaffected.

Testis, Epididymis, and Glans Penis Growth Parameters

In both Holtzman and Long Evans rats, in utero and lactational TCDD exposure has been shown to slightly although significantly decrease testis weight (Mably et al., 1992a; Bjerke et al., 1994b; Gray et al., 1995). However, when testis weight was normalized with respect to body weight, decreases were not usually statistically significant (Bjerke and Peterson, 1994; Roman et al., 1995). Compared to decreased testis weight, decreased epididymis and cauda epididymis weights in the Holtzman rat were more persistent and sensitive effects of in utero and lactational TCDD exposure, as weight decreases persisted as late as PND 120 with a maternal dose as low as 64 ng/kg (Mably et al., 1992c). Developmental TCDD exposure also reduced cauda epididymis weight in Long Evans rats (Gray et al., 1995). In Holtzman rats, in utero and lactational TCDD exposure decreased glans penis diameter and absolute weight but not length or relative weight on PND 63 (Bjerke and Peterson, 1994).

Sperm Numbers and Fertility

Another sensitive effect of in utero and lactational TCDD exposure in the male Holtzman rat is a decrease in DSP, which is a measure of mature testicular sperm number. Exposure to a range of TCDD doses (64–1000 ng/kg) resulted in dose-related decreases in DSP in pubertal (PND 49), postpubertal (PND 63), and sexually mature (PND 120) rats, with the magnitude of the reduction (26–43%) decreasing with age (Mably et al., 1992a). Statistically significant decreases were detected with a maternal TCDD dose as low as 64 ng/kg on PNDs 63 and 120, although 160 ng/kg was required to achieve statistical significance on PND 49. Similar decreases in DSP have been reported in two subsequent studies in Holtzman rats using a dose of 700 ng TCDD/kg (Bjerke and Peterson, 1994) or 1000 ng/kg (Sommer et al., 1996). In Long Evans rats exposed to 1000 ng TCDD/kg (Gray et al., 1995), DSP was reduced to 80% of control at 8–11 months of age, but this decrease was not statistically significant.

The mechanism by which TCDD decreases DSP has been investigated in Holtzman rats. Decreased DSP was not accompanied by decreased absolute paired testis weight (Mably et al., 1992a) or relative paired testis weight (Bjerke and Peterson, 1994), nor was it associated with decreased circulating follicle stimulating hormone (FSH) concentrations (Mably et al., 1992a). Histologically, spermatogenesis was qualitatively normal; there was no indication of a gross histological lesion nor any evidence of germ cell degeneration (D. Bjerke and K. Shinomiya, unpublished results). Flow cytometric analysis also failed to reveal any spermatogenic lesion, as the relative populations of tetraploid (4N), diploid (2N), and haploid (1N) cells were not altered by TCDD exposure. Although the amount of sperm produced by the testis can be directly correlated with Sertoli cell number and function, Sertoli cell proliferation rate and Sertoli cell number were not affected by in utero and lactational TCDD exposure (D. Bjerke and K. Shinomiya, unpublished results). TCDD exposure tended to decrease germ cell proliferation and ratios of early germ cells to Sertoli cells, suggesting that Sertoli cell function might be minimally impaired. However, these decreases were not statistically significant nor of a sufficient magnitude to fully explain TCDD-induced decreases in DSP.

Once mature sperm leave the testis, they travel through the efferent ducts to the epididymis and are stored in the cauda epididymis. As with DSP, significant decreases in CESN have been seen in the Holtzman rat on PNDs 63 and 120 with a maternal TCDD dose as low as 64 ng/kg (Mably et al., 1992a). This decrease in CESN was consistently more severe (by 8–25%) than the decrease in DSP (Mably et al., 1992a; Bjerke and Peterson, 1994; Sommer et al., 1996), and ejaculated sperm numbers were decreased even further (Sommer et al., 1996). The graded decline in sperm numbers as they travel from the testis through the excurrent duct system has also been observed in Long Evans rats aged 8–11 months, in which in utero and lactational TCDD exposure decreased DSP to 79% of control, CESN to 62% of control and ejaculated sperm numbers to 42% of control (Gray et al., 1995). While these results suggest that sperm transit rate through the excurrent duct system should be increased by TCDD exposure, this hypothesis has been carefully tested in Holtzman rats and ultimately rejected, ruling out the possibilities of sperm loss via spontaneous ejaculation or abnormal introduction into the urine. (Sommer et al., 1996).

Despite the fact that in utero and lactational TCDD exposure severely decreased CESN in Holtzman rats, the reproductive capability of these males was not different from control. TCDD-exposed males were just as successful as vehicle-exposed males in impregnating females, and resulting litters exhibited no difference in size or pup viability from birth through PND 21 (Mably et al., 1992a). This lack of effect on fertility of TCDD-exposed males is not surprising, considering that male rats ejaculate 10 times the number of sperm required for impregnation. In contrast, in adult Long Evans rats exposed to TCDD in utero and via lactation, fertility, as measured by the number of implantation sites in impregnated females, was decreased to 43% of control (Gray et al., 1995). The reason for this strain difference is not clear at this time.

Sexual Behavior and Regulation of LH Secretion

In addition to effects on development and function of male reproductive tract organs, in utero and lactational TCDD exposure also had effects on sexual behavior of male Holtzman rats. When mated with untreated, receptive females, parameters such as mount latency, intromission latency, ejaculation latency, and postejaculatory interval were increased in TCDD-exposed males in a dose-dependent manner, as were number of mounts preceding ejaculation (Mably et al., 1992b). In addition, the copulatory rate of these males, or the number of mounts and intromissions per minute, was significantly decreased by TCDD exposure in a dose-dependent manner. Taken together, these data signify a partial demasculinization of sexual behavior. In a more recent study in this rat strain, sexual behavior was again demasculinized, although fewer parameters assessed attained statistical significance (Bjerke et al., 1994a). Partial demasculinization of sexual behavior has also been observed in the Long Evans rat, in which ejaculation latency and number of mounts preceding ejaculation were significantly increased by in utero and lactational TCDD exposure, although mount latency and postejaculatory interval were not significantly affected (Gray et al., 1995).

Holtzman rats exposed to TCDD in utero and via lactation not only exhibited partial demasculinization of sexual behavior, but also exhibited partial feminization of sexual behavior. When castrated and primed with ovarian steroids, TCDD-exposed males displayed a higher frequency of lordosis when mounted by a stud male than did vehicle-exposed males (Mably et al., 1992b; Bjerke et al., 1994a; Bjerke and Peterson, 1994). Partial feminization was also exhibited in the regulation of LH secretion. When rats were castrated and primed with estrogen and progesterone, TCDD-exposed but not vehicle-exposed males exhibited a distinct plasma LH surge considered to be a female response pattern (Mably et al., 1992b). Feminization of sexual behavior was not observed in TCDD-exposed Long Evans rats (Gray et al., 1995) and therefore may represent another rat strain–specific response.

If TCDD-induced alterations in sexual behavior and the regulation of LH secretion in the male rat were to be attributed to interference with hormone-mediated imprinting of the CNS early in development, then biochemical and morphological indices of CNS sexual differentiation should have also been partially demasculinized and/ or partially feminized. For example, estrogen receptor concentrations in a number of brain regions involved in hormonal regulation of sexual behavior and gonadotropin secretion are sexually dimorphic, with concentrations being higher in females than in males (Brown et al., 1988). Incomplete masculinization and/or incomplete defeminization of the male CNS would be expected to result in increased estrogen receptor concentrations in these areas. However, there was no effect of in utero and lactational TCDD exposure on estrogen receptor concentrations in any of these sexually dimorphic regions (Bjerke et al., 1994a). In addition, there was no effect of TCDD exposure on the volume of the sexually dimorphic nucleus of the preoptic area (SDN-POA) (Bjerke et al., 1994a), the size of which positively correlates with available estrogens (aromatized from androgens) during the perinatal period, therefore being larger in males than in females (Gorski et al., 1978, 1980; Döhler et al., 1984). Similarly, there was no effect of TCDD exposure on the volume of the medial preoptic nucleus (Bjerke et al., 1994a), the size of which negatively correlates with available androgens during the perinatal period, therefore being larger in females than in males (Ito et al., 1986). These results indicate that in utero and lactational TCDD exposure causes partial demasculinization and feminization of sexual behavior that is not associated with morphological alterations in the CNS, again providing strong indirect evidence that the effects of TCDD exposure cannot be traced back to alterations in the perinatal hormonal environment. The mechanism by which in utero and lactational TCDD exposure alters male sexual behavior remains unknown, although it has been suggested (Gray et al., 1995) that the partial demasculinization of sexual behavior in the Long Evans rat is the result of alterations in penile sensitivity.

In Utero Versus Lactational TCDD Exposure

In order to assess the relative contributions of in utero TCDD exposure versus lactational TCDD exposure to developmental male reproductive toxicity, pregnant Holtzman rats were administered 1000 ng TCDD/kg or vehicle on GD 15, and standardized litters were either fostered to dams within the same treatment group or cross-fostered to dams in the opposite treatment group (Bjerke and Peterson, 1994). This produced four treatment groups of male offspring: no TCDD exposure, in utero TCDD exposure, lactational TCDD exposure, and in utero and lactational TCDD exposure. For some male reproductive endpoints, both the in utero route and the lactational route of TCDD exposure alone produced effects similar to those seen with combined in utero and lactational exposure. For example, ventral prostate weight, DNA content, and protein content; seminal vesicle weight, DNA content, and protein content; and CESN were decreased on PND 63 by both in utero exposure alone and lactational exposure alone. Interestingly, the combined in utero and lactational route did not further decrease seminal vesicle growth parameters or CESN, although it did further decrease ventral prostate growth parameters when compared to either in utero or lactational exposure alone. Preputial separation was delayed, DSP was reduced, and the diameter of the glans penis was reduced by in utero TCDD exposure but not by lactational TCDD exposure, whereas feminization of sexual behavior was observed in response to lactational exposure but not in utero exposure.

These results not only provide important information regarding the developmental time period during which the various parameters assessed may be affected by TCDD exposure, they also provide important information regarding the sensitivity of the developing male rat reproductive system to TCDD. The amount of TCDD to which the rat fetus was exposed was negligible compared to that to which the nursing pup was exposed. Yet most effects of in utero and lactational TCDD exposure on the developing male rat reproductive system were observed in response to in utero TCDD exposure alone, providing evidence that the developing male rat reproductive system is even more sensitive to TCDD exposure than was previously believed.

EFFECTS OF IN UTERO AND LACTATIONAL TCDD EXPOSURE ON THE MALE HAMSTER AND MOUSE REPRODUCTIVE SYSTEMS

In order to determine whether effects of in utero and lactational TCDD exposure were species-specific, Gray et al. (1995) and Theobald and Peterson (1997) assessed developmental male reproductive toxicity of TCDD in Syrian hamsters and in ICR mice, respectively; comparison of developmental reproductive effects of TCDD exposure in rats, hamsters, and mice is presented in Table 1. Pregnant hamsters were orally dosed with 2000 ng TCDD/kg on GD 11, whereas pregnant mice were orally dosed with 15,000–60,000 ng TCDD/kg on GD 14. In both cases, as with the rat, the time of dosing corresponded to the onset of testicular androgen production (Hoar and Monie, 1981; Greco et al., 1993). Lactational exposure continued until weaning on PND 24 (hamsters) or 21 (mice). As was seen in the rat, there was no effect of in utero and lactational TCDD exposure on circulating testosterone concentrations in either hamsters or mice when assessed between puberty and sexual maturity, and only some androgen-dependent parameters were affected by TCDD exposure. In the hamster, AGD on PNDs 3 and 10 was not decreased by TCDD exposure. And although preputial separation (complete by PND 47) was significantly delayed, flank gland development, an androgen-dependent process that occurs about the same time as preputial separation, was unaffected by TCDD exposure. In the mouse, neither anogenital distance (relative or absolute), testis descent, nor preputial separation was affected by TCDD exposure.

As in Long Evans rats, seminal vesicle weight was significantly reduced in adult hamsters. In contrast, effects of in utero and lactational TCDD exposure on the accessory sex organs of ICR mice were quite similar to those seen in Holtzman rats, with decreased ventral prostate weight being among the most sensitive effects. Coagulating gland weights were also significantly decreased by TCDD exposure, whereas seminal vesicle weights were unaffected. In the hamster, as in the Holtzman and Long Evans rat strains, in utero and lactational TCDD exposure decreased absolute testis and epididymis weights and resulted in a progressive loss of sperm as they travelled from the testis through the excurrent duct system. DSP in TCDD-exposed hamsters was 92% of control, while CESN was 68% of control and ejaculated sperm number was 35% of control. Sperm granulomas were present in the epididymides and/or testes of 25% of TCDD-exposed hamsters, suggesting that an immune response might contribute to the progressive sperm loss in this species. In the mouse, in utero and lactational TCDD exposure had no effect on testis or epididymis weights or on DSP, although epididymal sperm numbers were reduced to 70% of control. However, this effect was noted on PND 65 but did not persist beyond this time point.

Finally, in contrast to the rat, hamster masculine sexual behavior was not altered by in utero and lactational TCDD exposure. Neither ejaculation latency nor the number of mounts preceding ejaculation was increased by TCDD exposure. Effects on feminine sexual behavior and reproductive capability in the male hamster, and effects on masculine and feminine sexual behavior and reproductive capability in the male mouse, have not been evaluated.

It is important to note that, although the ICR mouse was relatively insensitive to TCDD-induced developmental male reproductive toxicity, in utero and lactational exposure of the male hamster to 2000 ng TCDD/kg maternal body weight produced similar effects on the reproductive system as in utero and lactational exposure of the male rat to 64–1000 ng TCDD/kg maternal body weight. The dose administered in the hamster study was only twice the maximum dose used in the rat studies, although the LD_{50} of TCDD in the hamster is more than 100 times greater than the LD_{50} of TCDD in the rat (Poland and Knutson, 1982). Although no dose response studies have been conducted in the hamster, it seems likely that the ED_{50} for observed effects in the hamster would be much less than 100 times the ED_{50} for these effects in the rat, implying that the relative resistance of the hamster to adult TCDD exposure cannot be extrapolated to developmental TCDD exposure.

EFFECTS OF IN UTERO AND LACTATIONAL PCB EXPOSURE ON THE MALE RODENT REPRODUCTIVE SYSTEM

Most studies in which effects of in utero and/or lactational PCB exposure on the male reproductive system have been assessed have used complex mixtures of PCB congeners such as Aroclors (produced by Monsanto) and Clophens (produced by Bayer). The use of PCB mixtures greatly complicates interpretation of results in terms of AhR-dependent and AhR-independent effects of PCBs because (1) these mixtures were contaminated with PCDFs, which are AhR agonists, and (2) of the 209 possible PCB congeners, only a small number are believed to cause toxicity, like TCDD, via an AhR-mediated mechanism. Those congeners that bind most strongly to the AhR and thus are most potent in producing AhR-mediated (TCDD-like) effects are the coplanar PCBs: 3,3′, 4,4′-tetrachlorobiphenyl (PCB 77); 3,3′, 4,4′,5-pentachlorobiphenyl (PCB 126); and 3,3′, 4,4′,5,5′-hexachlorobiphenyl (PCB 169). In addition, mono-*ortho* derivatives of the coplanar PCBs (i.e., those additionally substituted with chlorine in either the 2 or the 2′ position) have some, although much reduced, AhR agonist potency

(Safe, 1992). However, most PCB congeners do not bind to the AhR, and the contribution of AhR agonists to specific endpoints of Aroclor or Clophen toxicity is not fully understood.

Plasma Androgen Concentrations and External Indicators of Androgenic Status

Although most studies of developmental male reproductive toxicity of PCBs in rodents have focused solely on fertility, a few have addressed effects on plasma androgen concentrations, external indicators of androgenic status, sex organ growth, and spermatogenesis. Male offspring of Dunkin Hartley guinea pig dams exposed to 2–3 mg Clophen A50/kg/day between GDs 18 and 60 did not show any decrease in plasma testosterone concentration between PNDs 30 and 90 (Lundkvist, 1990). Similarly, male offspring of Holtzman rat dams administered a cumulative dose as high as 320 mg Aroclor 1254/ kg during lactation did not show decreased plasma testosterone concentrations as adults (Sager et al., 1991), and male offspring of Wistar rat dams administered 1.8 mg PCB 169/kg on GD 1 did not show decreased plasma testosterone or DHT concentrations on PND 100 (Smits-van Prooije and Waalkens-Berendsen, 1994). The latter study also reported a lack of effect on both absolute and relative AGD on PND 7, as well as a lack of effect on time to testis descent.

Accessory Sex Organ and Testis Weights

Although alterations in plasma androgen concentrations or external indicators of androgenic status have not been reported in response to in utero and/or lactational PCB exposure, decreases in androgen-dependent accessory sex organ weights have been noted. When lactating Holtzman rats were orally dosed with Aroclor 1254 between PNDs 1 and 9 (cumulative doses ranging from 48 to 384 mg/kg), both absolute and relative ventral prostate weight of offspring were significantly decreased in adulthood in a dose-dependent manner (Sager, 1983; Sager et al., 1991). Histologically, the ventral prostate epithelial cells were noticeably flattened and the ductal epithelium was less convoluted. Absolute and relative seminal vesicle weights were reduced at higher Aroclor 1254 doses only, and the percent decrease in relative weight (9–16%) was less than that observed for the ventral prostate (19–48%) (Sager, 1983; Sager et al., 1991). In contrast, male offspring born to Sprague-Dawley rat dams administered 30 mg/kg/day of Aroclor 1221, Aroclor 1242, or Aroclor 1260 between GDs 14 and 20 did not exhibit significantly decreased ventral prostate weights at 6 months of age (Gellert and Wilson, 1979). The difference in effect on ventral prostate weight in these two studies may reflect the different dosing periods (postnatal versus prenatal), different congener compositions of the Aroclor mixtures, and/or the degree of PCDF contamination of the Aroclor mixtures. In a cross-fostering study in which Wistar rat dams were administered a single dose of PCB 169 on GD 1 (Smits-van Prooije and Waalkens-Berendsen, 1994), combined in utero and lactational exposure of male offspring decreased absolute and relative ventral prostate weight in adulthood to 58% and 62% of control, respectively, although neither in utero nor lactational exposure alone was similarly effective. In contrast, in utero but not lactational or combined in utero and lactational PCB 169 exposure significantly decreased absolute seminal vesicle weight in adulthood to 83% of control, although relative seminal vesicle weight was not reduced by any exposure regimen.

Both absolute and relative testis weights were significantly increased in male Holtzman rat offspring born to dams administered cumulative doses of 192 and 384 mg Aroclor 1254/kg during lactation (Sager, 1983). Increased testis weight was also seen with in utero and lactational exposure of male Sprague-Dawley rats to a cumulative dose of 210 mg Aroclor 1260/kg maternal body weight (Gellert and Wilson, 1979) and with lactational exposure to a cumulative maternal dose as low as 42 mg Aroclor 1242 or 10 mg Aroclor 1254 (Cooke et al., 1996). These increases in testis weight were most likely due not to AhR agonists in the Aroclors but to other PCB congeners which alter thyroid hormone status, as transient hypothyroidism early in postnatal development has been shown to increase Sertoli cell number, sperm production, and testis size (Cooke et al., 1991a; Cooke and Meisami, 1991; van Haaster et al., 1992). In contrast, in utero exposure of Dunkin Hartley guinea pigs to 2–3 mg Clophen A50/kg maternal body weight/day between GDs 18 and 60 decreased both absolute and relative testis weights (Lundkvist, 1990).

Spermatogenesis

As would be expected, lactational exposure of male rats to Aroclor 1242 or Aroclor 1254 increased Sertoli cell proliferation rate and daily sperm production as a consequence of non-AhR agonist PCBs in the mixtures, which induced transient, neonatal hypothyroidism (Cooke et al., 1996). In contrast, in utero and lactational exposure to PCB 77, an AhR agonist, has been shown to transiently decrease spermatogenesis. Although the most advanced testicular cell type in seminiferous tubules was the same in C57/B1 mice born to dams administered either vehicle or a cumulative dose as low as 63 mg PCB 77/kg from before mating through lactation, the percentage of tubules displaying this most advanced cell type was significantly decreased in a dose-dependent manner (Ronnback and de Rooij, 1994). Spermatogenesis was suppressed on PNDs 14 and 28 but had qualitatively recovered by PND 56.

Fertility

The effect of in utero and/or lactational PCB mixture exposure on fertility of male rodents seems to vary with the type of Aroclor or Clophen used. In utero and lactational exposure of male Sprague-Dawley rats to a cumulative maternal dose of 210 mg/kg of Aroclor 1221, 1242, or 1260 had no effect on their ability to sire viable litters (Gellert and Wilson, 1979), and lactational exposure of male NMRI mice to a cumulative maternal dose of 200 mg Clophen A60/kg had no effect on the number of ova implanted when mated with control females (Kihlstrom et al., 1975). In contrast, when male Holtzman rats exposed via lactation to a cumulative maternal dose of 192 or 384 mg Aroclor 1254/kg were mated with untreated females (Sager, 1983), significant decreases in number of pregnancies, number of implantation sites per corpora lutea (reflecting preimplantation loss), and number of live pups per litter were observed. In addition, resorption rate was increased (reflecting postimplantation loss). This reduction in fertility of PCB-exposed males could not be attributed to reduced body weight gain because fertility of pair fed controls was not affected (Sager et al., 1987). Nor was reduced fertility accompanied by decreased DSP, decreased CESN, altered sperm morphology, impaired sperm motility, or decreased plasma FSH or testosterone concentrations (Sager et al., 1987, 1991). However, in control females mated to Aroclor

1254-exposed males, as many as 79% of ovulated eggs were unfertilized, polyspermic, or fragmented. When fertilization was successful, most fertilized ova failed to develop normally (Sager et al., 1987, 1991).

It seems unlikely that AhR agonists in Aroclor 1254 were responsible for decreasing fertility in lactationally exposed Holtzman rats in the absence of decreased DSP and CESN, because in utero and lactational TCDD exposure in this same strain significantly decreased DSP and CESN without affecting fertility (Mably et al., 1992a). However, in utero and lactational exposure to PCB 169 has been shown to alter mating behavior and cause infertility in male Wistar rats (Smits-van Prooije et al., 1993). When male offspring of dams administered 1.8 mg PCB 169/kg on GD 1 were paired with receptive, untreated females, only 6 of 14 pairings resulted in matings, and of the 6 matings, none resulted in pregnancy. In contrast, when male offspring of dams administered vehicle on GD 1 were paired with receptive, untreated females, 9 of 10 pairings resulted in matings, and of the 9 matings, 6 resulted in pregnancy. In a follow-up cross-fostering study (Smits-van Prooije and Waalkens-Berendsen, 1994), it was shown that either in utero or lactational PCB 169 exposure alone could decrease the reproductive success of males mated to untreated females, with most failures being the result of preimplantation loss. However, no effect on sperm number in the whole epididymis could be detected, and sperm motility and morphology appeared normal. Although most toxic effects of PCB 169 are believed to be mediated via the AhR, the role of the AhR in reducing fertility in response to in utero and/or lactational PCB 169 exposure is unclear at this time.

POSSIBLE MECHANISMS BY WHICH IN UTERO AND LACTATIONAL TCDD EXPOSURE ADVERSELY AFFECTS THE DEVELOPING MALE REPRODUCTIVE SYSTEM

TCDD as an Antiestrogen

TCDD has been shown to act as an antiestrogen in a number of model systems. For example, it inhibits estradiol-induced increases in uterine wet weight in mice and rats (Gallo et al., 1986; Romkes et al., 1987; Astroff and Safe, 1990) and interferes with estradiol-induced increases in rat uterine peroxidase activity (Astroff and Safe, 1990) and rat uterine epidermal growth factor receptor binding activity and mRNA level (Astroff et al., 1990). These effects are most likely due to TCDD-induced decreases in estrogen receptor protein levels and consequent decreases in binding activity, both of which have been observed in the uterus and liver (Romkes et al., 1987, DeVito et al., 1992). However, antiestrogenic effects of TCDD exposure may be species, strain, and/or age dependent, as TCDD does not inhibit estradiol-induced increases in uterine weight or keratinization of the vaginal epithelium in weanling female Sprague-Dawley rats (White et al., 1995). TCDD has also been reported to decrease estrogen receptor protein levels and estrogen-binding activity in cultured cells (Harris et al., 1990; Zacharewski et al., 1991). Increases in estradiol metabolism in a breast cancer cell line (MCF-7) have also been observed (Gierthy et al., 1988; Spink et al., 1990, 1992), although in vivo studies have in general failed to link increased estradiol metabolism with TCDD exposure (Shiverick and Muther, 1983; Namkung et al., 1985; DeVito et al., 1992).

The physiological role of estrogens in the development of the male reproductive system is not fully understood. Estrogen receptors have been detected in various organs of the developing male mouse reproductive tract by ^{3}[H]estradiol steroid autoradiography (Cooke et al., 1991b). Undifferentiated mesenchyme of the urogenital sinus (which gives rise to the prostate) and Wolffian duct (which gives rise to the seminal vesicles, epididymis and vas deferens) both exhibit estradiol binding sites as early as GD 16 that persist postnatally in differentiated stromal fibroblasts but not smooth muscle cells. In addition, epithelial estradiol binding sites are present in the epididymis prenatally and in the seminal vesicle and coagulating gland early postnatally, although the vas deferens and prostate epithelium do not exhibit estradiol binding sites as late as PND 10. Using immunohistochemistry, the mouse testis has been shown to express estrogen receptors in cells within the seminiferous tubules as well as in the interstitium as early as GD 13, although this expression seems to decline during late gestation (Greco et al., 1992). An important functional role for these receptors in the testis is implied by studies in estrogen receptor knockout (ERKO) mice, which are infertile and exhibit testes containing degenerating seminiferous tubules that produce only 10% of the amount of sperm produced by testes of wild-type mice (Lubahn et al., 1993; Eddy et al., 1996). Thus, all organs in the male reproductive tract express estrogen receptors in at least some cell types, but the testis is the only male reproductive organ in which the estrogen receptor has been shown to play an essential role in normal development. However, it should be noted that it is now known that at least two forms of the estrogen receptor exist (ERα and ERβ, Kuiper et al., 1996), and the role of ERβ in development of the male reproductive system has not been investigated. ERβ mRNA is very highly expressed in the adult rat prostate epithelium, with much lower levels in epididymis and testis (Kuiper et al., 1997).

The spectrum of effects of in utero and lactational TCDD exposure on the developing male reproductive system does not quite match the pattern of effects observed in response to perinatal antiestrogen administration (Table 2). Like perinatal TCDD exposure, neonatal exposure of male mice to the antiestrogen tamoxifen decreases testis, epididymis, and accessory sex organ weights (Iguchi and Hirokawa, 1986; Taguchi, 1987). However, unlike perinatal TCDD exposure, neonatal tamoxifen exposure causes sterility, testicular hypoplasia, and spermatogenic arrest, effects that appear to be permanent (Taguchi, 1987). Neonatal exposure of mice to another antiestrogen, keoxifene, also decreases testis, epididymis and seminal vesicle weights and results in spermatogenic arrest, although these effects are not permanent (Chou et al., 1992).

Perinatal antiestrogen exposure affects not only the developing male reproductive system but also the developing male CNS. Masculinization of the CNS, which results in male patterns of sexual behavior and gonadotropin regulation, is dependent on perinatal androgens available for aromatization to estrogens (McEwen et al., 1977; MacLusky and Naftolin, 1981) and can be gauged morphologically by the volume of the SDN-POA (Gorski et al., 1978, 1980; Döhler et al., 1984). Although in utero and lactational TCDD exposure induces partial demasculinization of sexual behavior, partial feminization of sexual behavior and partial feminization of the regulation of LH secretion, the volume of the SDN-POA is not decreased (Bjerke et al., 1994a). In contrast, while perenatal antiestrogen exposure demasculinizes and feminizes sexual behavior (McEwen et al., 1977; Matuszczyk and Larsson, 1995) and feminizes the regula-

Table 2 Comparison of Selected Effects of Developmental Exposure to TCDD, Antiestrogens, Antiandrogens, and 5α-Reductase Inhibitors[a]

	TCDD	Antiestrogen	Antiandrogen	5α-Reductase Inhibitor
External indicators of androgenic status				
Anogenital distance (relative)	No effect	—[b]	Decreased	Decreased
Preputial separation	Delayed	—	Delayed	Delayed
Sex organ weights (absolute)				
Ventral prostate	Decreased	Decreased	Decreased[c]	Decreased
Seminal vesicle	Decreased	Decreased	Decreased	Decreased[d]
Testis	Decreased	Decreased	Decreased	No effect
Epididymis	Decreased	Decreased	Decreased	No effect
Spermatogenesis				
Testicular histology	No effect	Altered[c]	Altered[f]	—
Testicular sperm production	Decreased	Decreased	Decreased	—
Cauda epididymal sperm number	Decreased	—	Decreased	—
Fertility	No effect	Sterility	Sterility	—
External genitalia	No effect	No effect	Hypospadias Cleft prepuce Feminization	Hypospadias Cleft prepuce Feminization
Nipple formation	No	No	Yes	Yes
Sexual behavior and CNS parameters				
Masculine behavior	Demasculinized	Demasculinized	Demasculinized	—
Feminine behavior	Feminized	Feminized	Feminized	—
SDN-POA volume	No effect	Decreased	No effect	—
Regulation of LH secretion	Feminized	Feminized	Feminized	—

[a]Data shown for developmental exposure to TCDD reflect effects in the Holtzman rat. Data shown for developmental exposure to antiestrogens, antiandrogens, and 5α-reductase inhibitors reflect effects in rodents from representative studies, as referenced in the text.
[b]Parameter not measured in studies referenced in the text.
[c]Antiandrogen administration frequently results in agenesis of the prostate.
[d]5α-Reductase inhibitor administration has only minimal effects on seminal vesicle weight.
[e]Testicular hypoplasia and spermatogenic arrest.
[f]Seminiferous tubule atrophy.

tion of LH secretion (MacLusky and Naftolin, 1981), these effects are accompanied by decreased volume of the SDN-POA (Döhler et al., 1984).

In summary, although TCDD is known to be antiestrogenic in several model systems, and some effects of in utero and lactational TCDD exposure are similar to those of developmental antiestrogen exposure, developmental male reproductive toxicity cannot be easily ascribed to the antiestrogenic action of TCDD. It should be noted that perinatal estrogen exposure has similar effects on the developing male reproductive system as perinatal antiestrogen exposure (for review, see Iguchi, 1992), but because TCDD has not been shown to be estrogenic in any model system, it is unlikely that an estrogenic mechanism could explain any aspect of the developmental male reproductive toxicity of TCDD.

TCDD as an Antiandrogen

In utero and lactational exposure of male rats to TCDD decreases accessory sex organ weights, delays preputial separation, decreases testicular sperm production and epididymal sperm storage, partially demasculinizes sexual behavior, partially feminizes sexual behavior, and partially feminizes the regulation of LH secretion (Mably et al., 1992a,b,c; Bjerke et al., 1994a,b; Bjerke and Peterson, 1994; Roman et al., 1995; Gray et al., 1995). The manifestation of these effects would be consistent with decreased testicular androgen production and/or circulating androgen concentrations, but neither of these parameters has been shown to be significantly affected, either perinatally or at later times, by in utero and lactational TCDD exposure (Mably et al,. 1992c; Chen et al., 1993; Roman et al., 1995; Gray et al., 1995). The possibility remains, however, that the androgenic deficiency-like syndrome which follows in utero and lactational TCDD exposure could be the result of interference with androgen action at the level of the androgen receptor.

Administration of an androgen receptor antagonist such as flutamide (Imperato-McGinley et al., 1992), the fungicide vinclozolin (Gray et al., 1994) or p,p'-DDE (a metabolite of DDT) (Kelce et al., 1995) to male rodents during perinatal development can result in decreased testis, epididymis, and accessory sex organ weights, delayed preputial separation, and decreased DSP and CESN (Table 2). However, with developmental antiandrogen exposure, the decreases in DSP and CESN stem from a high incidence of seminiferous tubule atrophy and, in some cases, a failure of the testis to descend (Gray et al., 1994), whereas with TCDD exposure, testicular histology remains qualitatively normal (D. Bjerke and K. Shinomiya, unpublished results). Effects of antiandrogen (Neumann et al., 1970; Döhler et al, 1986; Thornton et al., 1991) and TCDD exposure (Mably et al., 1992b) on sexual behavior, the volume of the SDN-POA, and the regulation of LH secretion are qualitatively similar. In contrast, antiandrogen exposure but not TCDD exposure decreases relative AGD, feminizes external genitalia, causes sterility, produces hypospadias with cleft prepuce, and results in nipple formation (Neumann et al., 1970; Jean-Faucher et al., 1985; Gray et al., 1994).

In some respects, effects of in utero and lactational TCDD exposure resemble those of in utero exposure to the 5α-reductase inhibitors (Table 2), finasteride, and 6-methylene-4-pregnene-3,20-dione (6-MP). 5α-Reductase is the enzyme that converts testosterone to DHT, the major androgen responsible for morphogenesis and growth of the prostate (Cunha et al., 1987), branching (but not initial outgrowth) of the SV (Shima et al., 1990), and masculinization of external genitalia (Imperato-McGinley et al., 1985). Ventral prostate DHT concentration is decreased by in utero and lactational TCDD exposure on PND 21 (Roman et al., 1995), and developmental exposure to either TCDD (Mably et al., 1992c; Bjerke et al., 1994b; Bjerke and Peterson, 1994; Roman et al., 1995) or finasteride (Clark et al., 1993) results in decreased ventral prostate weight and delayed preputial separation. In addition, both TCDD (Roman et al., 1998b) and 6-MP (Iguchi et al., 1991) decrease prostatic epithelial budding from the urogenital sinus. However, as with antiandrogen exposure, in utero exposure of male rats to finasteride results in feminization of external genitalia, hypospadias with cleft prepuce, and nipple formation (Clark et al., 1993), effects that are not observed in response to in utero and lactational TCDD exposure. In addition, 5α-reductase inhibition significantly decreases relative AGD, while TCDD exposure usually does not. These discrepancies cannot be explained by different windows of sensitivity for effects of 5α-reductase inhibition, as the TCDD exposure period (GD 15-PND 21) used in the above-described studies encompasses the window of sensitivity for most effects of finasteride exposure (GD 16–17). Nor can these discrepancies be explained by different dose-response curves for effects of 5α-reductase inhibition, because decreased AGD and nipple formation (not seen with TCDD exposure) are more sensitive endpoints of in utero finasteride exposure than delayed preputial separation (a result of TCDD exposure) (Clark et al., 1993). It is therefore unlikely that TCDD-induced inhibition of 5α-reductase activity could explain effects on prostate growth and preputial separation.

TCDD as a Modulator of Cell Proliferation and Differentiation

Although TCDD exposure does alter processes and behaviors that are known to be sex hormone-dependent, from the discussions above it is clear that effects of in utero and lactational TCDD exposure on the male reproductive system cannot be easily classified as antiestrogenic or antiandrogenic in nature. TCDD has been shown to cause tissue-specific hyperplasia and hypoplasia (Poland and Knutson, 1982) and induce apoptosis in thymocytes (Kamath et al., 1997). It has also been shown to interfere with differentiation processes in a number of organ systems including the epidermis (Greenlee et al., 1987), palate (Abbott and Birnbaum, 1989), thymus (Blaylock et al., 1992) and adipose tissue (Phillips et al., 1995; Brodie et al., 1996). By exposing male rats to TCDD between GD 15 and PND 21, a time period of active cell proliferation and differentiation in the male reproductive tract, it is possible that the lesions that ensue may be the result of interference with nonhormonal aspects of these processes. For example, prostatic budding and ductal morphogenesis are of course androgen-dependent but also involve important mesenchymal-epithelial interactions occurring downstream of androgen receptor action (Cunha et al., 1987) that TCDD might modulate. Western blot, immunohistochemical, and/or reverse transcriptase-polymerase chain reaction analyses have identified both the AhR and ARNT in the developing rat ventral and dorsolateral prostates (Roman and Peterson, 1998) as well as in many organs of the adult male rat reproductive system (testis, epididymis, vas deferens, ventral prostate, dorsolateral prostate, seminal vesicle) (Roman et al., 1998a) and have demonstrated that, except for the testis, each of these organs is responsive to TCDD exposure in terms of CYP1A1 induction, raising the possibility that adverse effects of in utero and lactational TCDD exposure on development and/or function of these organs could be the result of direct alterations in gene transcription.

EXTRAPOLATION OF DEVELOPMENTAL MALE REPRODUCTIVE TOXICITY OF TCDD IN LABORATORY RODENTS TO HUMAN HEALTH RISK ASSESSMENT: FUTURE RESEARCH NEEDS

TCDD is a potent male reproductive toxicant in rodents when exposure occurs during the very sensitive period of late fetal and early neonatal development. Adverse effects on the male reproductive system in response to in utero and lactational TCDD exposure were first described in the Holtzman rat (Mably et al., 1992a,b,c) and have since been characterized in the Long Evans rat and Syrian hamster (Gray et al., 1995) and ICR mouse (Theobald and Peterson, 1997). In general, similar effects occurred across rat strains and across species, although some species-specific effects have been observed (Table 1). Because developmental male reproductive effects are among the most sensitive endpoints of TCDD toxicity, they must be taken into consideration when assessing human health risk of TCDD exposure. However, in order to accurately assess the human health risk of TCDD exposure in terms of developmental male reproductive toxicity, several research needs must be met.

Determination of Toxic Equivalency Factors for TCDD-like Halogenated Aromatic Hydrocarbons with Respect to Developmental Male Reproductive Toxicity, and Evaluation of Developmental Male Reproductive Toxicity of Nonpersistent AhR Agonists

The relative potencies of PCDD, PCDF, and PCB congeners have traditionally been expressed as toxic equivalency factors (TEFs), which are assigned based on the ability of a congener to elicit AhR-mediated effects relative to TCDD and which can be used to convert the concentration of AhR agonists in a tissue sample to TEQs. In humans in Western industrialized countries, the average background body burden of AhR agonists is estimated to be 8–13 ng TEQs/kg (DeVito et al., 1995), whereas the LOAEL for the most sensitive developmental male reproductive effects of TCDD exposure is 64 ng/kg (Mably et al., 1992a,c). However, application of TEFs to the assessment of human health risk of TCDD-like compounds with respect to male reproductive toxicity is uncertain, as most TEFs are not based on reproductive endpoints. This uncertainty could be addressed by determining TEFs for developmental male reproductive endpoints following in utero and lactational exposure to those PCDD, PCDF, and PCB congeners that make the greatest contribution to the background body burden of TEQs in humans.

Persistent PCDD, PCDF, and PCB congeners are not the only xenobiotic AhR ligands to which humans are exposed. Other more readily metabolizable ligands include polycyclic aromatic hydrocarbons such as 3-methylcholanthrene and benzo(a)pyrene, which are by-products of cigarette smoke and combustion of fossil fuels, as well as indole-3-carbinol and related compounds, which are found in cruciferous vegetables of the Brassica genus (broccoli, cabbage, cauliflower). These ligands bind to the AhR and produce the same biochemical responses (e.g., induction of CYP1A1) as TCDD, but because they are metabolized, effects on gene transcription do not persist (Jellinck et al., 1993; Liu et al., 1994; Riddick et al., 1994). It has recently been shown that a single exposure to indole-3-carbinol on GD 15 did not produce a TCDD-like spectrum of effects on the male rat reproductive system; that is, no decreases in accessory sex organ weights or caudal epididymal sperm numbers were observed (Wilker et al., 1996).

However, other AhR ligands to which humans are frequently exposed have not been evaluated in this system. In addition, it is necessary to determine whether chronic exposure to metabolizable ligands causes developmental male reproductive toxicity, and whether this type of exposure might modulate the effects of exposure to the more persistent PCDD, PCDF, and PCB congeners.

Determination of the Role of the AhR in TCDD-Induced Developmental Male Reproductive Toxicity and Normal Male Reproductive System Development

Human health risk assessment, based on the very sensitive developmental male reproductive endpoints of TCDD exposure, needs to be validated with a better understanding of the mechanisms underlying these effects. Although most effects of TCDD exposure are believed to be initiated via an AhR-mediated mechanism (Okey et al., 1994), this has not been conclusively shown for developmental male reproductive effects. Therefore, it is important to show that developmental male reproductive effects considered in risk assessments are indeed AhR-mediated. By studying these responses in congenic mice that differ only at the *Ahr* locus (Birnbaum, 1986) insight into whether these effects are AhR-mediated could be obtained. Finally, the recently developed *Ahr$^{-/-}$* mice (Fernandez-Salguero et al., 1995; Schmidt et al., 1996) may be useful in determining whether developmental male reproductive toxicity of TCDD is AhR-mediated and will be invaluable in studying the role of the AhR in normal male reproductive system development.

More penetrating studies regarding the mechanistic basis of certain developmental male reproductive endpoints of TCDD exposure are warranted due to the sensitivity and reproducibility of these effects, as well as their potential relevance to human health. For example, inhibition of prostate development and decreased epididymal and ejaculated sperm counts are sensitive, reproducible effects of in utero and lactational TCDD exposure that could have human health relevance. A recent Institute of Medicine study (1996) found a limited association between exposure to Agent Orange (a defoliant used during the Vietnam war that was contaminated with TCDD) and prostate cancer. And although the function of the prostate is not necessary for life or fertility, it is the most diseased organ in the human male. Approximately 50% of men exhibit some degree of benign prostatic hyperplasia (BPH) by age 50, whereas 80% of men over 80 years old exhibit BPH at autopsy (Napalkov et al., 1995). In addition, prostate cancer is the most common male cancer in the United States and the second most common male cancer in many other countries (Boyle et al., 1995). The regulatory mechanisms involved in normal prostate growth regulation are not well understood, making identification of lesions involved in pathological prostate growth very difficult. Using in utero and lactational TCDD exposure as a model system, gene products involved in prostatic cell proliferation and differentiation that act downstream of androgen receptor ligand binding or independently of the androgen receptor signaling pathway may be identified, furthering the understanding of normal prostate development and perhaps the pathological states of BPH and cancer. With respect to sperm numbers, further study of the mechanism of action of TCDD is warranted as there has been increasing concern that sperm counts in human males may have declined over the past 50 years, and it has been postulated that environmental contaminants have contributed to this decline (Carlsen et al., 1992; Sharpe, 1993; Sharpe and Skakkeback, 1993; Olsen et al., 1995).

Collection of Epidemiological Data

There is no doubt that the male rodent reproductive system is very sensitive to developmental TCDD exposure. However, the majority of epidemiological data regarding effects of TCDD on the human male reproductive system have been gathered from cohorts exposed as adults. For example, adult occupational TCDD exposure has been associated with slight decreases in serum testosterone concentrations (Egeland et al., 1994), and, as mentioned above, a limited association between exposure to Agent Orange and prostate cancer has been demonstrated (Institute of Medicine, 1996). However, there is a paucity of epidemiological data linking developmental AhR agonist exposure to male reproductive effects in humans. Epidemiological studies of boys born to women who had ingested PCB- and PCDF-contaminated rice oil during pregnancy (Yu-Cheng incident in Taiwan, 1979) have shown that sexual maturation was not delayed and testicular and scrotal development was not altered. However, the exposed boys had significantly shorter penises (Guo et al., 1993). In contrast, penis diameter but not penis length has been shown to be decreased in male rats developmentally exposed to TCDD (Bjerke and Peterson, 1994). Because the rice oil ingested by these pregnant women was contaminated with PCDFs and PCBs, only some of which are AhR agonists, and quaterphenyls, which are not AhR agonists, observed effects in the prenatally exposed boys may have been contributed to by non-AhR agonists. Regardless of this difficulty in interpretation, though, the Yu-Cheng offspring are the best-characterized cohort of humans prenatally exposed to AhR agonists and as such provide an invaluable resource for future studies, which could include a wider breadth of developmental male reproductive endpoints.

In summary, in utero and lactational exposure to very low levels of TCDD has profound effects on reproductive system development and sexual behavior in male rodents. However, although the average background body burden of TEQs in humans is quite close to the LOAEL for the most sensitive developmental male reproductive effects of TCDD exposure in rats, it is difficult to say with any certainty whether human males are currently at risk from low-level TCDD exposure during prenatal and early postnatal development. By determining developmental male reproductive specific TEFs for TCDD-like halogenated aromatic hydrocarbons and determining whether nonpersistent AhR agonists can cause developmental male reproductive toxicity, a more accurate assessment of human exposure to compounds that might produce TCDD-like developmental male reproductive toxicity could be made. And by elucidating the role of the AhR in TCDD-induced developmental male reproductive toxicity, defining the biochemical and molecular mechanisms underlying the most sensitive TCDD-induced developmental male reproductive endpoints, and collecting human epidemiological data on male reproductive endpoints similar to those assessed in laboratory studies, the risk that in utero and lactational AhR agonist exposure poses to human male reproductive system development and function could be more accurately assessed.

ACKNOWLEDGMENTS

Portions of this research were supported by NIH grant ES01332. This article is Contribution 285, Environmental Toxicology Center, University of Wisconsin, Madison, WI 53706. BLR was supported by NIH Training Grant T32 ES07015 awarded to Environmental Toxicology Center, University of Wisconsin, Madison, WI 53706.

REFERENCES

Abbott, B. D., and Birnbaum, L. S. (1989). *Toxicol. Appl. Pharmacol.*, *99*: 276.

Anzick, S. L., Kononen, J., Walker, R. L., Azorsa, D. O., Tanner, M. N., Guan, X.-Y., Sauter, G., Kallioniemi, O.-P., Trent, J. M., and Meltzer, P. S. (1977). *Science*, 277: 965.

Astroff, B., and Safe, S. (1990). *Biochem. Pharmacol.*, *39*: 485.

Astroff, B., Rowlands, C., Dickerson, R., and Safe, S. (1990). *Mol. Cell. Endocrinol.*, *72*: 247.

Birnbaum, L. S. (1986). *Drug Metab. Dispos.*, *14*: 34.

Bjerke, D. L., Brown, T. J., MacLusky, N. J., Hochberg, R. B., and Peterson, R. E. (1994a). *Toxicol. Appl. Pharmacol.*, *127*: 258.

Bjerke, D. L., Sommer, R. J., Moore, R. W., and Peterson, R. E. (1994b). *Toxicol. Appl. Pharmacol.*, *127*: 250.

Bjerke, D. L., and Peterson, R. E. (1994). *Toxicol. Appl. Pharmacol.*, *127*: 241.

Blaylock, B. L., Holladay, S. D., Comment, C. E., Heindel, J. J., and Luster, M. I. (1992). *Toxicol. Appl. Pharmacol.*, *112*: 207.

Boyle, P., Maisonneuve, P., and Napalkov, P. (1995). *Urology*, *46*: 47.

Brodie, A. E., Azarenko, V. A., and Hu, C. Y. (1996). *Toxicol. Lett.*, *84*: 55.

Brown, T. J., Hochberg, R. B., Zielinski, J. E., and MacLusky, N. J. (1988). *Endocrinology*, *123*: 1761.

Burbach, K. M., Poland, A., and Bradfield, C. A. (1992). *Proc. Natl. Acad. Sci. USA*, *89*: 8185.

Carlsen, E., Giwercman, A., Keiding, N., and Skakkebaek, N. E. (1992). *Br. Med. J.*, *305*: 609.

Chen, S.-W., Roman, B. L, Saroya, S. Z., Shinomiya, K., Moore, R. W., and Peterson, R. E. (1993). *Toxicologist*, *13*: 104.

Choi, E. J., Toscano, D. G., Ryan, J. A., Riedel, N., and Toscano, W. A., Jr. (1991). *J. Biol. Chem.*, *266*: 9591.

Chou, Y.-C., Iguchi, T., and Bern, H. A. (1992). *Reprod. Toxicol.*, *6*: 439.

Chowdhury, A. K., and Steinberger, E. (1975). *Biol. Reprod.*, *12*: 609.

Chung, L. W. K., and Ferland-Raymond, G. (1975). *Endocrinology.*, *97*: 145.

Clark, R. L., Anderson, C. A., Prahalada, S., Robertson, R. T., Lochry, E. A., Leonard, Y. M., Stevens, J. L., and Hoberman, A. M. (1993). *Toxicol. Appl. Pharmacol.*, *119*: 34.

Cooke, P. S., and Meisami, E. (1991). *Endocrinology*, *129*: 237.

Cooke, P. S., Hess, R. A., Porcelli, J., and Meisami, E. (1991a). *Endocrinology*, *129*: 244.

Cooke, P. S., Young, P., Hess, R. A., and Cunha, G. R. (1991b). *Endocrinology*, *128*: 2874.

Cooke, P. S., Zhao, Y.-D., and Hansen, L. G. (1996). *Toxicol. Appl. Pharmacol.*, *136*: 112.

Crews, S. T., Thomas, J. B., and Goodman, C. S. (1988). *Cell, 52*: 143.

Cunha, G. R., Donjacour, A. A., Cooke, P. S., Mee, S., Bigsby, R. M., Higgins, S. J., and Sugimura, Y. (1987). *Endocr. Rev.*, *8*: 338.

DeVito, M. J., Thomas, T., Martin, E., Umbreit, T. H., and Gallo, M. A. (1992). *Toxicol. Appl. Pharmacol.*, *113*: 284.

DeVito, M. J., Birnbaum, L. S., Farland, W. H., and Gasiewicz, T. A. (1995). *Environ. Health Perspect.*, *103*: 820.

Döhler, K. D., Srivastava, S. S., Shryne, J. E., Jarzab, B., Sipos, A., and Gorski, R. A. (1984). *Neuroendocrinology*, *38*: 297.

Döhler, K. D., Coquelin, A., Davis, F., Hines, M., Shryne, J. E., Sickmoller, P. M., Jarzab, B., and Gorski, R. A. (1986). *Neuroendocrinology*, *42*: 443.

Dolwick, K. M., Swanson, H. I., and Bradfield, C. A. (1993). *Biochemistry*, *90*: 8566.

Eddy, E. M., Washburn, T. F., Bunch, D. O., Goulding, E. H., Gladen, B. C., Lubahn, D. B., and Korach, K. S. (1996). *Endocrinology, 137*: 4796.

Egeland, G. M. Sweeney, M. H., Fingerhut, M. A., Wille, K. K., Schnorr, T. M., and Halperin, W. E. (1994). *Amer. J. Epidemiol.*, *139*: 272.

Ema, M., Morita, M., Ikawa, S., Tanaka, M., Matsuda, Y., Gotoh, O., Saijoh, Y., Fujii, H., Hamada, H., Kikuchi, Y., and Fujii-Kuriyama, Y. (1996). *Mol. Cell. Biol.*, *16*: 5865.

Ema, M., Taya, S., Yokotani, N., Sogawa, K., Matsuda, Y., and Fujii-Kuriyama, Y. (1997). *Proc. Natl. Acad,. Sci. USA, 94*: 4273.

Feldman, S. C., and Bloch, E. (1978). *Endocrinology, 102*: 999.

Fernandez-Salguero, P., Pineau, T., Hilbert, D. M., McPhail, T., Lee, S. S. T., Kimura, S., Nebert, D. W., Rudikoff, S., Ward, J. M., and Gonzalez, F. J. (1995). *Science, 268*: 722.

Fernandez-Salguero, P. M., Hilbert, D. M., Rudikoff, S., Ward, J. M., and Gonzalez, F. J. (1996). *Toxicol. Appl. Pharmacol., 140*: 173.

Fernandez-Salguero, P., Pineau, T., Hilbert, D. M., McPhail, T., Lee, S. S. T., Kimura, S., Nebert, D. W., Rudikoff, S., Ward, J. M., and Gonzalez, F. J. (1995). *Science. 268*: 722.

Forsythe, J. A., Jiang, B.-H., Iyer, N. V., Agani, F., Leung, S. W., Koos, R. D., and Semenza, G. L. (1996). *Mol. Cell. Biol., 16*: 4604.

Gallo, M. A., Hesse, E. J., MacDonald, G. J., and Umbreit, T. H. (1986). *Toxicol. Lett., 32*: 123.

Gellert, R. J., and Wilson, C. (1979). *Environ. Res., 18*: 437.

Giannone, J. V., Okey, A. B., and Harper, P. A. (1995). *Can. J. Physiol. Pharmacol, 73*: 7.

Gierthy, J. F., Lincoln, D. W., Kampcik, S. J., Dickerman, H. W., Bradlow, H. L., Niwz, T., and Swaneck, G. E. (1988). *Biochem. Biophys. Res. Comm., 157*: 515.

Gorski, R. A., Gordon, J. H., Shryne, J. E., and Southam, A. M. (1978). *Brain Res., 148*: 333.

Gorski, R. A., Harlan, R. E., Jacobson, C. D., Shryne, J. E., and Southam, A. M. (1980). *J. Comp. Neurol., 193*: 529.

Gradin, K., McGuire, J., Wenger, R. H., Kvietikova, I., Whitelaw, M. L., Toftgård, R., Tora, L., Gassmann, M., and Poellinger, L. (1996). *Mol. Cell. Biol., 16*: 5221.

Gray, L. E., Jr., Ostby, J. S., and Kelce, W. R. (1994). *Toxicol. Appl. Pharmacol., 129*: 46.

Gray, L. E., Jr., Kelce, W. R., Monosson, E., Ostby, J. S., and Birnbaum, L. S. (1995). *Toxicol. Appl. Pharmacol., 131*: 108.

Greco, T. L., Furlow, J. D., Duello, T. M., and Gorski, J. (1992). *Endocrinology, 130*: 421.

Greco, T. L., Duello, T. M., and Gorski, J. (1993). *Endocr. Rev., 14*: 59.

Greenlee, W., Osborne, R., Dold, K., Hudson, L., Young, M., and Toscano, W. (1987). *Rev. Biochem. Toxicol., 8*: 1.

Guo, Y. L., Lai, T. J., Ju, S. H., Chen, Y. C., and Hsu, C. C. (1993). *Dioxin '93, 14*: 235.

Gustafsson, J.-Å., and Ingelman-Sundberg, M. (1979). *Biochem. Pharmacol., 28*: 497.

Harris, M., Zacharewski, T., and Safe, S. (1990). *Cancer Res., 50*: 3579.

Hirose, K., Morita, M., Ema, M., Mimura, J., Hamada, H., Fujii, H., Saijo, Y., Gotoh, O., Sogawa, K., and Fujii-Kuriyama, Y. (1996). *Mol. Cell. Biol., 16*: 1706.

Hoar, R. M., and Monie, I. W. (1981). *Developmental Toxicology* (C. Kimmel and J. Buelke-Sam, eds.). Raven Press, New York, p. 13.

Hoffman, E. C., Reyes, H., Chu, F.-F., Sander, F., Conley, L. H., Brooks, B. A., and Hankinson, O. (1991). *Science, 252*: 954.

Hogenesch, J. B., Cahn, W. K., Jackiw, V. H., Brown, R. C., Gu, Y.-Z., Pray-Grant, M., Perdew, G. H., and Bradfield, C. A. (1997). *J. Biol. Chem., 272*: 8581.

Iguchi, T. (1992). *Int. Rev. Cytol., 139*: 1.

Iguchi, T., and Hirokawa, M. (1986). *Proc. Jpn. Acad. Ser. B, 62*: 157.

Iguchi, T., Uesugi, Y., Takasugi, N., and Petrow, V. (1991). *J. Endocrinol., 128*: 395.

Imperato-McGinley, J., Binienda, Z., Arthur, A., Mininberg, D., Vaughan, E. D., and Quimby, F. (1985). *Endocrinology, 116*: 807.

Imperato-McGinley, J., Sanchez, R. S., Spencer, J. R., Yee, B., and Vaughan, E. D. (1992). *Endocrinology, 131*: 1149.

Institute of Medicine (1996). *Veterans and Agent Orange: Update 1996*, National Academy Press, Washington, D.C., p. 217.

Ito, S., Murakami, S., Yamanouchi, K., and Arai, Y. (1986). *Proc. Jpn. Acad. Ser. B., 62*: 408.

Jackson, R. R., Bargiello, T. A., Yun, S. H., and Young, M. W., (1986). *Nature, 320*: 185.

Jean-Faucher, C., Berger, M., DeTurckheim, M., Veyssiere, G., and Jean, C. (1985). *J. Endocrinol., 104*: 113.

Jellinck, P. H., Forkert, P. G., Riddick, D. S., Okey, A. B., Michnovicz, J. J., and Bradlow, H. L. (1993). *Biochem. Pharmacol.*, *45*: 1129.

Johnson, L., Dickerson, R., Safe, S. H., Nyberg, C. L., Lewis, R. P., and Welsh, T. H. (1992). *Toxicology*, *76*: 103.

Johnson, L., Wilker, C. E., Safe, S. H., Scott, B., Dean, D. D., and White, P. H. (1994). *Toxicology*, *89*: 49.

Kamath, A. B., Xu, H., Nagarkatti, P. S, and Nagarkatti, M. (1997). *Toxicol. Appl. Pharmacol.*, *142*: 367.

Kelce, W. R., Stone, C. R., Laws, S. C., Gray, L. E., Kemppainen, J. A., and Wilson, E. M. (1995). *Nature*, *375*: 581.

Khera, K. S., and Ruddick, J. A. (1973). *Chlorodioxins: Origin and Fate* (E. H. Blair, ed.). American Chemical Society, Washington, DC, p. 70.

Kihlstrom, J. E., Lundberg, C., Orberg, J., Danielsson, P. O., and Sydhoff, J. (1975). *Environ. Physiol. Biochem.*, *5*: 54.

King, D. P., Zhao, Y., Sangoram, A. M., Wilsbacher, L. D., Tanaka, M., Antoch, M. P., Steeves, T. D. L., Vitaterna, M. H., Kornhauser, J. M., Lowrey, P. L., Turek, F. W., Takahashi, J. S. (1997). *Cell, 89*, 641.

Kleeman, J. M., Moore, R. W., and Peterson, R. E. (1990). *Toxicol. Appl. Pharmacol.*, *106*: 112.

Korenbrot, C. C., Huhtaniemi, I. T., and Weiner, R. I. (1977). *Biol. Reprod.*, *17*: 298.

Kuiper, G. G. J. M., Carlsson, B., Grandien, K., Enmark, E., Häggblad, J., Nilsson, S., and Gustafsson, J.-Å. (1997). *Endocrinology, 138*: 863.

Kuiper, G. G. J. M., Enmark, E., Pelto-Huiko, M., Nilsson, S., and Gustafsson, J.-Å. (1996). *Proc. Natl. Acad. Sci. USA, 93*: 5925.

Li, X., Weber, L. W., and Rozman, K. K. (1995). *Fundam. Appl. Toxicol.*, *27*: 70.

Liu, H., Wormke, M., Safe, S. H., and Bjeldanes, L. F. (1994). *J. Natl. Cancer Inst.*, *86*: 1758.

Lubahn, D. B., Moyer, J. S., Golding, T. S., Couse, J. F., Korach, K. S., and Smithies, O. (1993). *Proc. Natl. Acad. Sci. USA*, *90*: 11162.

Lundkvist, U. (1990). *Toxicology.*, *61*: 249.

Mably, T. A., Bjerke, D. L., Moore, R. W., Gendron-Fitzpatrick, A., and Peterson, R. E. (1992a). *Toxicol. Appl. Pharmacol.*, *114*: 118.

Mably, T. A., Moore, R. W., Goy, R. W., and Peterson, R. E. (1992b). *Toxicol. Appl. Pharmacol.*, *114*: 108.

Mably, T. A., Moore, R. W., and Peterson, R. E. (1992c). *Toxicol. Appl. Pharmacol.*, *114*: 97.

MacLusky, N. J., and Naftolin, F. (1981). *Science*, *211*: 1294.

Matuszczyk, J. V., and Larsson, K. (1995). *Hormones Behav.*, *29*: 191.

McEwen, B. S., Lieberburg, I., Chaptal, C., and Krey, L. C. (1977). *Hormones Behav.*, *9*: 249.

Mebus, C. A., Reddy, V. R., and Piper, W. N. (1987). *Biochem. Pharmacol.*, *36*: 727.

Moore, R. W., Potter, C. L., Theobald, H. M., Robinson, J. A., and Peterson, R. E. (1985). *Toxicol. Appl. Pharmacol.*, *79*: 99.

Moore, R. W., Jefcoate, C. R., and Peterson, R. E. (1991). *Toxicol. Appl. Pharmacol.*, *109*: 85.

Nambu, J. R., Lewis, J. O., Wharton, K. A., Jr., and Crews, S. T. (1991). *Cell, 67*: 1157.

Namkung, M. J., Porubek, D. J., Nelson, S. D., and Juchau, M. R. (1985). *J. Steroid Biochem.*, *22*: 563.

Napalkov, P., Maisonneuve, P., and Boyle, P. (1995). *Urology*, *46*: 41.

Neumann, F., von Berswordt-Wallrabe, R., Elger, W., Steinbeck, H., Hahn, J. D., and Kramer, M. (1970). *Recent Prog. Horm. Res.*, *26*: 337.

Okey, A. B., Riddick, D. S., and Harper, P. A. (1994). *Toxicol. Lett.*, *70*: 1.

Olsen, G. W., Bodner, K. M., Ramlow, L. R., Ross, C. E., and Lipschultz, L. I. (1995). *Fertil. Steril.*, *63*: 887.

Orlowski, J., Bird, C. E., and Clark, A. F. (1988). *J. Endocrinol.*, *116*: 81.

Peterson, R. E., Theobald, H. M., and Kimmel, G. L. (1993). *Crit. Rev. Toxicol.*, *23*: 283.

Peterson, R. E., Cooke, P. S., Kelce, W. R., and Gray, L. E., Jr. (1997). *Comprehensive Toxicology, Vol. 10: Reproductive and Endocrine Toxicology* (K. Boekelheide, R. Chapin, P. Hoyer, and C. Harris, eds.). Pergamon, New York, p. 181.

Phillips, M., Enan, E., Liu, P. C. C., and Matsumura, F. (1995). *J. Cell Sci.*, *108*: 395.

Poland, A., and Glover, E. (1980). *Mol. Pharmacol.*, *17*: 86.

Poland A., and Knutson, J. C. (1982). *Ann. Rev. Pharmacol. Toxicol.*, *22*: 517.

Pollenz, R. S. (1996). *Mol. Pharmacol.*, *49*: 391.

Pollenz, R. S., Santostefano, M. J., Klett, E., Richardson, V. M., Necela, B., and Birnbaum, L. S. (1998). *Toxicol. Sci.*, in press.

Probst, M. R., Fan, C.-M., Tessier-Lavigne, M., and Hankinson, O. (1997). *J. Biol. Chem.*, *272*: 4451.

Rajfer, J., and Coffey, D. S. (1979). *Invest. Urol.*, *17*: 3.

Rajfer, J., and Walsh, P. C. (1977). *J. Urol.*, *118*: 985.

Reyes, H., Reisz-Porszasz, S., and Hankinson, O. (1992). *Science*, *256*: 1193.

Riddick, D. S., Huang, Y., Harper, P. A., and Okey, A. B. (1994). *J. Biol. Chem.*, *269*: 12118.

Roman, B. L., and Peterson, R. E. (1998). *Toxicol. Appl. Pharmacol.*, *150*: in press.

Roman, B. L., Pollenz, R. S., and Peterson, R. E. (1998a). *Toxicol. Appl. Pharmacol.*, *150*: in press.

Roman, B. L., Timms, B. G., Prins, G. S., and Peterson, R. E. (1998b). *Toxicol. Appl. Pharmacol.*, *150*: in press.

Roman, B. L., Sommer, R. J., Shinomiya, K., and Peterson, R. E. (1995). *Toxicol. Appl. Pharmacol.*, *134*: 241.

Romkes, M., Piskorska-Pliszczynska, J., and Safe, S. (1987). *Toxicol. Appl. Pharmacol.*, *87*: 306.

Rönnbäck, C., and de Rooij, D. G. (1994). *Pharmacol. Toxicol.*, *74*: 287.

Rowlands, J. C., McEwan, I. J., and Gustafsson, J.-Å. (1996). *Mol. Pharmacol.*, *50*: 528.

Rune, G. M., de Souza, P., Krowkw, R., Merker, H. J., and Neubert, D. (1991). *Histol. Histopathol.*, *6*: 459.

Safe, S. (1992). *Environ. Health Perspect.*, *100*: 259.

Sager, D. B. (1983). *Environ. Res.*, *31*: 76.

Sager, D. B., Shih-Schroeder, W., and Girard, D. (1987). *Environ. Contam. Toxicol.*, *38*: 946.

Sager, D., Girard, D., and Nelson, D. (1991). *Environ. Toxicol. Chem.*, *10*: 737.

Schmidt, J. V., and Bradfield, C. A. (1996). *Ann. Rev. Cell Dev. Biol.*, *12*: 55.

Schmidt, J. V., Su, G. H.-T., Reddy, J. K., Simon, M. C., and Bradfield, C. A. (1996). *Proc. Natl. Acad. Sci. USA*, *93*: 6731.

Semenza, G. L., and Wang, G. L. (1992). *Mol. Cell. Biol.*, *12*: 5447.

Sharpe, R. M. (1993). *J. Endocrinol.*, *136*: 357.

Sharpe, R. M., and Skakkebaek, N. E. (1993). *Lancet*, *341*: 1392.

Shima, H., Tsuji, M., Young, P., and Cunha, G. R. (1990). *Endocrinology*, *127*: 3222.

Shiverick, K. T., and Muther, T. F. (1983). *Biochem. Pharmacol.*, *32*: 991.

Smits-van Prooije, A. E., Lammers, J. H. C. M., Waalkens-Berendsen, D. H., Kulig, B. M., and Snoeij, N. J. (1993). *Chemosphere*, *27*: 395.

Smits-van Prooije, A. E., and Waalkens-Berendsen, D. H. (1994). TNO-report: Preliminary report study 1526.

Sommer, R. J., Ippolito, D. L., and Peterson, R. E. (1996). *Toxicol. Appl. Pharmacol*, *140*: 146.

Spink, D. C., Lincoln, D. W., Dickerman, H. W., and Gierthy, J. F. (1990). *Proc. Natl. Acad. Sci. USA*, *87*: 6917.

Spink, D. C., Eugster, H.-P., Lincoln, D. W., Schuetz, J. D., Schuetz, E. G., Johnson, J. A., Kaminsky, L. S., and Gierthy, J. F. (1992). *Arch. Biochem. Biophys.*, *293*: 342.

Startin, J. R. (1994). *Dioxins and Health* (A. Schecter, Ed.), Plenum Press, New York, p. 115.

Sutter, T. R., Guzman, K., Dold, K. M., and Greenlee, W. F. (1991). *Science*, *254*: 415.

Taguchi, O. (1987). *Biol. Reprod.*, *37*: 113.

Theobald, H. M., and Peterson, R. E. (1997). *Toxicol. Appl. Pharmacol.*, *145*: 124.

Thornton, J. E., Irving, S., and Goy, R. W. (1991). *Phys. Behav.*, *50*: 471.

Tian, H., McKnight, S. L., and Russell, D. W. (1997). *Genes Dev.*, *11*: 72.

van Haaster, L. H., de Jong, F. H., Docter, R., and de Rooij, D. G. (1992). *Endocrinology*, *131*: 1574.

Wang, G. L., Jiang, B.-H., Rue, E. A., and Semenza, G. L. (1995). *Proc. Natl. Acad. Sci. USA*, *92*: 5510.

Warren, D. W., Haltmeyer, G. C., and Eik-Nes, K. B. (1975). *Endocrinology*, *96*: 1226.

Warren, D. W., Huhtaniemi, I. T., Tapanainen, J., Dufau, M. L., and Catt, K. J. (1984). *Endocrinology*, *114*: 470.

White, T. E. K., Rucci, G., Liu, Z., and Gasiewicz, T. A. (1995). *Toxicol. Appl. Pharmacol.*, *133*: 313.

Whitelaw, M. L., Göttlicher, M., Gustafsson, J.-Å, and Poellinger, L. (1993). *EMBO J.*, *12*: 4169.

Wilk, R., Weizman, I., and Shilo, B.-Z. (1996). *Genes Dev.*, *10*: 93.

Wilker, C., Johnson, L., and Safe, S. (1996). *Toxicol. Appl. Pharmacol.*, *141*: 68

Wilson, J. D., and Lasnitzki, I. (1971). *Endocrinology*, *89*: 659.

Yao, T.-P., Ku, G., Zhou, N., Scully, R., and Livingston, D. M. (1996). *Proc. Natl. Acad. Sci. USA*, *93*:10626.

Zacharewski, T., Harris, M., and Safe, S. (1991). *Biochem. Pharmacol.*, *41*: 1931.

Zhou, Y.-D., Barnard, M., Tian, H., Ring, H. Z., Francke, U., Shelton, J., Richardson, J., Russell, D. W., and McKnight, S. L. (1997). *Proc. Natl. Acad. Sci. USA*, *94*: 713.

Zook, D. R., and Rappe, C. (1994). *Dioxins and Health* (A. Schecter, Ed.), Plenum Press, New York. p. 79.

25

Toxicity of Spermatogenesis and Its Detection

Richard M. Sharpe
MRC Reproductive Biology Unit, Edinburgh, Scotland

INTRODUCTION

Any scientist involved in the study of spermatogenesis is struck both by the beautiful complexity of this highly organized process and by its seeming fragility. In contrast, viewed from the perspective of fertility or sperm counts, spermatogenesis appears anything but fragile—robust even— and this would probably be the view of many pathologists in the pharmaceutical industry. The truth, as usual, is probably somewhere in between these two extreme perceptions, but the contrast has an important message to give us, namely, that the perceived susceptibility of spermatogenesis to disruption depends really on how it is being assessed. In humans, there are only isolated instances in which industrial chemicals have been shown to exert adverse effects on spermatogenesis, notably the case of the nematocide dibromochloropropane (DBCP), which causes catastrophic impairment of spermatogenesis (Whorton et al., 1979). Back in 1992, I wrote an article (Sharpe, 1992) that argued that people should not be misled into complacency by the lack of evidence that other industrial and environmental chemicals apparently had little impact on human male fertility or sperm production. My reasoning was that this absence of data was more a reflection of the absence of sensitive endpoints than anything else. The endpoints most commonly used for assessment of adverse reproductive effects of industrial chemical in humans (i.e., sperm counts, fertility) were only capable of detecting catastrophes; anything less would go undetected (Sharpe, 1992, 1993a). Almost on cue, the saga of "falling sperm counts" in the human male broke to emphasize my message (Carlsen et al., 1992). It is very much the same message that I hope to put over in this short chapter, though it is now modified by the "wisdom" of our new (and probably imperfect) perceptions about "new" pathways via which environmental chemicals may affect male reproductive function.

VULNERABILITY OF SPERMATOGENESIS

The process of spermatogenesis involves intense germ cell proliferation followed by meiosis and then a remarkable differentiation process whereby the "typical" round haploid spermatid is transferred into a spermatozoon, the smallest cell in the body (Sharpe, 1994). Once puberty is reached, spermatogenesis grinds on incessantly in a nonseasonal breeder such as the human. Indeed, a normal adult male would be expected to produce around 150–200 $\times$ 10^6 *new* spermatozoa per day, each of which will have taken 10 weeks to "manufacture." Each spermatozoon has to be more or less perfect morphologically, functionally, and genetically if it is to fulfill the task for which it was made. It has to survive ejaculation and its swim up the female reproductive tract, and at the end of this arduous trip (equivalent to a 35- to 40-mile trip for an adult man) it must successfully recognize, bind to, and then fertilize an oocyte if one is present. Anyone who sits back and thinks of the perils of this journey can only marvel at how well it works against seemingly impossible odds and obstacles. That it does succeed is all down to numbers—the production and ejaculation of hundreds of millions of spermatozoa. The very fact that each mammalian species, including humans, produces such vast numbers of spermatozoa (Sharpe, 1994) is shouting to us that fertilization is as vulnerable as it is vital, and that the vulnerability or weak link lies on the side of the male gamete. Therefore, to ensure fertilization there are two absolute requirements from the male side. First, spermatogenesis must generate huge numbers of spermatozoa. Second, a high proportion of these must be functionally and genetically competent. If this reasoning is accepted, then it follows that any factor that compromises *either* the number of sperm produced or their functional/genetic competence must be viewed as a potential threat to fertility.

PATHWAYS OF DISRUPTION OF SPERMATOGENESIS

In general all adverse effects on spermatogenesis occur via one of three routes. The first route would be as the result of some general toxicity, which does not primarily involve the hypothalamus, pituitary, or reproductive system but which leads, secondarily, to adverse effects on the process of spermatogenesis. For example, toxic effects on the liver that affect steroid metabolism or vitamin A metabolism would be likely to affect normal testicular function (Moore and Bullock, 1988; Kim and Wang, 1993). Similarly, any toxic effect that impaired the delivery of oxygen to the testis (e.g., effects on lung function, erythropoiesis, or erythrocyte function) or that raised core body temperature would very likely compromise normal spermatogenesis, as the testis is always poised on the brink of hypoxia (Setchell et al., 1994) and the process of sperm production is extremely heat sensitive (Mieusset and Bujan, 1995).

Second, spermatogenesis can be disrupted by any factor that alters hypothalamic-pituitary function such that FSH and/or LH secretion are altered. Grossly elevated levels of FSH, and especially LH, appear to be as detrimental as subnormal levels (Maddock and Nelson, 1952; Kerr and Sharpe, 1989; Sharpe, 1994). Many chemicals with steroidal or dopaminergic properties (either agonistic or antagonistic) can induce such changes (Sever and Hessol, 1985; Mattison et al., 1990; Morris et al., 1996).

Third, spermatogenesis can be disrupted by factors that directly affect functions of cells in the testis on which spermatogenesis depends. This could include factors that affect function of the vasculature of the testis (e.g., heavy metals) (Roe, 1964; Setchell et al., 1994), the production of testosterone by the Leydig cells (e.g., P$_{450}$ inhibitors, ethane dimethane sulfonate) (Lamb and Foster, 1988; Morris et al., 1996), the vital functions of the Sertoli cells (e.g., nitroaromatics, hexanedione) (Foster et al., 1986; Johnson et al., 1991; Richburg et al., 1994), or direct effects on germ cells (radiation, antimitotic drugs) (Meistrich, 1993, 1996; Hoyes and Morris, 1996). There are numerous examples of compounds that will affect one or more of these cell types (Sever and Hessol, 1985; Mattison et al., 1990; Morris et al., 1996), although discerning between primary and secondary effects of a compound can sometimes be quite difficult, especially when it comes to distinguishing whether the primary site of action is the Sertoli cell or the germ cells. For example, a compound that results primarily in reduced LH secretion will lead, secondarily, to reduced testosterone production by the Leydig cells, which, in turn, will lead to disruption of spermatogenesis (i.e., germ cell degeneration) because of inadequate effects of testosterone on the Sertoli cells (Sharpe, 1994). Only by measuring hormone levels at various times after exposure to the compound in question and relating this in detail to morphological changes in the testis will the sequence of cause and effect be established. In this instance, the primary site of action will probably be easy to detect because of the clues offered by altered hormone levels and the morphological "fingerprint" that this results in in the testis (Kerr et al. 1993). However, with many chemicals there are no such obvious clues. This brings me to the principal topic of this chapter, namely, how to detect adverse testicular effects of exogenous factors.

DETECTION OF ADVERSE CHANGES IN SPERMATOGENESIS

There are essentially two approaches that have been used to indicate whether or not a test chemical is able to exert an adverse effect on spermatogenesis: (1) evaluation of testicular morphology (= pathology) and (2) functional evaluation of spermatogenesis, i.e., assessment of fertility in mating trials. For regulatory purposes, detailed analysis of both of these endpoints is usually required (Lamb, 1988). Both of these approaches have particular, and different, strengths and weaknesses. For example, fertility assessment by mating trials will reveal whether or not *functionally* competent sperm are being produced in numbers sufficient to maintain normal fertility. In contrast, morphological evaluation of the testis will not detect a functional abnormality in developing germ cells or spermatozoa unless this also has a very pronounced morphological manifestation. However, fertility assessment provides little if any indication as to the nature of any defect in the sperm or as to when and how this defect has arisen. In contrast, good morphological evaluation of the testis provides quite detailed information on these issues (Russell et al., 1990), especially if assessment is made at various times after initiation of treatment. It is partly for these reasons that testicular morphology/pathology remains one of the frontline approaches in evaluating the potential safety of new chemicals/drugs (Lamb, 1988).

Abnormalities in Spermatogenesis/Testicular Morphology

As has already been indicated, evaluation of testicular histopathology does have limitations, which are often not well appreciated (Russell et al., 1990). In addition, recent developments concerning fetal/neonatal exposure to hormonally active chemicals has highlighted new and potentially important limitations to straightforward assessment of testicular morphology. These limitations are best illustrated by some examples (Fig. 1).

No pathologist has any problem recognizing when a chemical has caused severe impairment of spermatogenesis, manifest either by the loss of most of the germ cells (Fig. 1b) or by the presence of multiple degenerating germ cells and/or extensive vacuolation of the seminiferous epithelium (Fig. 1c). Such effects can be induced by a range of chemicals or nonchemical agents (e.g., elevated temperature, impaired fluid resorption, or obstruction of the excurrent ducts) (Ilio and Hess, 1994; Setchell et al., 1994; Mieusset and Bujan, 1995).

Stage-Dependent Germ Cell Degeneration

The situation is radically different when the lesions induced by chemicals or other factors are either more minor or are not obvious. For example, mild androgen insufficiency will probably result only in a small number of degenerating germ cells (Fig. 1d), which could easily be missed or misinterpreted as normal. Germ cell degeneration does occur in the normal testis, but it affects primarily spermatogonia and secondary spermatocytes at particular stages of the spermatogenic cycle (Kerr, 1992; Billig et al., 1995). In contrast, androgen withdrawal or insufficiency results in degeneration of a small percentage of preleptotene and pachytene spermatocytes and round and elongating spermatids and in a highly stage-specific manner (see Kerr et al., 1993; Sharpe, 1994). It is thought that the detection of such degenerative cells is the most sensitive indicator of androgen insufficiency and would be evident long before any significant change in blood levels of testosterone was demonstrable.

Effects on Sperm Release

Another induced abnormality of spermatogenesis that is extremely common but is often overlooked is the impairment of normal sperm release (Fig. 1e). The latter normally occurs late in stage VIII of the spermatogenic cycle, but exposure to a wide range of chemicals or other agents (e.g., heat) can interfere with this process, via unknown mechanisms, resulting in the abnormal retention of the sperm during stages IX–XIV or beyond (Russell et al., 1990; Russell, 1991); the retained sperm are eventually phagocytosed by the Sertoli cells. Although treatment-induced failure of sperm release is often accompanied by other manifestations, such as vacuolation of the basal epithelium and/or degenerating germ cells (Fig. 1c), in exceptional circumstances it can occur on its own (Fig. 1e; see Sharpe et al., 1995a) or as a prelude to more widespread and more obvious changes (Russell, 1991). In the latter situations, it is extremely easy to overlook as the testis will probably be of normal weight and testicular morphology will appear grossly normal; it is only on close inspection that the abnormal retention of sperm is detected (Fig. 1e). Retained sperm can give completely misleading reassurance if the normality of sperm production is being assessed by measuring daily sperm production (DSP). This entails enumeration of homogenization-resistant elongate spermatids in samples of testicular tissue (Blazak et al., 1993), but clearly this method will not distinguish between normal and retained sperm (Sharpe, 1994). Just because sperm are being made in normal numbers in the testis does not mean that there are normal sperm numbers in the ejaculate!

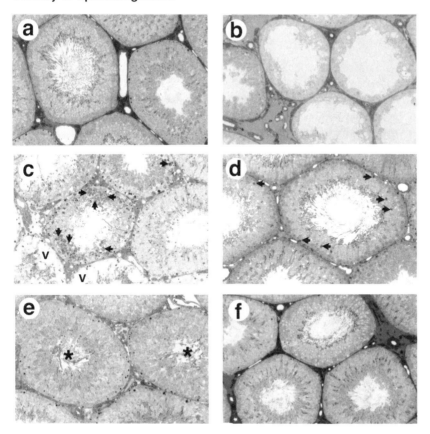

Figure 1 Treatment-induced histopathological changes in the testis of the adult rat, illustrating the range in presentation, the ease or difficulty of detecting such changes, and the relationship to testicular weight. (a) A control rat. (b) A rat treated neonatally with diethylstilbesterol (DES), which exhibits mainly Sertoli cell–only seminiferous tubules with only occasional tubules (left side of panel) containing germ cells. (c) A rat in which the scrotum had been heated to 43°C for 30 minutes (equivalent of a hot bath) one day previously. Note the extensive vacuolation (V) of the basal epithelium of some seminiferous tubules, the presence of multiple clusters of degenerating germ cells (arrows), and some retained spermatozoa (tubule in center). Note also that some seminiferous tubules (right of panel) appear morphologically near normal. (d) A rat treated 5 days previously with ethane dimethane sulfonate to destroy the Leydig cells and thus induce androgen withdrawal. Note that the seminiferous tubules appear completely normal except for the presence of a small number of degenerating pachytene spermatocytes/spermatids (arrows), which are restricted to seminiferous tubules at stages VII and VIII of the spermatogenic cycle. (e) Consequences of the failure of sperm release, shown here in a transgenic rat (Sharpe et al., 1995a). Note that the seminiferous tubules appear grossly normal except that many mature (step 19) elongate spermatids are retained (*stages X and XI) at stages of the spermatogenic cycle beyond stage VIII when they are normally released. (f) A rat treated neonatally with DES in which spermatogenesis is completely normal, although the diameter of the seminiferous tubules is probably reduced marginally. Testicular weight is often a quick and handy endpoint for assessing the degree of disruption of spermatogenesis caused by a particular treatment, but its limitations are illustrated by considering the (fixed) weights of the testes shown above: a, 1793 mg; b, 643; c, 1562; d, 1787; e, 1765; f, 1412.

Absence of Testicular Pathology: Windows of Sertoli Cell and Germ Cell Susceptibility

Finally, perhaps the biggest potential problem arises when the testis is morphologically completely normal after treatment (Fig. 1f), yet it is clear that an adverse effect has occurred because the testis is reduced significantly in size. Certainly in the past, such instances may have been passed off as "normal," especially if the decrease in testis weight was relatively small (10–20%). Such cases must now be viewed differently because of the realization that they probably reflect a specific adverse effect of treatment on the multiplication of Sertoli cells, which in the rat commences at around days 15–16 p.c. of gestation and continues up to days 15–18 of postnatal life (Sharpe, 1994; Sharpe et al., 1995b).

The reason why reduced Sertoli cell number leads to a reduction in testicular weight is as follows (Fig. 2). Each Sertoli cell supports a fixed number of germ cells through their development from spermatogonia into spermatozoa and, as the majority of the volume of the adult testis is made up of germ cells, then the greater the number of Sertoli cells, the greater will be the number of germ cells and the larger will be the testis (and vice versa) with corresponding changes in daily sperm production and probably in sperm counts. This relationship (Fig. 2) has been clearly demonstrated in a variety of natural and experimental situations (see Sharpe, 1993b; 1994a) involving both decreases and increases in Sertoli cell numbers (and testicular size) and can result from transient alterations in the blood levels of follicle-stimulating hormone, thyroid hormones, and probably other factors, such as estrogens and opiate peptides. Indeed, there is general agreement that most of the differences in testicular size and daily sperm production between different species, between strains of the same species, and between individuals of the same strain are due primarily to differences in Sertoli cell number (Sharpe, 1994). Perhaps the most remarkable aspect of this relationship is that it has been well established and well documented for a decade or more, but its implications

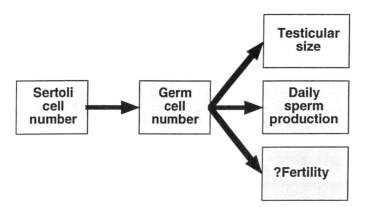

Figure 2 Diagrammatic representation of how Sertoli cell number, which is determined in neonatal life, subsequently determines total germ cell number and mass in adulthood and thus determines final testicular size and daily sperm production (DSP). The exact relationship between changes in DSP and fertility is unclear, although it would be expected that any reduction in DSP would lead to a reduction in average sperm counts in the ejaculate and thus cause a qualitative change in fertility, e.g., increased time to pregnancy in humans.

with respect to reproductive toxicology have not been appreciated until very recently. These observations are equally relevant to humans, for whom the most important period of Sertoli cell multiplication is probably also during fetal and early postnatal life (Cortes et al., 1987; Sharpe, 1994) and in whom the number of Sertoli cells and daily sperm production are closely linked in adulthood (Johnson et al. 1984).

The most important development to come from this "newfound" wisdom is the wider realization that it is not just Sertoli cell multiplication that has a limited and specific window of susceptibility, but that many other aspects of Sertoli cell and germ cell development and differentiation have similar windows (Fig. 3). The implications of this are considerable. Exposure to a chemical that perturbs some aspect of Sertoli cell or germ cell development during a very narrow, but highly specific, window of time can potentially induce an irreversible change, which may have long-term adverse consequences on spermatogenesis and fertility, though these may not be manifest at the time of the causative effect but only later in life, e.g., postpubertally in the case of spermatogenesis. Of course, such effects should still be detectable by *thorough* toxicological evaluation of a chemical, but this would have to include studies such as multigeneration tests to ensure not only that all potential windows of susceptibility were covered but also that the animals were then evaluated in adulthood. This has not always been the case up until now.

Whether a smaller, but morphologically normal, testis and the associated decrease in daily sperm production results in any change in fertility remains to be shown, as there are rather conflicting opinions as to whether decreased sperm counts equate to decreased fertility in rats. However, common sense tells us that a decrease in sperm production leading to a decrease in sperm counts in the ejaculate cannot improve fertility—it can only impair it, even if this is only a qualitative change such as reduced litter size in rats or increased time to achieve pregnancy in humans. When extrapolating from the

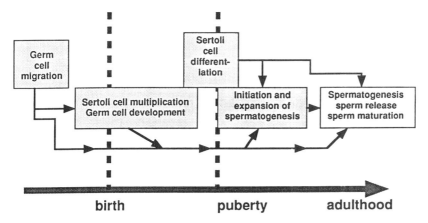

Figure 3 Diagrammatic illustration of developmental windows of susceptibility of spermatogenesis to disruption. Each of the grey-shaded boxes shows the approximate timing of a particular stage of development of either the germ cells and/or the Sertoli cells. Quantitatively and qualitatively normal spermatogenesis in adulthood (white box) depends on the successful completion of each of the earlier stages of development. There are well-documented examples in humans and animals of the consequences of the failure of each one of these steps, all of which lead to complete or partial loss of fertility.

rat to the human, there are other factors to be taken into consideration, the most important of which is the inherent difference in the efficiency of sperm production. In humans, sperm production is woefully inefficient when compared with the rat and other mammals (see Sharpe, 1994), and the quality of the sperm produced is also poor (only 30–65% of sperm are morphologically normal!) compared to other animals (>95% morphologically normal). There seems little doubt that this difference in efficiency of spermatogenesis largely explains the poor fecundity of humans compared with our fellow mammals. Thus, any further potential erosion of our sperm-producing capacity would have to be viewed with concern. A 10% decrease in sperm production in a laboratory rat or hamster would, theoretically, have little if any effect on fertility, whereas such a decrease in humans might be of significance.

LESSONS FOR THE FUTURE

The explosion of the world human population may have contributed to the view that little can go wrong with human fertility. This represents a handy "leaning post" for those who argue strongly that manmade chemicals have had negligible impact on any aspect of human reproduction over the past half-century. In global terms, this perception is probably difficult to argue against, but as any clinician will tell you, it is individual patients who matter, and the prevailing rates of subfertility are close to 10% of couples; for them there is probably no greater concern in their lives. In approximately half of these cases there is probably a male factor involved, though this may be an underestimate because our understanding and diagnosis of male factors is very inadequate. What contribution have manmade chemicals made to male infertility? The answer is that we do not know or, to use a more telling phrase, we are ignorant. This ignorance has been highlighted in a very public way over the past 2–3 years as claims and counterclaims have been made that human sperm counts are, or are not, falling (see Irvine, 1996). The truth on this matter still eludes us and probably will for some time to come. But something is happening out there to human male reproductive health, to judge by the number of pieces of corroborative (but indirect) evidence (Sharpe and Skakkebaek, 1993; Toppari et al., 1996), and we would be foolish to dismiss these as coincidences.

Our newfound wisdom tells us that it is entirely possible that exposure of the human male, either as a fetus or as an infant, to exogenous agents such as some chemicals has the potential to induce adverse changes that do not become manifest until adulthood when sperm are produced and fertility is eventually put to the test. If we are wise, we will invest in improving our understanding of these early programming events and of their susceptibility (or otherwise) to exogenous factors. We cannot fail to gain from such investment, for even if we are able to prove, in time, that this is an imagined rather than a real problem, this knowledge will be invaluable in all kinds of decision making. If the converse is true, then it is imperative that we know sooner rather than later because of the time bomb that might be ticking for some of the next generation.

At the same time, we should not forget the adult male. Our preoccupation with environmental chemicals may blind us to other, more straightforward factors that might impair sperm production in the adult male. A recent small study (Tiemessen et al., 1996), with a cross-over design showed nearly a trebling in motile sperm counts in men who wore loose underpants (boxer shorts) for 6 months compared with when they wore modern, tighter-fitting underpants for a similar period. This study needs to be repeated with some urgency, but if its findings are correct, then such effects, which are presumably attributable to increase in testicular temperature (Mieusset and Bujan, 1995), might dwarf any effects of chemicals.

Whether wearing tighter underwear affects male fertility remains unknown, although if there is a change it is probably qualitative. Indeed, we should also not lose sight of the fact that even in situations that are suboptimal, spermatogenesis does keep working in most men to the point that it enables them to father children. This is probably a reflection of the time and trouble that nature has invested in devising a successful and robust spermatogenic process. But this should not make us complacent. It seems certain that environmental pollution of our planet will get progressively worse with the burgeoning world population, and understanding of when and how this might affect sperm production will therefore become increasingly more important.

REFERENCES

Billig, H., Furuta, I., Rivier, C., Tapanainen, J., Parvinen, M., and Hsueh, A. J. W. (1995). *Endocrinology*, *136*:5.

Blazak, W. F., Treinen, K. A., and Juniewicz, P. E. (1993). *Male Reproductive Toxicology* (R. E. Chapin and J. J. Heindel, eds.). Academic Press, New York, p. 86.

Carlsen, E., Giwercman, A., Keiding, N., and Skakkebaek, N. E. (1992). *Br. Med. J.*, *305*: 609.

Cortes, D., Müller, J., and Skakkebaek, N. E. (1987). *Int. J. Androl.*, *10*: 589.

Hoyes, K. P., and Morris, I. D. (1996). *Int. J. Androl.*, *19*: 199.

Ilio, K. Y., and Hess R. A. (1994). *Micros. Res. Tech.*, *29*: 432.

Irvine, D. S. (1997). *Hum. Reprod.*, (in press).

Johnson, K. J., Hall, E. S., and Boekelheide, K. (1991). *Toxicol. Appl. Pharmacol.*, *111*: 432.

Johnson, L., Zane, R. S., Petty, C. S., and Neaves, W. B. (1984). *Biol. Reprod.*, *31*: 785.

Kerr, J. B. (1992). *J. Reprod. Fertil.*, *95*: 825.

Kerr, J. B., and Sharpe, R. M. (1989). *Cell Tiss. Res.*, *257*: 163.

Kerr, J. B., Millar, M., Maddocks, S., and Sharpe, R. M. (1993). *Anat. Rec.*, *235*: 547.

Kim, K. H., and Wang, Z. (1993). *The Sertoli Cell* (L. D. Russell and M. D. Griswold, eds.). Cache River Press, FL, p. 517.

Lamb, J. C., IV. (1988). *Physiology and Toxicology of Male Reproduction* (J. C. Lamb IV and P. M. D. Foster, eds.). Academic Press, New York, p. 137.

Lamb, J. C., IV, and Foster, P. M. D., eds. (1988). *Physiology and Toxicology of Male Reproduction*. Academic Press, New York.

Maddock, W. O., and Nelson, W. O. (1952). *J. Clin. Endocrinol. Metab.*, *12*: 985.

Mattison, D. R., Plowchalk, D. R., Meadows, M. J., Al-Juburi, A. Z., Gandy, J., and Malek, A. (1990). *Environ. Med.*, *74*: 391.

Meistrich, M. L. (1993). *Eur. J. Urol.*, *23*: 136.

Meistrich, M. L. (1996). *Male Gametes: Production and Quality* (S. Hamamah and R. Mieusset, eds.). Les Editions INSERM, Paris, p. 83.

Mieusset, R., and Bujan, L. (1995). *Int. J. Androl*, *18*: 169.

Moore, J. W., and Bulbrook, R. D. (1988). *Oxford Reviews of Reproductive Biology* (J. R. Clarke, ed.). Oxford University Press, Oxford, p. 180.

Morris, I. D., Hoyes, K. P., Taylor, M. F., and Woolveridge, I. (1996). *Male Gametes: Production and Quality* (S. Hamamah and R. Mieusset, eds.). Les Editions INSERM, Paris, p. 135.

Richburg, J. H., Redenbach, D. M., and Boekelheide, K. (1994). *Toxicol. Appl. Pharmacol.*, *128*: 302.

Roe, F. J. C. (1964). *Br. J. Cancer, 18*: 674.

Russell, L. D. (1991). *Int. J. Androl., 14*: 307.

Russell, L. D., Ettlin, R. A., Sinha-Hikim, A. P., and Clegg, E. D., eds. (1990). *Histological and Histopathological Evaluation of the Testis*. Cache River Press, FL.

Setchell, B. P., Maddocks, S., and Brooks, D. E. (1994). *The Physiology of Reproduction,* 2nd ed. (E. Knobil and J. D. Neill, eds.). Raven Press, New York, p. 1063.

Sever, L. E., and Hessol, N. A. (1985). In: Endocrine Toxicology (Eds. J. A. Thomas, K. S. Karach and J. A. McLachlan) Raven Press, New York, pp. 211–248.

Sharpe, R. M. (1992). *Chem. Ind.* (Feb 3): 88.

Sharpe, R. M. (1993a). *Reproductive Toxicology* (M. Richardson, ed.). VCH, Weinheim. p. 129.

Sharpe, R. M. (1993b). *J. Endocrinol., 136*: 357.

Sharpe, R. M. (1994). *The Physiology of Reproduction*, 2nd ed. (E. Knobil and J. D. Neill, eds.). Raven Press, New York, p. 1363.

Sharpe, R. M., and Skakkebaek, N. E. (1993). *Lancet, 341*: 1392.

Sharpe, R. M., Maguire, S. M., Saunders, P. T. K., Millar, M., Russell, L. D., Ganten, D., Bachmann, S., Mullins, L., and Mullins, J. J. (1995a). *Biol. Reprod., 53*: 214.

Sharpe, R. M., Fisher, J. S., Millar, M., Jobling, S., and Sumpter, J. P. (1995b). *Environ. Health Perspect., 103*: 1136.

Tiemessen, C. H. J., Evers, J. L. H., and Bots, R. S. G. M. (1996). *Lancet, 347*: 1844.

Toppari, J., Larsen, J. C., Christiansen, P., Giwercman, A., Grandjean, P., Guillette, L. J. Jr., Jégou, B., Jensen, T. K., Jouannet, P., Keiding, N., Leffers, H., McLachlan, J. A., Meyer, O., Müller, J., Rajpert-De Meyts, E., Scheike, T., Sharpe, R. M., Sumpter, J. S., and Skakkebaek, N. E. (1996). *Environ. Health Perspect., 104* (Suppl 4): 741.

Whorton, M. D., Milby, T. H., Krauss, R. M., and Stubbs, H. A. (1979). *J. Occup. Med., 21*: 161.

26

Gamete Toxicology: The Impact of New Technologies

Sally D. Perreault
*U. S. Environmental Protection Agency, Research Triangle Park,
North Carolina*

INTRODUCTION

Gamete toxicology, a subspecialty within the broad field of reproductive toxicology, is the study of how drugs and/or xenobiotics or their metabolites may alter the development and function of eggs and sperm. Many aspects of gamete biology can be evaluated after drug or toxicant exposure, including gamete production and release (numbers), structure (morphology), or function (ability to undergo normal transport, participate in fertilization, and support normal embryonic development). However, until recently, methods for evaluating the effects of toxicants on gametes were typically limited to routine histological or cytological observations during which sperm or eggs were counted and perhaps stained to evaluate external features. The past decade or two has produced numerous technological advances in cytology, molecular biology, and computer-enhanced measurements that are making it possible to probe gamete function to a far greater depth and specificity. These provide exciting opportunities for detecting toxicant-induced changes in gametes and for elucidating the cellular-molecular mechanisms by which toxicants may act.

This chapter will review recent progress in the use of several new technologies with respect to gamete physiology and toxicology and point out avenues for potential future applications. Therefore, the scope of this chapter is rather broad, and the intent is to provide the reader with departure points for promising research directions rather than a comprehensive review of each topic. It is recognized that many of these technologies are at an early stage of development and that some may prove more useful than others in the long term. Obviously, many of these are impractical to apply in routine reproductive toxicology test protocols (as reviewed in Zenick et al., 1994) and are too specific for use in screening compounds for reproductive toxicity. Rather, many of the more complex new technologies are more suited to mechanistic studies after routine tests

Disclaimer: This document has been reviewed in accordance with the U.S. Environmental Protection Agency policy and approved for publication. Mention of trade names or commercial products does not constitute endorsement or recommendation for use.

635

of reproductive performance demonstrate an adverse effect. Emphasis will be placed on rat gametes as the rat is the preferred test species for reproductive toxicology testing.

Spermatozoa have come under scrutiny since the invention of the microscope, and the assessment of human semen quality has been routine for clinical evaluation of fertility for many years (Keel, 1990). Likewise, human semen is readily available to serve as a biomarker of male reproductive effects for use in occupational, pharmacological, or epidemiological studies (Wyrobek et al., 1997). The evaluation of rodent sperm numbers and quality is a more recent practice in the field of reproductive toxicology. Nevertheless, a growing number of male reproductive toxicology studies and test protocols are including various sperm measures in addition to fertility assessments (Zenick et al., 1994), and norms for these measures are being established. In contrast, evaluation of oocytes for clinical assessment of fertility or as an outcome in an epidemiology study is simply not feasible, and most of our information about oocytes comes from animal studies (primarily in rodents). Nevertheless, the study of human oocytes has been markedly advanced as a consequence of the revolution in assisted reproductive technologies where at least those oocytes that failed to undergo fertilization are sometimes available for study. Rarely have morphological indicators of oocyte damage been used in reproductive toxicology, due largely to a lack of meaningful measures of oocyte "health." However, new probes specific for subcellular organelles are now making it possible to evaluate oocyte function in a far more meaningful fashion, and have great potential for use in evaluating the effects of toxicants on oocytes and zygotes.

COMPUTER VISION TECHNOLOGIES APPLIED TO SPERMATOZOA

Computer-Assisted Sperm Analysis for Evaluating Sperm Motion in Toxicology Studies

Background

Advances in photography and video micrography have recently made it possible to preserve images of sperm tracks and begin to evaluate their characteristics. The pioneering work of Makler (1978), who applied multiple exposure photography to objectify the analysis of sperm motility, and Katz and Overstreet (1981), who described a simple system for measuring sperm velocity from videotapes, set the stage for more sophisticated, computer-assisted modes of automating and perfecting this process. Once computer boards were integrated with video technology such that a series of video frames could be stored, it became possible to develop integrated systems that allowed tracking of the sperm from video frame to video frame, as well as storage of multiple tracks from a single field of view.

Today, computer-assisted sperm analysis (CASA) is a widely used and appreciated technology for measuring not only sperm velocity but numerous other features of the sperm track such as direction, progressiveness and pattern of motion. Figure 1 illustrates and defines the commonly reported CASA sperm track outcomes. Principles and limitations of CASA technology were reviewed recently by Boyers et al. (1989) and Mortimer (1990). The technology continues to advance with the availability of larger and faster computers so that more images can be processed in less time. Most recently, digital image storage has become an alternative for videotapes, and longer tracking capabilities are making it possible to monitor the behavior of individual sperm over longer periods of time (Moore and Akhondi, 1996).

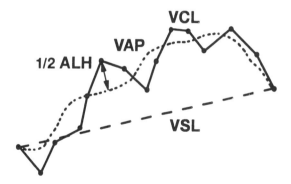

Figure 1 Diagram of a typical sperm track with definitions of computer-assisted sperm analysis (CASA) endpoints. Derived endpoints are defined as for the HTM-IVOS CASA instrument (Hamilton Thorne Research, Beverly, MA). VSL = Straight line velocity (distance from first to last point/time); VCL = curvilinear velocity (average point to point velocity); VAP = velocity of the average path (derived by mathematical smoothing of VCL); ALH = amplitude of lateral head displacement (maximum width of the head oscillation, derived from VCL and VAP); BCF = beat cross frequency (number of times track crosses the average path); STR = straightness (VSL/VAP × 100); LIN = linearity (VSL/VCL × 100).

The Rat Challenge

Working and Hurtt (1987) first reported the potential value of CASA-generated sperm motility and velocity measures for reproductive toxicology studies in the rat. These were evaluated within the context of other sperm measures such as epididymal sperm reserves and sperm morphology. As various labs began to evaluate the utility of existing CASA systems, it became clear that there were several problems unique to rat sperm. CASA had, to that date, been optimized for human sperm, which are relatively easy for computers to track since the nucleus appears as a bright dot under dark field optics. However, rat sperm are much larger than human sperm and their images under dark field optics are elongated because part of their tail is visible. Optimization of CASA for rat sperm entailed using an appropriate magnification (lower than for human sperm) (Working and Hurtt, 1987) and in some cases customizing the software to track the head region more accurately (Slott et al., 1993; Dostal et al., 1996). Most recently, attempts have been made to gate out tail (and other) artifacts by imaging only the sperm nucleus with the use of a DNA-specific dye and fluorescence microscopy (Farrell et al., 1996). This approach appears to be possible with rodent sperm as well (S. D. Perreault, unpublished observation). To date, good agreement between manually derived and CASA values has been obtained for rat spermatozoa with respect to the percentage of motile sperm and straight-line velocity (VSL) (Linder et al., 1992; Yeung et al., 1992; Slott et al., 1993), and for VSL, path (VAP), and curvilinear velocity (VCL) (Moore and Akhondi, 1996).

As more toxicology labs began to use CASA, it became increasingly clear that a number of methodological variables could affect the quality of rat sperm motion during the period of observation and create technical artifacts. The variables that need to be carefully controlled include the method of collection of the sample, the region of the epididymis or vas deferens sampled, the dilution medium used, the temperature at which the sample is analyzed, the type and depth of microscope chamber used, and the duration of tracking (Klinefelter et al., 1991; Slott et al., 1991; Toth et al., 1991a; Weir and Rumberger, 1995; Moore and Akhondi, 1996; Dostal et al., 1996).

CASA users from academia, industry, and government have convened several times to discuss the technology as it has been evolving, resolve common problems, and come to consensus with regard to the best methods to use for incorporating CASA data into reproductive toxicology studies. These workshops have focused on methods for the rat (Chapin et al., 1992), human (Schrader et al., 1992), and a combination of rat, rabbit, and dog (Seed et al., 1996). The reports conclude that each lab must specify and standardize its own procedures and establish the validity of its own measures within the context of the general guidance provided.

Interpretation of CASA Data in Reproductive Toxicology Studies

Data on sperm motility are most useful in studies where the effect of the chemical is on the *quality*, as opposed to the *quantity* of sperm produced. For example, when damage occurs at later stages of spermiogenesis in the testis or in epididymal sperm, sperm numbers may be normal, but sperm viability or motion characteristics, and therefore sperm function, may be impaired. Most toxicology testing protocols employ exposure intervals that are sufficiently long enough to encompass the entire process of spermatogenesis (Zenick et al., 1994). This insures that damage to early stages of germ cell development, such as death or genetic alteration of spermatogonia or spermatocytes, will be detectable as reduced fertility at the time of breeding. As discussed in the U.S. Environmental Protection Agency's *Reproductive Toxicity Risk Assessment Guidelines* (1996), if a testicular toxicant's primary effect is to produce testicular damage with germ cell depletion, then such endpoints as testicular histology and epididymal sperm reserves will usually provide more useful information, from the standpoint of risk assessment, than will sperm motility or morphology. Nevertheless, information on sperm motion and morphology may help describe/confirm the pathogenesis and consequences of the testicular effects. On the other hand, if infertility is found in the absence of testicular effects, then evaluation of sperm motility and viability is critical for understanding the underlying cause of the infertility and for determining whether the chemical is affecting reproductive cell function at doses below those that cause infertility. It now becomes important to understand the relationship between altered sperm motion and fertility. Specifically, how much does sperm velocity have to be altered before sperm function (fertility) is affected? Application of CASA technology has recently helped to shed light on this important question.

Several investigators have addressed this issue using a model compound that is thought to selectively affect sperm motion. The compound α-chlorohydrin (proposed many years ago as a potential male contraceptive) and its parent epichlorohydrin (a reactive chemical with broad industrial uses) produce rapid and reversible antifertility effects, coincident with altered sperm velocity, when administered orally to male rodents (reviewed by Jones, 1983; Perreault, 1997). By careful selection of dosages, a response can be generated such that the velocity of epididymal sperm is affected in a graded manner, while sperm numbers and viability remain normal. Males were then bred to untreated females to evaluate fertility and/or sperm fertilizing ability. In rats, both fertility and the percentage of fertilized ova (recovered the day after breeding) were significantly affected by epichlorohydrin in a dose-dependent manner (Toth et al., 1991b). Since fertility was correlated with the percentage of fertilized ova, the results suggest that the reduced fertility was due to fertilization failure (as opposed to early embryo death). Logistic regression models examining the relationship between CASA outcomes and fertility were more informative than linear models, and correlations between each fertility measure and each motion end point (except beat/cross frequency) were significant. Nevertheless, changes in the motion endpoints did not account for all of epichlorohydrin's antifertility effect.

In male hamsters a range of dosages of α-chlorohydrin, given for 4 days to isolated epididymal effects, produced a graded response of CASA outcomes in both epididymal and uterine sperm examined shortly after breeding (Slott et al., 1995, 1997). Both multivariate linear and logistic models examining the relationship between CASA outcomes and fertility demonstrated significant correlations between motion endpoints and both fertilizing ability (of eggs recovered shortly after breeding) and pregnancy outcome, with the logistic models being more predictive of fertility. However, the relationship was not linear. Rather, evaluation of the distribution of sperm velocity measures suggested the existence of a velocity threshold above which sperm are fertile. Presumably there must be sufficient numbers of sperm with sufficiently high velocity characteristics for the male to be fertile. Companion in vitro fertilization studies using epididymal sperm recovered from α-chlorohydrin–treated males were consistent with the in vivo data and suggested that there may be an effect on sperm capacitation as well as sperm motion.

These studies illustrate the value of CASA technology in reproductive toxicology studies. CASA technology made it practical for quantifying graded changes in sperm motion. This information supports the contention that test protocols that evaluate the quality (vigor and pattern) of sperm motion, as opposed to determining only the percentage of motile sperm, should provide a more comprehensive assessment of toxicant-induced effects on sperm and assist in the prediction of potential impacts on fertility. Results of statistical analyses based on the distributions of the CASA outcomes also indicated that changes in motion of *epididymal* sperm may not entirely explain changes in fertility. Rather, it may be important to consider changes in sperm motility that normally occur in the female tract and are important for sperm fertilizing ability.

Use of CASA to Evaluate Sperm "Hyperactivation"

Recent progress has been made in using CASA to monitor changes in sperm motion that accompany capacitation. "Hyperactivation" is a term used to describe a specialized form of motion that sperm of many species assume, usually while in the oviduct near the site of fertilization. First described by R. Yanagimachi in 1970, hyperactivation is characterized by highly vigorous but nonlinear motion. This motion, which has a "whiplash" or "figure 8" appearance, has been thoroughly described in hamster and human sperm using video tracings (Suarez, 1988, 1996; Mortimer and Swan, 1995). It occurs in concert with "capacitation" and is thought to be important for penetration of the oocyte investments, particularly the zona pellucida (reviewed by Olds-Clarke, 1996; Stauss et al., 1995). Therefore, hyperactivation may be a useful marker for normal sperm function. Attempts to identify hyperactivated sperm using multivariate analysis of CASA outcomes have demonstrated that information about both vigor (e.g., VCL) and pattern (e.g., ALH; LIN) is needed. By setting thresholds for these terms, sperm can be classified as progressive (or linear), intermediate, or hyperactivated. The ability of CASA to accurately measure sperm tracks is highly dependent upon framing rate, and the recent use of 60 Hz (frames per second) for CASA has enhanced its utility, especially for tracking very vigorous hyperactivated sperm. In addition, direct capture of digital (as opposed to video) images, and the analysis of longer tracks, appears to improve tracking efficiency (S. D. Perreault, unpublished observations).

Figure 2 shows representative CASA tracks, collected at 60 Hz, of cauda epididymal sperm from three rodent species: rat, mouse, and hamster. Mouse and hamster sperm were incubated under routine capacitating conditions, and both progressive and hyperactivated sperm tracks are shown. Clearly, hyperactivated tracks have higher VCL and ALH and lower VSL and LIN than progressive sperm. Approaches used for human sperm (reviewed in Mortimer and Swan, 1995) for setting reliable cutoffs for these endpoints are being applied to these rodent species in order to evaluate hyperactivation in toxicant treated or genetically altered rodent models (Neill and Olds-Clarke, 1987; Shivaji et al., 1995; Perreault et al., 1997). Although rat sperm apparently exhibit hyperactivated motion in the oviduct at the site of fertilization (Shalgi and Phillips, 1988), there have been no reports of induction of classical hyperactivated motility during in vitro fertilization. When in vitro conditions that support hyperactivated motility of rat sperm are discovered, this outcome may prove a useful adjunct in toxicology studies in this test species.

CASA Applied to Count Sperm

CASA motility assessments also include identification of immotile sperm so total sperm concentrations can be derived. However, conditions for monitoring motile sperm are not typically ideal for accurate counting of immotile sperm. Problems can be encountered in machine-identification of immotile sperm due to differences in brightness and debris in the sample. Furthermore, deeper chamber depths that are ideal for evaluating the motion of large rodent sperm may not provide optimal depth of focus for dead sperm lying on the bottom. Also, while images viewed on a dark field are ideal for tracking motile sperm, this optical system can make it difficult to see the sperm tail and therefore to confirm that the object being counted by the computer is, indeed, an intact sperm. These problems have recently been overcome by using a DNA-specific, fluorescent stain (IDENT, Hamilton Thorne Research) so that only the sperm nucleus is detected. If other nucleated cells are present in the preparation, they can be gated out on the basis of size or shape. This system was recently shown to provide results comparable to those using the hemacytometer as a "gold standard" for ejaculated human sperm (Zinaman et al., 1996) and rat cauda epididymal sperm (Strader et al., 1996). Significant savings in technician time and fatigue can be achieved using this method.

Image Analysis of Rat Sperm Morphology

Intact Sperm

Methods for computer-assisted image analysis of sperm head morphology were pioneered in human sperm that have a symmetrical shape (Katz et al., 1986; Moruzzi et al., 1988). As with CASA applied to sperm motion, careful attention to sample preparation and staining is essential for obtaining consistent results (Davis and Cravance, 1993). Application of automated methods to rat spermatozoa was more challenging due to the asymmetrical, hooked shape of the rat sperm nucleus and differences in head shape

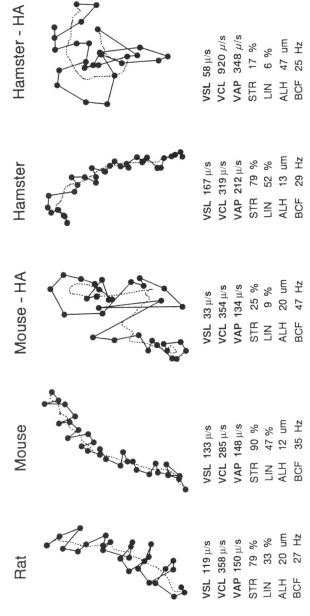

Rat	Mouse	Mouse - HA	Hamster	Hamster - HA
VSL 119 μ/s	VSL 133 μ/s	VSL 33 μ/s	VSL 167 μ/s	VSL 58 μ/s
VCL 358 μ/s	VCL 285 μ/s	VCL 354 μ/s	VCL 319 μ/s	VCL 920 μ/s
VAP 150 μ/s	VAP 148 μ/s	VAP 134 μ/s	VAP 212 μ/s	VAP 348 μ/s
STR 79 %	STR 90 %	STR 25 %	STR 79 %	STR 17 %
LIN 33 %	LIN 47 %	LIN 9 %	LIN 52 %	LIN 6 %
ALH 20 um	ALH 12 um	ALH 20 um	ALH 13 um	ALH 47 um
BCF 27 Hz	BCF 35 Hz	BCF 47 Hz	BCF 29 Hz	BCF 25 Hz

Figure 2 Representative CASA tracks of rodent epididymal spermatozoa made using the HTM-IVOS CASA instrument. Sperm were recorded shortly after collection and again after being incubated under capacitating conditions to illustrate hyperactivated (HA) motility (mouse and hamster). Sperm were tracked at 60 Hz (frames/second) and tracks were traced from the individual track screen (not to scale). Hyperactivated tracks exhibit higher VCL and ALH and lower LIN than progressive tracks.

dimensions with or without inclusion of the acrosome in the image. Recently, Davis et al. (1994) have incorporated measures of curvature as well as traditional dimensions (length, width) in an automated image analysis program to evaluate rat sperm head morphology. Using this system they have demonstrated several subpopulations of atypical sperm head shapes that result from treatment of rats with the testicular toxicant 1,3-dinitrobenzene. This technology may make it practical to evaluate large numbers of sperm and to document subtle changes in sperm head and/or nuclear shape that are associated with exposure to testicular toxicants. It will be important to demonstrate whether differences are due to loss of the acrosome and whether the abnormal shape affects fertilizing ability or reflects genetic defects in the sperm, as discussed later.

Sperm Nuclei Decondensed In Vitro

Mammalian sperm DNA is densely packaged by association with sperm-specific protamines (reviewed by Balhorn, 1989; Ward, 1993). The structural stability conferred upon sperm nuclei by this unique DNA packaging is enhanced by inter- and intramolecular protamine disulfide bonds, which form mainly during epididymal maturation. Sperm nuclei can be induced to decondense in vitro by incubation in a disulfide reducing agent together with a detergent, high salt, or an appropriate proteinase. The timing and extent of decondensation vary by species and may be altered within species by exposure to chemicals that cross-link (or break) DNA or protamine (reviewed by Perreault, 1997). A variety of in vitro decondensation assays have been proposed as potential tests for fertility and/or the genetic integrity of sperm (reviewed by Perreault, 1993).

Recently, image analysis was used to document differences in the extent of sperm chromatin decondensation in vitro produced in rat sperm by treatment with the chemotherapeutic agent cyclophosphamide. Decondensation of sperm from treated animals was less extensive, based on measurements of length and width, than that of controls. This information, together with biochemical measures of DNA, was used to demonstrate that the chemical induces both DNA cross-linking and breakage in sperm nuclei, an observation that may account for the male-mediated developmental effects of this drug (Qiu et al., 1995a,b). Similarly, Schwartz and Ricker (1997) used image analysis to show that cauda epididymal sperm from rats that have undergone surgical sympathectomy of the epididymis decondense to a lesser extent in vitro than sperm from controls. This treatment causes stasis of sperm in the epididymis, and although the sperm are able to fertilize eggs, the resulting pregnancy fails at an early stage, suggesting that the sperm nuclei may be genetically damaged (Ricker et al., 1997). Thus, fairly simple measures of nuclear stability to decondensing agents in vitro may prove a useful adjunct to more specific measures of DNA damage in assessing toxicant induced male-mediated infertility and/or adverse developmental effects.

Sperm Nuclei Decondensed in Frog Egg Extracts

An intriguing new system for monitoring DNA damage in sperm is based on early cell cycle research. A frog egg extract capable of decondensing frog sperm nuclei in vitro was shown to decondense mammalian sperm nuclei as well, as long as a disulfide reducing agent was included (Lohka and Masui, 1983). Unlike the simple in vitro decondensation assay described above that mimics only the first step in sperm chromatin processing, this system not only decondenses the sperm, but also supports remodeling of the decondensed sperm nuclei into pronucleuslike structures that are capable of DNA synthesis. Recently applied to evaluate human sperm, this human sperm activation assay (HSAA) was shown to correlate with male infertility and was proposed for use in evaluating chemically induced alterations in sperm nuclei in men exposed to environmental toxicants (Brown et al., 1995; Sawyer and Brown, 1995). This assay is being applied to rats (RSAA) in the context of toxicology studies (Sawyer and Brown, 1996). Decondensation is monitored and quantified using an image analysis system and DNA synthesis is assessed by autoradiography (incorporation of radioactive thymidine). Thus the coupling of computer vision to evaluate structural changes, with biochemical assays to evaluate function, may provide useful methods for elucidating toxicant-induced damage in sperm nuclei.

FLOW CYTOMETRIC MEASURES OF SPERM INTEGRITY

Sperm Viability and Acrosome Reaction

Like CASA, flow cytometry has the advantage of being able to measure fluorescence in thousands of cells in a short period of time. This technology has been applied to sperm to measure viability. For example, by simultaneously assessing rhodamine 123 (bright in the functional mitochondria of live cells) and ethidium bromide (staining the nuclei of dead cells only), flow cytometry can rapidly provide the percentage of live cells in a sample (Evenson et al., 1982). Similarly, carboxyfluorescein diacetate (bright in live cells) can be used in combination with propidium iodide (like ethidium bromide, bright in the nuclei of dead cells) (Garner et al., 1986) to evaluate sperm viability with flow cytometry. Recently this strategy was enhanced by using two nuclear dyes: propidium iodide to stain dead cells and a new membrane-permeant nuclear stain, SYBR-14, to stain live sperm (Garner and Johnson, 1995). This method has been validated for several species including human and mouse and has the advantage of identifying moribund sperm, which exhibit both colors. Although not yet used in rat toxicology studies, this technology would appear well suited to quantifying effects of chemicals in large numbers of animals as are used in toxicology testing.

In addition to tracking viability, flow cytometry can be used to monitor acrosome reaction in a sperm suspension. Since acrosome reaction is essential for fertilization (reviewed in Yanagimachi, 1994), it can be used as another biomarker of sperm function. Coupled with the viability stains mentioned above, information about capacitation and acrosome reaction can be obtained by adding an acrosome label. This makes it possible to identify the percentage of viable cells with intact or reacted (missing) acrosomes. For example, Graham et al. (1990) evaluated viability with propidium iodide and Rhodamine 123, and included a fluorescent lectin pisum sativum agglutinin (PSA) to tag the acrosome. Using this combination of labels, they employed flow cytometry to derive the percentage of live sperm that had undergone an acrosome reaction. (Viability was an important measure since sperm may lose their acrosomes secondary to cell death, so nonviable, acrosomeless sperm would not be reflective of physiological acrosome reaction.) To date, this flow cytometric approach has not been applied in toxicology, but at least one recent report (Oberlander et al., 1996) indicates that acrosome reaction can be detected in rat sperm using another fluorescence-based assay (the chlortetracycline assay).

Sperm Chromatin Structure Assay

While new technologies have emerged to evaluate sperm motion, viability, and acrosomal status, methods for evaluating the genetic integrity of the sperm nucleus have remained elusive. Yet it has been known for many years that a variety of toxicants can damage sperm DNA and that the condensed spermatids and epididymal spermatozoa are at a unique risk since they lack the ability to repair DNA damage (reviewed in Perreault, 1997). A promising flow cytometric test, the sperm chromatin structure assay (SCSA), evaluates the susceptibility of sperm DNA to heat- or acid-induced denaturation in situ (reviewed in Evenson and Jost, 1994). Sperm with intact sperm chromatin fluoresce green, while sperm with single-stranded DNA (after denaturation) fluoresce red when stained with the metachromatic dye acridine orange. By measuring the relative amounts of red and green in each sperm using flow cytometry, the extent of DNA denaturation can be quantified and is expressed as *alpha-t*. Alpha-t is defined as red fluorescence/(red + green). Several other endpoints can then be derived to describe the whole sample: mean and standard deviation of alpha-t and the percent cells outside the main population (COMP). COMP is particularly useful in toxicology studies since it provides a measure of the proportion of sperm with abnormal chromatin in a sample. For example, SCSA and especially COMP alpha-t were able to detect alternations in sperm chromatin shortly after exposure of mice to the alkylating agent methyl methanesulfonate (Evenson et al., 1993), a mutagen that produces dominant lethal effects. Since SCSA is easily evaluated in human sperm (fresh or frozen), it may be particularly suited as a biomarker of effect in large-scale epidemiology studies. Indeed, compared with other semen measures, SCSA outcomes were more stable over time within an individual man (Evenson et al., 1991). Furthermore, abnormal SCSA was recently shown to correlate with DNA damage detected in sperm using the comet assay (Evenson et al., 1997). This is a single cell agarose gel electrophoresis assay that detects small pieces of DNA (i.e., broken DNA) and can be applied to sperm cells (Singh et al., 1989).

TUNEL Applied to Sperm

Another assay that measures DNA breaks is the terminal deoxynucleotidyl transferase-mediated dUTP-biotin nick end-labeling assay (TUNEL) (Gavrieli et al., 1992). Based on labeling 3'-OH termini with biotinylated dUTP using exogenously supplied terminal transferase, this assay is being widely used as an indicator of apoptosis. The assay was recently applied to human sperm, and the biotinylated product was quantified by flow cytometry via linkage with fluoresceinated avidin (Gorczyca et al., 1993). The extent of DNA strand breakage measured with this assay was correlated with the extent of abnormal DNA determined with the SCSA, leading the authors to conclude that semen samples with abnormal SCSA probably have high levels of DNA damage and that sperm may undergo apoptosis. The association between the susceptibility of sperm DNA to in situ denaturation (abnormal SCSA) and the presence of DNA strand breaks (TUNEL) has been confirmed in humans, rams, bulls, and stallions to date (Sailer et al., 1995), making it likely that this assay could be applied to rodent sperm in toxicology studies. Interestingly, negative correlations were also found in human sperm between the percentage of sperm with fragmented DNA (again determined by TUNEL using flow sorting) and traditional measures of semen quality (motility, morphology, and sperm concentration) as well as the fertilization rate in vitro (Sun et al., 1997). This study also showed that smokers had a higher percentage of sperm with fragmented DNA. These observations suggest, again, that TUNEL and/or SCSA should be useful biomarkers of effect in human epidemiology or occupational studies.

ADVANCES IN METHODS FOR IN VITRO ASSESSMENT OF GAMETE FUNCTION

In Vitro Fertilization

While in vitro fertilization (IVF) methods are used for the treatment of human infertility, the application of IVF in reproductive toxicology studies in rodents has been limited, despite the availability of established methods (O'Brien et al., 1993; Perreault and Jeffay, 1993). IVF has been used to demonstrate deficits in sperm fertilizing ability after in vivo exposure to 1,3-dinitrobenzene (Holloway et al., 1990a) and ethylene glycol monomethyl ether (Holloway et al., 1990b) in the rat and α-chlorohydrin in the hamster (Slott et al., 1997). Other studies have evaluated effects of chemicals added to the culture medium during IVF, such as α-chlorohydrin in the rat (Woods and Garside, 1996) or polychlorinated or polybrominated biphenyls in the mouse (Kholkute et al., 1994a,b).

IVF has also been used as a means of evaluating sperm chromosomes. Once sperm have fertilized an oocyte and undergone decondensation and transformation into the male pronucleus, the zygote can be blocked at the first mitotic metaphase. Oocyte and sperm chromosomes can then be spread on slides for standard karyotype analysis. Indeed, human sperm chromosomes can be analyzed in this manner after fertilization of zona-free hamster eggs, and this technique has generated a large data base on the incidence of human chromosome aberrations, chromosome-specific aneuploidy and diploidy (reviewed by Martin, 1993). Although labor intensive, this method has shown that men exposed to radiotherapy and chemotherapy have an increased frequency of sperm chromosome abnormalities (reviewed by Martin, 1993). Using a similar technique, mouse or human sperm were exposed to antineoplastic agents in vitro and then allowed to fertilize mouse or hamster eggs (respectively) in vitro (Kamiguchi et al., 1985; Matsuda and Tobari, 1988). Again, eggs were examined at first metaphase to evaluate different types of chromosome aberrations, as well as to assess the repair capability of the eggs. Of course the method can be applied after in vivo fertilization as well. For example, male mice were treated with acrylamide, a germ cell mutagen, and bred to untreated females. Colchicine was given to arrest metaphase in vivo and the zygotes were recovered for karyotype analysis. Acrylamide induced sperm chromosome aberrations consistent with chromosome breakage as opposed to aneuploidy (Pacchierotti et al., 1994). Studies like this can also be conducted using IVF with sperm from treated males and oocytes from untreated females.

IVF can also be used to evaluate oocyte function directly (i.e., the ability to undergo normal fertilization and form a zygote capable of normal embryonic development). Indeed, the recent demonstration of complete oocyte development in vitro marks the way for expanded application of in vitro technology in gamete toxicology. Primordial follicles contained in newborn mouse ovaries were grown first in organ culture, and then as cumulus-oocyte complexes (Eppig and O'Brien, 1996). Once the cumulus-oocyte complexes became competent to mature, they were fertilized in vitro and developed to term after transfer into foster dams. This remarkable achievement should make it possible to explore the mechanisms of toxicity of such chemicals as benzo-*a*-pyrene (Mattison et al., 1983) and 4-vinylcyclohexene diepoxide (Springer et al., 1996), female reproducible toxicants that appear to target oocytes in primary or small preantral follicles.

Fluorescence Detection of Oocyte and Sperm Components During Fertilization

Advances in fluorescence immunohistochemistry have made it possible to detect alterations in subcellular components of oocytes and zygotes during IVF. For example, use of antitubulin antibodies to label the meiotic spindle and DNA-specific fluorochromes to label the egg chromosomes and sperm nucleus has revealed adverse effects of microtubule poisons on zygote formation. Schatten et al. (1985) demonstrated that colcemid (a derivative of colchicine) caused spindle disassembly with scattering of the metaphase chromosomes and failure of pronucleus formation in mouse oocytes undergoing IVF. Results with two other microtubule inhibitors, griseofulvin and nocodazole, were similar. These studies provided information not only about the importance of microtubule function during fertilization, but also demonstrated that exposure to exogenous chemicals during critical periods of oocyte maturation and fertilization can cause chromosome imbalance and arrest early development.

In vitro fertilization studies in the hamster also helped elucidate the mechanism by which the fungicide carbendazim (methyl 2-benzimidazolecarbamate, or MBC) causes early pregnancy loss after acute administration to females during oocyte maturation and at the time of fertilization. MBC exposure during IVF resulted in partial spindle dissolution, identified with antitubulin antibody immunofluorescence (Zuelke and Perreault, 1995). Nevertheless, the oocytes became penetrated by sperm and the sperm decondensed. However, spindle disruption resulted in two general zygotic phenotypes: either the oocytes contained scattered chromosomes and a decondensed sperm (meiotic arrest at metaphase II), or they exhibited multiple small female pronuclei (with or without the second polar body) along with a morphologically normal male pronucleus (Fig. 3). The latter phenotype was also observed in zygotes recovered from females treated with MBC just before fertilization, and a significant proportion of these zygotes were aneuploid (Zuelke and Perreault, 1995). IVF may be a practical means for comparing the relative potency of microtubule poisons (or other related chemicals). For example, effects of MBC on hamster IVF were compared with effects of the classic microtubule poisons colcemid and podophyllotoxin (Fig. 4). Again, the most common abnormality with MBC exposure was the occurrence of multiple small female pronuclei with normal male pronuclei. In contrast, colcemid and podophyllotin exposure usually resulted in zygotes arrested at meiosis II in which sperm decondensed but neither male nor female pronuclei formed (Fig. 4). These studies also showed that MBC was far less potent than colcemid or podophyllotoxin. Zygotes containing multiple female pronuclei were also reported in the mouse after performing IVF in the presence of nocodazole, a microtubule poison that is structurally similar to MBC (Maro et al., 1986).

Substances that alter the meiotic spindle would also be expected to have effects on oocyte maturation (i.e., the LH-induced progression of the oocyte from the germinal vesicle oocyte stage through the first meiotic division). Indeed, in vivo exposure of female hamsters to MBC shortly before the endogenous LH surge resulted in arrested

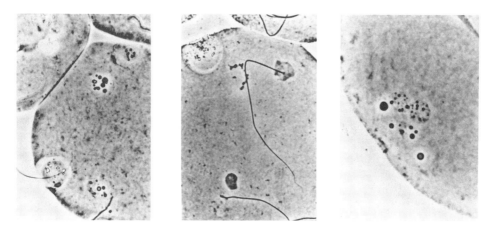

Figure 3 Effect of MBC on zygote formation during hamster IVF. (Left panel) A normal (control) hamster zygote with well-formed female pronucleus (top near second polar body) and male male pronucleus with associated sperm tail (bottom, near first polar body). (Middle panel). An MBC-exposed zygote illustrative of the arrested development phenotype. This oocyte was penetrated by two spermatozoa, and although they decondensed, they did not form male pronuclei. The oocyte chromosomes (top near first polar body) appear clumped. (Right panel) An MBC-exposed zygote illustrative of abnormal female pronucleus development. This zygote contains multiple small female pronuclei and lacks a second polar body.

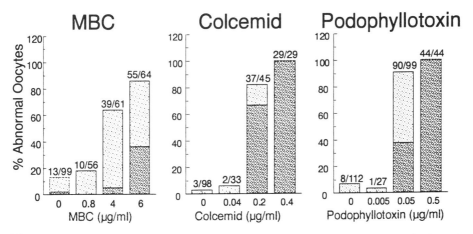

Figure 4 Abnormal in vitro fertilization of hamster oocytes in the presence of microtubule inhibitors. Hamster oocytes were inseminated with in vitro capacitated hamster sperm and cultured in the presence of inhibitors for 3.5 hours before being assessed for fertilization (as in Zuelke and Perreault, 1995). Bars indicate the percentage of eggs (combined from 3–5 experiments) exhibiting abnormal fertilization. Dark stippling indicates the arrested phenotype (eggs containing a decondensed sperm and clumped egg chromosomes as shown in Figure 3); light stippling indicates eggs containing a normal male pronucleus but two or more small female pronuclei (with or without a second polar body). While the classic microtubule inhibitors colcemid and podophyllotoxin tended to arrest meiosis, MBC, a reversible inhibitor of tubulin polymerization, was more likely to cause abnormal female pronuclei.

preimplantation embryo development and implantation failure, as well as early post-implantation loss (Jeffay et al., 1996). This toxicity was explored in subcellular detail by exposing mouse oocytes to MBC during in vitro oocyte maturation (Can and Albertini (1997). Immunofluorescence with confocal microscopy were used to visualize chromosomes, α-tubulin, and acetylated tubulin in the treated oocytes. The results suggested that the MBC appears to affect spindle pole components and may, therefore, interfere with centrosome function.

Spindle disruption was also detected by immunofluorescence microscopy in mature hamster oocytes exposed in vitro to the glutathione (GSH)–oxidizing agent diamide (Zuelke et al., 1997). This was associated with scattering of the metaphase II chromosomes. These results suggest that GSH is important in oocytes for spindle maintenance, as well as for sperm chromatin decondensation as demonstrated previously (reviewed by Perreault, 1992). When the diamide was washed out prior to insemination with capacitated sperm, the resulting zygotes exhibited a phenotype similar to those described above after MBC exposure and as illustrated in Figure 5. Presumably such zygotes result when the chromosomes lose their orderly arrangement on the metaphase plate and be-

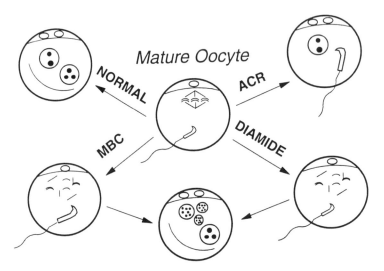

Figure 5 Diagram of abnormal zygote phenotypes seen in oocytes exposed to model compounds. The mature oocyte is arrested at meiotic metaphase II at the time of fertilization (center). The normal zygote (upper left) contains two large pronuclei along with the tail of the fertilizing sperm and exhibits two polar bodies. Exposure to MBC during fertilization (in vitro or in vivo) results in spindle defects with disruption of the metaphase plate (lower left). These zygotes usually complete meiosis and remodel the sperm nucleus into a normal male pronucleus. However, the spindle disruption leads to the formation of multiple small female pronuclei (lower center), often without extrusion of the second polar body (Zuelke and Perreault, 1995). In vitro exposure to diamide (lower right) also disrupts the spindle and inhibits sperm decondensation. Removal of the diamide allows recovery of male pronucleus formation; however, spindle regrowth is atypical and the zygotes frequently exhibit multiple female pronuclei (Zuelke et al., 1997). Exposure to acrylamide (ACR, upper right) during fertilization in vivo does not affect formation of the female pronucleus (or extrusion of the polar body), but results in delayed decondensation of the sperm chromatin and failure of normal male pronucleus formation. None of these abnormal phenotypes would be expected to be consistent with embryonic development to term.

come separated into groups, each of which subsequently forms its own nuclear envelope. Presumably such zygotes would also be predisposed to aneuploidy, which in rodent zygotes generally leads to early embryo death before or soon after implantation. However, some aneuploidies in human zygotes gives rise to full-term babies with severe deficits. Perhaps the less severely affected zygote with abnormal female pronuclei is a better predictor (biomarker) of serious developmental defects than is the more severely affected zygote that simply arrests prior to pronucleus formation.

Many in vivo studies using acute exposures to various environmental chemicals or drugs at the time of oocyte maturation have been shown to induce aneuploidy in mouse oocytes (reviewed in Mailhes and Marchetti, 1994), and many of these are believed to act through spindle disruption. Depending on the timing and dosage, such chemicals may simply arrest meiosis; indeed recent cell-cycle research indicates that resumption of meiosis in metaphase II oocytes (a process that requires inactivation of maturation promoting factor) is dependent upon the presence of a structurally intact spindle (Kubiak et al., 1993). While IVF and in vivo studies in rodents help shed light on the mechanisms of aneuploidy, the extent to which spindle poisons (or other environmental agents) contribute to aneuploidy in human populations remains to be determined (reviewed in Eichenlaub-Ritter, 1994).

As more probes for localization of specific sperm and egg components become available and their use increases in both static and living cells (reviewed by Simerly and Schatten, 1993), our understanding of toxicant action should be enhanced. Already, the combined use of fluorescent stains for tubulin and DNA (described above) are being used to help clinicians identify what went wrong in human eggs that failed to undergo normal fertilization in vitro (Asch et al., 1995). Evaluation of these eggs with laser scanning confocal microscopy provided superior and unprecedented resolution of fine, subcellular details. Furthermore, with digital imaging capabilities and the ability to superimpose multiple, pseudo-colored images, it becomes possible to visualize many subcellular structures in one photograph. For example, a recent report about the fate of the sperm tail during fertilization shows discretely labeled sperm nuclei (DNA stain), sperm midpiece (mitochondrial stain), sperm tail (α-tubulin or fibrous sheath stains), and oocyte microtubules all present in one image and each colored a different color (Sutovsky et al., 1996).

Fertilization by Sperm Microinjection

Intracytoplasmic sperm injection (ICSI) has made it possible to overcome some severe forms of male infertility (reviewed by Palermo et al., 1996). Technological advances in micromanipulation are making this procedure practical to include in many human IVF laboratories. By placing a sperm directly into the oocyte, sperm that may not be able to bind or transit the zona pellucida can still participate in zygote formation. This technique can be used in toxicology studies as well in order to specifically evaluate oocyte function. For example, we have used sperm injection to evaluate the ability of oocytes recovered from treated female hamsters to decondense a sperm nucleus. When female hamsters are exposed to acrylamide just before breeding to untreated males, their eggs can be fertilized but the fertilizing sperm does not develop into a normal male pronucleus (Fig. 5). To show that this is due to an inadequate oocyte and not to delayed fertilization or a sperm defect, we injected sperm nuclei into oocytes from treated females. The resulting zygotes showed the same phenotype as the in vivo fertilized zygotes, namely, a normal female pronucleus with a sperm arrested at the de-condensation stage (Perreault et al., 1995). We also used ICSI to show that the timing of sperm decondensation was related to the species-specific types of protamine present, and to the extent of protamine disulfide binding in the nucleus (Perreault et al., 1987, 1988). Based on these observations, we suggest that this assay would also be useful to specifically assess decondensation and male pronucleus formation by sperm recovered from males after exposure to DNA or protamine cross-linking agents. Furthermore, the method is proving useful in evaluating the chromosomes in morphologically abnormal human sperm after injection into mouse oocytes (Lee et al., 1996). Thus, this approach may be a valuable adjunct to assays that detect DNA damage in sperm.

FLUORESCENCE IN SITU HYBRIDIZATION TO EVALUATE SPERM AND EGG CHROMOSOMES

The development of chromosome-specific, repetitive DNA sequence probes has made it possible to identify particular chromosomes in interphase nuclei via fluorescence in situ hybridization (FISH). This approach was first applied to human sperm using a probe for chromosome Y (Guttenbach and Schmid, 1990; Wyrobek et al., 1990) and soon extended to include a probe for an autosome (Robbins et al., 1993) and then for both sex chromosomes (Wyrobek et al., 1994). With many labs now using this technique, a new level of understanding of human sperm aneuploidy is emerging (Williams et al., 1993). For example, it appears that paternal age influences aneuploidy, with older men more likely to exhibit sex chromosome aneuploidy (Robbins et al., 1995). This find-ing could influence study designs if this assay is used to detect toxicant-induced sperm aneuploidy. Advantages of using this method for toxicology studies over the IVF ap-proach are many: thousands of sperm can be evaluated so even rare events can be de-tected; all sperm can be evaluated, not just the ones that fertilize eggs; and aneuploidy can be distinguished from polyploidy. The method is now being used in mice (Wyrobek et al., 1995) and being developed for the rat (Robbins et al., 1995). It promises to further enhance our ability to detect toxicant-induced damage to sperm nuclei.

FISH with a chromosome-specific probe can also be applied in oocytes. An el-egant example is provided by a study evaluating the movement of the X chromosome during meiosis in the mouse (Hunt et al., 1995). Here, immunofluorescence staining of tubulin was used along with a DNA-specific dye and an FITC-labeled X-specific probe in order to study the X chromosome with respect to its position on the spindle. Resolution was enhanced by using confocal microscopy. Results indicate that oocytes appear to continue meiosis even when a univalent chromosome is present (Hunt et al., 1995). This situation is very unlike that in the male and may provide insight into the types of aneuploidy seen in females and the types of chemicals that might induce it.

CONCLUSION

New technologies continue to become available and innovation during the past decade has been substantial. These technologies are rapidly expanding our understanding of the basic biology and genetics of gametes and gamete function. Potential applications in the field of reproductive toxicology are widespread and hold great promise for the future both as screens to detect hazard and as in-depth tests to identify and elucidate mechanisms of toxicant action and of endogenous repair of damage.

REFERENCES

Aravindan, G. R., Bjordahl, J., Jost, L. K., and Evenson, D. P. (1997). *Exp. Cell Res., 235*: 231.

Asch, R., Simerly, C., Ord, T., Ord, V. A., and Schatten, G. (1995). *Human Reprod. 10*: 1897.

Balhorn, R. (1989). *Molecular Biology of Chromosome Function* (K. W. Adolph, ed.). Springer Verlag, New York, p. 366.

Boyers, S. P., Davis, R. O., and Katz, D. S. (1989). *Curr. Probl. Obstet. Gynecol. Fertil. 12*: 173.

Brown, D. B., Hayes, E. J., Uchida, T., and Nagamani, M. (1995). *Fertil. Steril., 64*: 612.

Can, A., and Albertini, D. F. (1997). *Mol. Reprod. Dev., 46*: 351.

Chapin, R. E., Filler, R. S., Gulati, D., Heindel, J. J., Katz, D. F., Mebus, C. A., Obasaju, F., Perreault, S. D., Russell, S. R., Schrader, S., Slott, V., Sokol, R. Z., and Toth, G. (1992). *Reprod. Toxicol. 6*: 267.

Davis, R. O., and Gravance, C. G. (1993). *Fertil. Steril., 59*, 412.

Davis, R. O., Gravance, C. G., Thal, D. M., and Miller, M. G. (1994). *Reprod. Toxicol., 8*: 521.

Dostal, L. A., Faber, C. K., and Zandee, J. (1996). *Reprod. Toxicol., 10*: 231.

Eichenlaub-Ritter, U. (1994). *Curr. Topics Dev. Biol., 29*: 281.

Eppig, J. J., and O'Brien, M. J. (1996). *Biol. Reprod., 54*: 197.

Evenson, D. P., and Jost, L. K. (1994). *Methods Cell Biol., 42*: 59.

Evenson, D. P., Darzynkiewicz, Z., and Melamed, M. R. (1982). *J. Histochem. Cytochem., 30*: 279.

Evenson, D. P., Jost, L. K., Baer, R. K., Turner, T. W., and Schrader, S. M. (1991). *Reprod. Toxicol., 5*: 115.

Evenson, D. P., Jost, L. K., and Baer, R. K. (1993). *Environ. Mol. Mutagen, 21*: 144.

Farrell, P. B., Foote, R. H., and Zinaman, M. J. (1996). *Fertil. Steril., 66*: 446.

Garner, D. L., and Johnson, L. A. (1995). *Biol. Reprod., 53*: 276.

Garner, D. L., Pinkel, D., Johnson, L. A., and Pace, M. M. (1986). *Biol. Reprod., 34*: 127.

Gavrieli, Y., Sherman, Y., and Ben-Sasson, S. A. (1992). *J. Cell. Biol., 119*: 493.

Gorczyca, W., Traganos, F., Jesionowskka, H., and Darzynkiewicz, Z. (1993). *Exp. Cell. Res., 207*: 202.

Graham, J. K., Kunze, E., and Hammerstedt, R. H. (1990). *Biol. Reprod., 43*: 55.

Guttenbach, M., and Schmid, M. (1990). *Am. J. Human Genet. 46*, 553.

Holloway, A. J., Moore, H. D. M., and Foster, P. D. M. (1990a). *Fundam. Appl. Toxicol., 14*: 113.

Holloway, A. J., Moore, H. D. M., and Foster, P. D. M. (1990b). *Reprod. Toxicol., 4*: 21.

Hunt. P., LeMaire, R., Embury, P., Sheean, L., and Mroz, K. (1995). *Human Mol. Genet., 4*: 2007.

Jeffay, S., Libbus, B., Barbee, R. R., and Perreault, S. D. (1996). *Reprod. Toxicol., 10*: 183.

Jones, A. R. (1983). *Aust. J. Biol. Sci., 36*: 333.

Kamiguchi, Y., Tateno, H., Iizawa, Y., and Mikamo, K. (1995). *Mutat. Res., 326*: 185.

Katz, D. F., and Overstreet, J. O. (1981). *Fertil. Steril., 35*: 188.

Katz, D. F., Overstreet, J. W., Samuels, S. J., Niswander, P. W., Bloom, T. D., and Lewis, E. L. (1986). *J. Androl., 7*: 203.

Keel, B. A. (1990). *Handbook of the Laboratory Diagnosis and Treatment of Infertility* (B. A. Keel and B. W. Webster, eds.). CRC Press, Boca Raton, FL, p. 27.

Kholkute, S. D., Rodriguez, J., and Dukelow, W. R. (1994a). *Reprod. Toxicol.*, *8*: 487.

Kholkute, S. D., Rodriguez, J., and Dukelow, W. R. (1994b). *Arch. Environ. Contam. Toxicol.*, *26*: 208.

Klinefelter, G. R., Gray, L. E., Jr., and Suarez, J. D. (1991). *Reprod. Toxicol.*, *5*: 39.

Kubiak, J. Z., Weber, M., de Pennart, H., Winston, N. J., and Maro, B. (1993). *EMBO J.*, *12*: 3773.

Lee, J. D., Kamiguchi, Y., and Yanagimachi, R. (1996). *Human Reprod.*, *11*: 1942.

Linder, R. E., Strader, L. F., Slott, V. L., and Suarez, J. D. (1992). *Reprod. Toxicol.*, *6*: 491.

Lohka, M., and Masui, Y. (1983). *Science*, *220*: 719.

Mailhes, J. B., and Marchetti, F. (1994). *Mutat. Res.*, *320*: 87.

Makler, A. (1978). *Fertil. Steril.*, *30*: 192.

Maro, B., Johnson, M. H., Webb, M., and Flach, G. (1986). *J. Embryol. Exp. Morph.*, *92*: 11.

Martin, R. H. (1993). *Reprod. Toxicol.*, *7*: 47.

Matsuda, Y., and Tobari, I. (1988). *Mutat. Res.*, *198*: 131.

Mattison, D. R., Shiromizu, K., and Nightingale, M. S. (1983). *Am. J. Ind. Med.*, *4*: 191.

Moore, H. D. M., and Akhondi, M. A. (1996). *J. Androl.*, *17*: 50.

Mortimer, D. (1990). *Handbook of the Laboratory Diagnosis and Treatment of Infertility* (B. A. Keel and B. W. Webster, eds.). CRC Press, Boca Raton, FL, p. 97.

Mortimer, S. T., and Swan, M. A. (1995). *Human Reprod.*, *10*: 873.

Moruzzi, J. F., Wyrobek, A. J., Mayall, B. H., and Gledhill, B. L. (1988). *Fertil. Steril.*, *58*: 763.

Neill, J. M., and Olds-Clarke, P. (1987). *Gamete Res.*, *18*: 121.

Oberlander, G., Yeung, C. H., and Cooper, T. G. (1996). *J. Reprod. Fertil.*, *106*: 231.

O'Brien, M. J., Wigglesworth, K., and Eppig, J. J. (1993). *Methods in Toxicology, Vol. 3* (J. J. Heindel and R. E. Chapin, eds.). Academic Press, Inc., New York, p. 128.

Olds-Clarke, P. (1996). *J. Androl.*, *17*: 183.

Pacchierotti, F., Tiveron, C., D'Archivio, M., Bassani, B., Cordelli, E., Leter, G., and Spano, M. (1994). *Mutat. Res.*, *309*: 273.

Palermo, G. D., Cohen, J., and Rosenwaks, M. D. (1996). *Fertil. Steril.*, *65*: 899.

Perreault, S. D. (1992). *Mutat. Res.*, *296*: 43.

Perreault, S. D. (1993). *Understanding Male Infertility: Basic and Clinical Approaches* (R. W. Whitcomb and B. R. Zirkin, eds.). Raven Press, New York, p. 267.

Perreault, S. D. (1997). *Comprehensive Toxicology*, Vol. 10 (I. G. Sipes, A. J. Gandolfi, and C. A. McQueen, eds.). Elsevier Science, New York, p. 165.

Perreault, S. D., and Jeffay, S. C. (1993). *Methods in Toxicology, Vol. 3* (J. J. Heindel and R. E. Chapin, eds.). Academic Press, Inc., Orlando, FL, p. 92.

Perreault, S. D., Naish, S. J., and Zirkin, B. R. (1987). *Biol. Reprod.*, *36*: 239.

Perreault, S. D., Barbee, R. R., Elstein, K. H., Zucker, R. M., and Keefer, C. L. (1988). *Biol. Reprod.*, *39*: 157.

Perreault, S. D., Jeffay, S. A., and Barbee, R. R. (1995). *Biol. Reprod.*, *52* (Suppl. 1): 97.

Perreault, S. D., Jeffay, S. C., and Katz, D. S. (1997). *J. Androl.*, *18* (Suppl 1): P-50.

Qiu, J., Hales, B. F., and Robaire, B. (1995a). *Biol. Reprod.*, *52*: 33.

Qiu, J., Hales, B. F., and Robaire, B. (1995b). *Biol. Reprod.*, *52*: 1465.

Ricker, D. D., Crone, J. K., Chamness, S. L., Klinefelter, G. R., and Chang, T. S. K. (1997). *J. Androl.*, *18*: 131.

Robbins, W. A., Segraves, R., Pinkel, D., and Wyrobek, A. J. (1993). *Am. J. Human Genet.*, *52*: 799.

Robbins, W. A., Baulch, J. E., Moore, D., Weier, H.-U., Blakey, D., and Wyrobek, A. J. (1995). *Reprod. Fertil. Dev.*, *7*: 799.

Sailer, B. L., Jost, L. K., and Evenson, D. P. (1995). *J. Androl.*, *16*: 80.

Sawyer, D. E., and Brown, D. B. (1995). *Reprod. Toxicol.*, *9*: 351.

Sawyer, D. E., and Brown, D. B. (1996). *Toxicologist*, *16*: 120.

Schatten, J., Simerly, C., and Schatten, H. (1985). *Proc. Natl. Acad. Sci. USA*, *82*: 4152.

Schrader, S. M., Chapin, R. E., Clegg, E. D., Davis, R. O., Fourcroy, J. L., Katz, D. F., Rothmann, S. A., Toth, G., Turner, T. W., and Zinaman, M. (1992). *Reprod. Toxicol.*, *6*: 275.

Schwartz, M., and Ricker, D. D. (1997). *J. Androl.*, *18* (Suppl. 1):P-31.

Seed, J., Chapin R. E., Clegg, E. D., Dostal, L. A., Foote, R. H., Hurtt, M. E., Klinefelter, G. R., Makris, S. L., Perreault, S. D., Schrader, S., Seyler, D., Sprando R., Treinen, K. A., Veeramachaneni, D. N. R., and Wise, L. D. (1996). *Reprod. Toxicol.*, *3*: 237.

Shalgi, R., and Phillips, D. M. (1988). *Biol. Reprod.*, *39*: 1207.

Shivaji, S., Peedicayil, J., and Devi, L. G. (1995). *Mol. Reprod. Dev.*, *42*: 233.

Simerly, C., and Schatten, G. (1993). *Methods Enzymol.*, *225*: 516.

Singh, N. P., Danner, D. B., Tice, R. R., McCoy, M. T., Collins, G. D., and Schneider, E. L. (1989). *Exp. Cell Res.*, *184*: 461.

Slott, V. L., Suarez, J. D., and Perreault, S. D. (1991). *Reprod. Toxicol.*, *5*: 449.

Slott, V. L., Suarez, J. D., Poss, P. M., Linder, R. E., and Perreault, S. D. (1993). *Fundam. Appl. Toxicol.*, *21*: 298.

Slott, V. L., Jeffay, S., Suarez, J. D., Barbee, R. R., and Perreault, S. D. (1995). *J. Androl.*, *16*: 523.

Slott, V. L., Jeffay, S. C., Dyer, C. J., Barbee, R. R., and Perreault, S. D. (1997). *J. Androl.*, *18* in press.

Springer, L. N., Flaws, J. A., Sipes, I. G., and Hoyer, P. B. (1996). *Reprod. Toxicol.*, *10*: 137.

Stauss C. R., Votta, T. J., and Suarez, S. S. (1995). *Biol. Reprod.*, *53*: 1280.

Strader, L. F., Linder, R. E., and Perrault, S. D. (1996). *Reprod. Toxicol.*, *10*: 529.

Suarez, S. S. (1988). *Gamete Res.*, *19*: 51.

Suarez, S. S. (1996). *J. Androl.*, *17*: 331.

Sun, I.-G., Jurisicova, A., and Casper, R. F. (1997). *Biol. Reprod.*, *56*: 602.

Sutovsky, P., Navara, C. S.,and Schatten, G. (1997). *Biol. Reprod.*, *55*: 1195.

Toth, G. P., Stober, J. A., George, E., Read, E. J., and Smith, M. K. (1991a). *Reprod. Toxicol.*, *5*: 487.

Toth, G. P., Stober, J. A., Zenick, H., Read, E. J., Christ, S. A., and Smith, M. K. (1991b). *J. Androl.*, *12*: 54.

U. S. Environmental Protection Agency. (1996). *Fed. Reg.*, *61*: 56274.

Ward, W. S. (1993). *Biol. Reprod.*, *48*: 1193.

Weir, P. J., and Rumberger, D. (1995). *Reprod. Toxicol.*, *9*: 327.

Williams, B. J., Ballenger, C. A., Malter, H. E., Bishop, F., Tucker, M., Zwingman, T. A., and Hassold, T. J. (1993). *Human Mol. Genet.*, *2*: 1929.

Woods, J., and Garside, D. A. (1996). *Reprod. Toxicol.*, *10*: 199.

Working, P. K., and Hurtt, M. E. (1987). *J. Androl.*, *8*: 330.

Wyrobek, A. J., Alhborn, T., Balhorn, R., Stanker, L., and Pinkel, D. (1990). *Mol. Reprod. Dev.*, *27*: 200.

Wyrobek, A. J., Robbins, W. A., Mehraein, Y., Pinkel, D., and Weier, H.-U. (1994). *Am. J. Med. Genet.*, *53*: 1.

Wyrobek, A. J., Lowe, X., Pinkel, D., and Bishop, J. (1995). *Mol. Reprod. Dev.*, *40*: 259.

Wyrobek, A. J., Schrader, S. M., Perreault, S. D., Fenster, L., Huszar, G., Katz, D. F., Osorio, A. M., Sublet, V., and Evenson, D. (1997). *Reprod. Toxicol.*, *11*: 243.

Yanagimachi, R. (1970). *J. Reprod. Fertil.*, *23*: 193.

Yanagimachi, R. (1994). *The Physiology of Reproduction*, 2nd ed. (E. Knobil and J. D. Neill, eds.). Raven Press, Ltd., New York, p. 189.

Yeung, C. H., Oberlander, G. and Cooper, T. G. (1992). *J. Reprod. Fertil.*, *96*: 427.

Zenick, H., Clegg, E. D., Perreault, S. D., Klinefelter, G. R., and Gray, L. E. (1994). *Principles and Methods of Toxicology*, 3rd ed. (A. W. Hayes, ed.). Raven Press, Ltd., New York, p. 937.

Zinaman, M. J., Uhler, M. L., Vertuno, E., Fisher, S. G., and Clegg, E. D. (1996). *J. Androl.*, *17*: 288.

Zuelke, K. A., and Perreault, S. D. (1995). *Mol. Reprod. Dev.*, *42*: 200.

Zuelke, K. A., Jones, D., and Perreault, S. D. (1997). *Biol. Reprod., 57,* in press.

27
Sertoli Cell Toxicants

Ling-Hong Li and Jerrold J. Heindel
National Institute of Environmental Health Sciences, Research Triangle Park, North Carolina

The Sertoli cell, originally described by the Italian physiologist Enrico Sertoli in 1865, is the nongerminal component of the seminiferous epithelium where mammalian spermatogenesis takes place. Sertoli cells play multiple roles in maintaining and regulating spermatogenesis, although the exact mechanisms remain to be determined. There is abundant evidence showing that the Sertoli cell is also the target of many testicular toxicants. The objective of this chapter is to provide a review of the current knowledge on the toxicities of some environmental agents in Sertoli cells. It will focus on the most studied Sertoli cell toxicities and thus will not be an exhaustive review of every agent shown to affect Sertoli cell functions. There will also be a brief review on the biology of Sertoli cells and experimental models applied in Sertoli cell toxicological research. For detailed information on the biology of spermatogenesis and Sertoli cells, as well as the toxicological methodology, we refer the readers to *The Sertoli Cell* by Russell and Griswold (1993) and reviews by Heindel and Treinen (1989), Lamb and Chapin (1993), Lamb (1993), Bardin et al. (1994), Jégou (1994), and Kierszenbaum (1994). Since the rat is the most frequently used animal in male reproductive toxicological studies, the Sertoli cell used in this chapter refers to rat Sertoli cells unless otherwise mentioned.

BIOLOGY OF THE SERTOLI CELL

Differentiation and Proliferation

It is generally believed that the precursors of the Sertoli cells are among the somatic cells of the sexually indifferent gonadal ridge. The primordial Sertoli cells appear on day 13 postfertilization and then aggregate to form seminiferous cords the following day in the rat embryo (Satoh, 1985). The mechanisms for the appearance of primordial Sertoli cells from their ancestors in the pregonadal mesonephros remain to be defined. A number of gene products, such as testis-determining factor (TDF), anti-Müllerian hormone (AMH), meiosis-preventing substance (MPS), FSH receptor, and inhibin have been shown to be involved in the appearance of primordial Sertoli cells (Pelliniemi et al., 1990; McLaren, 1991).

Following the initial appearance in the early stages of organogenesis, the differentiating Sertoli cells undergo rapid proliferation with a peak of cell division on day 20 of gestation. Thereafter, proliferation slows down and ceases in pups on day 15 after birth (Clermont and Perey, 1957; Steinberger and Steinberger, 1971; Orth, 1982). In normal adult rats, Sertoli cell numbers appear stable throughout life. Morphometric data also show that the Sertoli cell number per unit length of seminiferous tubules appears stable throughout the cycle of the seminiferous epithelium despite variations in the diameter of the tubules and the volume of the lumen. It is estimated that there are about 22×10^6 Sertoli cells per testis (Wing and Christensen, 1982).

Since the Sertoli cell population in the adult animal is determined during the perinatal period, as mentioned above, any disruption of the proliferative process will have a profound effect on the Sertoli cell number and, in turn, affect testicular size (Agelopoulou et al. 1984). Although the control mechanism of Sertoli cell mitosis during the perinatal period is not fully understood, there is considerable evidence that hormones such as thyroid hormone and FSH are key modulators. While FSH stimulates Sertoli cell proliferation, thyroid hormone directly promotes maturation and decreased Sertoli cell mitogenesis at the same time (Orth, 1982; Cooke, 1991).

Structure

Sertoli cells are large, stellate cells with numerous cytoplasmic processes with adjacent Sertoli cells and germ cells. At the light microscopic level, Sertoli cells can be distinguished from germ cells by their large polymorphous nucleus with a pale dusting chromatin structure and a large nucleolus. Ultrastructurally, Sertoli cells are characterized as tall (75–100 μm), irregularly columnar and stellate cells. The cells extend from the basal lamina toward the central point within the lumen of seminiferous tubules. They have numerous lateral and apical processes extending between and around germinal cells. Despite the difficulties in the morphometric measurement of Sertoli cells due to their complex structure, a stage V rat Sertoli cell has been reconstructed based on a series of electron microscopic photographs (reviewed by Russell, 1993). The reconstructed Sertoli cell shows a volume of about 6000 μm^3 and a surface area of about 12,000 μm^2. There are about 10 elongated spermatids extending deeply into the apical recesses of a stage V Sertoli cell. The numerical density of germ cells to Sertoli cells is about 16:1. The estimated ratio of volume density of germ cells to Sertoli cells is about 10:1. This ratio is less than the ratio of numerical density because the germinal elements are smaller than Sertoli cells (reviewed by Bardin et al., 1994).

Sertoli cells contain a well-developed network of cytoskeletons including all three classes of cytoskeletal proteins: actin filaments, intermediate filaments, and microtubules (Vogl et al., 1993). Actin filaments have both structural and motility-related functions in cells. In Sertoli cells, actin can be detected scattered throughout most of the cytoplasm but is concentrated in ectoplasmic specializations and tubulobulbar complexes (Vogl et al., 1989). The former is a complex formed by bundles of actin filaments sandwiched between the plasma membrane of the Sertoli cell and cisterns of smooth endoplasmic reticulum. This structure can be found in the basal regions of the seminiferous epithelium at sites of attachment between neighboring Sertoli cells and in the apical regions between Sertoli cells and germ cells, mainly elongated spermatids. The latter, tubulobulbar complex is a structure consisting of specialized projections that extend into Sertoli cells from neighboring spermatids (Russell, 1979). Sertoli cells also contain actin-associated proteins such as α-actinin, fimbrin, and vinculin (Jockusch and Isenberg, 1981; Grove and Vogl, 1989; Grove et al., 1990).

The functions proposed for cytoplasmic intermediate filaments include positioning the nucleus in cells, anchoring cells to each other and to extracellular matrix, and mechanically transducing signals between the cell surface and the nucleus (see reviews by Schliwa, 1986, and Klymkowsky et al., 1989). Intermediate filaments in normal Sertoli cells in mature mammals are of the vimentin type (Paranko et al., 1986; Mali et al., 1987; Amlani and Vogl, 1988). In a typical Sertoli cell, vimentin proteins are centered around the basally positioned nucleus. From there, they radiate to the periphery of the cell where they associate with the plasma membrane at sites of attachment both to adjacent cells and to the basal lamina.

Microtubules are elongate polymers composed of tubulin subunits. Like actin filaments, microtubules appear to have both structural and motility-related roles. In Sertoli cells, microtubules are concentrated in the supranuclear region, where they extend throughout the body and into apical stalks and processes that support developing spermatids. In these regions, microtubules are generally aligned parallel to the long axis of the Sertoli cell.

One of the most important functions of Sertoli cells is to form the blood-tubule barrier. The blood-tubule barrier is exclusively formed by inter–Sertoli cell tight junctions and hence is also called the Sertoli cell barrier. With this barrier, the seminiferous epithelium is structurally as well as functionally divided into two compartments: basal and adluminal (Dym and Fawcett, 1970; Russell, 1978). The basal compartment faces the interstitial space and houses the spermatogonia, preleptotene spermatocytes, and the peritubular face of the Sertoli cells. The adluminal compartment contains more advanced germ cells including primary and secondary spermatocytes and spermatids. There is an intermediate compartment formed for a relatively short period of time to allow clones of preleptotene spermatocytes to move from the basal to the adluminal compartment in stage VII–VIII tubules. It is thought that this barrier creates a specialized environmental essential for germ cell development. It also maintains this environment by exclusion of potentially harmful factors from the systemic circulation, and vice versa.

The structure of Sertoli cells undergoes cyclic changes during the cycle of spermatogenesis. For example, unlike the number of Sertoli cells per seminiferous tubule unit, which is constant along the tubule, Sertoli cell volume fluctuates in relation to the stages of the cycle (Bugge and Loen, 1986). The smallest volume is found at stage VII–VIII, whereas the largest volume is at stages X and XII–XIV of the cycle.

According to the early observations by Leblond and Clermont (1952), Sertoli cell nuclei can be categorized into two groups: the "parallel type" and the "perpendicular type." The former usually shows a flat or oval morphology parallel to the peritubular base of the cell and are mainly distributed in stage VIII (the time of sperm release). The latter is usually triangular or piriform with the apex pointed toward the lumen and mainly observed in stage IX–XIV tubules. It is thought that perinuclear microtubules and microfilaments play roles in shaping the nuclear morphology of Sertoli cells, though no convincing evidence is yet available to support this hypothesis (Suarez-Qian and Dym, 1992).

Both primary and secondary lysosomes have been well described in Sertoli cells. The number of lysosomes per Sertoli cell also changes dramatically during the cycle. From stage X to stage XIV, the number of lysosomes increases from 278 ± 23 to 556

± 8, respectively. After the heads and proximal part of the spermatid tails disengage from the Sertoli cell processes at stage VIII, the residual bodies are moved to the basal part of the Sertoli cell and are rapidly lysed by lysosomes at stages IX–X. Coupled with this activity, the secondary lysosomes merged with the residual bodies and the number of lysosomes dramatically decreased to 100 ± 24 (Morales et al., 1985, 1986).

Sertoli cells contain numerous lipid droplets. Following the phagocytosis at stage IX, the volume of lipid in the Sertoli cell droplets increased during stages IX–XIV (Asaaf and Clermont, 1981). A substantial increase in the amount of lipid material in Sertoli cells was also found when there is an increase in germ cell degeneration (Lacy et al., 1968). This evidence suggests that the lipid droplets may result from the lysis of residual bodies and/or degenerating germ cells.

Other Sertoli cell subcellular organelles such as mitochondria, rough and smooth endoplasmic reticulum, and Golgi apparatus also show cyclic variations (Kerr, 1988; Hermo et al,. 1991).

Function

The basic function of Sertoli cells is to create an appropriate microenvironment for germ cell development, although it is apparent that more detailed functions will be elucidated at the molecular level. The Sertoli cell performs its basic function through structural maintenance of an intact seminiferous epithelium and functional regulation of the biochemical environment in which the germinal cells successfully develop from undifferentiated spermatogonia into highly specialized spermatozoa. In the structural aspect, Sertoli cells (1) maintain the integrity of the seminiferous epithelium through their numerous attachments to other cells and acellular elements, (2) compartmentalize the epithelium with the tight junctions between Sertoli cells, (3) take part in the upward and probably also downward movement of germ cells within the epithelium, and (4) play an important role(s) in the process of spermiation (Russell, 1979).

Compared to their structural supporting roles in the process of spermatogenesis, the biochemical functions of Sertoli cells are more complicated. The major biochemical functions of Sertoli cells include delivery of nutrients to germ cells, secretion of fluid and proteins, steroidogenesis and steroid metabolism, phagocytosis, and serving as targets for hormones mainly follicle-stimulating hormone (FSH) and testosterone. To demonstrate the complexity of Sertoli cell functions, we briefly review here the biological roles of FSH in the seminiferous epithelium and the production of proteins by Sertoli cells.

Role of FSH

It is generally accepted that FSH receptors are present only in Sertoli cells, although there is limited evidence showing that FSH receptors may also be present on spermatogonia (Orth and Christensen, 1978). Like many other polypeptide hormones, FSH binds to the FSH receptor, which activates adenylate cyclase activity via the interaction between the receptor and a G-stimulator (Gs) protein. Activation of adenylate cyclase results in a large increase in cAMP levels and subsequently in cAMP-dependent kinase (PKA) activity. Increased PKA activity phosphorylates a variety of structural proteins such as vimentin, enzymes, and regulatory factors such as cAMP response element binding protein (CREB). The phosphorylated CREB then binds to a promotor which

contains a cAMP response element (CRE) and activates some genes such as those encoding for *c-fos* and other *trans*-acting factors. The biological sequences of this cascade varies with the developmental status of the animal and probably also with the cycle
of spermatogenesis in adult animals (reviewed by Griswold, 1993).

In prenatal and newborn animals, it is believed that FSH is of critical importance
in stimulating the proliferation of Sertoli cells (Orth, 1984). However, some recent
evidence does not support this conclusion. For example, when neonatal rats were treated
with 6-propyl-2-thiouracil (PTU) to induce hypothyroidism, decreased FSH levels were
observed during the treatment and into adulthood, but Sertoli cell number was considerably *increased* when treated rats developed into adulthood (Kirby et al., 1992). Nevertheless, in prepubertal rats, it appears that FSH is still required for the final process
in Sertoli cell maturation, such as formation of the inter–Sertoli cell tight junctions and
initiation of the first wave of spermatogenesis (Solari and Fritz, 1978; Fritz, 1978).

The action of FSH in adult rats is much more complicated than in neonatal and
immature rats. Although FSH receptors have been shown to be expressed at high levels in adult rats, a number of experiments have shown that FSH is not required for basic
spermatogenesis in the adult rat, but is necessary for maintaining quantitatively normal spermatogenesis (Yoon et al., 1990). For example, spermatogenesis is only qualitatively normal and total sperm production is significantly decreased in hypophysectomized rats that lack circulating FSH (Sharpe, 1987). In vitro, FSH prevents apoptosis
and stimulates DNA synthesis in cultured rat seminiferous tubules (Henriksen et al.,
1996). The effects of FSH on Sertoli cells seem to vary with different stages of the
spermatogenetic cycle (Parvinen, 1993). In fact, the final output of FSH action could
be considered as a balanced result of all the components of the cascade from FSH binding to its receptor to the final functioning of FSH effectors such as PKA-phosphory-
lated proteins. Cyclic changes in some of these components have been observed. For
example, FSH binding to its receptors gradually increases after stage VIII and reaches
the maximal level at stage I, after which it decreases again and become minimal at stages
VI–VII (Parvinen et al., 1980). In accordance with this observation, FSH receptor
numbers and mRNA levels showed a similar cyclic change (Heckert and Griswold,
1991). There are also a number of other factors that may affect the response of Sertoli
cells to FSH stimulation. For example, phosphodiesterases (PDEs), which degrade
cAMP to AMP, inhibitory Gi proteins, protein kinase C, calcium, and adenosine have
all been shown to interact with the FSH signal transduction system (Conti et al., 1984;
Monaco and Conti, 1986; Quirk and Reichert, 1988; Monaco et al., 1988; Grasso and
Reichert, 1989; Huhtaniemi et al., 1989).

Production of Proteins

The Sertoli cell secretes a large number of proteins including transport and binding
proteins, proteases and antiproteases, extracellular matrix proteins, as well as regulatory proteins such as growth factors (reviewed by Skinner, 1993). Examples of Sertoli
cell–secreted growth factors include insulinlike growth factors I and II (IGF-I and IGF-
II), epidermal growth factor (EGF), transforming growth factors a and b (TGF-a and
TGF-b), nerve growth factor (NGF), interleukin-1 (IL-1), and stem cell factor (SCF)
(see reviews by Heindel and Treinen, 1989; Skinner, 1993; Lamb, 1993; Packer et al.,
1995).Though it is generally recognized that the major function of a growth factor is
to regulate cell proliferation and differentiation within a tissue, the mechanisms whereby

the growth factors produced by Sertoli cells alter the testicular functions are not clear. As described above, the morphology and function of Sertoli cells differ at different stages of the cycle. Since the circulating gonadotropin and androgen levels remain relatively constant, it is presumed that local events involved in intercellular communications within and between testicular compartments are of particular significance. One hypothesis is that each step in spermatogenesis is regulated by a complex series of interactions involving growth factors (Lamb, 1993).

EXPERIMENTAL MODELS

In Vivo Experiments

The general approach used to elucidate the mechanism of action of a toxicant also applies to the male reproductive system. This approach has been discussed in detail by Chapin and Heindel (1993). First it is necessary to determine that an agent is toxic to the male reproductive system using in vivo whole animal studies. These studies can define the dose response, time course, reversibility, and effects of age for a toxic response, as well as provide some information on the likely target organ(s) or even cell type(s). This can be accomplished with breeding studies such as the conventional multigeneration study or continuous breeding study (Heindel et al., 1989; Gulati, et al., 1991). Once the agent has been identified as a male reproductive toxicant, a "pathogenesis" study is designed to decipher the target cell(s) (reviewed by Chapin and Heindel, 1993). Generally, this can be done by observing histological and/or biochemical endpoints at different time points after a single or multiple doses. Theoretically, if a chemical induces an early change (morphological or biochemical) in Sertoli cells shortly after an in vivo exposure, i.e., before changes are detected in other cell types, this chemical could be called a Sertoli cell toxicant. In practice, recognizing an agent as a Sertoli cell toxicant can be a difficult task, mainly due to the structural and functional complexity of the testis, and it is significantly affected by the observed endpoints and our current knowledge on the functions of Sertoli cells. For the purpose of description, one could classify the endpoints of Sertoli cell toxicity into two categories: biochemical and histological markers. One class of often-used biomarkers is the measurement of the concentrations of Sertoli cell–secreted proteins such as androgen-binding protein (ABP), inhibin, transferrin, etc., in serum or in the seminal plasma of the ejaculates (reviewed by Sharpe, 1992; Maddocks et al., 1992). The problem with these markers are that they are readily altered and are therefore not specific. For example, they could be altered by breakdown of the blood-tubule barrier, loss of polarity in the Sertoli cell transport and secretion pathways, altered communication among testicular cells, as well as breaches of the integrity of downstream structures such as the epididymis and vas deferens (reviewed by Linstrom et al., 1988; Sharpe, 1992). Evaluations of the hypothalamic-pituitary-testicular axis function are generally helpful in the initial assessment of suspected Sertoli cell toxicants. Data generated from these evaluations of biochemical markers could provide suggestive information about the target tissues, but they are generally difficult to use as an indicator of Sertoli cell functions. Therefore, as was discussed by Chapin and Heindel (1993), the histological demonstration of pathological changes, while not guaranteeing which cell type is the primary target, is currently the most satisfying demonstration of damage. In most cases, a preliminary idea of the mechanism causing damage can be generated by the morphological pattern of response. The histological endpoints frequently used to identify the Sertoli cell damage are described in the following sections.

Changes in Sertoli Cell Numbers

The Sertoli cell undergoes proliferation during a narrow perinatal period. Prepubertal rats treated with busulfan prenatally have reduced numbers of Sertoli cells with fewer FSH receptors (Viguier-Martinez et al., 1984). Reduced Sertoli cell numbers have also been observed in neonatal rats (2 and 3 weeks old), but not in adults, after exposure to di(2-ethylhexyl) phthalate (DEHP) (Dostal et al., 1988).

Neonatal hypothyroidism causes an extension of this proliferative period and results in an increase in the population of Sertoli cells in adult animals. Therefore, increased numbers of Sertoli cells creates an enlarged testis that produce more sperm than normal (Hess et al., 1993). Rats treated with hypothyroidism-inducing chemicals such as 6-propyl-2-thiouracil, Aroclor 1242, a polychlorinated biphenyl (PCB), during the neonatal period (0–25 days) showed a significant increase in testis weight and daily sperm production (Hess, et al., 1993; Cooke et al., 1996). In contrast, when a hyperthyroid condition is created by triiodothyronine injection during the neonatal period, the rate of Sertoli cell proliferation decreased and the proliferative activity stopped prematurely. Thus, decreased testis size and sperm production were observed in the adult animals (van Haaster et al., 1993). To date, there has been no report demonstrating that perinatal exposure to an environmental toxicant could yield fewer Sertoli cells in adult animals.

Vacuolization

Accumulation of various vacuoles is often an early response of Sertoli cells to a variety of insults such as exposures to testicular toxicants and testosterone withdrawal (Creasy et al., 1987; Kerr et al., 1993). At the light microscope level, at least two distinct forms of vacuoles can be observed. One form consists of multiple, small vacuoles and is often seen in the basal part of the epithelium. An example of this form is seen in rats after exposure to phthalates (Creasy et al., 1987). Another form of vacuoles consist of large, clear areas such as those seen with 2,5-hexanedione exposure (Chapin et al., 1983; Boekelheide, 1987a, 1988a). Though vacuolization of Sertoli cells is a common phenomenon and is an early response in the cascade of Sertoli cell damage, there are still questions about their localization as well as their mechanism of formation. While there are a number of reports showing they are within the cytoplasm of Sertoli cells and are probably formed by swelling and coalescence of intracellular membrane-bounded structures such as endoplasmic reticulum and vesicles (Creasy et al., 1983; Chapin et al., 1989), Kerr et al. (1993) showed that various vacuoles induced by testosterone withdrawal are intercellular and probably result from disruption of Sertoli cell junctional complexes.

Inhibited Spermiation

Spermiation is the process of sperm release from the seminiferous epithelium. This process takes place at stage VIII in normal rat testis. Although little is known about the factors that control this process, available evidence suggests that the Sertoli cell has a major role (Russell, 1991). Inhibited spermiation can be defined morphologically as the existence of late elongated spermatids at various levels within the epithelium at stages IX-XII tubules. According to Russell's observations (1991), about 50% of all the abnormal treatments/conditions he examined in common laboratory species showed inhibited spermiation. In fact, inhibited spermiation has been observed with most, if not all, of the known testicular toxicants such as boron (Ku et al., 1993), cadmium (Hew et al., 1993a,b), 1,3-dinitrobezene (DNB) (Hess et al., 1988b), 2,5-hexanedione (HD) (Chapin et al., 1983), and tri-o-cresyl-phosphate (TOCP) (Chapin et al., 1991). The mechanisms underlying the processes of inhibited spermiation remain to be determined. Several hypotheses have been proposed. One maintains that the inhibited spermiation results from Sertoli cell phagocytosis of late spermatids (Russell et al., 1990). The primary defect could either be in the spermatids (these abnormal spermatids being recognized and phagocytosed by Sertoli cells) or in Sertoli cells resulting in abnormal phagocytosis. If this hypothesis were true, one would expect to see no normal specializations between Sertoli cells and spermatids, and those retained spermatids should be embedded within membrane-bounded structures. Unfortunately, there is no evidence to differentiate these hypotheses.

There is also a question about whether the fluid movement within seminiferous lumen is important in spermiation. The concern arises from two observations. First, despite the long length of the seminiferous tubules, few released elongated spermatids are observed in the lumen of the seminiferous tubules at stages other than stages VIII–IX. This indicates that released sperm is immediately transferred from stage VIII tubules to the rete testis. In other words, the luminal fluid must move quickly. But this postulate is not supported by the early data showing that the rate of fluid movement along a single seminiferous tubule is only about $1\mu l/h$ (Hinton and Setchell, 1978; Setchell et al., 1978). Second, when stage VII–XIV seminiferous tubular segments are cultured with the two ends closed (see next section), there is no flow of luminal fluid. After culture for 24 hours, late elongated spermatids (step 19) can be frequently observed in stage IX–X seminiferous tubules (containing step 9–10 round spermatids), indicating almost completely inhibited spermiation (personal observation). Indeed, an association between a decrease in luminal fluid secretion and inhibited spermiation has been seen with exposures to phthalates and 2,5-hexanedione (Gray and Gangolli, 1986; Johnson et al,. 1991).

Apical Sloughing and Shedding

Germ cells are sloughed individually or as a group into the lumen of the seminiferous tubules, sometimes with fragments of attached Sertoli cells. Sloughed cells may be detected both within the seminiferous tubular lumen and the epididymal lumen (Russell et al,. 1981; Creasy et al., 1983; Boekelheide, 1988a). Since germ cell sloughing may result from the handling and cutting of the tissues during histological process, sloughed cells found in the epididymal lumen is the best evidence to indicate the occurrence of germ cell sloughing in seminiferous epithelium.

Disturbance of the Timing of Spermatogenesis

It is generally believed that the timing of spermatogenesis is under rigid genetic control with little or no change during germ cell development. This conclusion is mainly based on the fact that neither hormone treatments nor hormone withdrawal affect the timing of spermatogenesis (Clermont and Harvey, 1965). However, disturbances in the timing of spermatogenesis have been observed. For example, in vitamin A–deficient rats, young proliferative spermatogonia (Aa1 or A1) failed to divide into more mature spermatogonia (A2, etc.). It is thought that the germ cell is arrested at the differentiating A1 spermatogonia. Since the synchronization of spermatogenesis achieved in vitamin A–replenished rats is lost over the course of about 10 spermatogenetic cycles, there must be some heterogeneity in the duration of the spermatogenetic process (Morales and Griswold, 1987). It is established that the relative frequency of a spermatogenetic stage reflects the relative duration of that stage within the entire cycle of the seminiferous epithelium. While the relative frequency (percentage) of a given stage in normal rats is remarkably constant (Hess et al., 1990), several testicular toxicants have been shown to alter the frequencies of particular stages. For example, an obvious increase in stage XIV tubule has been reported in rats 24–48 hours after a single oral dose of 2-ME (Chapin et al., 1984; Creasy et al,. 1985), while a significant decrease in this stage was observed in rats treated with DNB (Hess et al., 1988b).

Germ Cell Degeneration

The normal structure and function of Sertoli cells and germ cells are mutually dependent. Germ cell degeneration is probably inevitable after severe damage in Sertoli cells. Additionally, germ cell degeneration is always associated with morphological changes in Sertoli cells. This mutually dependent complexity makes it extremely difficult to answer the question about which comes first, Sertoli cell damage or germ cell damage, when only morphological data are used. Nevertheless, germ cell degeneration is still the most frequently used morphological indicator to identify the stage and/or cell type preference of a toxic effect. While previous studies generally recognized germ cell death as necrosis, recent studies have shown that germ cells die by apoptosis regardless of the initiator of cell death, since internucleosomal DNA degradation, the hallmark of apoptosis, could be consistently observed in germ cell death induced by a number of testicular toxicants such as 2,5-hexendione (Blanchard et al., 1996), 2-methoxyethanol (2-ME) (Ku et al., 1995), mercury, and cadmium (Li et al., unpublished). One interesting phenomenon is the morphology of dying spermatocytes. While degenerating spermatocytes in rat testis usually show heavily stained cytoplasm with dispersed chromatin masses, a morphology typical of necrosis (Chapin et al., 1984), dying spermatocytes in guinea pigs (Ku et al,. 1994, 1995) and humans (Li et al., 1996) show a morphology typical of apoptosis (Wyllie et al., 1984). The mechanism(s) underlying this species difference remain to be answered.

In summary, though microscopic examination of testes provides limited information about Sertoli cell toxicities, careful characterization of a particular morphological pattern of response can often generate enough data to set up a working hypothesis for further investigation.

In Vitro Experiments

Once the Sertoli cell is identified as the probable early target cell type of an agent, in vitro models can be used to elucidate the mechanism(s) of action. Many in vitro model systems have been developed to evaluate Sertoli cell functions as well as the actions of the testicular toxicants on the Sertoli cells. Among them, Sertoli cell–enriched culture and Sertoli–germ cell co-culture have been successfully applied in toxicological studies.

Sertoli Cell–Enriched Culture

In this model, Sertoli cells are isolated from testis by enzymatic digestion and cultured in monolayer. In order to obtain a highly enriched Sertoli cell population, most studies have used immature animals as testis donors, particularly at an age shortly before the completion of the first cycle of meiotic division and the appearance of spermatids (e.g., 18- to 20-day-old rats) (Dorrington and Armstrong, 1975; Mather et al., 1982; Russell and Steinberger, 1989). As an alternative, Sertoli cell–enriched cultures can be prepared from 28-day-old rats. In either case, contaminating germ cells can be removed from the culture by hypotonic shock treatment (Galdieri et al., 1981; Wagle et al., 1986). Several biomarkers are frequently used to indicate the functions of cultured Sertoli cells in these models, in addition to morphological and/or viability data. Secretion of lactate, pyruvate, Sertoli cell–specific proteins (e.g., ABP, inhibin, etc.), and FSH-stimulated cAMP production have been used to elucidate the mechanisms of the Sertoli cell toxicities of several testicular toxicants such as DNB (Gray, 1988) and mono-2-ethylhexyl phthalate (MEHP) (Heindel and Chapin, 1989).

Sertoli–Germ Cell Co-Culture

In this model, Sertoli cells are prepared from 28-day-old rats and the attached germ cells are not removed from the cultures. Compared to Sertoli cell–enriched culture, this model can be used to examine the interaction(s) between Sertoli cells and germ cells, although the relationship between these two cell types in this model is largely different from that in vivo. One measure of toxicity using this model is the assay of germ cell detachment from the monolayer of Sertoli cells ("pop-off" assay) (Gray and Beamand, 1984). This endpoint is very similar to a part of the in vivo lesion where germ cells are shed from the epithelium. For example, treatment of Sertoli–term cell co-cultures with MEHP increased the germ cell detachment from the Sertoli cell monolayer, which partly mimicked the testicular damage induced by DEHP in vivo (Gray, 1986).

The advantage of these two models lies in their simplicity. They provide opportunities to look at the specific effects of toxicants (mechanism of actions) on relatively purified Sertoli cells. Once the action of a chemical on Sertoli cells is known, the toxicities of the parent chemicals, their metabolites, and structurally related chemicals can be screened. This will be useful for identifying toxic moieties and exploring possible mechanism(s) of action.

The disadvantages of the models also result from their simplicity—they lack intact structural and/or functional relationships between germ cells and Sertoli cells. While cultured Sertoli cells still retain some of their in vivo features, there is increasing evidence showing the difference between cultured Sertoli cells and Sertoli cells in vivo. For example, cultured Sertoli cells respond to cAMP stimulators differently from cultured seminiferous tubules and from Sertoli cells in vivo (Attramadal et al., 1984). This limitation could become critical when the target of a chemical in the testis is the structural and/or functional interactions between germ cells and Sertoli cells. An example is the Sertoli cell toxicity of 2-ME, a glycol ether that induces testicular atrophy and has been previously recognized as a germ cell toxicant (Chapin and Lamb, 1984). When enriched germ cell cultures were exposed to methoxyacetic acid (MAA, the active metabolite of 2-ME), germ cells did not degenerate (Gray et al., 1985; Ku and Chapin, 1994). Degeneration could only be found in spermatocytes attached to Sertoli cells. In contrast, the 2-ME–induced in vivo lesion could be faithfully mimicked in vitro in cultured rat seminiferous tubules that maintained the intact relationship between Sertoli cells and germ cells (Ku and Chapin, 1994).

Seminiferous Tubule Culture

A number of investigators have cultured testicular tissue chunks or individual seminiferous tubule segments isolated either from human or rat testes to investigate the dynamics and regulations of spermatogenesis (Steinberger et al., 1963; Chowdhury et al., 1975; Toppari et al., 1986; Seidl and Holstein, 1990; Nikular et al., 1990; McKinnell et al., 1995). Compared to the above-mentioned Sertoli cell co-cultures, seminiferous tubule cultures maintain the intact relationship between Sertoli cells and germ cells. Recently, Ku and Chapin (1994) used cultured seminiferous tubule cultures for their study on the testicular toxicity of 2-ME. They found that 2-ME–induced germ cell death shown in vivo could only be faithfully mimicked in vitro using cultured rat seminiferous tubules with MAA. These findings suggest that cultured seminiferous tubules may be a better in vitro model to investigate the mechanisms of the testicular toxicants. The advantage of this tubule culture model over the more disrupted Sertoli cell culture models is that it maintains the intact relationship between germ cells and Sertoli cells, which is believed to be important for the normal functions of these cells. Though mechanically isolated tubules have been cultured for more than 3 days (Toppari, et al., 1986), their overall morphology is generally destroyed after more than 48 hours in culture; optimal morphology was observed in 24-hour cultured seminiferous tubules isolated after a brief enzymatic digestion (Li et al., 1996a). Therefore, one of the limitations of the tubule culture model is the relatively short culture time. The model cannot be used for studies of agents that require more than 24-hour exposure. Nonetheless, the possible signal transduction pathways involved in 2-ME–induced spermatocyte apoptosis has been explored with cultured seminiferous tubules isolated from adult and 25-day-old rats and with cultured tissue chunks isolated from human testis (Li et al., 1996a, b) thereby showing its application.

REVIEWS OF THE SERTOLI CELL TOXICITIES OF KNOWN TESTICULAR TOXICANTS

In this section, rather than providing a detailed review of the Sertoli cell toxicities of all known testicular toxicants, we have chosen to concentrate on the five Sertoli cell toxicants that have been studied in greatest detail: cadmium, dinitrobenzene, glycol ethers, hexanedione, and phthalate esters. In each case, there is strong evidence that the Sertoli cell is the primary target. In addition, in vitro approaches have been used to give information on the mechanisms of actions of these agents. While in no case have definitive mechanisms been elucidated, these studies demonstrate the approaches that can be used to identify the possible targets and the information that can be used to develop a research plan to investigate the testicular toxicities of other agents.

Cadmium

Cadmium is a heavy metal that occurs widely in nature as a contaminant of zinc. It has no known physiological functions in vivo. Since trace amounts of cadmium are naturally present in water as well as plants and animals, humans are inevitably exposed to it through the food chains (reviewed in Steinberger, 1993). The air we breathe contains cadmium in the form of oxides and chlorides. In addition, it is commercially produced and used in electroplating, as a paint pigment, and in electrical batteries. Since it is nonbiodegradable and has a biological half-life of 200 days in rats (Webb, 1975), it is considered a cumulative toxicant. In contrast to the other toxicants reviewed here, cadmium toxicity in the testes has been known for over 75 years, since Alsberg and Schwartze (1919) reported a bluish discoloration of testes in animals given cadmium salts at 2.2 mg/kg.

Severe testicular toxicity can be induced in rats by a single s.c. dose of cadmium (Laskey et al., 1989 [2 mg/kg]; Kotsonis and Klaassen, 1977 [100 or 150 mg/kg]), but not after prolonged oral administration in food or water (Kotsonis and Klaasen, 1977; Zenick et al., 1982 [68.8 ppm]). Indeed, rats given 100 ppm cadmium in their drinking water for 12 weeks showed no testicular toxicity even though the concentration of cadmium in their testes was greater than that which caused testicular injury measured 14 days after a s.c. injection (Kotsonis and Klaasen, 1977, 1978). This suggests that it is not the concentration in the testis that is important for testicular toxicity, but the rate of administration, indicating a threshold for cadmium-induced testicular toxicity. Indeed, there is evidence that low doses of cadmium stimulate an increase in proteins (metal-lothionein and other cadmium-binding proteins) that protect against its toxicity; high doses probably overwhelm this defense mechanism (Waalkes et al., 1984; reviewed in Steinberger, 1993).

The initial testicular toxicity studies focused on high single doses at which cadmium caused increased permeability of the vascular endothelium. Blood vessels in the testis are particularly sensitive to this effect and are believed to be the primary reason for the testicular damage following high exposure to cadmium. Damage appears to consist of loss of the endothelial tight junctional barrier, which results in edema, increased fluid pressure, ischemia, and tissue necrosis (Steinberger, 1993; Hew et al., 1993a; Aoki and Hoffer, 1978). Thus the testicular toxicity of cadmium was thought to be secondary to this vascular damage. However, Lee and Dixon (1973) showed that 1 mg/kg of cadmium compromised spermatogenesis without noticeable alterations in the testicular vasculature. The first indication that the Sertoli cell might be the primary target of cadmium toxicity came from the work of Johnson (1969) and Setchell and Waites (1970), who showed that cadmium altered the blood-testis barrier. These in vivo results were confirmed in vitro using Sertoli cell monolayers in two compartment cultures (Janecki et al., 1992). Hew et al. (1993a,b) have proposed that the cadmium effect on Sertoli cell tight junctions (i.e., the blood-tubule barrier) may be due to its action on the actin filaments associated with these junctions.

Several experiments have been performed to histologically define the initial target for cadmium in the testis. In 1988, Gouveia looked for morphological damage to the testis after a single i.p. dose of 1 mg/rat of cadmium chloride to adult Wistar rats. Unfortunately this dose was an LD_{50} and therefore too high to be useful to detect a subtle initial lesion. They reported vascular leakage after 2 hours and major toxicity at the next time point of 24 hours. The high dose and the overwhelming preconception of vascular toxicity as the initial lesion precluded meaningful results. In 1993, Hew et al. used a lower i.p. dose of 1.0 mg/kg (~ 0.25 mg/rat), which caused no testicular vascular changes in Sprague-Dawley rats. The first effect, a failure of spermiation in stage IX tubules, was noted 24 hours postdosing. As time after treatment increased to 72 hours, failure of spermiation was seen in almost all tubules at stages IX–XII of spermatogenesis. They concluded that cadmium probably acts during early stage VIII of the seminiferous to induce failure of spermiation. Further, since Sertoli cells are ultimately involved in this process, they proposed that the Sertoli cells are the initial site of cadmium toxicity. Indeed, Hew et al. (1993b) showed that microfilament bundles in the basal region of the Sertoli cells were also disrupted by 24 hours and that the microfilaments were most severely altered in stage VIII of the seminiferous epithelium. Stages IV through VII did not show cadmium-induced microfilament changes indicating that cadmium, like most other testicular toxicants, is also a stage-specific toxicant. These data suggest a relationship between microfilaments, Sertoli cell tight junctions, and spermiation. Hew et al. (1993b) proposed that an as yet unidentified component of the tight junctions and spermiation process may be the actual target of cadmium in the testis. They have hypothesized that the sensitivity of some stages of the cycle of the seminiferous epithelium to cadmium toxicity may be due to decreased expression of hsp 27, a heat-shock protein, which is associated with microfilaments in Sertoli cells. Recently, Welsh et al. (1996) have shown that increased hsp 27 expression can confer resistance to cadmium toxicity. The next step could be to prepare hsp 27 transgenic mice to examine the direct role of this protein in the cadmium-induced testicular toxicity.

As with all the other Sertoli cell toxicants discussed, the testicular toxicity of cadmium is also age-related (Wong and Klaassen, 1980; Phelps and Laskey, 1989). The severity of cadmium-induced testicular damage increased with age for all variables measured even though the testicular concentrations of cadmium showed no differences with age. Utilizing testis weight as the endpoint, Wong and Klaassen (1980) reported that cadmium-induced testicular damage in rats occurred in three stages: testes in animals from birth to 2 weeks were insensitive, only slight degenerative changes occurred in testes from 3- to 5-week-old animals, and severe necrosis and ischemia occurred in the testes from animals 6 weeks of age and older. It has been proposed that differences in circulatory or vascular endothelium underlie these changes in age-related susceptibility. The possibility of altered levels of different isozymes of metallothionein or other cadmium-binding proteins with age has not been discounted. The proposal by Janecki et al. (1992) that the age-related toxicity may be related to the development of the blood-tubule barrier is also of merit, especially in light of the recent data showing cadmium effects on Sertoli cell microfilaments which are a part of the blood-tubule barrier (Hew et al., 1993a).

The cadmium-induced capillary toxicity which leads to necrosis and ischemia (i.e., the high-dose effects) can be prevented by the calmodulin inhibitors chlorpromazine, triflouperazine, and W7, suggesting that calmodulin may play a role in mediating the toxic effect of cadmium (Niewenhuis and Prozialeck, 1987). Since these agents tend to be biochemically nonselective, additional studies are needed to clarify these effects. Cadmium has also been shown to inhibit phosphodiesterase activity in vitro, probably via binding to calmodulin (Flick et al., 1987), lending further support to a role of calmodulin and calcium in the cadmium toxicity. Similarly, zinc, if given before or within 2 hours of cadmium exposure, is capable of reversing/inhibiting its deleterious effects on testicular vasculature, spermatogenesis, and fertility in the rat (Parizek, 1957; Webb, 1972; Saksena et al., 1983). This is probably because zinc competes with cadmium for sites on metal-binding proteins.

In summary, cadmium, at low doses, is a stage-specific Sertoli cell toxicant. The role of heat-shock proteins in the toxicity is intriguing as is the role of metallothionein and other binding proteins in protection against its testicular toxicity and the age-relatedness of its toxicity.

Dinitrobenzene

1,3-Dinitrobenzene (DNB) is a chemical intermediate in the manufacture of azodyes, explosives, and several organic syntheses including that of nitroanaline and phenylenediamine. It is acutely toxic to humans and animals. The estimated lethal dose range for humans is 5–50 mg/kg (Gleason et al., 1981) and the oral LD_{50} for rats is 83 mg/kg (Cody et al., 1981). The acute toxicity of DNB, like that of other nitroaromatic compounds, is due to cyanosis and hypoxia as a result of methemoglobinemia in humans and animals (Isihara et al., 1976; Watanabe et al., 1976).

As early as 1981 (Cody et al., 1981; Bond et al., 1981), there were reports that DNB could cause testicular toxicity in rats. These studies were confirmed and extended by Blackburn et al. (1985), Hess et al. (1985), and Linder et al. (1986). The latter reported that gavage of 1.5 mg/kg DNB to Sprague-Dawley rats decreased sperm production over a 12-week period. At 3 mg/kg/day, a dose without clinical signs of toxicity, there was a 60% decrease in testis weight, a 40% decrease in epididymal weight, and no motile sperm. These males had normal mating behavior but were infertile. These data were confirmed by Linder et al. in 1988 when they observed normal mating behavior and infertility 4–5 weeks after treatment with DNB (48 mg/kg).

Pathogenesis-type studies focusing on various endpoints were carried out to define both the earliest signs of toxicity and the testicular cell(s) most sensitive to the toxicity, i.e., the primary target cell in both Sprague-Dawley and Wistar rats (Foster et al., 1986a, 1987a; Blackburn et al., 1988; Hess et al., 1988; Rehnberg et al., 1988; Reader et al., 1991). Foster et al. (1986a) showed that 24 hours after a single dose of 25 mg/kg to sexually mature (60-day-old) Wistar-derived rats, there was cytoplasmic vacuolation in Sertoli cells, and retraction of the late pachytene spermatocytes occurred by 48 hours. These data led them to propose the Sertoli cell as the primary target but did not preclude direct toxicity to the germ cells. Blackburn et al. (1988), using this same system, showed that there was vacuolization and retraction of Sertoli cell cytoplasm in the region associated with primary spermatocytes, which already showed nuclear pyknosis and an increase in cytoplasmic eosinophilia as early as 12 hours after dosing with DNB. This toxicity was stage-dependent, occurring first in stages VIII–XI. By 24 hours, there was focal degeneration and loss of pachytene spermatocytes, disorientation of elongate spermatids and further vacuolization of Sertoli cell cytoplasm and vesiculation of the smooth endoplasmic reticulum in stages IX and X. Leydig cell morphology remained normal at all times (Blackburn et al., 1988). At about the same time Hess et al. (1988a,b) reported their pathogenesis study using 96-day-old Sprague-Dawley rats dosed with 48 mg/kg body weight (a dose that caused initial clinical signs of toxicity but no change in body weight). They showed that the first cells damaged at 24 hours postdosing were pachytene spermatocytes in stages VII–XIII and round spermatids. Sertoli cells contained large cytoplasmic vacuoles by 48 hours but not at 24 hours. By 48 hours, there was retention of step 19 spermatids in stage IX and X tu-bules, i.e., inhibited spermiation. Germ cells outside the blood-tubule barrier, i.e., spermatogonia and preleptotene spermatocytes, showed no detectable DNB toxicity. Hess et al. (1988a,b) concluded that these data could be explained by the existence of a metabolite formed by Sertoli cells that specifically targeted germ cells. This dose of 48 mg/kg was larger than the 25 mg/kg dose used by Foster et al. (1986a) and the animals used were slightly older and of the Sprague-Dawley vs. Wistar strain. Perhaps the larger dose resulted in too severe an initial toxicity to properly sort out the early timing relationship in the pathogenesis study, or there could be a strain difference since Blackburn et al. (1988) did not report early retention of step 19 spermatids in Wistar-derived rats. It is noteworthy that both studies reported a similar stage specificity to the toxicity.

Rehnberg et al. (1988), using a single dose of 32 mg/kg of DNB to 90-day-old Sprague-Dawley rats, reported a decrease in seminiferous tubule fluid after 3 hours and elevated levels of serum FSH after 2 weeks (confirmed by Reader et al., 1991) indica-tive of Sertoli cell toxicity. Pituitary and serum levels of LH were normal, as were the weights of the androgen-dependent organs (prostate and seminal vesicles), showing that the Leydig cell was not a target of DNB. Using flow cytometric analysis, Evenson et al. (1989) also showed that the first noticeable effect was a loss of pachytene sperma-tocytes one day after dosing Sprague-Dawley rats with 48 mg/kg DNB. Since this tech-nique only counts the number of cells based on ploidy, it could not look for early non-lethal signs of Sertoli cell toxicity and is therefore not an appropriate technique for a pathogenesis study. Thus multiple pathogenesis type experiments concluded that DNB toxicity was stage dependent and resulted in pachytene spermatocyte death and suggested to varying degrees that the Sertoli cell was the initial target cell.

The testicular toxicity of DNB is at least partially reversible. Linder et al. (1986) showed rapid regeneration of spermatogenesis in rats treated with DNB (6 mg/kg/day) for 12 weeks and allowed a recovery period of up to 5 months. By 10 days posttreat-ment, there was active spermatogenesis with the presence of spermatids in most tubules and complete recovery in three of four animals by 5 months. Similarly, Hess et al. (1988a) showed that after a single dose of 48 mg DNB/kg to adult Sprague-Dawley rats, spermatogenetic recovery could be detected by day 16 posttreatment. Recovery, however, was variable such that by 175 days posttreatment three males had 95% com-plete recovery, another had 97% atrophic tubules, and three more had 15–45% atro-phic tubules. The variability in the recovery of different tubules is an area where more attention is warranted.

The testicular toxicity of DNB is species and age-dependent. Mice are less sen-sitive than rats (Evanson et al., 1989a; Obasaju et al., 1991). Within each species pre-pubertal and pubertal animals are less sensitive to DNB toxicity than adults (Evenson et al., 1989a,; Linder et al., 1990b). Indeed 105-day-old animals are more resistant that 75-day-old animals (Linder et al., 1990). Hamsters appear insensitive to the tes-ticular toxicity of DNB (Obasaju et al., 1991). Metabolic differences could account for the insensitivity of the young animals as they eliminated DNB from blood more rap-idly than adult animals. However, there are only minor pharmacological differences that could account for the differential sensitivity of the 75-day-old rat and that of the 120-day-old rat (Brown et al., 1994). Clearly it is an area that is not resolved and needs more attention.

There are three isomers of DNB: 1,2-, 1,3-, and 1,4-DNB. These three isomers, 1,2-dinitrobezene, 1,3-dinitrobenzene (DNB), and 1,4-dinitrobenzene, all induce methemoglobinemia and anemia on prolonged exposure (Watanabe et al., 1976). However, 1,2 (*ortho*)- and 1,4 (*para*)-dinitrobenzene isomers are not testicular toxicants at doses where 1,3 (*meta*)-DNB shows testicular toxicity (Blackburn et al., 1988). 1,4- and 1,3-DNB seem to be of similar potency with regard to cyanosis and splenic enlargement, indicating different mechanisms for these two toxic effects versus the testicular effect (Blackburn et al., 1988) and that cyanosis is not the mechanism of the testicular toxicity to DNB. This is substantiated by data showing no testicular damage after prolonged elevation of methemoglobin levels in rats dosed with sodium nitrate (Bond et al., 1981). In 1961, Parker showed that nitroreduction in the gastrointestinal tract (anaerobically) and liver (aerobically) followed by conjugation is the major metabolic route for DNB. Indeed no parent compound is excreted in the urine (McEuen and Miller, 1991). The products of this nitroreduction, nitroanaline and nitroacetanilide, are not testicular toxicants in vivo (Cossum et al., 1986; McEuen and Miller, 1991) or in vitro in Sertoli germ cell cultures (Cave and Foster, 1990). Indeed nitroreduction is a detoxification route for testicular toxicity. When germ-free animals (to remove one site of nitroreduction) are dosed with DNB, the testicular toxicity is enhanced while the hematological effects are decreased providing additional evidence for the testicular toxicity being separate from the effects of methemoglobeinemia and for a testicular site of formation of an active metabolite. To this end, Ellis and Foster (1992) used testicular subfractions to show that the testis is capable of facilitating metabolism of DNB to nitrosonitrobenzene (NNB), nitrophenylhydroxylamine (NP), and nitroanaline (NA). Also, Cave and Foster (1990) reported that Sertoli–germ cell cocultures are capable of metabolizing DNB to NA and nitroacetanalide and MNB. Nitroanaline and nitroacetanalide were not toxic to the Sertoli–germ cell cultures, while NNB not only mimicked the effects of DNB but was more toxic (3–5 times) on an equimolar basis (Foster, 1989). Thus, at this point, NNB or a further metabolite of NNB appears to be the prime candidate for the proximal toxicant of DNB, and the testis appears to be the site (or at least one site) of its formation.

Foster et al. (1987a) have calculated that the peak blood and testis concentrations of DNB following a single oral dose of 50 mg/kg are 5×10^{-5} and 2×10^{-5} M, respectively. This information allows the use of in vitro systems to examine the mechanism of DNB toxicity using concentrations relevant to those found in vivo. Using relevant concentrations of DNB in Sertoli–germ cell cocultures, Foster et al. (1987a) were able to show a dose-related increase in vacuolation of Sertoli cells and detachment of germ cells accompanied by their phagocytosis by the Sertoli cells. These results are directly comparable to what is measured after in vivo administration and validate the model for use in mechanistic studies. The question is, what is the mechanism whereby DNB results in the release of viable germ cells and their subsequent death and phagocytosis by Sertoli cell since the Sertoli cell is postulated to the initial target. Several aspects of Sertoli cell function have been addressed. DNB stimulates pyruvate and lactate production with no effect on ATP levels or FSH-stimulated cAMP production (Foster, 1989). This initial work on Sertoli–germ cell cocultures has not been continued. Thus the mechanism of the Sertoli cell toxicity of DNB remains unclear, as is the questions of some direct germ cell toxicity. The lack of recent work in this area leaves the mechanisms of DNB testicular toxicity unclear.

Isolated tubule segments have been used as a model system and have provided further information on DNB toxicity. Allenby et al. (1991) have used this system to show that DNB increased inhibin secretion and that this effect matches an in vivo effect to increase in immunoreactive inhibin in testicular interstitial fluid 3 days posttreatment. McLaren et al. (1993) have used staged tubule preparations to show DNB causes stage-specific changes in methionine incorporation into secreted proteins. They have identified six proteins whose secretion is altered by DNB. Recent studies in our laboratory have found that DNB-induced germ cell death in cultured rat seminiferous tubules is apoptotic and could be prevented by calcium channel blockers such as nifedipine and verapamil (L.-H. Li and R. E. Chapin, unpublished).

Clearly, there are some data concerning the mechanism of the Sertoli cell and germ cell toxicity of DNB. Follow-up studies are needed to define the mechanism of the toxicity in Sertoli cells and to determine any direct effect on the germ cells themselves.

Glycol Ethers

Ethylene glycol ethers are a family of organic water-miscible solvents used in a wide variety of products including printing inks, textile dyes, epoxy resin coatings, and as an anti-icing additive in jet fuels. The commonly used glycol ethers are monoalkyl ethers of ethylene glycol, such as ethylene glycol monomethyl ether (EGME) or 2-methoxyethanol (2-ME), ethylene glycol monoethyl ether (EGEE) or 2-ethoxyethanol (2-EE), ethylene glycol monobutyl ether (EGBE) or 2-butoxyethanol (2-BE), etc. (reviewed in Oudiz and Zenick, 1986; Browning and Curry, 1994). The estimated annual U.S. production of glycol ethers reached 1 billion pounds in 1984, with 2-ME production estimated at 3.78 million kilograms (about 8.35 million pounds) in 1985 (Browning and Curry, 1994; HSDB, 1995). The National Institute of Occupational Safety and Health (NIOSH) estimated that 168,180 workers (about 70% of them male) were exposed to 2-ME in 1983 (HSDB, 1995). Glycol ethers with a short carbon chain cause a wide range of toxic effects in the reproductive, developmental, and hematopoietic systems. The developmental and reproductive toxicity of glycol ethers are inversely proportional to the length of the n-alkyl carbon substituent, with 2-ME being the most efficacious (Nagano et al., 1979). It has been shown that 2-ME targets a range of organs, including fetus, testis, thymus, and bone marrow, in a variety of species, including mouse, rat, guinea pig, rabbit, dog, as well as human (Hardin, 1983; Miller et al., 1983; Chapin et al., 1984; Creasy et al., 1985; Clarke et al., 1991; Holladay et al., 1994; Simialowicz et al., 1994; Ku et al., 1994a; Balasubramanian et al., 1995).

The ability of 2-ME to induce testicular damage was first demonstrated in rabbits in 1938 by Wiley et al. (1938). Nagano et al. (1979) used a variety of ethylene glycol monoalkyl ethers to investigate the testicular effects of those compounds in mice. They found that 2-ME and 2-EE were the most potent in their effects on the testis. The earliest effects on relative testis weight with 2-ME occurred after 5 weeks of oral treatment with 250 mg/kg/day dose level. The testicular toxicity of 2-ME was also demonstrated in mice, rats, and rabbits by the inhalation route of exposure (Miller et al., 1981, 1983). While these studies recognized that the testis is one of the major targets of 2-ME, they gave little indication of the nature of the early damage. In the 1980s, the National Toxicology Program at NIEHS performed a series of studies to evaluate the reproductive toxicity of glycol ethers (Chapin et al., 1984, 1985a,b; Chapin and Lamb, 1984; Lamb et al., 1984). Significant adverse effects on fertility were seen in both male and female CD-1 mice treated with 1% and 2% 2-ME in drinking water for 14 weeks. Testicular atrophy, decreased sperm motility, and increased abnormal sperm were noted in the treated males. Significantly decreased fertility was also observed in male F344

rats gavaged with 200 mg/kg/day for 5 days (Chapin et al., 1985a). Although the serum levels of testosterone, FSH, and LH have not been evaluated in treated animals,
epidemiological studies in 2-ME– and 2-EE–exposed workers did not show any change
in these parameters (Welch et al., 1988).

To define the target cells for 2-ME action, adult Sprague-Dawley rats were dosed
with 50–500 mg/kg body weight/day for 11 days (Foster et al., 1983; Creasy and Foster,
1984). Over the 11-day dosing period, 2-ME was found to produce testicular damage
at dose levels of and in excess of 100 mg/kg/day with a no-effect level at 50 mg/kg/
day. The first evidence of testicular damage following 2-ME treatment was degeneration in the spermatocytes. The degenerative changes at the light microscopic level consisted of general cellular shrinkage, increased cytoplasmic eosinophilia, and nuclear
pyknosis. These changes were restricted to spermatocytes at late phases of meiosis (primary spermatocytes at phases of late pachytene, diplotene, and dividing stages). The
susceptibility of spermatocytes to 2-ME was reported to be in the order of dividing
spermatocytes (Stage XIV) > early-pachytene spermatocytes (Stages I–III) > late-
pachytene spermatocytes (Stage IX–XIII) > mid-pachytene spermatocytes (Stage IV–
VIII). Leptotene/Zygotene spermatocytes and step I spermatids also showed degenerative changes, but only after prolonged dosing at high-dose levels. Sertoli cells and Leydig
cells appeared unaffected. Similar findings but in a shorter time frame were reported
by Chapin et al. (1984, 1985a,b) in their NTP studies using F344 rats. Chapin et al.
also showed that some indices of Sertoli cell function such as fluid production and
androgen-binding protein secretion were not affected by 2-ME, and they concluded that
late spermatocytes are targets for 2-ME while Sertoli cells are relatively unaffected. At
the electron microscopic level, while Foster et al. (1983) reported that "the most prominent changes 16 hr after a single oral dose of 500 mg/kg were mitochondrial swelling
and disruption, cytoplasmic vacuolation, and early condensation of nuclear chromatin"
in spermatocytes, Creasy et al. (1986) found that plasma membrane dissolution in 2-
ME–sensitive spermatocytes and vacuolation in Sertoli cell cytoplasm at 2-ME–sensitive stages were prominent early features of the lesion. However, both authors noted
that characterization of the 2-ME–induced lesion at an ultrastructural level was difficult because 2-ME–induced germ cell degeneration was a very rapid process; cells either appeared normal or in a stage of advanced necrosis. Therefore, Creasy et al. (1986)
concluded that the early occurrence of vacuolar changes in the Sertoli cells and their
association with the sensitive spermatocyte population may indicate a causal relationship, although ultrastructural studies provide no indication of the subcellular site of
damage.

There is a general consensus that metabolism of monoalkyl glycol ethers via alcohol and aldehyde dehydrogenases to alkoxyacetic acids is a prerequisite for development of their toxicity (Ghanayem et al., 1987, 1989; Foster et al., 1987b). The proposed metabolic pathway of 2-ME is 2-ME via alcohol dehydrogenase →
methoxyacetaldehyde via aldehyde dehydrogenase → methoxyacetic acid → inactive
glycine conjugates (Moss et al., 1985; Foster et al., 1986b). Indeed, methoxyacetic acid
(MAA) has been reported to be the proximate metabolite of 2-ME both in vivo and in
vitro. In vivo, methoxyacetic acid has been identified as a major urinary metabolite of
2-ME (Miller et al., 1983; Moss et al., 1985). MAA and 2-ME at equivalent doses
induced equal testicular damage (Foster et al., 1983). Furthermore, using ^{14}C-ME (250
mg/kg i.p.), Moss et al. (1985) showed that pretreatment with the alcohol dehydrogenase inhibitor pyrazole inhibited the conversion of 2-ME to MAA and at the same time

completely prevented the testicular toxicity of 2-ME. Also in this study, Moss et al. revealed a rapid conversion of 2-ME to MAA ($t_{1/2} = 0.6 \pm 0.03$ h) and a gradual clearance of radioactivity ($t_{1/2} = 19.7 \pm 2.3$ h). At a single i.p. dose of 250 mg/kg (which induced an obvious testicular lesion), the plasma concentrations of MAA ranged from ~3 mM to ~6 mM within 1–6 hours postdosing. Six hours after the dosing, the plasma concentration of MAA decreased rapidly, but a low level of MAA (ca. 1 mM) was still detectable 48 hours after the dosing. This study provided the basis for the choice of concentrations for subsequent in vitro studies.

Using in vitro Sertoli–germ cell cocultures, Gray et al. (1985) provided further evidence for MAA as the major metabolite for the testicular toxicity of 2-ME. In this study, 2-ME induced no morphological evidence of toxicity when added to the culture medium at concentrations up to 50 mM for 72 hours. In contrast, 2–10 mM MAA for 24–72 hours caused significant degeneration in pachytene and dividing spermatocytes, which were attached the monolayers of Sertoli cells. Consistent with this study, Beattie et al. (1984) reported that MAA, but not 2-ME, significantly decreased medium lactate concentrations and rates of lactate accumulation in the medium of primary Sertoli cell cultures.

Although it is generally believed that MAA is the ultimate toxic metabolite of 2-ME, the possible involvement of the other metabolites has not been completely excluded. Methoxyacetaldehyde (MALD), the intermediate metabolite of 2-ME, has been shown to produce the characteristic testicular lesion reported for MAA and 2-ME (Foster et al., 1986b). While the alcohol dehydrogenase inhibitor pyrazole is preventive against the testicular effects of 2-ME, pretreatment of animals with an aldehyde dehydrogenase inhibitor disulfiram did not inhibit the 2-ME-lesion (Foster et al., 1984; Moss et al., 1985). Further metabolism of MAA has also been proposed recently (Sumner and Fennell, 1993).

While it has been widely accepted that the germ cell, or more specifically, the late spermatocyte, is the prime target of 2-ME, a recent study by Ku and Chapin (1994b) revealed that Sertoli cells, instead of germ cells, may be the initial targets. By comparing the morphological germ cell death induced by MAA between in vivo and in vitro and between three in vitro models (enriched mixed germ cell cultures, Sertoli–germ cell co-cultures, and seminiferous tubule cultures), Ku and Chapin found that MAA-induced germ cell death in vivo was faithfully mimicked in vitro in cultured seminiferous tubules. However, germ cells in enriched mixed germ cell cultures and germ cells detached from the monolayer of Sertoli cells in the co-cultures showed no morphological degeneration. The authors also noted that germ cells that remained attached to the monolayers of Sertoli cells in the co-cultures did show morphological cell death. This study suggested that Sertoli cells that retain their intact relationship to germ cells are required for the expression of the testicular toxicity of 2-ME. In other words, either the Sertoli cell or the intact cell-to-cell communication between Sertoli cells and germ cells is the initial target. These data also showed that seminiferous tubule culture is the best model for in vitro studies on the testicular toxicity of 2-ME/MAA. The hypothesis that the Sertoli cell is the initial target of 2-ME/MAA is further supported by the evidence that lactate production of cultured Sertoli cells is inhibited by MAA (Beattie et al., 1984). However, addition of 5–10 mM lactate to the culture medium did not reduce the toxicity. Equally, removal of the pyruvate normally present in the medium did not enhance the toxicity, as judged morphologically (Gray et al., 1985). Therefore, Sertoli cells appear to be the initial targets of 2-ME/MAA.

In rats, degenerating spermatocytes induced by 2-ME are morphologically characterized with uniformly condensed chromatin or dispersed chromatin masses (karyorrhexis) followed by dissolution (karyolysis) with increased cytoplasmic eosinophilia, typical of necrosis (Chapin et al., 1984; Creasy et al., 1984). However, in guinea pig in vivo or in cultured human testicular tissues in vitro, 2-ME/MAA-induced degenerating spermatocytes showed marked chromatin condensation at the nuclear periphery, typical of apoptosis (Ku et al., 1995; Li et al., 1996). Furthermore, a pattern of DNA fragmentation consisting of 180–200 bp fragments (a ladder pattern) on agarose gel, the biochemical hallmark of apoptosis, was found in DNA samples isolated from 2-ME/MAA–treated testicular tissues in all the species (rats, guinea pigs, and humans). This indicates that 2-ME/MAA induces spermatocyte cell death by apoptosis, although degenerating spermatocytes showed different morphology in different species.

Two major hypotheses have been proposed for the mechanisms of the testicular toxicity of 2-ME. One is that 2-ME induces germ cell death by reducing the availability of purine bases, which consequently decreases RNA synthesis in spermatocytes (Mebus et al., 1989). This hypothesis is based on the following observations:

1. 2-ME targets tissues such as testis, embryos, bone marrow, and thymus, which undergo rapid proliferation (Simialowicz et al., 1991; Browning and Curry, 1994).
2. Simple physiological compounds such as serine, sarcosine, and acetate can prevent the testicular and embryotoxicity of 2-ME (Mebus et al., 1989; Mebus and Welsch, 1989).

It has been proposed that the ability of these compounds to attenuate the testicular toxicity of 2-ME may result from their ability to donate one-carbon units, which can be used in purine base biosynthesis. However, there is no evidence showing either that RNA synthesis is inhibited by 2-ME/MAA exposure or that the purine base biosynthesis is enhanced by these one-carbon unit suppliers. Furthermore, this hypothesis is based on the presumption that the germ cell is the initial target, which contradicts recent data suggesting that the Sertoli cell is actually the initial target, as mentioned above. This hypothesis has not been further explored since it was proposed in late 1980s.

The second hypothesis is that 2-ME induces germ cell death by interfering with one or several interregulating signal transduction pathways within either Sertoli cells or germ cells, which consequently disrupt the normal cell-to-cell communications between Sertoli cells and germ cells. This hypothesis is based on the evidence that intact relationship between Sertoli cells and germ cells is required for the expression of the testicular toxicity of 2-ME/MAA and that calcium channel blockers such as dihydropyridines, verapamil, diltiazem, and TMB-8, which are thought to inhibit calcium movement through plasma membranes, are preventive against the 2-ME/MAA–induced germ cell death both in vivo in rats and in vitro in cultured rat seminiferous tubules or human testicular tissues (Ghanayem and Chapin, 1990; Li et al., 1996). However, using indo-1 laser-scanning confocal microscope, an advanced technique to measure the intracellular free calcium concentration ($[Ca^{2+}]_i$), no detectable changes in $[Ca^{2+}]_i$ in sensitive spermatocytes were seen. Some new data support the second hypothesis by showing that molecules such as calmodulin and protein kinase C are possibly involved in the 2-ME/MAA–induced germ cell apoptosis (Li et al., 1997).

The question about the sensitivity of late spermatocytes to 2-ME/MAA is, in fact, also a question about mechanism. One possibility is that the difference in sensitivity between different stages may be due to the stage-dependent changes in the morphology and function of Sertoli cells, and also of the cell cycle of germ cells during meiosis, since late spermatocytes are those at the late phases of meiotic cell cycle. Indeed, a significant increase in the frequency of stage XIV tubules in rat testis in vivo has been independently noticed by Chapin et al. (1984) and Creasy et al. (1985). Using adult rat seminiferous tubule culture models, a recent in vitro study has found that MAA at concentrations of ≥ 0.08 mM also increased the frequency of stage XIV tubules (Li et al., 1997). While these data suggest a possible association between spermatocyte apoptosis and the meiotic cell cycle, the processes that control the meiotic cell cycle are unknown. The possible role(s) of G2/M-phase specific cyclin B–dependent p34cdc2 kinase in germ cell development at late phases of meiosis and also in the process of 2-ME–induced late spermatocyte apoptosis have been noticed but remain to be further investigated (Rhee and Wolgemuth, 1995; King and Cidlowski, 1995; L.-H. Li et al., unpublished). If the changes in germ cell meiotic cell cycle could result from initial changes in Sertoli cells, then the question would be: Is the stage-dependency of the toxicity due to stage-dependent sensitivity of germ cells at the different phases of meiotic cell cycle or to the stage-dependent sensitivity of Sertoli cells?

Thus, it is clear that MAA is the major toxic metabolite and that the 2-ME–induced spermatocyte cell death looks necrotic in rats, but it is really apoptotic in all the species observed. In addition, the 2-ME–induced germ cell degeneration is possibly mediated through initial Sertoli cell–based events such as changes in kinase activation/inactivation and disrupted cell cycle regulation. Dissecting the newly proposed pathways using both in vivo and in vitro models (such as seminiferous tubule cultures) will be the major task for the future.

n-Hexane

n-Hexane is a commonly used solvent. Its neurotoxicity, resulting in peripheral polyneuropathy, has been known for over 20 years (Allen et al., 1975; Korobkin, et al., 1975). *n*-Hexane is metabolically converted into aliphatic ketones and alcohols. The neurotoxicity correlates with the concentration of the diketone, 2,5-hexanedione (HD), although other congeners are also active (Boekelheide, 1990). The diketones result in pyrole formation. Subsequent oxidation and protein cross-linking are required for the neurotoxicity (Genter et al., 1987, 1988).

In vivo dosing of rats with n-hexane, usually in drinking water (O'Donaghue et al., 1978; Chapin et al., 1982), but also via subcutaneous injection (Spencer and Schaumburg, 1975), intraperitoneal injection (Di Vincenzo et al., 1976), or by gavage (Krasavage et al., 1980), results in testicular toxicity and neurotoxicity. Boekelheide (1987a) has been able to separate the testicular toxicity from the neurotoxicity, showing they are separate events.

Pathogenesis studies designed to determine the testicular target reported the Sertoli cell as the target cell. Sertoli cell injury, initially detected as vacuolization, leads to germ cell loss (Chapin et al., 1982, 1983). The Sertoli cell vacuoles are preferentially found in stages XII–XIV and I of the spermatogenetic cycle after 3 weeks of dosing with 1.0% HD to Fisher F344 rats (Chapin et al., 1983). Of the germ cells, the elongated spermatids are most susceptible to HD-induced toxicity. They were sloughed into the tubule lumen, a phenomenon consistent with that caused by microtubule-disrupting agents. It should be noted that both Sprague-Dawley and Fisher F344 rats have been used to study HD testicular toxicity. The Fisher rat, being an inbred strain, has a more homogeneous response to HD such that the toxicity occurs over a shorter period of time and is more irreversible (Boekelheide, 1988; Blanchard et al., 1996). Nonetheless, the testicular toxicity resulting from HD exposure is identical for the two strains of rats (Boekelheide, 1988; Johnson et al., 1991; Allard and Boekelheide, 1996). By analogy to the mechanism of the neurotoxicity, Boekelheide (1987a, 1988) has shown that HD is the active testicular toxicant of *n*-hexane exposure. He has also proposed that HD testicular toxicity results from alterations in Sertoli cell microtubules and that the altered microtubules result from pyrole-dependent cross-linking (Boekelheide, 1987b). In the case of HD intoxication, histopathological changes in the testis occurred at 2–3 weeks, while increased pyrole content and tubulin cross-linked dimers were detected at 2 weeks. This is the proper sequence for cause-and-effect relationship (but initially is only a correlation). These data also substantiate our initial premise that histopathological techniques can indeed define the target cell and also that biochemical changes can be expected to occur before the lesion is evident by the histopathological methods. In the ensuing years Dr. Boekelheide has singlehandedly pursued this problem of defining the mechanism of the Sertoli cell toxicity of HD.

Boekelheide's laboratory has shown that while general mechanisms of HD toxicity are similar in nervous system injury and testicular toxicity, differences in the biochemistry and physiology of these tissues result in differences in the specifics of the toxicity. For example, nervous system toxicity is dependent on the total accumulated dose, not the rate of exposure. Graham et al. (1982) showed that this is because there is no turnover of cytoskeleton elements in the axons, resulting in a continuous accumulation of pyrole-induced neurofilament cross-linking. In contrast, the testicular toxicity is rate-dependent and independent of total dose. This difference can be explained by the fact that the Sertoli cell target, microtubular proteins, constantly turn over. This was substantiated by the fact that the abnormal microtubule assembly and pyrole content returned to normal values 1–2 weeks following discontinuation of HD exposure. Thus, there is a threshold of altered microtubule assembly required for testicular toxicity because of the constant turnover of microtubules.

Before examining studies of the mechanism of HD toxicity, it is worth examining the results of other in vivo studies. HD toxicity is slow in onset. Even at the high dose of 1% HD in drinking water, testis weight did not decline until 7 weeks (2 weeks after termination of exposure) (Boekelheide, 1988). The first sign of toxicity was Sertoli cell vacuoles, which appeared at 3 weeks in F344 rats (Chapin et al., 1982, 1983) or at 4 weeks in CD rats (Boekelheide, 1988) followed by loss of elongated spermatids and spermatocytes with resulting disorganization of the germinal epithelium at 5 weeks and tubules consisting of only Sertoli cells and spermatogonia at 7 weeks and beyond. HD toxicity is largely irreversible. Boekelheide (1988a) exposed Sprague-Dawley rats to 1% 2,5-HD in drinking water for 5 weeks and then 17 weeks of recovery. A 17-week recovery period corresponds to eight cycles of the seminiferous epithelium. There was no increase in testis weight during this recovery period, although there was partial restoration of the germ cell population in two of five rats (Boekelheide and Hall, 1991). This lack of significant recovery occurred in the presence of normal numbers of stem cells. Indeed, there was minimal recovery of spermatogenesis up to 75 weeks after exposure to HD, indicating that there is a defect either in the stem cells present or in the Sertoli cells that prevents maintenance of mitotic daughter cells (Boekelheide and Hall, 1991). Thus the HD-exposed rat is a model of irreversible testicular injury. The atrophic testis resulting from HD exposure contains a reduced population of actively dividing spermatogenetic stem cell populations and residual type A spermatogonia with no postmitotic germ cells (Boekelheide and Hall, 1991; Allard et al., 1995). Thus, the failure of recovery of spermatogenesis after HD exposure probably can be explained by a combination of fewer stem cells and a block in spermatogenesis that results in death of maturing spermatogonia. Blanchard et al. (1996) have recently shown that the germ cell death following HD exposure in vivo is the result of increased apoptosis as defined by TUNEL staining, morphological analysis, and low molecular weight DNA ladder patterns. The earliest morphological signs of apoptosis appeared after 2 weeks of exposure to HD, thereby becoming one of the most sensitive toxicity endpoints. Changes in tubulin polymerization also occur at 2 weeks postdosing (Hall et al., 1991). While the mechanism of this block in spermatogenesis that leads to apoptosis is unknown, it has been hypothesized that it results from a lack of Sertoli cell paracrine factors needed for spermatogonial differentiation. This is clearly an attractive and testable hypothesis.

In vivo studies have also been used to show that HD alters the stage frequency of the spermatogenetic process (Chapin et al., 1983). More recently, these studies have been extended to show that HD altered the duration of the cycle of the seminiferous epithelium from 12.4 to 13.4 days (Rosiepen et al., 1995). Since the entire spermatogenetic process requires four spermatogenetic cycles, in the rat, the duration of spermatogenesis would be increased from 50 to 54 days during dosing with HD. This increased duration of spermatogenesis resulted from a reduced stage-dependent progression of germ cells from stages VII and VIII to stages IX and X. These results are significant for two reasons. First, they show that it is possible for toxicants to alter the timing of the spermatogenetic process, a process that appears to be highly species specific and tightly controlled. Second, they point to a possible mechanism of action of HD. The progression of spermatogenesis from stages VI to VII involves movement of elongated spermatids and creation of residual bodies, effects thought to be mediated by the Sertoli cell cytoskeleton. Thus, these in vivo data are also consistent with an effect of HD on Sertoli cell intermediate filaments and microtubules. With regard to HD effects on testicular tubulin (presumably mostly Sertoli cell tubulin), Boekelheide (1987b) reported that purified tubulin from testes of HD-treated animals assembled earlier and more rapidly than tubulin from control testes. This effect is consistent with an increased number of microtubule nucleation sites, which results in an increased number of shorter microtubules (Boekelheide, 1987a). A time-course study showed that the pyrole content of testes from rats treated with 1% HD for 22 weeks reached a maximal level at 2 weeks (first time point measured) then remained relatively constant (followed only for 8 of 22 weeks of exposure). Tubulin purified from testis showed a unique band on SDS-PAGE corresponding to a cross-linked tubulin dimer starting at 2 weeks (the earliest time point!), which like the pyrole content appeared to decline somewhat by 8 weeks of exposure. Biochemically this altered tubulin consisted of an early onset of microtubule assembly in vitro, resulting in a decreased nucleation time. The amount of the cross-linked tubulin dimer roughly correlated with the degree of assembly alteration (Boekelheide, 1988b). The HD cross-linked tubulin was evident on SDS-PSGE analysis as a band inconsistent with any multiple of tubulin dimer. Further analysis revealed that the cross-linked tubulin was an α-β heterodimer. This cross-linking of tubulin in the polymeric state suggests a fixation of the confirmation into that of the stable polymer in shorter segments. It is important to note that the microtubule (MT) assembly alterations produced by HD in vivo (measured in vitro) could be duplicated by in vitro incubation of monomeric tubulin with HD. The Boekelheide group has also examined the mechanism whereby HD alters microtubular associated events in addition to altered enucleation sites. For example, they have shown that the microtubule motor kinesin moves more slowly along microtubules exposed to HD and that there is an altered distribution of the microtubule motor cytoplasmic dynein after 3–4 weeks of exposure. These changes would be consistent with altered microtubular motor-dependent intracellular transport and/or positioning processes (Hall et al., 1995) and could explain the effects seen by Rosiepen et al. (1995) (i.e., altered duration of the cycle of the seminiferous epithelium).

In addition to microtubules, the Sertoli cell cytoskeleton also contains networks of intermediate and actin filaments. The intermediate filaments are concentrated around the nuclei and extend outward. They are associated with the Sertoli cell–germ cell ectoplasmic specializations that are thought to be involved in anchoring germ cells to the Sertoli cells and with the blood-tubule barrier. Actin filaments are also associated with ectoplasmic specializations and are thought to be important in anchoring step 19 spermatids to Sertoli cells. Using the usual dosing paradigm (1% HD in drinking water), Hall et al. (1991) showed that small foci of disorganized actin and vimentin (intermediate filament) of the tubulobulbar region of step 19 spermatids (3 weeks) preceded histopathological changes in Sertoli cells (i.e., vacuoles) (4 weeks). This suggested that a disruption of the tubulobulbar region (actin- and tubulin-containing structures) is one of the earliest consequences of HD treatment. It appears right after alterations in microtubule assembly (2 weeks). As with the microtubule effects of HD, the actin changes are also reported to be due to pyrole-dependent protein cross-linking (Hall et al., 1991).

In order to provide an in vitro model to examine mechanistic studies of the testicular lesion Hall et al. (1992) prepared and cultured Sertoli cells from HD exposed rats. The function of cultured Sertoli cells from HD-exposed rats was compared to that of cultures of Sertoli cells from 21-day-old control rats or cells from adult cryptorchid testes (a model of reversible testicular injury). A control of normal adult Sertoli cells was not used. The proper control for this experiment is unclear. Nonetheless, they did confirm that microtubules, actin, and vimentin filaments had an abnormal distribution in the Sertoli cells cultured from the HD-treated rats, at least as compared to the two controls utilized. An important question is how is the Sertoli cell toxicity, manifest as altered cytoskeleton, translated into germ cell toxicity? Indeed there is a hierarchy of sensitivity of germ cells to HD exposure where elongated spermatids are the most sensitive followed by the round spermatids. Johnson et al. (1991) and Richburg (1994), using two different techniques, proposed that HD action on microtubules translated into altered protein secretion and seminiferous tubule fluid secretion. Richburg et al. (1994) have shown that normal tubule fluid secretion requires an intact "microtubule-dependent membrane trafficking pathway." It is well known from studies in several tissues that microtubules are associated with the Golgi complex and the endoplasmic reticulum and that they function as tracks for organelle transport. Thus, it is plausible that an alteration in microtubules could result in disrupted tubule fluid formation. The time course of HD toxicity also fits with this hypothesis as HD exposure in vivo results in altered nucleation and assembly of tubulin in vitro as early as 2 weeks after exposure with altered secretion of tubule fluid noted after 3 weeks of exposure in testes with minimal histological evidence of damage (Richburg, 1994). Proof of the cause-and-effect relationship between altered microtubules and seminiferous tubule fluid production in

Sertoli cells and their relationship to the germ cell death after HD exposure awaits further experimentation. Nonetheless, this is an intriguing and testable hypothesis, which will undoubtedly lead to new discoveries of the biology of Sertoli cell function and spermatogenesis.

In summary, a picture is emerging of an initial effect of HD to cross-link cytoskeletal elements, thereby altering protein secretion and trafficking in the Sertoli cell. This causes altered Sertoli cell–germ cell contacts and loss of Sertoli cell paracrine support of germ cells, resulting in activation of the apoptotic pathway leading to germ cell death. Much remains to be determined but research is focused on the molecular events controlling apoptosis and microtubule dynamics and Sertoli cell paracrine support of germ cell such that continued progress can be expected. In addition, continued analysis of the irreversibility of this HD toxicity will provide information that may be translatable to other Sertoli cell toxicants such as MEHP which also results in irreversible toxicity.

Phthalate Esters

Esters of *o*-phthalic acid are used extensively in consumer products and medical devices as plasticizers to impart flexibility to plastic materials. Since they are not covalently bound to the plastic, they are able to leach into the environment, providing for widespread exposure (see review by Thomas and Thomas, 1984). Indeed, low but detectable concentrations of phthalates are found in virtually all human urine samples (Albro et al., 1984). While, in general, acute toxicity of most of the commonly encountered phthalates is low, quite high doses of some phthalates have been shown to be carcinogenic, to result in liver toxicity, and to cause reproductive or developmental toxicity (reviewed in Thomas and Thomas, 1984; Albro, 1986; Woodward, 1988).

Several phthalate esters including di(2-ethylhexyl) phthalate (DEHP) and dithexyl phthalate (DHP) are male reproductive toxicants in both rats and mice. The short-chain diethyl derivatives are not reproductive toxicants as measured in breeding studies (Lamb et al., 1987; Woodward, 1988; Heindel et al., 1989).

The testicular lesion produced by di(*n*-butyl)phthalate (DBP), dipentyl phthalate (DPP), DHP, and DEHP in immature rats is characterized by early sloughing of spermatids and spermatocytes and severe vacuolation of Sertoli cell cytoplasm (Cater et al., 1977; Foster et al., 1982; Fukuota et al., 1989). Further studies have shown that with DPP (and presumably for the other active phthalates), the Sertoli cell is the initial and primary target (Foster et al., 1982; Creasy et al., 1983; Creasy et al., 1987), with the result being a disturbance in Sertoli cell–germ cell interaction (Fukuota et al., 1989).

In vitro studies using Sertoli cell–germ cell cultures have shown that, in general, the monoesters of the active phthalates in vivo increase germ cell detachment in vitro, thereby mimicking one aspect of their in vivo actions (Gray and Beamand, 1984). In addition to the changes in the Sertoli cell identified by histological procedures, phthalate exposure in vitro also results in a reduction of seminiferous tubule fluid formation and secretion of androgen-binding protein. The work of Dostal et al. (1988) which showed decreased number of Sertoli cells after neonatal exposure to DEHP, further supports the Sertoli cell as the site of action of phthalate esters. This important finding has not been followed up.

Orally administered phthalate esters are rapidly hydrolyzed to the corresponding monoesters by nonspecific esterases primarily in the gut (Albro et al., 1973; Lake et al., 1977; Oishi and Hiraga, 1980a,b). Because the hydrolysis of the second ester linkage yielding phthalic acid occur much more slowly, the monoesters are the major metabolite and the proximal toxicant. Indeed, the testicular toxicity of DEHP can be mimicked in vivo by MEHP (mono-2-ethylhexyl phthalate), but not by MEHP-derived metabolites (Sjoberg et al, 1986). While similar studies have not been completed for all the phthalate esters toxic to the male reproductive system, the available data suggest that they cause a similar testicular lesion and that their toxicity is also due to the mono-substituted derivative (Cater et al., 1977; Fukuota et al., 1989). Additional evidence that MEHP is the proximal testicular toxicant of DEHP comes from data showing that Sertoli cells are incapable of further metabolism of MEHP while mimicking the in vivo toxicity (Gray and Beamand, 1984; Williams and Foster, 1988; Albro et al., 1989; Heindel et al., 1989).

The testicular toxicity of phthalates is age-dependent in in vivo studies but not in in vitro studies. Mature animals are less sensitive than immature animals (Gray and Gangolli, 1986; Oishi and Hiraga, 1986; Creasy et al., 1987). In fact, in neonatal rats, DEHP exposure will actually decrease the numbers of Sertoli cells per testis (Dostal et al., 1988). The F_1 males generated from F_0 males exposed to 1% DBP in drinking water in the Reproductive Assessment by Continuous Breeding (RACB) study showed significantly higher sensitivity to DBP than their parents (F_0) (NTP, 1989). In immature (3–5 weeks old) and young adult rats, DEHP causes severe Sertoli cell toxicity, while in adults (15 weeks of age), testicular toxicity of DEHP is minimal (reviewed by Gray and Gangolli, 1986). Similarly, Creasy et al. (1983, 1987) showed that immature rats (3–4 weeks of age) were more sensitive to DPP than were adults. In the immature animals, the toxicity was expressed evenly throughout the testis, resulting in Sertoli cell vacuolization and degeneration and partial exfoliation of the spermatid and spermatocyte population of all the tubules within 24 hours of dosing. However, in the adult, the effects were restricted to that subpopulation of tubules in Stages XI–XIV and II. This may be because the extent of absorption and hence the total systematic exposure to MEHP is higher in young rats. Also, when DEHP is given intravenously, the age-related change in sensitivity is lost (Sjoberg et al., 1986). The hypothesis that the age-dependent toxicity is due to the age-related changes in metabolism has been substantiated by in vitro experiments showing that the MEHP effects on the Sertoli cell (increased lactate, decreased FSH stimulated cAMP production, altered FSH binding to its membrane receptor) are not age dependent (Heindel and Powell, 1992; Grasso et al., 1993).

In addition to the in vivo age-related toxicity to phthalate esters, there is also species specificity. DGP produces testicular toxicity in the mouse and no toxicity in the hamster (Oishi and Hiraga, 1980a, b; Gray et al., 1982). Hamsters are also insensitive to the testicular toxicity of DEHP. Similar to the age-related change in sensitivity, the species differences appear to be due to differences in metabolism and disposition. MEHP in vivo is a testicular toxicant in the hamster (Gray et al., 1982), and in vitro it is a hamster Sertoli cell toxicant (J. J. Heindel, unpublished observations).

Thus, in vivo experiments backed up by in vitro data have shown clearly which phthalates are testicular toxicants and identified the proximal toxicants and the primary cellular target. In vitro studies using primary Sertoli cell cultures have been utilized to define the mode of action. Indeed, MEHP has been shown to cause disturbances in many Sertoli cell functions using these in vitro techniques. The challenge is to decipher the pathway(s) of toxicity and to develop a logical overall hypothesis and to then test this hypothesis in vivo.

Stimulation of lactate secretion has been proposed as part of the action of phthalate monoesters on Sertoli cell–germ cell function (Chapin et al., 1988; Moss et al., 1988; Williams and Foster, 1989). This stimulation of lactate production by MEHP in vitro is evident in Sertoli cells cultured from animals 13–28 days old (Heindel and Powell, 1992). Indeed, the IC_{50} for inhibition of cAMP accumulation and the ED_{50} for stimulation of lactate secretion are similar for all age groups, suggesting the possibility that these two effects may be somehow related. The stimulation of lactate production is also apparent and unrelated to age with monopentyl phthalate but not with the nontoxic methyl, ethyl, and propyl side-chain phthalates. There is not effect of monobutyl phthalate on lactate production at any age, which agrees with the data of Moss et al. (1988), who used Sertoli cells from 28-day-old rats. Whether this implies that stimulation of lactate production is not part of the general mechanism of action of phthalate esters or that this lack of stimulation is because monobutyl phthalate is the least toxic phthalate, as far as Sertoli cell toxicity is concerned, is not clear.

In the liver, MEHP and MBP have been shown to inhibit mitochondrial respiration by altering the permeability of the inner membrane and by inhibiting succinate dehydrogenase activity in a noncompetitive manner. DPP was not tested (Melnick and Schiller, 1982). Thus, there is precedence for an effect of phthalate monoesters to alter energy metabolism. In vivo, DPP was shown histochemically to reduce succinate dehydrogenase activity in mitochondria of Sertoli cells (Foster et al., 1982; Gangolli, 1982). Chapin et al. (1988) reported that MEHP inhibited Sertoli cell succinate dehydrogenase activity in a mixed manner. They were, however, unable to detect changes in mitochondrial membrane potential as was shown for liver mitochondria (Inouye et al., 1978; Melnick and Schiller, 1982). Additionally, hepatic mitochondria swell when exposed to specific phthalate esters (Melnick and Schiller, 1982; Ohyama, 1987). The literature on Sertoli cell mitochondria is contradictory, with some reports showing mitochondrial condensation after in vivo exposure to phthalates (Foster et al., 1982), while others show mitochondrial hypertrophy after phthalate exposure in vivo or in vitro (Creasy et al., 1983, 1987). MEHP is the only phthalate monoester that results in reduced Sertoli cell ATP levels, an effect that only occurred in Sertoli cells from immature animals (Heindel and Powell, 1992). Thus, while the data suggest that at least some of the phthalate esters (MEHP, DPP) may alter energy metabolism, the effects are not as impressive as those reported for liver. More information is needed, perhaps with more sensitive probes of mitochondrial function, to ascertain the importance of mitochondria as a site of action of the active phthalate esters. In addition, the relationship between these metabolic alterations and the mechanism of toxicity is not yet clear.

The fact that the initial testicular lesion in DPP-treated adult rats is restricted to tubules in the successive stages XI–XIV, I, and II of the spermatogenetic cycle (Creasy et al., 1987), i.e., the stages with the highest FSH responsiveness (Parvinen, 1982), prompted several investigators to examine an effect of phthalates, specifically MEHP, on FSH stimulation of cAMP accumulation in Sertoli cell cultures. Indeed, MEHP specifically inhibits FSH-stimulated cAMP accumulation in cultured Sertoli cells with no effect on cAMP accumulation stimulated by isoproterenol, prostaglandin E, forskolin, or cholera toxin (Lloyd and Foster, 1988; Heindel and Chapin, 1989; Heindel and Powell, 1992). This inhibition is time and dose dependent and is only partial, usually to 40–60% of control values. This MEHP-induced inhibition was not affected by incubation in the presence of methylisobutylxanthine, a phosphodiesterase inhibitor, suggesting that MEHP does not stimulate the breakdown of cAMP. Furthermore, inhibition occurs in the presence of pertussis toxin, suggesting that MEHP action is independent of the adenylate cyclase inhibitory pathway. While MEHP is the most potent phthalate monoester at inhibiting the FSH-stimulated cAMP response in cultured Sertoli cells, monopentyl phthalate (MPP) and monobutyl phthalate (MBP) also inhibit this response, but only in cells prepared from older animals (>36 days of age). The methyl, ethyl, and propyl monophthalate esters have no effect on FSH-stimulated cAMP accumulation and are not testicular toxicants in vivo (Heindel and Powell, 1992). It is important to note that the phthalate concentrations in vitro that resulted in decreased FSH-stimulated cAMP accumulation (10^{-6}–10^{-4} M) are in the same range as those found in testes (2×10^{-4} M) of rats given a toxic dose of DEHP of 2 g/kg (Tanoka et al., 1975).

To further define the site of action of MEHP on the FSH receptor signal transduction system, Grasso et al. (1993) incubated membranes from cultured Sertoli cells previously exposed to MEHP with ^{125}I-hFSH to monitor FSH binding. They showed that MEHP inhibited FSH binding when preincubated with Sertoli cells in culture but not when added simultaneously with ^{125}I-hFSH to the purified membrane preparation. Attenuation of FSH binding was evident after a 3-hour preincubation with 100 μM MEHP (18%) and was maximal after 15–24 hours of preincubation (70–90%). Preincubation of Sertoli cells for 24 hours with 100 μM DEHP had no effect on FSH binding. Half-maximal inhibition occurred at approximately 0.1 μM MEHP. Scatchard analysis indicated a fourfold decrease in FSH affinity with no change in receptor concentration. Exposure of Sertoli cells to MEHP was age-independent over the range of 18–45 days. The order of potency and age specificity for the effects of MEHP, MBP, and MPP on inhibition of FSH binding to purified rat Sertoli cell membranes was MEHP > MPP > MBP, similar to that previously shown for inhibition of FSH-stimulated cAMP accumulation in cultured Sertoli cells (Heindel and Powell, 1992). The order of potency and degree of inhibition of FSH-stimulated cAMP accumulation in cultured Sertoli cells is slightly different from that seen in vivo (DPP > DEHP > DBP) (Foster et al., 1980). This may be related to differences in metabolism and disposition in vivo. Perhaps more importantly, the nontoxic monomethyl phthalate had no effect on either FSH binding or FSH-stimulated cAMP accumulation, indicating that these in vitro actions have the proper phthalate specifically for an important in vivo mechanism. Moreover, the action of MEHP is likely not at the FSH receptor but on the GTP-binding protein that couples the receptor to the catalytic unit. The mechanism whereby MEHP alters the interaction of the FSH receptor with the GTP-binding protein, presumably Gs, is still unclear.

On the basis of the in vivo and the in vitro data, we conclude that the ability of certain phthalate esters to reduce both FSH binding and subsequent cAMP accumulation is likely to be at least a part of the mechanism responsible for their testicular toxicity. These effects cannot be solely responsible for phthalate testicular toxicity, as MEHP also inhibits Sertoli cell estradiol and inhibin secretion at a post-cAMP site (Heindel et al., 1990). In addition, the relationship of the stimulation of lactate secretion and the inhibition of mitochondrial function to the effects on FSH-stimulated cAMP production, as well as their role in the ultimate Sertoli cell toxicity and subsequent germ cell death, is unclear.

Recent studies by Richburg and Boekelheide (1996) suggest that the collapse of vimentin filaments in Sertoli cells by MEHP leads to a loss of Sertoli–germ cell contacts. They have developed a hypothesis that relates changes in Sertoli cell vimentin, release of germ cells from contact with Sertoli cells, and an alteration in the Fas-associated apoptotic pathway in germ cells. This is a testable hypothesis that shows how actions on Sertoli cells at a molecular level can lead to germ cell toxicity. It will be important to relate these events to the effects already shown for phthalates on Sertoli cell function to determine if there is a single initiating event leading to a cascade of effects or multiple initial interactions. Thus, while there is much data, an integrated understanding of the mechanism of the reproductive toxicity of phthalates remains to be developed and will require further examination of their effects on Sertoli cell and germ cell function. It would be helpful to determine, for example, how phthalates interact specifically with the G_S protein coupled to the FSH receptor and what happens to the FSH receptor/G_S/membrane interaction such that the phthalate specifically is altered with increasing animal age. Since it is known that protein kinase C interferes with FSH-cAMP-PKA pathway, the possible involvement of PKC should also be taken into consideration, though previous preliminary data showed that PKC is probably not involved in MEHP-inhibited granulosa cell functions (Treinen and Heindel, 1992). Answers to these questions will also help our understanding of Sertoli cell function. One possibility is that the hydrophobic MEHP molecule targets membranes, cellular and mitochondrial, and FSH receptor–mediated effects and vimentin by affecting their attachment to or activity in the membranes.

Boekelheide (1990) proposed that "cell polarity, membrane integrity, membrane pumps, regulatory fluid secretion and intracellular volume control and intercellular or extracellular matrix adhesive properties are several fundamental and potentially unique characteristics of early postnatal Sertoli cells which could be usefully studied in the search for the mechanism of phthalate-induced testicular injury." We agree that these fields are worthy of attention. However, over the past 5 years there has been minimal activity in this area such that these techniques and some new molecular tools are still waiting to be used to finalize the mechanism of phthalate-induced testicular toxicity.

FUTURE DIRECTIONS

Considering the fundamental roles of the Sertoli cell in spermatogenesis, it is not surprising that many reproductive toxicologists believe that the Sertoli cell is the initial target for most testicular toxicants. While identification of the initial target of a testicular toxicant is still a time-consuming task, it is relatively straightforward compared to definition of mechanistic pathways. Although some very common phenomena such as inhibited spermiation, vacuolization, disturbed timing of spermatogenesis, and germ cell death have been shown to be caused by several Sertoli cell toxicants, the mechanisms underlying these changes remain unknown. One of reasons for the slow progress in this field is lack of validated models with sensitive and reliable mechanistic endpoints. Validation of established models and biomarkers through interlaboratory collaboration is needed to further our knowledge in this field. While it is important to put more effort into developing in vitro models that most closely mimic the in vivo situation, it will also be very helpful to apply transgenic tools to clarify the cause-effect relationship and to decipher the possible involvement of certain pathways in the process of damage. Another important question to answer is whether environmental toxicants affect the development of Sertoli cells, especially when exposure is during the perinatal period. With the rapid progress in the field of molecular biology, we should be able to answer such questions in the not-too-distant future.

ACKNOWLEDGMENTS

The authors gratefully acknowledge Drs. Robert E. Chapin and Barbara J. Davis for their critical comments on the manuscript.

REFERENCES

Agelopoulou, R., Magre, S., Patsavoudi, E., and Jost, A. (1984). *J. Embryol. Exp. Morphol.*, *83*:15.

Albro, P. W. (1986). *Rev. Biochem. Toxicol.*, *8*: 73.

Albro, P. W., Thomas, R., and Fishbi, L. (1973). *J. Chromatogr.*, *76*: 321.

Albro, P. W., Jordan, S., Corbett, J. T., and Schroeder, J. L. (1984). *Anal. Chem.*, *56*: 247.

Albro, P. W., Chapin, R. E., Corbett, J. T., Schroeder, J., and Phelps, J. L. (1989). *Toxicol. Appl. Pharmacol.*, *100*: 193.

Allard, E. K., and Boekelheide, K. (1996). *Toxicol. Appl. Pharmacol.*, *137*: 149.

Allard, E. K., Hall, S. J., and Boekelheide, K. (1995). *Biol. Reprod.*, *53*: 186.

Allen, N., Mendell, J. R., Billmaier, D. J., Fontaine, R. E., and O'Neill, J. (1975). *Arch. Neurol.*, *32*: 209.

Allenby, G., Foster, P. M. D., and Sharpe, R. M. (1991). *Fundam. Appl. Pharmacol.*, *16*: 710.

Alsberg, C. L., and Schwartze, E. W. (1919). *J. Pharmacol.*, *13*: 504.

Amlani, S., and Vogl, A. W. (1988). *Anat. Rec.*, *220*: 143.

Aoki, A., and Hoffer, A. P. (1978). *Biol. Reprod.*, *18*: 579.

Assaf, A. A., and Clermont, Y. (1981). *Anat. Rec.*, *19*: 12a.

Attramadal, H., Jahnsen, T., Le Gac, F., and Hansson, V. (1984). *Regulation of Target Cell Responsiveness* (K. W. McKerns, A. Aakvaag, and V. Hansson, eds.). Plenum Press, New York, p. 3.

Balasubramanian, H., Campbell, G. A., and Moslen, M. T. (1994). *Toxicologist*, *14*: 298.

Bardin, C. W., Cheng, C. Y., Mustow, N. A., and Gunsalus, G. L. (1994). *Physiology and Reproduction*, 2nd ed. (E. Knobil and J. D. Neil, eds.). Raven Press, New York, p. 1291.

Beattie, P. J., Welsh, M. J., and Brabec, M. J. (1984). *Toxicol. Appl. Pharmacol.*, *76*: 56.

Blackburn, D. M., Lloyd, S. C., Gray, A. J., and Foster, P. M. D. (1985). *Toxicologist*, *5*: 121.

Blackburn, D. M., Gray, A. J., Lloyd, S. C., Sheard, C. M., and Foster, P. M. D. (1988). *Toxicol. Appl. Pharmacol.*, *92*: 54.

Blanchard, K. T., Allard, A. K., and Boekelheide, K. (1996). *Toxicol. Appl. Pharmacol.*, *137*: 141.

Boekelheide K. (1987a). *Toxicol. Appl. Pharmacol.*, *88*: 370.

Boekelheide K. (1987b). *Toxicol. Appl. Pharmacol.*, *88*: 383.

Boekelheide K. (1988a). *Toxicol. Appl. Pharmacol.*, *92*: 18.

Boekelheide K. (1988b). *Toxicol. Appl. Pharmacol.*, *92*: 28.

Boekelheide K. (1990). *Sertoli Cell* (L. D. Russell and G. D. Griswold, ed.), Cache River Press, Clearwater, FL, p. 551.

Boekelheide K., and Eveleth, J. (1988). *Toxicol. Appl. Pharmacol.*, *94*: 76.

Boekelheide K., and Hall, S. J. (1991). *J. Androl.*, *12*: 18.

Bond, J. A., Chism, J. P., Rickert, D. E., and Popp, J. A. (1981). *Fund. Appl. Pharmacol.*, *1*: 389.

Brown, C. D., Forman, C. L., McEuen, S. F., and Miller, M. G. (1994). *Fundam. Appl. Toxicol.*, *23*: 439.

Browning, R. G., and Curry, S. C. (1994). *Human Exp. Toxicol.*, *13*: 325.

Bugge, H. P., and Ploen, L. (1986). *J. Reprod. Fertil.*, *76*: 39.

Cater, B. R., Cook, M. W., Gangolli, S. D., and Grasso, P. (1977). *Toxicol. Appl. Pharmacol.*, *41*: 609.

Cave, D. A., and Foster, P. M. D. (1990). *Fundam. Appl. Toxicol.*, *14*: 199.

Chapin, R. E., Norton, R. M., Popp, J. A., and Bus, J. S. (1982). *Toxicol. Appl. Pharmacol.*, *62*: 262.

Chapin, R. E., and Heindel, J. J. (1993). *Methods in Toxicoloy, Vol. 3 Part A: Male Reproductive Toxicology*. Academic Press, San Diego.

Chapin, R. E., and Lamb, IV, J. C. (1984). *Environ. Health Perspect.*, *57*: 219.

Chapin, R. E., Morgan, K. T. and Bus, J. S. (1983). *Exp. Mol. Pathol.*, *38*: 149.

Chapin, R. E,, Dutton, S. L., Ross, M. D., Sumrell, B. M., and Lamb, J. C. (1984). *J. Androl.*, *5*: 369.

Chapin, R. E., Dutton, S. L., Ross, M. D., and Lamb, IV, J. C. (1985a). *Fundam. Appl. Toxicol.*, *5*: 182.

Chapin, R. E., Dutton, S. L., Ross, M. D., Swaisgood, R. R., and Lamb, IV, J. C. (1985b). *Fundam. Appl. Toxicol.*, *5*: 515.

Chapin, R. E., Gray, T. J. B., Phelps, J. L. and Dutton, S. L. (1988). *Toxicol. Appl. Pharmacol.*, *92*: 467.

Chapin, R. E., Phelps, J. L., Burka, L. T., Abou-Donia, M. B., and Heindel, J. J. (1991). *Toxicol. Appl. Pharmacol.*, *108*: 194.

Chowdhury, A. K., Steinberger, A., and Steinberger, E. (1975). *Andrologia*, *7*: 297.

Clarke, D. O., Mebus, C. A., Miller, F. J., and Welsch, F. (1991). *Toxicol. Appl. Pharmacol.*, *110*: 514.

Clermont, Y., and Harvey, S. C. (1965). *Endocrinology*, *76*: 80.

Clermont, Y., and Perey, B. (1957). *Am. J. Anat.*, *100*: 241.

Cody, T. E., Witherup, S., Hastings, L., Stemmer, K., and Christian, R. T. (1981). *J. Toxicol. Environ. Health*, *7*: 829.

Conti, M., Monaco, L, Toscano, M. V., and Stefanini, M. (1984). *INSERM*, *123*: 187.

Cooke, P. S. (1991). *Ann. NY Acad. Sci.*, *637*: 122.

Cooke, P. S., Zhao, Y.-D., and Hansen, L. G. (1996). *Toxicol. Appl. Pharmacol.*, *136*: 112.

Cossum, P. A., Rickert, D. E., and Working, P. K. (1986). *Pharmacologist*, *28*: 178.

Creasy, D. M. and Foster, P. M. D. (1984). *Exp. Mol. Pathol.*, *40*: 169.

Creasy, D. M., Foster, J. R., and Foster, P. M. D. (1983). *J. Pathol.*, *139*: 309.

Creasy, D. M., Glynn, J. C., Gray, T. J. B., and Butler, W. H. (1985). *Exp. Mol. Pathol.*, *43*: 321.

Creasy, D. M., Beech, L. M., Gray, T. J. B., and Butler, W. H. (1986). *Exp. Mol. Pathol.*, *45*: 311.

Creasy, D. M., Beech, L. M., Gray, T. B., and Butler, W. H. (1987). *Exp. Mol. Pathol.*, *46*: 357.

Di Vincenzo, G. D., Kaplan, C. J., and Dedinas, J. (1976). *Toxicol. Appl. Pharmacol.*, *36*: 511.

Dorrington, J. H., and Amstrong, D. T. (1975). *Proc. Natl. Acad. Sci. USA*, *72*: 2677.

Dostal, L. A., Chapin, R. E., Stefanski, F. A., Harris, M. W.,and Schwetz, B. A. (1988). *Toxicol. Appl. Pharmacol.*, *895*: 104.

Dym, M., and Fawcett, D. W. (1970). *Biol. Reprod.*, *3*: 308.

Ellis, M. K., and Foster, P. M. D. (1992). *Toxicol. Lett.*, *62*: 201.

Evenson, D. P., Janca, F. C., Baer, R. K., Jost, L. K., and Karabinus, D. S. (1989a). *J. Toxicol. Environ. Health*, *28*: 67.

Evenson, D. P., Janca, F. C., Jost, L. K., Baer, R. K., and Karabinus, D. S. (1989b). *J. Toxicol. Environ. Health*, *28*: 81.

Flik, G., van de Winkel, G. J., Part, P., Bonga, S. E. W., and Lock, R. A. C. (1987). *Arch. Toxicol.*, *59*: 353.

Foster, P. M. D. (1989). *Arch. Toxicol.*, *13*: 3.

Foster, P. M. D., Thomas, L. V., Cook, M. W., and Gangoli, S. D. (1980). *Toxicol. Appl. Pharmacol.*, *54*: 392.

Foster, P. M. D., Foster, J. R., Cook, M. W., Thomas, L. V., and Gangoli, S. D. (1982). *Toxicol. Appl. Pharmacol.*, *63*: 120.

Foster, P. M. D., Creasy, D. M., Foster, J. R., Thomas, L. V., Cook, M. W., and Gangoli, S. D. (1983). *Toxicol. Appl. Pharmacol.*, *69*: 385.

Foster, P. M. D., Creasy, D. M., Foster, J. R., and Gray, T. J. M. (1984). *Environ. Health. Perspect.*, *57*: 207.

Foster, P. M. D., Sheard, C. M., and Lloyd, S. C. (1986a). *Excerpta Med. Int. Congr.*, *716*: 281.

Foster, P. M. D., Blackburn, D. M., Moore, R. B., and Lloyd, S. C. (1986b). *Toxicol. Lett.*, *32*: 73.

Foster, P. M. D., Lloyd, S. C., and Prout, M. S. (1987a). *Toxicol. In Vitro*, *1*: 31.

Foster, P. M. D., Lloyd, S. C., and Blackburn, D. M. (1987b). *Toxicology*, *43*: 17.

Fritz, I. (1978). *Biochemical Actions of Hormones* (G. Litwack, ed.). Academic Press, New York, p. 249.

Fukuota, M., Tanimoto, T., Zhou, Y., Kawaski, N., Tanaka, A., Ikemoto, I., and Machida, T. (1989). *J. Appl. Toxicol.*, *9*: 277.

Galdieri, M., Ziparo, E., Palomibi, F., Russo, M. A., and Stefanini, M. (1981). *J. Androl.*, *2*: 249.

Gangolli, S. D. (1982). *Environ. Health Perspect.*, *45*: 77.

Genter, M. B., Szakal-Quin, G., Anderson, C. W., Anthony, D. C., and Graham, D. G. (1987a). *Toxicol. Appl. Pharmacol.*, *87*: 351.

Genter, M. B., Clair, M. B., Amarnath, V., Moody, M. A., Anthony, D. C., Anderson, C. W., and Graham, D. G. (1987b). *Chem. Res. Toxicol.*, *1*: 179.

Ghanayem, B. I., and Chapin, R. E. (1990). *Exp. Mol. Pathol.*, *52*: 279.

Ghanayem, B. I., Burka, L. T., Mattews, H. B. (1987). *Toxicol. Appl. Pharmacol.*, *242*: 222.

Ghanayem, B. I., Burka, L. T., and Matthews, H. B. (1989). *Chem.-Biol. Interact.*, *70*: 339.

Gleason, M. N., Gosselin, R. E., Hodge, H. C., and Smith, R. P. (1969). *Clinical Toxicology of Commercial Products*, 3d ed. Williams & Wilkins, Baltimore, p. 58.

Gouveia, M. A. (1988). *Andrologia*, *20*: 225.

Graham, D. G., Anthony, D. C., and Boekelheide, K. (1982). *Toxicol. Appl. Pharmacol.*, *64*: 415.

Grasso, P., and Recichert, L. J. (1989). *Endocrinology*, *125*: 3029.

Grasso, P., Heindel, J. J., Powell, C., and Reichert, L., Jr. (1993). *Biol. Reprod.*, *48*: 454.

Gray, T. J. B. (1986). *Food. Chem. Toxicol.*, *6/7*: 601.

Gray, T. J. B. (1988). *Physiology and Toxicology of Male Reproduction* (J. C. Lamb IV and P. M. D. Foster, eds.). Academic Press, San Diego, p. 225.

Gray T. J. B., and Beamand, J. A. (1984). *Food. Chem. Toxicol.*, *22*: 123.

Gray T. J. B., and Gangolli, S. D. (1986). *Environ. Health Perspect.*, *65*: 229.

Gray, T. J. B., Rowland, I. R., Foster, P. M. D., and Gangoli, S. D. (1982). *Toxicol. Lett.*, *11*: 142.

Gray, T. J. B., Moss, E. J., Creasy, D. M., and Gangolli, S. D. (1985). *Toxicol. Appl. Pharmacol.*, *79*: 490.

Griswold, G. D. (1993). *Sertoli Cell* (L. D. Russell and G. D. Griswold, eds.). Cache River Press, Clearwater, FL, p. 493.

Grove, B. D., and Vogl, A. W. (1989). *J. Cell. Sci.*, *93*: 309.

Grove, B. D., Pfeiffer, D. C., Allen, S., and Vogl, A. W. (1990). *Am. J. Anat.*, *188*: 44.

Gulati, D. K., Hope, E., Teague, J., and Chapin, R. E. (1991). *Fundam. Appl. Toxicol.*, *17*: 270.

Hall, E. S., Eveleth, J., and Boekelheide, K. (1991). *Toxicol. Appl. Pharmacol.*, *111*: 443.

Hall, E. S., Hall, S. J., and Boekelheide, K. (1992). *Toxicol. Appl. Pharmacol.*, *117*: 99.

Hall, E. S., Hall, S. J., and Boekelheide, K. (1995). *Toxicol. Appl. Pharmacol.*, *24*: 173.

Hardin, B. D. (1983). *Toxicology*, *27*: 91.

Heckert, L. L., and Griswold, M. D. (1991). *Mol. Endocrinol.*, *5*: 670.

Heindel, J. J. and Chapin, R. E. (1989). *Toxicol. Appl. Pharmacol.*, *97*: 377.

Heindel, J. J., and Powell, C. (1992). *Toxicol. Appl. Pharmacol.*, *115*: 116.

Heindel, J. J., and Treinen, K. A. (1989). *Toxicol. Pathol.*, *17*: 411.

Heindel, J. J., Gulati, D. K., Mounce, R. C., Russel, S. R., and Lamb, J. C. (1989). *Fund. Appl. Toxicol.*, *12*: 508.

Heindel, J. J., Powell, C., and Culler, M. D. (1990). *Biol. Reprod.*, *42*: 66.

Henriksén, K., Kangasniemi, M., Parvinen, M., Kaipia, A., and Hakovirta, H. (1996). *Endocrinology*, *137*: 2141.

Hermo, L., Oko, R., and Hecht, N. B. (1991). *Anat. Rec.*, *229*: 3150.

Hess, R. A., Linder, R. E., and Strader, L. (1985). *Toxicologist*, *5*: abstraact 554.

Hess, R. A., Linder, R. E., Strader, L. F., and Perreault, S. D. (1988a). *J. Androl.*, *9*: 317.

Hess, R. A., Linder, R. E., Strader, L. F., and Perreault, S. D. (1988b). *J. Androl.*, *9*: 327.

Hess, R. A., Schaeffer, D. J., Eroschenko, V. P., and Keen, J. E. (1990). *Biol. Reprod.*, *43*: 517.

Hess, R. A., Cooke, P. S., Bunick, D., and Kirby, J. D. (1993). *Endocrinology*, *132*: 2607.

Hew, K., Ericson, W. A., and Welsh, M. J. (1993a). *Toxicol. Appl. Pharmacol.*, *121*: 15.

Hew, K., Heath, G. L., Jiwa, A. H., and Welsh, M. J. (1993b). *Biol. Reprod.*, *49*: 840.

Hinton, B. T., and Setchell, B. P. (1978). *J. Physiol. (London)*, *284*: 16P.

Holladay, S. D., Comment, C. E., Kwon, J., and Luster, M. I. (1994). *Toxicol. Appl. Pharmacol.*, *129*: 53.

HSDB. (1995). Hazardous Substances Data Bank: 2-methoxyethanol.

Huhtaniemi, I., Nikula, H., and Parvinen, M. (1989). *Mol. Cell Endocrinol.*, *62*: 89.

Inouye, B., Ogino, Y., Ishida, T., Ogata, M., and Utsumi, K. (1978). *Toxicol. Appl. Pharmacol.*, *43*: 189.

Ishihara, N., Kanaya, A., and Ikeda, M. (1976). *Int. Arch. Occup. Environ. Health*, *36*: 161.

Janecki, A., Jakubowiak, A., and Steinberger, A. (1992). *Toxicol. Appl. Pharmacol.*, *112*: 51.

Jégou, B. (1994). *Cell Biol. Toxicol.*, *8*: 49.

Jockusch, B. M., and Isenberg, G. (1981). *Proc. Natl. Acad. Sci. USA*, *78*: 3005.

Johnson, K. J., Hall, E. S., and Boekelheide, K. (1991). *Toxicol. Appl. Pharmacol.*, *111*: 432.

Johnson, M. H. (1969). *J. Reprod. Fertil.*, *19*: 551.

Kerr, J. B. (1988). *Anat. Embryol.*, *179*: 191.

Kerr, J. B., Savage, G. N., Millar, M., and Sharpe, R. M. (1993). *Cell. Tissue Res.*, *274*: 153.

Kierszenbaum, A. L. (1994). *Endocr. Rev.*, *15*: 116.

King, K. L., and Cidlowski, J. A. (1995). *J. Cell. Biochem.*, *58*: 175.

Kirby, J. D., Jetton, A. E., Cooke, P. S., Hess, R. A., Bunick, D., Ackland, J., Turek, F., and Schwartz, N. B. (1992). *Endocrinology*, *131*: 559.

Klymkowsky, M. W., Bachant, J. B., and Domingo A. (1989). *Cell. Motil. Cytoskel.*, *14*: 309.

Korobkin, R., Asbury, A. K., Sumner, A. J., and Nielsen, S. L. (1975). *Arch. Neurol.*, *32*: 158.

Kotsonis, F. N., and Klaassen, C. D. (1977). *Toxicol. Appl. Pharmacol.*, *41*: 667.

Kotsonis, F. N., and Klaassen, C. D. (1978). *Toxicol. Appl. Pharmacol.*, *46*: 39.

Krasavage, W. J., O'Donoghue, J. L., DiVincenzo, G., and Terhaar, C. J. (1980). *Toxicol. Appl. Pharmacol.*, *52*; 433.

Ku, W. W., and Chapin, R. E. (1994). *Toxicol. In Vitro*, *8*: 1191.

Ku, W. W., Chapin, R. E., Wine, R. N., and Gladen, B. C. (1993). *Reprod. Toxicol.*, *7*: 305.

Ku, W. W., Ghanayem, B. I., Chapin, R. E., and Wine, R. N. (1994). *Exp. Mol. Pathol.*, *61*; 119.

Ku, W. W., Wine, R. N., Chae, B. Y., Ghanayem, B. I., and Chapin, R. E. (1995). *Toxicol. Appl. Pharmacol.*, *134*: 100.

Lacy, D., Vinson, G. P., Collins, P., Bell, J., Fryson, P., Pudney, J., and Pettitt, A. J. (1968). *Proc. 3rd. Int. Cong. End.*, p. 1019.

Lake, B. G., Phillips, J. C., Linnell, J. C., and Gangoli, S. D. (1977). *Toxicol. Appl. Pharmacol.*, 88: 239.

Lamb, D. J. (1993). *J. Urol.*, *150*: 583.

Lamb, IV, J. C., and Chapin, R. E. (1993). *Reprod. Toxicol.*, *7*: 17.

Lamb, IV, J. C., Gulati, D. K., Russell, V. S., Hommel, L., and Sabharwal, P. S. (1984). *Environ. Health Perspect.*, *57*: 85.

Lamb, J. C., IV, Chapin, R. E., Teague, J., Lawton, A. D., and Reel, J. R. (1987). *Toxicol. Appl. Toxicol.*, 88: 255.

Laskey, J. W., Rehnberg, G. L., Laws, S. C., and Hein, J. F. (1986). *J. Toxicol. Environ. Health*, *19*: 393.

Leblond, C. P., and Clermont, Y. (1952). *Ann. NY Acad. Sci.*, *55*: 548.

Lee, I., and Dixon, R. L. (1973). *J. Pharmacol. Exp. Ther.*, *187*: 641.

Li, L.-H., Wine, R. N., and Chapin, R. E. (1997). *J. Androl.*, *17*: 153.

Li, L.-H., Wine, R. N., Miller, D. S., Reece, J. M., and Chapin, R. E. (1996b). *Toxicol. Appl. Pharmacol.*, *144*: 105.

Linder, R. E., Hess, R. A., and Strader, L. F. (1986). *J. Toxicol. Environ. Health*, 19: 477.

Linder, R. E., Hess, R. A., Perreault, S. D., Strader, L. F., and Barbee, R. R. (1988). *J. Androl.*, *9*: 317.

Linder, R. E., Strader, L. F., Barbee, R. R., Rehnberg, G. L., and Perreault, S. D. (1990). *Fundam. Appl. Toxicol.*, *14*: 284.

Linstrom, P., Harris, M., Ross, M., Lamb IV, J. C., and Chapin, R. E. (1988). *Fundam. Appl. Toxicol.*, *11*: 528.

Lloyd, S. C., and Foster, P. M. D. (1988). *Toxicol. Appl. Pharmacol.*, 95: 484.

Maddocks, S., Kerr, J. B., Allenby, G., and Sharpe, R. M. (1992). *J. Endocrinol.*, *132*: 439.

Mali, P., Virtanen, I., and Parvinen, M. (1987). *Andrologia*, *19*: 644.

Mather, J. P., Zhuang, L., Perez-Infante, V., and Philips, D. M. (1982). *Ann. NY Acad. Sci.*, *383*: 44.

McEven, S. F., and Miller, M. G. (1991). *Drug Metab. Dispos.*, *19*: 661.

McKinnell, C., Brackenbury, E. T., Qureshi, S. J., Hargreave, T. B., and Sharpe, R. M. (1995). *Int. J. Androl.*, *18*: 103.

McLaren, A. (1991). *Nature*, *351*: 96.

McLaren, T. T., Foster, P. D. M., and Sharpe, R. M. (1993a). *Fundam. Appl. Toxicol.*, *21*: 384.

McLaren, T. T., Foster, P. M. D., and Sharpe, R. M. (1993b). *Int. J. Androl.*, *16*: 370.

Mebus, C. A., and Welsch, F. (1989). *Toxicol. Appl. Pharmacol.*, 99: 98.

Mebus, C. A., Welsch, F., and Working, P. K. (1989). *Toxicol. Appl. Pharmacol.*, 99: 110.

Melnick, R. L., and Schiller, C. M. (1982). *Environ. Health Perspect.*, 45: 51.

Miller, R. R., Ayres, J. A., Young, J. T., and McKenna, M. J. (1983). *Fundam. Appl. Toxicol.*, *3*: 49.

Miller, R. R., Ayres, J. A., Calhoun, L. L., Young, J. T., and McKenna, M. J. (1981). *Toxicol. Appl. Pharmacol.*, *61*: 368.

Monaco, L., and Conti, M. (1986). *Biol. Reprod.*, *35*: 258.

Monaco, L., Adamo, S., and Conti, M. (1988). *Endocrinology*, *123*: 2032.

Morales, C. R., and Griswold, M. D. (1987). *Endocrinology*, *121*: 432.

Morales, C. R., Hermo, L., and Clermont, Y. (1985). *Am. J. Anat.*, *173*: 203.

Morales, C. R., Clermont, Y., and Nadler, N. (1986). *Biol. Reprod.*, *34*: 207.

Mori, H., and Christensen, A. K. (1980). *J. Cell Biol.*, *84*: 340.

Moss, E. J., Thomas, L. V., Cook, M. W., Walters, D. G., Foster, P. M. D., Creasy, D. M., and Gray, T. J. B. (1985). *Toxicol. Appl. Pharmacol.*, *79*: 480.

Moss, E. J., Cook, M. W., Thomas, L. V., and Gray, T. J. B. (1988). *Toxicol. Lett.*, *40*: 77.

Nagano, K., Nakayama, E., Kogano, M., Oobayashi, H., Adachi, H., and Yamada, T. (1979). *Jpn. J. Ind. Health.*, *21*: 29.

National Toxicology Program. (1992). *Final Report on the Reproductive Toxicity of Di-n-butyl Phthalate in Sprague-Dawley Rats*. Report #T-oo35C, PB92111996.

Niewenhuis, R. J., and Prozialeck, W. C. (1987). *Biol. Reprod.*, *37*: 127.

Nikular, H., Vihko, M., and Huhtaniemi, I. (1990). *Mol. Cell. Endocrinol.*, *70*: 247.

Obasaju, M. F., Katz, D. F., and Miller, M. G. (1991). *Fundam. Appl. Toxicol.*, *16*: 257.

O'Donoghue, J. L., Krasavage, W. J., and Terhaar, C. J. (1978). *Toxicol. Appl. Pharamcol.*, *45*: 269.

Ohyama, T. (1987). *J. Biochem.*, *82*: 9.

Oishi, S., and Hiraga, K. (1980a). *Toxicol. Appl. Pharmacol.*, *53*: 35.

Oishi, S., and Hiraga, K. (1980b). *Toxicol. Lett.*, *5*: 413.

Oishi, S., and Hiraga, K. (1986). *Toxicology*, *15*: 197.

Orth, J. M. (1982). *Anat. Rec.*, *203*: 485.

Orth, J. M. (1984). *Endocrinology*, *115*: 1248.

Orth, J. M., and Christensen, A. K. (1978). *Endocrinology*, *103*: 1944.

Oudiz, D., and Zenick, H. (1986). *Toxicol. Appl. Pharmacol.*, *84*: 576.

Paranko, J., Kallajoki, M., Pelliniemi, L. J., Lehto, V. P., and Vietanen, I. (1986). *Dev. Biol.*, *117*: 35.

Parizek, J. (1957). *J. Endocrinol.*, *15*: 56.

Parker, A. I., Besmer, P., and Bachvarova, R. F. (1995). *Mol. Reprod. Dev.*, *42*: 303.

Parker, D. V. (1961). *Biochem. J.*, *78*: 262.

Parvinen, M. (1982). *Endocr. Rev.*, *3*: 404.

Parvinen, M. (1993). *Sertoli Cell* (L. D. Russell and G. D. Griswold, eds.). Cache River Press, Clearwater, FL, p. 331.

Parvinen, M., Marana, R., Robertson, D. M., Hansson, V., and Ritzén, E. M. (1980). *Testicular Development, Structure and Function* (A. Steinberger and E. Steinberger, eds.). Raven Press, New York, p. 425.

Pelliniemi, L. J., Frojdman, K., and Paranko, J. (1993). *Sertoli Cell* (L. D. Russell and G. D. Griswold, eds.). Cache River Press, Clearwater, FL, p. 87.

Phelps, P. V., and Laskey, J. W. (1989). *J. Toxicol. Environ. Health*, *27*: 95.

Quirk, S. M., and Reichert, L. J. (1988). *Endocrinology*, *123*: 230.

Reader, S. C. J., Shingles, C., and Stonard, M. D. (1991). *Fundam. Appl. Toxicol.*, *16*: 61.

Rehnberg, G. L., Lender, R. E., Goldman, J. M., Hein, J. F., McElroy, W. K., and Cooper, R. L. (1988). *Toxicol. Appl. Pharmacol.*, *95*: 255.

Rhee, K., and Wolgemuth, D. J. (1995). *Dev. Dyn.*, *204*: 406.

Richburg, J. H., and Boekelheide, K. (1996). *Toxicol. Appl. Pharmacol. 137*: 42

Richburg, J. H., Redenbach, D. M., and Boekelheide, K. (1994). *Toxicol. Appl. Pharmacol.*, *128*: 302.

Rosiepen, G., Chapin, R. E., and Weinbauer, G. F. (1995). *J. Androl.*, *16*: 127.

Russell, L. D. (1978). *Anat. Rec.*, *190*: 90.

Russell, L. D. (1979). *Anat. Rec.*, *194*: 233.

Russell, L. D. (1991). *Int. J. Androl.*, *14*: 307.

Russell, L. D. (1993). *Sertoli Cell* (L. D. Russell and G. D. Griswold, eds.), Cache River Press, Clearwater, FL, p. 2.

Russell, L. D., and Griswold, G. D., eds. (1993). *Sertoli Cell*. Cache River Press, Clearwater, FL.

Russell, L. D., and Stenberger, A. (1989). *Biol. Reprod.* 41:571.

Russell, L. D., Malone, J. P., and MacCurd, D. S. (1981). *Tissue Cell, 13*: 349.

Russell, L. D., Ettlin, R. A., Hikim, A. P. S., and Clegg, E. D. (1990). *Histological and Histopathological Evaluation of the Testis.* Cache River Press, Clearwater, FL, p. 213.

Saksena, S., White, M. J., Mertzlufft, J., and Lau, I. (1983). *Contraception, 27*: 521.

Satoh, M. (1985). *J. Anat., 143*: 17.

Schliwa, M. (1986). *The Cytoskeleton, An Introductory Survey.* Springer-Verlag, New York.

Seidl, K., and Holstein, A.-F. (1990). *Cell Tiss. Res., 261*: 539.

Setchell, B. P., and Waites, G. M. H. (1970). *J. Endocrinol., 47*: 841.

Setchell, B. P., Davies, R. V., Gladwell, R. T., Hinton, R. T., Main, S. J., Pilsworth, L., and Waites, G. M. H. (1978). *Ann. Biol. Anim. Biochim. Biophys., 18*: 623.

Sharpe, R. M. (1987). *J. Endocrinol., 113*: 1.

Sharpe, R. M. (1992). *Int. J. Androl., 15*: 201.

Simialowicz, R. J., Riddle, M. M., and Williams, W. C. (1994). *Int. J. Immunopharmacol., 16*: 695.

Sjoberg, P., Lindquist, N. G., and Ploen, L. (1986). *Environ. Health Perspect., 65*: 237.

Skinner, M. K. (1993). *Sertoli Cell* (L. D. Russell and G. D. Griswold, eds.). Cache River Press, Clearwater, FL, p. 237.

Solari, A. J., and Fritz, J. B. (1978). *Biol. Reprod., 18*: 329.

Spencer, P. S., and Schaumburg, H. H. (1975). *J. Neurol. Neurosurg. Psychiatry, 38*: 771.

Steinberger, A. (1993). *Handbook of Hazardous Materials.* Academic, New York, p. 113.

Steinberger, A., and Steinberger, E. (1971). *Biol. Reprod., 4*: 84.

Steinberger, A., Steinberger, E., and Perloff, W. H. (1963). *Fed. Proc., 22*: 372.

Suarez-Quian, C. A., and Dym, M. (1992). *Microsc. Res. Tech., 20*: 219.

Sumner, S. C. J., and Fennell, T. R. (1993). *Toxicol. Appl. Pharmacol., 120*: 162.

Tanaka, A., Adachi, T., Takahashi, T., and Yamaha, T. (1975). *Toxicology, 4*: 253.

Thomas, J. A., and Thomas, M. J. (1984). *CRC Crit. Rev. Toxicol., 13*: 283.

Toppari, J., Mali, P., and Eerola, E. (1986). *J. Histochem. Cytochem., 34*: 1029.

Treinen, K., and Heindel, J. J. (1992). *Reprod. Toxicol., 6*: 143.

van Haaster, L. H, de Jong, F. H., Docter, R., and de Rooij, D. G. (1993). *Endocrinology, 133*: 755.

Viguier-Martinez, M. C., Hochereau-de Reviers, M. T., Barenton, B., and Perreau, C. (1984). *J. Reprod. Fertil., 70*: 67.

Vogl, A. W. (1989). *Int. Rev. Cyt., 119*: 1.

Vogl, A. W., Pfeiffer, D. C., Redenbach, D. M., and Grove, B. D. (1993). *Sertoli Cell* (L. D. Russell and G. D. Griswold, eds.). Cache River Press, Clearwater, FL, p. 39.

Waalkes, M. P., Chernoff, S. B., and Klaassen, C. D. (1984). *Biochem. J., 220*: 811.

Wagle, J. R., Heindel, J. J., Sanborn, B. M., and Steinberger, A. (1986). *In Vitro, 22*: 325.

Watanabe, T., Ishihara, N., and Ikeda, M. (1976). *Int. Arch. Occup. Environ. Health, 37*: 157.

Webb, M. (1972). *Biochem. Pharmacol., 21*: 2767.

Webb, M. (1975). *Br. Med. Bull., 31*: 246.

Welch, L., S., Schrader, S. M., Turner, T. W., and Cullen, M. R. (1988). *Am. J. Ind. Med., 14*: 509.

Welsch, F., Sleet, R. B., and Green, J. A. (1987). *J. Biochem. Toxicol., 2*: 225.

Welsh, M. J., Wu, W., Parvinen, M., and Gilmont, R. R. (1996). *Biol. Reprod., 56*: 141.

Wiley, F. H., Hueper, W. C., Bergen, D. S., and Blood, F. R. (1938). *J. Ind. Hyg. Toxicol., 20*: 269.

Williams, J., and Foster, P. M. D. (1988). *Toxicol. Appl. Pharmacol., 94*: 160.

Wing, T.-Y., and Christensen, A. K. (1982). *Am. J. Anat., 165*: 13.

Wong, K. L., and Klaassen, C. D. (1980). *Toxicol. Appl. Pharmacol., 55*: 456.

Woodward, K. N. (1988). *Phthalate Esters: Toxicity and Metabolism*. Boca Raton, FL, CRC Press: *1*: 141–168.

Wyllie, A. H., Morris, R. G., Smith, A. L., and Dunlop, D. (1984). *J. Pathol., 142*: 67.

Yoon, D. J., Reggiardio, D., and David, R. (1990). *J. Endocrinol., 125*: 293.

Zenick, H., Hastings, L., Goldsmith, M., and Neiwenhius, R. J. (1982). *J. Toxicol. Environ. Health, 9*: 377.

Index